AN ATLAS OF

ILLINOIS

FISHES

AN ATLAS OF
ILLINOIS
FISHES

150 Years of Change

BRIAN A. METZKE, BROOKS M. BURR, LEON C. HINZ JR.,
LAWRENCE M. PAGE, AND CHRISTOPHER A. TAYLOR

**UNIVERSITY OF
ILLINOIS PRESS**
Urbana, Chicago, and Springfield
www.press.uillinois.edu

Library of Congress Cataloging-in-Publication Data
Names: Metzke, Brian A., 1980– author. | Burr, Brooks M., author.
   | Hinz, Leon C., author. | Page, Lawrence M., author. | Taylor,
   Christopher Alan, 1969– author.
Title: An atlas of Illinois fishes: 150 years of change / Brian A.
   Metzke, Brooks M. Burr, Leon C. Hinz Jr., Lawrence M. Page,
   Christopher A. Taylor.
Description: Urbana: University of Illinois Press, [2022] | Includes
   bibliographical references and index.
Identifiers: LCCN 2021038205 (print) | LCCN 2021038206
   (ebook) | ISBN 9780252044144 (hardback) | ISBN
   9780252053085 (ebook)
Subjects: LCSH: Fishes—Illinois--Identification. | Fishes—
   Illinois—Maps. | Fishes—Illinois—History.
Classification: LCC QL628.I3 M48 2022 (print) | LCC QL628.I3
   (ebook) | DDC 597.09773—dc23
LC record available at https://lccn.loc.gov/2021038205
LC ebook record available at https://lccn.loc.gov/2021038206

CONTENTS

155 Southern Redbelly Dace, *Chrosomus erythrogaster*

156 Redside Dace, *Clinostomus elongatus*

157 Lake Chub, *Couesius plumbeus*

158 Red Shiner, *Cyprinella lutrensis*

159 Spotfin Shiner, *Cyprinella spiloptera*

160 Blacktail Shiner, *Cyprinella venusta*

161 Steelcolor Shiner, *Cyprinella whipplei*

162 Silverjaw Minnow, *Ericymba buccata*

163 Gravel Chub, *Erimystax x-punctatus*

164 Western Silvery Minnow, *Hybognathus argyritis*

165 Brassy Minnow, *Hybognathus hankinsoni*

166 Cypress Minnow, *Hybognathus hayi*

167 Mississippi Silvery Minnow, *Hybognathus nuchalis*

168 Plains Minnow, *Hybognathus placitus*

169 Bigeye Chub, *Hybopsis amblops*

170 Pallid Shiner, *Hybopsis amnis*

171 Striped Shiner, *Luxilus chrysocephalus*

172 Common Shiner, *Luxilus cornutus*

173 Bleeding Shiner, *Luxilus zonatus*

174 Scarlet Shiner, *Lythrurus fasciolaris*

175 Ribbon Shiner, *Lythrurus fumeus*

176 Redfin Shiner, *Lythrurus umbratilis*

177 Sturgeon Chub, *Macrhybopsis gelida*

178 Shoal Chub, *Macrhybopsis hyostoma*

179 Sicklefin Chub, *Macrhybopsis meeki*

180 Silver Chub, *Macrhybopsis storeriana*

181 Hornyhead Chub, *Nocomis biguttatus*

182 River Chub, *Nocomis micropogon*

183 Golden Shiner, *Notemigonus crysoleucas*

184 Pugnose Shiner, *Notropis anogenus*

185 Emerald Shiner, *Notropis atherinoides*

ACKNOWLEDGMENTS

An undertaking of this magnitude requires contributions from more than the authors. We are extremely grateful to the individuals who have assisted us in completing this project, and if we have left out any names, we apologize and say thank you.

Funding was generously provided by the Illinois Department of Natural Resources, Division of Natural Heritage. We especially thank Ann Marie Holtrop and Jenny Skufca for appreciating the value of this project and committing to its completion. Andrew Hulin and Don Bricker of the Illinois Department of Natural Resources provided database and Graphic Information System technical assistance. Illinois Natural History Survey technicians Candice Shea, Morgan Oschwald, and Samantha Carey contributed greatly in compiling records and generating distribution maps.

Many individuals produced records and supporting data on behalf of their agencies and institutions: Jeannie Barnes, Robert Hrabik, Tara Kieninger, Mike McClelland, Nerissa McClelland, Greg Sass, Vic Santucci, Matt Short, Frank Veraldi, and Greg Whitledge. We thank the museum curators and staff who accessed and shared information and answered questions about specimen records: Mariangeles Arce and Mark Sabaj, Academy of Natural Sciences of Drexel University; David Eisenhour, Morehead State University; Caleb McMahon and Philip Willink, Field Museum of Natural History; Jeremy Tiemann, Illinois Natural History Survey; and Jeffrey Williams, Smithsonian National Museum of Natural History.

Illustrations in the identification keys came from many sources. Most were created or modified by Karen Fiorino, and she deserves special thanks. Partial funding to support her work was supplied by Roy Heidinger, Southern Illinois University Carbondale. Permissions to use previously published illustrations were graciously provided by the following individuals and institutions: Noel Burkhead; Kansas Museum of Natural History; Alice Prickett; Missouri Department of Conservation; Henry Robison; the Royal Ontario Museum; Karl Scheidegger; Jo-Ellen Trecartin; and the families of Edward Menhinick, Charles Schwartz, and Milton Trautman. Layout and editorial assistance for the key was kindly provided by Jennifer Davis and Danielle Ruffatto, Prairie Research Institute, University of Illinois at Urbana-Champaign. We also are grateful to Julie McMahon for designing the cover of this book.

Numerous photographers graciously donated their images for use in the species accounts. These include Robert Criswell, Kandis Elliot, Collin Gallagher, John Lyons, David Neely, Zachary Randall, Konrad Schmidt, Nate Tessler, Matthew Thomas, and Brian Zimmerman.

We sincerely thank all of the many individuals who assisted in the field collection of specimens and who have spent time in the field with the authors over the past 50 years.

Blackside Darter, *Percina maculata.* Watercolor by L. Hart in S. A. Forbes and R. E. Richardson's Fishes of Illinois (1909).

AN ATLAS OF

ILLINOIS

FISHES

Shorthead Redhorse, *Moxostoma macrolepidotum.* Watercolor by L. Hart in S. A. Forbes and R. E. Richardson's Fishes of Illinois (1909).

PURPOSE AND SCOPE

Over forty years have passed since Philip W. Smith published his *Fishes of Illinois*, and it has been over 110 years since Stephen A. Forbes and Robert E. Richardson published their book of the same title. In their times these books were viewed as preeminent treatises of Illinois fishes and have been used by generations of ichthyologists, ecologists, and naturalists. Much has changed since these books, including species introductions and extirpations, species-distribution changes, taxonomic changes, implementation of standardized fish-survey programs that greatly expanded distributional information, digitization of species occurrence records that enhanced access to data, and proliferation of personal computers and software that improved the pace of data analysis. These changes implore a contemporary treatment of Illinois fishes.

Forbes and Richardson envisioned the purpose of their *Fishes of Illinois* (1909) "to furnish those interested in Illinois fishes a reliable guide to a knowledge of species, [and] a careful account of their local and general distribution and of their relations to their environment," and a similar goal is aspired to here. This book provides a resource for identifying Illinois fishes and documenting their distributional status and trends. It is composed in such a way that both scientists and naturalists may appreciate its contents. This book is organized into 3 sections: a brief characterization of Illinois and review of ichthyological history to provide context for the diversity of Illinois fishes, keys for identifying Illinois fishes with supporting morphological illustrations, and family and species accounts with distribution maps and representative photos. Those readers wishing to supplement the information found in this book may reference Peterson's *Field Guide to Freshwater Fishes of North America North of Mexico* (Page & Burr 2011) and other state-specific treatments of fishes, such as the *Fishes of Missouri* (Pflieger 1997) or *A Naturalist's Guide to the Fishes of Ohio* (Rice & Zimmerman 2019). Angling information for waterbodies and species in Illinois can be found on the I Fish Illinois website. Conservation status of Illinois fishes may be found on the Illinois Endangered Species Protection Board website and the Illinois Wildlife Action Plan website.

A Brief History of Illinois

The diversity of Illinois fishes is a result of biogeographical history, landscape and climate diversity, and human activities. A chronological characterization of Illinois's geology, landcover, hydrology, and conservation provides readers with context for evaluating species distributions.

The ichthyofaunal history of Illinois dates back millions of years. The fish fossil record of Illinois is summarized to provide readers with an appreciation of prehistoric diversity of the Illinois ichthyofauna and aquatic habitats. Archeological evidence of fish consumption dates back approximately 10,000 years; the written history of Illinois fishes begins in the mid-1700s, and these records provide some insight into Illinois ichthyofauna prior to large-scale harvest and landscape impacts. Focused scientific surveys of Illinois fishes began in the mid-1800s and continue today. A treatment of scientific survey history, including significant contributors, survey methods, and survey effort, provides readers context for the information presented herein. Recognition of those whose work preceded this book is in reverence of their contributions to the study of fishes in Illinois.

Identifying Illinois Fishes

Forbes and Richardson (1909, 1920) included a key to the families of Illinois fishes, and Smith (1979) to the species, but the "Key to the Families of Illinois Fishes" provided in this book is the first in Illinois to include complementary illustrations for species. A family key precedes keys to species. The key uses morphological characteristics to distinguish alternatives of dichotomous couplets. Representative photographs found in species accounts also aid species identification. Readers can find illustrations and descriptions of morphological characteristics in the section "Fish Morphology" and the glossary.

Accounts of Illinois Fishes

Nearly 12 million individual fish collected during 166 years of surveys from approximately 15,200 unique localities were synthesized into almost 325,000 species occurrence records and illustrated in 215 distribution maps for 217 species.[1] Species accounts

---

1. As this book was going to press, the Tippecanoe Darter (*Etheostoma tippecanoe*) and Streamline Chub *Erimystax dissimilis*) were recorded in the Vermilion River, Vermilion County, from Danville downstream to the Vermilion River's confluence with the Wabash River. This discovery brings the total number of fish species in Illinois to 219.

Figure 1. Boat electrofishing unit. A generator produces power for the electrical field, which is projected in front of the boat. Netters in the fore of the boat collect stunned fish. Photo by S. Pescitelli.

Figure 2. Electric seine. A generator delivers power to a wire between the two brail poles of the seine, creating an electrical field around the seine. The seine is pulled through the water, and fish passing through the field are stunned and collected by netters. Photo by B. Metzke.

provide readers with morphological descriptions and brief status assessments of distributions. Records plotted on distribution maps are divided into 3 eras (Forbes and Richardson, Smith, and contemporary) to illustrate distribution trends. Readers should exercise caution when evaluating species distributions as there are limitations inherent to occurrence records (e.g., detection inefficiencies during surveys, incomplete coverage of survey effort) (figures 20–27).

150 Years of Change

The first inventories of Illinois fishes were published in the mid-1800s (Jordan 1878, Kennicott 1855, Nelson 1876), and the first statewide treatment of fishes was published in 1909 (Forbes & Richardson 1909). Since then the human population of Illinois has increased tenfold and large-scale anthropogenic landscape and hydrologic alterations have occurred. Species invasions have altered native-species distributions and community composition (e.g., Chick et al. 2019, Janssen & Jude 2001, McClelland et al. 2012). By the 1870s stocking of sport and food fishes was carried out by individuals and government agencies (Forbes & Richardson 1909, Illinois State Fish Commission 1884), and stocking and accidental releases have occurred continuously since then. These changes, which are mostly a result of human activities, have produced a fish community that is dynamic in space and time.

Burr and Page (2009) predicted additional introductions of non-native fish species (i.e., those deliberately or accidentally introduced or those expanding their range into Illinois) to Illinois and the extirpation of approximately 5 species per decade. Forbes and Richardson (1909) identified 1 non-native species, but by the 1970s there were 13 (Smith 1979), 22 by 1990 (Burr 1991), and 24 now. Smith (1979) identified 9 extirpated species, and Burr (1991) identified 12; however, only 7 species are presumed contemporarily extirpated. Stocking (Alligator Gar, Muskellunge) and contemporary rediscovery (Bigeye Chub, Harlequin Darter, Northern Madtom) are responsible for the improvement in the number of extirpated species. The prophecies of Burr and Page (2009) have proven partially correct; novel

non-native species continue to arrive and spread through Illinois's waters (e.g., Black Carp), but increasing diversity of native species offers some reason for optimism. The degree to which we prioritize mitigation of detrimental human activities while allocating resources to conservation of our species and systems will determine the future of native fishes in Illinois.

In plotting occurrence records by three temporal eras spanning more than 150 years, the species account maps in this atlas allow for a rapid visualization of distributional changes for Illinois fishes. Interpreting the causes of some changes is a far more difficult task. The introduction and spread of non-native fishes and the large-scale alteration of the landscape have clearly impacted some native-fish distributions. However, other factors may either partially or wholly account for other observed changes, including the dramatic increase in sampling sites and sampling frequency over the past 40 years (see this book's section "Methods for Mapping Species Distributions"), and changes in the sampling gear. Early biologists sampled fishes with seines and traps, while by the mid-1900s the fish toxicant rotenone and boat-mounted electrofishing gear (figure 1) also began to be employed for lake, river, and stream sampling. By 1987 the use of rotenone had been mostly discontinued, and many Illinois agencies began using electric seines (figure 2) in addition to traditional seines for wadeable river and stream sampling. Gear types vary in capture efficiency across the range of habitat types and fish body size and habitat preferences. For example, efficiency of electric seines in detecting species richness ranges from 0.25 to 0.95, and efficiency increases with fish body size (Bayley and Dowling 1990). In many cases, in-depth analyses of species-specific biological factors, anthropogenic changes, and climatological data will need to be conducted to elucidate distributional changes.

This book synthesizes one of the most extensive state-specific biological datasets in the United States. It captures the diversity and status of Illinois fishes at a single point in a history that spans millions of years. We hope it will be used as a benchmark from which to document, measure, and study the ever-changing composition of Illinois fish communities and the habitats that support them.

SETTING THE SCENE

Illinois encompasses approximately 58,000 mi.^{2} (150,000 km^2), spans approximately 380 mi. (615 km) north to south, and is divided into 102 counties (figure 3). The Mississippi River forms the western border, and the Ohio and Wabash Rivers make up much of the southern and eastern boundaries of Illinois, while the state is diagonally bisected by the Illinois River running southwest from near Chicago to Alton (figure 4). Twenty-four large rivers flow into the Mississippi, Wabash, and Ohio Rivers (figure 4). We divided the Illinois riverscape into 31 major basins to facilitate description of species distributions (figure 5). Illinois contains more than 2,400 lakes and impoundments of more than 10 ac. (0.04 km^2) and 21 of at least 1,000 ac. (4.05 km^2; figure 6). Approximately 1 million acres (4,047 km^2) of Lake Michigan are within Illinois (figure 3).

The geologic, topographic, climatic, and anthropogenic history of Illinois has shaped the past and present diversity and distribution of the state's ichthyofauna. A chronological summary of geology, landcover, and conservation and management of fishes provides context for patterns described in "Illinois Family and Species Accounts."

Early Period (Pre-1850s)

Surficial Geology

While the geology of Illinois continues to be investigated, general surface conditions have been well described (Kolata & Nimz 2010). The topographic surface of the state was formed through the transport of materials by wind, water, and ice. With the exception of the unglaciated areas in the southernmost and northwestern portions of Illinois, most of the surface of the state is covered with glacial drift, which was deposited during at least 6 glacial episodes over the past 2.5 million years. Alluvium and glacial outwash occur along the major stream courses (Kolata 2010). Lake Michigan is at the northeastern edge of the state, and large rivers form approximately two-thirds of the remaining state boundary.

The Illinoian glaciation (300,000 to 125,000 years ago) advanced south to within 18.6 mi. (30 km) of the Mississippi Embayment (Willman & Frye 1970), and late Wisconsin Episode (25,000 to 10,000 years ago) glaciers advanced only into the northeastern portion of Illinois (Hansel & McKay 2010). Retreating glaciers left behind shallow lakes and wetlands in a poorly drained low-relief landscape. Over time many of these low-relief landforms were buried under windblown silt (loess) and are concealed by current surface features (Hansel & McKay 2010). Streams in these areas (figure 7;

figure 11) formed relatively recently (< 100,000 years ago), and fish assemblages were formed through postglacial colonization events (Page 1991). These reinvasions of glaciated regions were mostly from the south and occurred through newly established drainage systems that crossed older basin divides. Other possible reinvasion routes may have included the upper Missouri River basin or the Driftless Area of northwestern Illinois (Page 1985).

The Driftless Area, which includes a portion of northwestern Illinois, remained unglaciated throughout the Pleistocene. This area contains much-higher topographic diversity than the more recently glaciated regions in most of the rest of the state (Page 1985). Streams in this area (figure 8) and those in the unglaciated southern areas of the state (i.e., Shawnee Hills area; figure 9) are characterized by higher gradients, bedrock near the surface, and distinctive aquatic communities (Page 1991).

South of the unglaciated Shawnee Hills in Illinois lies the northern terminus of the Mississippi Embayment. These low-gradient streams, cypress swamps, including the northernmost Bald Cypress–Tupelo swamp in the United States, and large-river backwaters resemble aquatic habitats found in the Gulf Coastal Region of the south-central United States (figure 10). Like the Shawnee Hills area, the Mississippi Embayment is home to distinctive aquatic communities.

Soils and Groundwater

Soils are the result of the weathering of geological materials and include inorganic components and living and dead organic matter. Glacial, alluvial, and wind-borne deposits make up the parent materials for soils in most of the northern two-thirds of Illinois. Sixty-three percent of the soils have developed in loess, with the remainder in glacial till (12%), alluvium (12%), outwash (8%), and bedrock (5%) (Fehrenbacher et al. 1984). These loess soils have high moisture and nutrient storage capacity (Fehrenbacher et al. 1984), making them a highly productive growing medium that easily erodes.

As a result of Illinois's geological history, groundwater resources are unevenly distributed across the state. About 50% of the state has bedrock or sand and gravel aquifers within 50 feet of the surface (Berg 2010). Productive aquifers occur in areas of thick, saturated deposits of sand and gravel within buried bedrock valleys and near the surface in many parts of central and northwestern Illinois (Larson & Herzog 2010). In areas where aquifers intersect with the surface, groundwater discharge into streams sustains flow during dry periods (Larson & Herzog 2010).

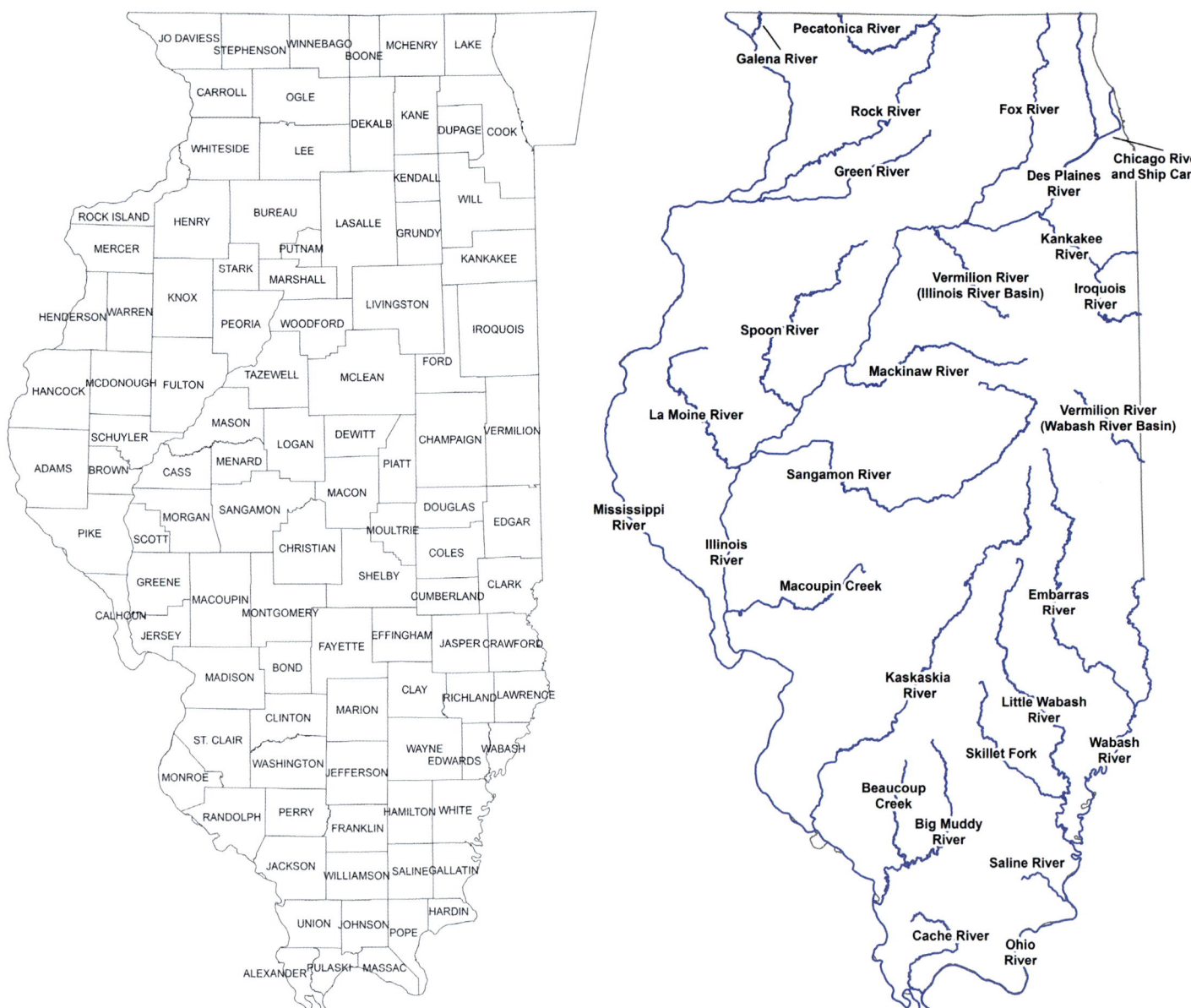

Figure 3. Illinois state and county boundaries.

Figure 4. Large rivers of Illinois.

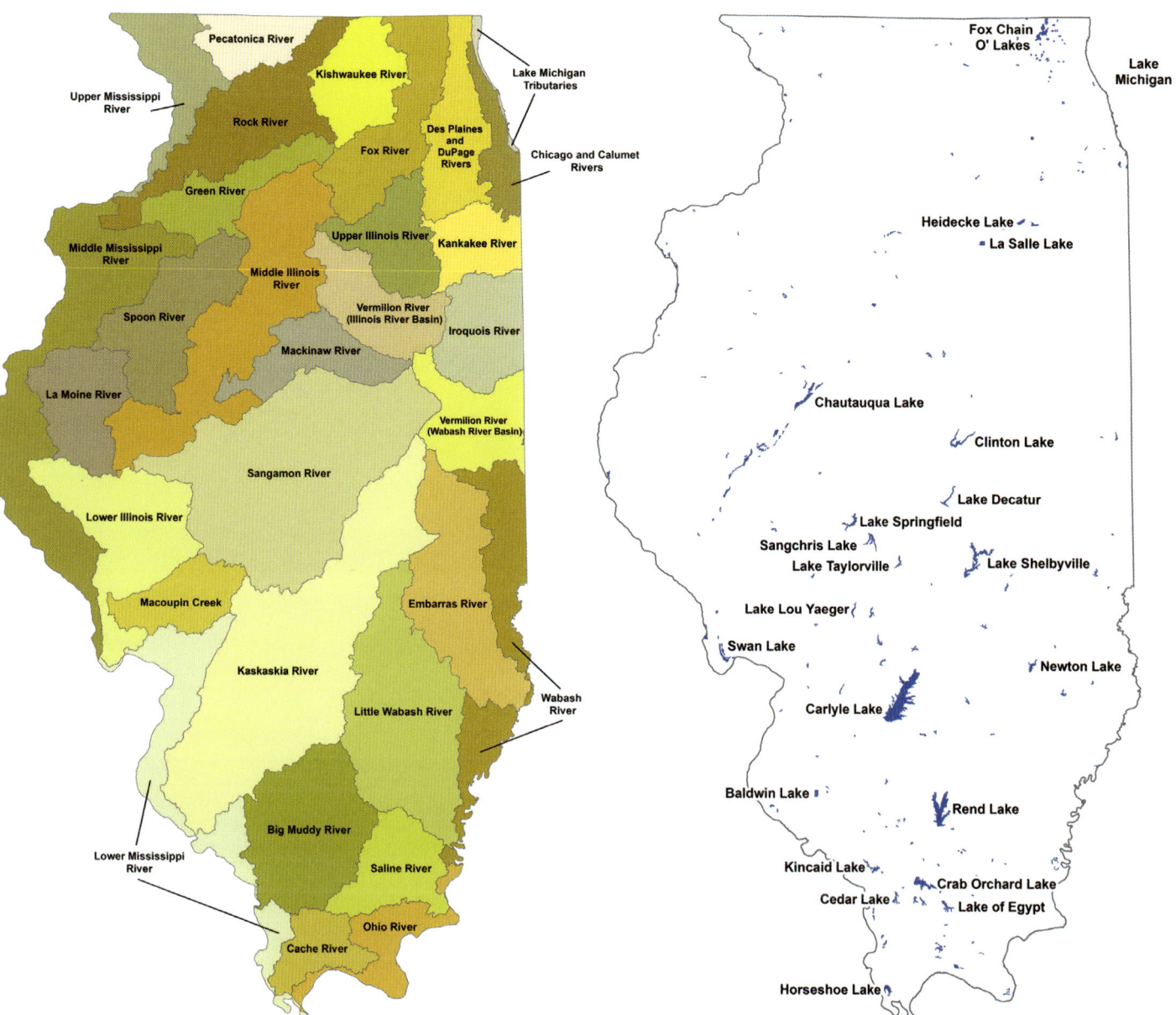

Figure 5. Major basins of Illinois.

Figure 6. Lakes and impoundments of Illinois (those > 100 acres are shown, those > 1,000 acres are labeled).

Figure 7. Mackinaw River, McLean County. This low-gradient stream is characteristic of the glaciated region of central Illinois. Photo by M. Jeffords.

Figure 8. Apple River, Apple River Canyon State Park, Jo Daviess County. This stream is characteristic of the Driftless Area of northwestern Illinois. Photo by M. Jeffords.

Figure 9. Bay Creek, Pope County. This stream is characteristic of the Shawnee Hills area of southern Illinois. Photo by M. Jeffords.

Figure 10. A cypress swamp in the Section 8 Woods Nature Preserve, Pulaski County. This is representative of swamps in the Mississippi Embayment region of extreme southern Illinois. Photo by M. Jeffords.

Figure 11. Mackinaw River, McLean County. This landscape is characteristic of Illinois's highly altered agricultural regions. Photo by M. Jeffords.

Landcover

Early descriptions by European settlers suggest the Illinois landscape was dominated by tallgrass prairie, upland and bottomland forests, and water, including large rivers, backwater lakes, and wetlands (Iverson 1991). Prairies were mainly in the northern two-thirds of the state, and forests were located in the south and along waterways. Wetlands, including swamps, bogs, and wet-prairies, are believed to have occupied about 24% of the total land area in Illinois prior to the 1840s (Taylor et al. 2009). Forested areas, concentrated along river corridors and southern and western Illinois, were approximately 41% of the land cover of early 1800s Illinois.

Forbes and Richardson Era (1853–1909)

Landcover

Stephen A. Forbes and Robert E. Richardson began the first systematic survey of Illinois ichthyofauna in the late 1800s, and by this time the landscape of Illinois had been dramatically altered. During this period, lands containing natural vegetation continued to be developed for agricultural and urban uses. The state's landcover was greatly changed as areas with natural prairie and forest vegetation were transformed into farmland and converted into wood products (Barnhardt 2010). Illinois's landscape transition continued with forests being reduced to 22% of what was estimated to have been

present in the early 1800s (Iverson 1991) and prairies reduced to small remnants. Smith (1979) indicated, "By the beginning of the 20th Century, most of Illinois had been denuded of native vegetation."

From 1818, when Illinois achieved statehood, to 1900, the human population in Illinois had increased by over 4.8 million people and included over 1.6 million people in Chicago (US Census Bureau 1900). Major engineering efforts greatly modified Illinois waterways to support this increased population throughout this era.

Wetland drainage activities became widespread in the mid- to late 1800s (Prince 1997) with tile drainage and levees reducing the functioning area of floodplains on the Illinois River up to 75% or more (Thompson 2002). The formation of drainage districts was precipitated by passage of the Illinois Farm Drainage Act of 1879, which furthered the development of agricultural lands. Drainage districts are governing bodies formed by local landowners to improve drainage for agricultural, mining, or sanitary purposes and are authorized to construct and maintain ditches, levees, and other drainage structures. Incorporation of Illinois agricultural lands into drainage districts peaked between 1900 and 1909, and nearly 10% of Illinois land area was within drainage districts by the end of the Forbes and Richardson era (McCorvie & Lant 1993).

Reservoirs were constructed throughout the state to supply water for domestic, commercial, and recreational needs by damming streams and permanently flooding adjacent lands resulting in the loss of a variety of aquatic habitats (Smith 1979). The earliest Illinois reservoirs were constructed in the 1850s, and by the end of the Forbes and Richardson era, more than 40 were complete (United

States Army Corps of Engineers 2020a). Water from Lake Michigan was diverted into the Illinois River through the Sanitary and Ship Canal, increasing flows and raising river stages by up to 1.2 meters (Thompson 2002). Untreated sewage and industrial wastes from Chicago were also flushed through the canal rather than being emptied into Lake Michigan (Havera 1989). This diversion altered patterns of flow and water quality in the Illinois River.

Statewide annual mean temperature ranged from approximately 50.9 °F (10.5 °C) to 51.9 °F (11.1 °C), and mean precipitation ranged from 34 in. (86.6 cm) to 42.5 in. (108 cm) during the Forbes and Richardson era (Illinois State Climatologist Office 2020).

Status of Illinois Fishes

Commercial fishery operations focused on large native species were relatively common. Burr (1991) reported that of the 142 presently valid species recorded by Forbes and Richardson (1909) from Illinois waters, only 1 was a non-native species (Common Carp). By 1920 Common Carp dominated the Illinois River fishery (Forbes & Richardson 1920).

Smith Era (1910–1978)

Landcover

The trends observed in the previous era continued into the late twentieth century. Nearly 1,100 reservoirs were constructed during this era, including most of Illinois's largest reservoirs, like Lake Shelbyville, Rend Lake, and Carlyle Lake (United States Army Corps of Engineers 2020a). Smaller rivers also were dammed for flood control and water supply as the human population of the state increased. Lock-and-dam systems installed to improve navigation along the Illinois River raised river stages by several feet (Thompson 2002). These efforts included 6 dams constructed in the 1930s to create a 9-foot-deep commercial navigation channel (Havera 1989). Dams raised Illinois River flood stages resulting in communities along the river initiating additional levee and pump-station construction to prevent flooding and provide further draining of the floodplain for agriculture (Miner & Miller 2010). Similar lock-and-dam projects occurred on the Ohio and Mississippi Rivers bordering Illinois. Four such structures were completed in the Illinois portion of the Ohio River in the 1920s (United States Army Corps of Engineers 2020b). The earliest locks-and-dams in the Mississippi River were operational in the early 1900s, and by 1953 there were 15 in the Illinois portion of the Mississippi.

Further expansion of drain tiling lowered water tables and increased the amount of land suitable for agriculture while continuing to alter the way water moved and was stored on the landscape. Stream channels were modified to handle increased runoff (figure 12), and associated sediment loads resulted in dramatically altered flow regimes and bed substrates (Webb 2002).

Marshes, forested floodplain lakes, and prairies continued to be converted into farmland and other highly managed uses. By the 1930s over 30% of all Illinois's farmland was drained (McCorvie & Lant 1993). Drainage records from Champaign County mirror statewide patterns, where by 1900, 36% of the county had drainage improvements, and most marshes and natural ponds were removed to convert the county's marshy prairies to well-drained farmland; by 1960 this had increased to 82% (Larimore & Smith 1963). Agricultural production intensified through the use of heavy machinery, agricultural chemicals (i.e., fertilizers, herbicides, insecticides), and new plant varieties. These changes led to the loss or modification of many aquatic habitats. Smith (1979) noted that decreased soil water-storage capacity due to sediment loss led to greatly increased flow variability even in once relatively stable streams.

Status of Illinois Fishes

Landscape modifications continued to shift the availability and characteristics of aquatic habitats throughout Illinois. Intensive agriculture, deforestation, and development continued to degrade stream quality (Page & Jeffords 1991). Stream conditions shifted during this era toward wider and shallower wetted channels, decreases in gravel substrates with concurrent increases in silt and sand, and less aquatic vegetation (Larimore & Smith 1963). These changes decreased habitat availability for many fishes but increased suitable habitat for species that were tolerant to these conditions (Burr 1991).

During this era there were 199 fish species known from Illinois including 13 non-native species (Smith 1971). Seventy native species exhibited patterns of range reduction or extirpation since the Forbes and Richardson era (Smith 1971). Some non-native species had been introduced to provide sportfishing opportunities (e.g., Brown Trout, Pacific Salmon) or assist with aquaculture (e.g., Grass Carp) while others were accidental introductions (e.g., Alewife, Rainbow Smelt, Sea Lamprey). Seven factors were considered responsible for the loss of native species from Illinois waters: siltation, drainage of wetlands, desiccation during drought, species interactions, pollution, loss of habitat from dams and impoundments, and increased water temperatures (Smith 1971).

Passage of environmental protections, such as the Clean Water Act (Federal Water Pollution Control Act of 1972, 33 U.S.C. §1251 et seq.) and the Endangered Species Act of 1973 (16 U.S.C. §1531 et seq.), as well as the establishment of the United States Environmental Protection Agency in 1970 (Exec. Order No. 110.2). occurred during this era. Efforts associated with these actions began to improve conditions for Illinois fishes as implementation occurred.

Smith (1979) noted that at the time of establishment of the Endangered Species Act, no Illinois fish species qualified for listing as endangered, and only 1 species (Lake Sturgeon) qualified as threatened. Despite this he continued with a brief discussion of rarity and restricted distributions of some Illinois fishes and listed 6 species as "seriously decimated" in Illinois (Lake Sturgeon, Alligator Gar, Cisco, Pallid Shiner, Bluehead Shiner, Bluebreast Darter) and 13 species as "decimated to a somewhat lesser degree" (Lake Trout, Pugnose Shiner, Blacknose Shiner, Blackchin Shiner, Blacktail Shiner, Largescale Stoneroller, River Redhorse, Northern Madtom, Banded Killifish, Bantam Sunfish, Western Sand Darter, Eastern Sand Darter, Harlequin Darter). Twenty-one other species were noted to

Figure 12. Channelized stream in the Kaskaskia River basin in central Illinois. Note the flowing drainage tile on the right bank. Photo by B. Metzke.

"have long been either rare or extremely restricted in distribution" but were not considered to be threatened at that time (Least Brook Lamprey, Northern Brook Lamprey, Pallid Sturgeon, Lake Chub, Alabama Shad, Lake Whitefish, Sturgeon Chub, Sicklefin Chub, Bigeye Shiner, Ironcolor Shiner, Weed Shiner, River Chub, Longnose Sucker, Shawnee Hills Cavefish, Northern Studfish, Slimy Sculpin, Fourhorn Sculpin, Spotted Sunfish, Banded Pygmy Sunfish, Starhead Topminnow, Cypress Darter).

Contemporary Era (1979–2019)

Landcover

The contemporary landscape is dominated by human uses focused on agricultural, residential, and industrial purposes. Illinois land cover comprises approximately 64.8% agriculture, 17.1% woodland, 13.3% urban, and 2.5% wetland. Eleven metropolitan areas have a population greater than 100,000. Modification of the landscape has continued into the contemporary era with land conversion now targeting areas currently developed as agriculture, rather than natural lands, for residential and suburban uses (Pierce et al. 2014). Many aquatic habitats have been lost or degraded despite having large quantities of water available on the landscape (Burr 1991) as indicated by less than 1% of the state being currently maintained in a high-quality natural state (White 1978).

Wetlands currently occur on approximately 2.5% of the total land area in Illinois as a result of over 7.9 million acres (31,970 km²) being drained since the 1800s (Miner & Miller 2010). Most wetlands were converted into farmland, as discussed earlier. This loss has occurred throughout the state with nearly half of counties losing more than 90% of their wetland area. Floodplain forest wetlands make up the majority of existing wetlands in Illinois despite many having been cleared, leveed, ditched, or hydrologically modified (Miner & Miller 2010).

Annual mean daily temperature ranges from 57.7 °F (14.3 °C) in southern Illinois to 47.1 °F (8.4 °C) in northern Illinois (Illinois State Water Survey 2011). The northern portion of the state averages 33 in. (84 cm) of precipitation each year, which increases to over 48 in. (122 cm) in southern Illinois (Illinois State Water Survey 2011).

Status of Illinois Fishes

Human activities are as much a factor in the contemporary composition of fish assemblages in Illinois's waters as is their evolutionary and geological history (Page 1991). Major threats recognized early in this era, such as siltation, drainage of floodplain lakes, introductions of exotic species, pollution, and channelization (Havera 1989, Page 1991, Smith 1979), are still problematic although the importance of additional threats are now being recognized (e.g., climate change, pharmaceuticals).

Smith (1979) speculated that the most damaging factors for the fish fauna in Illinois would be "excessive siltation, additional stream impoundments, and conversion of more large rivers into barge canals." While siltation continues to be a major problem in many Illinois streams, the development of large impoundments and conversion of rivers to barge canals have not generally occurred.

Smith did not foresee the magnitude of 2 problems that are now considered primary stressors to freshwater fishes: non-native species and climate change.

Burr (1991) summarized broad differences in the Illinois fish fauna as described by Forbes and Richardson (1909), Smith (1979), and more recent knowledge through 1990. Many species were observed to have experienced range reductions, although most sport-fishes appeared to have had their ranges expanded via purposeful introductions. The total number of Illinois fish species in 1990 was 209 including 187 native species and 22 non-native species (Burr 1991). Despite being largely unknown in Illinois during the Smith era, Sharpbellies (i.e., Bighead Carp and Silver Carp) were specifically mentioned as becoming widespread in big rivers and reservoirs (Burr 1991). Currently there are 24 established non-native species in Illinois. Non-native species recorded since Smith (1979) are White Perch, Striped Mullet, Inland Silverside, Bleeding Shiner, Threespine Stickleback, Black Carp, Silver Carp, Bighead Carp, Round Goby, and Oriental Weatherfish.

Retzer (2005) found an average loss of over 8 species per basin over a 100-year period when comparing fish species richness in 7 Illinois river basins driven largely by reduced numbers of minnow species (Leuciscidae). However, between the 1980s and 1990s, 5 basins exhibited an increase in species richness driven by suckers, sunfishes, and darters, presumably as a result of improved water quality (Retzer 2005).

In 2020 there were 19 endangered and 17 threatened fish species in Illinois (Illinois Endangered Species Protection Board 2020). All 6 of the species considered "seriously decimated" and 10 of 13 species considered "decimated to a somewhat lesser degree" in the Smith era are currently either state-listed or considered extirpated in Illinois. Thirteen of the 22 species that were identified as rare or with restricted distributions in Smith (1979) are currently state-listed as threatened (5 species) or endangered (8 species). Seven species are considered extirpated: Ohio Lamprey, Bluehead Shiner, Rosefin Shiner, Channel Darter, Gilt Darter, Stargazing Darter, and Spoonhead Sculpin. Smith (1979) listed Blackfin Cisco, Greater Redhorse, Cypress Minnow, Muskellunge, and Crystal Darter as extirpated, but contemporary records confirm their presence in Illinois, although Blackfin Cisco has not been recorded in Illinois since 1979.

A HISTORICAL PERSPECTIVE OF ILLINOIS FISHES

Fish Fossil Record

The ichthyofaunal history of Illinois dates back millions of years. Between 500 and 325 million years ago (mya), Illinois was covered by an expansive tropical ocean (Wiggers 1997). Glaciation and erosion have resulted in a sparse fish fossil record from that period, yet paleontological discoveries offer some insight into the prehistoric diversity of Illinois fishes. The fossil fish collection at the Field Museum of Natural History in Chicago contains more than 2,500 specimens originating from 35 Illinois counties (Field Museum of Natural History 2019), although most were recovered from northeastern Illinois (Grundy, Kankakee, Vermilion, and Will Counties). Specimens represent cartilaginous and bony fishes, including at least 46 families of sharks, lampreys, hagfishes, lobe-finned fishes, ray-finned fishes, and armored fishes from both marine and freshwater environments. The earliest specimens date to the Middle Devonian Period (~398–383 mya), but the majority are from the Pennsylvanian Period (~323–299 mya). These discoveries give a glimpse into the diversity of Illinois's prehistoric fishes and allow for inferences regarding characteristics of prehistoric aquatic environments.

19th-Century Exploitation of Fishes

Archeological findings and written accounts from European explorers and settlers provide a glimpse into the fishes of Illinois prior to focused scientific surveys. Some of the earliest zooarchaeological records in Illinois come from the American Bottom region along the Mississippi River near what is now the East St. Louis metropolitan area (Madison and St. Clair Counties). Exploitation of fishes by indigenous peoples in this region began in the Late Archaic Period as early as 8000 BCE and extended through the abandonment of the region in the 1500s CE (Kelly & Cross 1984, McElrath et al. 1984). More than 30 taxa of fishes have been identified from excavation sites in the American Bottom, including sturgeons (Acipenseridae), gars (Lepisosteidae), Bowfin, shads (Clupeidae), buffalos and carpsuckers (Catostomidae), catfishes (Ictaluridae), pikes (Esocidae), sunfishes (Centrarchidae), Walleye, and Freshwater Drum (Kelly 1997, Kelly & Cross 1984). At least 30 species with a composition similar to that of the American Bottom were identified from excavation sites in the lower Illinois River valley (Parmalee et al. 1972, Styles 1981). Of note, Northern Pike and River Redhorse were found at these sites, but neither species has been recorded in this portion of the Illinois River in the history of scientific surveys. Inventories at both the American Bottom and lower Illinois River valley sites indicate fishes were the most frequently exploited faunal resource. Indigenous peoples captured fishes using nets, traps, and hooks made of bone. Early European explorers recorded consumption of catfishes and Freshwater Drum by the indigenous inhabitants of the Mississippi River valley in Illinois (Carlander 1954). In the 1830s a record of catch at the mouth of the Illinois River included what are likely basses (*Micropterus* spp. and *Morone* spp.) and catfishes (*Ameiurus* spp.) (Carlander 1954). Sportfishing and fish market reports from the Mississippi River in the mid-1800s list gars (likely *Lepisosteus* spp.), what may have been Walleye or Yellow Perch, and Buffalo (*Ictiobus* spp.) as commonly harvested (Carlander 1954).

Lake Michigan has provided sustentive opportunities for centuries (Tanner 1987). In the second half of the 19th century, Lake Michigan supported a robust Lake Whitefish commercial fishery, and much of the catch was distributed to interior markets through the Chicago transportation network (Bogue 2000); however, overfishing and mortality caused by the non-native parasitic Sea Lamprey decimated the lower Lake Michigan population by the 1930s (Baldwin et al. 2006), and both Forbes & Richardson (1909) and Smith (1979) considered the species extirpated or nearly so. Lake Sturgeon and Lake Trout have exhibited a similar pattern of prevalence and decline in southern Lake Michigan due to exploitation and introductions of non-native species (Baldwin et al. 2006, Bogue 2000). The Lake Sturgeon is extirpated from the Illinois portion of Lake Michigan, while the Lake Trout remains in the lake, at least in part due to frequent stocking.

Ichthyofaunal Surveys

The rich history of ichthyological investigations in Illinois has been reviewed, in part, by Bennett (1958), Smith (1979), Burr (1991, 1997), Sabaj et al. (1997), and Burr and Page (2009). The history of fisheries science in Illinois was reviewed by Wahl et al. (2009) and is mentioned only briefly here.

By the time the Illinois Natural History Society, Illinois's legislatively sanctioned natural history museum and research institution, was established in 1858, about three-fourths of the Illinois fish fauna had been named and described by such distinguished naturalists as Samuel L. Mitchill (1764–1831), Charles A. Lesueur (1778–1846), Constantine Samuel Rafinesque (1783–1840), Jared P. Kirtland

Table 1. Fish species originally discovered and described from Illinois waters

Current scientific name	Original scientific name	Common name	Author	Year described	Type locality
Cottus ricei	*Cottopsis ricei*	Spoonhead Sculpin	Nelson	1876	Lake Michigan, Evanston, Cook County
Crystallaria asprella	*Pleurolepis asprellus*	Crystal Darter	Jordan	1878	tributary of Mississippi River, Warsaw, Hancock County
Etheostoma asprigene	*Poecilichthys asprigenis*	Mud Darter	Forbes in Jordan	1878	near Pekin, Tazewell County
Etheostoma caeruleum	*Etheostoma caerulea*	Rainbow Darter	Storer	1845	Fox River
Etheostoma kennicotti	*Catonotus kennicotti*	Stripetail Darter	Putnam	1863	Union County
Forbesichthys papilliferus	*Chologaster papilliferus*	Shawnee Hills Cavefish	Forbes	1882	spring, Union County
Fundulus dispar	*Zygonectes dispar*	Starhead Topminnow	Agassiz	1854	near East St. Louis, river unknown
Hybognathus nuchalis	*Hybognathus nuchalis*	Mississippi Silvery Minnow	Agassiz	1855	Mississippi River, Quincy, Adams County
Ichthyomyzon castaneus	*Ichthyomyzon castaneus*	Chestnut Lamprey	Girard	1858	Near Galena, Jo Daviess County
Lepomis symmetricus	*Lepomis symmetricus*	Bantam Sunfish	Forbes in Jordan & Gilbert	1883	Illinois River, Pekin, Tazewell County
Notropis anogenus	*Notropis anogenus*	Pugnose Shiner	Forbes	1885	Fox River, McHenry, McHenry County
Notropis nubilus	*Alburnops nubilus*	Ozark Minnow	Forbes in Jordan	1878	Rock River, Ogle County
Noturus exilis	*Noturus exilis*	Slender Madtom	Nelson	1876	Mackinaw Creek, McLean County
Percina phoxocephala	*Etheostoma phoxocephalum*	Slenderhead Darter	Nelson	1876	Illinois River and its tributaries
Scaphirhynchus albus	*Parascaphirhynchus albus*	Pallid Sturgeon	Forbes & Richardson	1905	Mississippi River, Grafton, Jersey County

(1793–1877), David H. Storer (1804–1891), Louis Agassiz (1807–1873), Charles F. Girard (1822–1895), and Frederic W. Putnam (1839–1914). Fifteen of these species were first discovered and described from Illinois (table 1).

The first regional list of Illinois fishes, which treated 30 species of the Chicago area, was published in 1855 by Robert Kennicott (1835–1866) (figure 13). Comprehensive catalogs of fishes of the entire state were composed in 1876 by Edward W. Nelson (1855–1934) (figure 14), in 1878 by David Starr Jordan (1851–1931) (figure 15), in 1884 by Stephen A. Forbes (1844–1930) (figure 16), and in 1903 by Thomas Large (birth and death dates unknown). Thomas L. Hankinson (1876–1935) was active in life-history investigations of fishes from 1902 to 1919 during his tenure at Eastern Illinois State Normal College (now Eastern Illinois University).

Intensive ichthyological surveys began with Forbes, the first director of the Illinois State Laboratory of Natural History (until 1877 the Illinois Natural History Society), initially located in Normal and moved to Urbana-Champaign in 1885. Forbes had entertained the idea of a seminal work on Illinois fishes for many years (Croker 2001), especially after establishing a biological field station on the Illinois River at Havana in 1894. Peak survey effort occurred between 1899 and 1901 when horse-drawn wagon parties were sent to approximately 475 localities representing 93 counties in Illinois, resulting in collection of more than 200,000 specimens (Forbes & Richardson 1909). Methods of collecting used by investigators at this time were mostly nets, seines, traps, and angling. Teams of collectors were also sent to surrounding states for surveys of regions adjacent to Illinois. Forbes solicited the aid of high school teachers by requesting they collect fishes in local waters and send them for identification. He would return an identified, synoptic set of fishes for teaching purposes.

Forbes published 24 papers on fishes and ecology in the late 1800s. Among them were 10 written between 1878 and 1888 that summarized his work on the food of fishes, his classic *The Lake as a Microcosm* (Forbes 1887), which is credited by many with initiating the field of ecology, and a paper using a statistical approach

Figure 13. Robert Kennicott (ca. 1860).

Figure 14. Edward W. Nelson (ca. 1920).

Figure 15. David Starr Jordan (1892).

Figure 16. Stephen A. Forbes (1880).

Figure 17. Phillip W. Smith (1959).

to Illinois fish distribution (Forbes 1907). Croker (2001) considered Forbes to be the founder of American ecology and the most significant contributor to that field in the late 19th and early 20th centuries. Forbes and Robert Earl Richardson (1877–1935) also published several works on fish ecology and studies of the Illinois River, including on its pollution. Forbes provided original descriptions of 5 species of fishes: Ozark Minnow (Forbes *in* Jordan 1878), Mud Darter (Forbes *in* Jordan 1878), Bantam Sunfish (Forbes *in* Jordan & Gilbert 1883), Pugnose Shiner (Forbes 1885), and Pallid Sturgeon (Forbes & Richardson 1905). The Pallid Sturgeon is the last species described with a type locality in Illinois and is presently recognized as federally endangered.

Thirty years of survey and cataloging effort materialized as *The Fishes of Illinois*, authored by Forbes and his assistant and colleague, Richardson. Although no publication date is given in the volume and was assumed by many subsequent authors, including Smith (1979), to be 1908, Sabaj et al. (1997) clarified its proper date as 1909. A separate atlas of 103 maps, 98 of which depict species distributions, accompanied the volume. A total of 142 currently recognized species were recorded by Forbes and Richardson, and only 1, Common Carp, was an established non-native species. About 20,000 specimens used to develop distribution maps in *The Fishes of Illinois* were deposited in the collection of the Illinois State Laboratory of Natural History. At the time, *The Fishes of Illinois* was considered by many to be the best regional publication on fishes in North America and was the first to use dot-point distribution maps. Exceptionally skillful

watercolor images of 52 species, some never published in color, were crafted by Lydia (Hart) Green and Charlotte M. Pinkerton and made the book even more appealing. Most copies of the book were burned in a warehouse fire, and a second edition was produced in 1920. Few copies of the 1909 edition remain. Unfortunately, after the death of both Forbes and Richardson, their fish collections housed at the Illinois Natural History Survey (INHS; until 1917 Illinois State Laboratory of Natural History) fell into mismanagement, and many valuable specimens were discarded.

Although no statewide survey efforts occurred in the 4 decades following *The Fishes of Illinois* (Forbes & Richardson 1909), several ichthyologists completed regional surveys or contributed to inventories of Illinois fishes. Seth Meek (1859–1914) and Samuel Hildebrand (1883–1949) published a study in 1910 on the fishes of the Chicago region (117 species recognized), and D. J. O'Donnell published a list in 1935 of Illinois fishes, which added a few species to the known fauna of the state. Many collections made during the 1940s by Aden C. Bauman (1911–1947), a student of Carl L. Hubbs (1894–1979) at the University of Michigan, contributed many significant records (e.g., Cypress Minnow), particularly from the southern half of the state. Bauman's collections are at the University of Michigan Museum of Zoology.

In the 1950s Philip W. Smith (1921–1986) (figure 17), author of *The Amphibians and Reptiles of Illinois* (1961), initiated a statewide survey of Illinois fishes. This task provided an opportunity for comparing contemporaneous distributional data with the classic work of Forbes and Richardson (1909, 1920). The bulk of Smith's fieldwork began in the summer of 1962 and continued until the mid-1970s. During this time Smith coauthored an account of the fishes of Champaign County (Larimore & Smith 1963), an annotated list of Illinois fishes (Smith 1965), an assessment of Illinois streams based on fish distribution data (Smith 1971), and a key to Illinois fishes (Smith 1973). After nearly 20 years of surveying, Smith published the second *Fishes of Illinois* in 1979. The book summarizes the identifying features, biology, and distribution of the Illinois fish fauna and included color plates of 34 species by Alice P. Prickett. Only 2 color paintings in Smith (1979) repeated illustrations that Forbes and Richardson used. Smith's distribution maps distinguish pre-1909 records (Forbes & Richardson 1909) from post-1950 records, providing visual evidence of changes in distribution of individual species over time. Smith's book was reprinted with a new cover in 2002.

Smith and his associates found 199 species in Illinois (a late discovery, Inland Silverside, first recorded in 1978, brought the number to 200), made over 3,100 collections from over 2,000 localities from all 102 counties, and deposited as vouchers about 400,000 specimens at the INHS. When he compared his records with those of Forbes and Richardson (1909), Smith (1971) found that about 70 fishes clearly showed decreases in their distribution, with 9 having been extirpated from the state. Smith also recorded 13 established non-native species in Illinois.

Smith relied heavily on Illinois Department of Conservation (now the Illinois Department of Natural Resources, IDNR) surveys from which specimens were turned over for identification and deposition

at the INHS. Department of Conservation biologists were of great assistance in making collections along the shore of Lake Michigan and
in the bordering Mississippi and Ohio Rivers (Smith 1986). William C.
Starrett (1912–1971), director of the Stephen A. Forbes Biological
Station at Havana from 1949 to 1971, conducted numerous surveys
along the Illinois River and deposited fish specimens at the INHS.
Smith also made efforts to survey areas of states bordering Illinois
to detect species of potential occurrence but not yet recorded in
Illinois. Collecting methods used during Smith's surveys were seines
of small mesh size (⅛ inch), boat electrofishers, and early models
of backpack electrofishers. Ichthyotoxins (e.g., rotenone) were used
infrequently, and the electric seine was used beginning in the 1960s.

Since the publication of Smith (1979), state fish biologists have
continued to survey the Illinois ichthyofauna. Particularly active have
been ichthyologists and fish biologists from the state's universities,
the INHS, the IDNR, the Illinois Environmental Protection Agency,
federal agencies, local governments, and privately owned consulting
firms. Additional discoveries of non-native species (e.g., Sharpbellies), native species previously unreported (e.g., Taillight Shiner), and
the expansion into Illinois of more southern species (e.g., Striped
Mullet) emphasize the dynamic nature of the Illinois ichthyofauna
and the importance of continued sampling of fishes even in presumably well-surveyed areas. Long-term monitoring stations are
present on several of the large rivers in Illinois, and there has been
regular surveying of the Illinois River mainstem since the Forbes and
Richardson era. Among the most widespread methods for sampling
today are the Missouri Trawl, a variety of nets, electric seines, and
backpack, barge, and boat-mounted electrofishers.

FISH MORPHOLOGY

Figures 18 and 19 illustrate morphological terms, structures, counts, and measurements used to identify fishes (see this book's glossary for definitions of terms). Measurements are always made in a straight line (not along a body contour) and expressed as a proportion of another measurement (e.g., total length) to facilitate comparison. For further instruction on how to perform counts and measurements, see *Fishes of the Great Lakes Region* (Hubbs & Lagler 2004).

Fishes have median fins (dorsal, caudal, and anal) and paired fins (pectoral and pelvic). In some fishes (e.g., most minnows), the dorsal fin is supported entirely by flexible, segmented "soft" rays. In others, only inflexible spiny rays ("spines") occur in the anterior portion of the fin. The spiny part of the fin may be contiguous with or separated from the soft-rayed part. When the front section is separated or nearly separated from the soft-rayed part (e.g., in darters), the fish is said to have 2 dorsal fins. Spiny rays are referred to herein as "spines," and soft rays are referred to simply as "rays." The sunfish (at top) in figure 18 has 10 spines and 11 soft rays in the dorsal fin; the catfish has 1 large spine and 6 rays.

Likewise, the anal fin may have only soft rays or have spines (usually only 1–3) preceding the rays. Because of the marked variation in morphology, counts of anal-fin rays are made differently in different groups of fishes. In most (e.g., sunfishes), there is an obvious break between spines and rays in the anal fin, just as there is in the dorsal fin. In others (e.g., catfishes), the small (unbranched) rays at the beginning of the anal fin increase gradually in size (and become branched), and all rays are counted. In all fishes, the last anal ray is counted as 1 even if its branches are separated to the fin base (i.e., if the branches of the ray look as if they will join just below the surface, they probably do and are counted as 1 ray). The sunfish in figure 18 has 3 spines and 10 rays in the anal fin; the catfish has no spines and 32 rays in the anal fin.

The mouth is described as *terminal* if it opens at the front end of the head with the upper and lower jaws being equal or nearly equal in length. It is *upturned* if it opens above that point and *subterminal* if it opens on the underside of the head.

Lateral-scale count (or lateral-line scale count) begins just behind the head and continues along the lateral line (or along midside if the lateral line is absent) to the origin of the caudal fin (which is located by bending the caudal fin to either side and noting the crease between body and caudal fin). Scales on the caudal fin are not included in lateral-scale counts even if they are pored. Scales above and below the lateral line begin at the origin of the dorsal or anal fin, respectively, and continue diagonally to the lateral line (but do not include the lateral-line scale). A transverse scale count is a continuation of the count of scales below the lateral line diagonally upward (including the lateral-line scale) to the dorsal fin. Scales around the caudal peduncle are those around the narrowest part. Predorsal scales are those along the midline of the nape from the rear of the head to the dorsal-fin origin. The number of rakers on the first gill arch can be seen in most fishes by lifting the opercle, the largest bone in the gill cover (figure 19). Gill rakers vary greatly in shape and number and are used in some groups to separate otherwise morphologically similar species. The gill-raker count is the total for the entire first arch (one closest to the gill cover) unless upper or lower limb only is specified. Branchiostegal rays are long, thin bones supporting the gill membranes on the underside of the head; all (short and long) are counted. The count given is the total number of rays on one side.

To examine pharyngeal teeth, it is necessary to remove the pharyngeal arch (which lies just posterior to the last gill arch) by lifting the gill cover (cut along the ventral side of the gill cover if necessary to loosen it from the body) and removing the arch. A scalpel or strong forceps is inserted between the arch and the bone behind it, starting at the upper angle of the gill opening and cutting down along the bone. The flesh holding the upper and lower ends of the arch in position is cut, and the arch is lifted out. Attached flesh is removed to expose the teeth. A pharyngeal arch may have 0, 1, or 2 rows of teeth on it, and counts are given for the minor and major rows on the left side followed by the counts for the major and minor rows on the right side. For example, the Weed Shiner has 2 rows of teeth on each arch, and the count is given as 2,4-4,2.

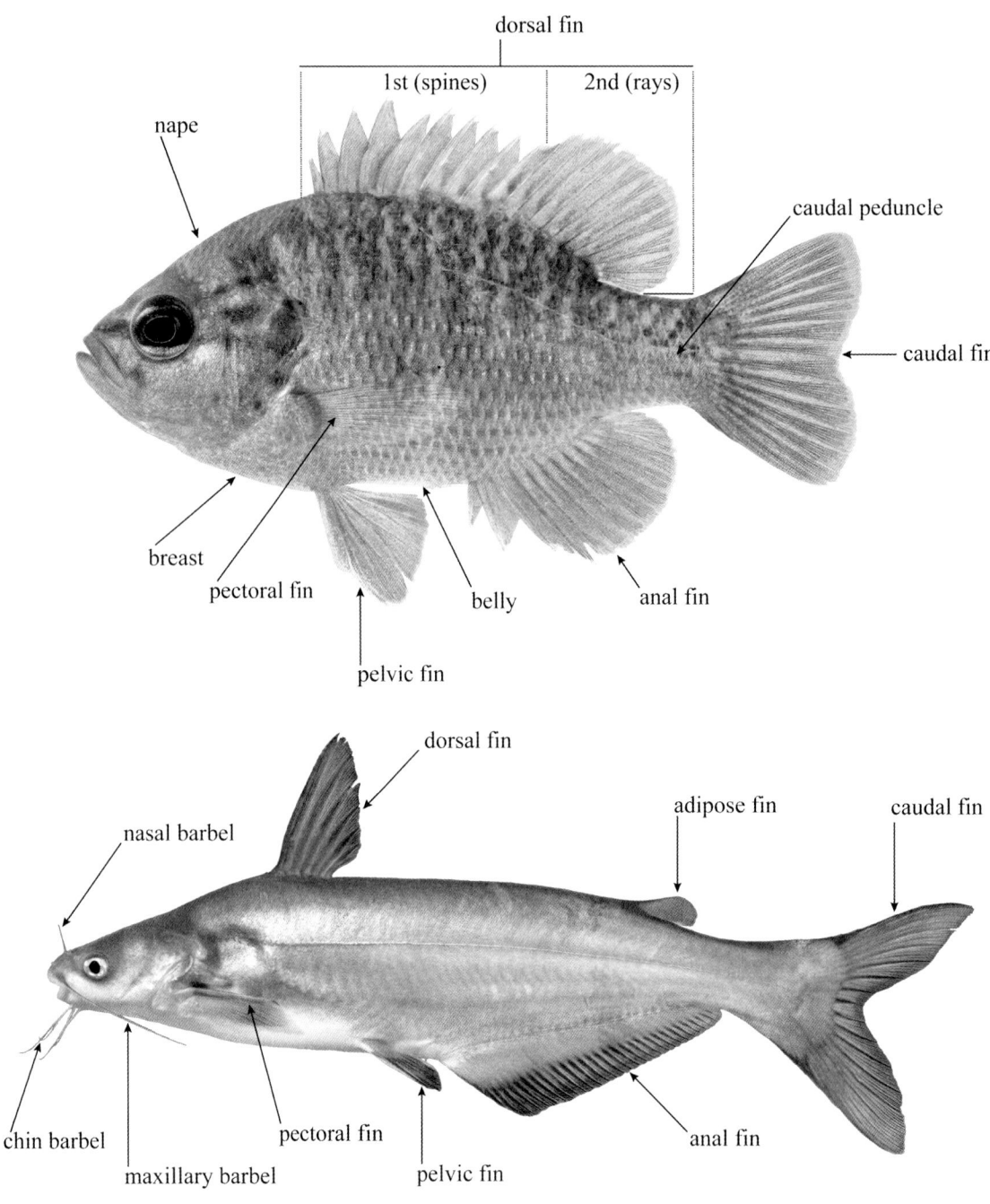

Figure 18. Terms used in the morphological descriptions of fishes.

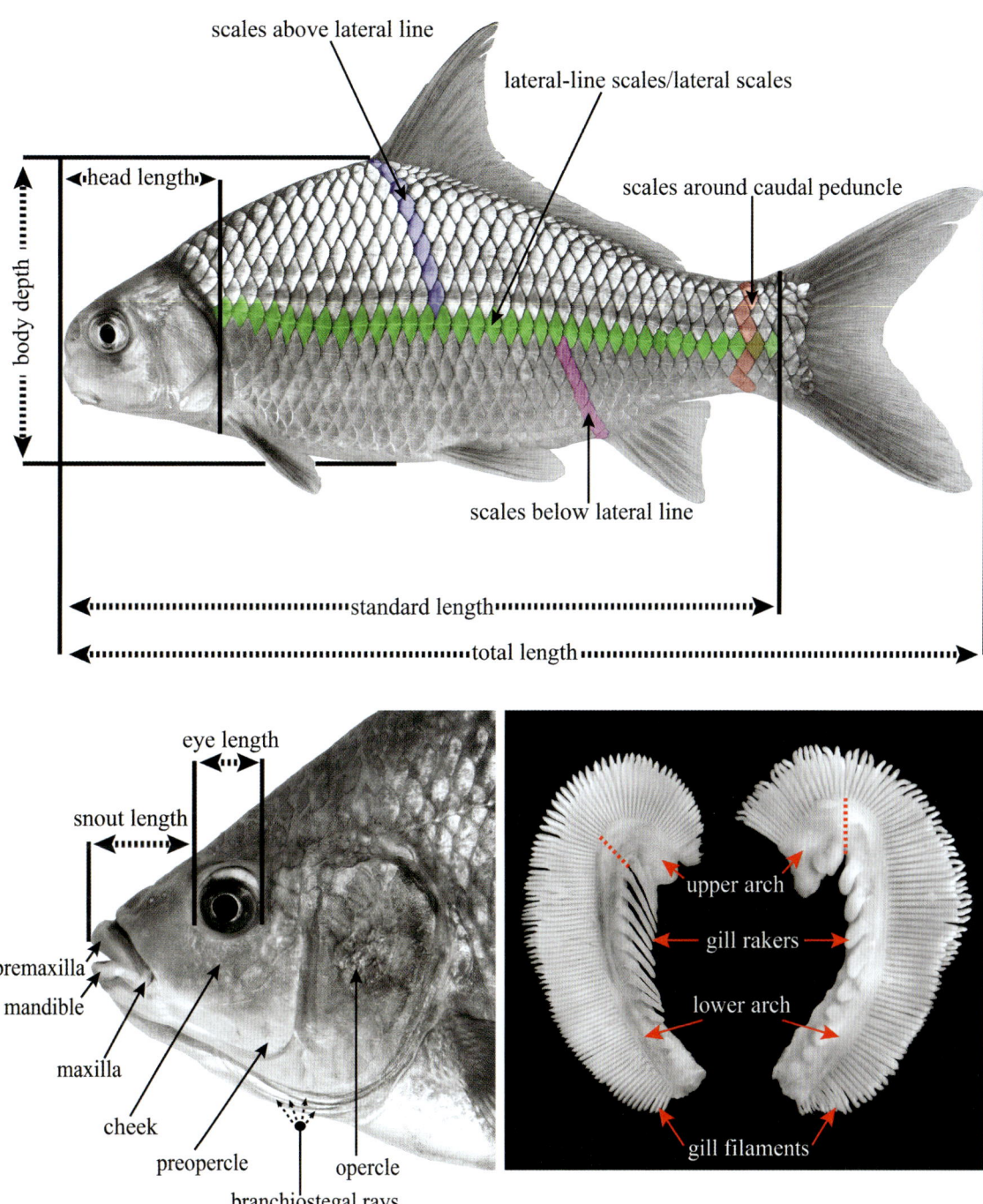

scales above lateral line

lateral-line scales/lateral scales

scales around caudal peduncle

head length

body depth

scales below lateral line

standard length

total length

eye length

snout length

premaxilla

mandible

maxilla

cheek

preopercle

opercle

branchiostegal rays

upper arch

gill rakers

lower arch

gill filaments

Figure 19. Terms, counts, and measurements used in morphological descriptions of fishes.

ORGANIZATION OF ILLINOIS FAMILY AND SPECIES ACCOUNTS

Family Accounts

Accounts for Illinois's 33 families of fishes precede associated species accounts and provide a brief overview of morphology, behavior, distribution, and diversity. Family names and arrangement follow "Eschmeyer's Catalog of Fishes" (Fricke et al. 2020). Common names of families follow Page et al. (2011), except Moronidae, which follows Fricke et al. (2020); the newly elevated cypriniform families, Leuciscidae and Xenocyprididae (Tan & Armbruster 2018), and their common names follow general use.

Species Accounts

Accounts for the 217 species treated in this book are arranged alphabetically by genus within families and by species within genera. Scientific names follow Fricke et al. (2020), with the exception of most sculpins, which were moved from *Cottus* to *Uranidea* by Kinziger et al. (2005).

Photographs

Lateral views of live individuals are shown for most species. Efforts were made to obtain images that best represent the life coloration observed in live individuals found in Illinois, but in a few cases it was necessary to use images of preserved specimens. More than one photograph is included for some species that have alternative coloration patterns in breeding individuals or in juveniles and adults. The names of the waterbody, county, and state where each photograph was taken are provided in the section "Photo Credits and Locations."

Maps

A map illustrates the distribution of each species in Illinois over 3 major time periods. Yellow squares denote records prior to 1910 (Forbes and Richardson era), orange dots are records from 1910 to 1978 (Smith era), and red dots are records after 1978 (contemporary era). More information on data used in preparation of maps is provided in the section "Methods for Mapping Species Distributions."

Identification

Each account provides a summary of morphological characteristics useful in identification. Most accounts provide a description of body shape, mouth position, fins, color, and other morphological characteristics that augment information in the dichotomous identification keys. Differences in coloration between sexes, breeding and non-breeding adults, and juveniles and adults are described when considered important for identification. Maximum total length is estimated, although Illinois specimens may not achieve this size.

Similar Species

Distinguishing characteristics are provided for other species in Illinois that may be confused with the species being described. Readers should refer to the accounts for these other species and to the dichotomous identification keys for assistance in differentiating similar species.

Distribution in Illinois

The contemporary and historical distributions of each species in Illinois are summarized with changes in distribution highlighted. Comments from previous studies regarding species distribution, in particular, those of Smith (1979) and Forbes and Richardson (1909), are included when insightful for understanding changes in distribution or frequency of occurrence.

Remarks

Relevant remarks regarding taxonomy and ecology are provided for some species, including taxonomic uncertainties, subspecies designations, presence of subspecies within Illinois, notable color or morphological variation, and hybridization with other Illinois fishes. Native distributions for non-native species and likely dates and routes of introduction are also discussed.

METHODS FOR MAPPING SPECIES DISTRIBUTIONS

A goal of this book is to document the spatiotemporal distributions of Illinois fishes. Species occurrence records (each with species name, locality, and date) were compiled from a variety of relevant and accessible sources into a single, comprehensive mapping database for development of distribution maps. Characteristics of source databases and mapping and validation procedures are described in the subsequent subsections.

Record Sources and Characteristics

The Long-Term Survey and Assessment of Large River Fishes in Illinois, or the Long-Term Electrofishing (LTEF) (figure 20) program, was initiated in 1959 to assess the Illinois and Des Plaines Rivers (Sparks & Starrett 1975). Since 1959 approximately 26 localities have been surveyed annually using 15-minute to 1-hour boat electrofishing runs (Pegg & McClelland 2004). In 2010 the LTEF program was expanded to include an additional 20 localities each on the Illinois, Des Plaines, Mississippi, Ohio, Wabash, Kankakee, and Iroquois Rivers (Michaels et al. 2011). At these additional localities, 15-minute boat electrofishing runs were conducted thrice annually in summer and fall. Gill and hoop nets were deployed at a subset of these additional localities.

The Long-Term Resource Monitoring Program (LTRMP) (figure 21) has surveyed 3 reaches of the Mississippi River and 1 reach of the Illinois River since 1989 (Ratcliff et al. 2014). Survey reaches range from 50 km to 110 km. Fishes are surveyed using 15-minute boat electrofishing runs, nets, and trawling thrice annually at 31 fixed and randomly selected localities within each of the 4 reaches.

The IDNR began recording results of fish surveys in 1952 (figure 22). Surveys between the 1950s and 1970s mainly focused on sportfishes in rivers and larger creeks. By the 1980s the construction of numerous dams and the implementation of standardized survey programs resulted in an expansion of the IDNR's surveys to more frequently include creeks, impoundments, and lakes, including Lake Michigan. The IDNR employs electrofishing, active and passive netting and, rarely, rotenone to survey fish assemblages. Many surveys are community based; however, targeted sportfish surveys also are conducted.

Vouchered-specimen holdings of 24 institutions were accessed from 2012 to 2014 by contacting curators or collection managers or by compiling fish records georeferenced under a National Science Foundation–funded project to georeference United States fish

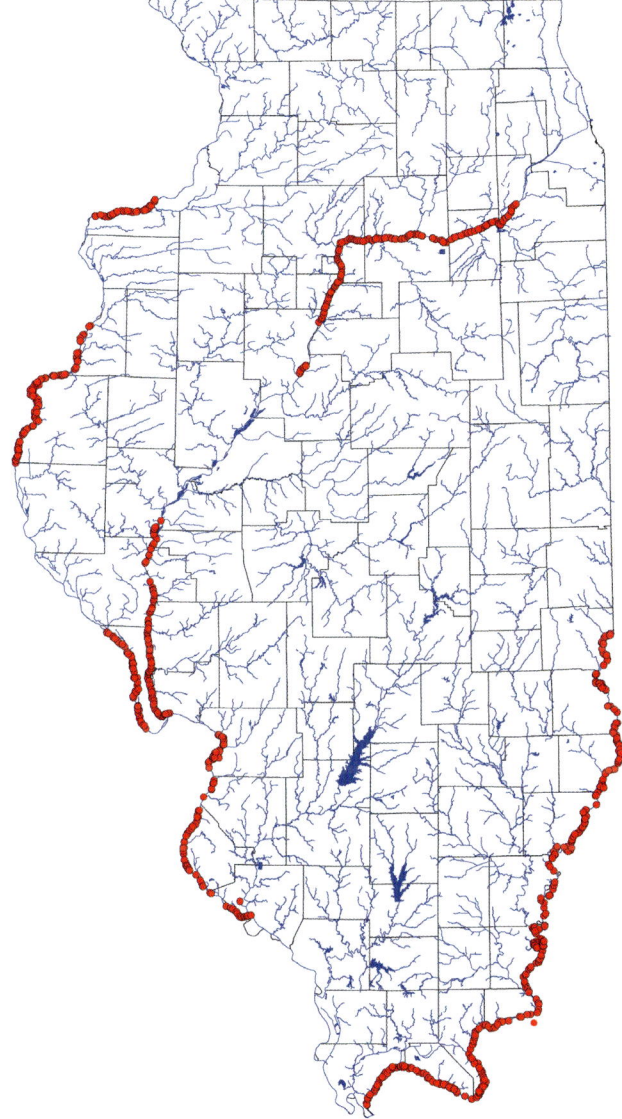

Figure 20. Long-term electrofishing (LTEF) records.

collections (table 2; figure 23). The INHS, Southern Illinois University Carbondale, and the Field Museum of Natural History contributed approximately 93% of institution records used in distribution maps. The INHS fish collection contributed the majority with more than 52,000 records. INHS specimens were collected using a variety of methods in both community and targeted surveys, and although

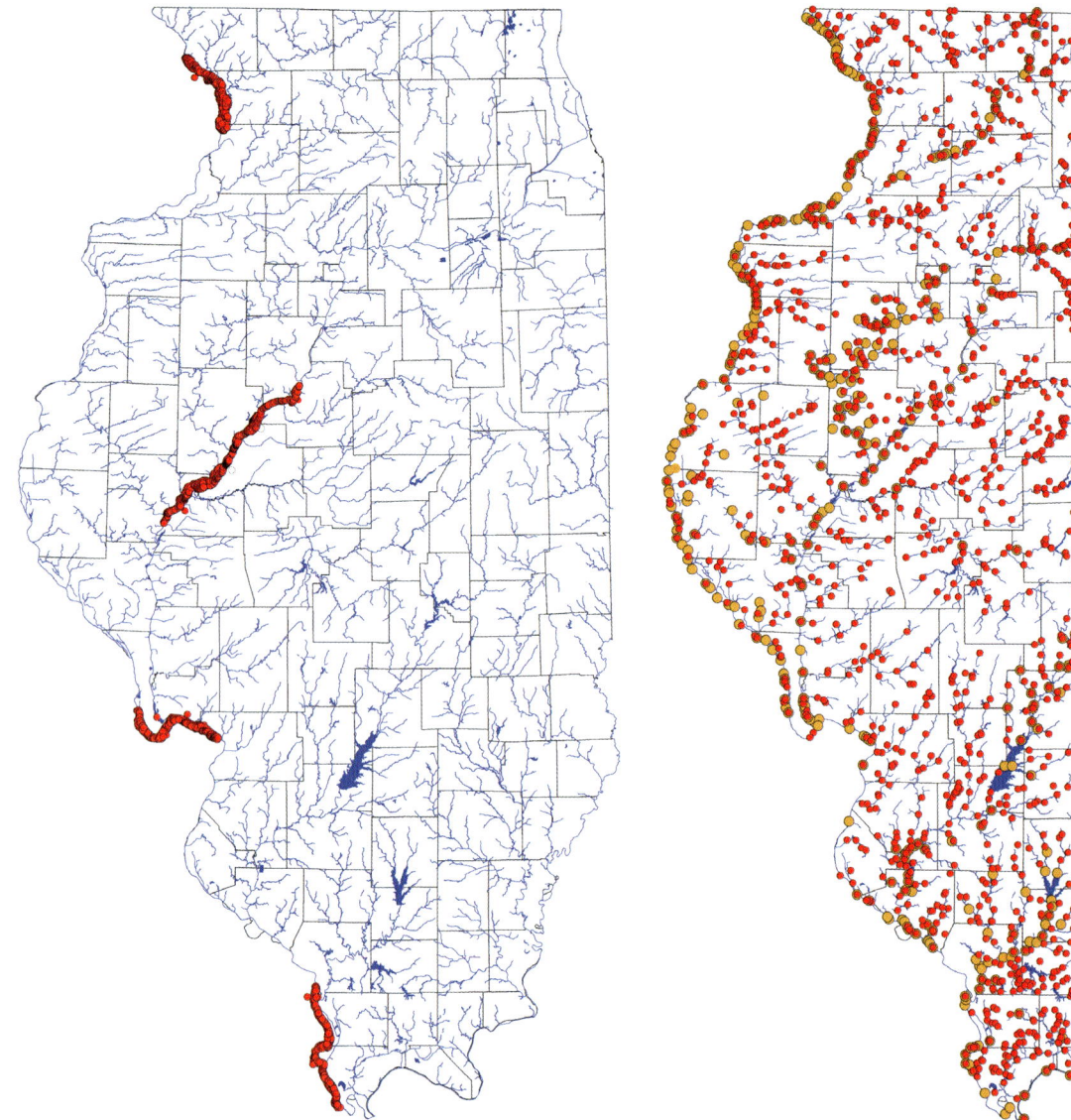

Figure 21. Long-term Resource Monitoring Program (LTRMP) records.

Figure 22. Illinois Department of Natural Resources (IDNR) records.

many specimens were collected by INHS staff, other specimens were donated to the collection by outside agencies and institutions. Many records provided by Southern Illinois University are from Illinois's southernmost counties, an area with unique aquatic habitats and species and underrepresented by most survey programs. Records from the Field Museum are concentrated in the northeastern quarter of the state.

The IDNR Natural Heritage Database (i.e., Biotics) (figure 24) is a Geographic Information System (GIS)-based tool for storing and mapping Illinois threatened and endangered species occurrence records. Records are submitted by local and statewide agencies, universities, conservation groups, and individuals. Accuracy of locality information for provisional submissions is ensured before being accepted into the database. Many Biotics entries are redundant with those of the other sources, but given the breadth of survey efforts and localities represented by Biotics records, these records filled some spatiotemporal gaps left by standardized surveys.

Mapping Process

Source database records were received in 2012 and 2013, and so the most recent record-years available ranged from 2010 to 2013 (table 3); however, more recent records that represent significant change in distributions of species were also included in the comprehensive mapping database and are noted in this book's section "Illinois Family and Species Accounts." Post-2013 records were accessed and added from the iDigBio web portal or directly from the individual that collected the specimen. More than 11.8 million individuals from 269 species and hybrids (table 3) were retrieved from sources. Only those species known to be or likely reproducing in Illinois and native species believed to be extirpated were included in the comprehensive mapping database. Hybrids were excluded. Consequently, accounts for 217 species are included in this book. The number of records for a species ranged from 1 (7 species) to 19,240 (Gizzard Shad) with a mean of 1,504 and median of 138.

Table 2. Institutional records of Illinois fishes

Institution	No. records
Academy of Natural Sciences of Drexel University	81
Canadian Museum of Nature	13
Chicago Academy of Sciences	87
Field Museum of Natural History	8,503
Harvard University Museum of Comparative Zoology	26
Illinois Natural History Survey	52,148
Kansas University Biodiversity Institute and Natural History Museum	128
Los Angeles County Museum	6
Mississippi Museum of Natural Science	19
Natural History Museum of Sweden	2
North Carolina Museum of Natural Sciences	42
Ohio State University Museum	1,844
Oklahoma Museum of Natural History	1
Royal Ontario Museum	2
Southern Illinois University Carbondale	11,678
Texas A&M University Biodiversity Research and Teaching Museum	14
Texas Natural History Collections	15
Tulane University	205
United States National Museum	35
University of Colorado Museum of Natural History	1
University of Florida, Florida Museum of Natural History	1,480
University of Michigan Museum of Zoology	1,128
University of Tennessee	209
Yale Peabody Museum of Natural History	162

Note: These records were incorporated into the comprehensive mapping database.

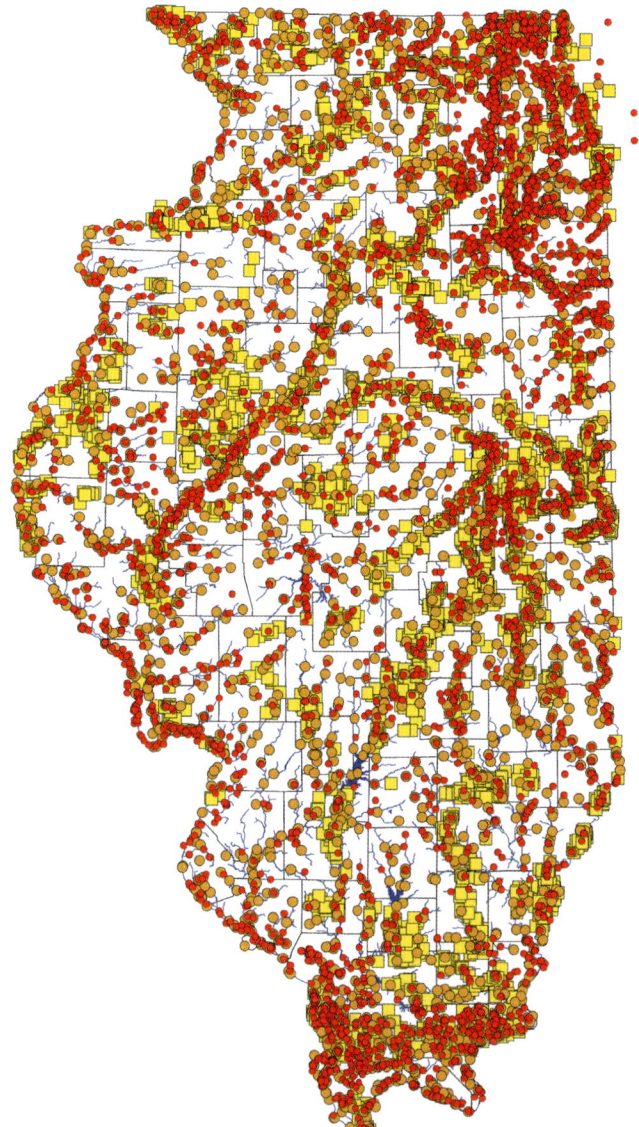

Figure 23. Institution specimen records.

To facilitate comparison with previous works describing distributions of Illinois fishes and to provide sufficient resolution for broad assessments of spatiotemporal distribution patterns, records were divided into 3 time periods or eras: those records collected before 1910 (i.e., Forbes and Richardson era), those collected between 1910 and 1978 (i.e., Smith era), and those collected after 1978 (i.e., contemporary era). The Forbes and Richardson era roughly corresponds with the survey, cataloging, and publication efforts of Forbes and Richardson (1909) (figure 25). The Smith era (figure 26) documents species distributions between Forbes and Richardson (1909) and Smith (1979). The contemporary era (figure 27) captures records between 1979 and the most recent record-year used from source databases (table 3).

Georeferenced collection localities were available for all LTEF, LTRMP, and Biotics records, but for IDNR and institutional records not georeferenced, coordinates were estimated using reference localities or georeferencing procedures. Coordinates were assigned to IDNR records either by using coordinates from synonymous reference localities (e.g., Illinois Environmental Protection Agency stations, United States Army Corps of Engineers river miles) or manual georeferencing using given verbal descriptions of sampling sites.

Coordinates could be estimated for approximately 93% of IDNR records. For institutional lots cataloged before GPS technology was available or for those where coordinates were not designated, coordinates were estimated using computer-assisted georeferencing (i.e., GEOlocate) (Rios & Bart 2010) or manually following Wieczorek et al. (2004). An estimate of uncertainty was determined for computer-assisted cases, and those with uncertainty values that equated to greater than 0.62 mi. (1.0 km) were excluded from the comprehensive mapping database. Many of the specimens from the Forbes and Richardson era in the INHS fish collection were destroyed in the 1930s, and so Forbes and Richardson (1909) and Smith (1979) were used to identify localities for those records and estimate coordinates.

Distribution maps were created in ArcMap 10.6 (Environmental Systems Research Institute [ESRI] 2017). Draft maps were reviewed to identify and remove or correct erroneous records. When possible the collector of a questionable LTEF, LTRMP, or IDNR record was con-

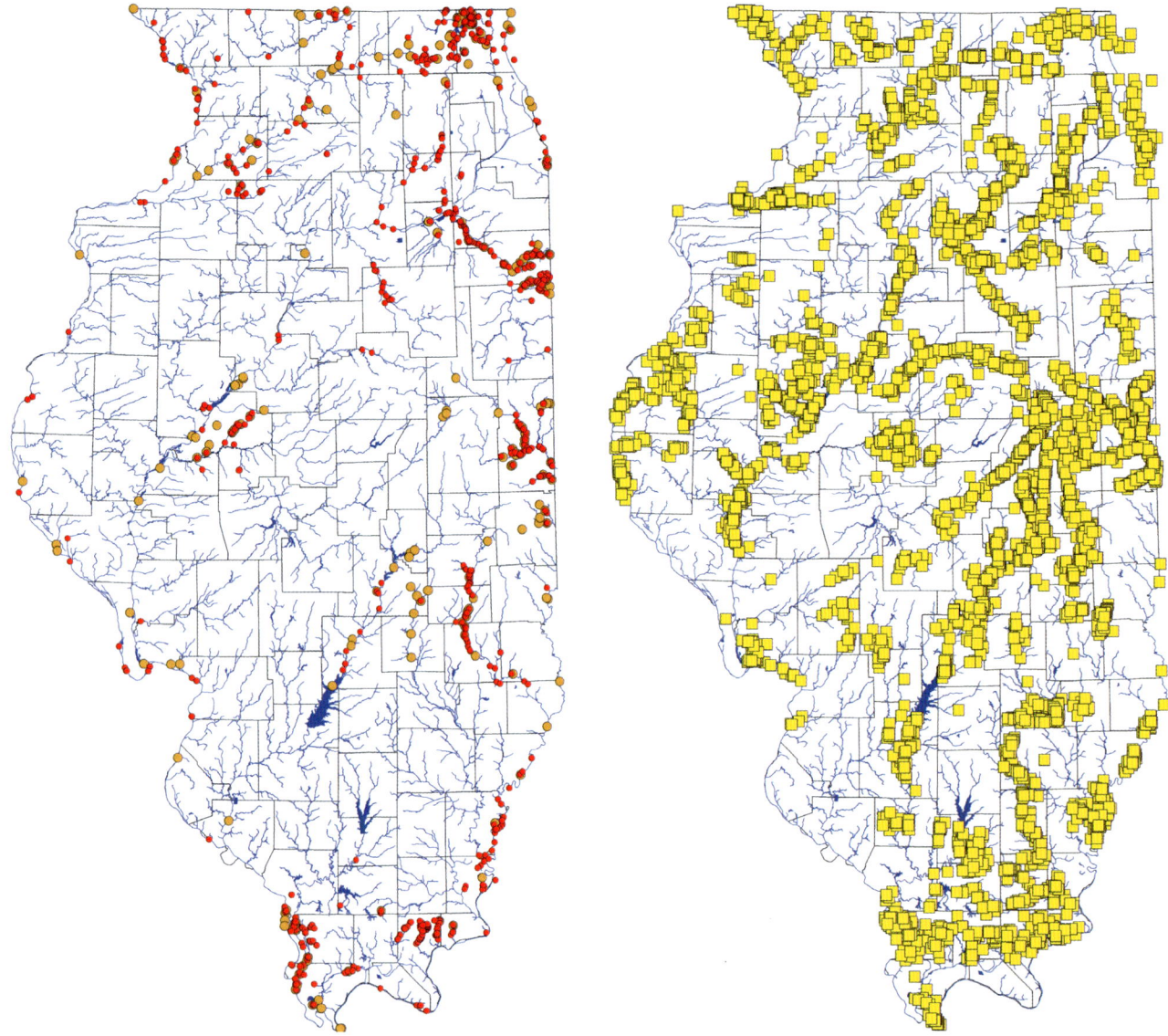

Figure 24. Illinois Department of Natural Resources Natural Heritage Database (i.e., Biotics) records.

Figure 25. Forbes and Richardson era (before 1910) records.

Table 3. Characteristics of source databases used to develop the comprehensive mapping database.

Source database	Date range*	Source database contents	Waterbody types	No. individual fish	No. records
Biotics	1882–2010	State-listed species occurrence records	Creeks, rivers, impoundments, lakes, etc.	—	1,553
IDNR	1952–2012	Community survey and targeted survey records	Creeks, rivers, impoundments, lakes, etc.	9,880,701	263,623
Institutional collections	1853–2013	Specimens	Creeks, rivers, impoundments, lakes, etc.	886,417	77,829
LTEF	1959–2012	Community survey records	Large rivers	381,934	199,228
LTRMP	1989–2010	Community survey records	Large rivers	769,181	164,895

* Spatiotemporal distribution of records is illustrated in figures 20–27. Noteworthy records documented more recently than the latest year in this range were incorporated into the comprehensive mapping database.

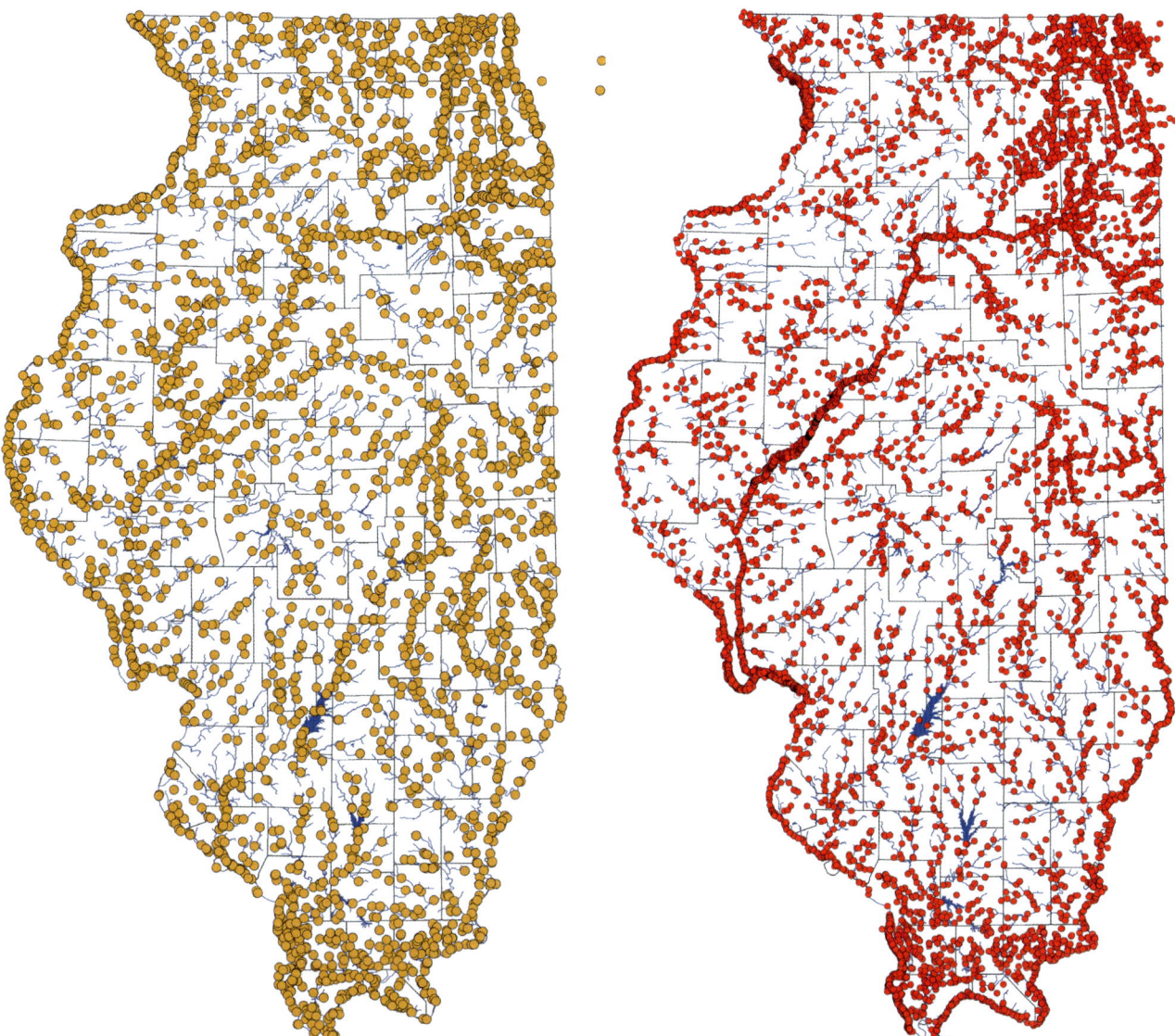

Figure 26. Smith era (1910–1978) records.

Figure 27. Contemporary era (after 1978) records.

tacted to provide validating evidence or identify the cause of error. If no corroborating evidence for a questionable LTEF, LTRMP, or IDNR record could be supplied, the record was retained only if museum records were present from a locality that supported the questionable record's validity. Questionable museum records were validated by examining the associated specimen(s). Following this validation procedure, 324,955 records were included in the comprehensive mapping database.

KEY TO THE FAMILIES
OF ILLINOIS FISHES

1a Jaws absent (mouth an oval suction cup or hood-like); pectoral and pelvic fins absent; 7 pore-like gill openings in a line behind head; single median nostril...... **Page 41**

Lampreys (Petromyzontidae)

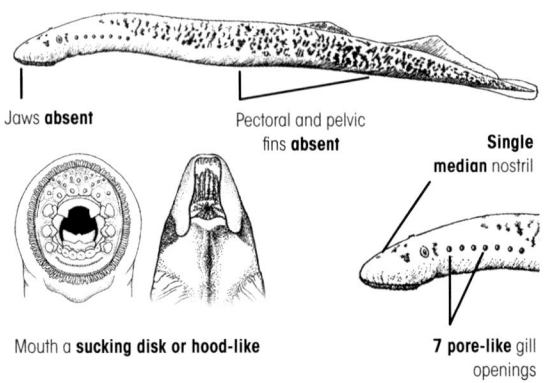

Jaws **absent**

Pectoral and pelvic fins **absent**

Single median nostril

Mouth a **sucking disk or hood-like**

7 pore-like gill openings

1b Articulated jaws present; mouth not an oval suction cup; pectoral fins present; pelvic fins usually present; one gill opening on each side of head; 2 pairs of nostrils, one on each side of head. **2**

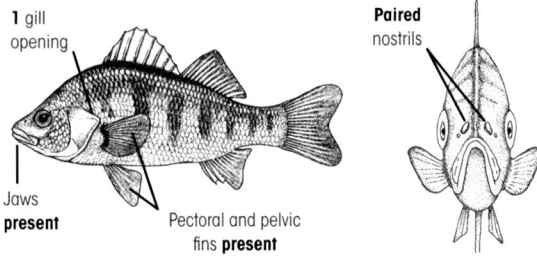

1 gill opening

Paired nostrils

Jaws **present**

Pectoral and pelvic fins **present**

2a Caudal fin deeply forked with backbone turned upward and extending nearly to tip of upper lobe of fin (heterocercal tail). **3**

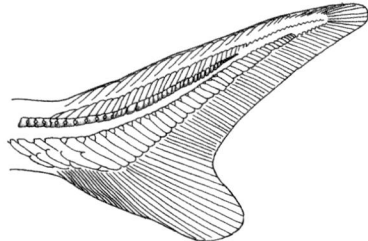

2b Caudal fin forked or unforked, but with backbone not turned upward and extending into upper lobe of fin (homocercal tail) **4**

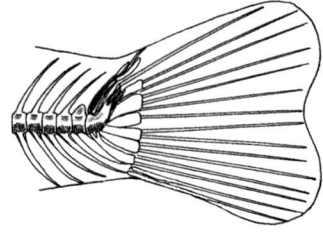

3a Snout long, shaped like a canoe paddle; gill cover long, flexible, pointed at end; body smooth without bony plates. . . . . **Page 59**

Paddlefishes (Polyodontidae)

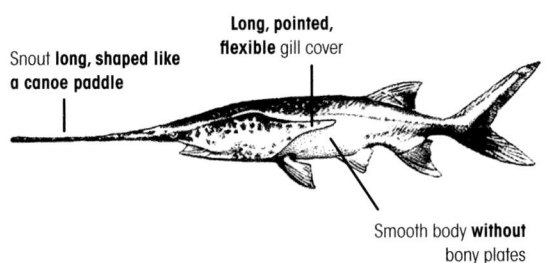

Snout **long, shaped like a canoe paddle**

Long, pointed, flexible gill cover

Smooth body **without** bony plates

3b Snout shovel-shaped or conical; gill cover short, rounded at end; body with rows of bony plates . . . . **Page 53**

Sturgeons (Acipenseridae)

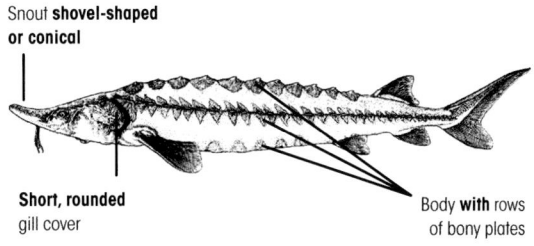

Snout **shovel-shaped or conical**

Short, rounded gill cover

Body **with** rows of bony plates

4a Jaws long with long, sharp teeth; body with a continuous sheath of hard, plate-like, diamond-shaped scales . . . . **Page 61**

Gars (Lepisosteidae)

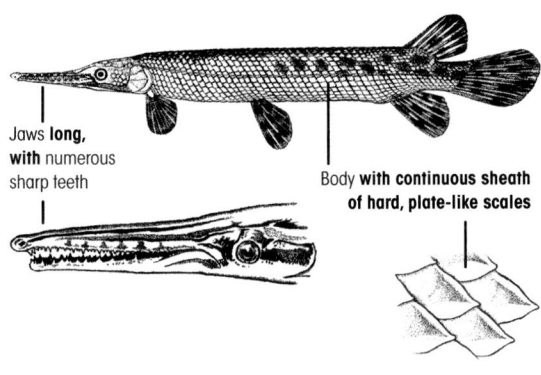

Jaws **long, with** numerous sharp teeth

Body **with continuous sheath of hard, plate-like scales**

4b Jaws short, without long, sharp teeth; body without scales or with flexible scales that overlap **5**

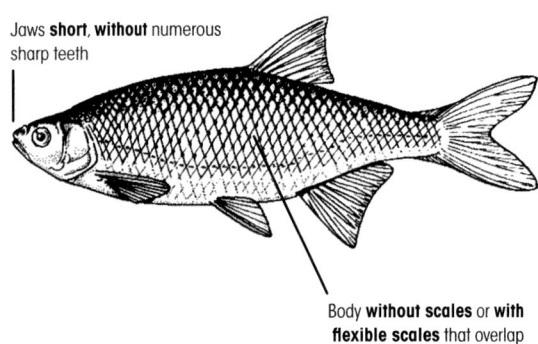

Jaws **short, without** numerous sharp teeth

Body **without scales** or with **flexible scales** that overlap

5a Adipose fin present, sometimes as a low, fleshy ridge . . .**6**

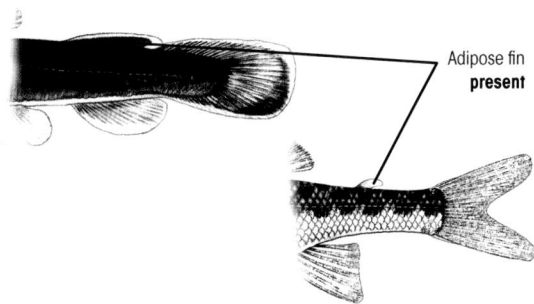

Adipose fin **present**

5b Adipose fin absent **9**

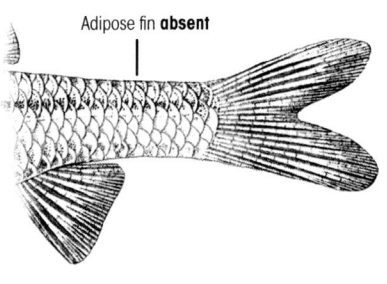

Adipose fin **absent**

6a Eight large barbels or "whiskers" around mouth; skin smooth, scales absent; pectoral fin with a strong, sharp spine at front. . . . .**Page 216**

6b No barbels or "whiskers" around mouth; scales present (small and easily overlooked in trouts and smelts); pectoral fin without a spine. **7**

North American Catfishes (Ictaluridae)

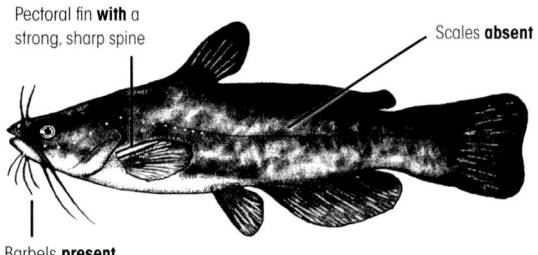

Pectoral fin **with** a strong, sharp spine

Scales **absent**

Barbels **present**

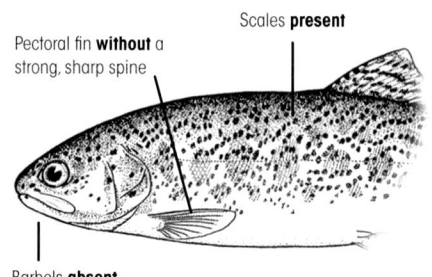

Scales **present**

Pectoral fin **without** a strong, sharp spine

Barbels **absent**

7a Mouth small, not extending behind front of eye; scales rough-edged (ctenoid); front edge (origin) of pelvic fin below middle of pectoral fin. . . . . **Page 264**

7b Mouth large, extending behind front of eye; scales smooth-edged (cycloid); front edge (origin) of pelvic fin far behind end of pectoral fin **8**

Trout-perches (Percopsidae)

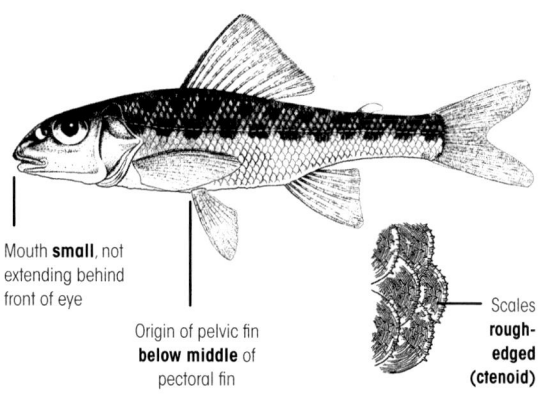

Mouth **small**, not extending behind front of eye

Origin of pelvic fin **below middle** of pectoral fin

Scales **rough-edged (ctenoid)**

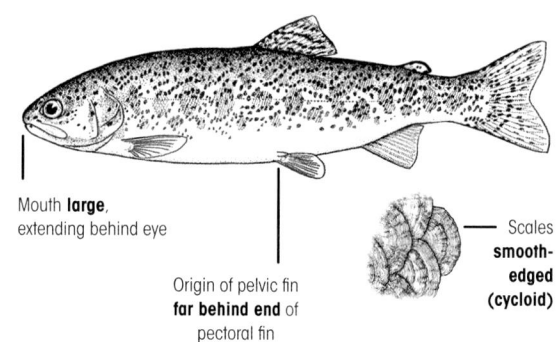

Mouth **large**, extending behind eye

Origin of pelvic fin **far behind end** of pectoral fin

Scales **smooth-edged (cycloid)**

8a Pelvic axillary process present (may be small and blunt in *Salvelinus*); scales small, with usually 70 or more in lateral series; teeth not conspicuously enlarged on jaws. . . . . **Page 245**

Trouts, Salmons, Whitefishes (Salmonidae)

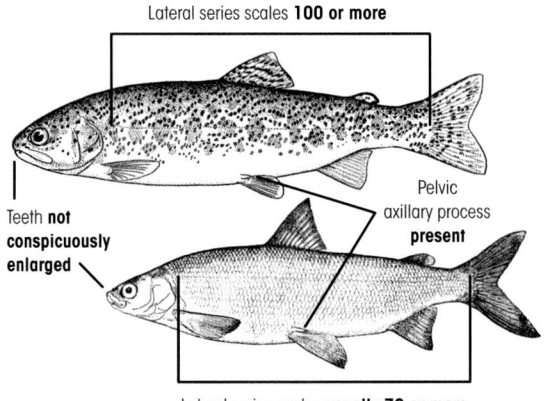

Lateral series scales **100 or more**

Teeth **not conspicuously enlarged**

Pelvic axillary process **present**

Lateral series scales **usually 70 or more**

8b Pelvic axillary process absent; scales relatively large, with about 61–70 in lateral series; teeth conspicuously enlarged on jaws . . . . **Page 262**

Smelts (Osmeridae)

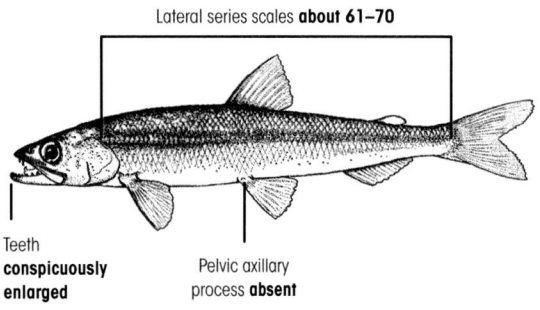

Lateral series scales **about 61–70**

Teeth **conspicuously enlarged**

Pelvic axillary process **absent**

9a Pelvic fins absent; scales present, but so small that body appears naked.. . . . . **10**

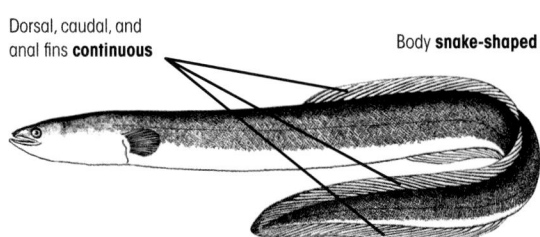

Pelvic fins **absent**

Scales **present**, but so small that body appears naked

9b Pelvic fins present; scales present or absent **11**

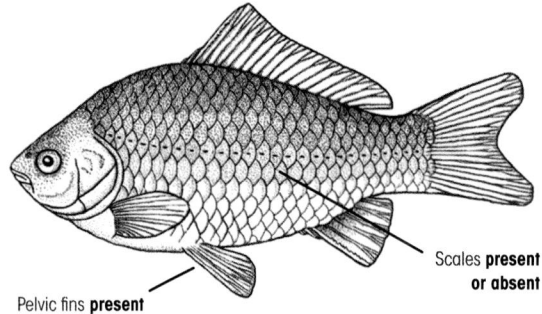

Scales **present or absent**

Pelvic fins **present**

10a Body snake-shaped; dorsal, caudal, and anal fins continuous. . . . . **Page 70**

Freshwater Eels (Anguillidae)

Dorsal, caudal, and anal fins **continuous**

Body **snake-shaped**

10b Body not snake-shaped; dorsal, caudal, and anal fins separate. . . . . **Page 268**

Cavefishes (Amblyopsidae)

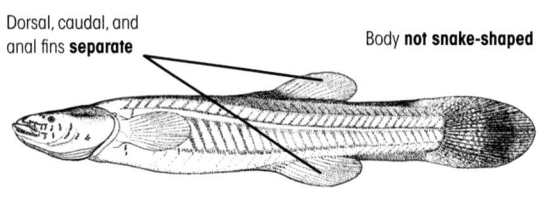

Dorsal, caudal, and anal fins **separate**

Body **not snake-shaped**

11a Anus far in front of anal fin, ahead of pelvic fins except in young, where positioned at lesser distances in front of anal fin. . . . . **Page 266**

11b Anus just in front of anal fin **12**

Pirate Perches (Aphredoderidae)

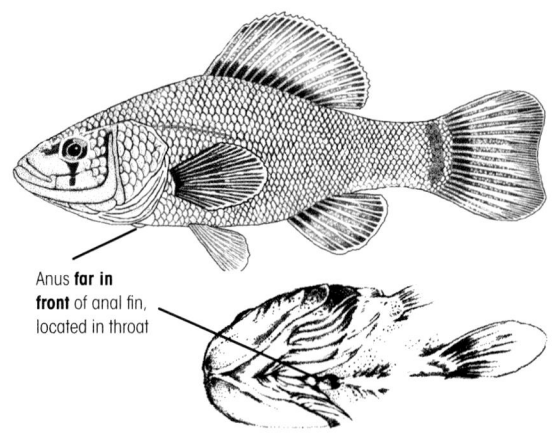

Anus **far in front** of anal fin, located in throat

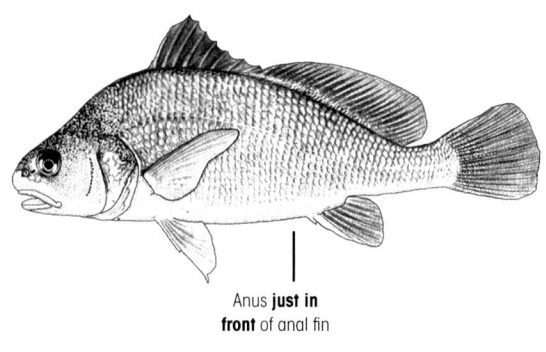

Anus **just in front** of anal fin

12a Two to 9 "free" dorsal spines, not connected to one another by membranes. . . . . **Page 376**

12b Dorsal spines absent or present, when present, not "free," connected to one another by membrane **13**

Sticklebacks (Gasterosteidae)

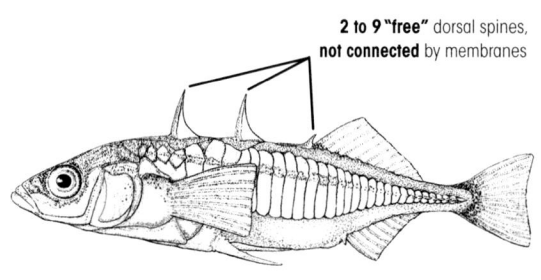

2 to 9 "free" dorsal spines, **not connected** by membranes

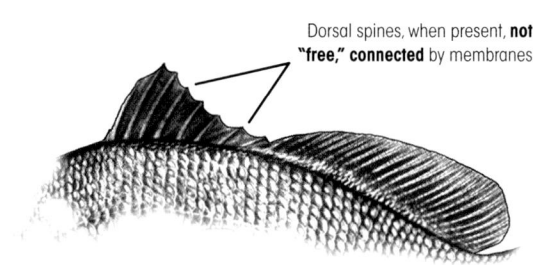

Dorsal spines, when present, **not "free," connected** by membranes

13a Dorsal fin single, without spines or with one stout saw-toothed spine **14**

Dorsal fin **single, without spines**

Dorsal fin **single, with one stout saw-toothed spine**

13b Dorsal fin divided into two distinct parts, or single and with four or more stiff spines **26**

Dorsal fin **divided** into two distinct parts

Dorsal fin **single and with four or more stiff spines**

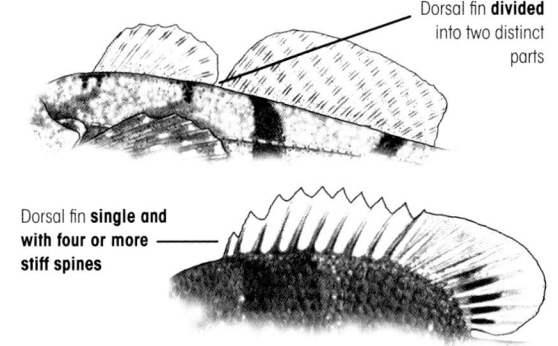

14a Caudal fin emarginate or forked **15**

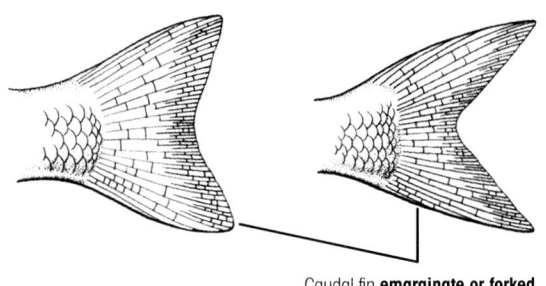

Caudal fin **emarginate or forked**

14b Caudal fin rounded or truncate **16**

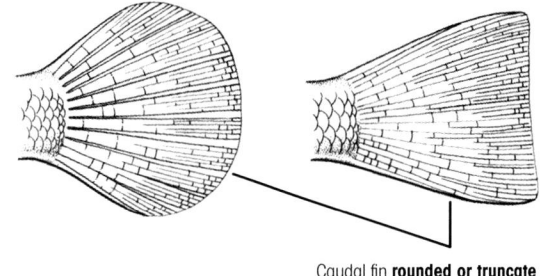

Caudal fin **rounded or truncate**

15a Snout shaped like a duck's bill; scales present on side of head; scales small, more than 95 in lateral series. ......
....**Page 238**

Pikes (Esocidae) in part

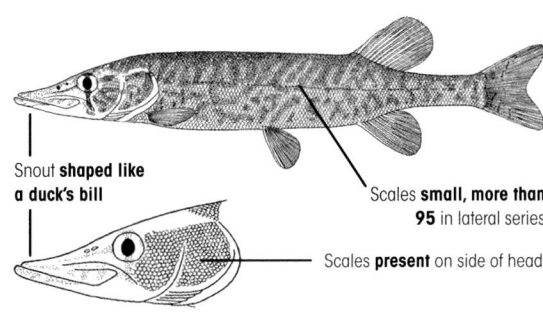

Snout **shaped like a duck's bill**

Scales **small, more than 95** in lateral series

Scales **present** on side of head

15b Snout rounded or pointed, not shaped like a duck's bill; scales absent on side of head; scales large, fewer than 95 in lateral series. **17**

Snout **rounded or pointed**, not shaped like a duck's bill

Scales **large, fewer than 95** in lateral series

Scales **absent** on side of head

16a Dorsal fin long (with 42–53 rays), its base more than half of total length; nostrils tubular; gular plate (large bony plate on underside of head) present. . . . . **Page 68**

Bowfins (Amiidae)

16b Dorsal fin short (with fewer than 16 rays), its base much less than half of total length; nostrils not tubular; gular plate absent. **23**

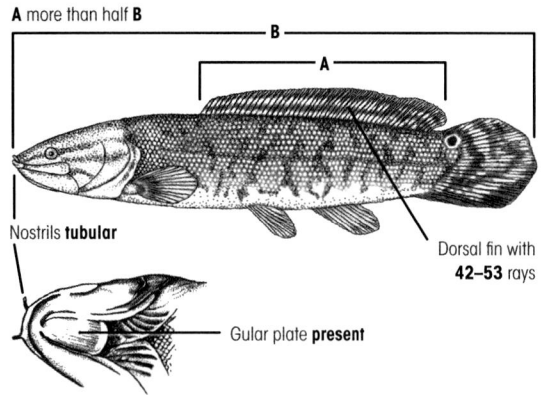

A more than half B

B

A

Nostrils **tubular**

Dorsal fin with **42–53** rays

Gular plate **present**

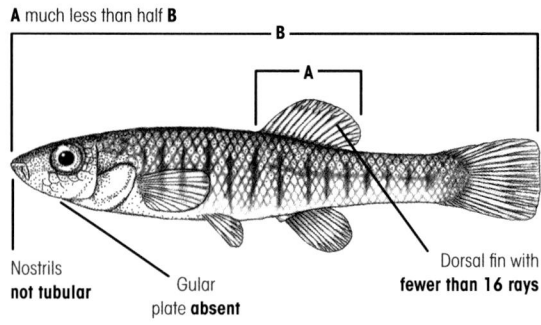

A much less than half B

B

A

Nostrils **not tubular**

Gular plate **absent**

Dorsal fin with **fewer than 16 rays**

17a Pelvic axillary process (small flap-like projection at upper margin of pelvic fin base) present; gill opening extending forward on throat to beneath eye; lower margins of gill covers overlapping on midline of throat; principal rays of anal fin 17 or more. **18**

17b Pelvic axillary process absent; gill opening not extending forward on throat to beneath eye; lower margins of gill covers not overlapping on midline of throat; principal rays of anal fin 16 or fewer **19**

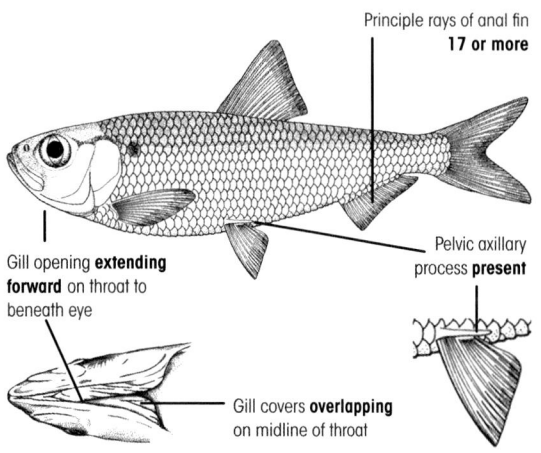

Principle rays of anal fin **17 or more**

Pelvic axillary process **present**

Gill opening **extending forward** on throat to beneath eye

Gill covers **overlapping** on midline of throat

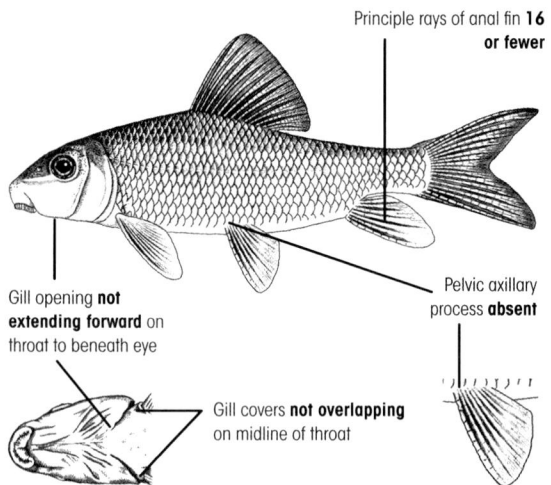

Principle rays of anal fin **16 or fewer**

Pelvic axillary process **absent**

Gill opening **not extending forward** on throat to beneath eye

Gill covers **not overlapping** on midline of throat

18a Midline of belly with sharp, saw-toothed projections; dorsal fin far forward of anal fin; lateral line absent . . . . **Page 76**

Shads and Herrings (Clupeidae)

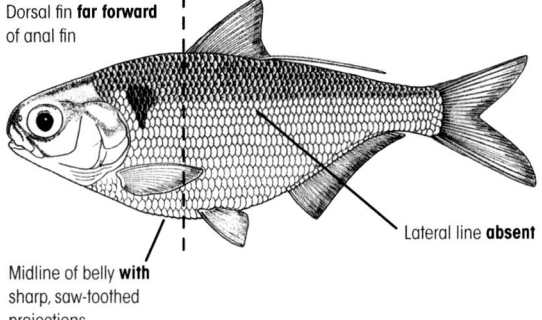

Dorsal fin **far forward** of anal fin

Lateral line **absent**

Midline of belly **with** sharp, saw-toothed projections

18b Midline of belly without sharp, saw-toothed projections; dorsal fin at least partly over anal fin; lateral line present . . . . **Page 72**

Mooneyes (Hiodontidae)

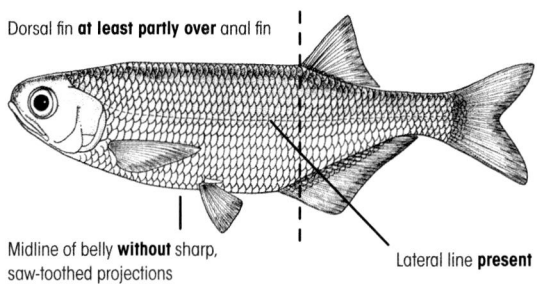

Dorsal fin **at least partly over** anal fin

Midline of belly **without** sharp, saw-toothed projections

Lateral line **present**

19a Dorsal and anal fins each with a stout, saw-toothed spine at front . . . . **Page 113**

Barbs and Carps (Cyprinidae)

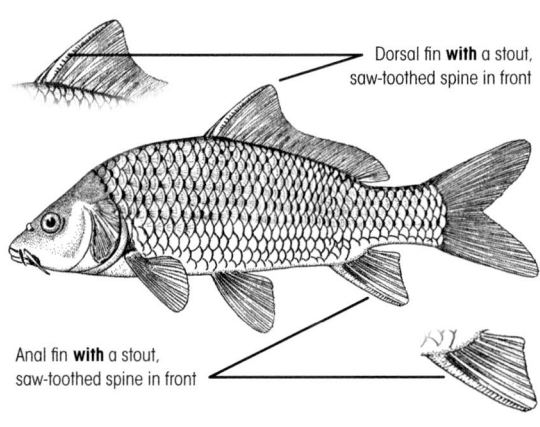

Dorsal fin **with** a stout, saw-toothed spine in front

Anal fin **with** a stout, saw-toothed spine in front

19b Dorsal and anal fins without stout, saw-toothed spines **20**

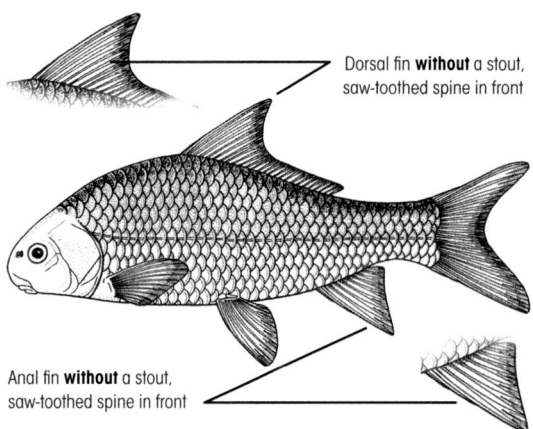

Dorsal fin **without** a stout, saw-toothed spine in front

Anal fin **without** a stout, saw-toothed spine in front

20a Dorsal fin with 8 (in 2 species 9 or 10) principal rays; length A divided by B less than 2.5; lips usually without bumps or plications; branched caudal fin rays 17; pharyngeal (throat) teeth in 1 or 2 rows, with 6 or fewer teeth in major row. **21**

20b Dorsal fin with 10 or more principal rays, or if only 9, then lateral line absent or reduced to a few pores; length A divided by B more than 2.5; lips usually thick with bumps or plications; branched caudal fin rays 16; pharyngeal (throat) teeth in 1 row of 10 or more teeth . . . . . **Page 84**

Suckers (Catostomidae)

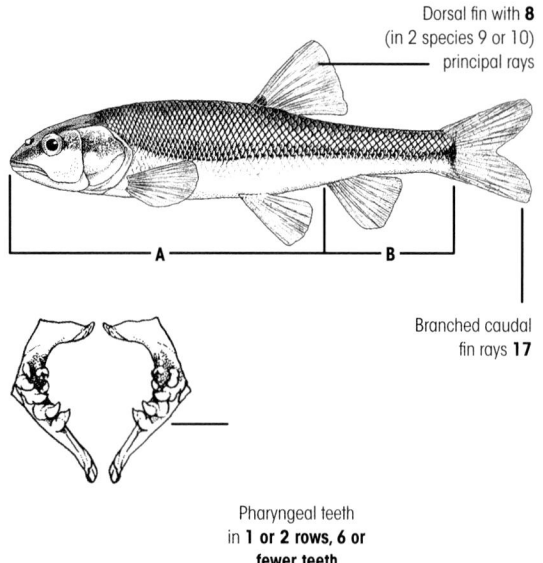

Dorsal fin with **8** (in 2 species 9 or 10) principal rays

Branched caudal fin rays **17**

Pharyngeal teeth in **1 or 2 rows, 6 or fewer teeth**

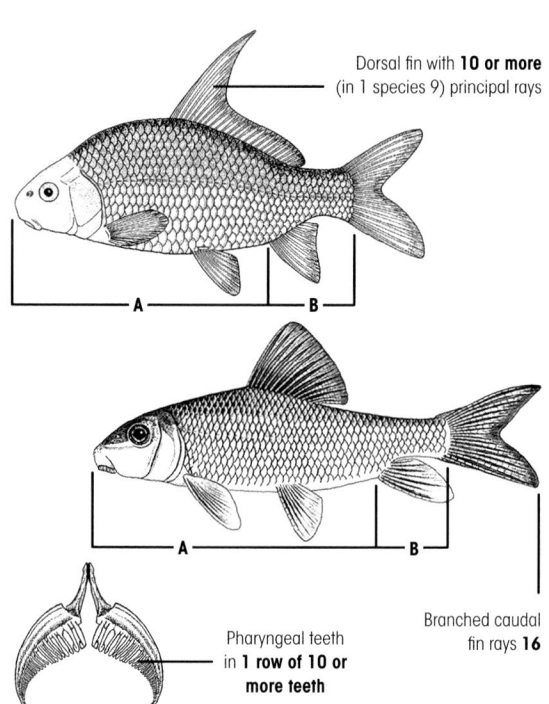

Dorsal fin with **10 or more** (in 1 species 9) principal rays

Pharyngeal teeth in **1 row of 10 or more teeth**

Branched caudal fin rays **16**

21a Distance from front of anal fin base to base of caudal fin (A) going 3 times or more into distance from front of anal fin base to tip of snout (B) . . . . **Page 117**

21b Distance from front of anal fin base to base of caudal fin (A) going 2.5 times or less into distance from front of anal fin base to tip of snout (B) **22**

Sharpbellies (Xenocyprididae) in part

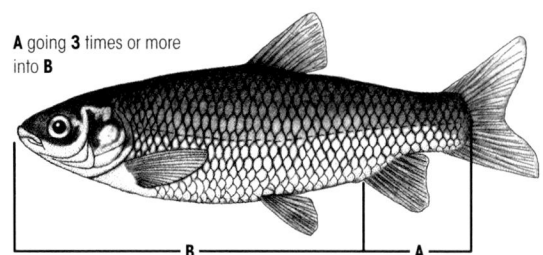

A going **3** times or more into **B**

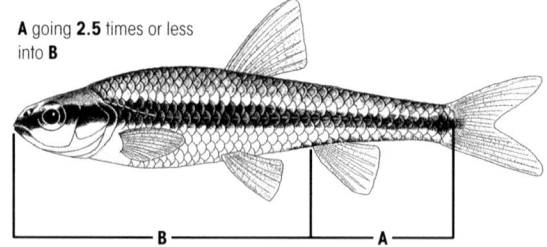

A going **2.5** times or less into **B**

22a Eye low on head, below central axis of body; pectoral fin long, extending well beyond front of pelvic fin; scales small, 85 or more in lateral line . . . . **Page 117**

Sharpbellies (Xenocyprididae) in part

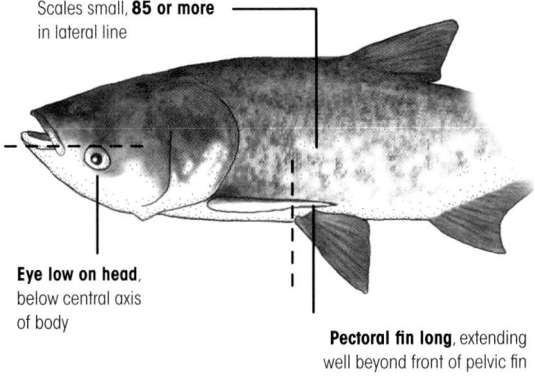

Scales small, **85 or more** in lateral line

Eye low on head, below central axis of body

Pectoral fin long, extending well beyond front of pelvic fin

22b Eye higher on head, not below central axis of body; pectoral fin short, rarely extending to front of pelvic fin; scales larger (except *Chrosomus erythrogaster*), usually 85 or fewer in lateral line. . . . . **Page 124**

Minnows (Leuciscidae)

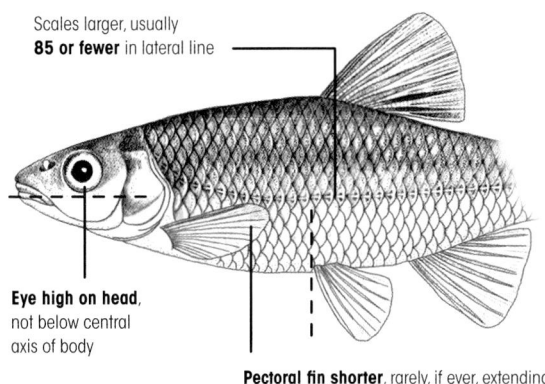

Scales larger, usually **85 or fewer** in lateral line

Eye high on head, not below central axis of body

Pectoral fin shorter, rarely, if ever, extending beyond front of pelvic fin

23a Five or 6 pairs (or 10–12 single) of relatively large barbels or "whiskers" around mouth; body shape elongate; tiny scales, more than 100 in a lateral series . . . . . **Page 214**

Spined Loaches (Cobitidae)

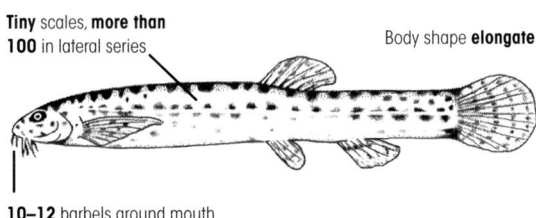

Tiny scales, **more than 100** in lateral series

Body shape **elongate**

10–12 barbels around mouth

23b Barbels absent; body shape not elongate; scales relatively large, less than 55 in a lateral series **24**

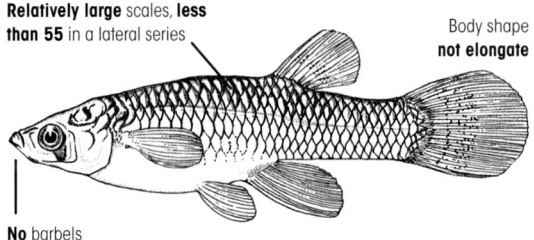

Relatively large scales, **less than 55** in a lateral series

Body shape **not elongate**

No barbels

24a Most of length of pelvic fin behind anterior base (origin) of dorsal fin; mouth terminal, groove of upper lip not continuous over snout . . . .**Page 238**

Mudminnows (Esocidae) in part

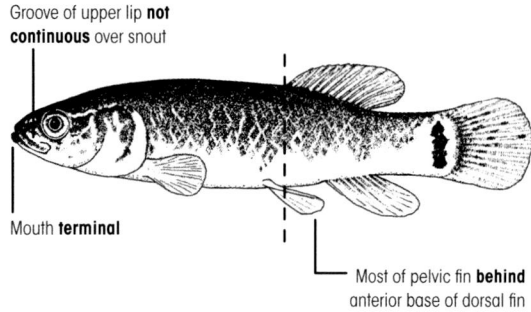

Groove of upper lip **not continuous** over snout

Mouth **terminal**

Most of pelvic fin **behind** anterior base of dorsal fin

24b Most or all of length of pelvic fin lying in advance of anterior base (origin) of dorsal fin; mouth superior, groove of upper lip continuous over snout **25**

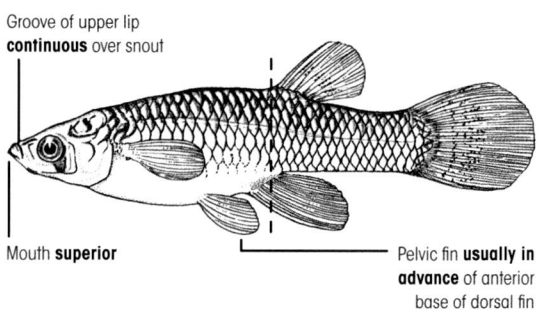

Groove of upper lip **continuous** over snout

Mouth **superior**

Pelvic fin **usually in advance** of anterior base of dorsal fin

25a Dorsal fin base almost entirely over anal fin base; scales in lateral series usually more than 30; anal fin of male not slender and rod-like; third ray of anal fin branched . . . . . .
. . . .**Page 278**

Topminnows (Fundulidae)

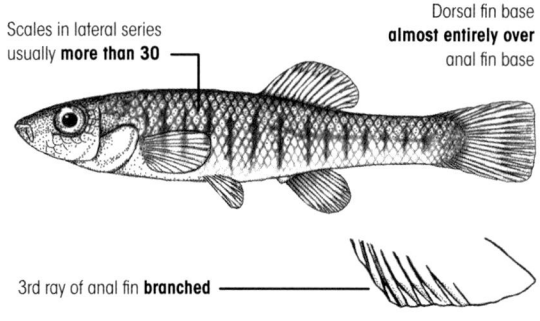

Scales in lateral series usually **more than 30**

Dorsal fin base **almost entirely over** anal fin base

3rd ray of anal fin **branched**

25b Dorsal fin base almost entirely (females and immatures) or entirely (males) behind anal fin base; scales in lateral series usually 30 or fewer; anal fin of male slender and rod-like (modified into intromittent organ); third ray of anal fin unbranched. . . . .**Page 286**

Livebearers (Poeciliidae)

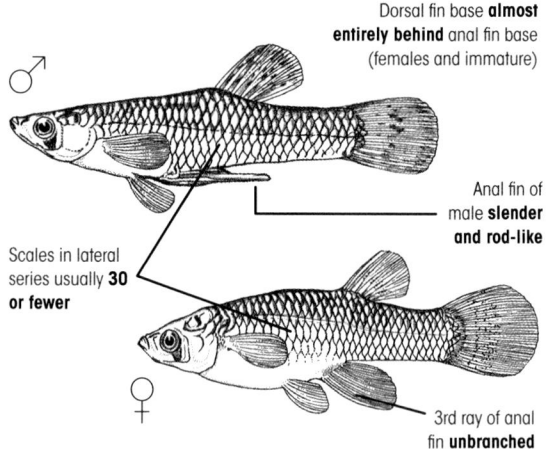

Dorsal fin base **almost entirely behind** anal fin base (females and immature)

Anal fin of male **slender and rod-like**

Scales in lateral series usually **30 or fewer**

3rd ray of anal fin **unbranched**

26a One median barbel on chin; soft dorsal fin long, with 60 or more rays . . . .**Page 270**

26b No median barbel on chin; soft dorsal fin short, with 35 or fewer rays **27**

Cuskfishes (Lotidae)

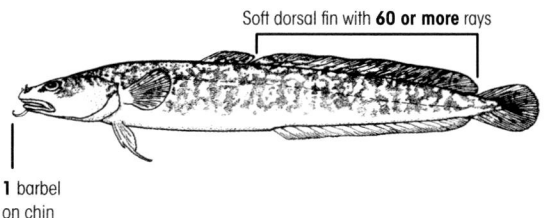

Soft dorsal fin with **60 or more** rays

1 barbel
on chin

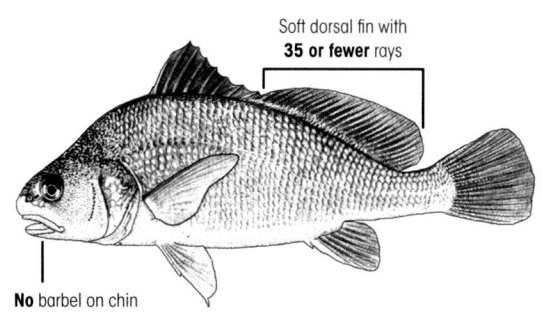

Soft dorsal fin with
35 or fewer rays

No barbel on chin

27a Pelvic fins fused to one another, forming a cone or sucking disc on belly . . . .**Page 272**

27b Pelvic fins separate and paired, normal with rays and membranes obvious **28**

Gobies (Gobiidae)

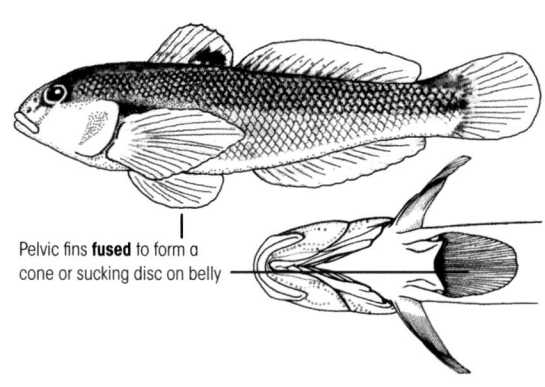

Pelvic fins **fused** to form a
cone or sucking disc on belly

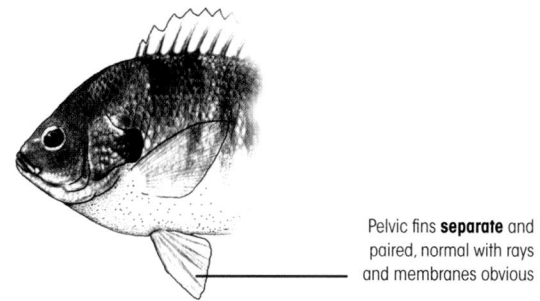

Pelvic fins **separate** and
paired, normal with rays
and membranes obvious

28a Spinous dorsal fin clearly separated from soft dorsal fin and with 3 to 5 thin spines; pectoral fin base near upper end of gill opening **29**

28b Spinous dorsal fin connected to soft dorsal fin, or, if separate, with 6 or more stout spines; pectoral fin base far below upper end of gill opening **30**

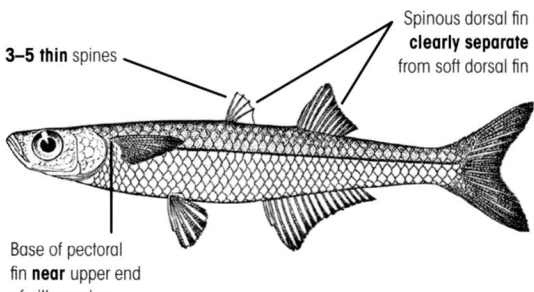

Spinous dorsal fin
clearly separate
from soft dorsal fin

3–5 thin spines

Base of pectoral
fin **near** upper end
of gill opening

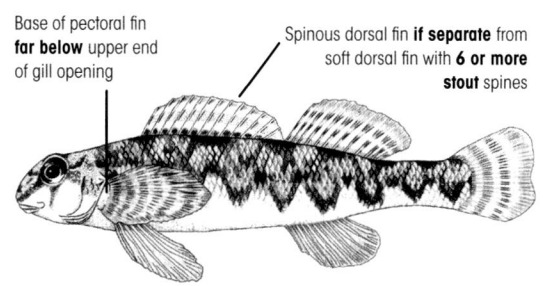

Base of pectoral fin
far below upper end
of gill opening

Spinous dorsal fin **if separate** from
soft dorsal fin with **6 or more
stout** spines

29a Anal fin with 1 spine; eye near middle of head; adipose eyelid absent . . . . **Page 274**

New World Silversides (Atherinopsidae)

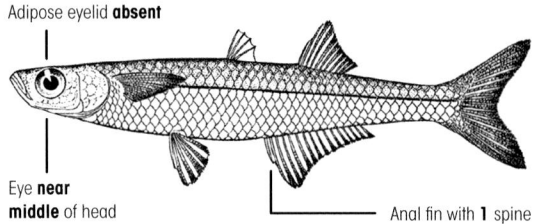

Adipose eyelid **absent**

Eye **near middle** of head

Anal fin with **1** spine

29b Anal fin with 2 (young) or 3 spines; eye distinctly anterior to middle of head; adipose eyelid present . . . . **Page 288**

Mullets (Mugilidae)

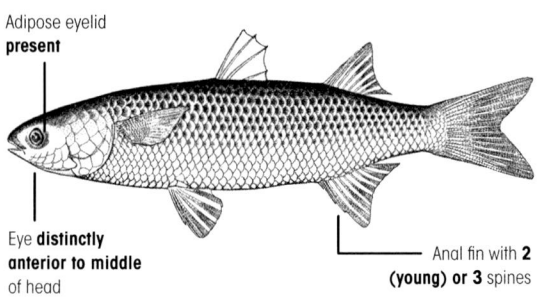

Adipose eyelid **present**

Eye **distinctly anterior to middle** of head

Anal fin with **2 (young) or 3** spines

30a Body without scales, with prickles only; pelvic fin with one thin spine and 3 or 4 soft rays; anal fin spines absent; spines of dorsal fin flexible, superficially resembling soft rays . . . . **Page 381**

Sculpins (Cottidae)

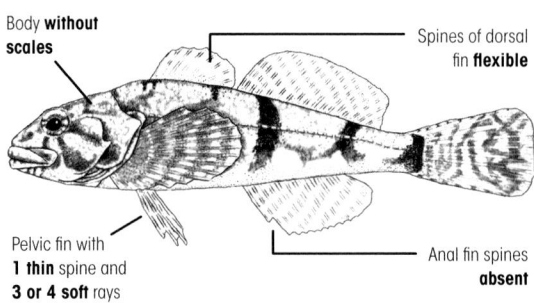

Body **without scales**

Spines of dorsal fin **flexible**

Pelvic fin with **1 thin** spine and **3 or 4 soft** rays

Anal fin spines **absent**

30b Body at least partly scaled; pelvic fin with 1 thin spine and 5 soft rays; anal fin spines present; spines of dorsal fin stiff. **31**

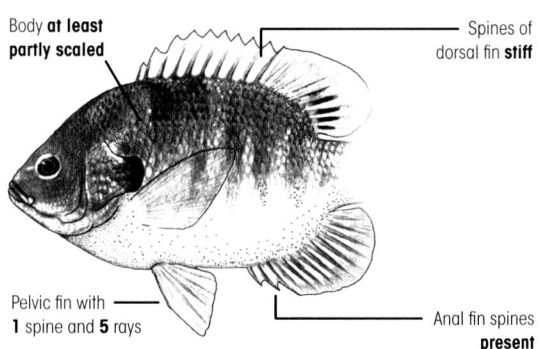

Body **at least partly scaled**

Spines of dorsal fin **stiff**

Pelvic fin with **1 spine and 5 rays**

Anal fin spines **present**

31a Anal fin spines 3 or more **32**

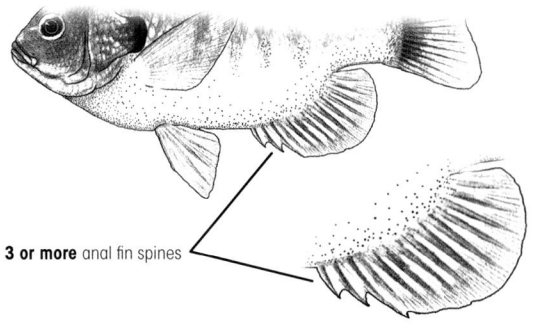

3 or more anal fin spines

31b Anal fin spines 1 or 2 . . . . **33**

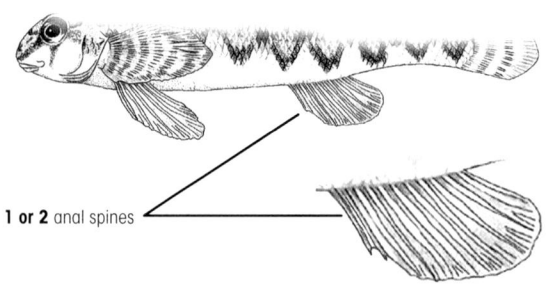

1 or 2 anal spines

32a Lateral line absent; total dorsal fin spines and rays 15 or fewer; maximum size less than 45 mm total length . . . .**Page 318**

32b Lateral line present, sometimes incomplete but with a few pored scales anteriorly; total dorsal fin spines and rays 16 or more; maximum size variable. **34**

Pygmy Sunfishes (Elassomatidae)

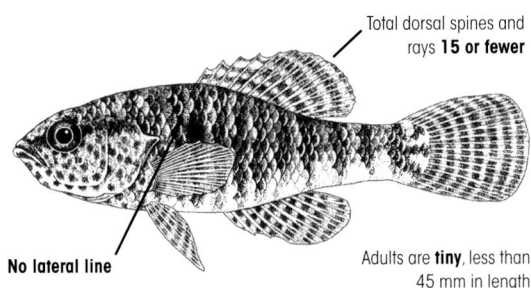

Total dorsal spines and rays **15 or fewer**

No lateral line

Adults are **tiny**, less than 45 mm in length

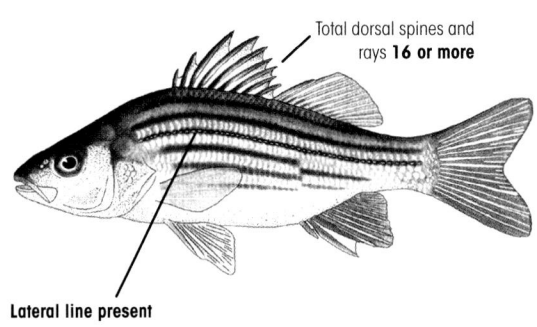

Total dorsal spines and rays **16 or more**

Lateral line present

33a Soft dorsal fin with 23 or more rays; lateral line extending to end of caudal fin; 2nd spine of anal fin stout and much longer than 1st. . . . . **Page 327**

33b Soft dorsal fin with 23 or fewer rays; lateral line not extending to end of caudal fin; 2nd spine of anal fin slender and not much longer than 1st. . . . . **Page 329**

Drums and Croakers (Sciaenidae)

Darters and Perches (Percidae)

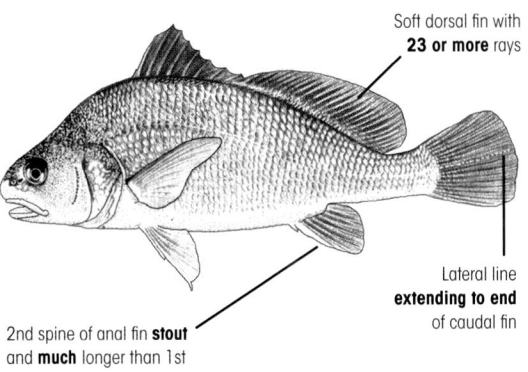

Soft dorsal fin with **23 or more** rays

Lateral line **extending to end** of caudal fin

2nd spine of anal fin **stout** and **much** longer than 1st

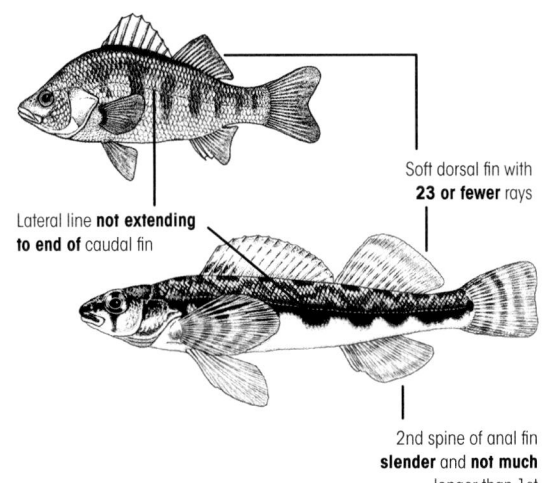

Soft dorsal fin with **23 or fewer** rays

Lateral line **not extending to end of** caudal fin

2nd spine of anal fin **slender** and **not much** longer than 1st

34a Spinous dorsal and soft dorsal fins separate or only slightly connected; a sharp spine near back of gill cover; margin of preopercle (bone just ahead of gill cover) strongly saw-toothed (serrate) . . . .**Page 320**

White Basses (Moronidae)

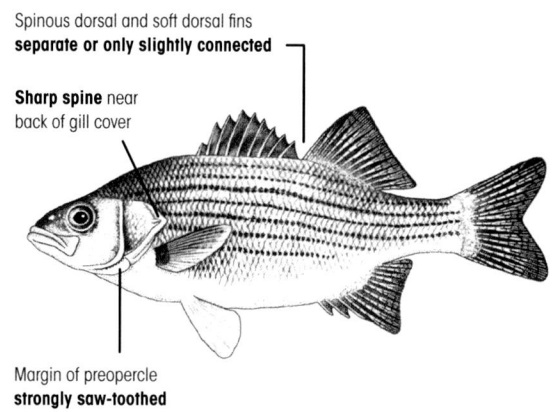

Spinous dorsal and soft dorsal fins
separate or only slightly connected

Sharp spine near
back of gill cover

Margin of preopercle
strongly saw-toothed

34b Spinous dorsal and soft dorsal fins well connected with, at most, a deep notch between them; no sharp spine near back of gill cover; margin of preopercle smooth, weakly saw-toothed in a few species . . . .**Page 290**

Sunfishes and Black Basses (Centrarchidae)

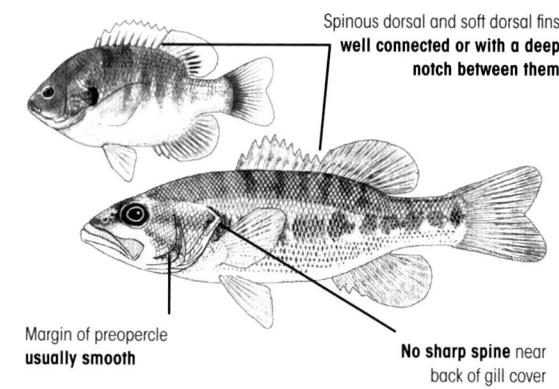

Spinous dorsal and soft dorsal fins
**well connected or with a deep
notch between them**

Margin of preopercle
usually smooth

No sharp spine near
back of gill cover

LAMPREYS

Family Petromyzontidae Bonaparte 1831

Lampreys are ancient, eel-like fishes that lack jaws, scales, paired fins, and bone. Adults have a cartilaginous skeleton, 1 median nostril, 7 pairs of pore-like gill openings, 1 or 2 dorsal fins continuous with the caudal fin, and an oral disc with rasping teeth on the tongue. Ammocoetes lack most characteristics used to identify adults; however, myomere counts are constant throughout life, and the counts given in species accounts can be used to identify ammocoetes as well as adults.

Lampreys excavate pits in stream riffles (rarely in wave-swept areas of lakes) to be used as spawning sites by removing stones with their suction-disc mouths and by fanning out fine particles with vibrations of the body. Eggs hatch into blind larvae called ammocoetes, which later metamorphose into adults. Larvae may last 3–8 (or, rarely, more) years, living in mud- or sand-bottomed pools and feeding by filtering microorganisms from water. Some species, the so-called brook or nonparasitic lampreys, do not feed as adults and spawn the spring following metamorphosis. Other species are parasitic and feed by attaching to and rasping a hole in the side of a large fish. Adults of several parasitic species migrate to the ocean but must return to fresh water to spawn.

The Sea Lamprey, *Petromyzon marinus*, may have played a role in the decline of commercial fishing in Lake Michigan, but the other 3 parasitic riverine species have had a negligible effect, if any at all, on fish populations in Illinois. Ammocoetes are used as bait by anglers, and adult lampreys have been marketed in Chicago in the distant past but have received little attention as a human food source in North America. Of the 23 species known in North America, 7 are recorded from Illinois.

KEY TO THE LAMPREYS (Petromyzontidae)

1a Mouth a horseshoe-shaped hood without tongue and teeth; eyes poorly developed and covered by skin; gill openings connected by a continuous horizontal groove . . . Ammocoetes (larval or immature stage of lamprey) Identification to species extremely difficult.

1b Mouth a circular sucking disc lined with teeth (teeth poorly developed in three species); eyes well developed, not covered by skin; gill openings "free", not connected by a horizontal groove.............................**2**

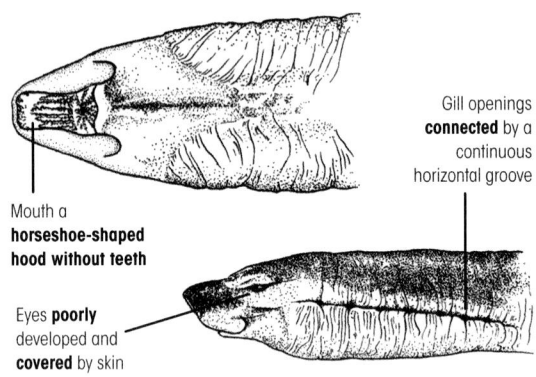

Mouth a **horseshoe-shaped hood without teeth**

Eyes **poorly** developed and **covered** by skin

Gill openings **connected** by a continuous horizontal groove

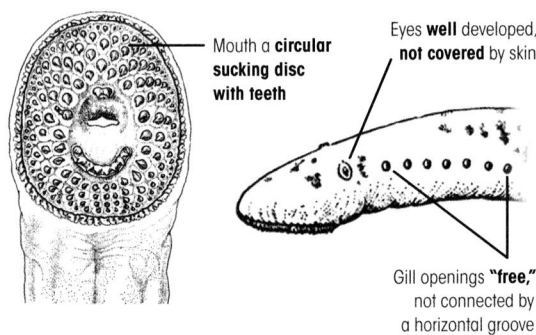

Mouth a **circular sucking disc with teeth**

Eyes **well** developed, **not covered** by skin

Gill openings **"free,"** not connected by a horizontal groove

2a Dorsal fin sometimes slightly notched but never divided into two distinct parts **3**

2b Dorsal fin divided into two distinct parts **4**

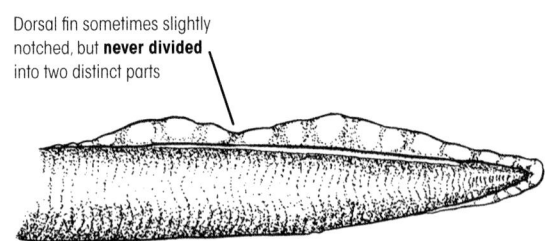

Dorsal fin sometimes slightly notched, but **never divided** into two distinct parts

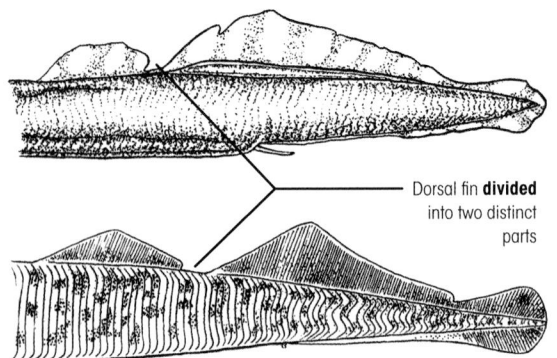

Dorsal fin **divided** into two distinct parts

LAMPREYS (Petromyzontidae)

42

3a Most disc teeth poorly developed, especially near outer margins of disc; sucking disc, when expanded, narrower than head; total length never more than 7 inches (17.8 cm); all disc teeth of innermost circle with 1 point; lateral line organs on body not darkened . . . . **Page 48**

Northern Brook Lamprey (*Ichthyomyzon fossor*)

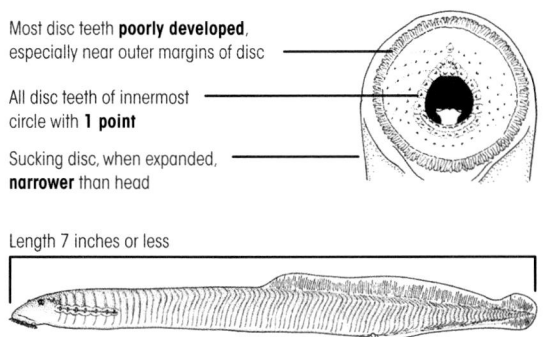

Most disc teeth **poorly developed**, especially near outer margins of disc

All disc teeth of innermost circle with **1 point**

Sucking disc, when expanded, **narrower** than head

Length 7 inches or less

3b All disc teeth well developed; sucking disc, when expanded, wider than head; total length often more than 8 inches (20.3 cm); all disc teeth of innermost circle with either 1 or 2 points; lateral line organs on body darkened, appearing as black dots. **5**

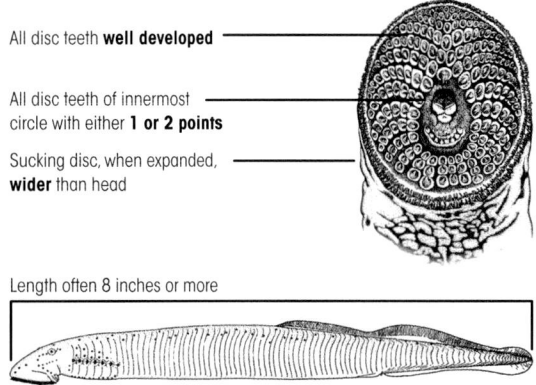

All disc teeth **well developed**

All disc teeth of innermost circle with either **1 or 2 points**

Sucking disc, when expanded, **wider** than head

Length often 8 inches or more

4a Pair of disc teeth at top of innermost circle (supraoral teeth) close together; sucking disc when expanded, wider than head; all disc teeth well developed, gradually diminishing in size near outer margins of disc; body mottled or marbled with black. . . . . **Page 52**

Sea Lamprey (*Petromyzon marinus*)

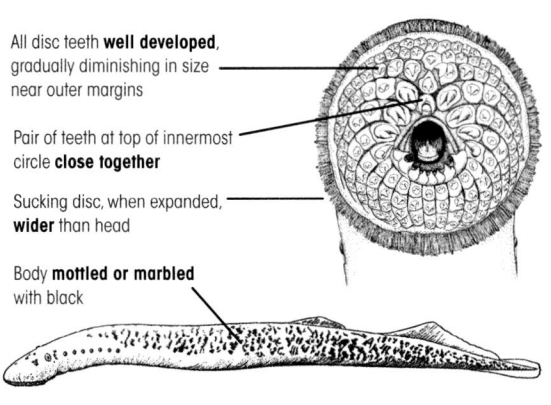

All disc teeth **well developed**, gradually diminishing in size near outer margins

Pair of teeth at top of innermost circle **close together**

Sucking disc, when expanded, **wider** than head

Body **mottled or marbled** with black

4b Pair of disc teeth at top of innermost circle widely separated; sucking disc, when expanded, narrower than head; most disc teeth poorly developed, especially near outer margins of disc; body not mottled or marbled with black **7**

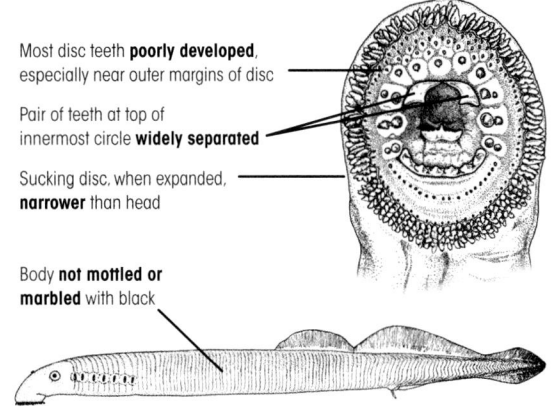

Most disc teeth **poorly developed**, especially near outer margins of disc

Pair of teeth at top of innermost circle **widely separated**

Sucking disc, when expanded, **narrower** than head

Body **not mottled or marbled** with black

5a All disc teeth of innermost circle with 1 point; myomeres between last gill opening and vent usually 49–52 . . . . . .
. . . . **Page 49**

5b All disc teeth of innermost circle with 2 points; myomeres between last gill opening and vent usually 54–62 . . . . . .
. . . .**6**

Silver Lamprey (*Ichthyomyzon unicuspis*)

Myomeres between last gill opening and vent usually **49–52**

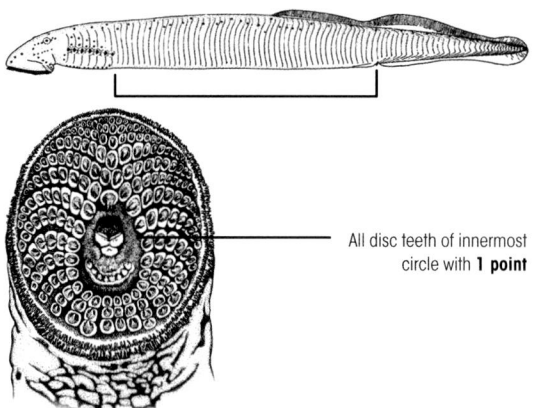

All disc teeth of innermost circle with **1 point**

Myomeres between last gill opening and vent usually **54–62**

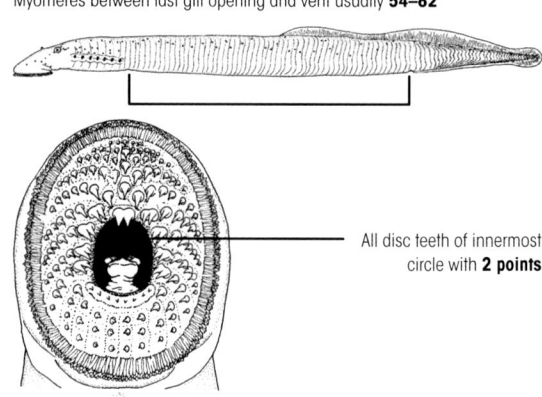

All disc teeth of innermost circle with **2 points**

6a Myomeres between last gill opening and vent usually 54–56; transverse lingual lamina with usually 36–53 tooth points (denticles) . . . . **Page 47**

6b Myomeres between last gill opening and vent usually 58–62; transverse lingual lamina with usually 15–32 tooth points (denticles) . . . . **Page 46**

Chestnut Lamprey (*Ichthyomyzon castaneus*)

Myomeres between last gill opening and vent usually **54–56**

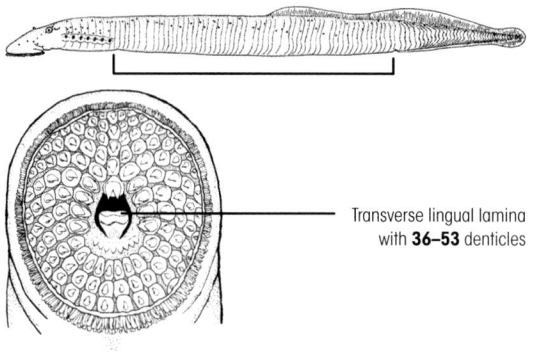

Transverse lingual lamina with **36–53** denticles

Ohio Lamprey (*Ichthyomyzon bdellium*)

Myomeres between last gill opening and vent usually **58–62**

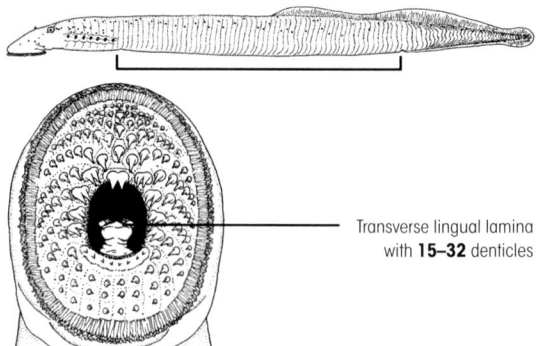

Transverse lingual lamina with **15–32** denticles

7a Myomeres between last gill opening and vent usually 54–60; disc teeth poorly developed above and below innermost row of teeth . . . . **Page 50**

Least Brook Lamprey (*Lampetra aepyptera*)

Myomeres between last gill opening and vent usually **54–60**

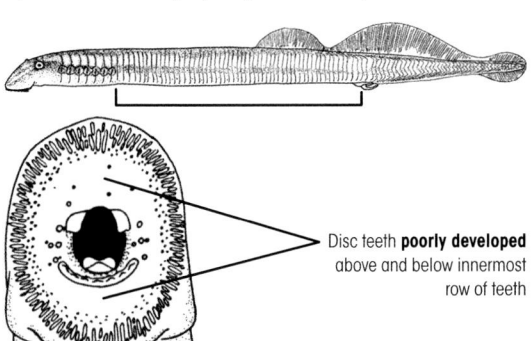

Disc teeth **poorly developed** above and below innermost row of teeth

7b Myomeres between last gill opening and vent usually 66–73; disc teeth moderately developed above and below innermost row of teeth . . . . **Page 51**

American Brook Lamprey (*Lethenteron appendix*)

Myomeres between last gill opening and vent usually **66–73**

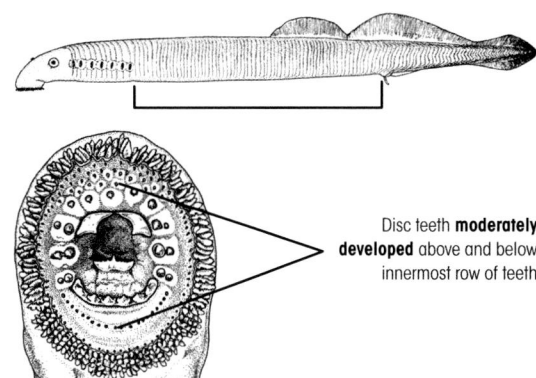

Disc teeth **moderately developed** above and below innermost row of teeth

OHIO LAMPREY *Ichthyomyzon bdellium* (Jordan 1885)

IDENTIFICATION The Ohio Lamprey is an eel-like fish that lacks jaws and paired fins and has 1 slightly notched dorsal fin, 1 median nostril, 7 pairs of pore-like gill openings, and an oral disc with rasping teeth on the tongue. The expanded oral disc is as wide or wider than the head and has sharp, well-developed disc teeth that are bicuspid. There are usually 56–62 myomeres along the trunk of the body; distinct black pigment is on the lateral-line pores. Adults are yellow or tan on the back and side with white to light olive-yellow below; the fins are olive-yellow. The species is parasitic. To 12 in. (30 cm).

SIMILAR SPECIES The Chestnut Lamprey, *I. castaneus*, usually has 51–56 myomeres on the trunk of the body. The Silver Lamprey, *I. unicuspis*, has sharp unicuspid teeth in the oral disc and usually 49–52 myomeres on the trunk of the body.

HABITAT Known only from large rivers in Illinois, almost certainly attached to other fishes. Ammocoetes in other parts of the species range live near debris in muddy pools and backwaters.

DISTRIBUTION IN ILLINOIS There are 4 records from the Forbes and Richardson era for the Ohio Lamprey, 2 from the Wabash River and 2 from the Embarras River. Smith (1979) noted that the specimens on which the records are based were taken before 1918, and only 1 individual was identifiable; the others were poorly preserved or had intermediate characteristics. The nearly identical Chestnut Lamprey and Silver Lamprey are the only parasitic species found in the Wabash River basin today. The Ohio Lamprey, if properly identified, has been extirpated from Illinois for at least a century.

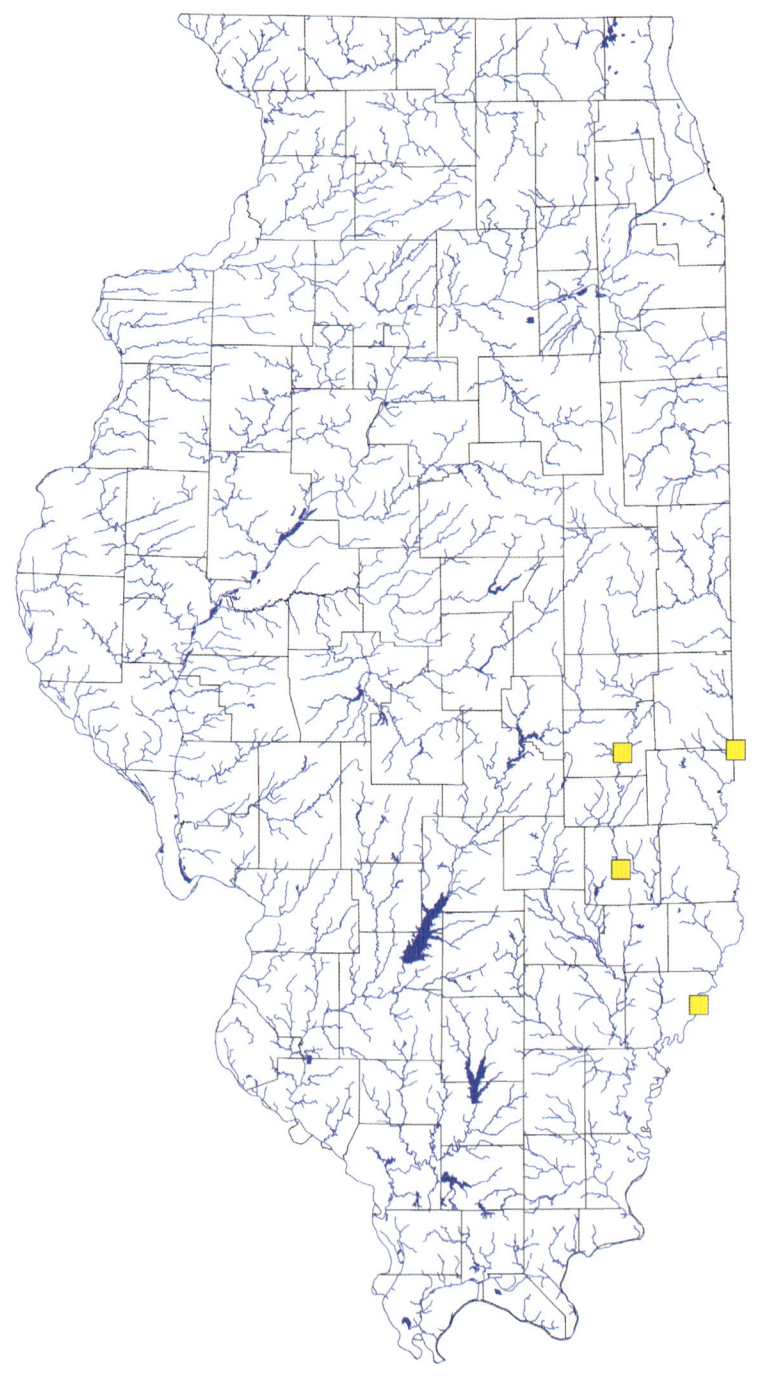

CHESTNUT LAMPREY *Ichthyomyzon castaneus* Girard 1858

IDENTIFICATION The Chestnut Lamprey is an eel-like fish that lacks jaws and paired fins and has 1 slightly notched dorsal fin, 1 median nostril, 7 pairs of pore-like gill openings, and an oral disc with rasping teeth on the tongue. The expanded oral disc is as wide or wider than the head and has large, sharp disc teeth that are invariably bicuspid. There are usually 51–56 myomeres on the trunk of the body; distinct black pigment is on the lateral-line pores. Adults are yellow or tan on the back and side with white to light olive-yellow below; the fins are olive-yellow. The species is parasitic. To 15 in. (38 cm).

SIMILAR SPECIES The Silver Lamprey, *I. unicuspis*, has sharp unicuspid teeth in the oral disc and usually 49–52 myomeres on the trunk of the body. The Ohio Lamprey, *I. bdellium*, usually has 56–62 myomeres on the trunk of the body.

HABITAT The Chestnut Lamprey is most often found attached to large fishes in bordering rivers and streams. At capture they may detach from their host but leave an obvious round scar. Adults are occasionally found ascending streams, probably migrating to their spawning sites. Ammocoetes occupy sand- and silt-bottomed pools and backwaters.

DISTRIBUTION IN ILLINOIS Smith (1979) discussed the difficulty in determining the early distribution of the Chestnut Lamprey in Illinois, due to conflicting reports, in part related to taxonomic confusion between the Chestnut and Silver Lampreys. The Chestnut Lamprey has essentially the same distribution in the contemporary era as in the Smith era, although it recently has been captured in Lake Michigan.

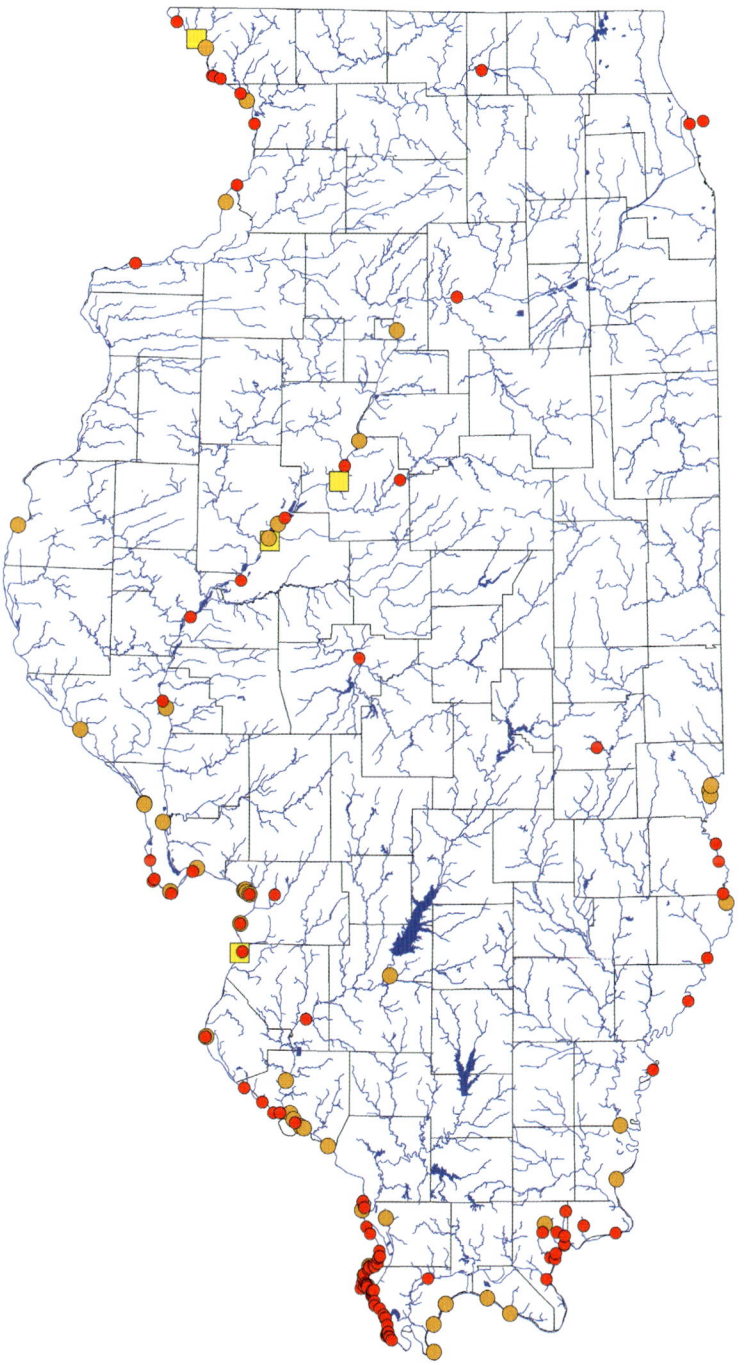

NORTHERN BROOK LAMPREY *Ichthyomyzon fossor* Reighard & Cummins 1916

IDENTIFICATION The Northern Brook Lamprey is an eel-like fish that lacks jaws and paired fins and has 1 slightly notched dorsal fin, 1 median nostril, 7 pairs of pore-like gill openings, and an oral disc with rasping teeth on the tongue. The expanded oral disc is narrower than the head and has very small, blunt disc teeth. There are usually 50–52 myomeres on the trunk of the body, and there is no black pigment on the lateral-line pores. Adults are dark gray to brown and often have a pale line along their back. The lower body and fins are pale gray, yellow, or silver-white. The species is nonparasitic. To 6¾ in. (17 cm).

SIMILAR SPECIES The parasitic species of adult *Ichthyomyzon* have large, sharp oral-disk teeth, black pigment on their lateral-line pores, and no pale line along their back.

HABITAT Adults of the Northern Brook Lamprey occupy creeks and small rivers with clean, clear rocky runs and slow runs above riffles. Ammocoetes drift downstream and live near debris in muddy pools and backwaters.

DISTRIBUTION IN ILLINOIS Forbes and Richardson (1909) did not record the Northern Brook Lamprey in Illinois, and Smith (1979) found it only in the Kankakee River in Kankakee County, where the species persists in the contemporary era.

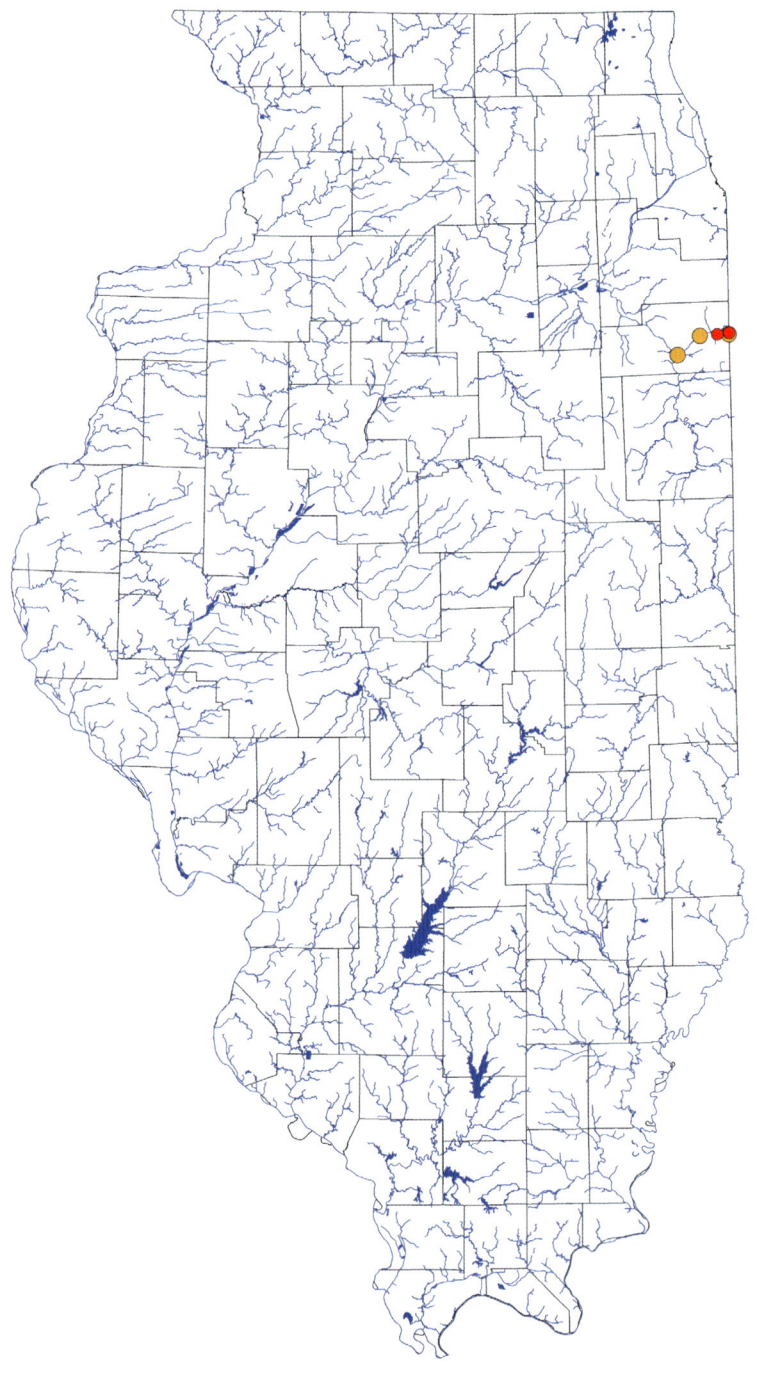

SILVER LAMPREY *Ichthyomyzon unicuspis* Hubbs & Trautman 1937

IDENTIFICATION The Silver Lamprey is an eel-like fish that lacks jaws and paired fins and has 1 slightly notched dorsal fin, 1 median nostril, 7 pairs of pore-like gill openings, and an oral disc with rasping teeth on the tongue. The expanded oral disc is as wide or wider than the head and has large, sharp disc teeth that are invariably unicuspid. There are usually 49–52 myomeres on the trunk of the body, and distinct black pigment is on the lateral-line pores. Adults are gray-brown to yellow-tan on the back and side with light yellow or tan below; the fins are yellow-gray. The species is parasitic. To 15½ in. (39 cm).

SIMILAR SPECIES The Chestnut Lamprey, *I. castaneus*, and Ohio Lamprey, *I. bdellium*, have bicuspid circumoral teeth and usually more than 51 myomeres on the trunk of the body. The Northern Brook Lamprey, *I. fossor*, has blunt disc teeth, a pale line along the back, and no black pigment on the lateral-line pores.

HABITAT Adults are usually found attached to other fishes in large rivers, lakes, and impoundments. Adults migrate upriver to spawn in gravel riffles and runs. Ammocoetes inhabit sandy and muddy pools and backwaters.

DISTRIBUTION IN ILLINOIS Forbes and Richardson (1909) and Smith (1979) recorded the Silver Lamprey throughout the state in large rivers. In the contemporary era, it is much less widespread and frequently captured only in the Mississippi River in northwestern Illinois, the Vermilion River in Vermilion County, and in the Wabash River. It appears to be extirpated from Lake Michigan, the Illinois River, and the Ohio River and is rarely encountered elsewhere.

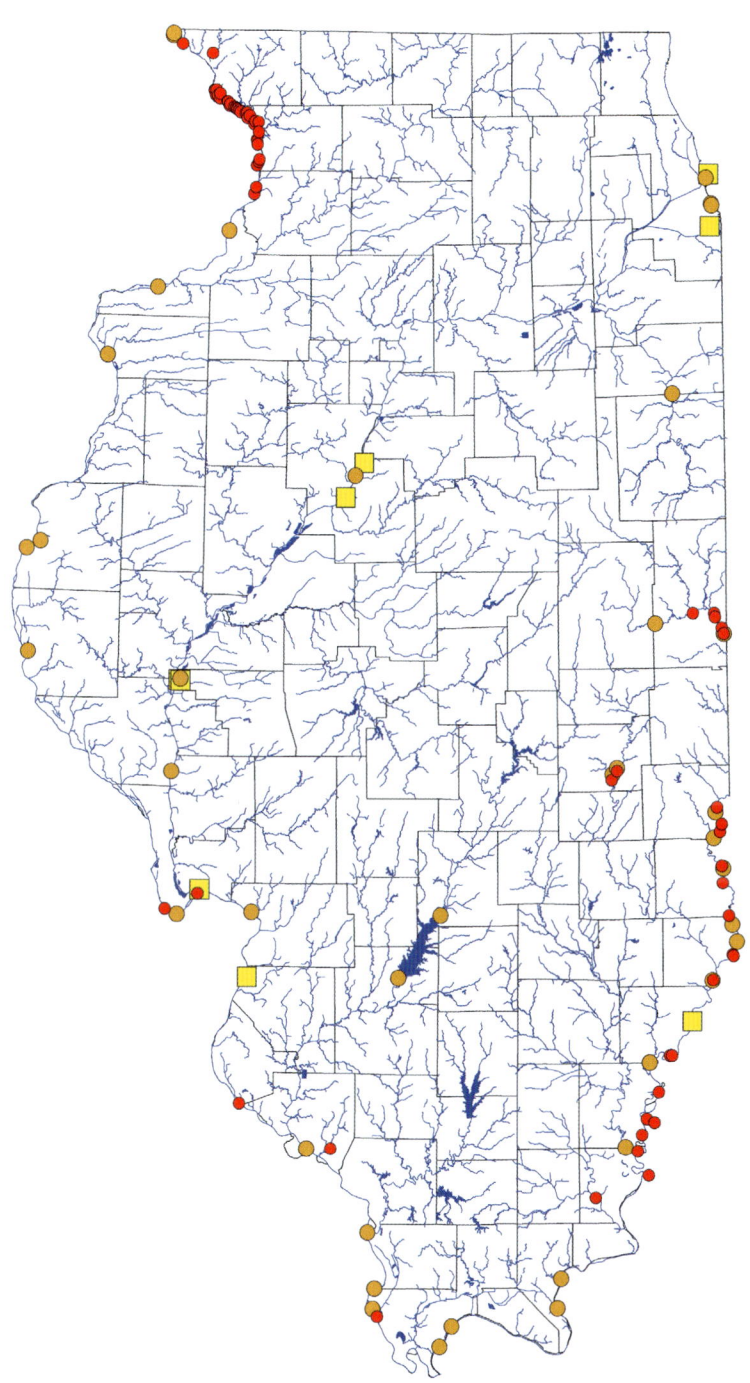

LEAST BROOK LAMPREY *Lampetra aepyptera* (Abbott 1860)

IDENTIFICATION The Least Brook Lamprey is an eel-like fish that lacks jaws and paired fins and has 2 dorsal fins narrowly connected at the base, 1 median nostril, 7 pairs of pore-like gill openings, and an oral disc with rasping teeth on the tongue. The expanded oral disc is narrower than the head and has blunt, extremely degenerate disc teeth. There are usually 52–59 myomeres on the trunk of the body. There is no black pigment on the lateral-line pores. Adults are light tan to silver-gray on their back and sides. The lower body is yellow or white, and the dorsal and caudal fins are yellow or gray. The breeding adult has mottled gray-brown on the back and a black stripe on the side of the body through the eye and at the base of the first dorsal fin. There are dusky black edges on the dorsal fins, and a gold stripe runs from the caudal fin through the middle of the dorsal fins. The caudal fin is dark tipped. The species is nonparasitic. To 7 in. (18 cm).

SIMILAR SPECIES The American Brook Lamprey, *Lethenteron appendix*, has 67–73 myomeres on the trunk of its body.

HABITAT Adults are found in gravel riffles and runs of spring-fed creeks and small rivers. The ammocoetes float downstream and bury themselves tail first in quiet pools and backwaters near debris.

DISTRIBUTION IN ILLINOIS The Least Brook Lamprey was first discovered in Illinois in 1952 in Sugar Creek in Williamson County in southeastern Illinois, and contemporary-era records show the species to be widespread in tributaries of the Ohio River in southeastern Illinois. Although recorded as recently as 2010 in Sugar Creek—the only known locality in the Saline River basin—several more recent surveys suggest that the species has disappeared from Sugar Creek.

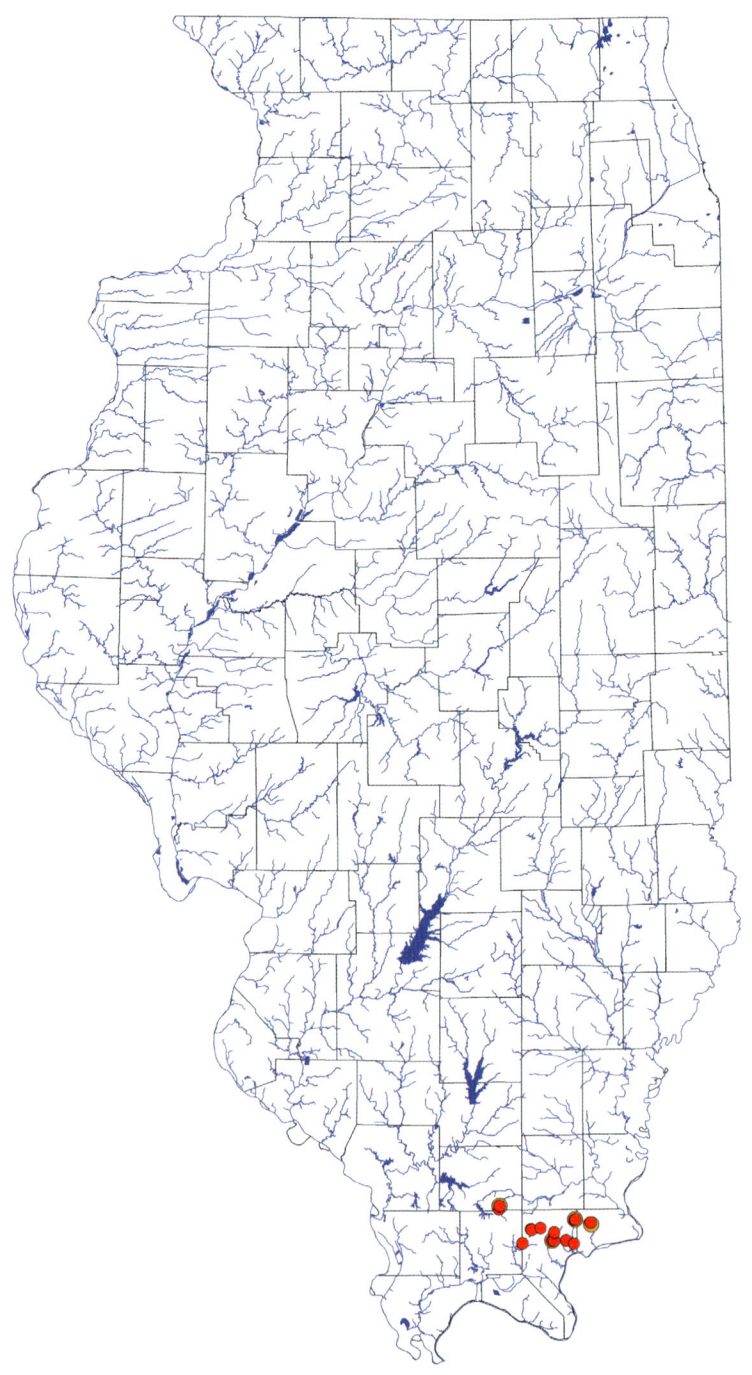

LAMPREYS (Petromyzontidae)

50

AMERICAN BROOK LAMPREY *Lethenteron appendix* (DeKay 1842)

IDENTIFICATION The American Brook Lamprey is an eel-like fish that lacks jaws and paired fins and has 2 dorsal fins narrowly connected at the base, 1 median nostril, 7 pairs of pore-like gill openings, and an oral disc with rasping teeth on the tongue. The expanded oral disc is narrower than the head and usually has blunt disc teeth. There are usually 67–73 myomeres along the trunk of the body. There is no black on the lateral-line pores. Adults are lead gray to slate blue on their back and side and dusky to silver-white below. The dorsal fins are yellow, and there is a dark gray to black blotch on the caudal fin. The breeding adult is olive-green or pink-purple to shiny black on the back and side; there is a black stripe at the base of the dorsal fins. The species is nonparasitic. To 13¾ in. (35 cm).

SIMILAR SPECIES The Least Brook Lamprey, *Lampetra aepyptera*, has extremely degenerate disc teeth and usually has 52–59 myomeres along the trunk of the body.

HABITAT Adults of the American Brook Lamprey occupy gravel- and sand-bottomed riffles and runs of creeks and small to medium rivers. Ammocoetes live near debris in muddy pools and backwaters.

DISTRIBUTION IN ILLINOIS Forbes and Richardson (1909) recorded the American Brook Lamprey only in Lake Michigan, where it has not been found since. Smith (1979) found the species at highly disjunct localities in the Mississippi, Illinois, and Wabash River basins. It persists in those same areas in the contemporary era and has been found in a few other localities, all in the northern two-thirds of the state.

REMARKS Forbes and Richardson (1909) treated this species as *Lampetra wilderi*, and Smith (1979) treated it as *Lampetra lamottei*.

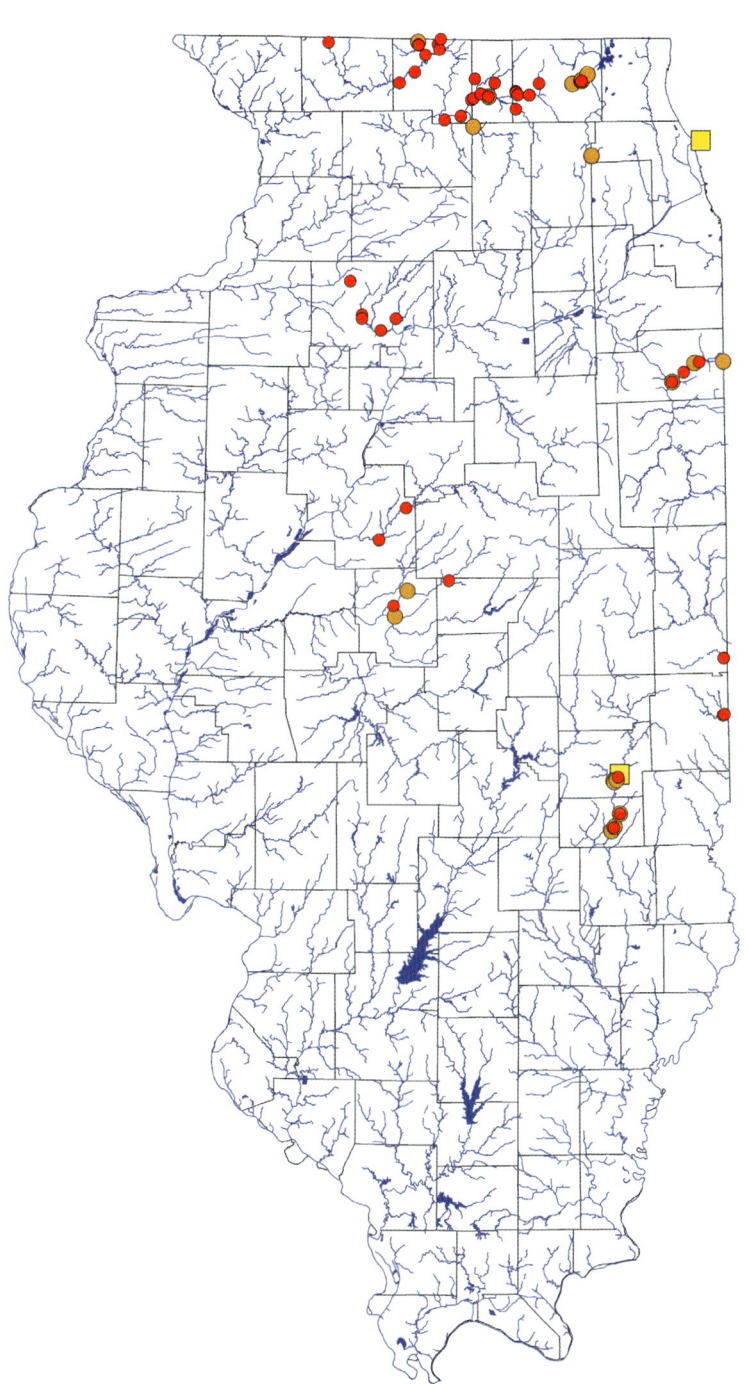

LAMPREYS (Petromyzontidae)

51

SEA LAMPREY
Petromyzon marinus Linnaeus 1758

IDENTIFICATION The Sea Lamprey is an eel-like fish that lacks jaws and paired fins and has 2 widely separated dorsal fins, a round or spatulate caudal fin, 1 median nostril, 7 pairs of pore-like gill openings, and an oral disc with rasping teeth on the tongue. The expanded oral disc is as wide or wider than the head and has large, sharp disc teeth. There are usually 66–75 myomeres on the trunk of the body. There is no black pigment on the lateral-line pores. Adults have prominent black mottling on a blue-gray to olive-brown back, sides, and dorsal-fin bases. The lower body is cream, yellow-white, or light brown. The breeding male has a prominent ropelike ridge on the nape. The species is parasitic. To 47 in. (120 cm); landlocked individuals rarely exceed 25 in. (64 cm).

SIMILAR SPECIES The Silver Lamprey, *Ichthyomyzon unicuspis*, and Chestnut Lamprey, *I. castaneus*, have 1 slightly notched dorsal fin, black pigment on the lateral-line pores, usually fewer than 56 trunk myomeres, and no prominent black mottling on the body.

HABITAT The Sea Lamprey is landlocked in most of the Great Lakes, including the Illinois portion of Lake Michigan. Because adults are parasitic, they attach to other fish species to feed and can be carried almost anywhere in Lake Michigan.

DISTRIBUTION IN ILLINOIS The Sea Lamprey was first recorded in the Illinois part of Lake Michigan in 1950. It had gained access to the Great Lakes from Atlantic Slope basins where it was native through the Erie Barge and Welland Canals. Efforts to eradicate the Sea Lamprey through use of lampricides in tributaries where Sea Lampreys migrate to spawn may have been successful, at least in Illinois. The last recorded observation of the species in Lake Michigan was in 2004. The only record for the species in Illinois other than for Lake Michigan was in the DuPage River in DuPage County in 1957.

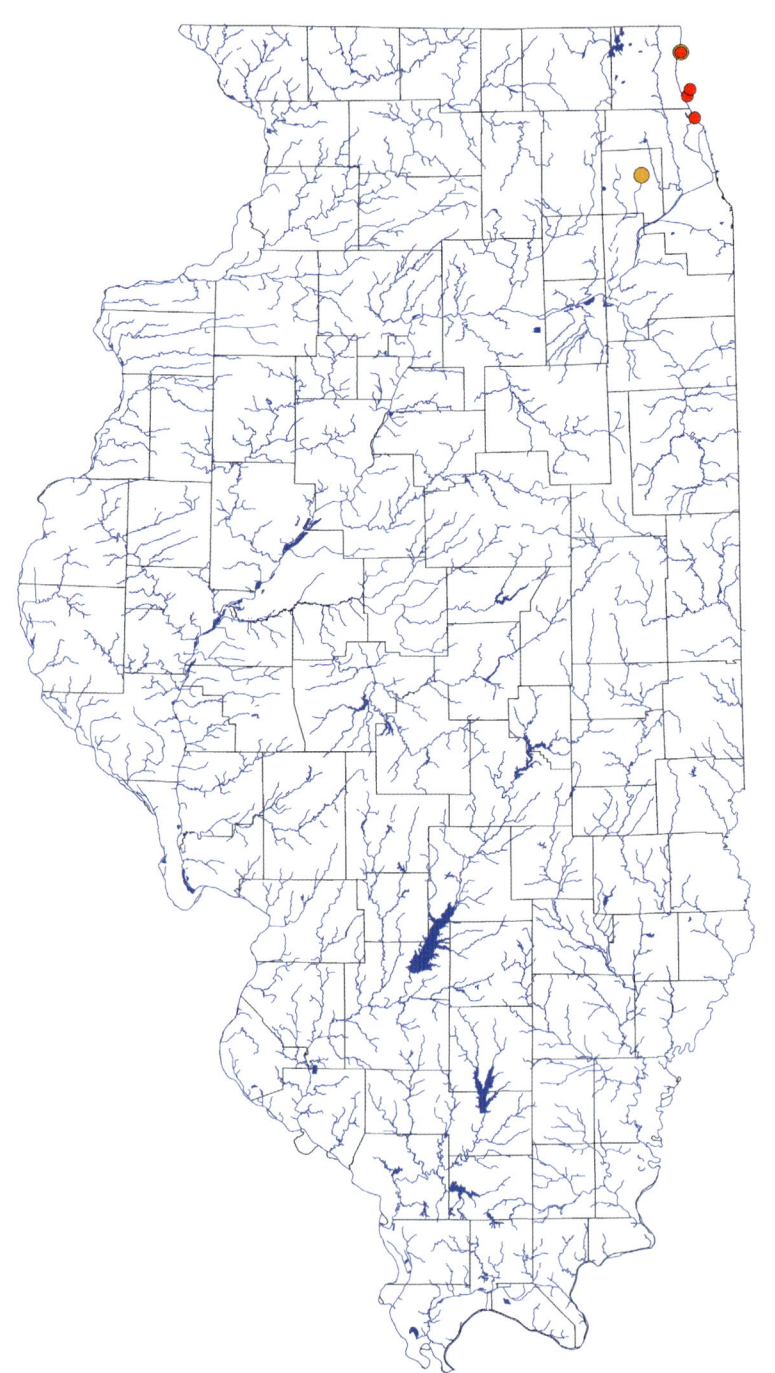

STURGEONS

Family Acipenseridae Bonaparte 1831

The family Acipenseridae is a group of large-bodied ancient fishes that have a row of bony plates (scutes) along the back, 2 rows of scutes on the side of the body, and dorsal and anal fins far back on the body. Their skeleton is composed mostly of cartilage except for ossified bones in the skull, and they have a spiral valve in the intestine, heterocercal caudal fins, and flattened heads with the mouth underneath. Their body morphology is adapted for life on the bottom of rivers and lakes. Members of the family are found in North America, Europe, and Asia, and most occur in fresh water, although some species spend most of their lives in salt water and only return to fresh water to spawn. The eggs of several species are used for caviar. Eight species are known from North America, 3 of which occur in Illinois.

1a Lower lip with 2 smooth lobes; barbels not fringed; caudal peduncle not completely covered with bony plates; length of caudal peduncle (A) less than distance from origin of pelvic fin to posterior edge of anal fin (B); upper lobe of tail fin without a long filament . . . . **Page 56**

Lake Sturgeon (*Acipenser fulvescens*)

1b Lower lip with 4 fringed lobes; barbels fringed; caudal peduncle completely covered with bony plates; length of caudal peduncle (A) about equal in length as distance from origin of pelvic fin to posterior edge of anal fin (B); upper lobe of tail fin with a long filament (frequently broken off) **2**

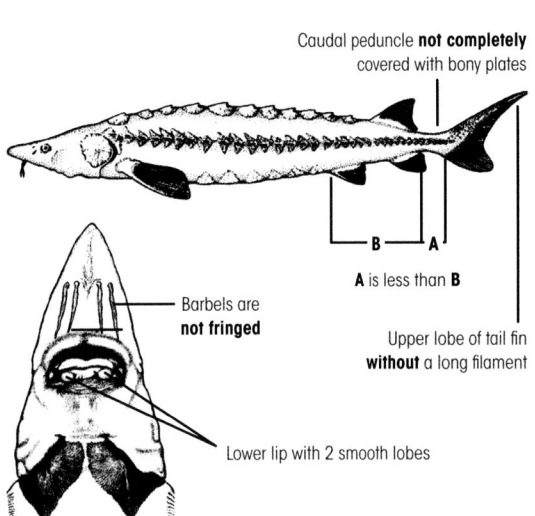

Caudal peduncle **not completely** covered with bony plates

Barbels are **not fringed**

A is less than **B**

Upper lobe of tail fin **without** a long filament

Lower lip with 2 smooth lobes

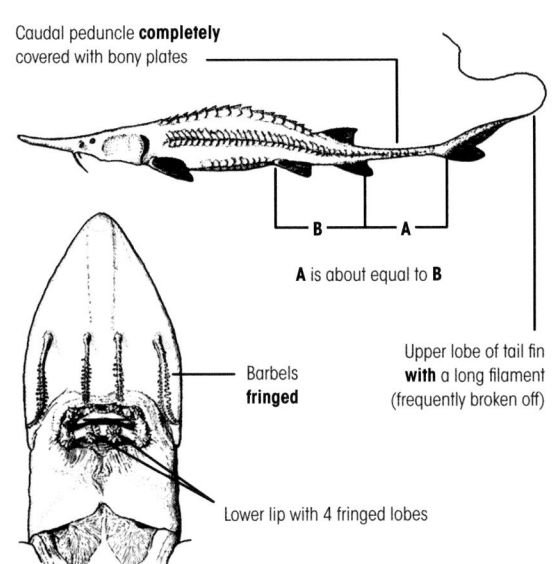

Caudal peduncle **completely** covered with bony plates

A is about equal to **B**

Barbels **fringed**

Upper lobe of tail fin **with** a long filament (frequently broken off)

Lower lip with 4 fringed lobes

2a Bases of outer barbels in line with or ahead of bases of inner barbels; length of inner barbel (A) going less than 6 times into head length (B); anal fin rays 23 or fewer; dorsal fin rays 36 or fewer; belly covered with small bony plates, except in juveniles . . . . **Page 58**

Shovelnose Sturgeon (*Scaphirhynchus platorynchus*)

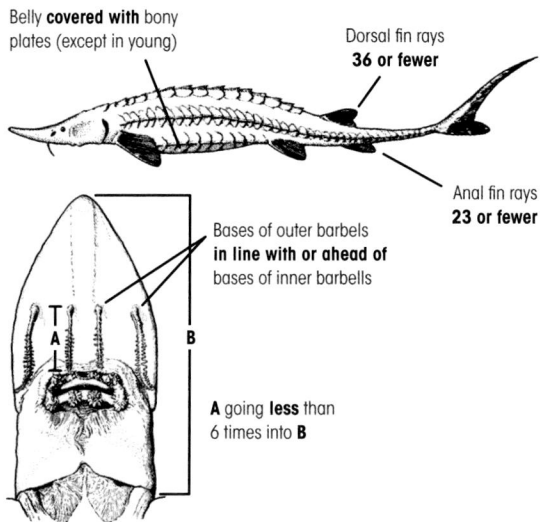

Belly **covered with** bony plates (except in young)

Dorsal fin rays **36 or fewer**

Anal fin rays **23 or fewer**

Bases of outer barbels **in line with or ahead of** bases of inner barbells

A going **less** than 6 times into **B**

2b Bases of outer barbels usually behind bases of inner barbels; length of inner barbel (A) going more than 6 times into head length (B); anal fin rays 24 or more; dorsal fin rays 37 or more; belly without bony plates . . . . . **Page 57**

Pallid Sturgeon (*Scaphirhynchus albus*)

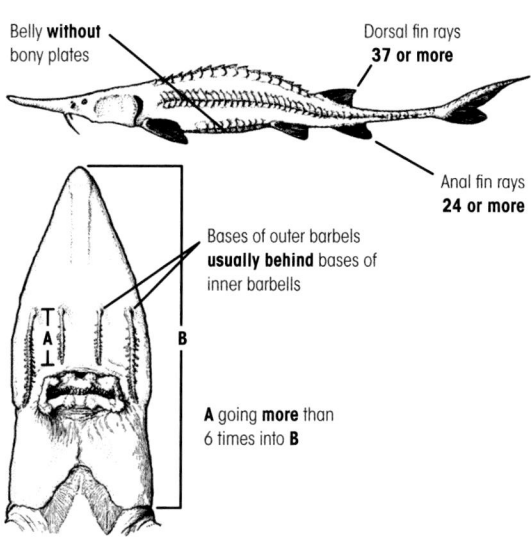

Belly **without** bony plates

Dorsal fin rays **37 or more**

Anal fin rays **24 or more**

Bases of outer barbels **usually behind** bases of inner barbells

A going **more** than 6 times into **B**

LAKE STURGEON *Acipenser fulvescens* Rafinesque 1817

IDENTIFICATION The Lake Sturgeon has a large, elongate, dorsoventrally flattened body, a large, flattened head and pointed snout, a rather flat back, a short caudal peduncle, and a caudal fin without a long filament extending from the upper lobe. Individuals are olive-brown to gray above and on the side with scutes of the same color and cream to white below. Juveniles are usually mottled above and on the side with dark blotches of various size. The fins are dark gray or brown. To 9 ft. (2.7 m).

SIMILAR SPECIES The Pallid Sturgeon, *Scaphirhynchus albus*, and Shovelnose Sturgeon, *S. platorynchus*, have an arched back, long, narrow caudal peduncle, and thin filament extending out from the upper lobe of the caudal fin.

HABITAT The Lake Sturgeon occurs on the bottom of large rivers and shallow areas of large lakes. It is usually encountered over substrates of mud, sand, and gravel.

DISTRIBUTION IN ILLINOIS The Lake Sturgeon is rare in Illinois, with few records from the Forbes and Richardson era to the present. It has been recorded from widely separated localities in the Mississippi and Illinois River basins and Lake Michigan, but it has not been recorded from the upper Illinois River or Lake Michigan since the Forbes and Richardson era. The species was reported to be abundant in Lake Michigan and present in large rivers of the state prior to 1880 (Nelson 1876, Jordan 1878) but rare in those waterbodies by 1909 (Forbes & Richardson 1909). Smith (1979) reported but didn't plot a record for the species in the lower Rock River collected in 1934 and noted that the species was still infrequently encountered in that river.

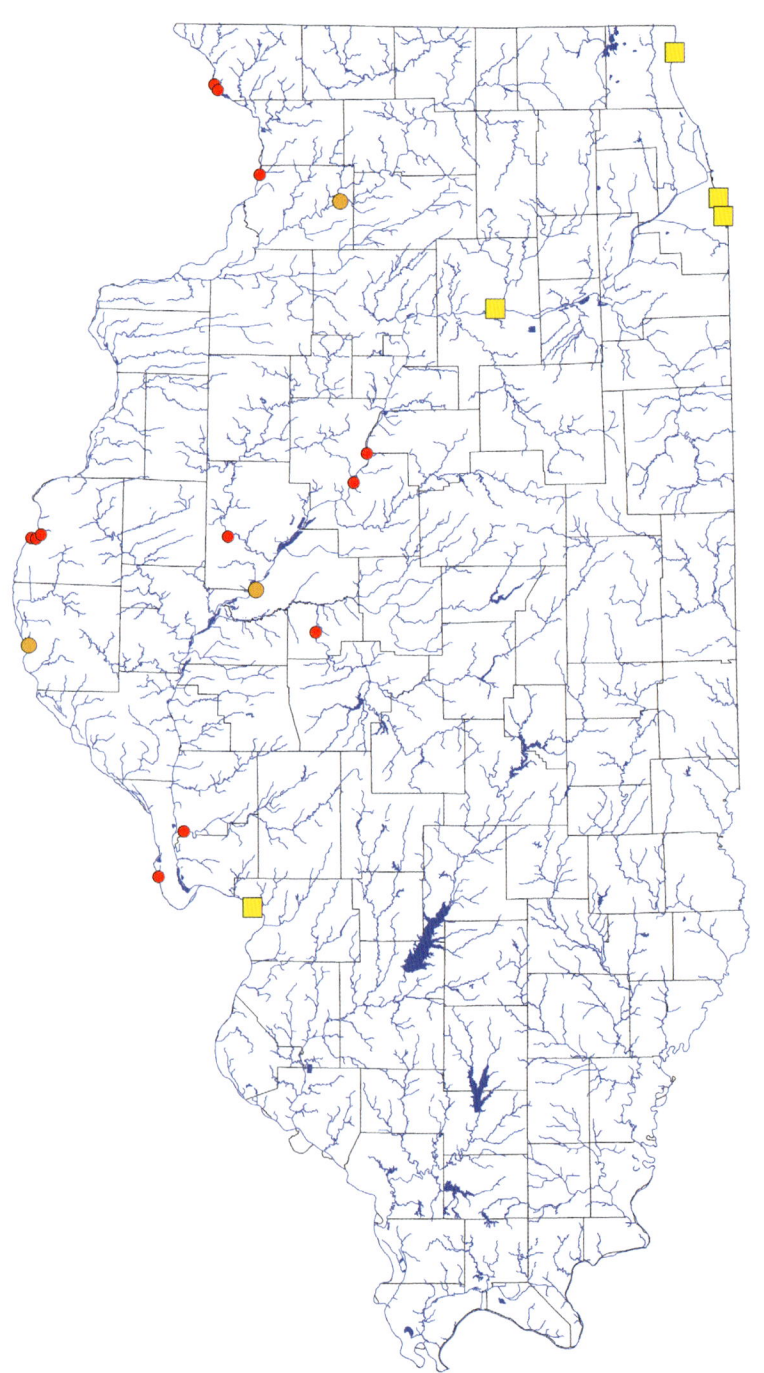

PALLID STURGEON *Scaphirhynchus albus* (Forbes & Richardson 1905)

IDENTIFICATION The Pallid Sturgeon has a large, elongate, dorsoventrally flattened body, a large, flattened head and pointed snout, no scutes on the belly, an arched back, a long, thin caudal peduncle, and a caudal fin with a long filament extending from the upper lobe. The bases of the outer barbels on the bottom of the snout are usually behind the bases of the inner barbels, and the inner barbels are half the length of the outer barbels. Individuals are light gray-brown above and on the side and white below. The fins are light gray to brown. To 6 ft. (1.83 m).

SIMILAR SPECIES The Lake Sturgeon, *Acipenser fulvescens*, has a short caudal peduncle and relatively flat back and lacks a long filament extending from the upper lobe of the caudal fin. The Shovelnose Sturgeon, *S. platorynchus*, has scutes on the belly, the bases of the outer barbels on the bottom of the snout are usually in line with or ahead of the bases of the inner barbels, and the inner barbels are the same length of the outer barbels.

HABITAT The Pallid Sturgeon occurs on the bottom of the main channel or embayments of large, turbid rivers. It is usually encountered in areas with flow over substrates of mud, sand, and gravel.

DISTRIBUTION IN ILLINOIS The Pallid Sturgeon is found only sporadically in the Mississippi River between the mouths of the Missouri and Ohio Rivers.

REMARKS The species is very rare and currently listed as an endangered species under the federal Endangered Species Act (United States Fish and Wildlife Service 2021). Culturing, recovery, and restocking efforts have been underway for 25 years, and restocked individuals can be found in the Mississippi River near St. Louis. The species has been reported to hybridize with the Shovelnose Sturgeon, *S. platorynchus* (Tranah et al. 2004), and in some cases genetic analysis is needed to confirm the identification of Pallid Sturgeon.

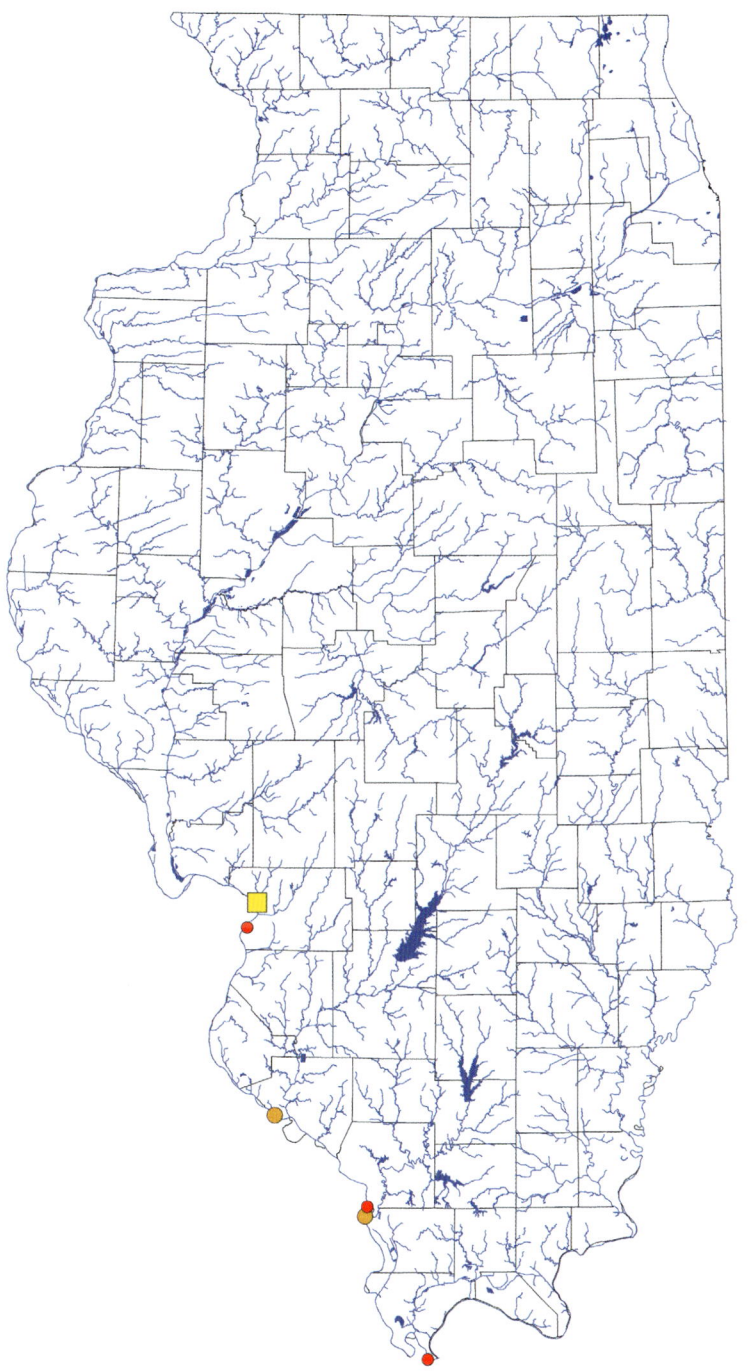

STURGEONS (Acipenseridae)

57

SHOVELNOSE STURGEON   *Scaphirhynchus platorynchus* (Rafinesque 1820)

IDENTIFICATION The Shovelnose Sturgeon has a large, elongate, dorsoventrally flattened body, a large, flattened head and pointed snout, scutes on the belly of all but young individuals, an arched back, a long, thin caudal peduncle, and a caudal fin with a long filament extending from the upper lobe. The bases of the outer barbels on the bottom of the snout are usually in line with or in front of the bases of the inner barbels, and the inner barbels are nearly the same length of the outer barbels. Individuals are light brown to dusky gray above and on the side and white below. The fins are light brown. To 43 in. (1.1 m).

SIMILAR SPECIES The Lake Sturgeon, *Acipenser fulvescens*, has a short caudal peduncle, relatively flat back, and lacks a long filament extending from the upper lobe of the caudal fin. The Pallid Sturgeon, *S. albus*, lacks scutes on the belly, the bases of the outer barbels on the bottom of the snout are usually below the bases of the inner barbels, and the inner barbels are half the length of the outer barbels.

HABITAT The Shovelnose Sturgeon occurs on the bottom of the main channel or embayments of large, turbid rivers. It is usually encountered in areas with flow over substrates of mud, sand, and gravel.

DISTRIBUTION IN ILLINOIS Forbes and Richardson (1909) discussed the presence of the Shovelnose Sturgeon in Illinois in the Mississippi and Illinois Rivers, including its decline in the Illinois River, but provided no specific localities. In the contemporary era it is found in the Mississippi, lower Illinois, and Wabash Rivers and is noticeably absent from the Ohio River. Smith (1979) also recorded the species in these rivers and from the extreme lower reaches of the Vermilion, Kaskaskia, and Big Muddy Rivers.

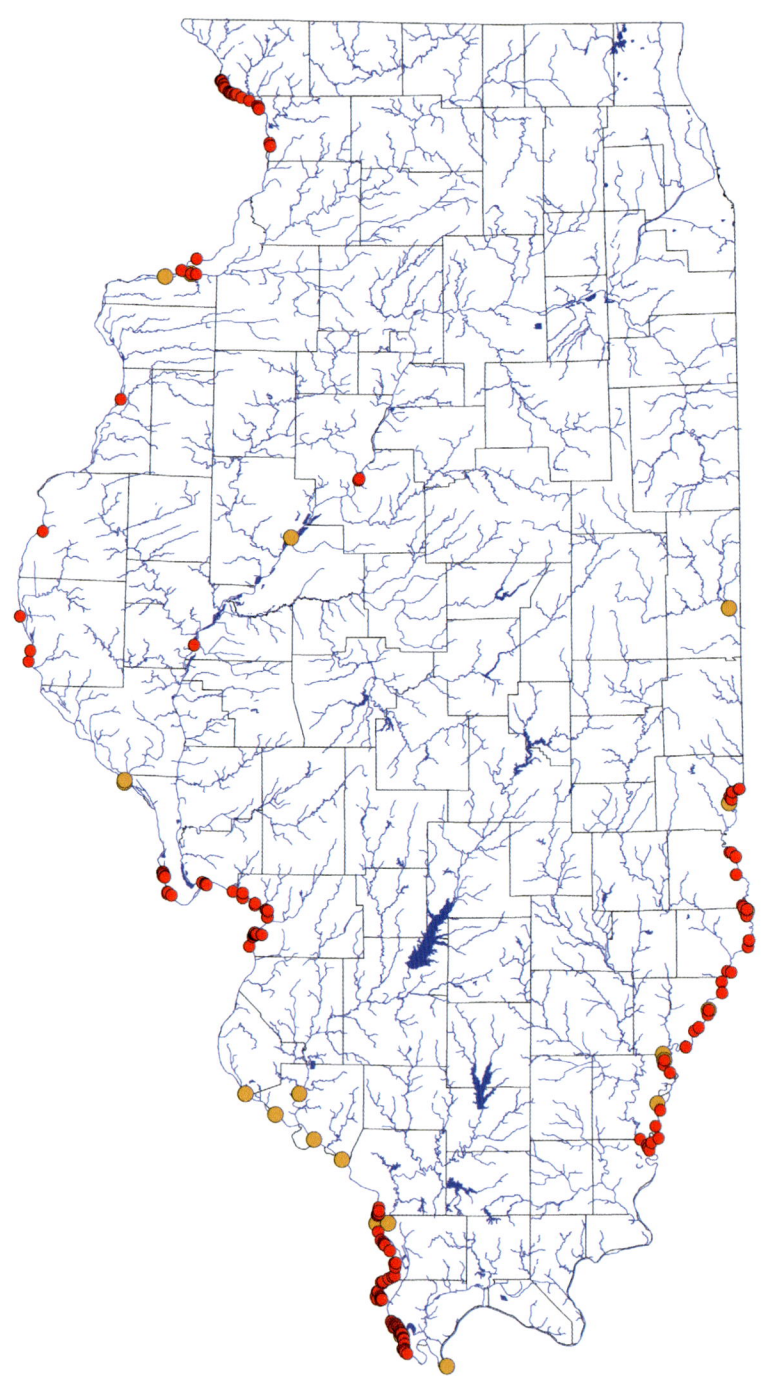

PADDLEFISHES

Family Polyodontidae Bonaparte 1835

Paddlefishes are a group of ancient fishes represented today by a single genus and species, *Polyodon spathula*. Until recently a second species, the Chinese Paddlefish, *Psephurus gladius*, occurred in the lower Yangtze River basin. That species is now thought to be extinct. Paddlefish have a long, paddle-shaped snout, a large mouth used for filter feeding, a spiral valve intestine, a skeleton composed of cartilage, and a heterocercal tail. The paddle-shaped snout is covered with taste buds that help the species locate pelagic food items. The Paddlefish can reach in excess of 59 in. (1.5 m) and is recreationally harvested in some localities. Its eggs are eaten as caviar.

PADDLEFISH *Polyodon spathula* (Walbaum 1792)

IDENTIFICATION The Paddlefish has a large, elongate body, a very distinctive paddle-shaped snout, an elongate flap on the rear of the gill covering, and a heterocercal tail. Individuals are gray to blue-gray above and on the side, with both the top and sides often mottled with small, dark spots and white below. Fins are gray to grayish-brown. To 87 in. (2.2 m).

SIMILAR SPECIES No other fish in Illinois has a long, paddle-shaped snout and heterocercal tail.

HABITAT The Paddlefish occurs in slower-flowing sections of large rivers, floodplain lakes, and impoundments. The species is usually encountered in waters deeper than 4 ft. (1.2 m).

DISTRIBUTION IN ILLINOIS Forbes and Richardson (1909) reported the Paddlefish from the Mississippi and Illinois Rivers. In the Smith era it was recorded from these same rivers and from the Wabash, Kaskaskia, and Big Muddy Rivers. Forbes and Richardson (1909) and Smith (1979) discussed the decline of this species in Illinois, and it is apparent that in the contemporary era the species is much less widespread in the upper Mississippi River and probably in the Illinois River than it was in previous eras.

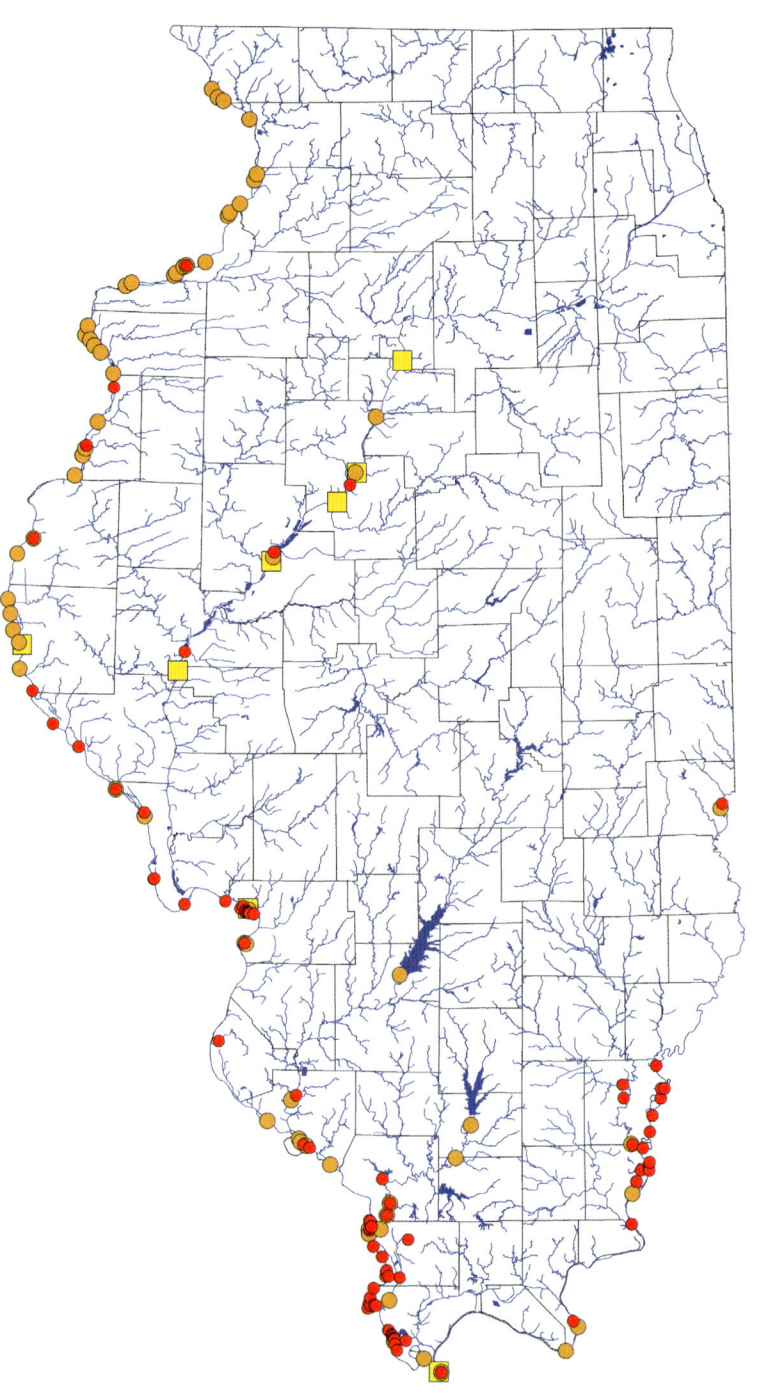

GARS

Family Lepisosteidae Bonaparte 1835

Gars are a group of distinctive fishes easily recognized by their long, sharply toothed jaws, heavily armored bodies with diamond-shaped and nonoverlapping ganoid scales, and dorsal and anal fins far back on the body. All gars are large, and 1 species exceeds 10 ft. (3 m) in length. Gars have a spiral valve in the intestine to aid in digestion, a lunglike gas bladder used to breathe in poorly oxygenated water, and an abbreviated (rounded externally) heterocercal caudal fin. Small individuals (< 10 in., or 25 cm) have a fleshy filament extending above the caudal fin. Very small individuals have an adhesive pad on the snout, which they use to cling to vegetation. As adults, gars feed on fishes that they catch by lying quietly and ambushing with a rapid lunge forward or with a sideways swipe of their long snout. The 7 extant species of gars are native to Cuba, North America, and Central America, where they inhabit lowland lakes, swamps, and quiet areas of large creeks and rivers; 2 species occasionally enter brackish and marine environments. Four species occur in Illinois.

KEY TO THE GARS (Lepisosteidae)

1a Snout long and narrow, its least width (A) going more than 10 times into its length (B); width of upper jaw at nostrils (C) less than eye diameter (D) . . . . **Page 66**

Longnose Gar (*Lepisosteus osseus*)

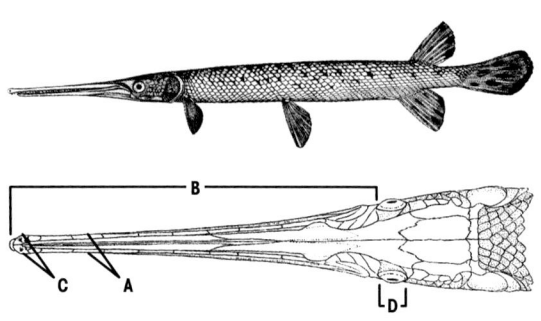

A going more than 10 times into **B**
C less than **D**

1b Snout short and broad, its least width (A) going less than 10 times into its length (B); width of upper jaw at nostrils (C) greater than eye diameter. **2**

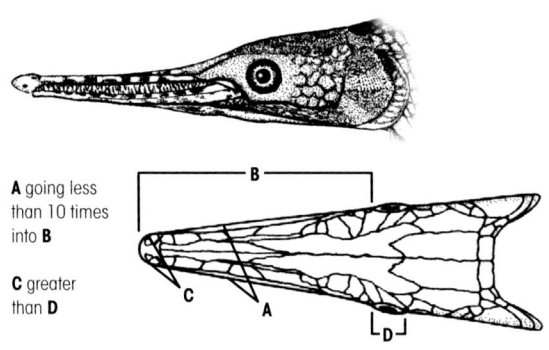

A going less than 10 times into **B**

C greater than **D**

2a Snout very short and broad, its least width (A) going 4.5 or fewer times into its length (B); distance from tip of snout to corner of mouth (C) shorter than rest of head (D); gill rakers on 1st gill arch 60 or more; young with a pale stripe along top of head and back . . . . **Page 64**

Alligator Gar (*Atractosteus spatula*)

C shorter than **D**

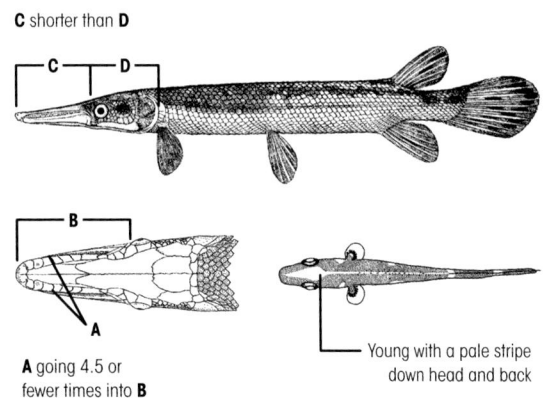

A going 4.5 or fewer times into **B**

Young with a pale stripe down head and back

2b Snout longer and not as broad, its least width (A) going 5 or more times into its length (B); distance from tip of snout to corner of mouth (C) longer than rest of head (D); gill rakers on 1st gill arch 30 or fewer; young lack a pale stripe on top of head and back **3**

C longer than **D**

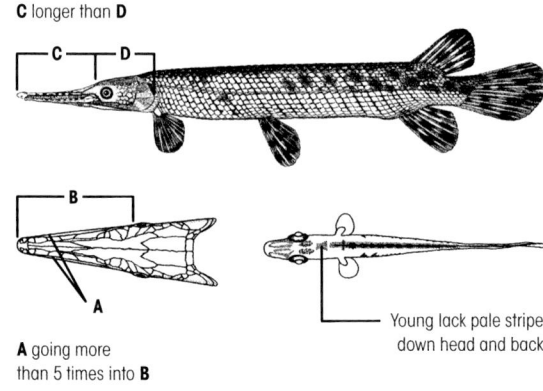

A going more than 5 times into **B**

Young lack pale stripe down head and back

3a Top of head, sides of lower jaw, pectoral fins, and pelvic fins with dark irregularly placed blotches; scales in lateral series (including small scale at base of caudal fin) usually 54–58 . . . . **Page 65**

Spotted Gar (*Lepisosteus oculatus*)

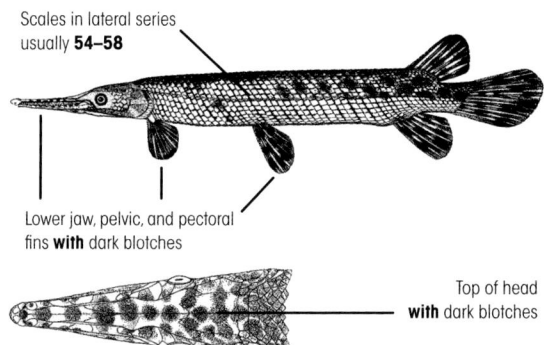

Scales in lateral series usually **54–58**

Lower jaw, pelvic, and pectoral fins **with** dark blotches

Top of head **with** dark blotches

3b Top of head, sides of lower jaw, pectoral fins and pelvic fins without dark blotches; scales in lateral series usually 60–64 . . . . **Page 67**

Shortnose Gar (*Lepisosteus platostomus*)

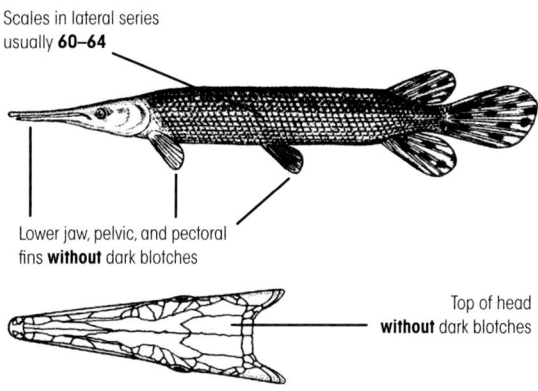

Scales in lateral series usually **60–64**

Lower jaw, pelvic, and pectoral fins **without** dark blotches

Top of head **without** dark blotches

ALLIGATOR GAR *Atractosteus spatula* (Lacepède 1803)

IDENTIFICATION The Alligator Gar is one of the largest fish in North America, with individuals reported to reach 12 ft. (3.7 m) in length. It is the largest fish in Illinois. It is an elongate fish with a short, broad snout, an upper jaw that is shorter than the rest of the head, and 2 rows of teeth on the upper jaw. Individuals are dark olive-brown (sometimes black) on the back and upper side and white to pale yellow on the lower side and belly. The sides may occasionally be spotted. Fins are dark brown, with dark spots on the median fins. Juveniles have a light stripe along the back from the tip of the snout to upper base of the caudal fin. There are 58–62 lateral scales and 59–66 rakers on the first gill arch. To 12 ft. (3.7 m).

SIMILAR SPECIES The Spotted Gar, *Lepisosteus oculatus*, Shortnose Gar, *L. platostomus*, and Long-nose gar, *L. osseus*, have the upper jaw longer than the rest of the head and fewer than 40 rakers on the first gill arch.

HABITAT The Alligator Gar occurs in swamps, sloughs, and sluggish backwaters of large rivers.

DISTRIBUTION IN ILLINOIS The Alligator Gar was known in the Forbes and Richardson era from 1 locality in the Big Muddy River and from reports of the species from the Mississippi above St. Louis and the lower Illinois River. Smith (1979) reported it from several southern Illinois localities, based mostly on anglers' reports and photographs. The species was probably never common in Illinois and was likely extirpated by the 1980s. In 2010 the Illinois Department of Natural Resources began a culture and reintroduction program, and the species was stocked in the lower Illinois and Kaskaskia Rivers. All contemporary records for the Alligator Gar, which include records through 2018, are likely a result of that effort.

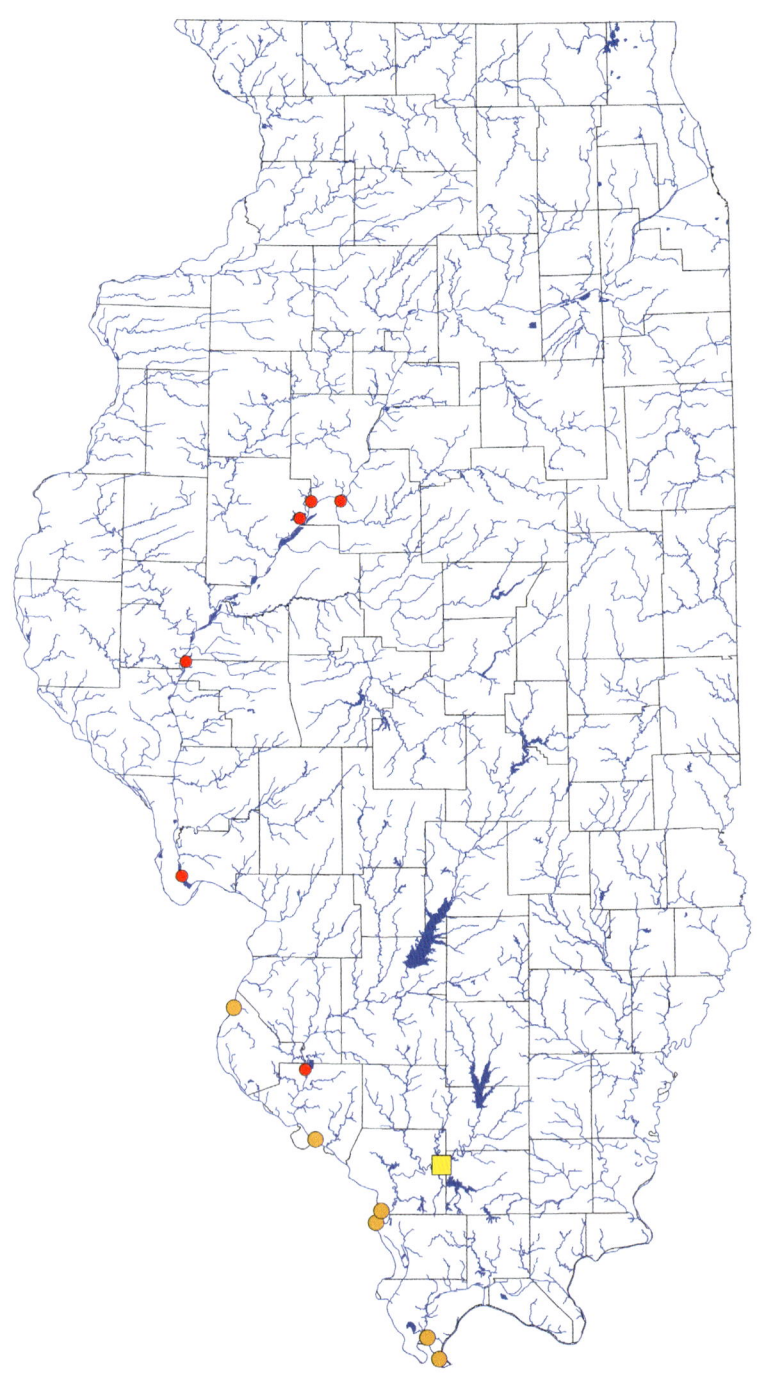

SPOTTED GAR *Lepisosteus oculatus* Winchell 1864

IDENTIFICATION The Spotted Gar is an elongate fish that has a long snout with its upper jaw longer than the rest of its head and with 1 row of teeth. It is olive-brown to black on the back and upper side and white to pale yellow on the lower side and belly. It has many large, dark olive-brown to black spots on the body, head, and all fins. Juveniles have dark stripes along the back and side. There are 54–58 lateral scales and 15–24 rakers on the first gill arch. To 44 in. (1.1 m).

SIMILAR SPECIES The Alligator Gar, *Atractosteus spatula*, has an upper jaw shorter than the rest of the head and more than 50 rakers on the first gill arch. The Longnose gar, *L. osseus*, lacks large, dark spots on the top of the head and snout and has a very long, thin snout that is twice as long as the rest of the head. The Shortnose Gar, *L. platostomus*, lacks large, olive or black spots on top of the head and snout and has 59–65 lateral scales.

HABITAT The Spotted Gar occurs in clear, quiet pools and backwaters of small to large rivers, swamps, and sloughs. The species is most frequently encountered near vegetation.

DISTRIBUTION IN ILLINOIS The Spotted Gar was recorded in the Forbes and Richardson era from only a handful of scattered localities in the Green, Illinois, and Mississippi Rivers. In the Smith era the species was reported from the middle Illinois River and in scattered localities in southern Illinois. Contemporary-era records show the species now to be more widespread, especially in the southern one-third of the state.

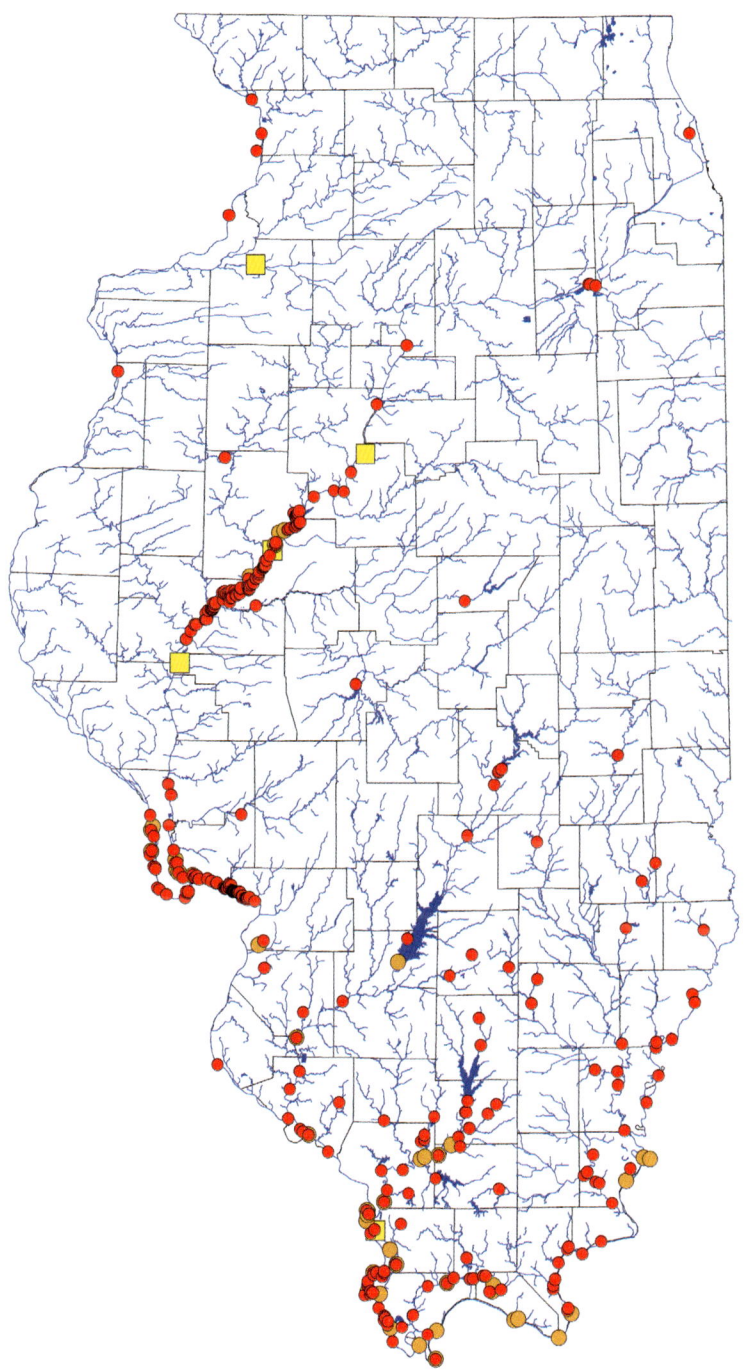

LONGNOSE GAR   *Lepisosteus osseus* (Linnaeus 1758)

IDENTIFICATION The Longnose Gar is an elongate fish that has a very long, thin snout with its upper jaw twice the length of the rest of the head and with 1 row of teeth. The back and upper side are olive-brown, and the lower side and belly are silvery-white to yellow. A wide, dark stripe is sometimes present along the side. Dark spots are present on the median fins and on the body in individuals from clear water. Juveniles have a narrow brown stripe along the back. There are 57–63 lateral scales and 14–31 rakers on the first gill arch. To 72 in. (1.8 m).

SIMILAR SPECIES The Alligator Gar, *Atractosteus spatula*, has an upper jaw shorter than the rest of the head and more than 50 rakers on the first gill arch. The Spotted Gar, *L. oculatus*, and Shortnose Gar, *L. platostomus*, have snouts that are longer than the head but not as long as the Longnose snout.

HABITAT The Longnose Gar is found most often near woody debris and vegetation in quiet areas of creeks and small to large rivers but also can be found in flowing water.

DISTRIBUTION IN ILLINOIS The Longnose Gar occurs statewide in large creeks and rivers. Its range in Illinois has not changed since the time of Forbes and Richardson except, as noted by Smith (1979), it no longer occurs in Lake Michigan.

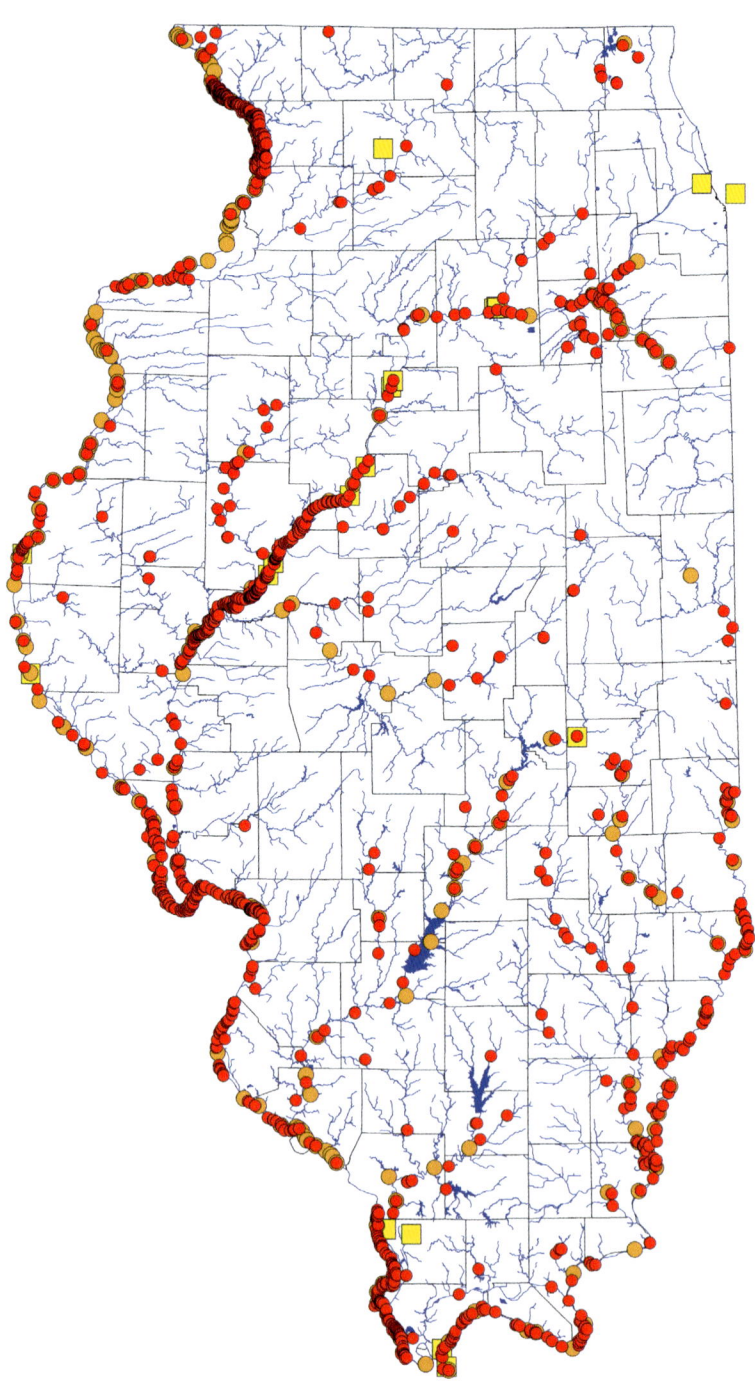

SHORTNOSE GAR *Lepisosteus platostomus* Rafinesque 1820

IDENTIFICATION The Shortnose Gar is an elongate fish that has a short, broad snout with an upper jaw slightly longer than the rest of the head and with 1 row of teeth. The back and upper side are olive or brown, and the underside is white to light silver. The median fins have black spots, but the paired fins usually lack spots. Some black spots may be present on the head of individuals found in clear water. Juveniles have fairly broad, dark brown stripes along the back and side. There are 60–64 lateral scales and 16–25 rakers on the first gill arch. To 33 in. (83 cm).

SIMILAR SPECIES The Alligator Gar, *Atractosteus spatula*, has an upper jaw shorter than the rest of the head and more than 50 rakers on the first gill arch. The Spotted Gar, *L. oculatus*, has large, dark olive or black spots on the top of the head and snout and 53–59 lateral scales. The Longnose Gar, *L. osseus*, has a very long snout going more than twice the length of the head.

HABITAT The Shortnose Gar occurs in quiet areas of creeks and small to large rivers and in lowland lakes and swamps. The species is frequently encountered near vegetation.

DISTRIBUTION IN ILLINOIS The Shortnose Gar has essentially the same distribution in the contemporary era as it had during the Smith era, although it is more widespread in the Illinois River basin in central Illinois. It is absent from much of northern and east-central Illinois.

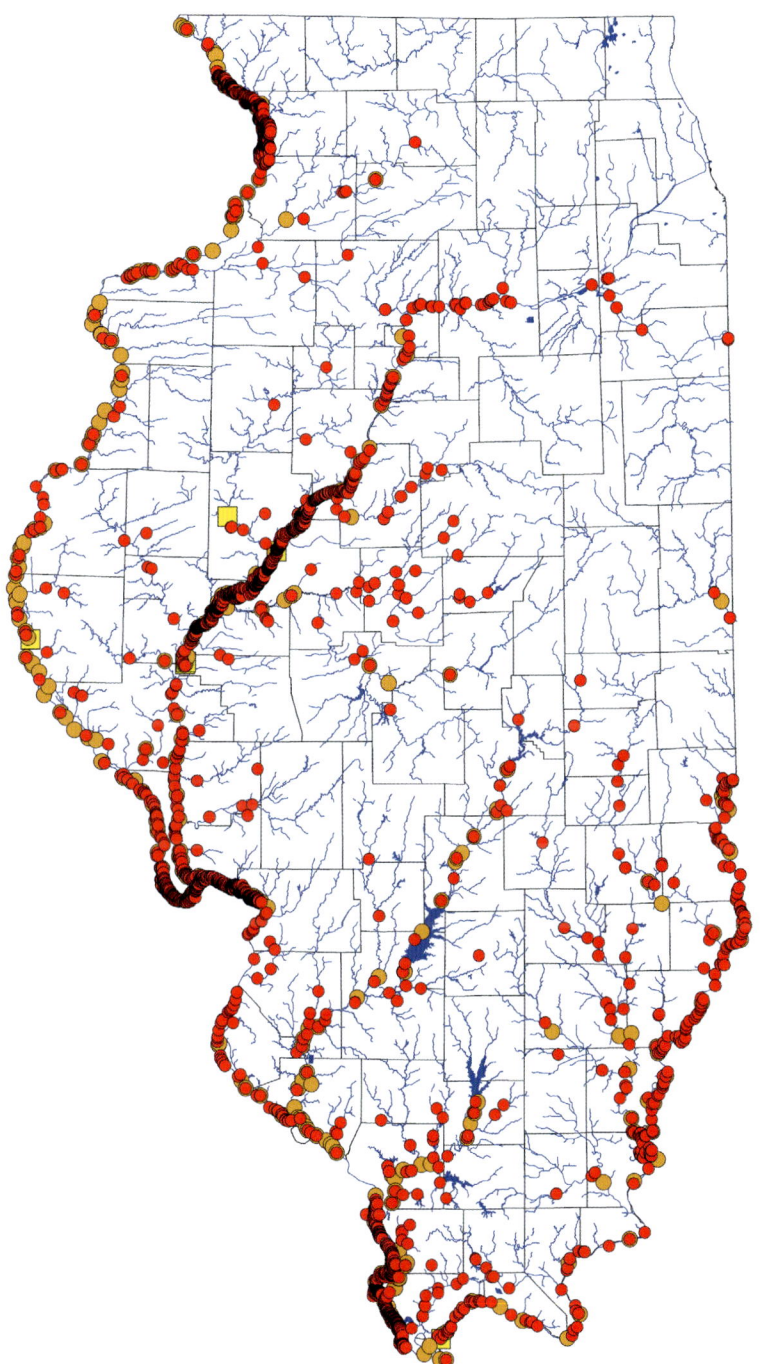

BOWFINS

Family Amiidae Bonaparte 1831

The family Amiidae is a group of ancient fishes represented today by a single genus and species. Fossils of members of the family have been found on all continents except Antarctica (Grande & Bemis 1999), and the 1 extant member is found only in the eastern United States and Canada. That 1 extant species, *Amia calva*, is distinctive in that it possesses several primitive characters, such as a gular plate on the throat, an abbreviated heterocercal caudal fin, and a lung-like gas bladder connected to its stomach. This latter feature, in addition to the rigid structures of the gill filaments and lamellae that prevent their collapse (Daxobeck et al. 1981), allows Bowfin to survive long periods of exposure to air.

BOWFIN *Amia calva* Linnaeus 1766

IDENTIFICATION The Bowfin has a long, nearly cylindrical body, a large head and large, terminal mouth, and a long dorsal fin that is more than half the length of the body. It has a large, bony gular plate on the throat, tubular nostrils, and rounded pectoral, pelvic, and caudal fins. Individuals are mottled olive above and on the upper side and cream-yellow to pale green below. Juveniles and smaller adults can have dark bands or converging stripes and a yellow to orange halo around a prominent black spot near the base of upper caudal fin rays. Adults have a less distinctive dark spot at the base of the upper caudal fin rays. The dorsal and caudal fins have black bands. During the breeding season males have brilliant turquoise-green lips, throat, belly, and ventral fins. To 43 in. (1.1 m).

SIMILAR SPECIES No other fish in Illinois has a long, cylindrical body with a dark caudal spot and a bony gular plate on the throat. The young of the species superficially resemble the Central Mudminnow, *Umbra limi*, but that species lacks the long dorsal fin and gular plate.

HABITAT The Bowfin occurs in swamps, sloughs, lakes, impoundments, and pools and backwaters of creeks and rivers of all sizes. It is usually encountered near vegetation.

DISTRIBUTION IN ILLINOIS The Bowfin's distribution in Illinois has changed little since the Forbes and Richardson era. It is most commonly encountered in low-gradient habitats across the southern one-third of the state and in the pools and backwaters of large rivers.

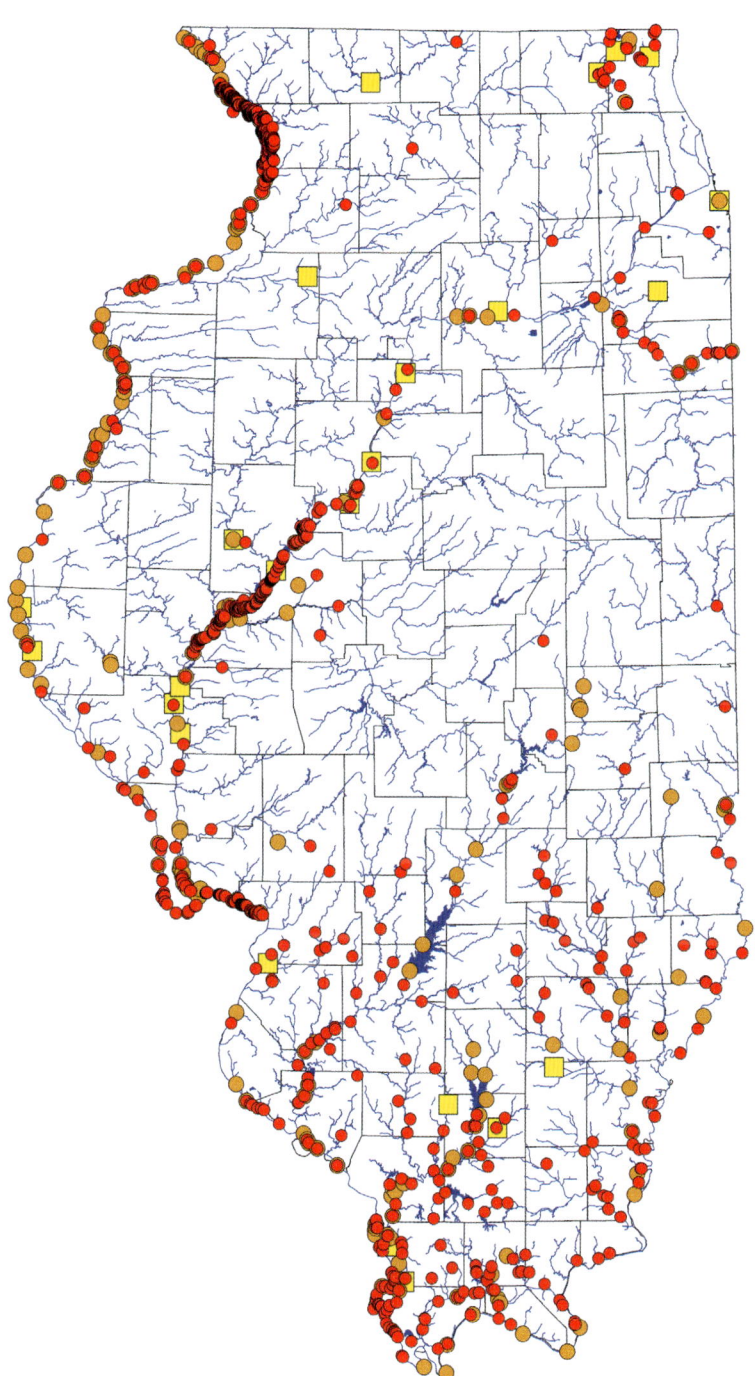

MOONEYES

Family Hiodontidae Valenciennes 1847

An exclusively North American family containing only 2 living species. Mooneyes are silvery to golden fishes that resemble shad (Clupeidae) but have a lateral line and an untoothed keel along the belly. They are strongly compressed, large-eyed fishes with 1 dorsal fin, no spines in the fins, no scales on the head, cycloid scales on the body, an adipose eyelid, abdominal pelvic fins, an axillary process at the base of the pelvic fin, and a deeply forked caudal fin. Mooneyes feed on aquatic invertebrates and small fishes. Eggs of Mooneyes are semi-buoyant and drift downstream following an upstream migration and spawning by adults.

KEY TO THE MOONEYES (Hiodontidae)

1a Front of dorsal fin even with or slightly behind front of anal fin; dorsal fin with 9–10 principal rays; maxillary long, extending past middle of eye; keel on midline of belly extending forward from vent nearly to pectoral fin bases. . . . . **Page 74**

1b Front of dorsal fin distinctly in front of anal fin; dorsal fin with 11–12 principal rays; maxillary shorter, not extending past middle of eye; keel on midline of belly extending forward from vent only as far as pelvic fin bases. . . . . **Page 75**

Goldeye (*Hiodon alosoides*)

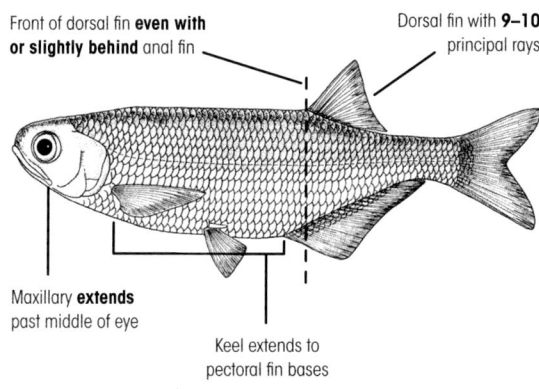

Front of dorsal fin **even with or slightly behind** anal fin

Dorsal fin with **9–10** principal rays

Maxillary **extends** past middle of eye

Keel extends to pectoral fin bases

Mooneye (*Hiodon tergisus*)

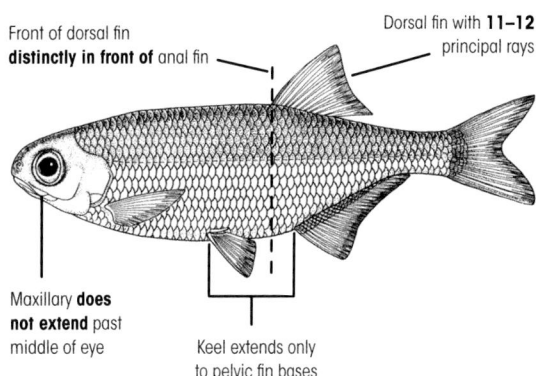

Front of dorsal fin **distinctly in front of** anal fin

Dorsal fin with **11–12** principal rays

Maxillary **does not extend** past middle of eye

Keel extends only to pelvic fin bases

GOLDEYE *Hiodon alosoides* (Rafinesque 1818)

IDENTIFICATION The Goldeye is deep and strongly compressed with an untoothed keel along the belly from the pectoral fin base to the anal fin. The back and upper side are silvery blue-green, the lower side is silver-white, and fins are clear or dusky. The snout is blunt and rounded, and the mouth is large, reaching behind the pupil of the eye. The eye is yellow-gold. The origin of the dorsal fin is opposite or behind the origin of the anal fin. There are 57–62 lateral scales, usually 9–10 dorsal rays, and 29–34 anal rays. To 20 in. (51 cm).

SIMILAR SPECIES The Mooneye, *H. tergisus*, has the origin of the dorsal fin in front of the anal-fin origin, the fleshy keel along the belly extending from the pelvic-fin base to the anal fin, usually 11–12 dorsal rays, and 26–29 anal rays.

HABITAT The Goldeye lives in the open water of pools and channels of medium to large rivers.

DISTRIBUTION IN ILLINOIS With the exception of the Rock and Fox River basins, the Goldeye occurs essentially statewide in medium to large rivers. It is most frequently encountered in the largest, most turbid rivers, including the Illinois River and the Mississippi River below the mouth of the Illinois River. It appears to be less widespread in the contemporary era than it was in the Smith era in the upper Mississippi, Kaskaskia, and Big Muddy Rivers, all of which have impoundments that probably interfere with migratory habits of the Goldeye (McPhail & Lindsey 1970, Smith 1979).

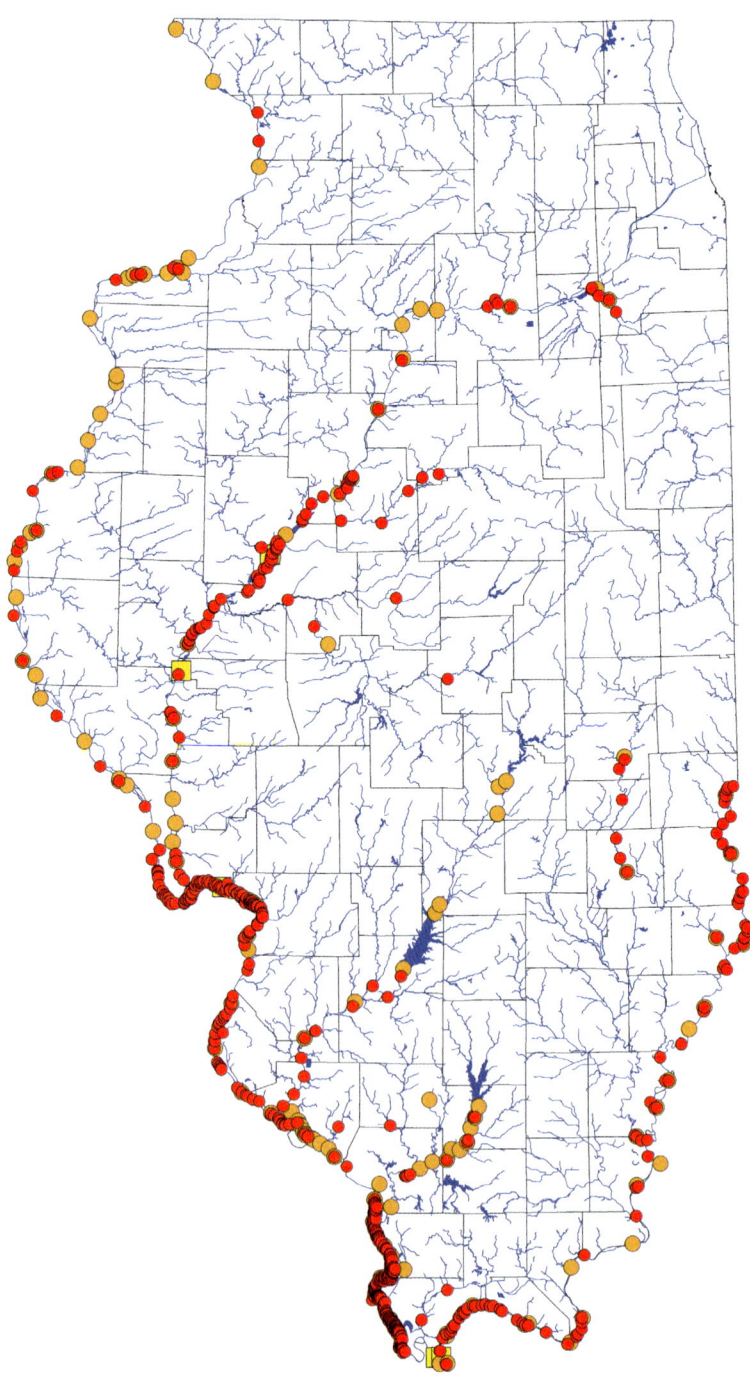

MOONEYE *Hiodon tergisus* Lesueur 1818

IDENTIFICATION The Mooneye is deep and strongly compressed with an untoothed keel along the belly from the pelvic-fin base to the anal fin. The eye is large and silver-gold. The origin of the dorsal fin is in front of the origin of the anal fin. The snout is blunt and rounded, and the mouth is large, extending to below the pupil of the eye. The back and upper side are silver with a blue-green tint, the lower side is silver-white, and fins are clear or slightly dusky. There are 52–57 lateral scales, usually 11–12 dorsal rays, and 26–29 anal rays. To 19 in. (48 cm).

SIMILAR SPECIES The Goldeye, *H. alosoides*, has the origin of the dorsal fin opposite or behind the origin of the anal fin, the fleshy keel along the belly extending from the pectoral fin base to the anal fin, usually 9–10 dorsal rays, and 29–34 anal rays.

HABITAT The Mooneye lives in the open water of pools and backwaters of medium to large rivers, lakes, and impoundments.

DISTRIBUTION IN ILLINOIS Like the Goldeye, *H. alosoides*, the Mooneye is found statewide in medium to large rivers except for the Fox River. However, it is more frequently recorded than the Goldeye in the upper Mississippi River and less commonly found in the Wabash and Ohio Rivers. Like the Goldeye, it appears to be less widespread in the contemporary era in the Kaskaskia and Big Muddy Rivers than it was in the Smith era, presumably because of the presence of large impoundments.

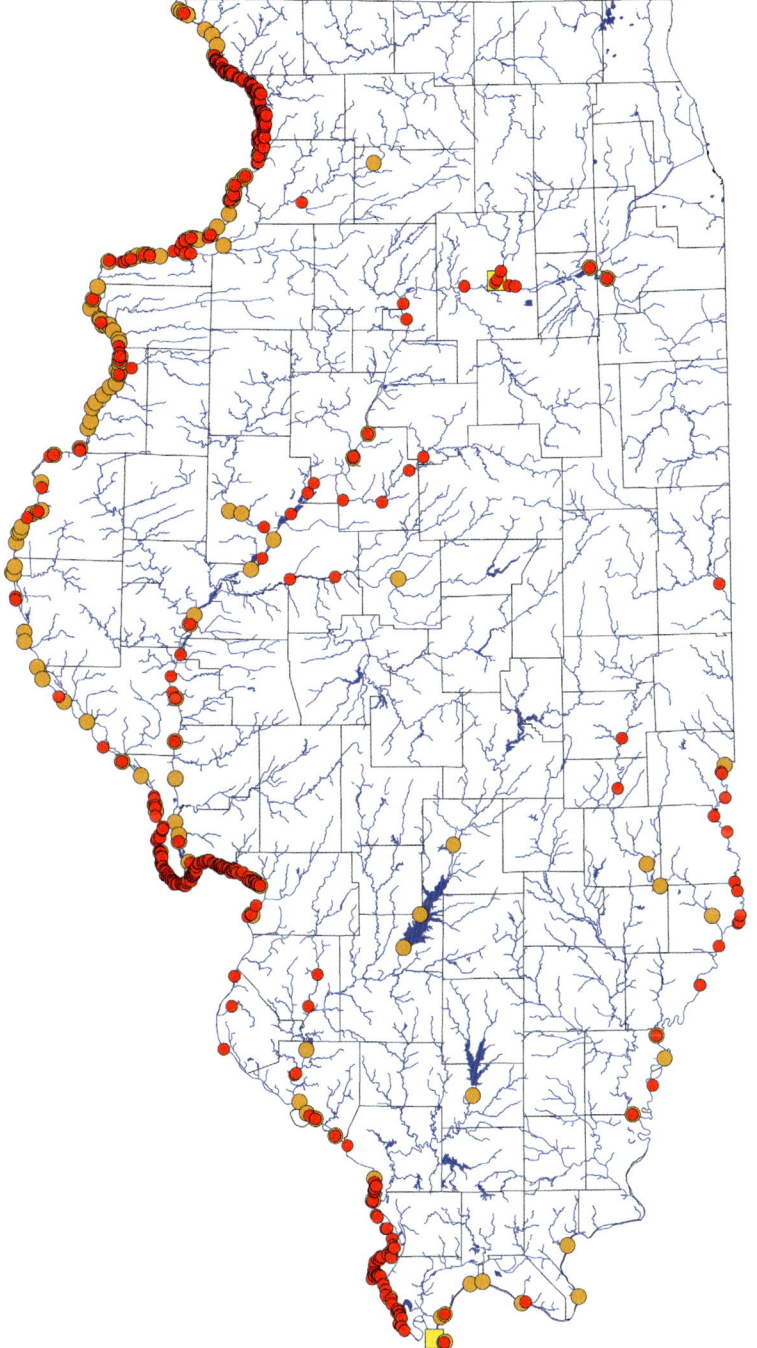

MOONEYES (Hiodontidae)

SHADS AND HERRINGS

Family Clupeidae Cuvier 1816

Clupeids are mostly silvery fishes with a strongly compressed body, no scales on the head, 1 dorsal fin, a conspicuous adipose eyelid, abdominal pelvic fins, an axillary process just above the base of the pelvic fin, no adipose fin, and no spines in the fins. Most have no lateral line; a few species have a short lateral line. The belly has sharply pointed scales, creating a saw-toothed edge. Most herrings and shads are marine, but a few species live only in fresh water, and some marine species frequently enter fresh water. Clupeids typically swim in schools, and most feed on plankton. Included among the 216 species of clupeids are sardines, menhaden, shads, and other commercially harvested and economically important fishes. Five species are found in Illinois, of which 4 are native and 1, the Alewife, has been introduced.

KEY TO THE SHAD AND HERRINGS (Clupeidae)

1a Last ray of dorsal fin elongated into a long, slender filament; principal rays of dorsal fin usually 15 or fewer; median line in front of dorsal fin unscaled **2**

1b Last ray of dorsal fin not elongated, similar in length to preceding ray; principal rays of dorsal fin usually 16 or more; median line in front of dorsal fin scaled **3**

Median line in front of dorsal fin unscaled | Principal rays of dorsal fin **15 or fewer** | Last ray of dorsal fin **elongated** into a long, slender filament

Median line in front of dorsal fin scaled | Principal rays of dorsal fin **16 or more** | Last ray of dorsal fin **not elongated** into a long, slender filament

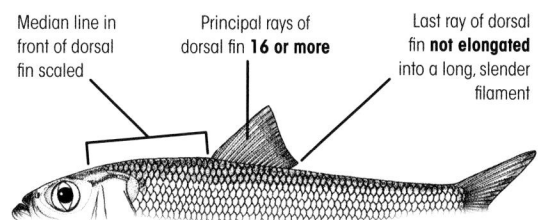

2a Tip of lower jaw projecting beyond tip of snout; upper jaw without a notch on ventral margin; rays in anal fin usually 20–25; scales in lateral series usually 41–48; caudal fin yellow in life . . . .**Page 83**

2b Tip of lower jaw not projecting beyond tip of snout; upper jaw with notch present on ventral margin; rays in anal fin usually 29–35; scales in lateral series usually 52–70; caudal fin dusky or black in life . . . .**Page 82**

Threadfin Shad (*Dorosoma petenense*)

Upper jaw **without** a notch on ventral margin | Caudal fin **yellow** in life

Tip of lower jaw **projecting** beyond tip of snout | Rays in anal fin usually **20–25**

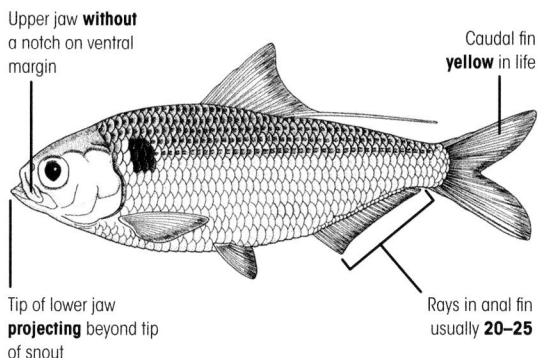

Gizzard Shad (*Dorosoma cepedianum*)

Upper jaw **with** a notch on ventral margin | Caudal fin **dusky or black** in life

Tip of lower jaw **not projecting** beyond tip of snout | Rays in anal fin usually **29–35**

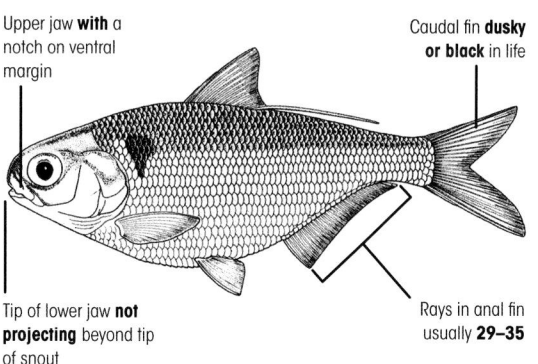

3a Tip of lower jaw projecting far beyond tip of snout when jaws are closed and viewed from the side; gill rakers on lower limb of first arch usually 20–24. . . . .**Page 80**

Skipjack Herring (*Alosa chrysochloris*)

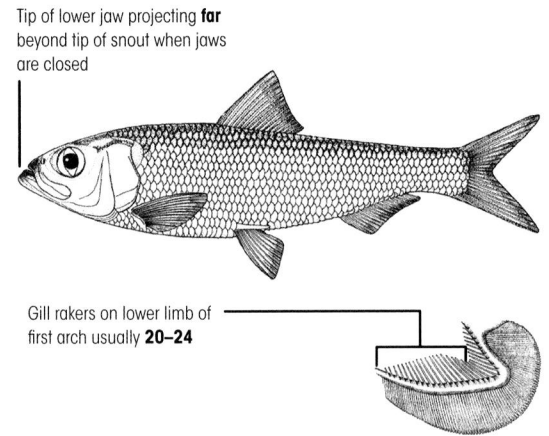

Tip of lower jaw projecting **far** beyond tip of snout when jaws are closed

Gill rakers on lower limb of first arch usually **20–24**

3b Tip of lower jaw projecting little, if any, beyond tip of snout when jaws are closed and viewed from the side; gill rakers on lower limb of first arch usually 35 or more. **4**

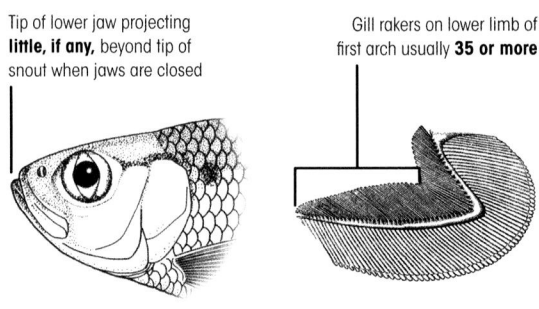

Tip of lower jaw projecting **little, if any,** beyond tip of snout when jaws are closed

Gill rakers on lower limb of first arch usually **35 or more**

4a Mouth sharply upturned (about 45° to horizontal); scales in lateral series usually 42–54; gill rakers on lower limb of first arch usually 39–41 . . . . **Page 81**

Alewife (*Alosa pseudoharengus*)

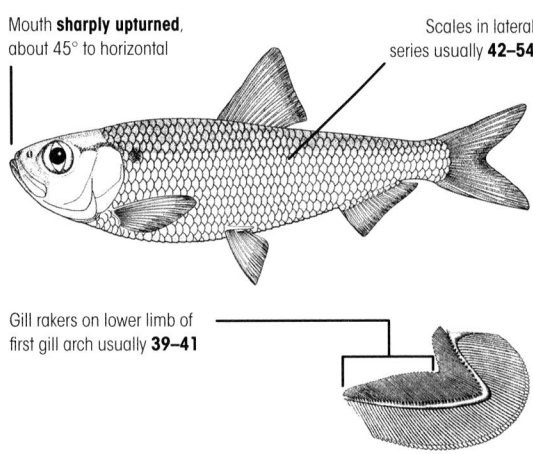

Mouth **sharply upturned**, about 45° to horizontal

Scales in lateral series usually **42–54**

Gill rakers on lower limb of first gill arch usually **39–41**

4b Mouth nearly horizontal, not sharply upturned; scales in lateral series usually 55–60; gill rakers on lower limb of first arch usually 42–48 . . . .**Page 79**

Alabama Shad (*Alosa alabamae*)

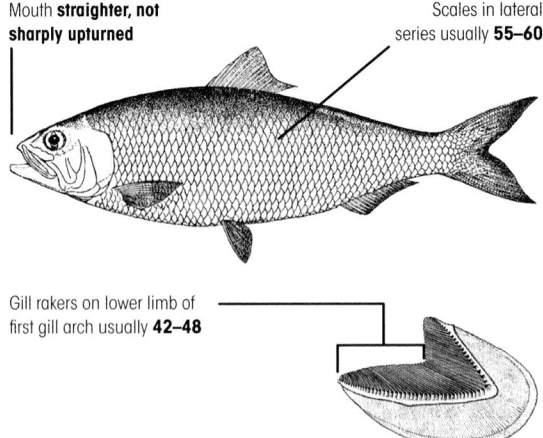

Mouth **straighter, not sharply upturned**

Scales in lateral series usually **55–60**

Gill rakers on lower limb of first gill arch usually **42–48**

ALABAMA SHAD *Alosa alabamae* Jordan & Evermann 1896

IDENTIFICATION The Alabama Shad has a strongly compressed body with a saw-toothed keel along the belly. The snout is moderately pointed, the mouth is terminal, and the lower jaw is equal to or projects only slightly beyond the upper jaw. The cheek is decidedly deeper than long, and the mouth is less oblique than in other shad in Illinois. Adults lack teeth on the jaws. The back and upper side of the body are silver with a blue-green sheen, and the lower side is silver-white. A blue-black spot near the upper edge of the gill cover usually is followed by 1 or 2 rows of smaller spots. The dorsal and caudal fins are dusky, and there is a black edge on the caudal fin; other fins are clear to light green. There are 42–48 rakers on the lower limb of the first gill arch. To 20¼ in. (51 cm).

SIMILAR SPECIES The Skipjack Herring, *A. chrysochloris*, has the lower jaw projecting beyond the upper jaw, teeth on the lower jaw, and a strongly oblique mouth, and its cheek is longer than or about equal to its depth. There is no blue-black spot near the upper edge of the gill cover, and there are 20–24 rakers on the lower limb of the first gill arch. The Alewife, *A. pseudoharengus*, has a more oblique mouth, teeth on the lower jaw, the cheek longer than or about equal to its depth, and 39–41 rakers on the lower limb of the first gill arch.

HABITAT The Alabama Shad has been found only in turbid water along the shoreline of the Mississippi River over sand and gravel substrates.

DISTRIBUTION IN ILLINOIS The Alabama Shad is an anadromous species that migrates far upstream to spawn. It was first recorded from the upper Mississippi River by Coker (1930) at Keokuk, Iowa, and then by Smith (1979) from Monroe County in 1962. Two contemporary-era records for the Mississippi River are based on juveniles that presumably were migrating south to the Gulf coast.

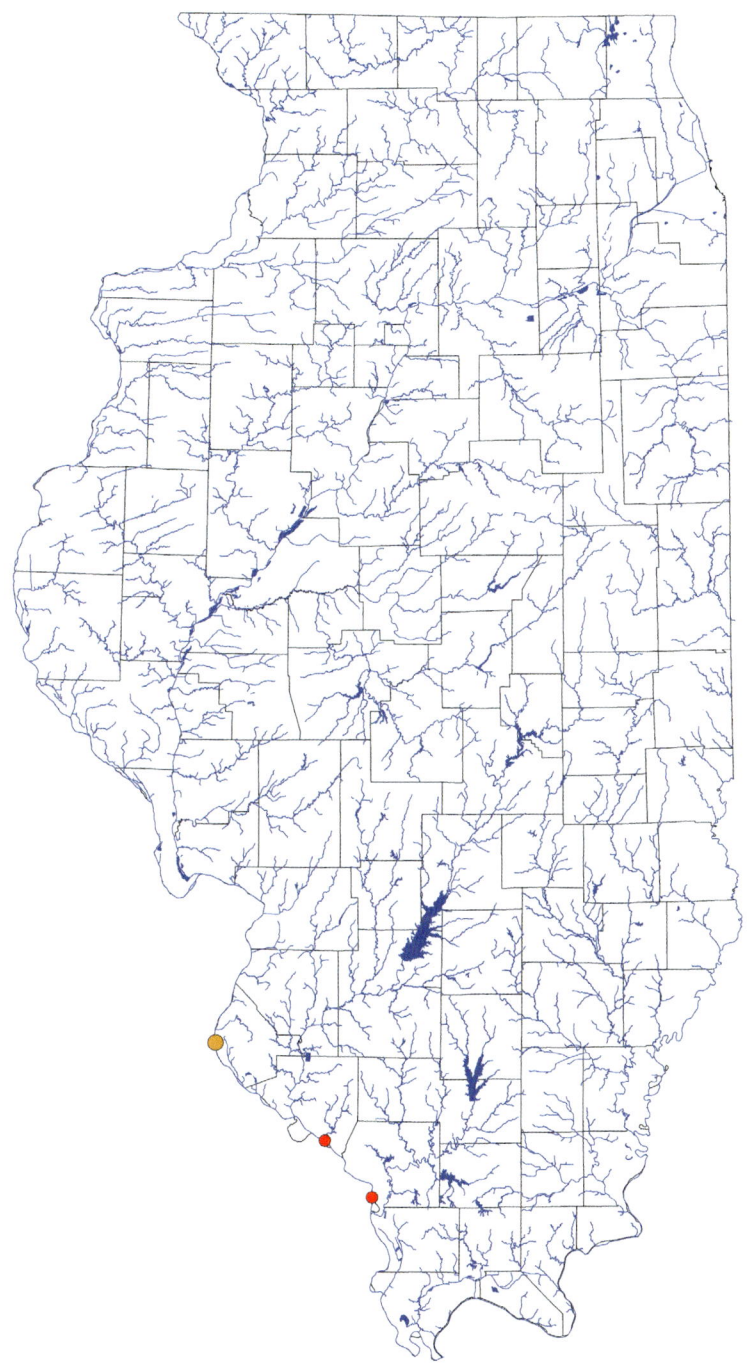

SKIPJACK HERRING *Alosa chrysochloris* (Rafinesque 1820)

IDENTIFICATION The Skipjack Herring has a strongly compressed body with a saw-toothed keel along the belly. The snout is moderately pointed, the mouth is terminal, and the lower jaw projects beyond the upper jaw. The mouth is strongly oblique (about 45° to the horizontal), and the cheek is longer than or about equal to its depth. Teeth are present on the lower jaw. The back and upper side of the body are silver with a blue-green sheen that ends abruptly, rather than gradually shading into the silver on the side of the body. The lower side is silver-white. The dorsal and caudal fins are dusky, often with a black edge; other fins are clear. There are 20–24 rakers on the lower limb of the first gill arch. To 21 in. (53 cm).

SIMILAR SPECIES The Alabama Shad, *A. alabamae,* has the lower jaw equal to or projecting only slightly beyond the upper jaw, the cheek deeper than long, a less-oblique mouth, no teeth on the jaws, a blue-black spot near the upper edge of the gill cover usually followed by 1 or 2 rows of smaller spots, and 42–48 rakers on the lower limb of the first gill arch. The Alewife, *A. pseudoharengus,* has the lower jaw equal to or projecting only slightly beyond the upper jaw, usually a blue-black spot near the upper edge of the gill cover, thin dusky stripes on the upper side, and 39–41 rakers on the lower limb of the first gill arch.

HABITAT The Skipjack Herring lives over sand and gravel in runs and flowing pools of medium to large rivers. It also is found in reservoirs.

DISTRIBUTION IN ILLINOIS Although Forbes and Richardson (1909) mentioned records for the Skipjack Herring from the Rock, Illinois, and Mississippi Rivers, precise localities or vouchers are available for only 1 record, from the Illinois River at Peoria. Smith (1979) found the Skipjack Herring to be widespread in the Wabash and Ohio Rivers and less frequently encountered in the Illinois and Mississippi Rivers. In the contemporary era, it is captured in all of these rivers and a few of their tributaries but is less widespread in the upper Mississippi River than in other large rivers.

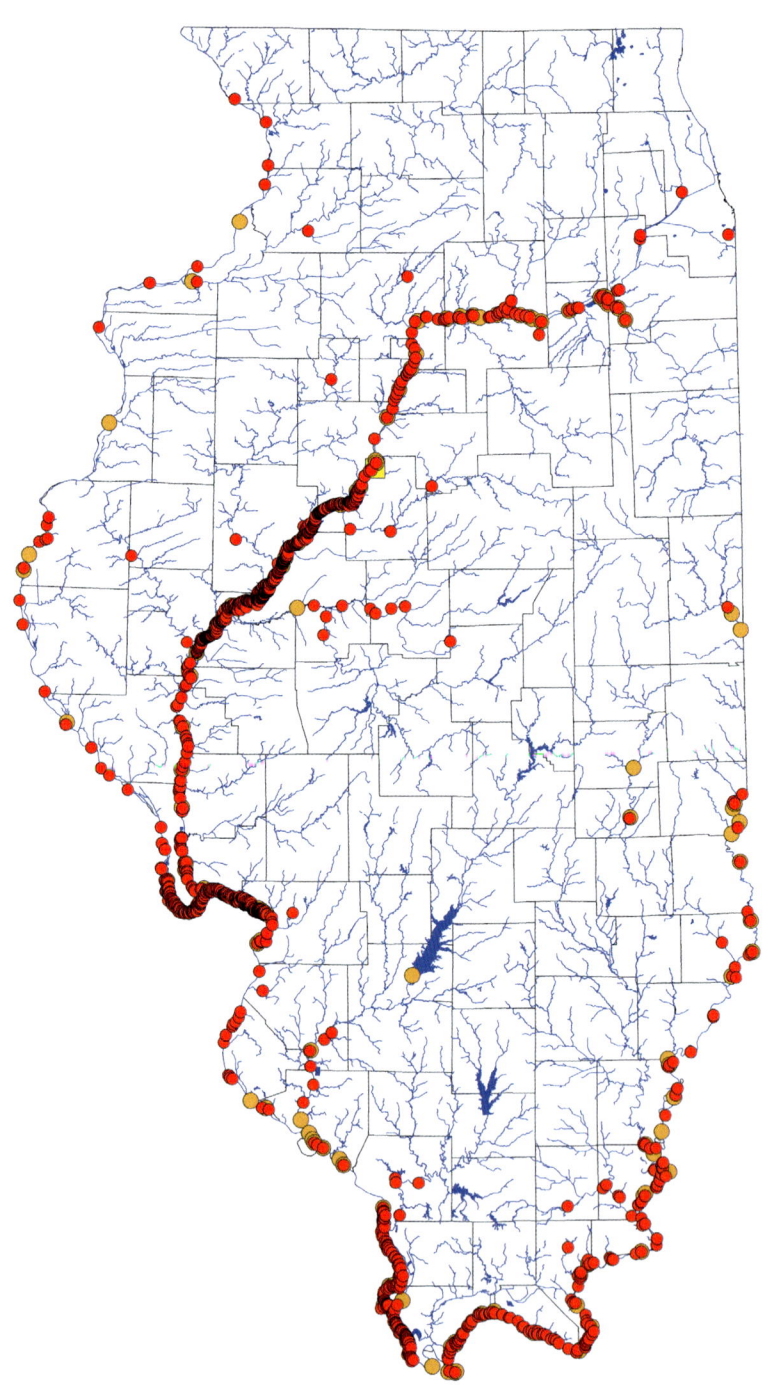

ALEWIFE *Alosa pseudoharengus* (Wilson 1811)

IDENTIFICATION The Alewife has a strongly compressed body with a saw-toothed keel along the belly. The snout is moderately pointed, the mouth is terminal, and the lower jaw is equal to or projects only slightly beyond the upper jaw. The mouth is strongly oblique (about 45° to the horizontal), and the cheek is longer than or about equal to its depth. Teeth are present on the lower jaw. The back and upper side of the body are silver with a blue-green sheen, and the lower side is silver-white. There is usually a blue-black spot near the upper edge of the gill cover and often thin dusky stripes on the upper side. Median fins are clear to dusky green or yellow; paired fins are clear. There are 39–41 rakers on the lower limb of the first gill arch. To 15 in. (38 cm).

SIMILAR SPECIES The Skipjack Herring, *A. chrysochloris*, has the lower jaw projecting beyond the upper and 20–24 rakers on the lower limb of the first gill arch, no blue-black spot near the upper edge of the gill cover, and no thin dusky stripes on the upper side of the body. The Alabama Shad, *A. alabamae*, has the cheek deeper than long, a less-oblique mouth, no teeth on the lower jaw, and 42–48 rakers on the lower limb of the first gill arch.

HABITAT The Alewife lives in the open water of Lake Michigan.

DISTRIBUTION IN ILLINOIS The Alewife is native to the Atlantic Coast and Lake Ontario. It had gained access to the western Great Lakes through the Welland Canal by the 1930s and has been in Lake Michigan since at least 1949 (Smith 1979). It became extremely common in the Illinois portion of the lake in the 1960s, and some individuals were recorded in the upper Illinois River. The species is less widespread in the contemporary era in Lake Michigan, and there are no contemporary records for the Illinois River basin.

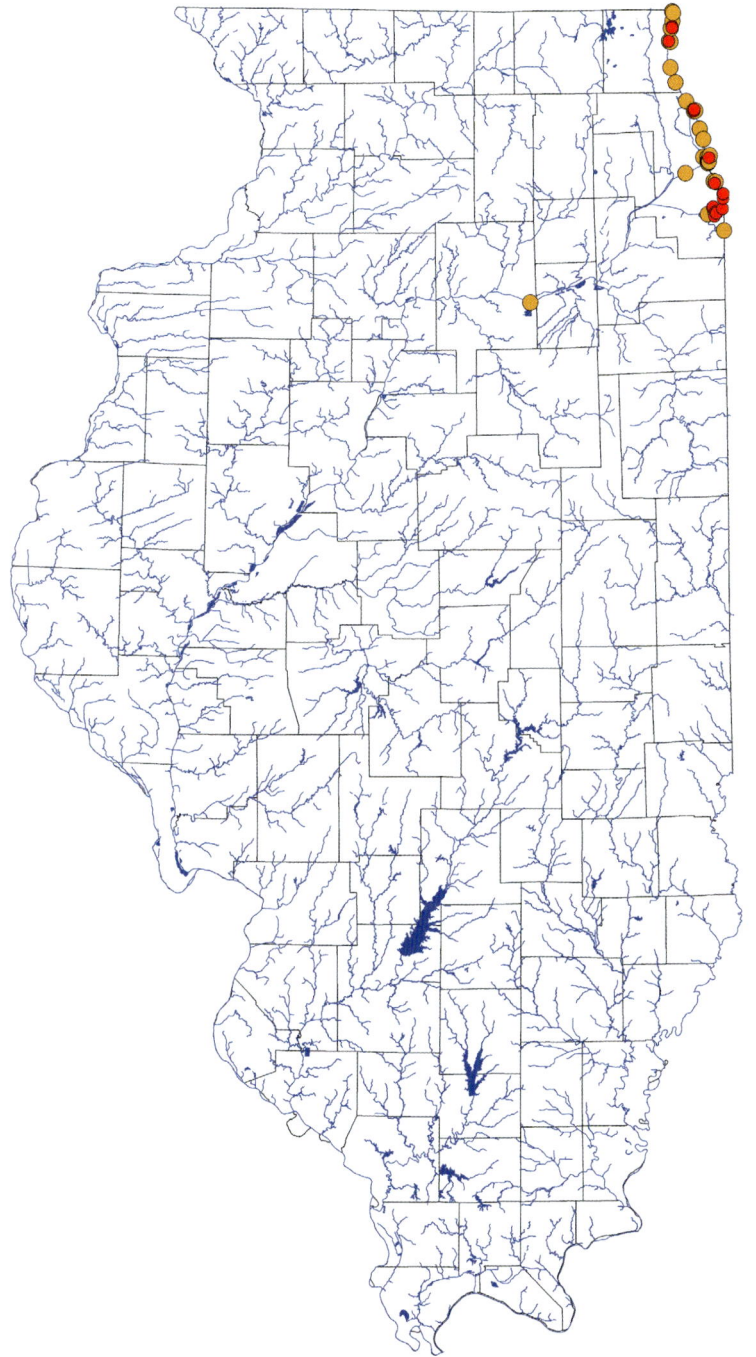

GIZZARD SHAD *Dorosoma cepedianum* (Lesueur 1818)

IDENTIFICATION The Gizzard Shad has a strongly compressed body with a saw-toothed keel along the belly, and a long, whiplike last dorsal fin ray. The snout is blunt, the mouth is subterminal, and there is a deep notch at the center of the upper jaw. The back and upper side of the body are silver, often with a blue-green sheen and usually 6–8 dusky black stripes. The lower side is silver-white. Juveniles and small adults have a large, purple-blue spot near the upper edge of the gill cover; the spot is faint or absent in large individuals. The fins are clear to dusky black. There are no black specks on the chin or floor of the mouth. There is no lateral line but 52–70 lateral scales, 10–13 dorsal rays, and 25–36 anal rays. To 20½ in. (52 cm).

SIMILAR SPECIES The Threadfin Shad, *D. petenense*, has a projecting lower jaw, a more pointed snout, yellow fins, black specks on the chin and floor of the mouth, and 40–48 lateral scales.

HABITAT The Gizzard Shad is found in open water of lakes, impoundments, and pools and runs of small to large rivers.

DISTRIBUTION IN ILLINOIS In earlier eras the Gizzard Shad was nearly statewide, although it was absent in northeastern Illinois, except for 1 record for Lake Michigan, and much of northern Illinois. In the contemporary era it is widespread throughout Illinois.

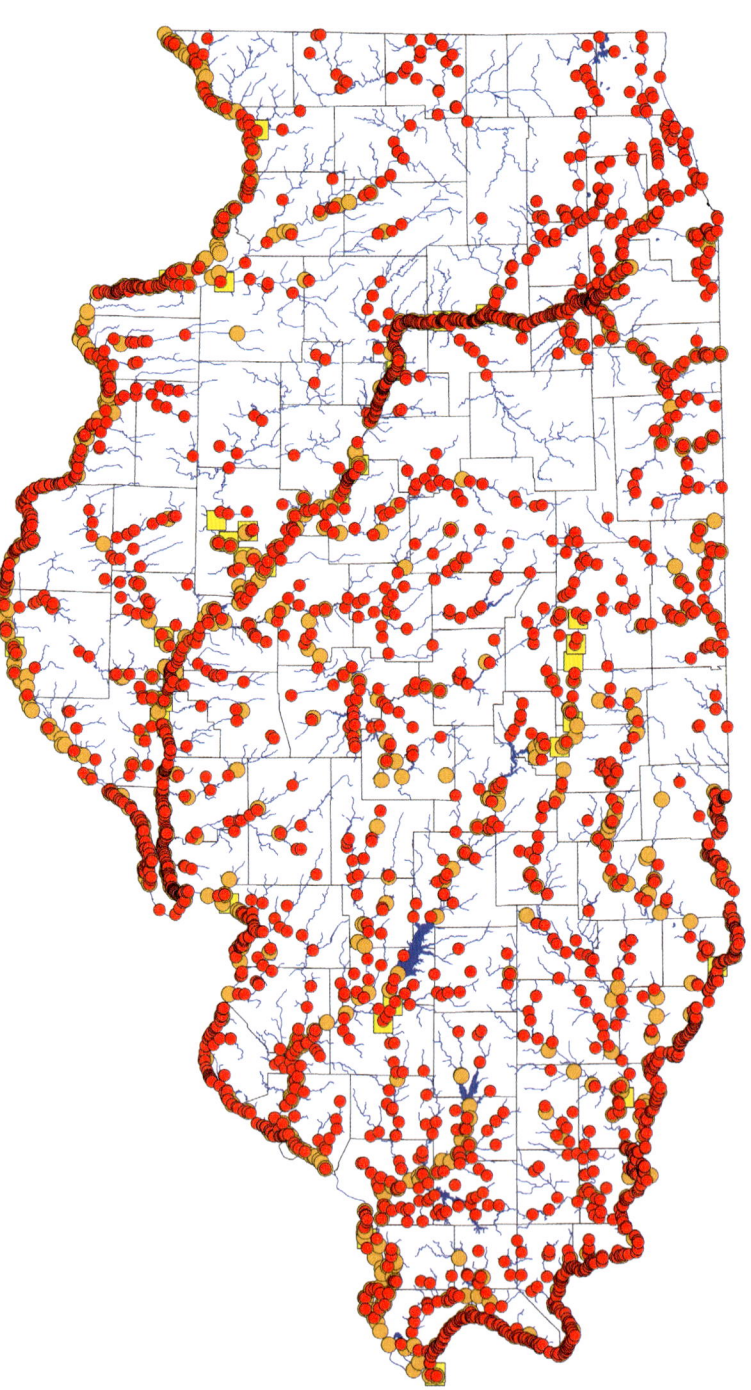

82

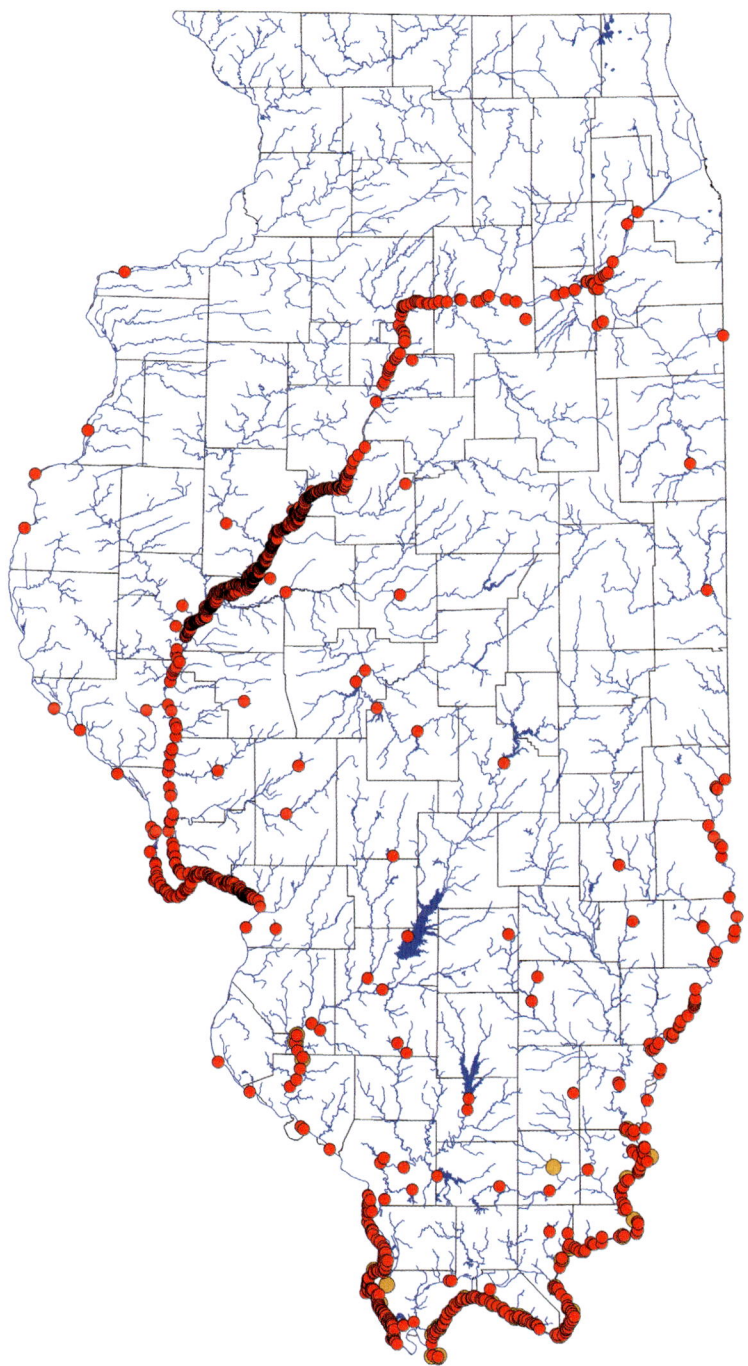

THREADFIN SHAD *Dorosoma petenense* (Günther 1867)

IDENTIFICATION The Threadfin Shad has a strongly compressed body, a saw-toothed keel along the belly, and a long, whiplike last dorsal fin ray. The snout is moderately pointed, the mouth is terminal, and the lower jaw projects beyond the upper jaw. The back and upper side of the body are silver with a blue-green sheen and usually 5–8 rows of small, black spots (sometimes forming continuous stripes) on the upper side of the body. There is a large, purple-blue spot near the upper edge of the gill cover. The lower side is silver-white. The fins are clear to dusky yellow. There are large, black specks on the chin and floor of the mouth. There is no lateral line, but there are 40–48 lateral scales, 11–14 dorsal rays, and 17–27 anal rays. To 9 in. (23 cm).

SIMILAR SPECIES The Gizzard Shad, *D. cepedianum*, has a blunt snout, a subterminal mouth, 52–70 lateral scales, and no black specks on the chin or floor of the mouth.

HABITAT The Threadfin Shad lives in lakes, backwaters, and pools of small to large rivers and usually is found in open water over sand or mud.

DISTRIBUTION IN ILLINOIS The Threadfin Shad was confined to Gulf Coast basins until the 1940s when it began to appear in the north, including in impoundments in Kentucky and Tennessee. The first specimens recorded for Illinois were collected in 1957 in the Ohio River basin. By the time Smith (1979) had finished his surveys, there were records for the Ohio, lower Wabash, lower Mississippi, and lower Kaskaskia River basins. This species has greatly expanded its range since then into the upper Mississippi River, the Illinois River basin, and scattered localities throughout central and southern Illinois. Some records for impoundments may be the result of stocking.

SHADS AND HERRINGS (Clupeidae)

83

SUCKERS

Family Catostomidae Agassiz 1850

Suckers are medium to large (to 40 in., 100 cm), deep-bodied fishes, with large, thick lips lined with papillae or plicae. They have 16 or more comblike or molarlike teeth in a single row on the pharyngeal arches but no teeth on the jaws, cycloid scales on the body, and no scales on the head. They have 1 dorsal fin with 9 or more rays, abdominal pelvic fins, and the anal fin far back on the body. Their fins lack spines. There are 82 species of suckers in North America, 19 of which occur in Illinois. One species in North America, the Longnose Sucker, *Catostomus catostomus*, also occurs in Siberia, and 1 species is endemic to China.

The sucker mouth is subterminal in most species, and the large lips are used to extract and ingest invertebrates and detritus from stream and lake beds. Because of the abundance and large size reached by most suckers, they often account for the largest biomass in streams and lakes and have been commercially important in Illinois.

KEY TO THE SUCKERS (Catostomidae)

1a Dorsal fin long, its margin deeply concave with 22 or more principal rays **2**

1b Dorsal fin shorter, its margin rounded, straight, or slightly concave, with 17 or fewer principal rays **8**

Dorsal fin **long**, its margin **deeply concave**, with **22 or more** principal rays

Dorsal fin **shorter**, its margin **straight, or slightly concave**, with **17 or fewer** principal rays

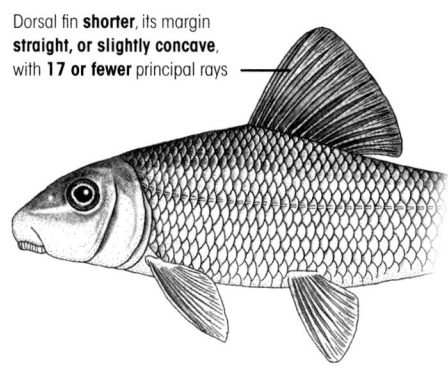

2a Snout length (A) greater than distance from back of eye to rear margin of gill cover (B); body elongate; lateral line scales 53 or more; lips papillose (with wart-like projections) . . . .**Page 99**

Blue Sucker (*Cycleptus elongatus*)

2b Snout length (A) less than distance from back of eye to rear margin of gill cover (B); body moderately deep; lateral line scales 42 or fewer; lips smooth or with shallow plicae (striations) **3**

A greater than **B**

Lateral line scales **53 or more**

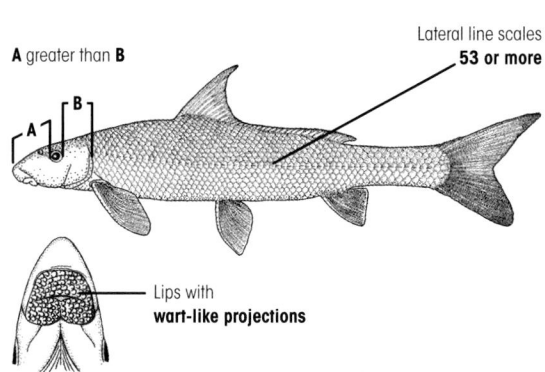

Lips with **wart-like projections**

A less than **B**

Lateral line scales **42 or fewer**

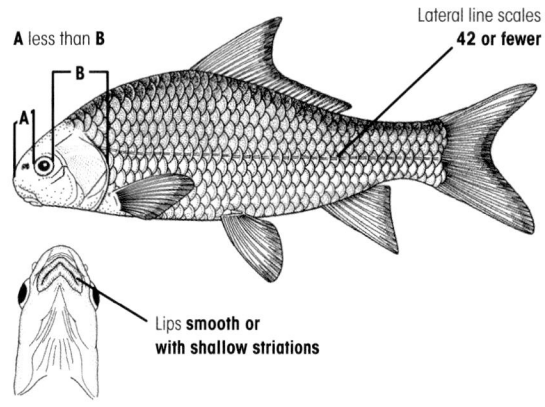

Lips **smooth or with shallow striations**

SUCKERS (Catostomidae)

85

3a Body grayish, blackish, or greenish; subopercle (bone at lower angle of gill cover) broadest at middle, its outer margin evenly rounded; distance A less than B; pelvic fins densely speckled with black; intestinal coiling with elongated loops; roof of skull without an anterior fontanelle (opening in front part of skull); anal fin rays usually 8 or more, rarely 7 . . . . **4**

Buffalofishes (*Ictiobus*)

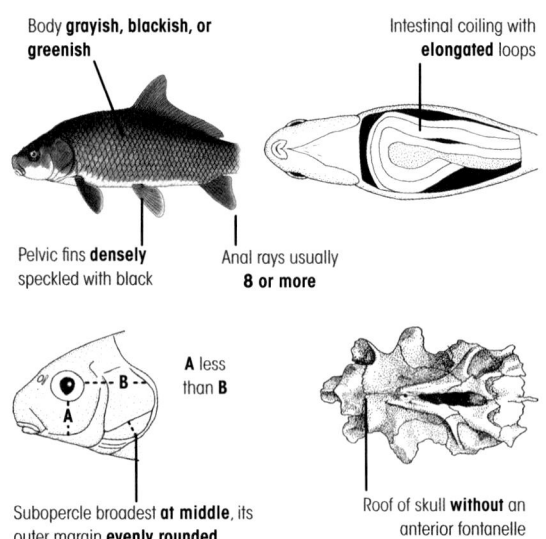

Body **grayish, blackish, or greenish**

Intestinal coiling with **elongated** loops

Pelvic fins **densely** speckled with black

Anal rays usually **8 or more**

A less than B

Subopercle broadest **at middle**, its outer margin **evenly rounded**

Roof of skull **without** an anterior fontanelle

3b Body silvery; subopercle bone broadest below middle, its outer margin somewhat angular; distance A about equal to B; pelvic fins scarcely or not at all speckled with black; intestinal coiling with circular loops; roof of skull with anterior fontanelle; anal fin rays usually 7 . . . . **6**

Carpsuckers (*Carpiodes*)

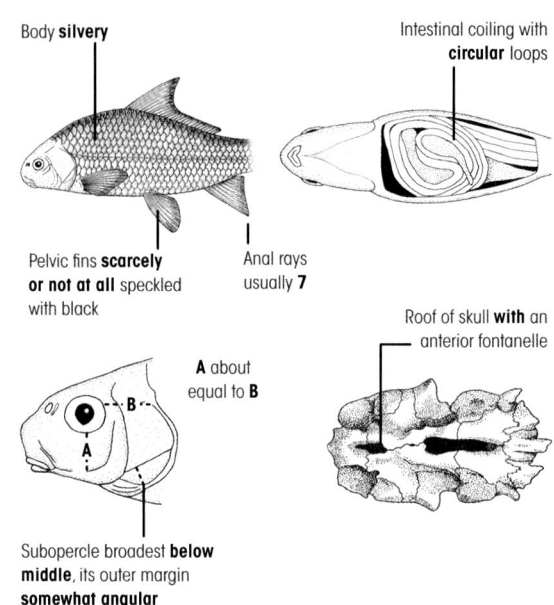

Body **silvery**

Intestinal coiling with **circular** loops

Pelvic fins **scarcely or not at all** speckled with black

Anal rays usually **7**

A about equal to **B**

Roof of skull **with** an anterior fontanelle

Subopercle broadest **below middle**, its outer margin **somewhat angular**

4a Front of upper lip above or about level with lower margin of eye; length of upper jaw (A) nearly equal to snout length (B); upper lip thin, only shallowly grooved; gill rakers on first arch 40–60 . . . . **Page 104**

Bigmouth Buffalo (*Ictiobus cyprinellus*)

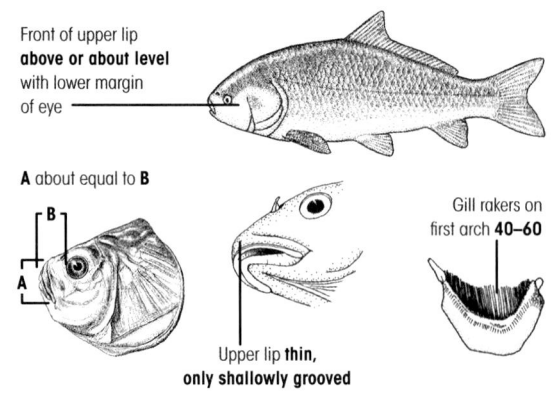

Front of upper lip **above or about level** with lower margin of eye

A about equal to **B**

Upper lip **thin, only shallowly grooved**

Gill rakers on first arch **40–60**

4b Front of upper lip far below lower margin of eye; length of upper jaw (A) much less than snout length (B); upper lip thicker, more deeply grooved; gill rakers on first arch 38 or fewer **5**

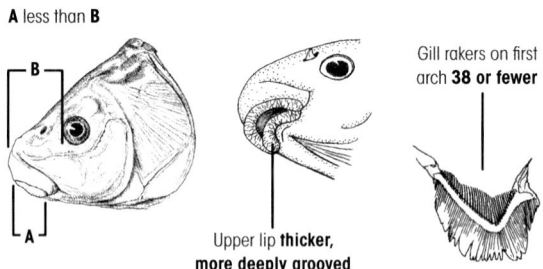

A less than **B**

Gill rakers on first arch **38 or fewer**

Upper lip **thicker, more deeply grooved**

5a Body depth (A) usually going more than 2.9 times into standard length (B); body surface in front of dorsal fin rounded or only weakly ridged; eye diameter (C) equal to or less than distance from fleshy posterior edge of upper jaw to fleshy anterior tip of lower jaw in adults (D) . . . . **Page 105**

Black Buffalo (*Ictiobus niger*)

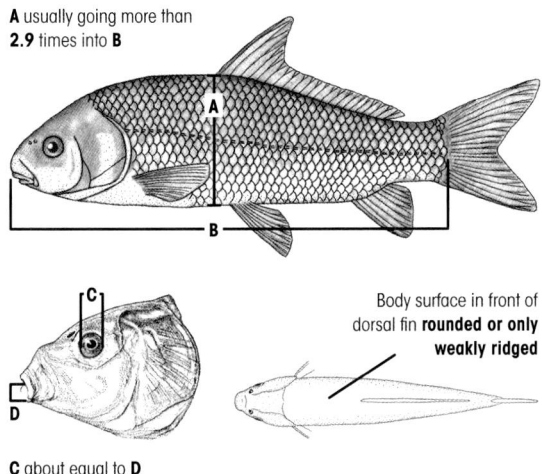

A usually going more than **2.9** times into **B**

Body surface in front of dorsal fin **rounded or only weakly ridged**

C about equal to **D**

5b Body depth (A) usually going 2.7 times or less into standard length (B); body surface in front of dorsal fin strongly ridged; eye diameter (C) usually greater than distance from fleshy posterior edge of upper jaw to fleshy anterior tip of lower jaw in adults (D). . . . .**Page 103**

Smallmouth Buffalo (*Ictiobus bubalus*)

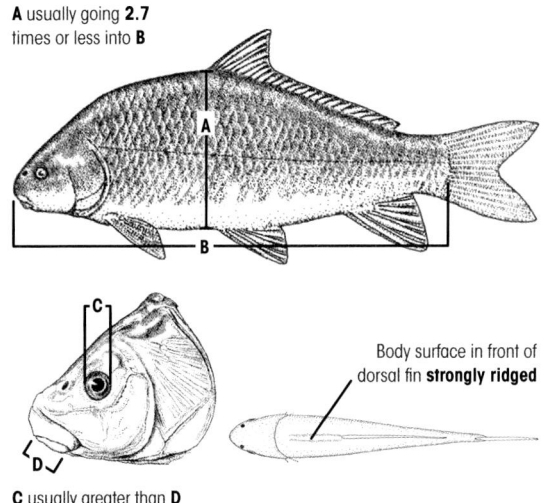

A usually going **2.7** times or less into **B**

Body surface in front of dorsal fin **strongly ridged**

C usually greater than **D**

6a No nipple-like projection at middle of lower lip; snout long, its length (A) about equal to distance from back of eye to upper end of gill cover (B); lateral line scales usually 37–40 . . . . **Page 95**

Quillback (*Carpiodes cyprinus*)

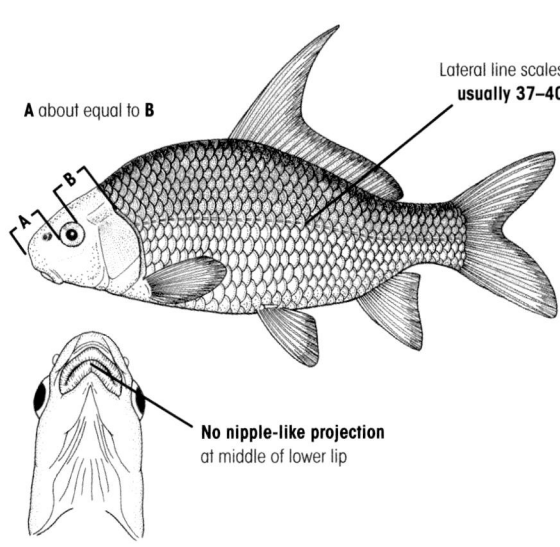

A about equal to **B**

Lateral line scales **usually 37–40**

No nipple-like projection at middle of lower lip

6b A nipple-like projection at middle of lower lip; snout shorter, its length (A) less than distance from back of eye to upper end of gill cover (B); lateral line scales usually 33–37 . . . .**7**

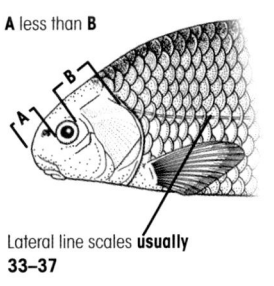

A less than **B**

Lateral line scales **usually 33–37**

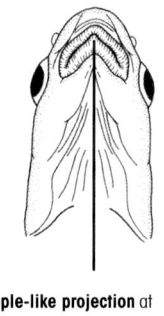

Nipple-like projection at middle of lower lip

7a First principal ray of dorsal fin very long, reaching to or beyond back of fin; body deeper, its depth (A) usually going 2.6 times or less into standard length (B) . . . . . . **Page 96**

Highfin Carpsucker (*Carpiodes velifer*)

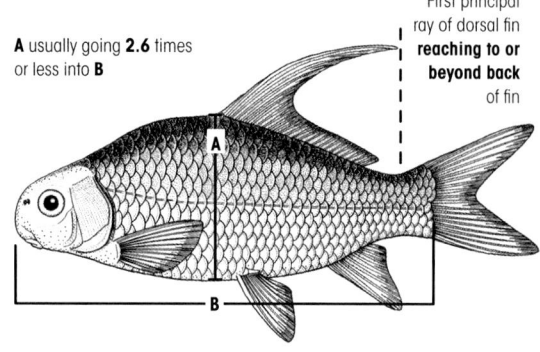

A usually going **2.6** times or less into **B**

First principal ray of dorsal fin **reaching to or beyond back** of fin

7b First principal ray of dorsal fin shorter, not reaching much beyond middle of fin; body more slender, its depth (A) usually going more than 2.6 times into standard length (B) . . . . **Page 94**

River Carpsucker (*Carpiodes carpio*)

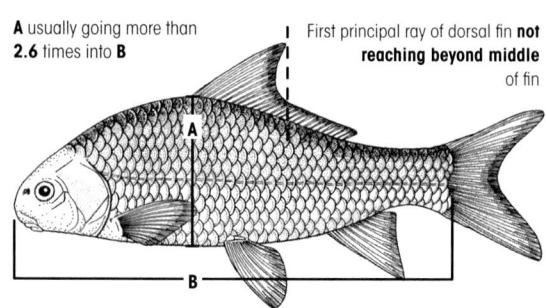

A usually going more than **2.6** times into **B**

First principal ray of dorsal fin **not reaching beyond middle** of fin

8a Head between eyes broad and strongly concave; head squarish in cross section; back and upper side with 4–6 prominent dark crossbars . . . . **Page 102**

Northern Hog Sucker (*Hypentelium nigricans*)

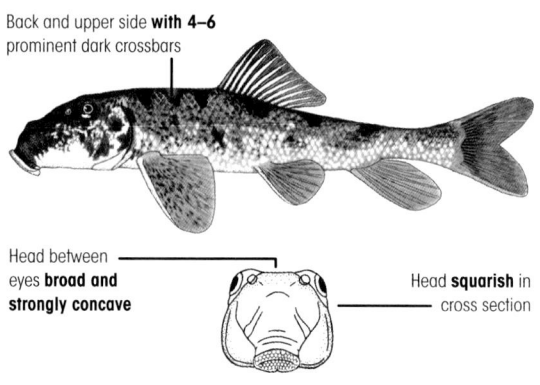

Back and upper side **with 4–6** prominent dark crossbars

Head between eyes **broad and strongly concave**

Head **squarish** in cross section

8b Head between eye narrower and convex; head rounded in cross section; back and upper side without prominent dark crossbars **9**

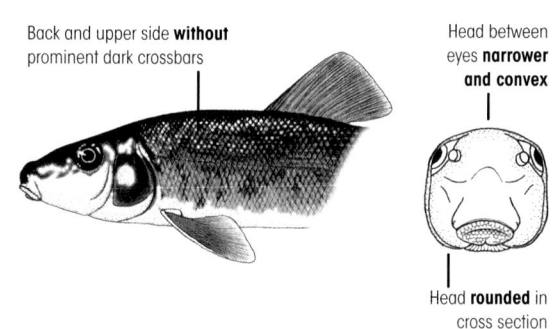

Back and upper side **without** prominent dark crossbars

Head between eyes **narrower and convex**

Head **rounded** in cross section

9a Scales small, especially toward head with usually 53 or more in lateral line; lips with numerous small wart-like projections (papillose) **10**

9b Scales larger, with usually 50 or fewer in lateral line; lips with numerous plicae (parallel folds), rarely papillose **11**

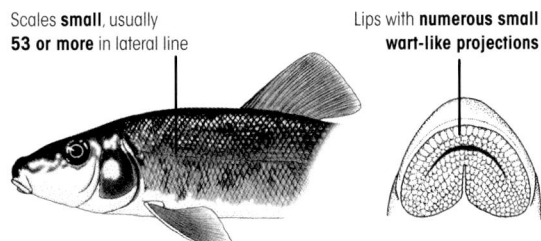

Scales **small**, usually **53 or more** in lateral line

Lips with **numerous small wart-like projections**

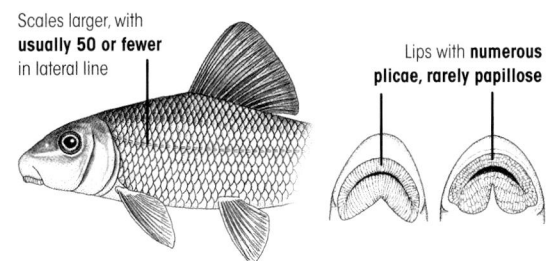

Scales larger, with **usually 50 or fewer** in lateral line

Lips with **numerous plicae, rarely papillose**

10a Lateral line scales 53–74; caudal peducle scale rows 20–27; snout rounded, projecting only slightly beyond upper lip. . . . .**Page 98**

10b Lateral line scales 95–120; caudal peducle scale rows 26–34; snout bulbous, projecting well beyond upper lip . . . .**Page 97**

White Sucker (*Catostomus commersonii*)

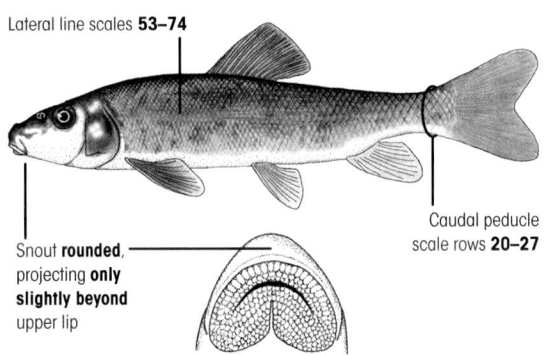

Lateral line scales **53–74**

Snout **rounded**, projecting **only slightly beyond** upper lip

Caudal peducle scale rows **20–27**

Longnose Sucker (*Catostomus catostomus*)

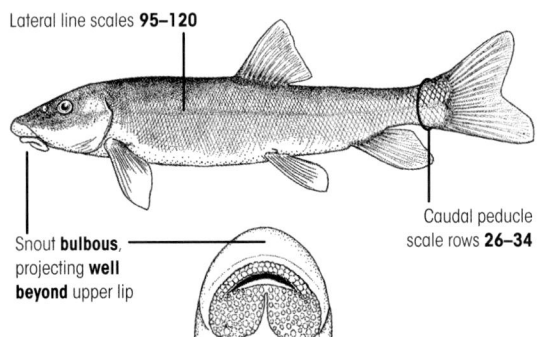

Lateral line scales **95–120**

Snout **bulbous**, projecting **well beyond** upper lip

Caudal peducle scale rows **26–34**

11a Lateral line complete to base of caudal fin; air bladder with 3 chambers **12**

11b Lateral line absent or incomplete, with only a few scales pored at front; air bladder with 2 chambers **17**

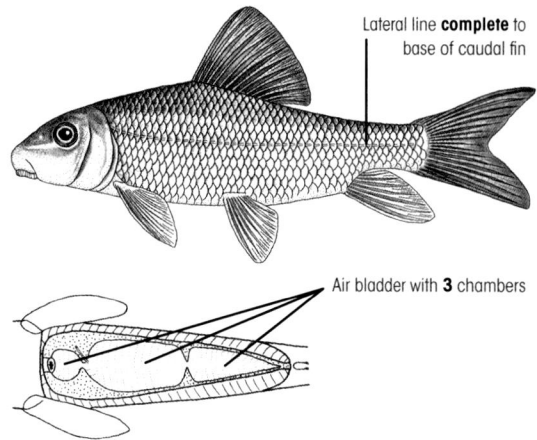

Lateral line **complete** to base of caudal fin

Air bladder with **3** chambers

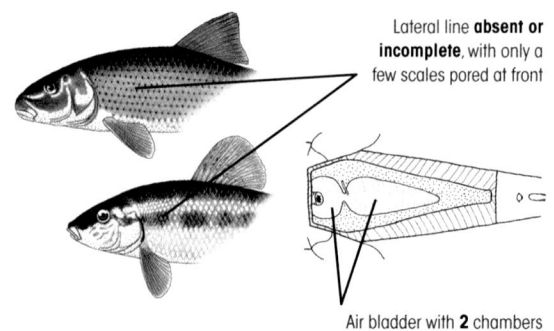

Lateral line **absent or incomplete**, with only a few scales pored at front

Air bladder with **2** chambers

12a Dorsal fin rays usually 14–16; rear margin of lower lip forming an acute angle; lower lip with parallel folds (plicae) broken into wart-like (papillae) structures . . . .
. . . . **Page 107**

12b Dorsal fin rays usually 12–13, rarely 14; rear margin of lower lip nearly straight, or forming a slight to moderate angle; lower lip with plicae only **13**

Silver Redhorse (*Moxostoma anisurum*)

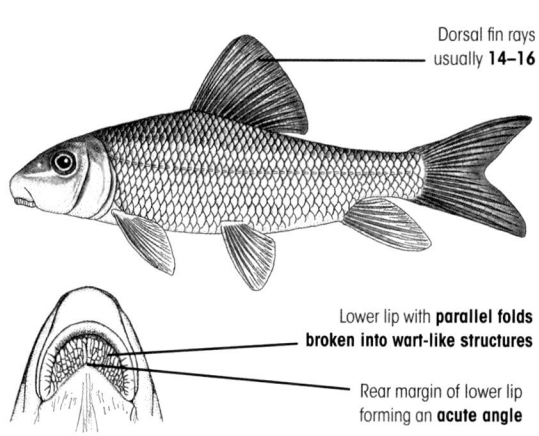

Dorsal fin rays usually **14–16**

Lower lip with **parallel folds broken into wart-like structures**

Rear margin of lower lip forming an **acute angle**

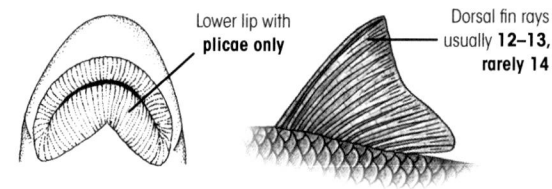

Lower lip with **plicae only**

Dorsal fin rays usually **12–13, rarely 14**

Rear margin of lower lip nearly straight, or forming a **slight to moderate angle**

13a Caudal peduncle scale rows usually 15–16 . . . **Page 112**

13b Caudal peduncle scale rows usually 12–13 . . . . **14**

Greater Redhorse (*Moxostoma valenciennesi*)

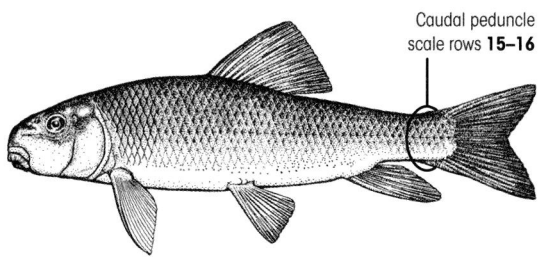

Caudal peduncle
scale rows **15–16**

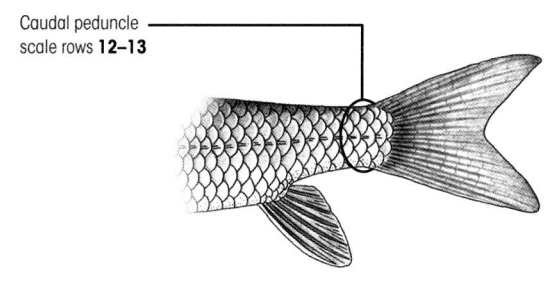

Caudal peduncle
scale rows **12–13**

14a Scales of back and upper sides each with a crescent-shaped dark spot at base; caudal fin in live or freshly preserved specimens bright red **15**

14b Scales of back and upper sides without dark spots at base; caudal fin in life or freshly preserved specimens slate colored. **16**

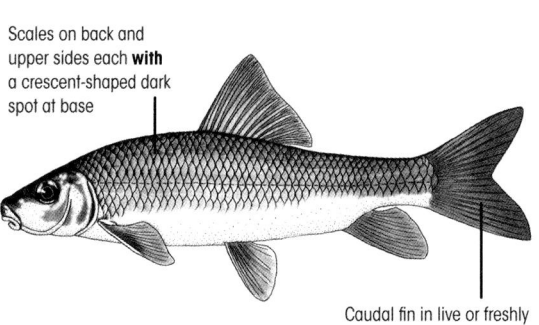

Scales on back and
upper sides each **with**
a crescent-shaped dark
spot at base

Caudal fin in live or freshly
preserved specimens **bright red**

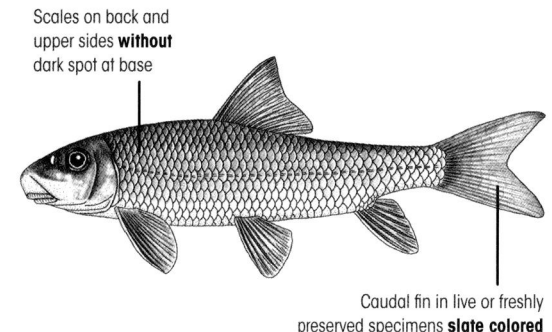

Scales on back and
upper sides **without**
dark spot at base

Caudal fin in live or freshly
preserved specimens **slate colored**

15a Rear margin of lower lip nearly straight; head short, its length (A) going 4.5 times or more into standard length (B); pharyngeal arch thin, with slender teeth in comb-like series . . . . **Page 111**

15b Rear margin of lower lip forming a definite V-shaped angle; head longer, its length (A) going fewer than 4.5 times into standard length (B); pharyngeal arch thick, with molar-like teeth . . . . **Page 108**

Shorthead Redhorse (*Moxostoma macrolepidotum*)

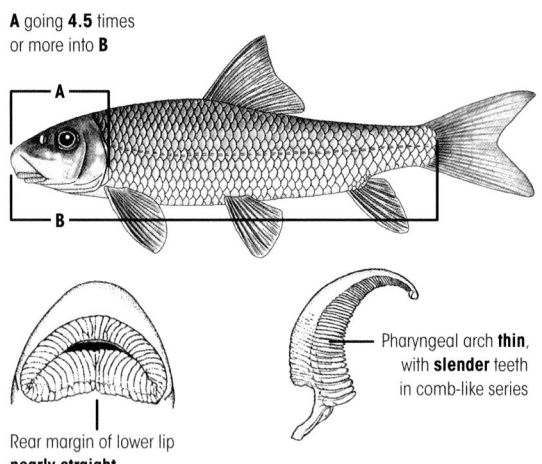

A going **4.5** times or more into **B**

Pharyngeal arch **thin**, with **slender** teeth in comb-like series

Rear margin of lower lip **nearly straight**

River Redhorse (*Moxostoma carinatum*)

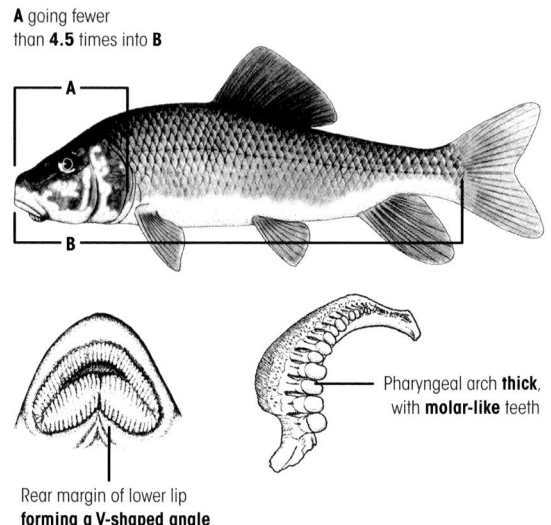

A going fewer than **4.5** times into **B**

Pharyngeal arch **thick**, with **molar-like** teeth

Rear margin of lower lip **forming a V-shaped angle**

16a Lateral line scales usually 44–47; caudal peduncle slender, its least depth (A) going 2.0 times or more into distance from base of caudal fin to front of anal fin base (B); pelvic fin rays usually 10, often 9 or 11; breeding males with tubercles on snout only . . . . **Page 109**

16b Lateral line scales usually 40–43; caudal peduncle deeper, its least depth (A) going fewer than 2.0 times into distance from base of caudal fin to front of anal fin base (B); pelvic fin rays usually 9, often 8 or 10; breeding males with tubercles on head . . . . **Page 110**

Black Redhorse (*Moxostoma duquesnei*)

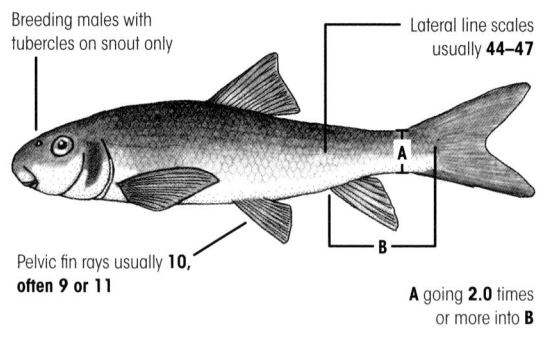

Breeding males with tubercles on snout only

Lateral line scales usually **44–47**

Pelvic fin rays usually **10, often 9 or 11**

A going **2.0** times or more into **B**

Golden Redhorse (*Moxostoma erythrurum*)

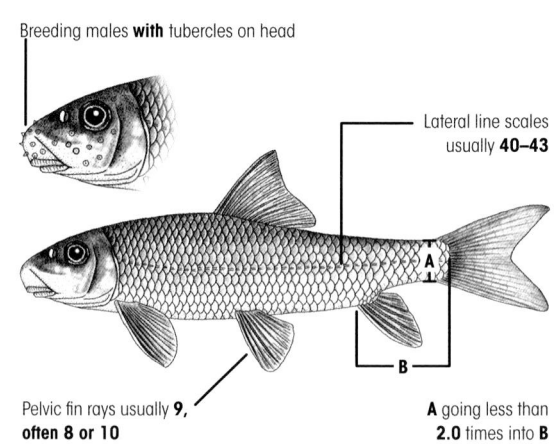

Breeding males **with** tubercles on head

Lateral line scales usually **40–43**

Pelvic fin rays usually **9, often 8 or 10**

A going less than **2.0** times into **B**

17a Scales on sides with dark spots at bases, forming interrupted parallel lines; outer margin of dorsal fin straight or concave, and with a black blotch at edge in young; body depth (A) going 4.0 or more times into standard length (B) .... **Page 106**

17b Scales on sides without dark spots at bases, but sides may have single lateral stripe, blotches, or uniform coloration; outer margin of dorsal fin convex or rounded and never with a black blotch; body depth (A) going fewer than 4.0 times into standard length (B) **18**

Spotted Sucker (*Minytrema melanops*)

Scales on sides **with** dark spots at bases, forming interrupted parallel lines

Outer margin of dorsal fin **straight or concave, with** a black blotch in young

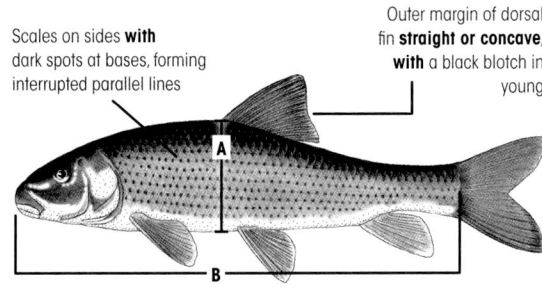

A going **4.0** times or more into **B**

Scales on sides **without** dark spots at bases, but sides **may have single lateral stripe**

Outer margin of dorsal fin **convex or rounded, never** with a black blotch

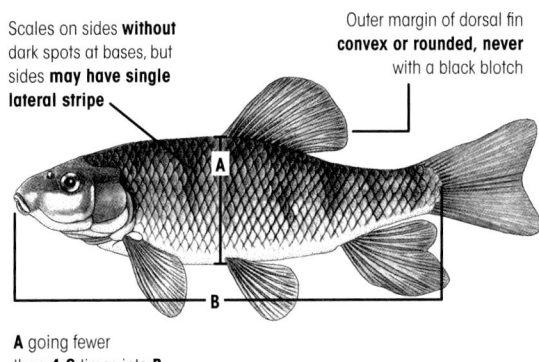

A going fewer than **4.0** times into **B**

18a Scales in lateral series usually 34–39; dark stripe along midside continuous in young and subadults (often indistinct in adults) .... **Page 101**

18b Scales in lateral series usually 40–45; dark stripe along midside in young only, broken into 5–8 confluent blotches in subadults (often indistinct in adults). ...... **Page 100**

Lake Chubsucker (*Erimyzon sucetta*)

Scales in lateral series usually **34–39**

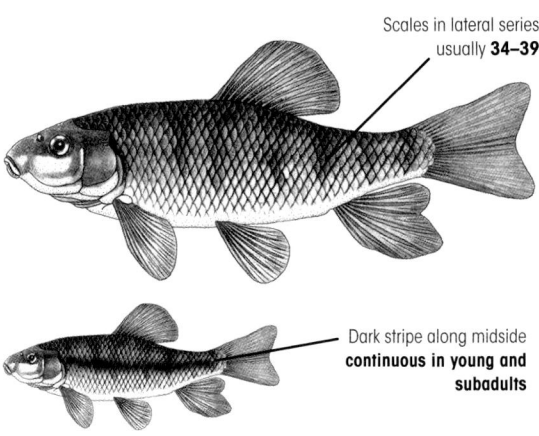

Dark stripe along midside **continuous in young and subadults**

Western Creek Chubsucker (*Erimyzon claviformis*)

Scales in lateral series usually **40–45**

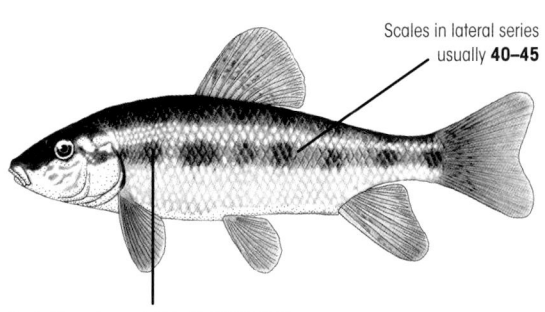

Dark stripe along midside **in young only**, broken into 5–8 confluent blotches in subadults

RIVER CARPSUCKER *Carpiodes carpio* (Rafinesque 1820)

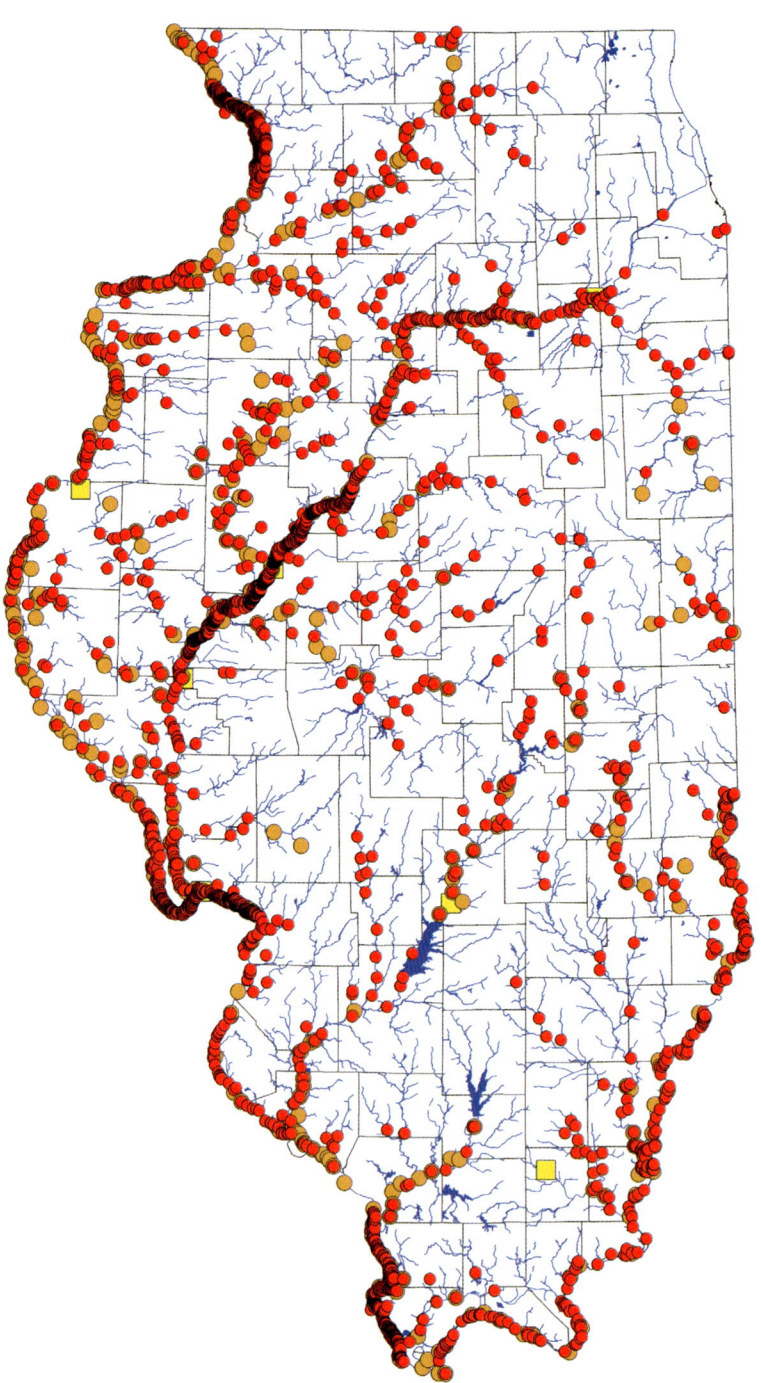

IDENTIFICATION The River Carpsucker has a long, falcate dorsal fin with 23–30 rays, a deep, laterally compressed body, a short, rounded snout, and a large papilla at the middle of the lower lip. The first dorsal ray is long but usually does not reach beyond the middle of the dorsal fin. The halves of the lower lip meet obliquely, not nearly a right angle. The length of the snout is less than the distance from the back of the eye to the upper end of the gill opening, and the upper jaw extends beyond the front of the eye. The subopercle is triangular and widest below its midpoint. The back and side of the body are silver-gray to bronze, and the lower side and belly are silver-white. Median fins are dusky gray, and paired fins are clear white to pink-orange. There are usually 33–37 lateral scales. To 25 in. (64 cm).

SIMILAR SPECIES The Quillback, *C. cyprinus*, lacks a large papilla on the lower lip, and the halves of the lower lip meet at nearly a right angle. The Highfin Carpsucker, *C. velifer*, has a blunt snout, and a long first dorsal ray reaching to or beyond the rear of the dorsal fin. Buffalos, *Ictiobus*, have a dusky black side of the body and dusky black pelvic fins, no large papilla on the lower lip, and a subopercle that is semicircular and widest at its midpoint.

HABITAT The River Carpsucker is found in pools and backwaters of small to large rivers, less often in creeks and lakes.

DISTRIBUTION IN ILLINOIS The River Carpsucker occurs statewide although it is sporadic and rare in the northeastern part. This is the same general distribution as the species had in earlier surveys.

REMARKS Juveniles of species of *Carpiodes* are difficult to separate from one another, and some records on the map for the River Carpsucker and other species of *Carpiodes* may be in error as identification of unvouchered records cannot be confirmed.

QUILLBACK   *Carpiodes cyprinus* (Lesueur 1817)

IDENTIFICATION The Quillback has a long, fal-cate dorsal fin with 23–30 rays, a deep, laterally compressed body, a short, rounded snout, and no large papilla on the lower lip. The long first dorsal ray reaches beyond the middle but does not reach to the rear of the dorsal fin. The halves of the lower lip meet at nearly a right angle. The length of the snout is subequal to the distance from the back of the eye to the upper end of the gill opening, and the upper jaw does not extend beyond the front of the eye. The subopercle is triangular and widest below its midpoint. The back and side of the body are silver-gray to bronze, and the lower side and belly are silver-white. Median fins are dusky gray, and paired fins are clear white to orange. There are usually 36–37 lateral scales. To 26 in. (66 cm).

SIMILAR SPECIES The Highfin Carpsucker, *C. velifer* has a blunt snout, a large papilla at the middle of the lower lip, and a long first dorsal ray reaching to or beyond the rear of the dorsal fin; the halves of the lower lip meet obliquely, not approaching a right angle. The River Carpsucker, *C. carpio*, has a blunt snout and a large papilla at the middle of the lower lip; the halves of the lower lip meet obliquely, not approaching a right angle. Buffalos, *Ictiobus*, have a dusky black side of the body and dusky black pelvic fins, and a subopercle that is semicircular and widest at its midpoint.

HABITAT The Quillback occurs in flowing pools and backwaters of creeks and small to large rivers, and can occasionally be found in lakes.

DISTRIBUTION IN ILLINOIS The Quillback occurs statewide but is much more frequently encountered in the northern half of the state. Given the greater collecting effort in recent decades, the species appears to be slightly less widespread in the contemporary era than in earlier eras.

REMARKS Juveniles of species of *Carpiodes* are difficult to separate from one another, and some records on the map for the Quillback may be in error as identification of unvouchered records cannot be confirmed.

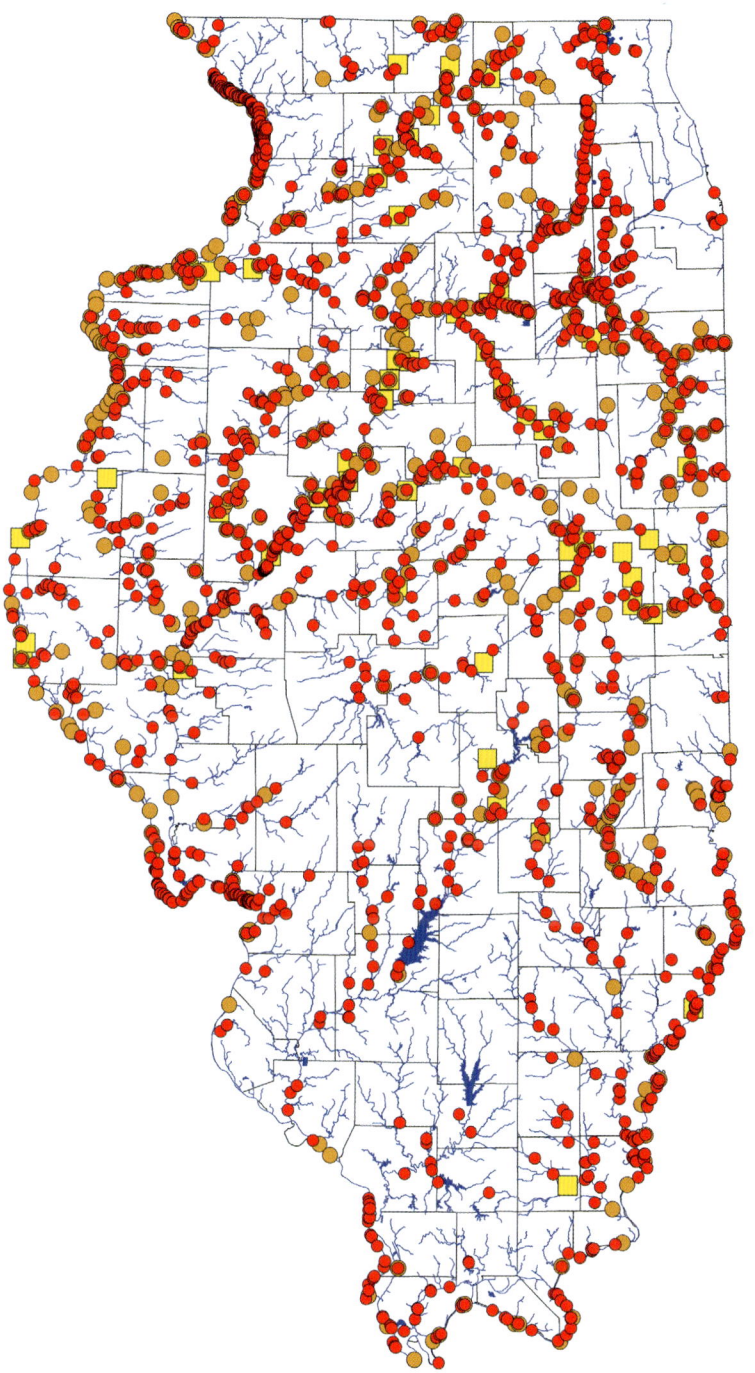

HIGHFIN CARPSUCKER *Carpiodes velifer* (Rafinesque 1820)

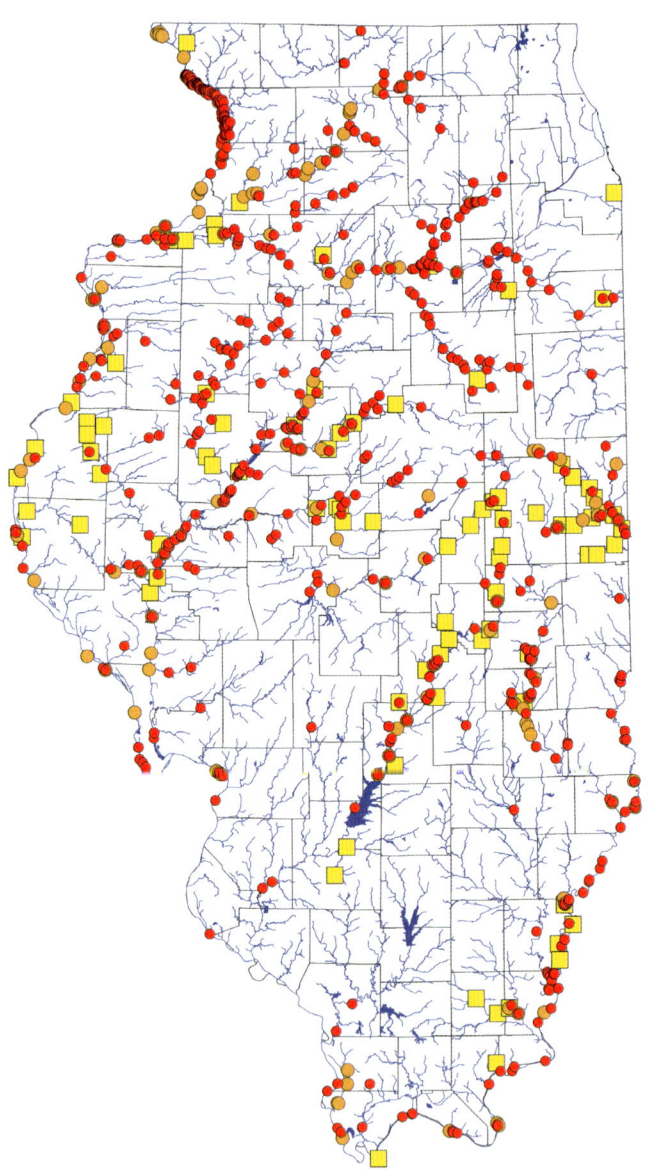

IDENTIFICATION The Highfin Carpsucker has a long, falcate dorsal fin with 23–30 rays, a deep, laterally compressed and highly arched body, a blunt—almost perpendicular—snout, and a large papilla at the middle of the lower lip. The long first dorsal ray reaches well beyond the middle of the dorsal fin—often reaching to or beyond the rear of the fin. The halves of the lower lip meet obliquely, not approaching a right angle. The subopercle is triangular and widest below its midpoint. The back and side of the body are silver-gray to bronze, and the lower side and belly are silver-white. Median fins are dusky gray, and paired fins are clear white to orange. There are usually 33–36 lateral scales. To 19½ in. (50 cm).

SIMILAR SPECIES The River Carpsucker, *C. carpio*, has a rounded snout, and a shorter first dorsal ray that does not reach well beyond the middle of the dorsal fin. The Quillback, *C. cyprinus*, has a shorter first dorsal ray that does not reach well beyond the middle of the dorsal fin and lacks a large papilla at the middle of the lower lip, and the halves of the lower lip meet at nearly a right angle. Buffalos, *Ictiobus*, have a dusky black side of the body and dusky black pelvic fins, no large papilla on the lower lip, and a subopercle that is semicircular and widest at its midpoint.

HABITAT The Highfin Carpsucker occurs in flowing pools and backwaters of small to large rivers. It usually is found in areas with gravel or gravel and sand substrates.

DISTRIBUTION IN ILLINOIS The Highfin Carpsucker occurs statewide but is most frequently encountered in central and north-central portions of the state. Smith (1979) commented on the decline of this species in Illinois, noting that Forbes and Richardson (1909) had reported the Highfin Carpsucker to be the "most generally distributed and most abundant carpsucker in the state." The distribution of the species in the contemporary era is essentially the same as that during the Smith era.

REMARKS Juveniles of species of *Carpiodes* are difficult to separate from one another, and some records on the map for the Highfin Carpsucker may be in error as identification of unvouchered records cannot be confirmed.

LONGNOSE SUCKER *Catostomus catostomus* (Forster 1773)

IDENTIFICATION The Longnose Sucker has a long, slender body, a bulbous snout that projects well beyond the upper lip, and papillose lips on a subterminal mouth. The lower lip, with a deep median notch, is about twice as thick as the upper lip. The back and upper side of the body are dark olive or gray, the side is dusky olive-gray that abruptly changes to white on the lower side and belly. Median fins are clear to dusky, and paired fins are clear to amber-pink. Juveniles and some adults have 3 dark blotches on the side of the body. During the breeding season, males are nearly black on the back and on upper side and have a rosy or red stripe along the midside. Large females are green-gold to copper-brown on the upper side, fading to white, yellow, or pink on the underside and may have a maroon stripe along the side. The scales are small with 90–120 along the lateral line, and there are usually 9–11 dorsal rays. To 25 in. (64 cm).

SIMILAR SPECIES The White Sucker, *C. commersonii*, has 53–74 lateral scales and a less-bulbous snout not extending well beyond the upper lip.

HABITAT The Longnose Sucker occurs in large, deep cold-water lakes, often near rocky shorelines or breakwaters, and in lower reaches of tributaries feeding the lakes.

DISTRIBUTION IN ILLINOIS The Longnose Sucker is restricted in Illinois to Lake Michigan as first recorded by Smith (1979). Contemporary-era records indicate that it is widespread in the lake.

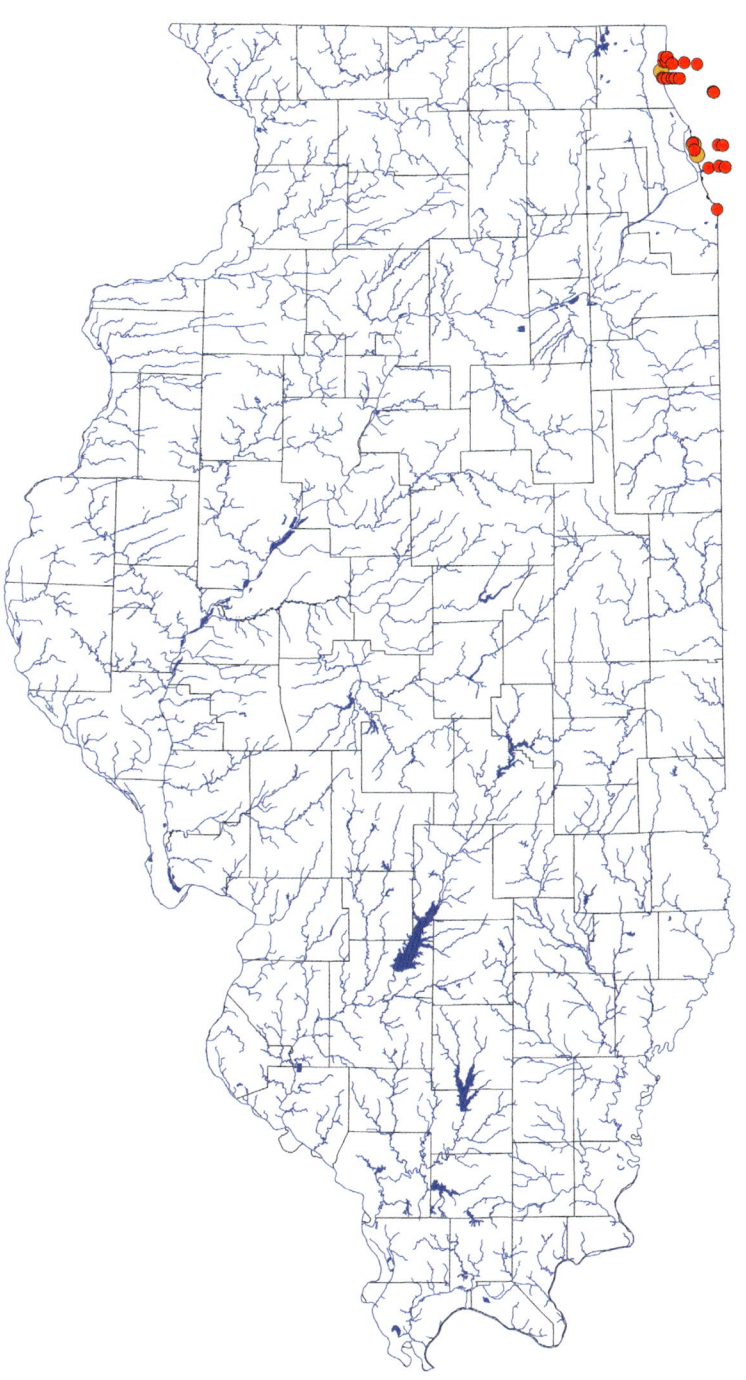

WHITE SUCKER *Catostomus commersonii* (Lacepède 1803)

IDENTIFICATION The White Sucker has a long, slender body, a rounded snout that slightly extends beyond the upper lip, and papillose lips on a subterminal mouth. The lower lip, with a deep median notch, is about twice as thick as the upper lip. The back and upper side of the body are bronze to slate-gray, often with scales darkly outlined. The side is dusky yellow to olive, the lower side and belly are white, and fins are clear to dusky. Juveniles and some adults have 3 dark blotches on the side of the body. During the breeding season, males have a light red or gold stripe along the side. The scales are small with 53–74 along the lateral line, and there are usually 10–12 dorsal rays. To 25 in. (64 cm).

SIMILAR SPECIES The Longnose Sucker, *C. catostomus*, has 90–120 lateral scales and a bulbous snout that projects well beyond the upper lip.

HABITAT The White Sucker is found in every type of aquatic habitat in Illinois, although it is most frequently encountered over sand or gravel in creeks and small to medium rivers and rarely is found in the largest rivers.

DISTRIBUTION IN ILLINOIS The White Sucker is the most widespread sucker in Illinois and occurs statewide. Smith (1979) noted that the species was the least frequently encountered in south-central Illinois, where streams tend to be less rocky and have more clay or muddy substrates. In the contemporary era the species occurs more widely throughout that region, although it remains somewhat less widespread there than elsewhere.

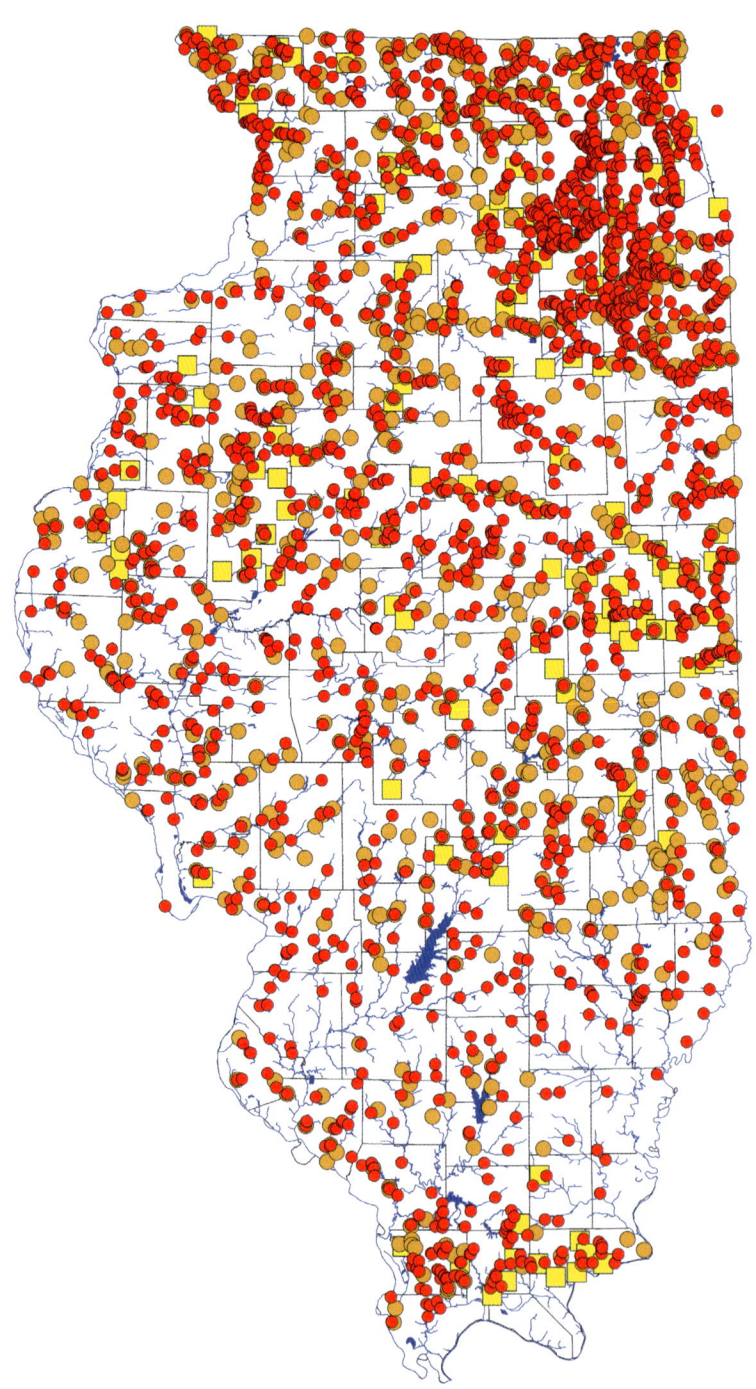

BLUE SUCKER *Cycleptus elongatus* (Lesueur 1817)

IDENTIFICATION The Blue Sucker has a small head, a long, falcate dorsal fin with 28–35 rays, a long, slender body that is deepest at the nape and tapers to a long, slender caudal peduncle and a rounded snout that extends beyond the small, subterminal mouth. Lips are fleshy and covered with many small papillae. The back and upper side of the body are olive-blue or blue-gray, the lower side and belly are light blue, and the fins are dark blue-gray. During the breeding season, males are blue-black and have small, white tubercles on the head, body, and fins. Females are light blue with fewer tubercles. There are 53–58 lateral scales and 7 anal rays. To 39 in. (99 cm).

SIMILAR SPECIES No other sucker in Illinois has an elongate blue body.

HABITAT The Blue Sucker occurs over bedrock and gravel substrates in fast-water channels of medium to large rivers.

DISTRIBUTION IN ILLINOIS The Blue Sucker is found in the Mississippi, Ohio, and Wabash Rivers and less often in smaller interior rivers. Smith (1979) found essentially this same distribution, although the species seems to be more widespread in the contemporary era in the Wabash River and parts of the Mississippi River. Forbes and Richardson (1909) recorded Blue Suckers only in the Mississippi and Rock Rivers.

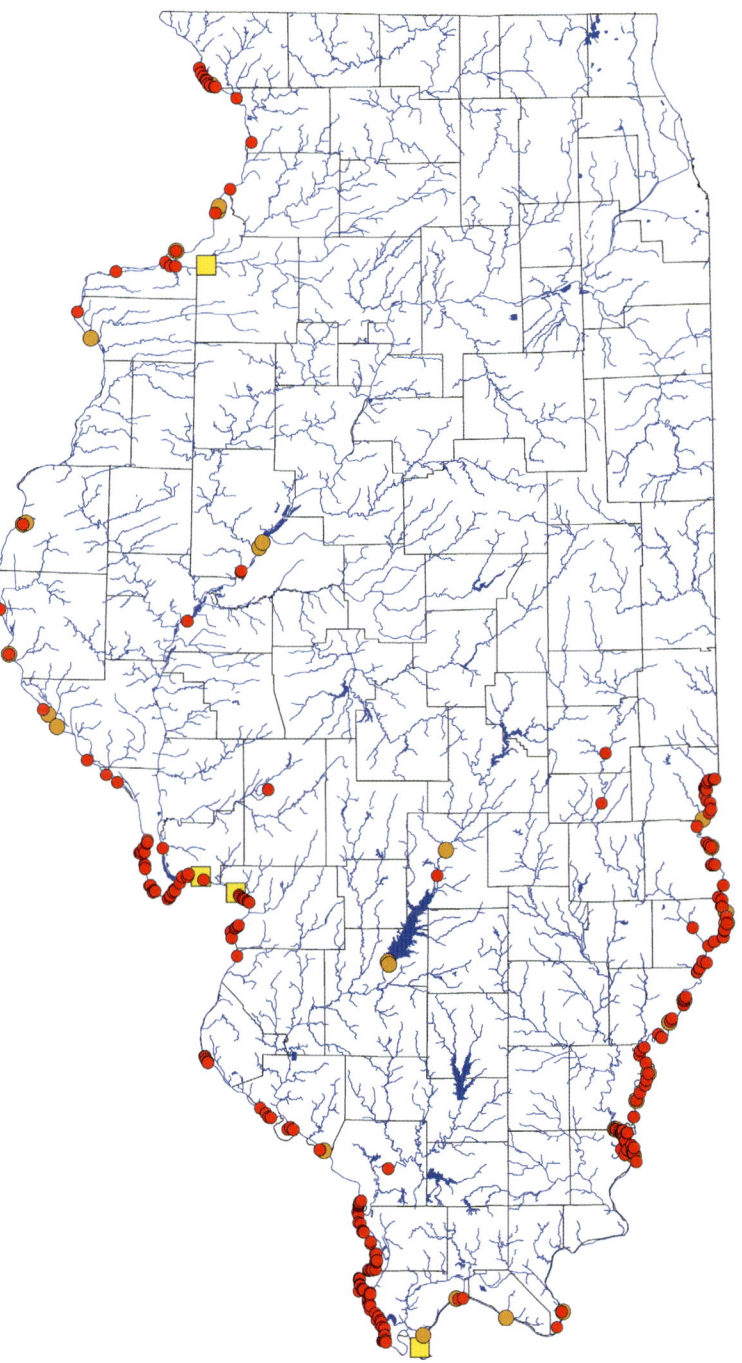

SUCKERS (Catostomidae)

99

WESTERN CREEK CHUBSUCKER *Erimyzon claviformis* (Girard 1856)

IDENTIFICATION The Western Creek Chubsucker has a thick body, a dorsal fin with a rounded edge, no lateral line, a small, sub-terminal mouth, and thick, plicate lips. The halves of the lower lip meet at nearly a right angle. The back and upper side of the body are golden olive-brown with dark-edged scales, the lower side and belly are yellow-olive to white, and fins are olive-gray to yellow-orange. There often is a series of large, black blotches along the back and another along the midside. Juveniles have an amber or red caudal fin and a broad, yellow-green stripe over 5–8 confluent dark blotches along the side from the tip of the snout to the base of the caudal fin. Males have a bilobed anal fin; during the breeding season they are dark brown with a yellow-pink breast and belly with light yellow median fins, orange paired fins, and 3 large tubercles on each side of the snout. There are 37–45 lateral scales, 9–11 (usually 10) dorsal rays, and 7 anal rays. To 9 in. (23 cm).

SIMILAR SPECIES The Lake Chubsucker, *E. sucetta*, has a deeper, chubbier body, 34–39 lateral scales, and 10–13, usually 11–12, dorsal rays. Juvenile Lake Chubsuckers have an intense dark stripe along the side of the body that extends onto the caudal fin.

HABITAT The Western Creek Chubsucker occurs in low-gradient headwaters and pools and backwaters of creeks and small rivers. It is often found over mud or sand near vegetation and woody debris.

DISTRIBUTION IN ILLINOIS As it was in the Smith era, the Western Creek Chubsucker is widespread throughout eastern and southern Illinois except in the Iroquois and Kankakee River basins. Forbes and Richardson (1909) found the species to be much more widespread, including in tributaries of the Mississippi and lower Illinois Rivers and in Lake Michigan. The species has disappeared in neighboring Wisconsin, where it was last collected in the state in 1928 in the Des Plaines River, a tributary of the upper Illinois River (Becker 1983).

REMARKS The Western Creek Chubsucker was referred to as the Creek Chubsucker, *E. oblongus*, in previous eras. *Erimyzon oblongus*, now the Eastern Creek Chubsucker, refers to populations on the Atlantic Slope and in the Lake Ontario basin of New York.

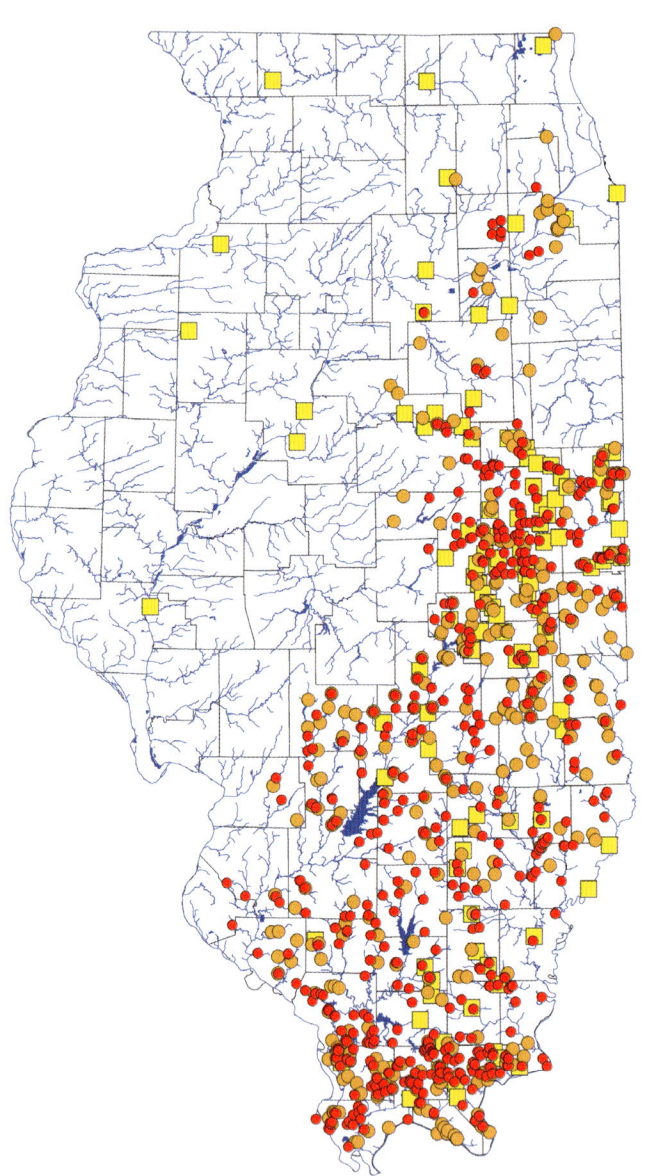

LAKE CHUBSUCKER *Erimyzon sucetta* (Lacepède 1803)

IDENTIFICATION The Lake Chubsucker has a thick body—somewhat arched and deepest at the origin of the dorsal fin, a dorsal fin with a rounded edge, no lateral line, a small, subterminal mouth, and thick plicate lips. The halves of the lower lip meet at nearly a right angle. The back and upper side of the body are yellow-brown with dark-edged scales, the lower side and belly are yellow-olive to white, and fins are olive-gray to dusky yellow-orange. There often is a series of large, black blotches along the back and another along the midside. Juveniles have an intense dark stripe along the side of the body that extends onto an amber or red caudal fin. Males have a bilobed anal fin and during the breeding season are golden brown with yellow to orange paired fins and 2–3 large tubercles on each side of the snout. There are 34–39 lateral scales, 10–13 (usually 11–12) dorsal rays, and 7 anal rays. To 16 in. (41 cm).

SIMILAR SPECIES The Western Creek Chubsucker, *E. claviformis*, has a slightly shallower body, 37–45 lateral scales, and 9–11, usually 10, dorsal rays. Juvenile Western Creek Chubsuckers do not have an intense dark stripe along the side of the body that extends onto the caudal fin.

HABITAT The Lake Chubsucker occurs in clear, well-vegetated sandy and muddy lakes, ponds, swamps, and quiet pools and backwaters of creeks and small rivers.

DISTRIBUTION IN ILLINOIS The Lake Chubsucker has a sporadic distribution across Illinois that is similar to that during the Smith era. It is most frequently encountered in the glacial lakes in northeastern Illinois and in the small areas of the Kankakee and Illinois River basins that have thick deposits of sand. It has disappeared from the Kaskaskia and Pecatonica Rivers, where it was recorded by Forbes and Richardson (1909), and from extreme southwestern Illinois, where it was found in the 1930s and 1940s (Smith 1979).

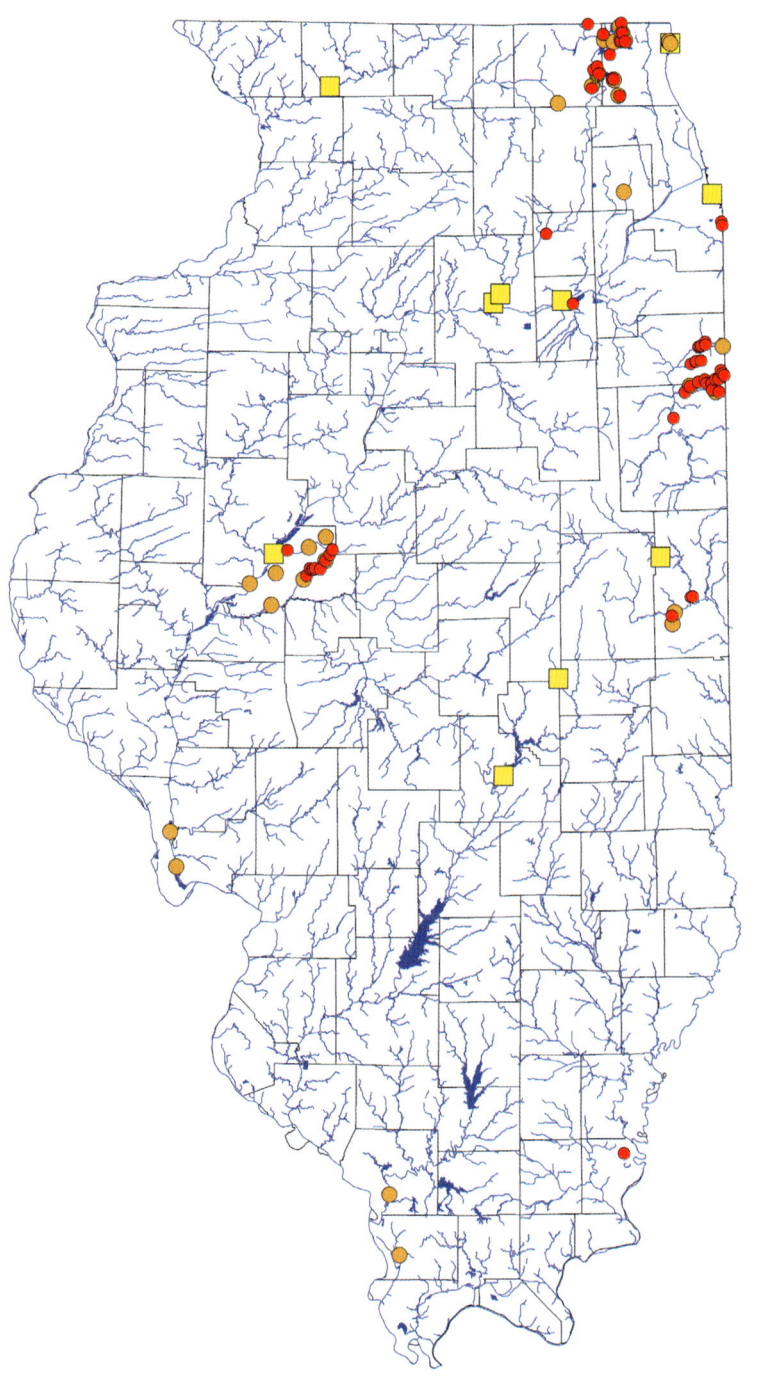

NORTHERN HOG SUCKER *Hypentelium nigricans* (Lesueur 1817)

IDENTIFICATION The Northern Hog Sucker has a large, wide head, a thick body that tapers to a long and slender caudal peduncle, and a large, subterminal mouth with fleshy lips covered with large papillae. The top of the head is flat in juveniles and becomes concave in large individuals. The back and upper side of the body are olive-bronze to red-brown with 3–6 dark brown saddles. The side of the body is light brown and often has pale yellow stripes along the scale rows. The lower side and belly are pale yellow or white. Large individuals have a blue-black snout, olive to light orange fins, and often a black edge on the dorsal and caudal fins. The lateral line is complete with 44–54 scales, and there are usually 11 dorsal rays. To 24 in. (61 cm).

SIMILAR SPECIES No other sucker in Illinois has a thick body that tapers to a long and slender caudal peduncle, a large head that is concave between the eyes, or brown saddles on the back and upper side of the body.

HABITAT The Northern Hog Sucker occurs in clear rocky riffles and runs in creeks and small to medium rivers with pebble, gravel, or sandy substrates.

DISTRIBUTION IN ILLINOIS The Northern Hog Sucker is widespread in northern and east-central Illinois, less so in west-central Illinois, and isolated in clear rocky streams in Hardin and Pope Counties in southeastern Illinois. It is absent in much of south-central and southern Illinois and from many small tributaries of the Mississippi River. This is essentially the same distribution found in earlier eras, except the species previously occurred in clear streams in Union County in southwestern Illinois and in the Chicago metropolitan region.

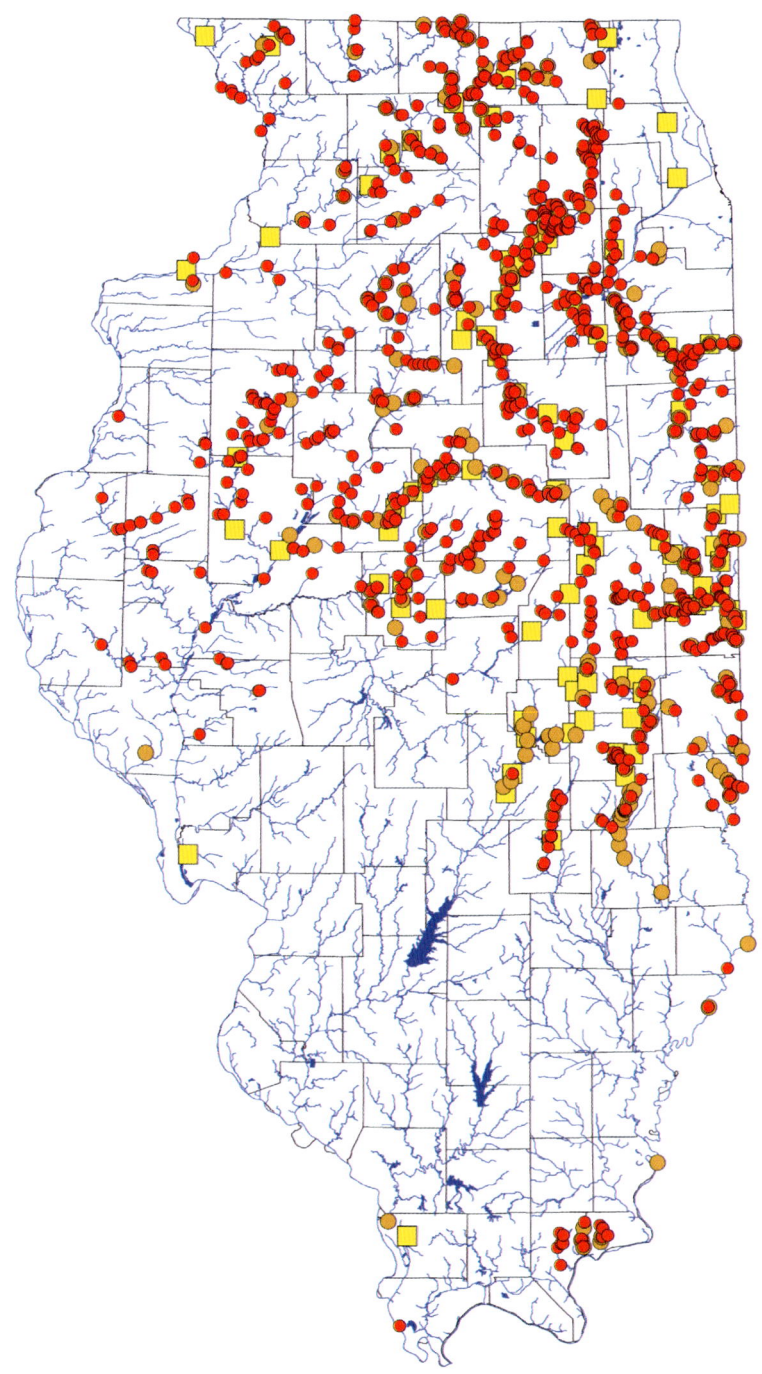

SMALLMOUTH BUFFALO *Ictiobus bubalus* (Rafinesque 1818)

IDENTIFICATION The Smallmouth Buffalo has a long, falcate dorsal fin with 24–31 rays, a deep body, a small head, and a nearly horizontal subterminal mouth. The front of the upper lip is well below the lower edge of the eye. The thick upper lip has distinct grooves. Large individuals have a moderately keeled nape. The subopercle is semicircular and widest at its midpoint. The back and upper side of the body are gray-black to olive-bronze, the side is black to dusky olive, the lower side and belly are dusky white to light yellow, and the fins are olive to black. There are 35 or fewer rakers on the first gill arch and usually 36–37 lateral scales, 10 pelvic rays, and 9 anal rays. To 31 in. (78 cm).

SIMILAR SPECIES The Black Buffalo, *I. niger*, has a shallower body with a more rounded nape, a more oblique mouth, and a more ovoid head. The Bigmouth Buffalo, *I. cyprinellus*, has a sharply oblique terminal mouth, an upper lip that is level with the lower edge of the eye, a rounded nape, shallow grooves on the upper lip, and 40 or more gill rakers on the first gill arch. Carpsuckers, *Carpiodes*, have silver sides of the body, clear-white or orange pelvic fins, and a subopercle that is triangular and widest below its midpoint.

HABITAT The Smallmouth Buffalo occurs in flowing pools and the main channels of small to large rivers, lakes, and impoundments.

DISTRIBUTION IN ILLINOIS Forbes and Richardson (1909) and Smith (1979) found the Smallmouth Buffalo to occur statewide but mostly restricted to large rivers. The species's distribution in the contemporary era is essentially the same, although it appears to be somewhat more frequently encountered in smaller streams (but see Remarks).

REMARKS Juveniles of species of *Ictiobus* are difficult to separate from one another, and some records on the map for the Smallmouth Buffalo and other species of *Ictiobus* may be in error as identification of the unvouchered records cannot be confirmed. Also, Bart et al. (2010) documented widespread hybridization among all 3 species of *Ictiobus* that occur in Illinois, further complicating the interpretation of distributions.

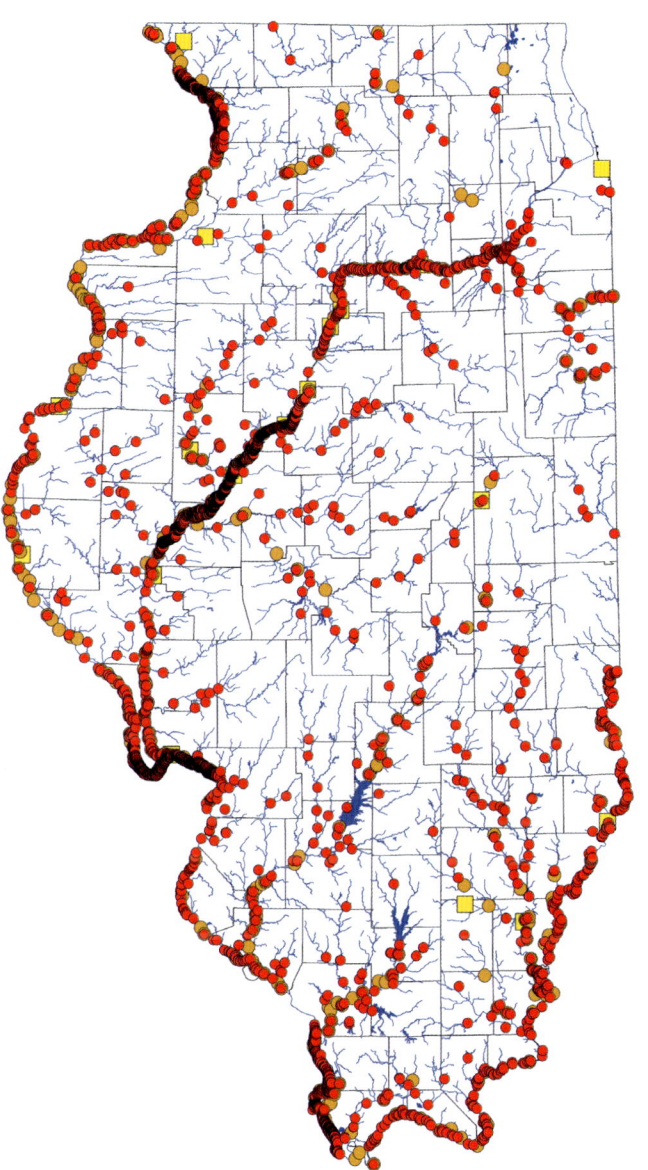

BIGMOUTH BUFFALO *Ictiobus cyprinellus* (Valenciennes 1844)

IDENTIFICATION The Bigmouth Buffalo has a long, falcate dorsal fin with 24–31 rays, a deep thick body, a large head, and an oblique, terminal mouth. The front of the upper lip is level with the lower edge of the eye. The upper lip has shallow grooves. Large individuals have a rounded nape. The subopercle is semicircular and widest at its midpoint. The back and upper side of the body are gray-black to olive-bronze with a green to copper sheen, the side is black to dusky olive, the lower side and belly are dusky white to light yellow, and the fins are brown to black. There are 40 or more rakers on the first gill arch and usually 35–36 lateral scales, 10–11 pelvic rays, and 8–9 anal rays. To 40 in. (100 cm).

SIMILAR SPECIES The Smallmouth Buffalo, *I. bubalus* has a nearly horizontal subterminal mouth, an upper lip that is well below the lower edge of the eye, distinct grooves on the upper lip, a moderately keeled nape, and 35 or fewer rakers on the first gill arch. The Black Buffalo, *I. niger*, has a shallower body, a more subterminal and less-oblique mouth, and distinct grooves on the upper lip, and the front of the upper lip is well below the lower edge of the eye. Carpsuckers, *Carpiodes*, have silver sides of the body, clear white or orange pelvic fins, and a subopercle that is triangular and widest below its midpoint.

HABITAT The Bigmouth Buffalo occurs in flowing pools, backwaters, and the main channels of small to large rivers, lakes, and impoundments.

DISTRIBUTION IN ILLINOIS Both Forbes and Richardson (1909) and Smith (1979) found the Bigmouth Buffalo to occur statewide in medium to large rivers and less often in small rivers. The species's distribution in the contemporary era is essentially the same, although it appears to be less frequently encountered in the upper Mississippi River and somewhat more frequently found in southern Illinois.

REMARKS Juveniles of species of *Ictiobus* are difficult to separate from one another, and some records on the map for the Bigmouth Buffalo and other species of *Ictiobus* may be in error as identification of the unvouchered records cannot be confirmed. Also, Bart et al. (2010) documented widespread hybridization among all 3 species of *Ictiobus* that occur in Illinois, further complicating the interpretation of distributions.

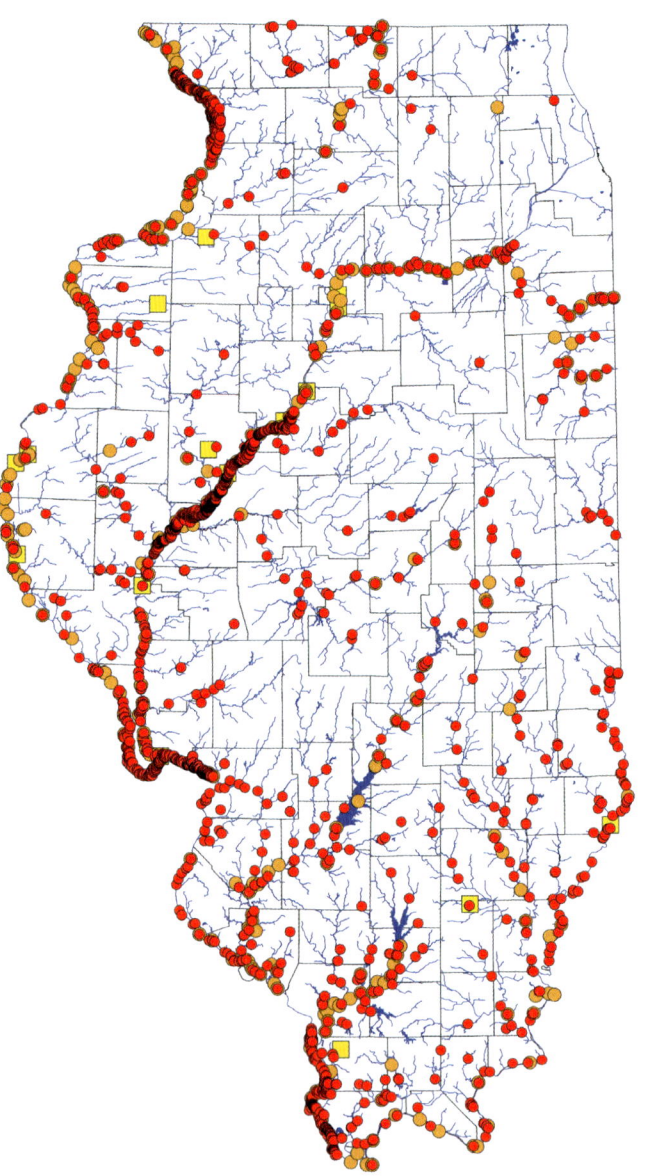

BLACK BUFFALO *Ictiobus niger* (Rafinesque 1819)

IDENTIFICATION The Black Buffalo has a long, falcate dorsal fin with 24–31 rays, a deep, thick body, a large, ovoid head, and a slightly oblique terminal mouth. The front of the upper lip is well below the lower edge of the eye. The thick upper lip has shallow grooves. Large individuals have a rounded nape. The subopercle is semicircular and widest at its midpoint. The back and upper side of the body are gray-black to olive-bronze, the side is black to dusky olive, the lower side and belly are dusky white to light yellow, and the fins are olive to black. There are 35 or fewer rakers on the first gill arch and usually 37–39 lateral scales, 9–11 pelvic rays, and 9 anal rays. To 37 in. (93 cm).

SIMILAR SPECIES The Smallmouth Buffalo, *I. bubalus*, has a nearly horizontal subterminal mouth, a deeper body, and a moderately keeled nape. The Bigmouth Buffalo, *I. cyprinellus*, has a sharply oblique terminal mouth, an upper lip that is level with the lower edge of the eye, shallow grooves on the upper lip, and 40 or more gill rakers on the first gill arch. Carpsuckers, *Carpiodes*, have silver sides of the body, clear white or orange pelvic fins, and a subopercle that is triangular and widest below its midpoint.

HABITAT The Black Buffalo occurs in flowing pools, backwaters, and the main channels of small to large rivers, lakes, and impoundments.

DISTRIBUTION IN ILLINOIS Smith (1979) found the Black Buffalo to occur statewide but mostly restricted to large rivers and less frequently recorded than other species of *Ictiobus*. Forbes and Richardson (1909) had no records for the Black Buffalo from east-central or southern Illinois. In the contemporary era the species occurs statewide and appears to be somewhat more frequently encountered in smaller streams than in previous surveys; however, it remains the least widespread species of *Ictiobus* in Illinois.

REMARKS Juveniles of species of *Ictiobus* are difficult to separate from one another, and some records on the map for the Black Buffalo and other species of *Ictiobus* may be in error as identification of the unvouchered records cannot be confirmed. Also, Bart et al. (2010) documented widespread hybridization among all 3 species of *Ictiobus* that occur in Illinois, further complicating the interpretation of distributions.

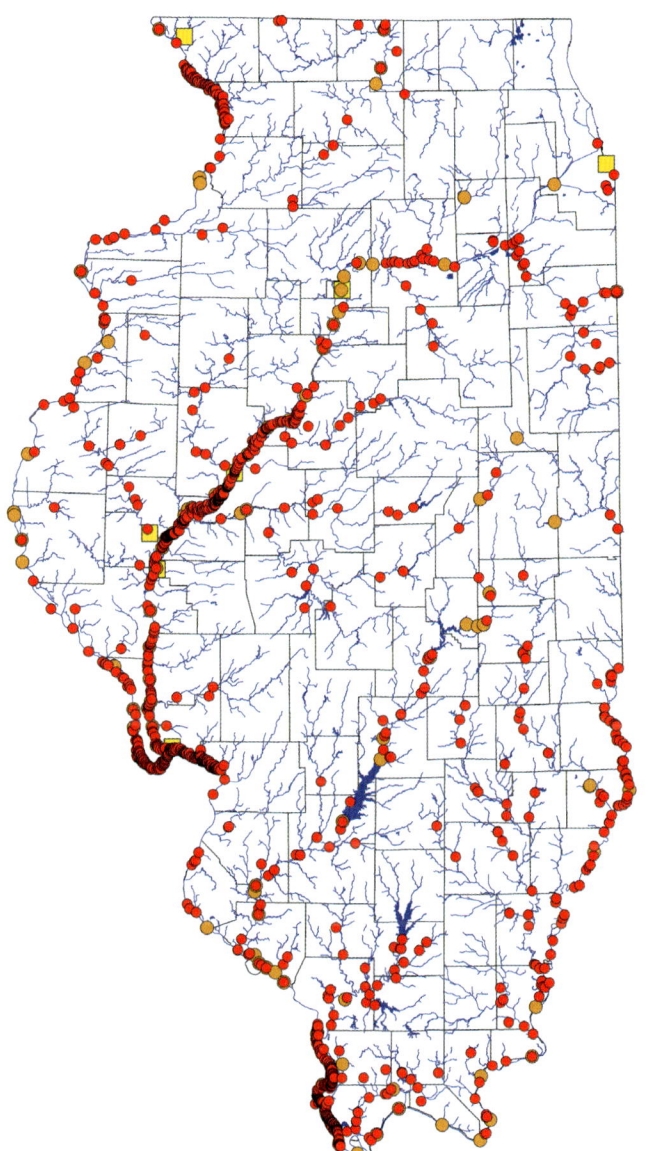

SPOTTED SUCKER *Minytrema melanops* (Rafinesque 1820)

IDENTIFICATION The Spotted Sucker has a slender body, a small mouth with thin (relative to other suckers) plicate lips, a U-shaped lower lip, and 8–12 parallel rows of dark brown to black spots on the back and side of the body. The lateral line is absent or present on only a few scales anteriorly, and the dorsal fin has a straight or concave edge. The back and upper side are olive to silver-gray, the side is silver-yellow to light brown, the underside is white, median fins are clear olive to orange, and paired fins are clear to dusky orange. The edges of the dorsal fin and lower lobe of the caudal fin are black. Juveniles have pink median fins. There are usually 42–47 lateral scales and 12 dorsal rays. To 19½ in. (50 cm).

SIMILAR SPECIES Redhorses, *Moxostoma*, lack 8–12 parallel rows of dark spots on the side of the body and have a lateral line and thicker lips. Chubsuckers, *Erimyzon*, have a deeper and less-slender body and a rounded dorsal fin.

HABITAT The Spotted Sucker occurs in flowing pools and runs of small creeks to large rivers with firm clay, sand, or gravel substrates and in impoundments.

DISTRIBUTION IN ILLINOIS The Spotted Sucker is generally distributed throughout Illinois, including in the northeast, where it was not recorded in earlier eras. It is widespread in the Wabash River basin of east-central Illinois, in the upper Mississippi River, and in much of southern Illinois, but it occurs sporadically in much of northern and central Illinois. Records from the Forbes and Richardson era suggest that the species was more widespread, especially in the Illinois River basin.

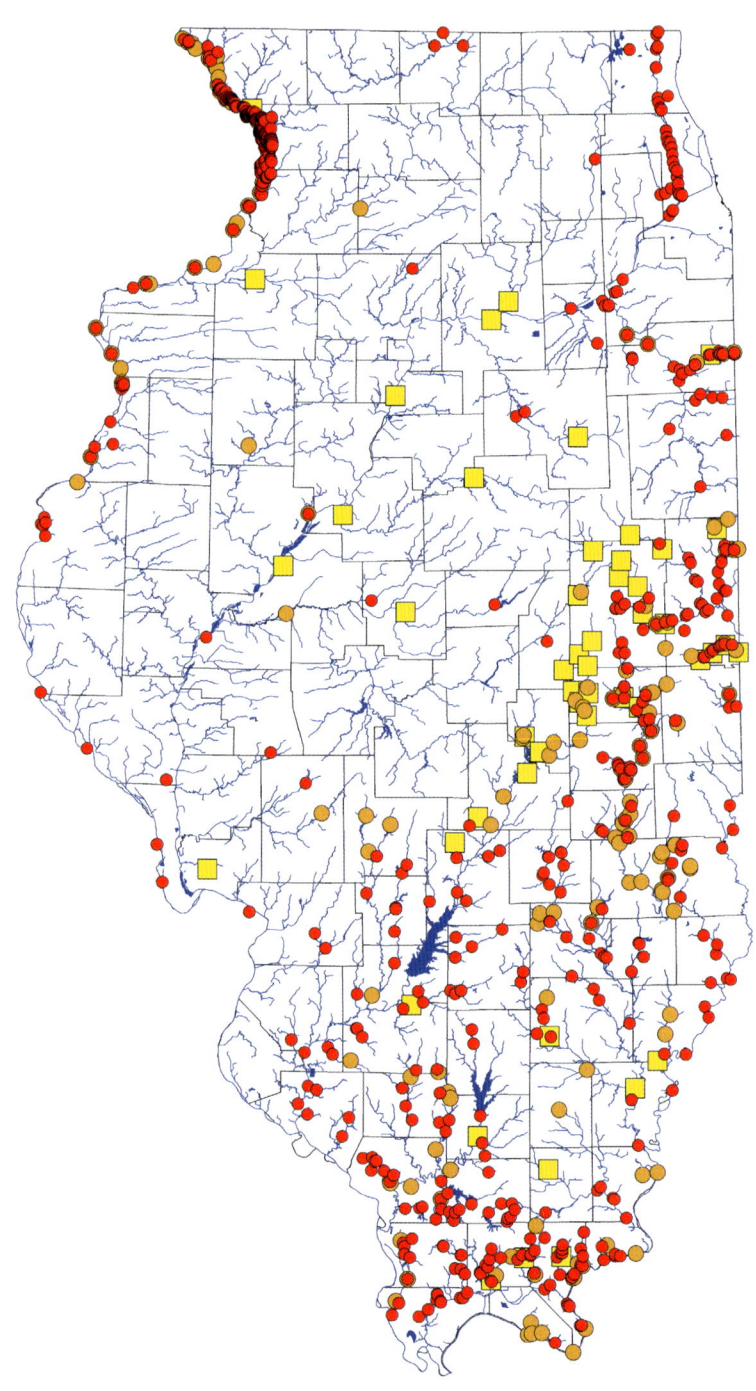

SILVER REDHORSE *Moxostoma anisurum* (Rafinesque 1820)

IDENTIFICATION The Silver Redhorse has a relatively deep, thick body that tapers to a slender caudal peduncle, a large head, and a dorsal fin with a straight or slightly convex edge and 14–16 rays. The lower lip is distinctly bilobed with an acutely V-shaped rear edge. The lips are plicate with small papillae. The back and side are silvery blue-green to brassy yellow, the underside is white to yellow, the dorsal fin is clear to dusky gray, and other fins are clear, dusky, or orange. Scales lack crescent-shaped black spots. There are usually 40–42 lateral scales and 12 scale rows around the caudal peduncle. To 28 in. (71 cm).

SIMILAR SPECIES The Shorthead Redhorse, *M. macrolepidotum*, River Redhorse, *M. carinatum*, and Greater Redhorse, *M. valenciennesi*. have crescent-shaped dark spots on the anterior edges of scales on the side and a pink or red caudal fin, and lack a deep V-shaped rear edge on the lower lip. The Golden Redhorse, *M. erythrurum*, and Black Redhorse, *M. duquesnei*, have a dorsal fin with a concave edge and lack a bilobed lower lip with a deep V-shaped rear edge.

HABITAT The Silver Redhorse occurs in deep pools and runs of creeks to large rivers.

DISTRIBUTION IN ILLINOIS The Silver Redhorse is generally distributed across the northern half of Illinois and in the northern tributaries of the Wabash River. The same general distribution was found in earlier eras, although the species was not recorded from the Des Plaines and Fox River basins, and Forbes and Richardson (1909) had no records for the Wabash River basin other than 1 record in the Vermilion River (Wabash) basin in Champaign County.

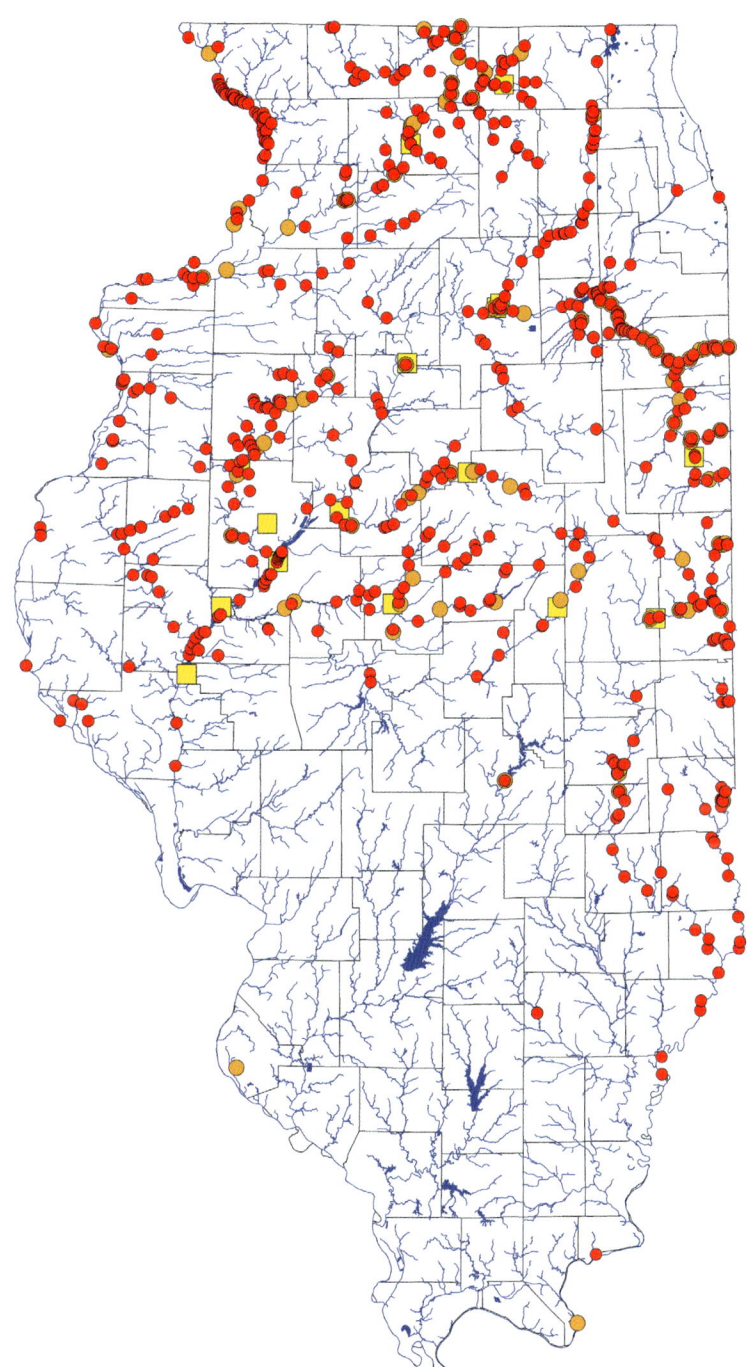

RIVER REDHORSE *Moxostoma carinatum* (Cope 1870)

IDENTIFICATION The River Redhorse has a relatively deep, thick body that tapers to a slender caudal peduncle, a large head that goes 4 times or less into the standard length, a dorsal fin with a straight or slightly concave edge and 12–13 rays, and a distinctly thickened lower lip with a broad U-shaped rear edge. The lips are plicate and without small papillae. Stout, molar-like teeth are present on the pharyngeal arch. The back of the body is olive-bronze, the side is pale yellow-silver to brassy with crescent-shaped dark spots on the anterior edges of the scales. The underside is white or yellow. The dorsal fin is reddish slate, anal and paired fins are yellow to orange, and the caudal fin is red. Breeding males have a dark stripe along the side and a bright red caudal fin. There are 42–44 lateral scales and 12–13 scale rows around the caudal peduncle. To 30 in. (77 cm).

SIMILAR SPECIES The Shorthead Redhorse, *M. macrolepidotum*, has a short head that goes more than 4 times into its standard length and small papillae on the rear of the lower lip. The Greater Redhorse, *M. valenciennesi*, has 15–16 scale rows around the caudal peduncle and lacks molar-like teeth on the pharyngeal arch. The Silver Redhorse, *M. anisurum*, the Golden Redhorse, *M. erythrurum*, and Black Redhorse, *M. duquesnei*, lack dark spots on the anterior edges of their scales and a pink or red caudal fin.

HABITAT The River Redhorse occurs in rocky pools and runs of small to medium rivers.

DISTRIBUTION IN ILLINOIS The River Redhorse occurs in most of the upper Illinois River basin, with the exception of the Des Plaines River, and in the Vermilion River basin in east-central Illinois. The species appears to be more widespread in the upper Illinois River basin than during the Smith era, during which it was found only in the Fox and Kankakee Rivers. The large number of new records suggests an in-state population increase or the increased use of boat electrofishing, which is the most effective method of capturing species of *Moxostoma* (Burr et al. 1996). Forbes and Richardson (1909) recorded the species at 1 locality each in the Rock River in Ogle County and the Kaskaskia River in Douglas County. The species has not been recorded from these 2 basins since then (Retzer & Kowalik 2002).

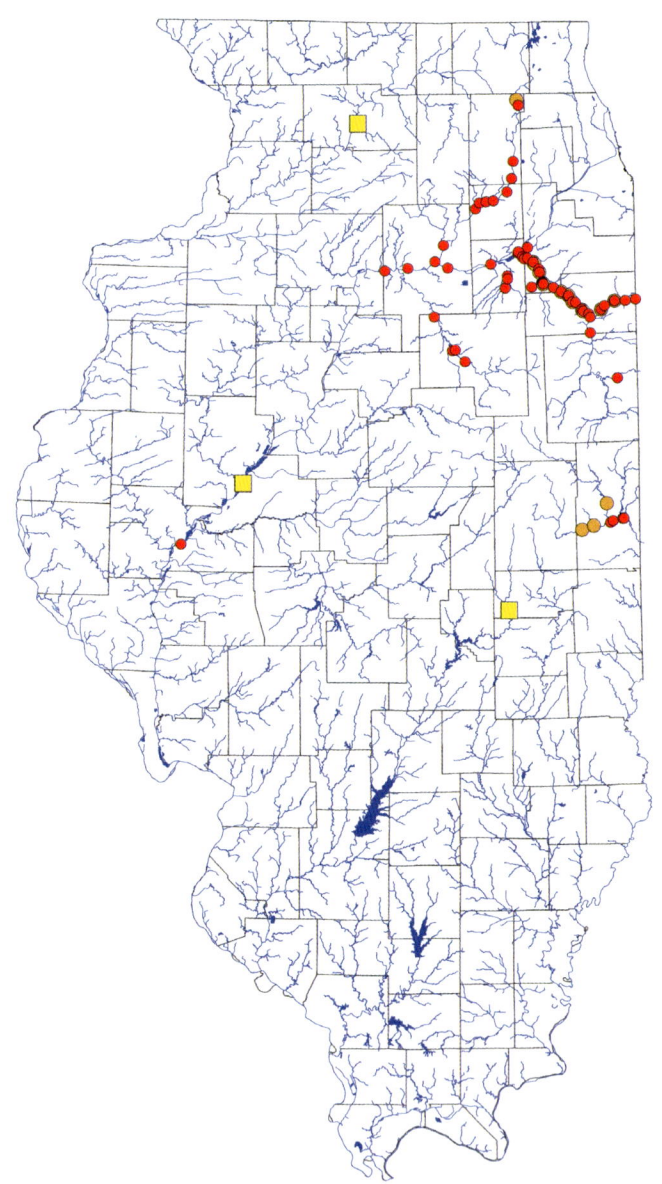

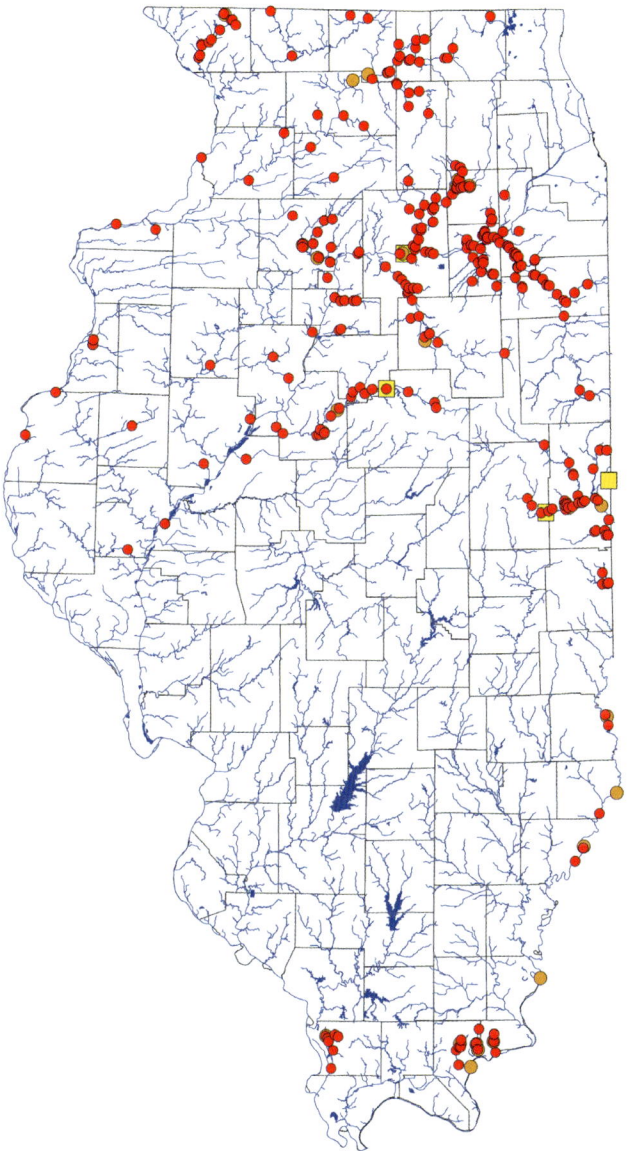

BLACK REDHORSE *Moxostoma duquesnei* (Lesueur 1817)

IDENTIFICATION The Black Redhorse has a relatively long, slender body that tapers to a slender caudal peduncle, a moderately large head that goes 4 times or less into the standard length, a dorsal fin with a straight or slightly concave edge and 12–14 rays, and a lower lip with a broad V-shaped rear edge. The lips are plicate and without small papillae. The back is light gray to dusky olive, the side is gold to brassy, and the underside is white to yellow. The anal and paired fins are orange, and the dorsal and caudal fins are dusky to slate. Breeding males have a pink-orange stripe along the side, pink-orange anal and paired fins, and small tubercles on the snout. Scales lack crescent-shaped black spots. There are 44–47 lateral scales and 12–14 scale rows around a narrow caudal peduncle. To 20 in. (51 cm).

SIMILAR SPECIES The Shorthead Redhorse, *M. macrolepidotum*, has a short head that goes more than 4 times into the standard length, small papillae on the rear of the lower lip, a pink or red caudal fin, and crescent-shaped dark spots on the anterior edges of the scales. The Greater Redhorse, *M. valenciennesi*, and River Redhorse, *M. carinatum*, have a pink or red caudal fin and crescent-shaped dark spots on the anterior edges of their scales. The Silver Redhorse, *M. anisurum*, has a bilobed lower lip with an acutely V-shaped rear edge. The Golden Redhorse, *M. erythrurum*, has 39–43 lateral scales and a deeper caudal peduncle.

HABITAT The Black Redhorse occurs in pools and runs of high-gradient creeks and small to medium rivers, usually over substrates of mud, sand, or gravel.

DISTRIBUTION IN ILLINOIS The Black Redhorse occurs in all basins in the northern half of Illinois, except the Des Plaines River, the Wabash River, and streams draining the western and eastern extremes of the Shawnee Hills in southern Illinois. It is noticeably absent from the Sangamon, Embarras, Little Wabash, Kaskaskia, and Big Muddy River basins of central and southern Illinois, most likely due to its preference for high-gradient creeks and rivers. The species appears more widespread in the contemporary era in the western half of northern Illinois than it was in previous eras.

GOLDEN REDHORSE *Moxostoma erythrurum* (Rafinesque 1818)

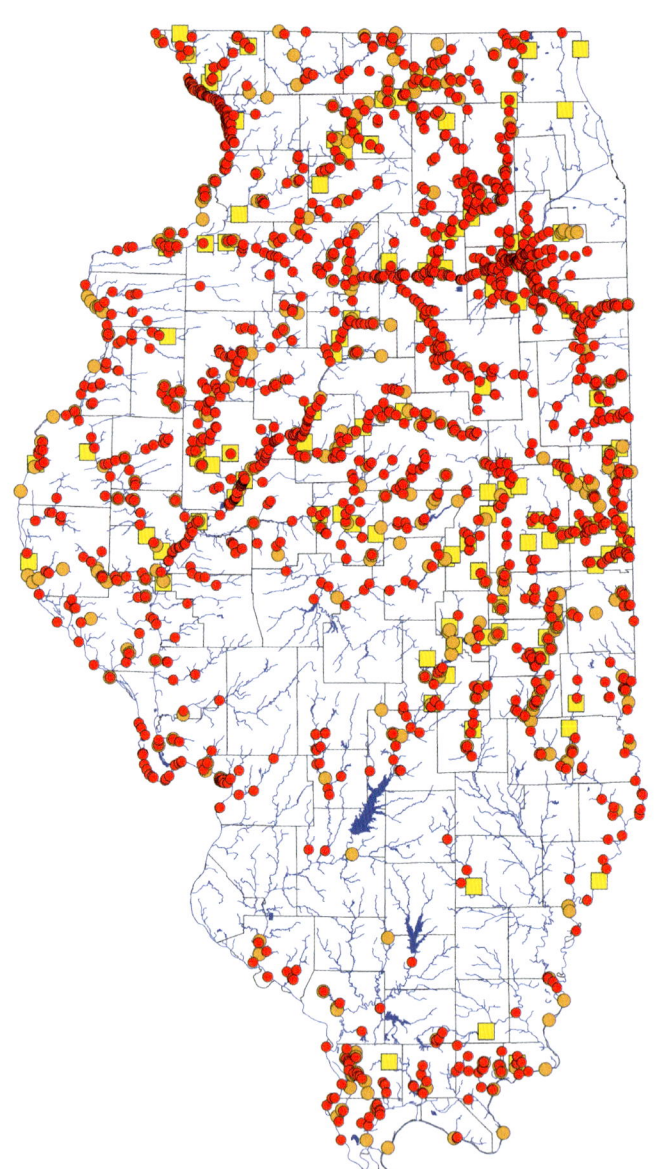

IDENTIFICATION The Golden Redhorse has a relatively deep body that tapers to a slender caudal peduncle, a moderately large head that goes 4 times or less into the standard length, a dorsal fin with a straight or slightly concave edge and 12–14 rays, and a lower lip with a broad V-shaped rear edge. The lips are plicate and without small papillae. The back is brassy brown to dusky olive, the side is light yellow to brassy, and the underside is white to yellow. The anal and paired fins are yellow to orange, and the dorsal and caudal fins are dusky to slate. Breeding males have a dark stripe along the side, bright pink or bright orange anal and paired fins, and usually large tubercles on the snout. Scales lack crescent-shaped black spots. There are 39–43 lateral scales and 12 scale rows around the caudal peduncle. To 30½ in. (78 cm).

SIMILAR SPECIES The Shorthead Redhorse, *M. macrolepidotum*, has a short head that goes more than 4 times into the standard length, small papillae on rear of the lower lip, a pink or red caudal fin, and crescent-shaped dark spots on the anterior edges of the scales. The Greater Redhorse, *M. valenciennesi*, and River Redhorse, *M. carinatum*, have a pink or red caudal fin and crescent-shaped dark spots on the anterior edges of their scales. The Silver Redhorse, *M. anisurum*, has a bilobed lower lip with an acutely V-shaped rear edge. The Black Redhorse, *M. duquesnei*, has 44–47 lateral scales and a narrower caudal peduncle.

HABITAT The Golden Redhorse lives in pools, runs, and deep riffles of creeks and small to large rivers. It is usually found over gravel or mixed sand and gravel and occasionally is captured in impoundments.

DISTRIBUTION IN ILLINOIS Forbes and Richardson (1909) and Smith (1979) found the Golden Redhorse to occur throughout much of Illinois but absent from most of the Mississippi River tributaries in southwestern Illinois and rare in the Wabash River tributaries below the Embarras River. In the contemporary era the species essentially shows the same distribution but is somewhat more widespread in southwestern Illinois and the Wabash River basin. The Golden Redhorse has not been found in the Lake Michigan basin since the Forbes and Richardson era.

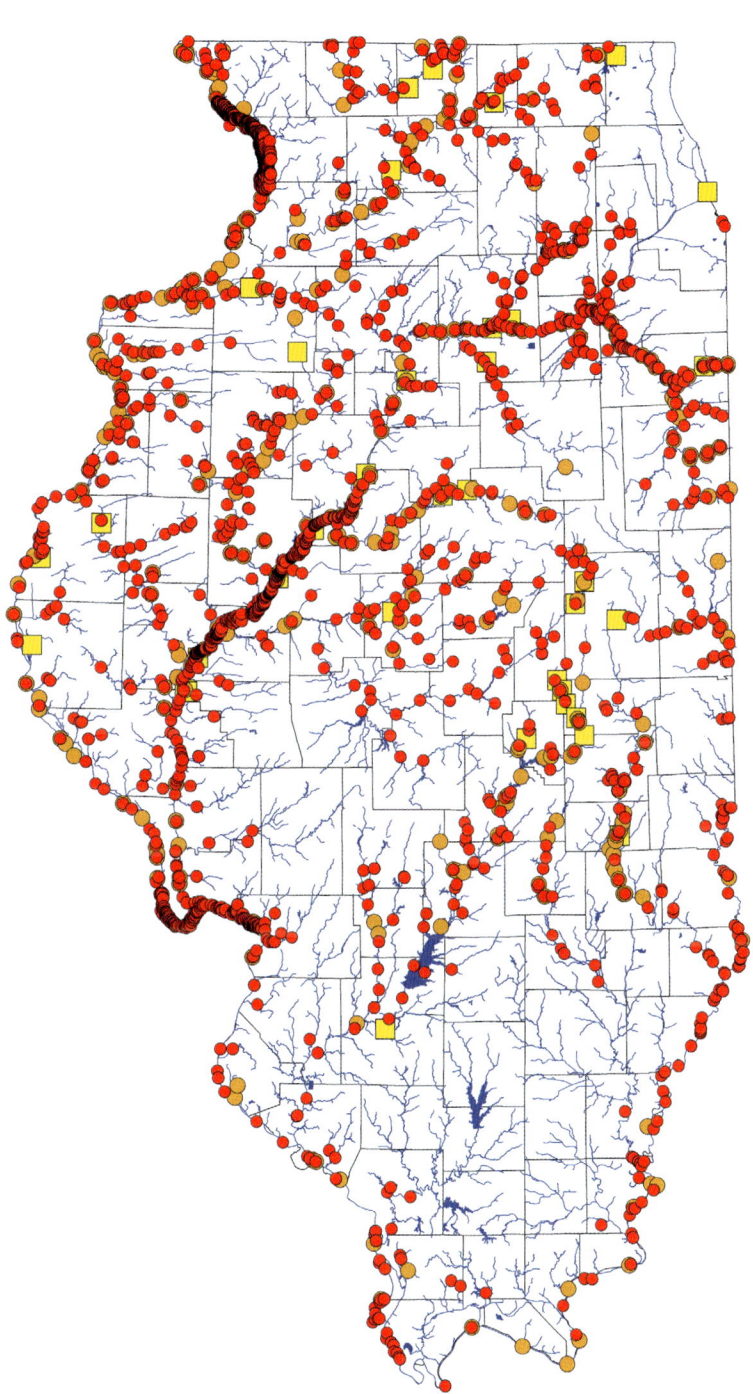

SHORTHEAD REDHORSE   *Moxostoma macrolepidotum* (Lesueur 1817)

IDENTIFICATION The Shorthead Redhorse has a relatively elongate body that tapers to a slender caudal peduncle, a small head that goes 4.2 times or more into the standard length, a dorsal fin with a straight or slightly concave edge and 12–13 rays, and a lower lip with a broad V-shaped to nearly straight rear edge. The lower lip is plicate in the forward half and with small papillae on the rear half. The back is olive to tan, the side is olive-yellow with a brassy or silver sheen, and the underside is white to yellow. The scales on the side of the body have crescent-shaped dark spots on the anterior edges. The anal and paired fins are light pink to reddish-orange, and the dorsal and caudal fins are pale pink to deep red. Breeding males have small tubercles on the snout and caudal and anal fins. There are 42–44 lateral scales and 12–13 scale rows around the caudal peduncle. To 29½ in. (75 cm).

SIMILAR SPECIES Other Redhorse species in Illinois have large heads that go 4 times or less into the standard length. The Silver Redhorse, *M. anisurum*, Black Redhorse, *M. duquesnei*, and Golden Redhorse, *M. erythrurum*, lack pink or red caudal fins and dark spots on the anterior edges of their scales.

HABITAT The Shorthead Redhorse is found over rocky substrates in flowing pools, runs, and riffles of creeks and small to large rivers.

DISTRIBUTION IN ILLINOIS As found in previous eras, the Shorthead Redhorse is generally distributed in the northern two-thirds of the state but is nearly absent from the Big Muddy, Skillet Fork, and Saline River basins in southern Illinois. It also is absent in the upper Des Plaines River basin and rare in Lake Michigan.

GREATER REDHORSE *Moxostoma valenciennesi* Jordan 1885

IDENTIFICATION The Greater Redhorse has a relatively deep, thick body that tapers to a slender caudal peduncle, a large head that goes 4 times or less into the standard length, a dorsal fin with a straight or slightly convex edge and 13–14 rays, and a distinctly thickened lower lip with a broad V-shaped rear edge. The lips are plicate and without small papillae. The back and side of the body are bronze to copper, and the underside is silver-yellow to white. There are crescent-shaped dark spots on the anterior edges of scales on the back and side. The dorsal fin is reddish slate, anal and paired fins are yellow to reddish orange, and the caudal fin is red. Breeding males have small tubercles on the head and scales, and large tubercles on the anal fin and lower lobe of the caudal fin. There are 42–45 lateral scales and 15–16 scale rows around the caudal peduncle. To 31½ in. (80 cm).

SIMILAR SPECIES The Shorthead Redhorse, *M. macrolepidotum*, has a short head that goes more than 4 times into the standard length and small papillae on the rear of the lower lip. The River Redhorse, *M. carinatum*, has 12–13 scale rows around the caudal peduncle and has molar-like teeth on the pharyngeal arch. The Silver Redhorse, *M. anisurum*, the Golden Redhorse, *M. erythrurum*, and the Black Redhorse, *M. duquesnei*, lack crescent-shaped dark spots on the anterior edges of their scales and a pink or red caudal fin.

HABITAT The Greater Redhorse occurs in sandy to rocky pools and runs of creeks and small to large rivers. Larger adults are more frequently encountered in medium to large rivers. The species also occurs in lakes in other states.

DISTRIBUTION IN ILLINOIS The Greater Redhorse occurs in a relatively small area of the upper Illinois River basin in the lower Fox and Vermilion River basins, in Aux Sable Creek, and in the Illinois River. Smith (1979) reported a single specimen collected in 1901 from Salt Creek of the Des Plaines River basin that had been mixed in with a Forbes and Richardson collection of Golden Redhorse, *M. erythrurum*. The 78-year absence of records for the Greater Redhorse caused Smith (1979) to assume that the species had been extirpated from Illinois. Given the distance to extant populations of the Greater Redhorse, recolonization in the upper Illinois River in recent decades is unlikely. Rather, the Greater Redhorse was missed during sampling through the Smith era. The large number of new records in rivers that had been sampled in previous surveys sug-

gests an in-state population increase (Retzer & Kowalik 2002) or the increased use of boat electrofishing, which is the most effective method of capturing species of *Moxostoma* (Burr et al. 1996). A similar increase has occurred in the Fox River in Wisconsin (Lyons et al. 2000).

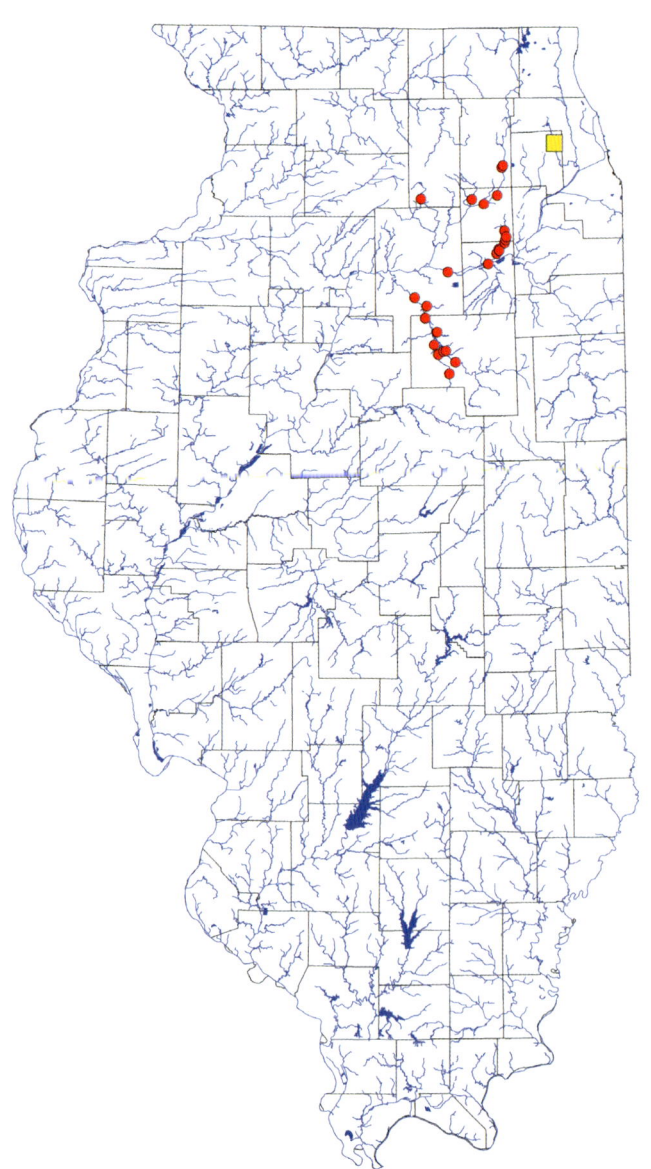

BARBS AND CARPS

Family Cyprinidae Rafinesque 1815

Carps lack teeth on the jaws but share with minnows and suckers the presence of pharyngeal teeth that are used to grind food against the roof of the mouth. They have 1 dorsal fin, no adipose fin, abdominal pelvic fins, and the anal fin far back on the body. Most species have 2 pairs of barbels, and pharyngeal teeth are usually in 2 or 3 rows. Males of many species become brightly colored, and some develop tubercles during the breeding season. Cyprinids are native to Eurasia and Africa, where they, along with leuciscids (minnows), usually are the most diverse and abundant fishes in streams and lakes. Cyprinids comprise a huge diversity of forms, including the Common Carp (*Cyprinus carpio*) and many popular aquarium fishes, such as the Goldfish (*Carassius auratus*), Cherry Barb (*Puntius titteya*), Red-Tailed Shark (*Epalzeorhynchos bicolor*), and Tiger Barb (*Puntigrus tetrazona*). The family contains about 1,700 species of which 2, the Common Carp and Goldfish, have been introduced into North America and are reproducing in Illinois.

1a   Upper jaw with 2 fleshy barbels on each side; lateral line scales usually 32–41 . . . . **Page 116**

1b   Upper jaw without barbels; lateral line scales usually 25–31 . . . . **Page 115**

Common Carp (*Cyprinus carpio*)

Goldfish (*Carassius auratus*)

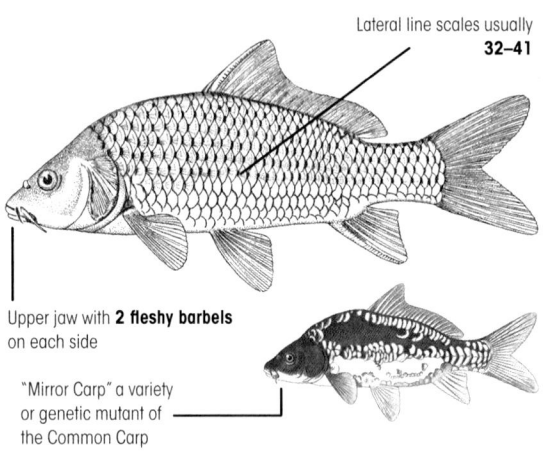

Lateral line scales usually **32–41**

Upper jaw with **2 fleshy barbels** on each side

"Mirror Carp" a variety or genetic mutant of the Common Carp

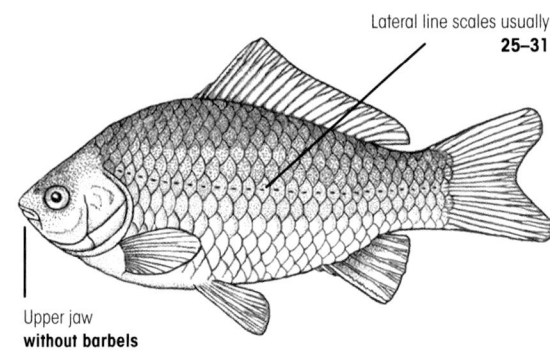

Lateral line scales usually **25–31**

Upper jaw **without barbels**

GOLDFISH

Carassius auratus (Linnaeus 1758)

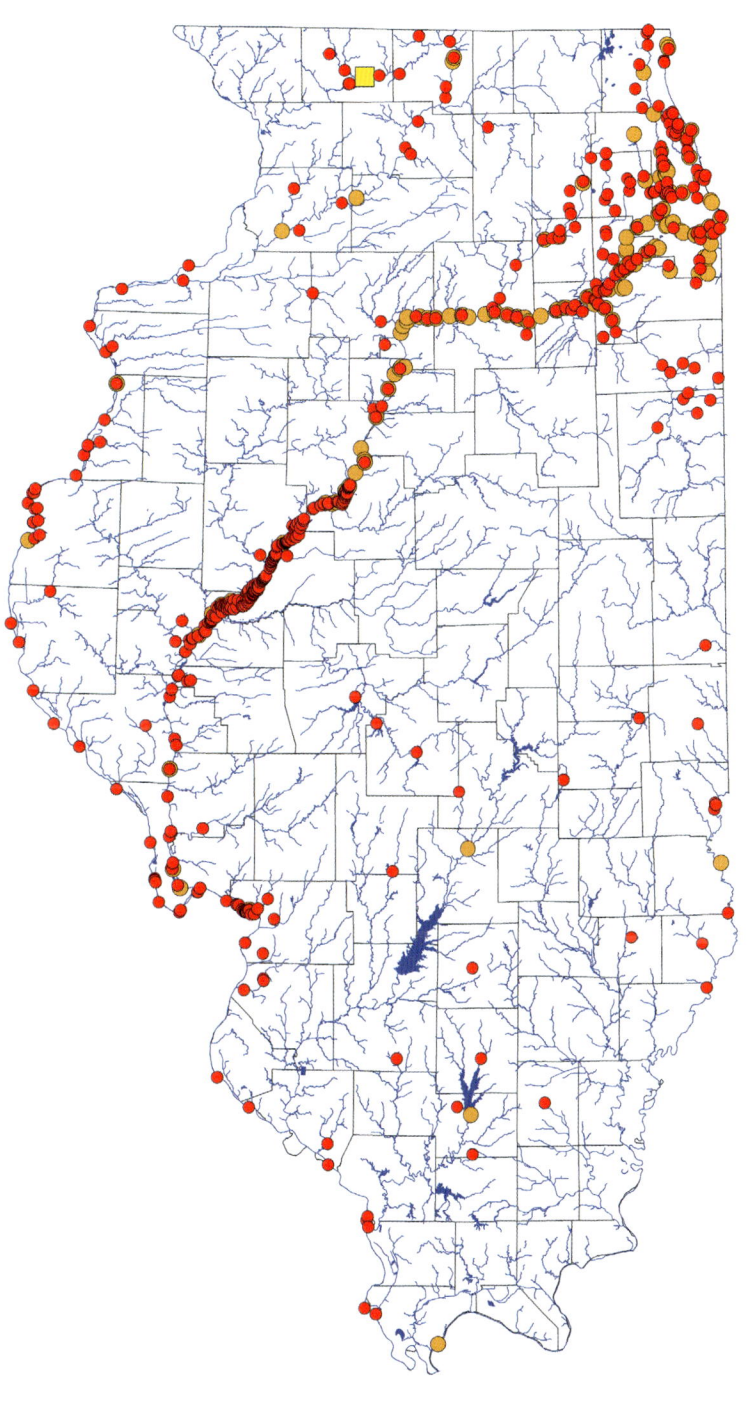

IDENTIFICATION The Goldfish has large scales with 25–31 along the lateral line and a long dorsal fin with 15–21 rays. There is a stout, saw-toothed spine and 2 smaller spines at the front of the dorsal and anal fins. The body is deep and thick, the mouth is terminal, and the caudal fin is large. The back and side of the body are gray-green with a brassy sheen, and the underside is white to yellow. The dorsal and caudal fins are gray to brown (see Remarks). There are 5–6 anal rays, and 0,4-4,0 pharyngeal teeth. To 16 in. (41 cm).

SIMILAR SPECIES The Common Carp, *Cyprinus carpio*, has 2 pairs of barbels, 32–38 lateral scales, and 1,1,3-3,1,1 pharyngeal teeth.

HABITAT The Goldfish lives in shallow, muddy pools and backwaters of sluggish rivers, ponds, and lakes. It usually is found in turbid water and often near vegetation. It is more tolerant than most fishes of some forms of pollution.

DISTRIBUTION IN ILLINOIS Smith (1979) stated that the first Goldfish to appear in Illinois were probably "escapees from ponds where they were kept for ornamental purposes." The species has spread rapidly and is abundant in the Chicago region and throughout the Illinois River. The huge floods of 1993 and 1994 allowed Goldfish to move into the Mississippi River, where it was found frequently both above and below the mouth of the Missouri River. It now occurs throughout Illinois, and an occasional individual may be seen almost anywhere in the state.

REMARKS The Goldfish, native to Asia, was first introduced into the United States in the late 1600s. Recently released Goldfish with "pet store" colors (gold, red, white, blue, black) may be encountered in the wild; however, reproducing populations quickly revert to natural cryptic colors. Goldfish readily hybridize with the Common Carp in Illinois, and in polluted areas, hybrids (which are fertile) outnumber the parent species.

BARBS AND CARPS (Cyprinidae)

115

COMMON CARP

Cyprinus carpio Linnaeus 1758

IDENTIFICATION The Common Carp has 2 large barbels, the rear barbel much longer, on each side of the upper jaw. The dorsal fin is long with 17–21 rays. There are a stout, saw-toothed spine and 2 smaller spines at the front of the dorsal and anal fins. The body is deep, thick, and strongly arched to the dorsal fin, which is somewhat flattened below. The mouth is terminal on young individuals and more subterminal on adults. The back and side of the body are gray (juveniles) to brassy green (adults), and the scales on the back and upper side are dark edged, with a black spot at the base of each scale. The underside is white to yellow, and fins are clear to dusky gray. Large individuals have bright red-orange caudal and anal fins. There are 32–38 lateral scales, 5–6 anal rays, and 1,1,3-3,1,1 pharyngeal teeth. To 4 ft. (1.2 m).

SIMILAR SPECIES The Goldfish, *Carassius auratus*, lacks barbels and dark-edged scales, has 25–31 lateral scales and 0,4-4,0 pharyngeal teeth.

HABITAT The Common Carp lives in muddy pools or along edges of small to large rivers and is common in lakes and ponds. It often is found in impoundments and turbid, sluggish streams with organic matter.

DISTRIBUTION IN ILLINOIS There were few records for the Common Carp during the Forbes and Richardson (1909) era, but Smith (1979) found the species to be statewide in occurrence. The Common Carp can be found almost anywhere in the state.

REMARKS The Common Carp, native to Eurasia, was first introduced to North America in 1831 and now is widely distributed in southern Canada, most of the United States, and northern Mexico. Individuals known as the "mirror carp" (with few, enlarged scales) or "leather carp" (scaleless) are fairly frequently encountered.

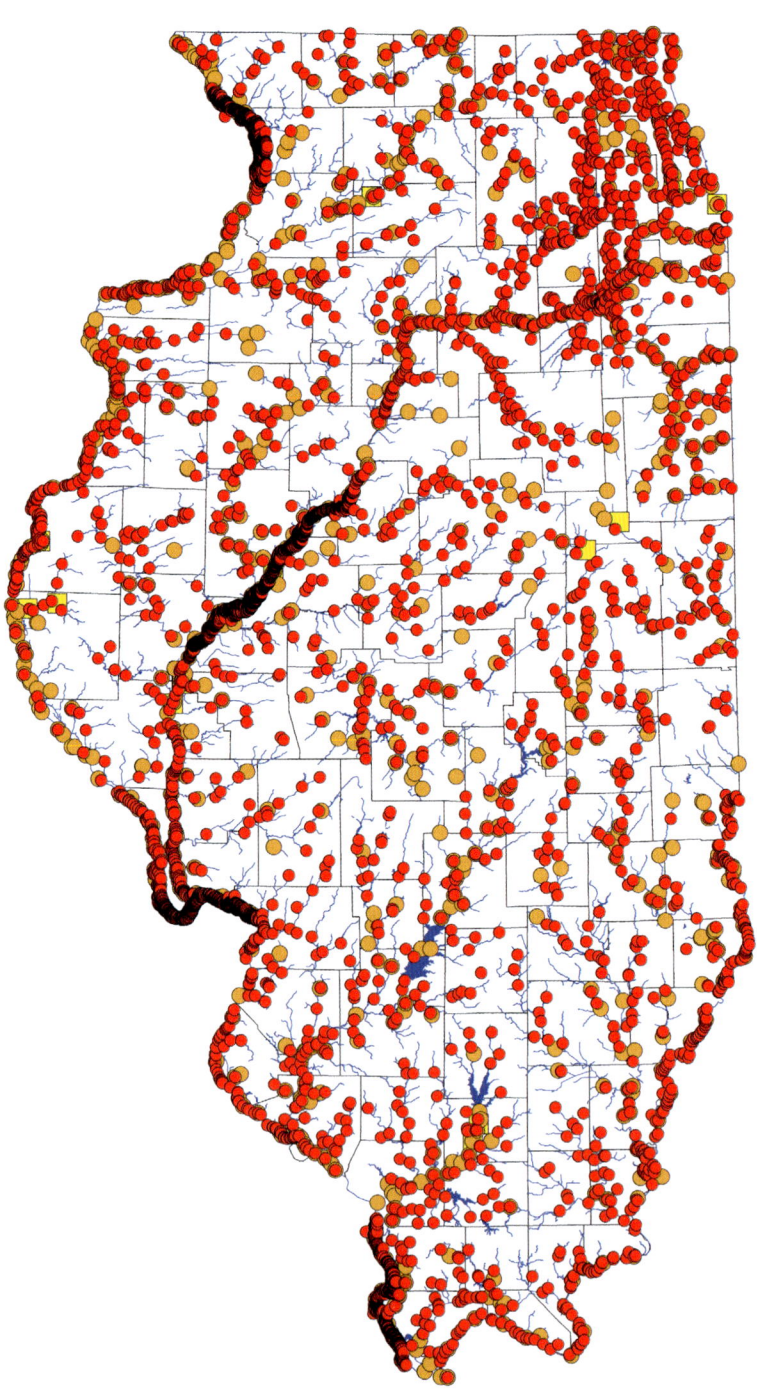

SHARPBELLIES

Family Xenocyprididae Günther 1868

Like minnows (Leuciscidae), members of Xenocyprididae were until recently (Tan & Armbruster 2018) placed in Cyprinidae. Osteological and molecular data rather than external morphology have been used to distinguish xenocyprids from cyprinids (barbs and carps) and leuciscids (minnows), but all 4 species of xenocyprids present in Illinois can be readily distinguished from native minnows by their much-larger size. All reach at least 4 ft. (1.2 m), whereas the largest minnows native to Illinois reach only 12½ in. (32 cm). Xenocyprids lack teeth on the jaws but share with cyprinids and leuciscids the presence of pharyngeal teeth that are used to grind food against the roof of the mouth. They have 1 dorsal fin, no adipose fin, and abdominal pelvic fins, and the anal fin is far back on the body. Four of the 160 species, all of which are native to Asia, have been introduced into North America and are reproducing in Illinois.

KEY TO THE SHARPBELLIES (Xenocyprididae)

1a Eye low on head, below central axis of body; pectoral fin long, extending well beyond front of pelvic fin; scales small, 85 or more in lateral line **2**

1b Eye higher on head, not below central axis of body; pectoral fin short, rarely extending to front of pelvic fin; scales larger, usually 85 or fewer in lateral line **3**

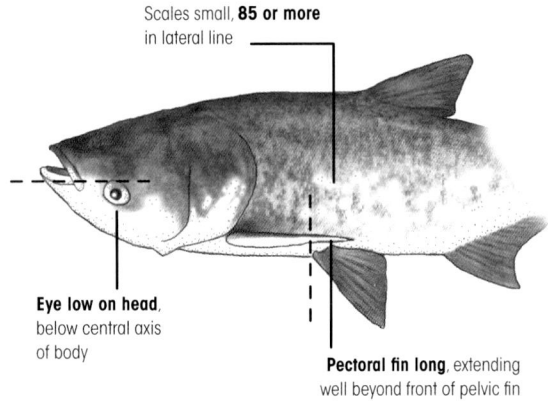

Scales small, **85 or more** in lateral line

Eye low on head, below central axis of body

Pectoral fin long, extending well beyond front of pelvic fin

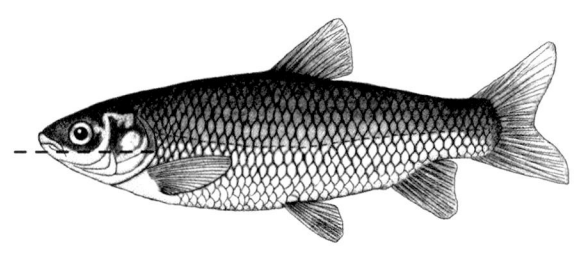

2a Belly with a smooth midventral keel (ridge) extending from front of anal fin to front of pelvic fins; side with scattered dark blotches; gill rakers long and slender . . . .**Page 122**

2b Belly with a smooth midventral keel extending from front of anal fin to under gill covers; side silvery, without scattered dark blotches; gill rakers a comb-like mass. . . .**Page 121**

Bighead Carp (*Hypophthalmichthys nobilis*)

Silver Carp (*Hypophthalmichthys molitrix*)

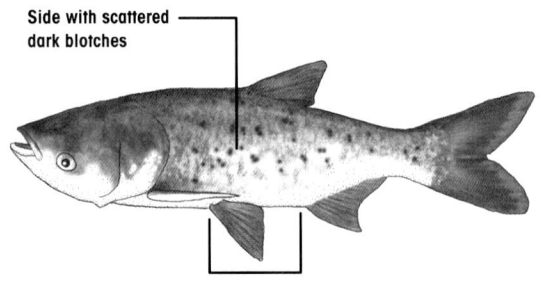

Side with scattered dark blotches

Belly with midventral keel extending from front of anal fin to front of pelvic fins

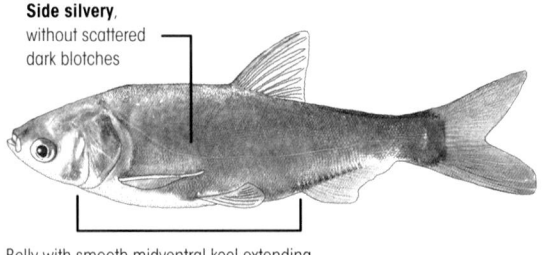

Side silvery, without scattered dark blotches

Belly with smooth midventral keel extending from front of anal fin to under gill covers

3a Scales in lateral line usually 34–38; throat teeth in 2 rows with prominent parallel grooves; scales brassy or olive with darkened edges . . . .**Page 120**

3b Scales in lateral line usually 39–45; throat teeth in 1 or 2 rows, molar-like in appearance, without prominent grooves; scales silvery-black without darkened edges. . .**Page 123**

Grass Carp (*Ctenopharyngodon idella*)

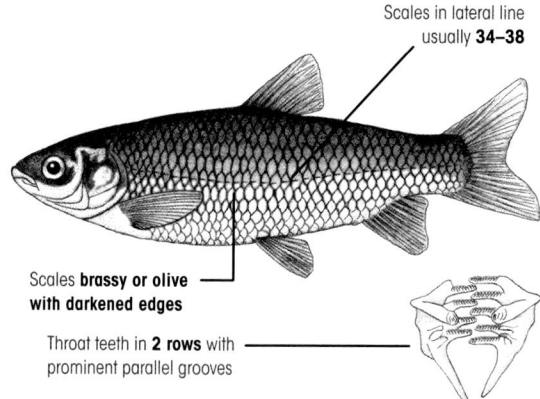

Scales in lateral line usually **34–38**

Scales **brassy or olive with darkened edges**

Throat teeth in **2 rows** with prominent parallel grooves

Black Carp (*Mylopharyngodon piceus*)

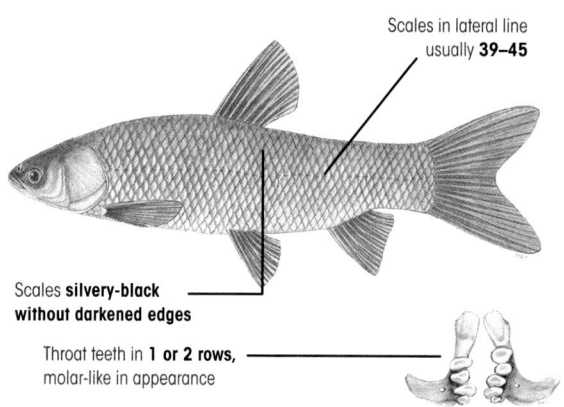

Scales in lateral line usually **39–45**

Scales **silvery-black without darkened edges**

Throat teeth in **1 or 2 rows,** molar-like in appearance

GRASS CARP   *Ctenopharyngodon idella* (Valenciennes 1844)

IDENTIFICATION The Grass Carp has a slender (especially the adult) and fairly compressed body, a wide head, and a terminal mouth. Scales are large with 34–45 along the lateral line; each scale is dark edged with a black spot at the base. The dorsal-fin origin is in front of the pelvic-fin origin. The caudal peduncle is deep, and the caudal fin is large. The back and upper side of the body are gray to brassy green, the underside is white to yellow, and fins are clear to gray-brown. There are 7 dorsal rays, 8 anal rays, and 12–16 rakers on the first gill arch. The pharyngeal teeth are 2,5-4,2 or 2,4-4,2, elongate with prominent parallel grooves on the grinding surfaces, and often hooked at the tip. To 5 ft. (1.5 m).

SIMILAR SPECIES The Black Carp, *Mylopharyngodon piceus*, is much darker, usually blue-gray or black on the back and upper side of the body, and has dark, almost black fins and usually 18–21 rakers on the first gill arch. The pharyngeal teeth are 0,4-5,0, massive, molar-like, and smooth.

HABITAT The Grass Carp has been introduced into lakes and ponds and now is also found in the pools and backwaters of large rivers.

DISTRIBUTION IN ILLINOIS The first Grass Carp reported in Illinois was caught by commercial fishers in the Mississippi River in 1971 (Smith 1979). Since then it has been introduced into farm ponds, small impoundments, and some of the glacial lakes in northeastern Illinois. Contemporary-era records indicate a tremendous expansion of its range, especially in the Illinois, lower Mississippi, Ohio, and Wabash Rivers.

REMARKS The Grass Carp, native to east Asia, was introduced into Arkansas and Alabama in 1963 and now is found in almost every state. It was originally introduced to control aquatic weeds in lakes and ponds.

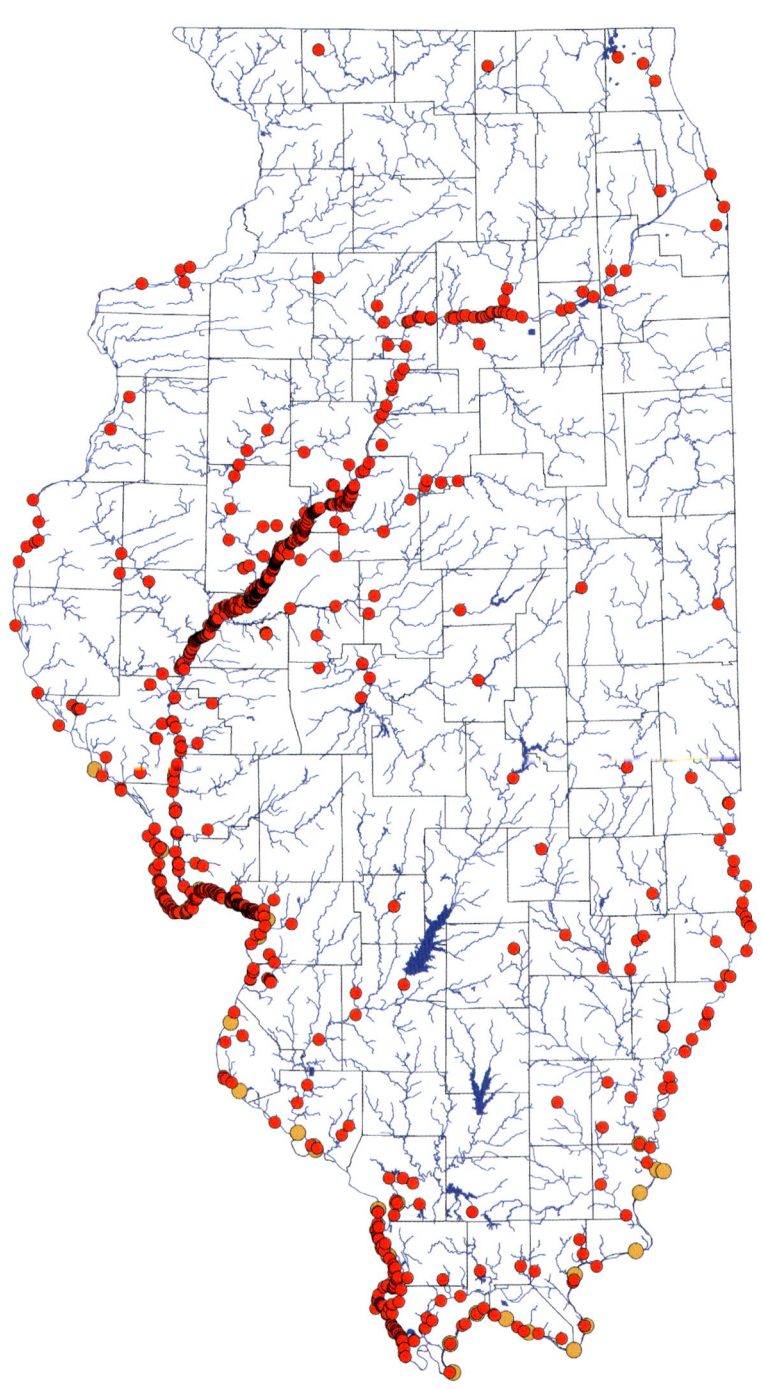

SHARPBELLIES (Xenocyprididae)

120

SILVER CARP *Hypophthalmichthys molitrix* (Valenciennes 1844)

IDENTIFICATION The Silver Carp has a deep, laterally compressed body, a large, terminal mouth, and a keel on the belly from the junction of the branchiostegal membranes to the anus. The eye sits below the middle of the head and is directed downward. The dorsal-fin origin is about even with or slightly behind the pelvic-fin origin. The back and upper side of the body are olive to gray, the side is silver to bronze, and the underside is white. Scales are small with 91–124 along the lateral line. There are 8 dorsal rays, 11–14 anal rays, and 0,4-4,0 pharyngeal teeth. The first gill arch has over 100 rakers fused into a spongelike filtering apparatus. To 4 ft. (1.2 m).

SIMILAR SPECIES The Bighead Carp, *Hypophthalmichthys nobilis*, has a keel from the pelvic-fin origin to the anus and unfused rakers on the first gill arch; adults have irregular gray-black blotches on the body.

HABITAT The Silver Carp is abundant in the main channels of large rivers and in the open waters of reservoirs.

DISTRIBUTION IN ILLINOIS The Silver Carp first appeared in Illinois in the 1980s. It now is abundant in the Illinois, lower Mississippi, Ohio, and Wabash Rivers and in some reservoirs in central and southern Illinois. Records through 2018 are illustrated on the species-distribution map.

REMARKS The Silver Carp is native to eastern China and was first introduced into the United States in 1973. It escaped from flooded culture ponds and now occurs in at least 16 states. It is established in the middle and lower Mississippi River basin from Iowa to Louisiana and is abundant where established. The species feeds mainly on phytoplankton and was introduced to improve water quality in aquaculture ponds. Silver Carp are known to jump into boats with outboard motors, sometimes seriously injuring humans.

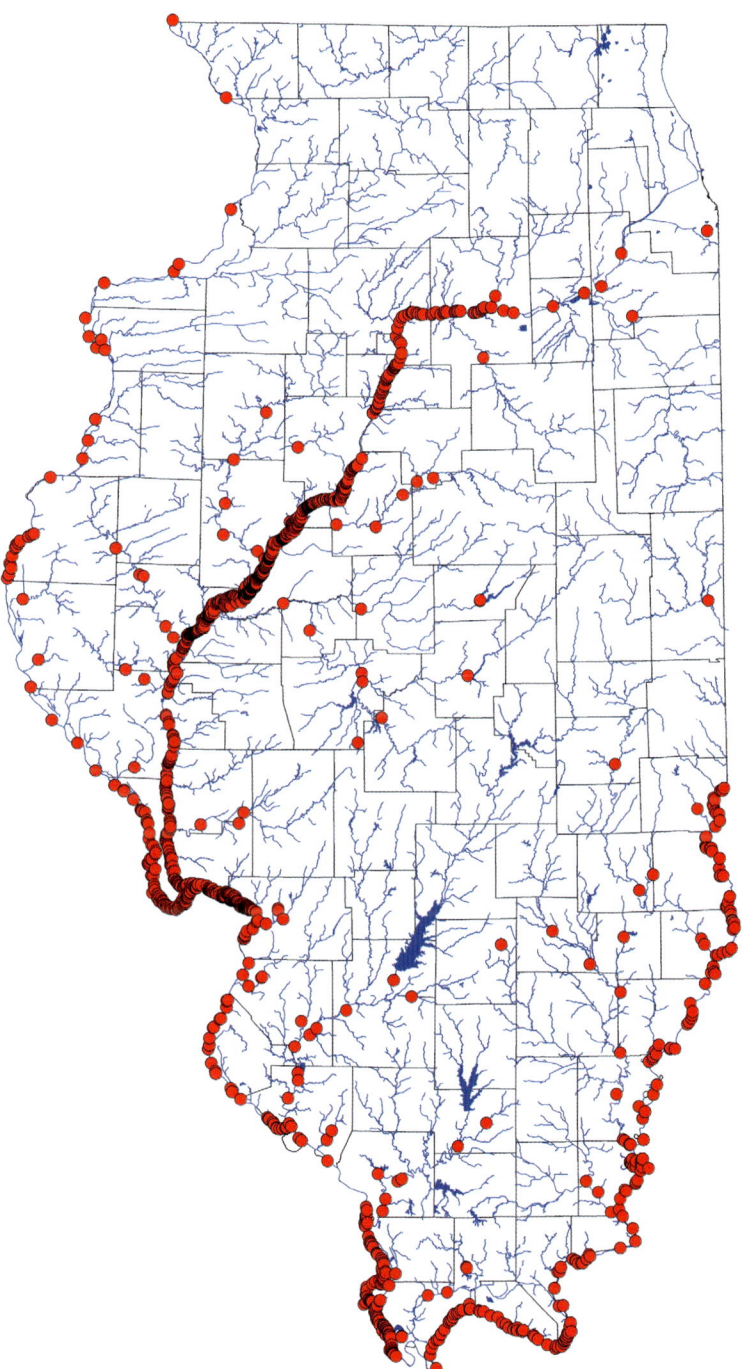

BIGHEAD CARP *Hypophthalmichthys nobilis* (Richardson 1845)

IDENTIFICATION The Bighead Carp has a deep, laterally compressed body, a large, terminal mouth, and a keel on the belly from the base of the pelvic fins to the anus. The eye sits below the middle of the head and is directed downward. The dorsal-fin origin is behind the pelvic-fin origin. The back and upper side of the body are dark gray, the side is silvery with dark blotches (in adults), and the lower side and belly are dull white. Juveniles are silver until about 2 months of age when irregular gray-black blotches develop on the side. Scales are small with 99–120 along the lateral line. There are 8 dorsal rays, usually 13–14 anal rays, and 0,4-4,0 pharyngeal teeth. The first gill arch has about 130 unfused rakers. To 5 ft. (1.5 m).

SIMILAR SPECIES The Silver Carp, *Hypophthalmichthys molitrix*, has a keel from the junction of the branchiostegal membranes to the anus, fused rakers on the first gill arch, and lacks gray-black blotches on the side of the body.

HABITAT The Bighead Carp occurs in the main channels of large rivers, backwaters, floodplain lakes, reservoirs, and ponds.

DISTRIBUTION IN ILLINOIS The Bighead Carp first appeared in Illinois in the 1980s. It now is abundant in the Illinois River and in parts of the Mississippi River and occurs elsewhere sporadically, including some reservoirs in central and southern Illinois. Records through 2018 are illustrated on the species-distribution map.

REMARKS The Bighead Carp was introduced from its native range in east Asia into aquaculture ponds in Arkansas in 1972. It is now established in the Missouri, Mississippi, and Ohio River basins. The Bighead Carp feeds mainly on zooplankton and was introduced to improve water quality in aquaculture facilities.

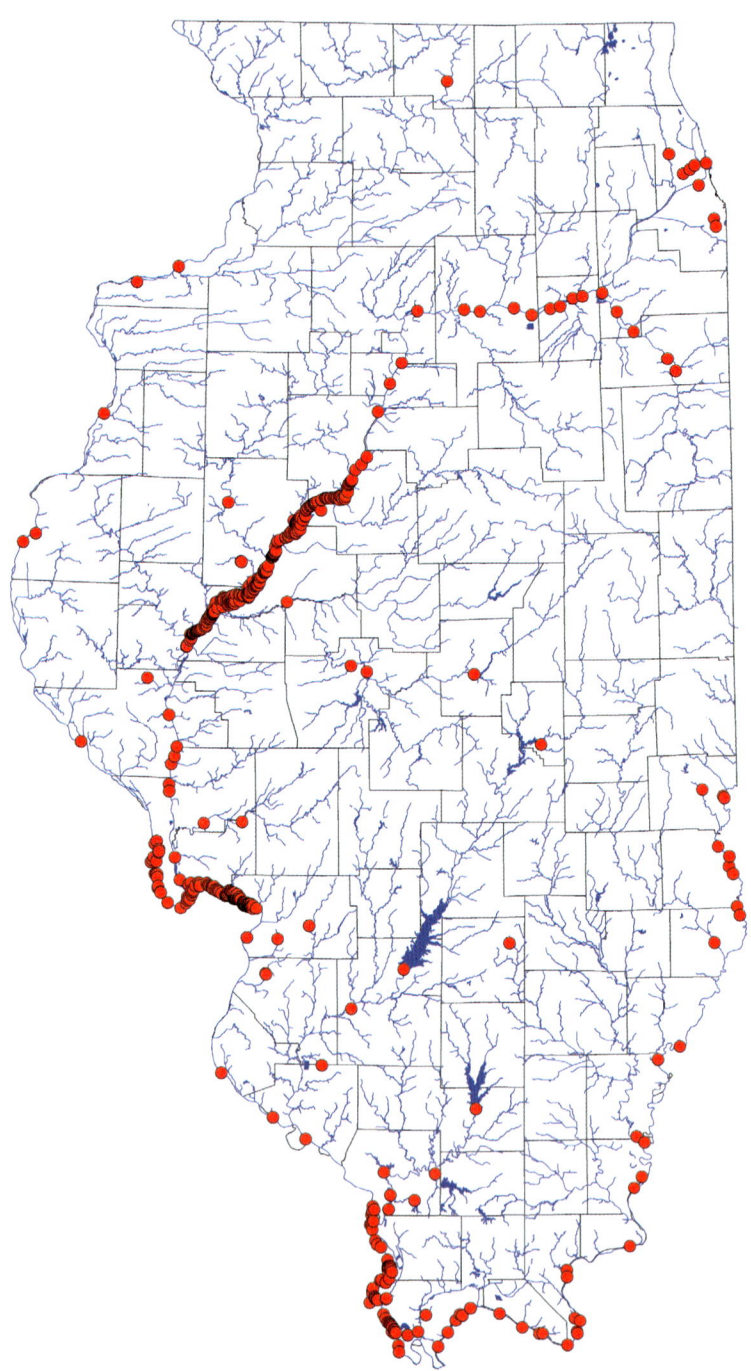

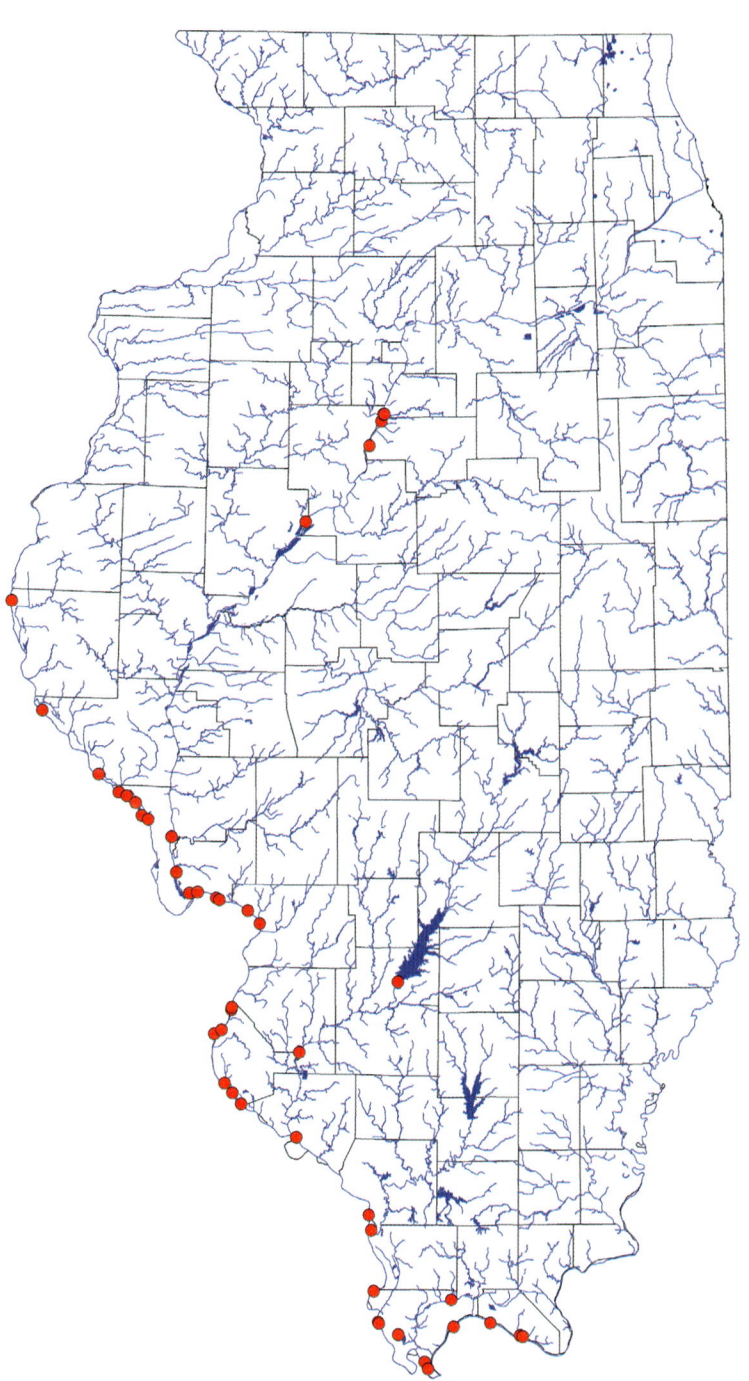

BLACK CARP *Mylopharyngodon piceus* (Richardson 1846)

IDENTIFICATION The Black Carp has a wide head, the mouth is terminal, the caudal peduncle is short and relatively deep, and the caudal fin is large. The scales are large, with usually 39–46 along the lateral line. The body is somewhat slender (especially in the adult) and fairly compressed. The dorsal-fin origin is in front of the pelvic-fin origin. The body is blue-gray or black on the back and side. All fins are dark gray, almost black. There are 7–9 dorsal rays, 7–9 anal rays, and usually 18–21 rakers on the first gill arch. Pharyngeal teeth are 0,4-5,0 (rarely, with 1 or 2 small, minor-row teeth), massive, molar-like, and smooth. To 6½ ft. (2 m).

SIMILAR SPECIES The Grass Carp, *Ctenopharyngodon idella*, is gray to brassy green on the back and upper side of the body and has clear to gray-brown fins and usually 12–16 rakers on the first gill arch. The pharyngeal teeth are 2,5-4,2 or 2,4-4,2, elongate, and have prominent parallel grooves on the grinding surfaces.

HABITAT The Black Carp has been found in flood-plain lakes, backwaters, and the main channels of large rivers.

DISTRIBUTION IN ILLINOIS The first Black Carp in Illinois was captured in March 2003 in Horseshoe Lake, Alexander County, by a fisher using trammel nets. Recent records are from the Mississippi, Illinois, lower Kaskaskia, Cache, and Ohio Rivers. Similar to other Sharpbellies, the Black Carp will almost certainly spread throughout the large rivers of Illinois. Records through 2018 are illustrated on the species-distribution map.

REMARKS The Black Carp was introduced from its native range in east Asia into the United States in the 1970s to control mollusks in lakes and ponds.

MINNOWS

Family Leuciscidae Bonaparte 1835

Minnows lack teeth on the jaws but share with suckers, barbs, and carps, the presence of pharyngeal teeth that are used to grind food against the roof of the mouth. Minnows have 1 dorsal fin, no adipose fin, abdominal pelvic fins, and the anal fin far back on the body. Males of many species become brightly colored, and some develop tubercles during the breeding season. Most species lack barbels, and pharyngeal teeth are in 1 or 2 rows. Leuciscids are native to North America, Eurasia, and northern Africa. In most streams and lakes in North America, minnows are the most diverse and abundant fishes, and they are as varied ecologically as they are morphologically. Most live in midwater, but species of *Campostoma* and *Rhinichthys* are benthic, and most are insectivores, but species of *Campostoma* and *Hybognathus* feed on algae. Some minnows, including species of *Nocomis* and *Luxilus*, have complex breeding strategies that involve interspecific construction and maintenance of nests. The family contains about 680 species, of which 61 are native to Illinois.

1a Belly behind pelvic fins with a keel or low ridge. . . . .
. . . . **Page 183**

1b Belly behind pelvic fins without a keel or ridge **2**

Golden Shiner (*Notemigonus crysoleucas*)

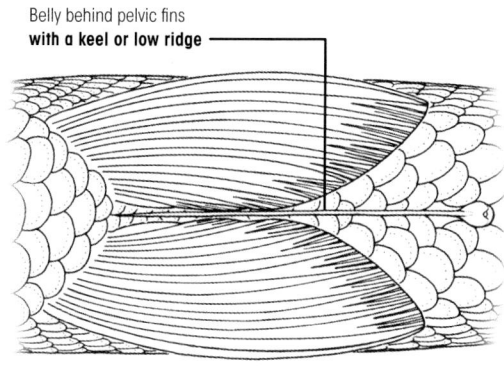

Belly behind pelvic fins
with a keel or low ridge

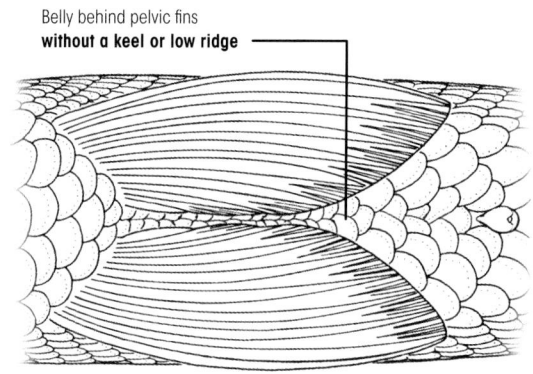

Belly behind pelvic fins
without a keel or low ridge

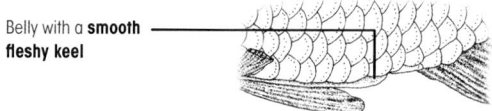

Belly with a **smooth
fleshy keel**

2a Dorsal fin rays 9 or 10 . . . .**3**

2b Dorsal fin rays 8 . . . .**4**

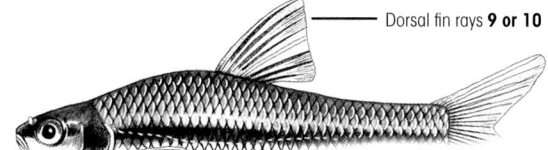

Dorsal fin rays **9 or 10**

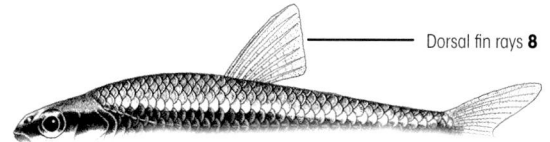

Dorsal fin rays **8**

3a First 4 and last 3 dorsal fin rays outlined with melano-phores, rays 5 and 6 lack pigment; anal fin rays 8; lateral line present at least to front of anal fin . . . . **Page 204**

3b All dorsal fin rays outlined with pigment; anal fin rays 9 or 10; lateral line absent, or with very few pored scales near gill cover . . . . **Page 210**

Pugnose Minnow (*Opsopoeodus emiliae*)

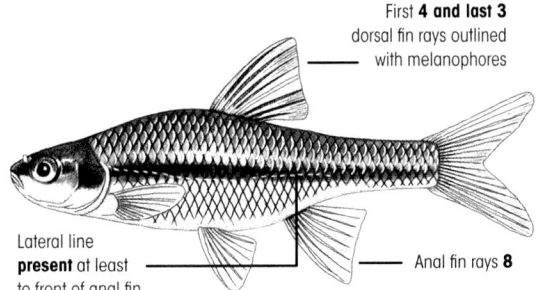

First **4 and last 3** dorsal fin rays outlined with melanophores

Lateral line **present** at least to front of anal fin

Anal fin rays **8**

Bluehead Shiner (*Pteronotropis hubbsi*)

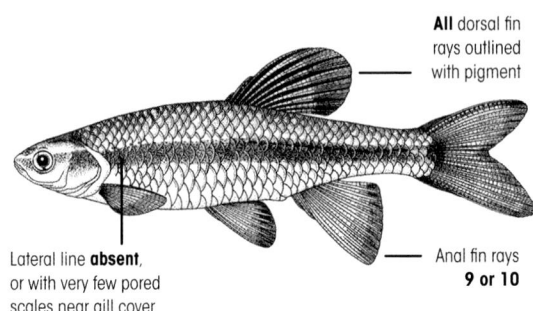

All dorsal fin rays outlined with pigment

Lateral line **absent**, or with very few pored scales near gill cover

Anal fin rays **9 or 10**

4a Side with 2 dusky or black stripes, separated by a broad light or yellow stripe; lateral line scales 70 or more; scales very small, scarcely visible without magnification . . . .**Page 155**

4b Side usually with 1 dusky stripe or without a dusky stripe; lateral line scales 65 or fewer; scales larger, plainly visible without magnification. **5**

Southern Redbelly Dace (*Chrosomus erythrogaster*)

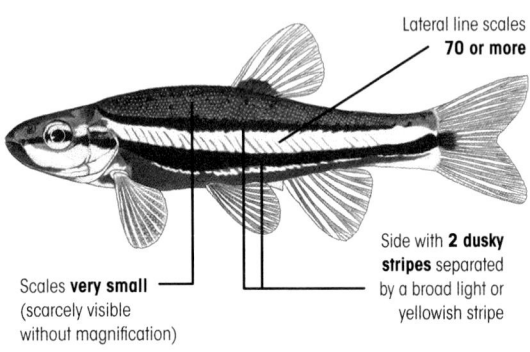

Lateral line scales **70 or more**

Side with **2 dusky stripes** separated by a broad light or yellowish stripe

Scales **very small** (scarcely visible without magnification)

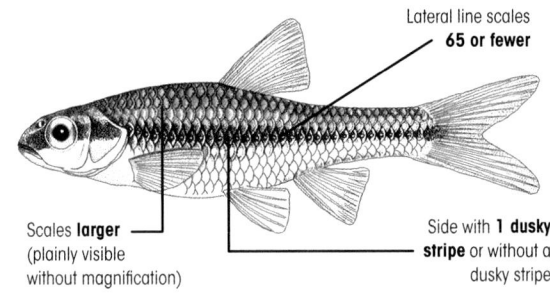

Lateral line scales **65 or fewer**

Scales **larger** (plainly visible without magnification)

Side with **1 dusky stripe** or without a dusky stripe

5a Area ahead of dorsal fin with scales much smaller than those on upper side; short ray at front of dorsal fin somewhat thickened and separated from first principal ray by a membrane **6**

5b Area ahead of dorsal fin with scales about same size as those on upper side; short ray at front of dorsal fin thin and tightly bound to first principal ray **8**

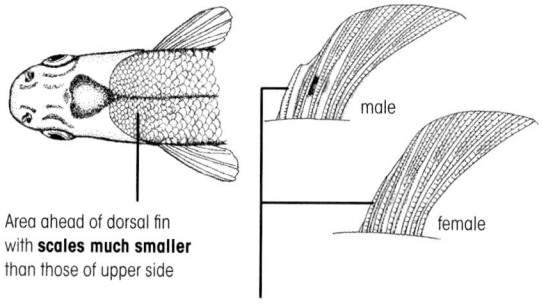

Area ahead of dorsal fin with **scales much smaller** than those of upper side

male

female

Short ray at front of dorsal **fin somewhat thickened and separated from first principal ray** by a membrane

Short ray at front of dorsal **fin thin and tightly bound** to first principal ray

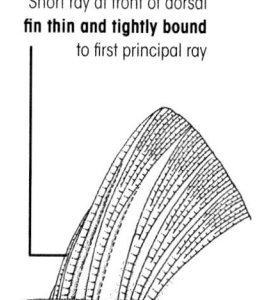

Area ahead of dorsal fin with **scales about the same size** as those of upper side

---

6a Intestine short, with a single S-shaped loop; lining of body cavity silvery, a dark, crescent-shaped mark on snout between nostril and upper lip; large eye directed somewhat upwardly on upper half of head . . . . **Page 208**

6b Intestine long, with several loops; lining of body cavity black; no dark, crescent-shaped mark on snout; smaller eye not directed upwardly, lower on side of head **7**

Bullhead Minnow (*Pimephales vigilax*)

Head of breeding male blackened and with tubercles

Intestine **short**, with a single S-shaped loop

Eye smaller, lower on head, not directed upwardly

Intestine **long**, with several loops

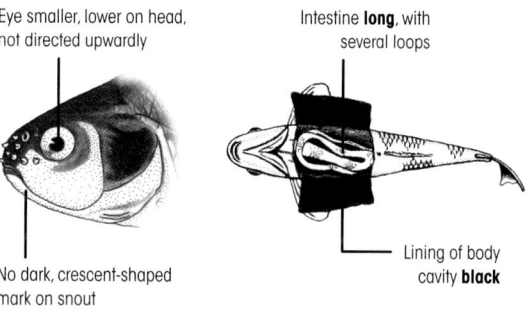

No dark, crescent-shaped mark on snout

Lining of body cavity **black**

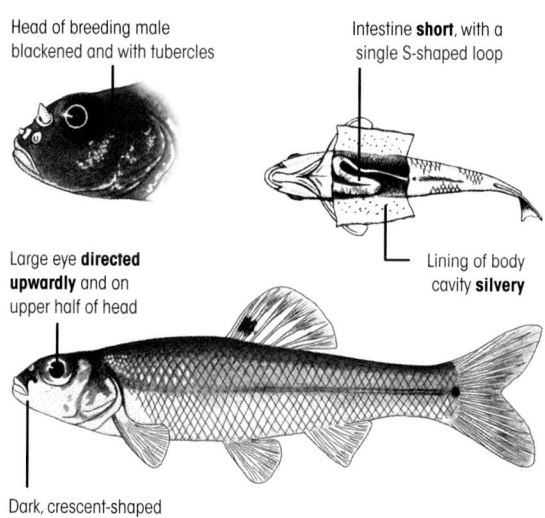

Large eye **directed upwardly** and on upper half of head

Lining of body cavity **silvery**

Dark, crescent-shaped mark on snout

7a Lateral line incomplete, not extending to base of caudal fin; predorsal scale rows usually 23 or more; body deep, its depth (A) going fewer than 4 times into standard length (B); herringbone lines on upper side; mouth oblique, not overhung by snout . . . . **Page 207**

Fathead Minnow (*Pimephales promelas*)

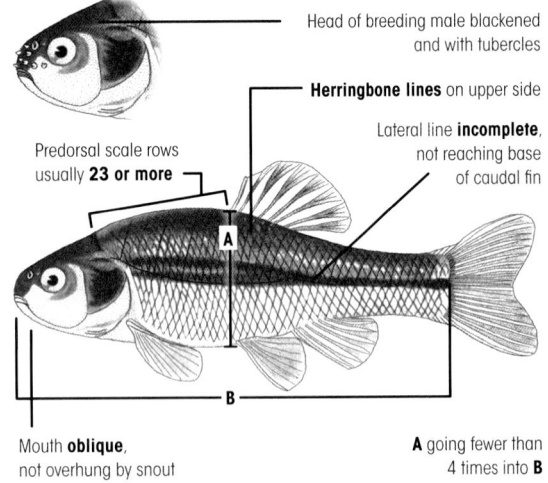

Head of breeding male blackened and with tubercles

Herringbone lines on upper side

Lateral line **incomplete**, not reaching base of caudal fin

Predorsal scale rows usually **23 or more**

Mouth **oblique**, not overhung by snout

A going fewer than 4 times into **B**

7b Lateral line complete to base of caudal fin; predorsal scale rows usually 22 or fewer; body more slender, its depth (A) going 4 or more times into standard length (B); no herringbone lines on upper side; mouth horizontal, overhung by rounded snout . . . . **Page 206**

Bluntnose Minnow (*Pimephales notatus*)

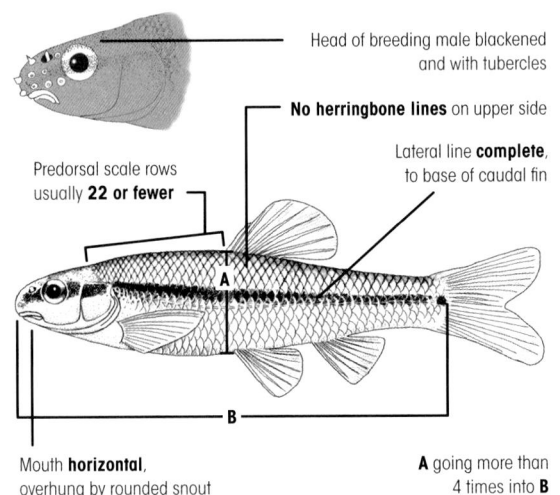

Head of breeding male blackened and with tubercles

No herringbone lines on upper side

Lateral line **complete**, to base of caudal fin

Predorsal scale rows usually **22 or fewer**

Mouth **horizontal**, overhung by rounded snout

A going more than 4 times into **B**

8a A small conical barbel at corner of mouth **9**

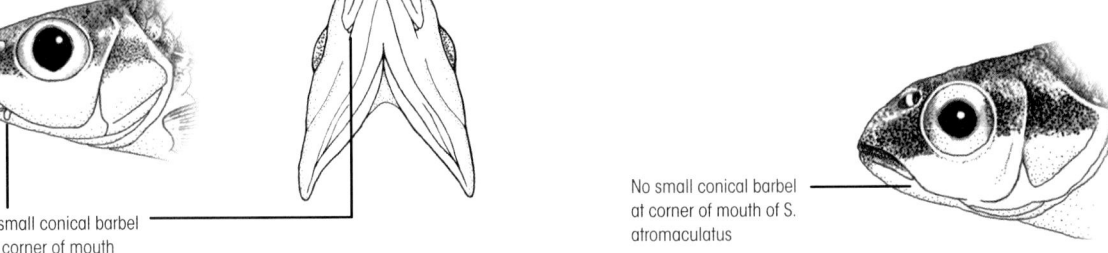

A small conical barbel at corner of mouth

8b No barbel at corner of mouth (a small flap-like barbel in groove above upper lip just forward of corner of mouth in *Semotilus atromaculatus* but not in plain sight) **20**

No small conical barbel at corner of mouth of S. atromaculatus

9a No groove between upper lip and tip of snout; upper jaw bound to snout by fleshy frenum. **10**

9b A distinct groove between upper lip and tip of snout; upper jaw free from snout, no frenum present **11**

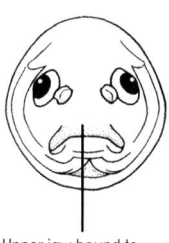 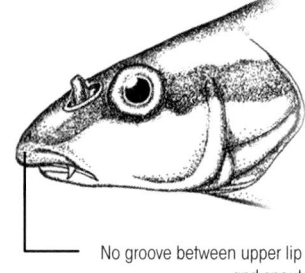

Upper jaw bound to snout by fleshy frenum

No groove between upper lip and snout

Distinct groove between upper lip and tip of snout; upper jaw free from snout, no frenum present

---

10a Eye diameter (A) greater than distance from tip of snout to anterior tip of lower jaw (B) . . . . **Page 212**

10b Eye diameter (A) less than or equal to distance from tip of snout to anterior tip of lower jaw (B). . . . . **Page 211**

Western Blacknose Dace (*Rhinichthys obtusus*)

Longnose Dace (*Rhinichthys cataractae*)

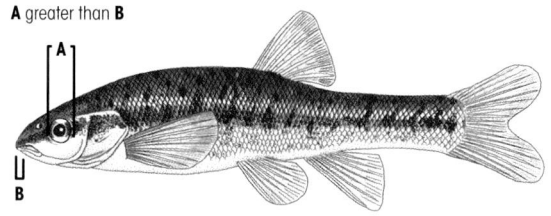

A greater than **B**

A less than or equal to **B**

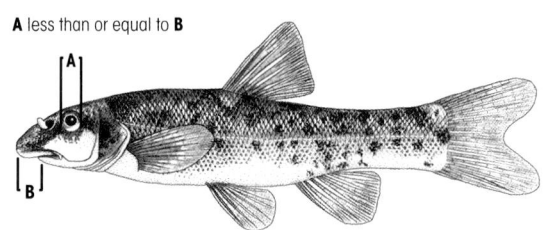

11a Snout projecting only slightly beyond upper lip; total radii (lines on scales radiating from center of scale toward scale margin) on scales from upper side of body usually 15 or more **12**

11b Snout projecting well beyond upper lip; total radii on scales near upper side of body usually 14 or fewer. **14**

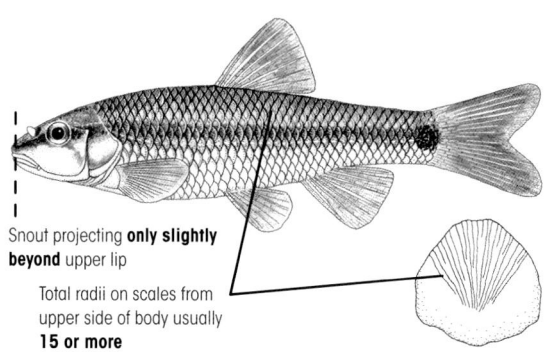

Snout projecting **only slightly beyond** upper lip

Total radii on scales from upper side of body usually **15 or more**

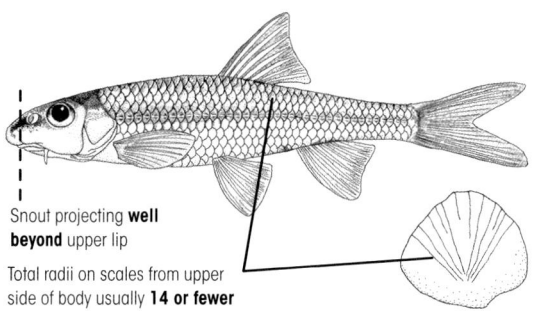

Snout projecting **well beyond** upper lip

Total radii on scales from upper side of body usually **14 or fewer**

12a Lateral line scales usually 58–70; anal fin rays 8; front of dorsal fin slightly behind front of pelvic fins . . . **Page 157**

12b Lateral line scales usually 38–45; anal fin rays 7; front of dorsal fin about equal to or slightly in front of pelvic fins. **13**

Lake Chub (*Couesius plumbeus*)

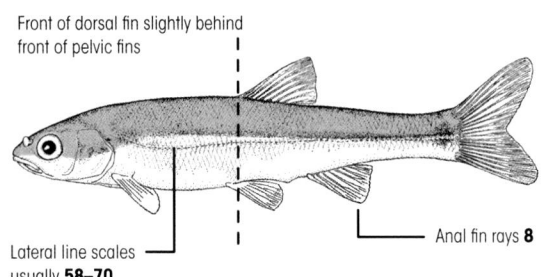

Front of dorsal fin slightly behind front of pelvic fins

Lateral line scales usually **58–70**

Anal fin rays **8**

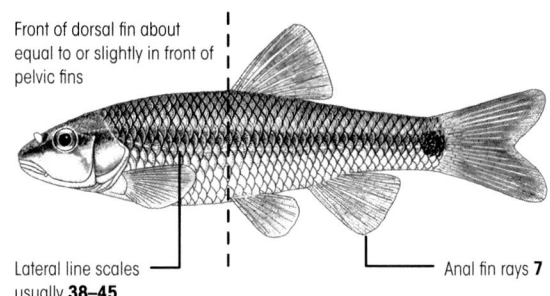

Front of dorsal fin about equal to or slightly in front of pelvic fins

Lateral line scales usually **38–45**

Anal fin rays **7**

13a Spot at base of caudal fin large and prominent; distance from front of eye to tip of snout (A) going 1.3 or more times into distance from front of eye to back edge of gill cover (B); breeding male with tubercles confined to top of head; prominent red or brassy spot behind eye; caudal fin of young usually red or orange . . . . **Page 181**

13b Spot at base of caudal fin, if present, small and pale; distance from front of eye to tip of snout (A) going 1.2 or fewer times into distance from front of eye to top origin of gill cover (B); breeding male with tubercles extending onto snout; no red spot behind eye; caudal fin of young usually slate-gray . . . . **Page 182**

Hornyhead Chub (*Nocomis biguttatus*)

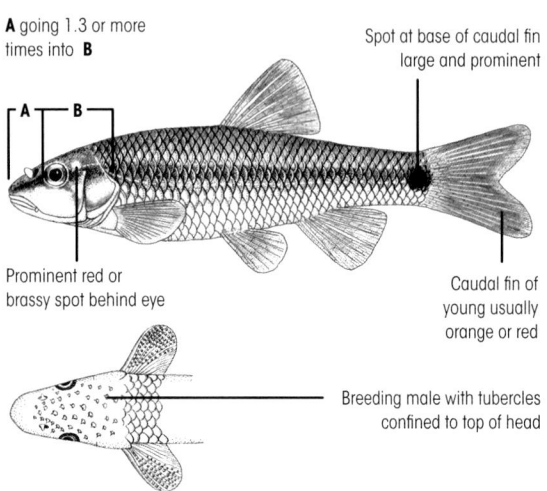

A going 1.3 or more times into **B**

Spot at base of caudal fin large and prominent

Prominent red or brassy spot behind eye

Caudal fin of young usually orange or red

Breeding male with tubercles confined to top of head

River Chub (*Nocomis micropogon*)

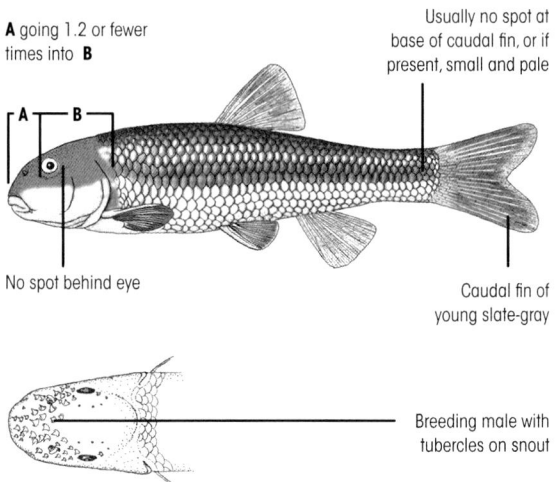

A going 1.2 or fewer times into **B**

Usually no spot at base of caudal fin, or if present, small and pale

No spot behind eye

Caudal fin of young slate-gray

Breeding male with tubercles on snout

14a Side of body with numerous dark X-, V- and Y-shaped markings or speckles **15**

14b Side of body with a continuous dark stripe extending forward from base of caudal fin to tip of snout or side uniformly colored **16**

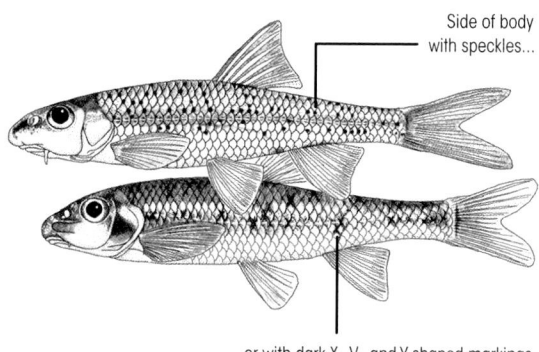

Side of body with speckles...

...or with dark X-, V-, and Y-shaped markings

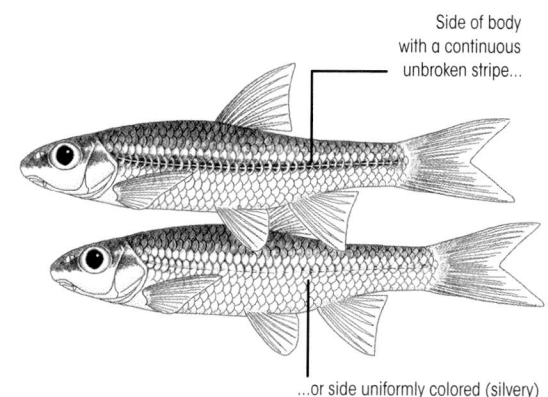

Side of body with a continuous unbroken stripe...

...or side uniformly colored (silvery)

15a Side of body with numerous small, roundish dark speckles; upper jaw extending past front of eye; eye smaller, its diameter (A) less than distance from back of eye to upper end of gill cover (B) . . . .**Page 178**

Shoal Chub (*Macrhybopsis hyostoma*)

15b Side of body with numerous X-, V-, and Y-shaped markings; upper jaw not extending past front of eye; eye larger, its diameter (A) about equal to distance from back of eye to upper end of gill cover (B) . . . .**Page 163**

Gravel Chub (*Erimystax x-punctatus*)

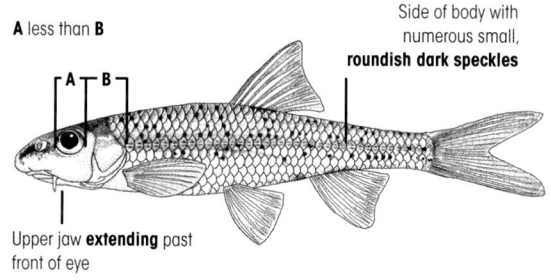

A less than **B**

Side of body with numerous small, **roundish dark speckles**

Upper jaw **extending** past front of eye

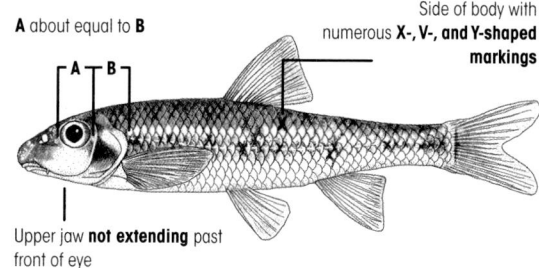

A about equal to **B**

Side of body with numerous **X-, V-, and Y-shaped markings**

Upper jaw **not extending** past front of eye

16a Eye small, its diameter (A) going more than 4.5 times into head length (B); lateral line scales usually more than 40 .
. . . .**17**

16b Eye large, its diameter (A) going fewer than 4 times into head length (B); lateral line scales usually fewer than 40 .
. . . .**19**

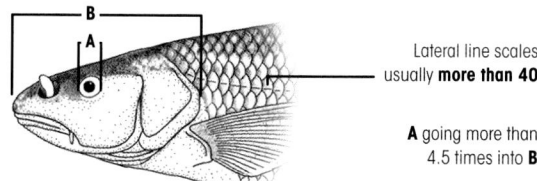

Lateral line scales usually **more than 40**

A going more than 4.5 times into **B**

Lateral line scales usually **fewer than 40**

A going fewer than 4 times into **B**

17a Snout length (A) about equal to distance from back of eye to rear margin of gill cover (B); body scales with distinct keels; usually 39–44 lateral line scales ...... **Page 177**

17b Snout length (A) less than distance from back of eye to rear margin of gill cover (B); body scales without keels; usually 45–60 lateral line scales **18**

Sturgeon Chub (*Macrhybopsis gelida*)

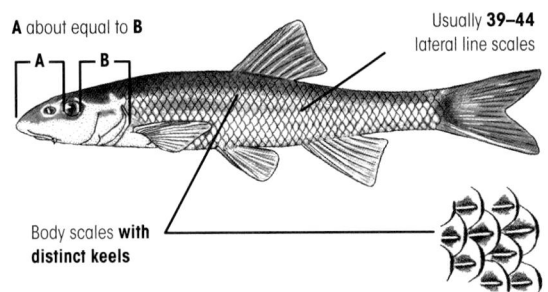

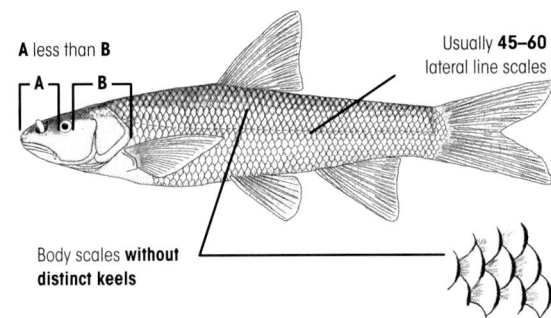

18a Head flattened, its width (A) greater than its depth (B); breast scaled; tip of pectoral fin not reaching behind base of pelvic fin in adults .... **Page 209**

18b Head rounded, its width (A) less than its depth (B); breast not scaled; tip of pectoral fin extending behind base of pelvic fin in adults .... **Page 179**

Flathead Chub (*Platygobio gracilis*)

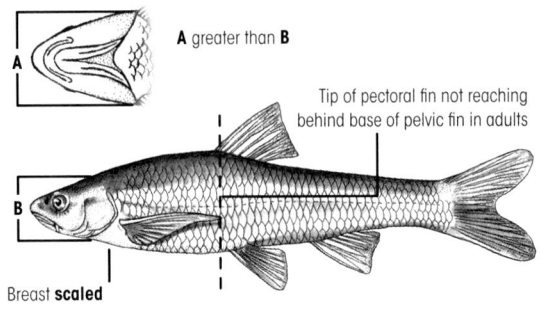

Sicklefin Chub (*Macrhybopsis meeki*)

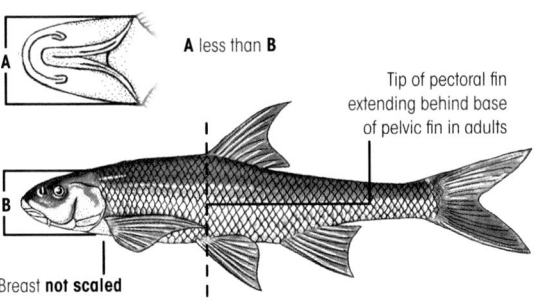

19a Stripe along midside dark and well developed; distance from tip of snout to front of dorsal fin (A) equal to or greater than distance from front of dorsal fin base to base of caudal fin (B) . . . .**Page 169**

19b No distinct stripe along midside (body silvery); distance from tip of snout to front of dorsal fin (A) much less than distance from front of dorsal fin to base of caudal fin (B) . . . .**Page 180**

Bigeye Chub (*Hybopsis amblops*)

A about equal to or greater than **B**

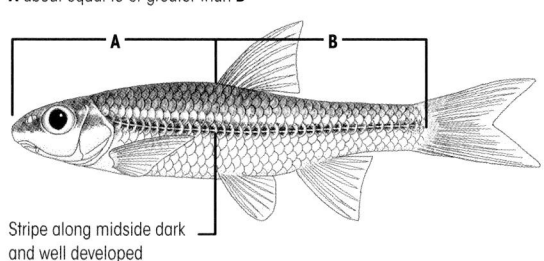

Stripe along midside dark and well developed

Silver Chub (*Macrhybopsis storeriana*)

A less than **B**

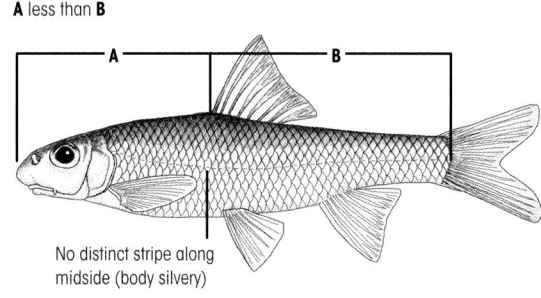

No distinct stripe along midside (body silvery)

20a Mouth large, its length from tip of lower lip to corner close (A) to or more than 1/2 head length (B); front half of adult side red . . . .**Page 156**

20b Mouth smaller, its length from tip of lower lip top corner (A) about 1/3 of head length (B) front half of adult side lacks red. **21**

Redside Dace (*Clinostomus elongatus*)

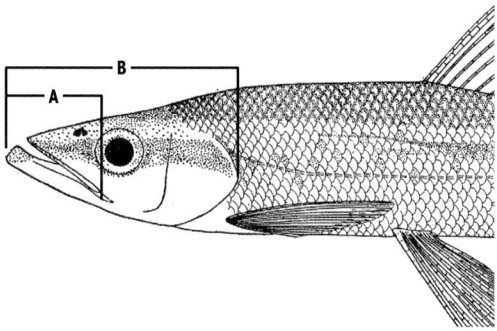

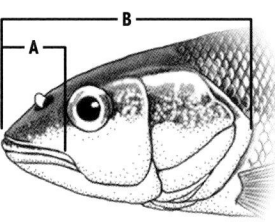

21a Intestine long, much more than twice the standard length of fish, with several loops (often visible through body wall of preserved specimens); lining of body cavity black. **22**

21b Intestine short, less than twice the standard length of fish, with a single s-shaped loop; lining of body cavity silvery in most species, with or without dark speckles **29**

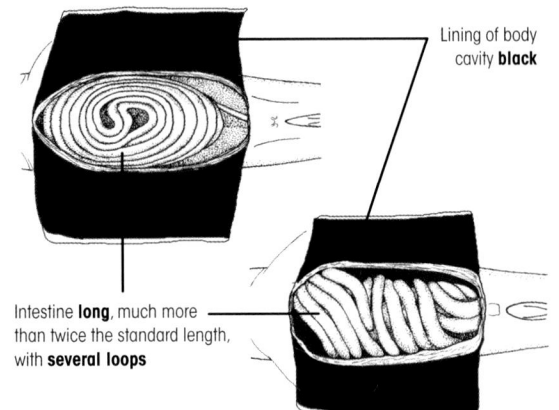

Lining of body cavity **black**

Intestine **long**, much more than twice the standard length, with **several loops**

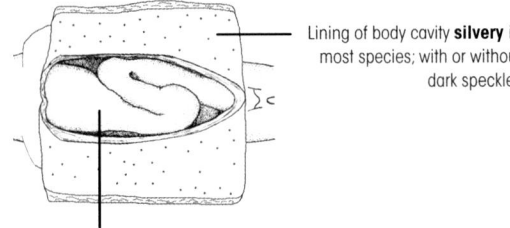

Lining of body cavity **silvery** in most species; with or without dark speckles

Intestine **short**, less than twice the standard length of fish, with a **single s-shaped loop**

22a Lower jaw with a hard, shelf-like extension, separated from lower lip by a groove; anal fin rays 7 . . . . **23**

22b Lower jaw without a hard, shelf-like extension; anal fin rays 8 . . . . **24**

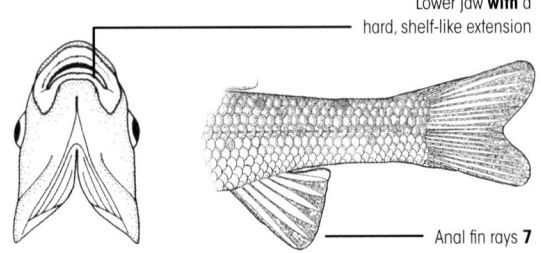

Lower jaw **with** a hard, shelf-like extension

Anal fin rays **7**

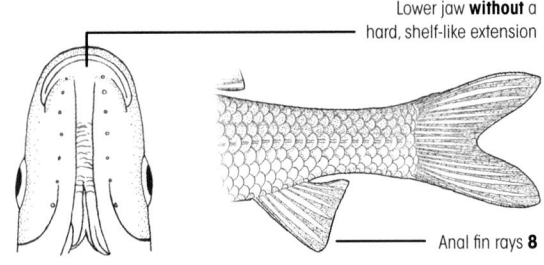

Lower jaw **without** a hard, shelf-like extension

Anal fin rays **8**

23a Number of scale rows around body just in front of dorsal fin usually 31–36; least width of skull between eyes (A) about equal to distance from back of eye to upper end of gill cover (B); nuptial male without tubercles beside each nostril. . . . .**Page 154**

Largescale Stoneroller (*Campostoma oligolepis*)

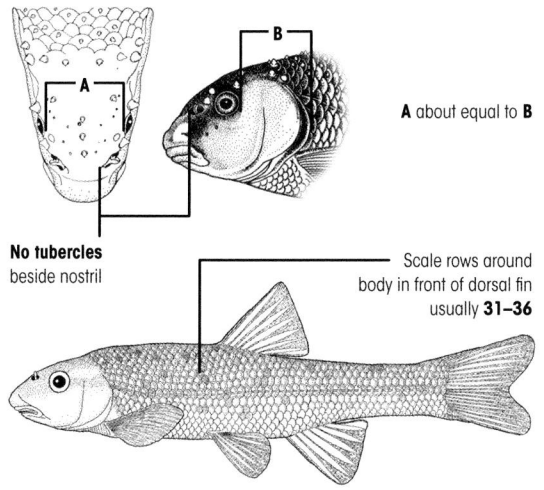

A about equal to **B**

No tubercles beside nostril

Scale rows around body in front of dorsal fin usually **31–36**

23b Number of scale rows around body just in front of dorsal fin usually 37–48; least width of skull between eyes (A) usually less than distance from back of eye to upper end of gill cover (B); nuptial male with 1–3 tubercles beside each nostril. . . . .**Page 153**

Central Stoneroller (*Campostoma anomalum*)

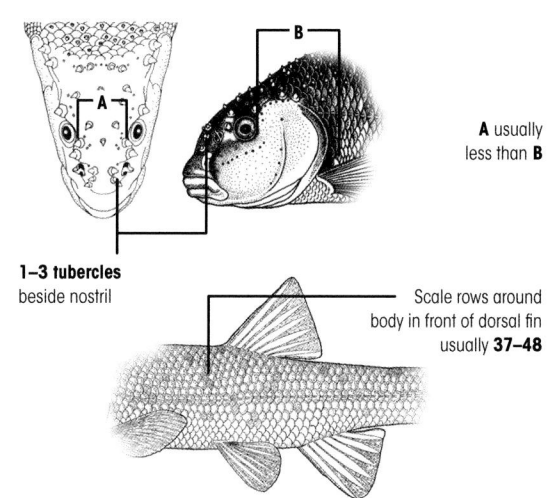

A usually less than **B**

1–3 tubercles beside nostril

Scale rows around body in front of dorsal fin usually **37–48**

24a Front of dorsal fin about equal with pelvic fin base; eye larger, its diameter (A) about equal to distance from back margin of eye to upper end of gill cover (B); stripe along midside prominent, extending forward to tip of snout . . . .
. . . .**Page 195**

Ozark Minnow (*Notropis nubilus*)

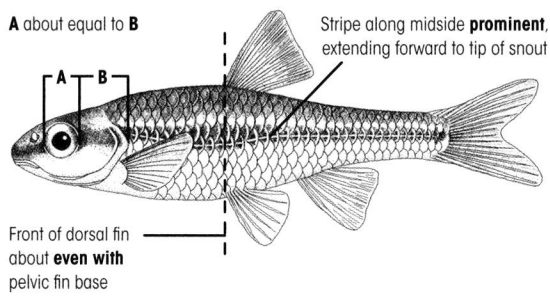

A about equal to **B**

Stripe along midside **prominent**, extending forward to tip of snout

Front of dorsal fin about **even with** pelvic fin base

24b Front of dorsal fin ahead of pelvic fin base; eye smaller, its diameter (A) less than distance from back margin of eye to upper end of gill cover (B); stripe along midside absent or, if present, not extending forward to tip of snout . . . **25**

Hybognathus

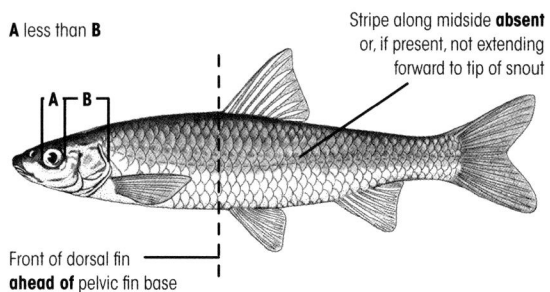

A less than **B**

Stripe along midside **absent** or, if present, not extending forward to tip of snout

Front of dorsal fin **ahead of** pelvic fin base

25a Dorsal fin rounded at outer margin, its first principal ray shorter than the second and third; total radii (lines on scales radiating from near center of scale toward scale margin) about 20; color in life yellowish on side. . . . .**Page 165**

Brassy Minnow (*Hybognathus hankinsoni*)

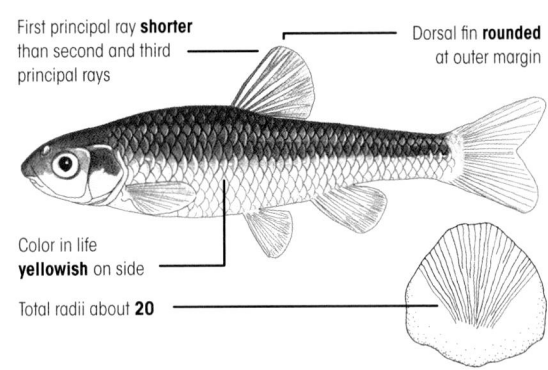

First principal ray **shorter** than second and third principal rays

Dorsal fin **rounded** at outer margin

Color in life **yellowish** on side

Total radii about **20**

25b Dorsal fin pointed at outer margin, its first principal ray as long or longer than the second or third; total radii about 10; color in life silvery on side. **26**

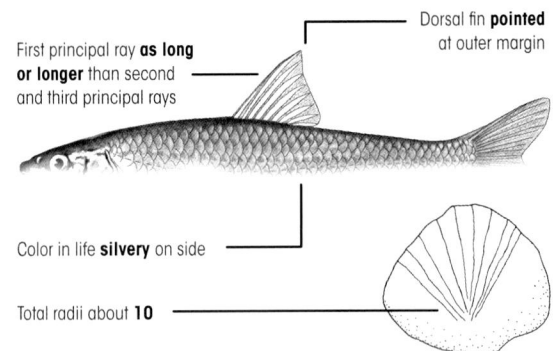

First principal ray **as long or longer** than second and third principal rays

Dorsal fin **pointed** at outer margin

Color in life **silvery** on side

Total radii about **10**

26a Snout length (A) less than or equal to eye diameter (B); front of upper lip level with bottom of eye (sometimes to middle of eye); scales on upper side dark-edged, forming a distinct diamond-shaped pattern. . . . .**Page 166**

Cypress Minnow (*Hybognathus hayi*)

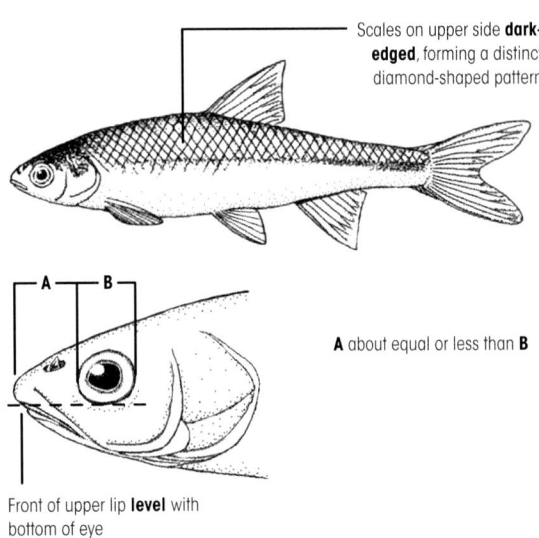

Scales on upper side **dark-edged**, forming a distinct diamond-shaped pattern

A about equal or less than **B**

Front of upper lip **level** with bottom of eye

26b Snout length (A) greater than eye diameter (B); front of upper lip usually below bottom of eye; scales on upper side not prominently dark-edged, not forming a diamond-shaped pattern. **27**

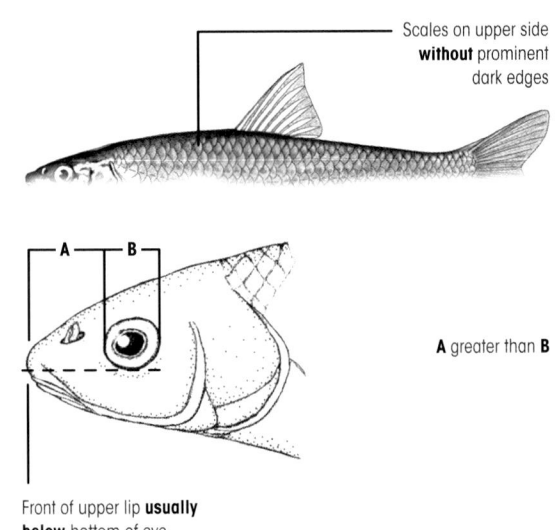

Scales on upper side **without** prominent dark edges

A greater than **B**

Front of upper lip **usually below** bottom of eye

27a Basioccipital process (backward extension of bone at lower rear margin of skull) narrow and peglike . . . .
. . . . **Page 168**

Plains Minnow (*Hybognathus placitus*)

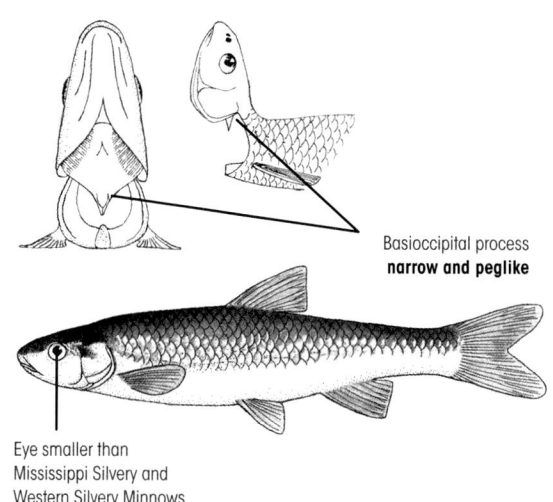

Basioccipital process
narrow and peglike

Eye smaller than
Mississippi Silvery and
Western Silvery Minnows

27b Basioccipital process broad and bladelike **28**

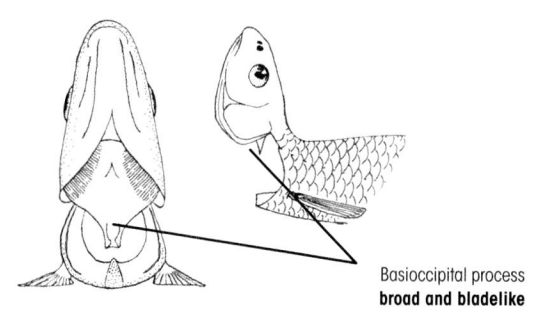

Basioccipital process
broad and bladelike

28a Diameter of eye (A) greater than width of mouth (B); rear margin of basioccipital process deeply concave . . . .
. . . . **Page 167**

Mississippi Silvery Minnow
(*Hybognathus nuchalis*)

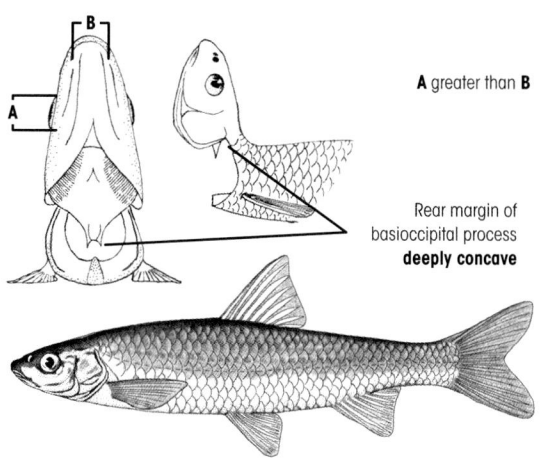

A greater than **B**

Rear margin of
basioccipital process
deeply concave

28b Diameter of eye (A) less than width of mouth (B); rear margin of basioccipital process straight or only slightly concave . . . .**Page 164**

Western Silvery Minnow
(*Hybognathus argyritis*)

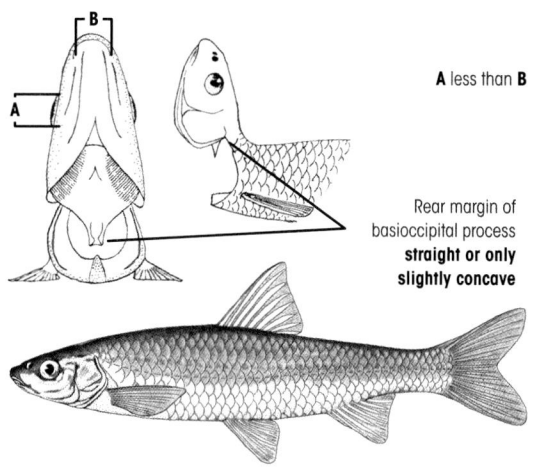

A less than **B**

Rear margin of
basioccipital process
**straight or only
slightly concave**

29a Mouth suckerlike; lower lip with a prominent lobe on each side . . . . **Page 205**

29b Mouth not suckerlike, lower lip without a prominent lobe on each side **30**

Suckermouth Minnow (*Phenacobius mirabilis*)

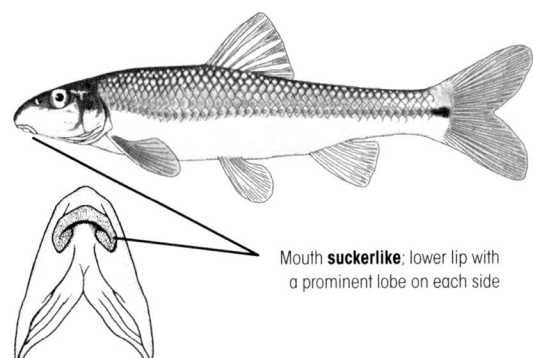

Mouth **suckerlike**; lower lip with a prominent lobe on each side

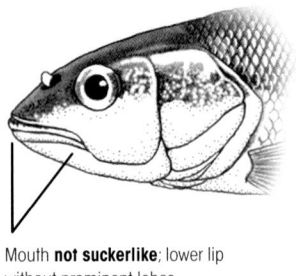

Mouth **not suckerlike**; lower lip without prominent lobes

30a A small flap-like barbel in groove above upper lip near corner of mouth; upper lip wider at midline than on either side; lateral line scales usually more than 48 . . . .
. . . . **Page 213**

30b No flap-like barbel in groove above upper lip; upper lip nearly uniform in width; lateral line scales usually fewer than 48 . . . .**31**

Creek Chub (*Semotilus atromaculatus*)

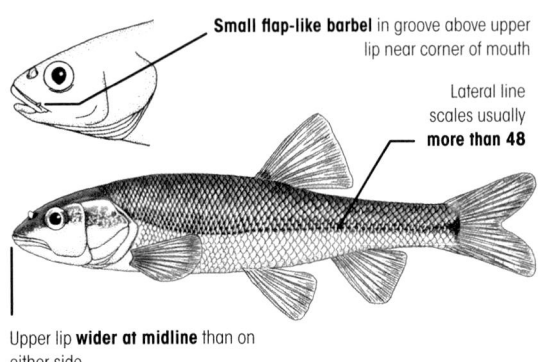

Small flap-like barbel in groove above upper lip near corner of mouth

Lateral line scales usually **more than 48**

Upper lip **wider at midline** than on either side

No flap-like barbel in groove above upper lip

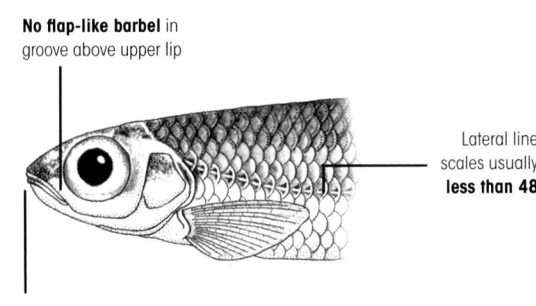

Lateral line scales usually **less than 48**

Upper lip **nearly uniform** in width

31a Undersurface of head distinctly flattened; cavernous lateral line chambers on lower cheek and undersurface of head; eye directed upward . . . .**Page 162**

Silverjaw Minnow (*Ericymba buccata*)

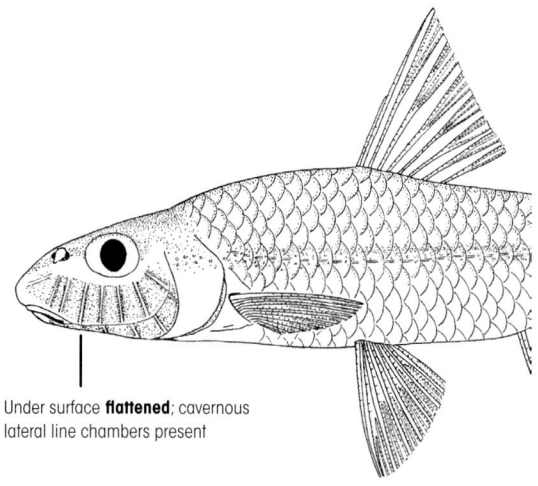

Under surface **flattened**; cavernous lateral line chambers present

31b Undersurface of head not distinctly flattened in most species; no cavernous lateral line chambers on cheeks or head; eye directed upward in only one species (*Notropis dorsalis*) **32**

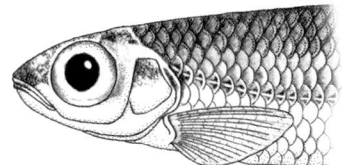

32a Base of caudal fin with a prominent black spot as large or larger than pupil of eye and roundish or squarish in outline **33**

Black spot **prominent, as large or larger** than pupil of eye and roundish or squarish in outline

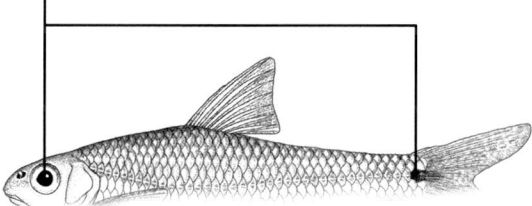

32b Base of caudal fin without a black spot or with spot much smaller than pupil of eye and often triangular or squarish in outline **35**

Black spot **much smaller** than pupil of eye and often triangular or squarish in outline or...

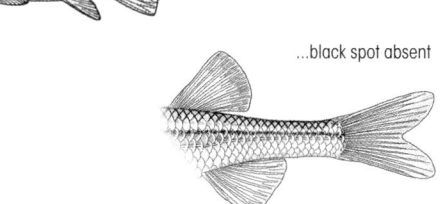

...black spot absent

33a Body elongate, its depth (A) going more than 5 times into standard length (B); small black spot above and below caudal spot; midside with dark stripe from base of caudal fin to tip of snout ....**Page 194**

Taillight Shiner (*Notropis maculatus*)

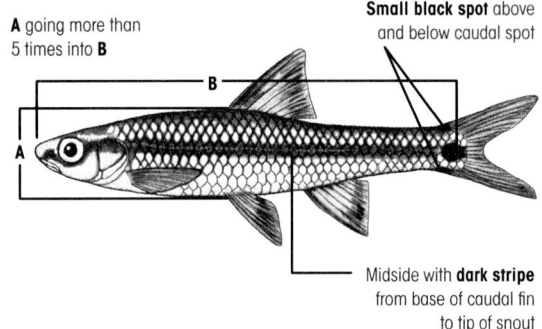

A going more than
5 times into **B**

Small black spot above
and below caudal spot

Midside with **dark stripe**
from base of caudal fin
to tip of snout

33b Body deeper, its depth (A) going less than 5 times into standard length (B); no small black spot above and below caudal spot; midside without dark stripe or with stripe only on back half of body, becoming indistinct toward head **34**

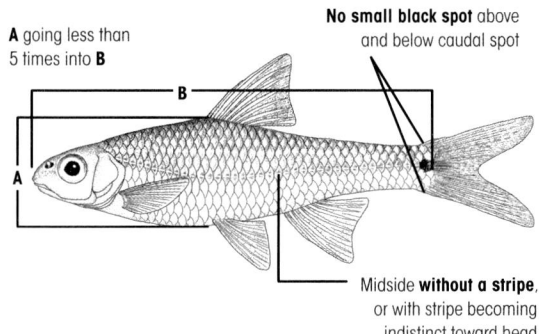

A going less than
5 times into **B**

No small black spot above
and below caudal spot

Midside **without a stripe**,
or with stripe becoming
indistinct toward head

34a Dorsal fin with dark speckles on some membranes between fin rays, forming a dusky blotch in back part of fin of adults; distance from tip of snout to front of dorsal fin (A) greater than distance from front of dorsal fin to base of caudal fin (B) ....**Page 160**

Blacktail Shiner (*Cyprinella venusta*)

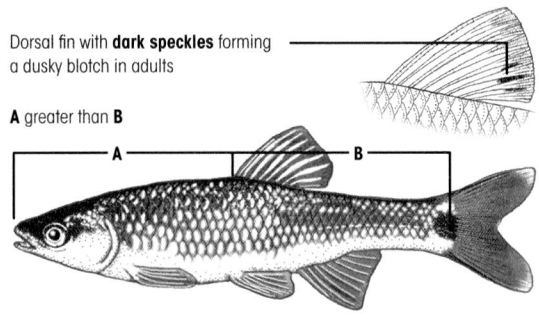

Dorsal fin with **dark speckles** forming
a dusky blotch in adults

A greater than **B**

34b Dorsal fin with dark speckles absent or confined to margins of fin rays; distance from tip of snout to front of dorsal fin (A) less than distance from front of dorsal fin to base of caudal fin (B) **Page193**

Spottail Shiner (*Notropis hudsonius*)

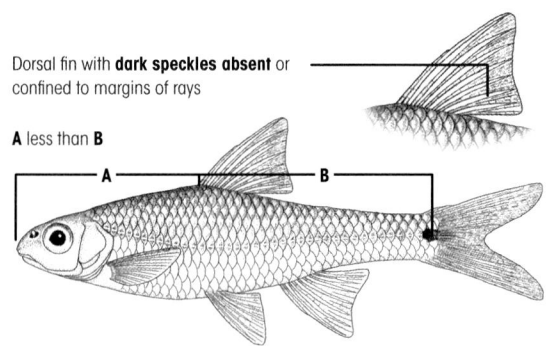

Dorsal fin with **dark speckles absent** or
confined to margins of rays

A less than **B**

35a Front of dorsal fin far behind pelvic fin base (near middle of pelvic fin) **36**

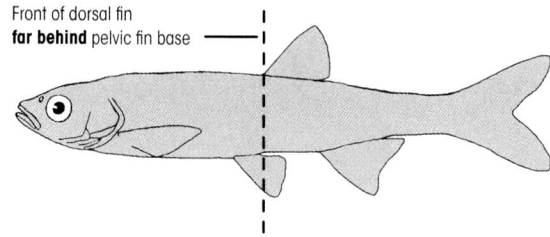

Front of dorsal fin
far behind pelvic fin base

35b Front of dorsal fin about even with, ahead of, or slightly behind pelvic fin. **40**

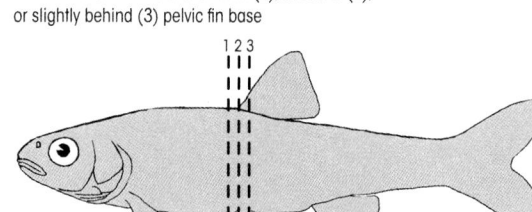

Front of dorsal fin about even with (2), ahead of (1),
or slightly behind (3) pelvic fin base

1 2 3

36a Predorsal scales (scales along midline of back in front of dorsal fin) small and crowded in 21 or more rows, and distinctly smaller than postdorsal scales (scale rows behind dorsal fin). **37**

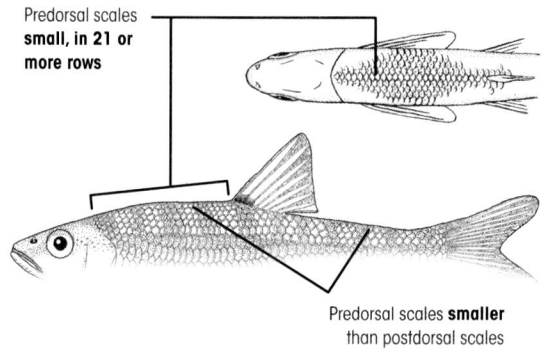

Predorsal scales
**small, in 21 or
more rows**

Predorsal scales **smaller**
than postdorsal scales

36b Predorsal scales larger, in 20 or fewer rows, and about the same size as postdorsal scales **39**

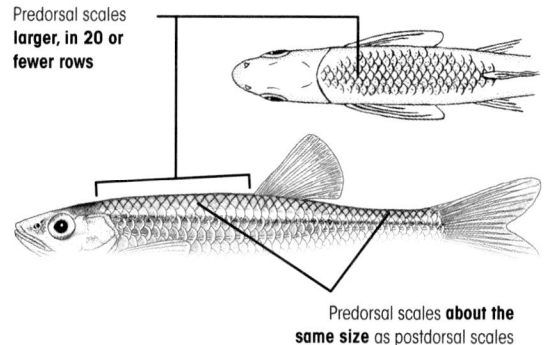

Predorsal scales
**larger, in 20 or
fewer rows**

Predorsal scales **about the
same size** as postdorsal scales

37a Black blotch at base of first few rays and membranes of dorsal fin; anal fin rays usually 9–11; breeding males have red fins. **38**

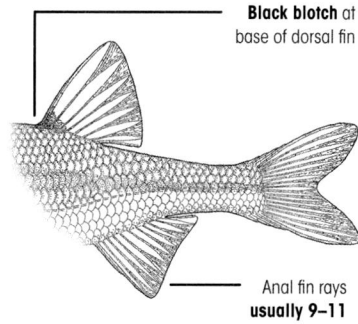

Black blotch at
base of dorsal fin

Anal fin rays
usually 9–11

37b No black blotch at base of dorsal fin; anal fin rays usually 11–12; breeding males have yellow or plain fins. . . . .
. . . .**Page 175**

Ribbon Shiner (*Lythrurus fumeus*)

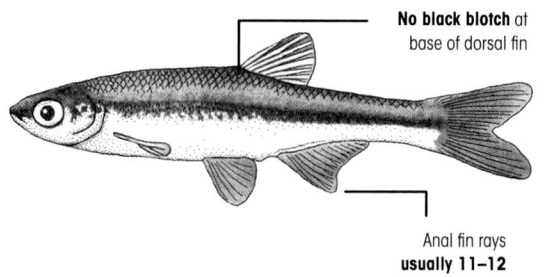

No black blotch at
base of dorsal fin

Anal fin rays
usually 11–12

38a Anal fin rays 10–12, modally 11; body deep, its depth (A) going fewer than 4.5 times into standard length (B); usually 15–17 scales around caudal peduncle; nuptial male with breeding tubercles below eye and 2 rows of tubercles along lower jaw . . . .**Page 176**

Redfin Shiner (*Lythrurus umbratilis*)

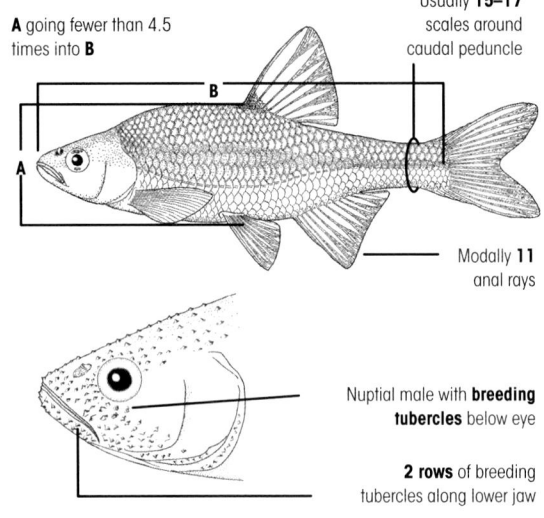

A going fewer than 4.5 times into **B**

Usually **15–17** scales around caudal peduncle

Modally **11** anal rays

Nuptial male with **breeding tubercles** below eye

2 rows of breeding tubercles along lower jaw

38b Anal fin rays 9–11, modally 10; body more slender, its depth (A) going 4.5 times or more into standard length (B); usually 13–16 scales around caudal peduncle; nuptial males without breeding tubercles below eye and 1 row of tubercles along lower jaw. . . . .**Page 174**

Scarlet Shiner (*Lythrurus fasciolaris*)

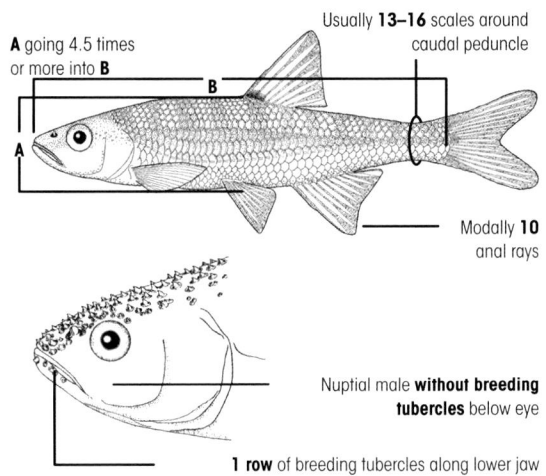

A going 4.5 times or more into **B**

Usually **13–16** scales around caudal peduncle

Modally **10** anal rays

Nuptial male **without breeding tubercles** below eye

1 row of breeding tubercles along lower jaw

39a Anal fin margin straight when expanded; tip of dorsal fin rounded; anterior lateral-line pores outlined above and below (stitched) with melanophores (tiny black specks); dark stripe along midline of back distinct . . . .
. . . . **Page 198 (Rosyface), Page 196 (Carmine)**

Rosyface Shiner (*Notropis rubellus*)
Carmine Shiner (*Notropis percobromus*)

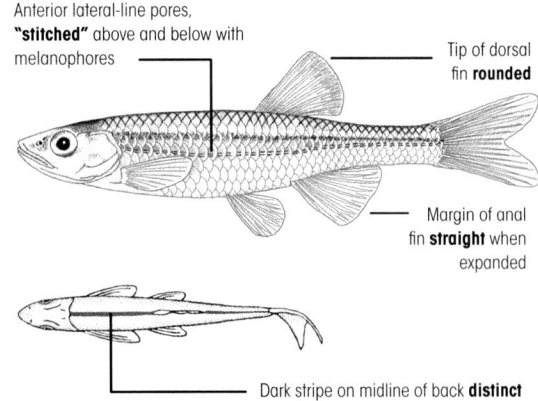

Anterior lateral-line pores, **"stitched"** above and below with melanophores

Tip of dorsal fin **rounded**

Margin of anal fin **straight** when expanded

Dark stripe on midline of back **distinct**

39b Anal fin margin concave when expanded; tip of dorsal fin pointed; anterior lateral-line pores not distinctly outlined above and below by melanophores; dark stripe along midline of back faint consisting mostly of small, separate dots . . . .**Page 185**

Emerald Shiner (*Notropis atherinoides*)

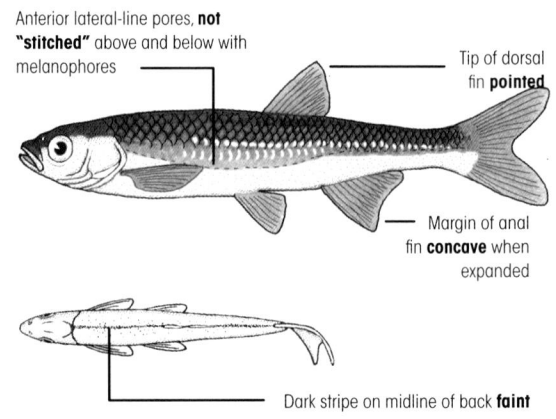

Anterior lateral-line pores, **not "stitched"** above and below with melanophores

Tip of dorsal fin **pointed**

Margin of anal fin **concave** when expanded

Dark stripe on midline of back **faint**

40a Anal fin rays 10–11; 2 dark crescent marks between nostrils. . . . . **Page 197**

40b Anal fin rays 9 or less; no dark crescent marks between nostrils. **41**

Silver Shiner (*Notropis photogenis*)

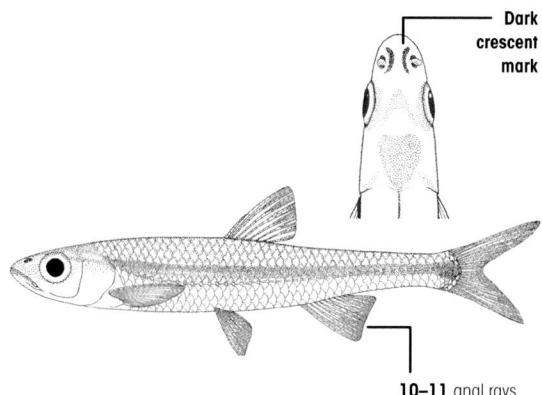

Dark crescent mark

10–11 anal rays

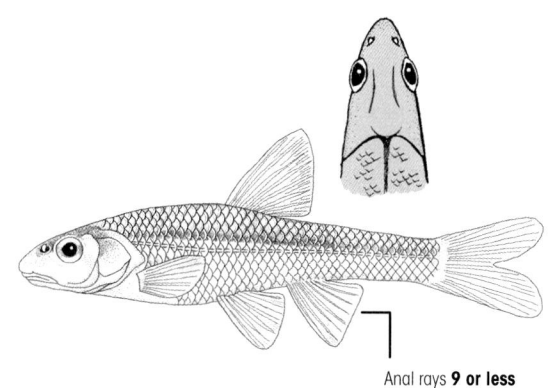

Anal rays **9 or less**

41a Scales on anterior part of body dark-edged, forming a diamond-shaped pattern; dorsal fin with dark pigment or speckles rather uniformly distributed or forming a distinct blotch in back part of fin. **42**

41b Scales on anterior part of body not forming a diamond-shaped pattern; dorsal fin with dark pigment and speckles absent or confined to margins of rays. **44**

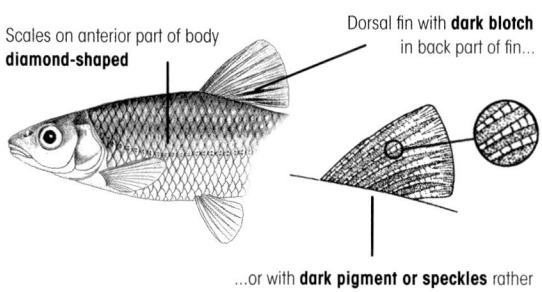

Scales on anterior part of body **diamond-shaped**

Dorsal fin with **dark blotch** in back part of fin...

...or with **dark pigment or speckles** rather uniformly distributed

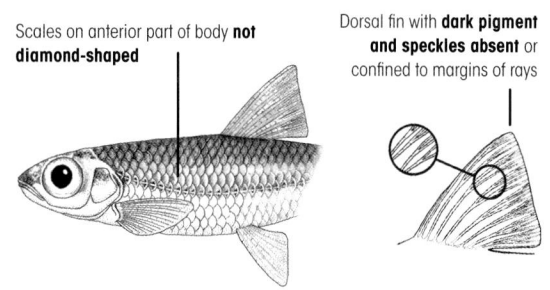

Scales on anterior part of body **not diamond-shaped**

Dorsal fin with **dark pigment and speckles absent** or confined to margins of rays

42a Dorsal fin with dark pigment rather uniformly distributed on membranes, not forming a distinct blotch in back part of fin; body depth (A) usually going less than 3.5 times into standard length (B) . . . . **Page 158**

Red Shiner (*Cyprinella lutrensis*)

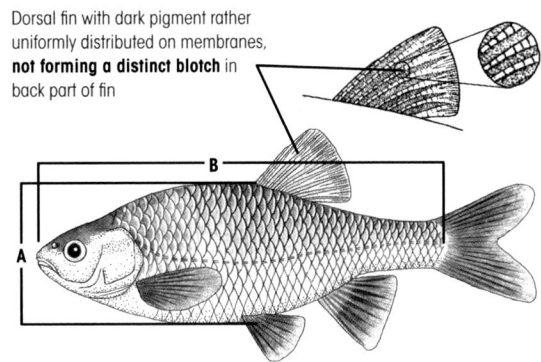

Dorsal fin with dark pigment rather uniformly distributed on membranes, **not forming a distinct blotch** in back part of fin

A usually going less than 3.5 times into **B**

42b Dorsal fin with dark pigment concentrated on membranes in back part of fin, forming a distinct black blotch in adults; body depth (A) usually going more than 3.5 times into standard length (B) **43**

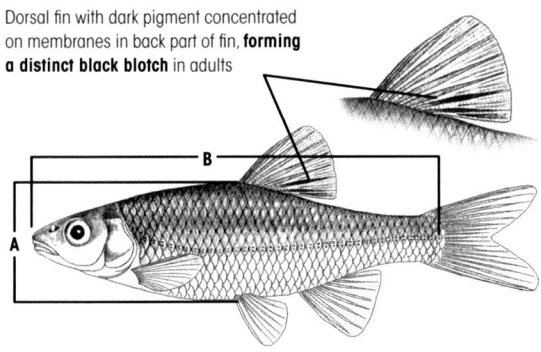

Dorsal fin with dark pigment concentrated on membranes in back part of fin, **forming a distinct black blotch** in adults

A usually going more than 3.5 times into **B**

43a Anal fin rays usually 8; first 2 or 3 membranes of dorsal fin with dark pigment absent, confined to margins of rays (except in breeding males); dark stripe on side of caudal peducle narrow and prominent, sharply defined above, and centered below pores of lateral line . . . . . . **Page 159**

Spotfin Shiner (*Cyprinella spiloptera*)

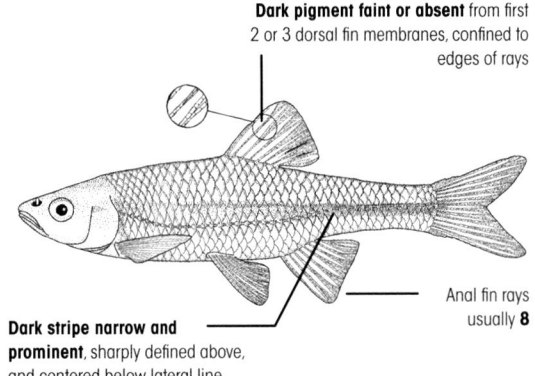

Dark pigment faint or absent from first 2 or 3 dorsal fin membranes, confined to edges of rays

Dark stripe narrow and prominent, sharply defined above, and centered below lateral line

Anal fin rays usually **8**

43b Anal fin rays usually 9; first 2 or 3 membranes of dorsal fin with dark pigment present (except in small young); dark stripe on side of caudal peduncle broad and faint, poorly defined above, and centered on pores of lateral line . . . . **Page 161**

Steelcolor Shiner (*Cyprinella whipplei*)

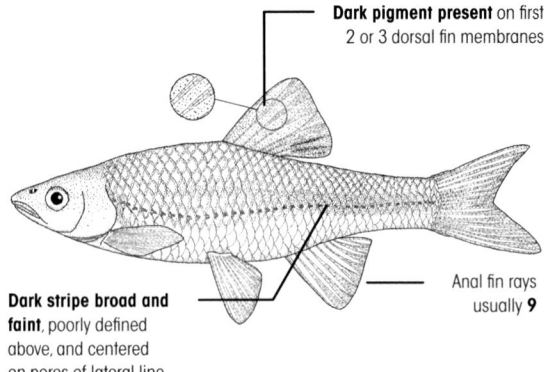

Dark pigment present on first 2 or 3 dorsal fin membranes

Dark stripe broad and faint, poorly defined above, and centered on pores of lateral line

Anal fin rays usually **9**

44a Dark stripe on midline of back sharply defined in front of dorsal fin, not consisting of separate black specks, and frequently as wide or wider than dorsal fin base **45**

44b Dark stripe on midline of back rather poorly defined in front of dorsal fin, consisting of separate dark specks or reduced to a thin line narrower than dorsal fin base **52**

Dark stripe on midline of back **sharply defined** in front of dorsal fin, not consisting of separate black specks, and frequently as wide or wider than dorsal fin base

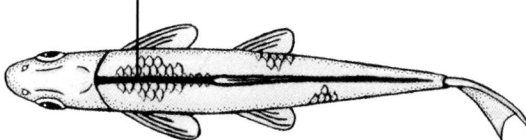

Dark stripe on midline of back **rather poorly defined** in front of dorsal fin, consisting of separate dark specks or...

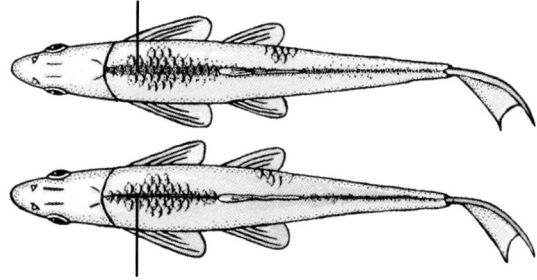

...reduced to a thin line narrower than dorsal fin base

45a Anal fin rays usually 9 to 11; lining of body cavity uniformly black **46**

45b Anal fin rays usually 7 or 8; lining of body cavity silvery, with or without dark speckles **48**

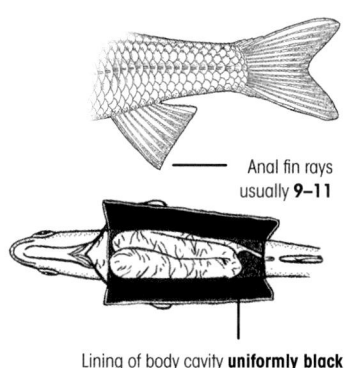

Anal fin rays usually **9–11**

Lining of body cavity **uniformly black**

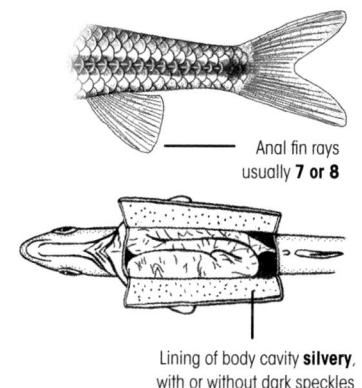

Anal fin rays usually **7 or 8**

Lining of body cavity **silvery**, with or without dark speckles

46a Stripe along midside dark and abruptly narrower behind gill cover; rear margin of gill cover with wide, dark bar; fins, mouth, and gill cover of breeding males with bright red highlights. . . . .**Page 173**

Bleeding Shiner (*Luxilus zonatus*)

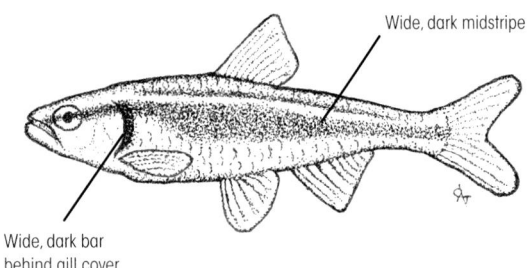

Wide, dark midstripe

Wide, dark bar behind gill cover

46b Sides of body without dark stripe; rear margin of gill cover without wide, dark bar; fins, mouth, and gill cover without bright red highlights, color, if present, pink **47**

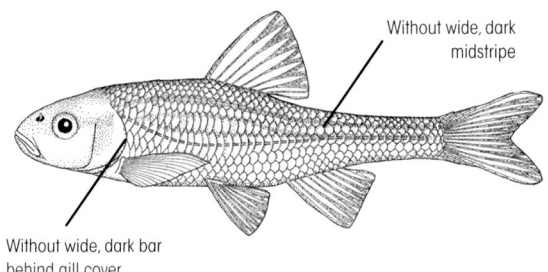

Without wide, dark midstripe

Without wide, dark bar behind gill cover

47a Chin sprinkled with dusky pigment; upper surface of body with dark lines, those from opposite sides of body converging behind dorsal fin forming V-shaped markings; predorsal scales usually 22 or less; scale rows around body in front of dorsal fin usually 26–29 . . . . . **Page 171**

Striped Shiner (*Luxilus chrysocephalus*)

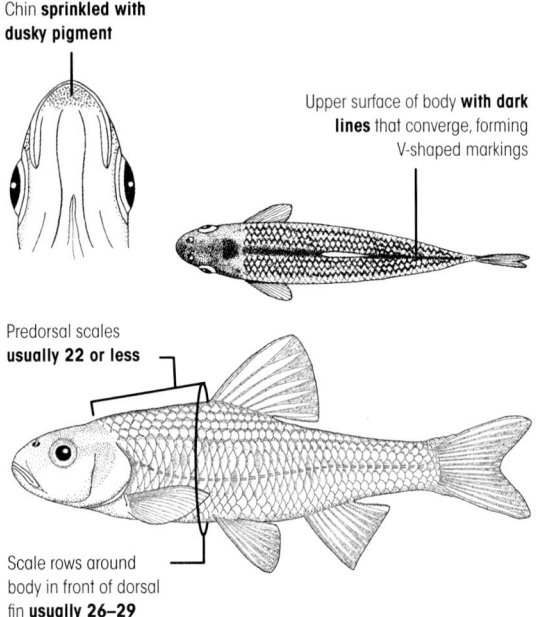

Chin **sprinkled with dusky pigment**

Upper surface of body **with dark lines** that converge, forming V-shaped markings

Predorsal scales **usually 22 or less**

Scale rows around body in front of dorsal fin **usually 26–29**

47b Chin not sprinkled with dusky pigment; upper surface of body without dark lines that converge behind dorsal fin; predorsal scales usually 23 or more; scale rows around body in front of dorsal fin usually 30–35 . . . .**Page 172**

Common Shiner (*Luxilus cornutus*)

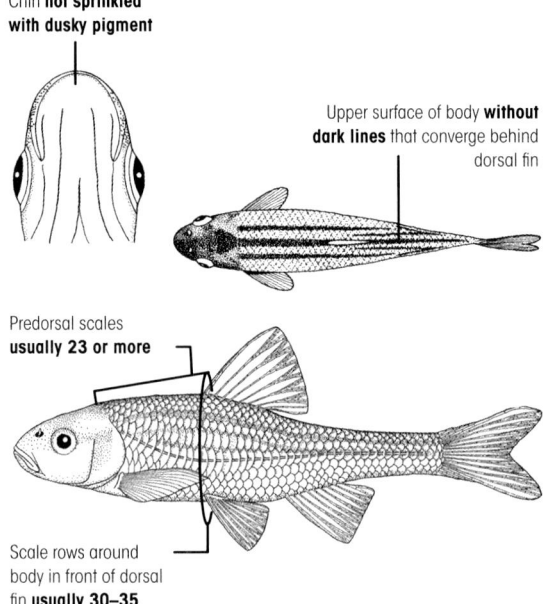

Chin **not sprinkled with dusky pigment**

Upper surface of body **without dark lines** that converge behind dorsal fin

Predorsal scales **usually 23 or more**

Scale rows around body in front of dorsal fin **usually 30–35**

48a Eyes directed upward, the lower margins of pupils usually visible when fish is viewed from above; lower surface of head distinctly flattened; mouth large, length of upper jaw (A) much greater than eye diameter (B) . . . . . .**Page 190**

48b Eyes directed to sides, the lower margins of pupils not visible when fish is viewed from above; lower surface of head not noticibly flattened; mouth smaller, length of upper jaw (A) not greater than eye diameter **49**

Bigmouth Shiner (*Notropis dorsalis*)

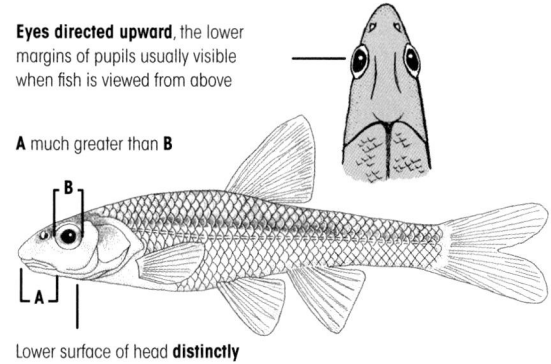

Eyes directed upward, the lower margins of pupils usually visible when fish is viewed from above

A much greater than **B**

B

A

Lower surface of head **distinctly flattened**

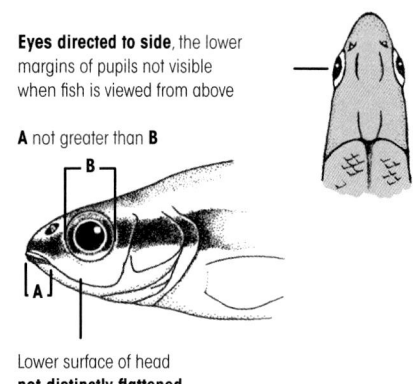

Eyes directed to side, the lower margins of pupils not visible when fish is viewed from above

A not greater than **B**

B

A

Lower surface of head **not distinctly flattened**

49a Dark stripe along midside prominent, extending forward onto snout **50**

49b Dark stripe along midside absent, or becoming indistinct toward head, usually ending under dorsal fin **51**

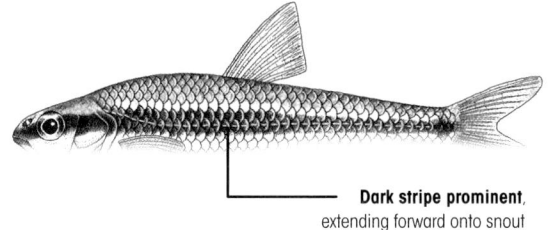

Dark stripe prominent, extending forward onto snout

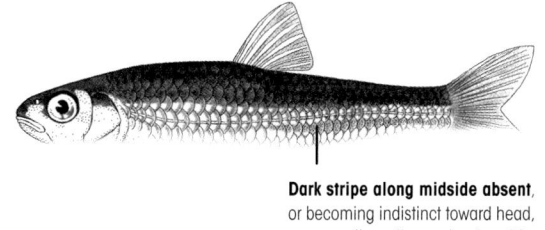

Dark stripe along midside absent, or becoming indistinct toward head, usually ending under dorsal fin

50a Anal fin rays usually 7; several scales below dark stripe along midside outlined with black pigment; inside of mouth not sprinkled with black pigment . . . . . . **Page 201**

Weed Shiner (*Notropis texanus*)

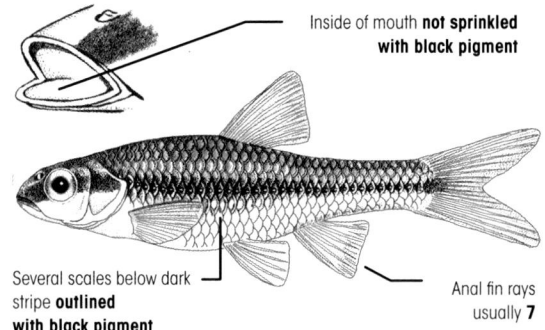

Inside of mouth **not sprinkled with black pigment**

Several scales below dark stripe **outlined with black pigment**

Anal fin rays usually **7**

50b Anal fin rays usually 8; scales below dark stripe along midside not outlined with black pigment; inside of mouth sprinkled with black pigment . . . . **Page 189**

Ironcolor Shiner (*Notropis chalybaeus*)

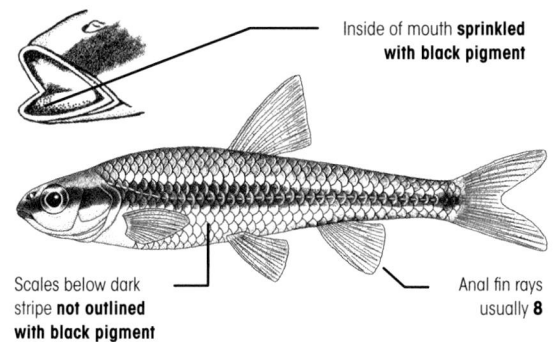

Inside of mouth **sprinkled with black pigment**

Scales below dark stripe **not outlined with black pigment**

Anal fin rays usually **8**

51a Stripe along midline of back not expanded just in front of dorsal fin; lateral-line pores not outlined above and below (stitched) with melanophores (tiny black specks); mouth larger, length of upper jaw (A) about equal to eye diameter (B) . . . . **Page 186**

River Shiner (*Notropis blennius*)

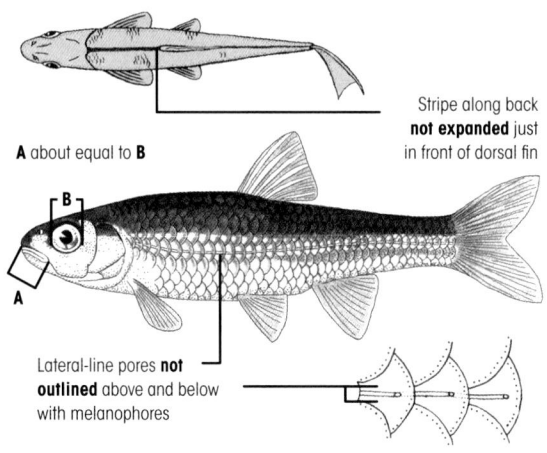

A about equal to **B**

Stripe along back **not expanded** just in front of dorsal fin

Lateral-line pores **not outlined** above and below with melanophores

51b Stripe along midline of back expanded into a wedge-shaped blotch in front of dorsal fin; lateral-line pores outlined above and below with melanophores; mouth smaller, length of upper jaw (A) less than eye diameter (B) . . . . **Page 200**

Sand Shiner (*Notropis stramineus*)

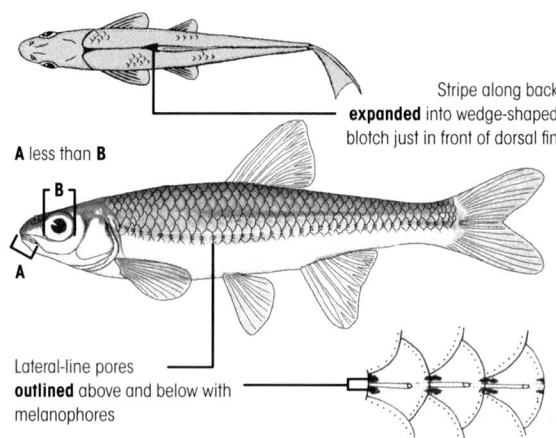

A less than **B**

Stripe along back **expanded** into wedge-shaped blotch just in front of dorsal fin

Lateral-line pores **outlined** above and below with melanophores

52a Stripe along midside dark and prominent, extending forward on head to tip of snout.................... **53**

Stripe along midside **dark and prominent**, extending
forward on head to tip of snout

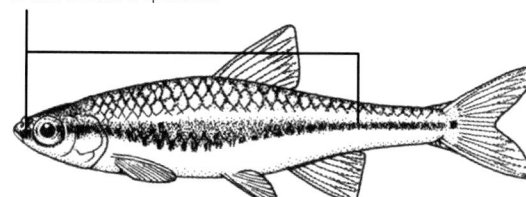

52b Stripe along midside indistinct or absent, not extending forward on head to tip of snout.................... **57**

Stripe along midside **indistinct or absent**,
not extending forward on head to tip of snout

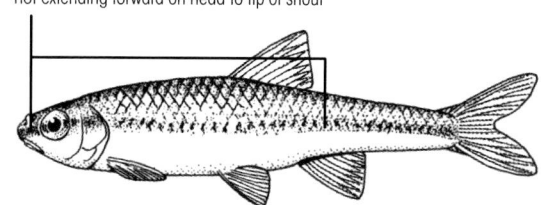

53a Lower lip and chin without black pigment; mouth nearly horizontal................................ **54**

Mouth **nearly
horizontal**

Lower lip
and chin **without
black pigment**

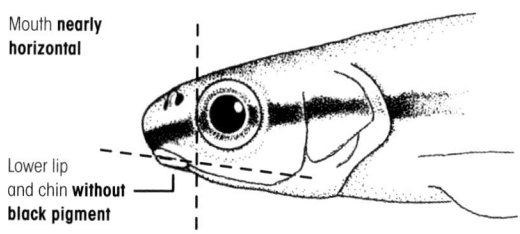

53b Lower lip and chin with black pigment; mouth oblique **55**

Mouth **oblique**

Lower lip
and chin **with
black pigment**

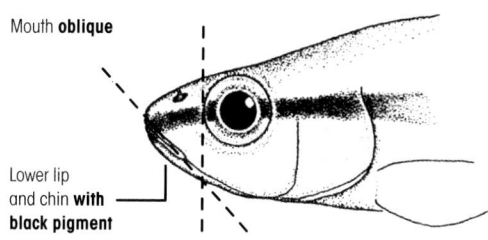

54a Anterior lateral series scales with black crescent-shaped marks, the tips of which point backwards; lateral line incomplete............................**Page 192**

Blacknose Shiner (*Notropis heterolepis*)

Anterior lateral series scales **with
black crescent-shaped marks**, the
tips of which point backwards

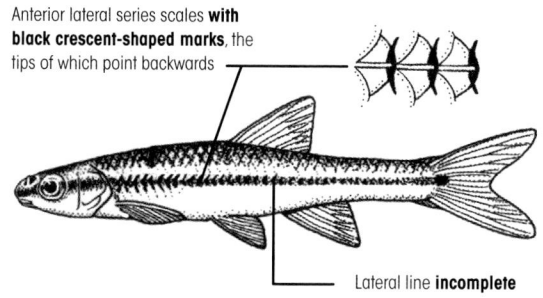

Lateral line **incomplete**

54b Anterior lateral series scales without black crescent-shaped marks; lateral line complete.........**Page 170**

Pallid Shiner (*Hybopsis amnis*) (in part)

Anterior lateral series scales
**without black crescent-shaped
marks**, often with little
black pigment

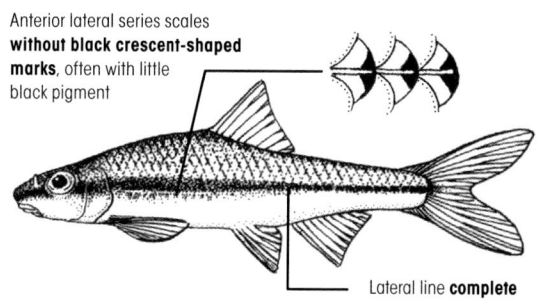

Lateral line **complete**

MINNOWS (Leuciscidae)

55a Mouth nearly vertical (about 80° with horizontal); mouth small, upper jaw extending only to below anterior nostril . . . .**Page 184**

55b Mouth oblique (about 45° with horizontal); mouth larger, upper jaw extending at least to posterior nostril **56**

Pugnose Shiner (*Notropis anogenus*)

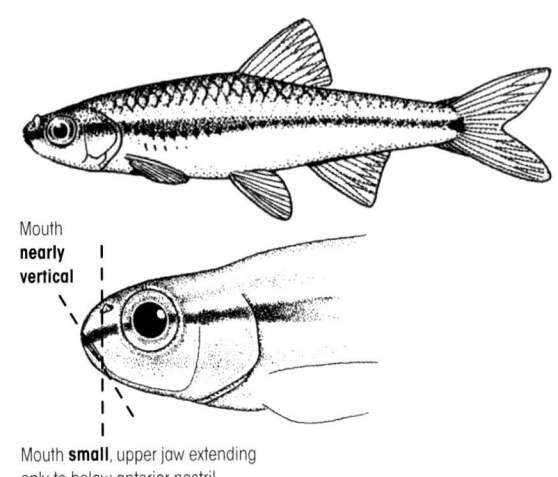

Mouth **nearly vertical**

Mouth **small**, upper jaw extending only to below anterior nostril

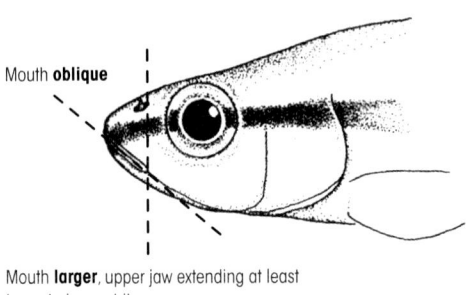

Mouth **oblique**

Mouth **larger**, upper jaw extending at least to posterior nostril

56a Lining of body cavity silvery; lateral line incomplete; black stripe along midside often with a zigzag appearance . . . .**Page 191**

56b Lining of body cavity uniformly black; lateral line complete; black stripe along midside with a stitched appearance. . . . .**Page 187**

Blackchin Shiner (*Notropis heterodon*)

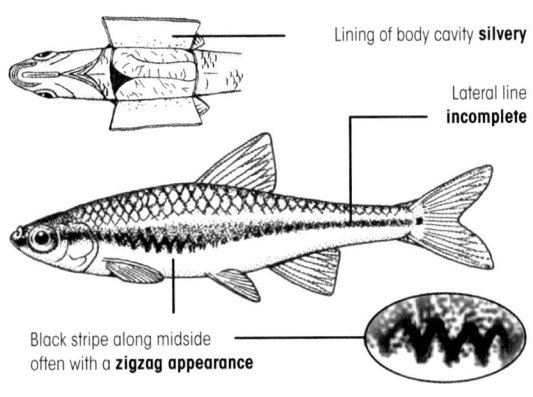

Lining of body cavity **silvery**

Lateral line **incomplete**

Black stripe along midside often with a **zigzag appearance**

Bigeye Shiner (*Notropis boops*)

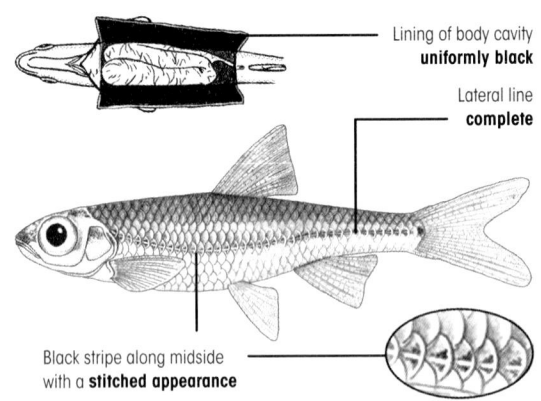

Lining of body cavity **uniformly black**

Lateral line **complete**

Black stripe along midside with a **stitched appearance**

57a Scales in anterior part of lateral line with width of exposed margins (A) going 2 times or less into their depth (B). **58**

57b Scales in anterior part of lateral line with width of exposed margins (A) going more than 2 times into their depth (B). **59**

A going 2 times or less into B

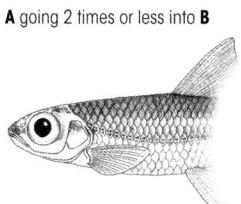 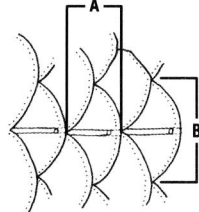

A going more than 2 times into B

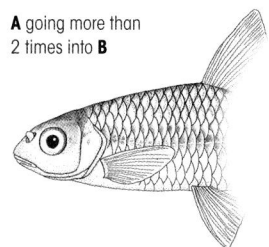 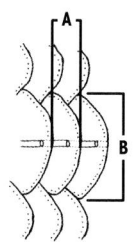

58a Front of upper lip on a level with lower margin of eye; mouth small, upper jaw nearly horizontal and not extending to front of eye; snout projecting distinctly beyond upper lip; pelvic fin rays usually 8. . . . .**Page 170**

Pallid Shiner (*Hybopsis amnis*) (in part)

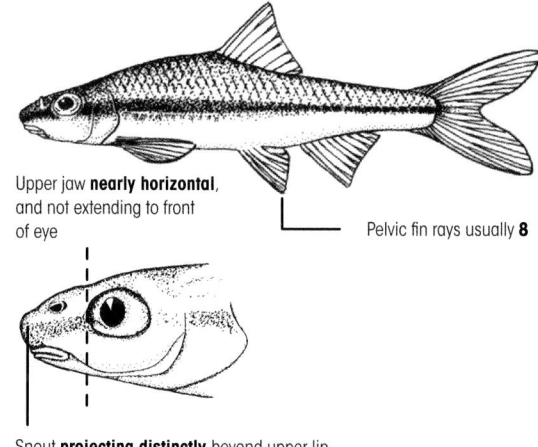

Upper jaw **nearly horizontal**, and not extending to front of eye

Pelvic fin rays usually **8**

Snout **projecting distinctly** beyond upper lip

58b Front of upper lip on a level with center of eye; mouth larger, upper jaw oblique and extending to or beyond front of eye; snout scarcely projecting beyond lip; pelvic fin rays usually 9 . . . .**Page 199**

Silverband Shiner (*Notropis shumardi*)

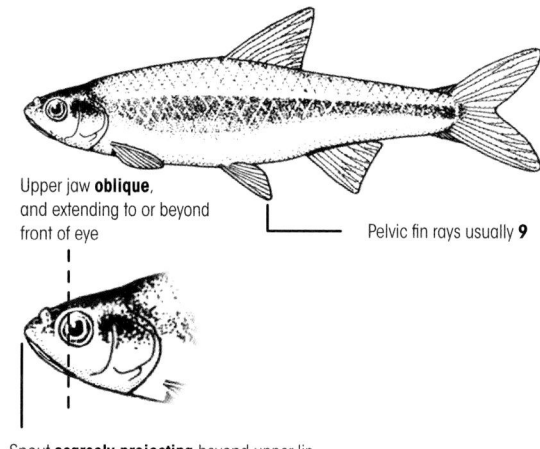

Upper jaw **oblique**, and extending to or beyond front of eye

Pelvic fin rays usually **9**

Snout **scarcely projecting** beyond upper lip

59a Body coloration very pallid; dusky stripe along midside absent or reduced to a few faint speckles; lining of body cavity silvery; sensory canal beneath eye absent or greatly reduced; tips of pelvic fins reaching to or beyond front of anal fin..............................**Page 188**

59b Body coloration darker; dusky stripe along midside developed on back half of body; lining of body cavity silvery with dark speckles well developed; sensory canal beneath eye complete; tips of pelvic fins not reaching front of anal fin............................. **60**

Ghost Shiner (*Notropis buchanani*)

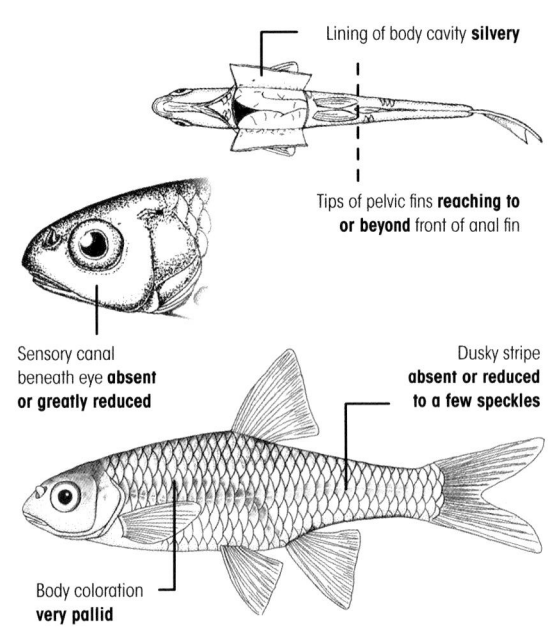

Lining of body cavity **silvery**

Tips of pelvic fins **reaching to or beyond** front of anal fin

Sensory canal beneath eye **absent or greatly reduced**

Dusky stripe **absent or reduced to a few speckles**

Body coloration **very pallid**

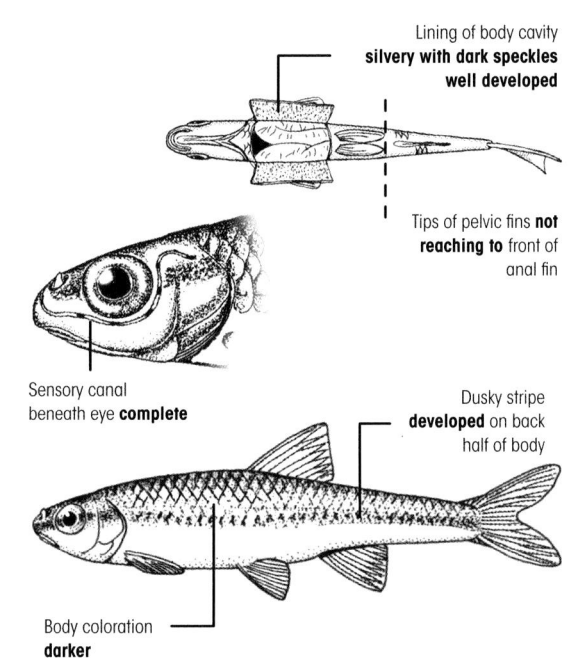

Lining of body cavity **silvery with dark speckles well developed**

Tips of pelvic fins **not reaching to** front of anal fin

Sensory canal beneath eye **complete**

Dusky stripe **developed** on back half of body

Body coloration **darker**

60a Dark stripe on back behind dorsal fin absent or weakly developed; lateral line scales usually 35–37 ....
.....................................**Page 202**

Mimic Shiner (*Notropis volucellus*)

60b Dark stripe on back behind dorsal fin fairly well developed; lateral line scales usually 33–35 ....**Page 203**

Channel Shiner (*Notropis wickliffi*)

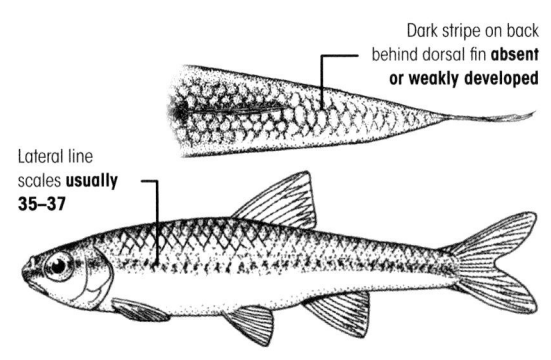

Dark stripe on back behind dorsal fin **absent or weakly developed**

Lateral line scales **usually 35–37**

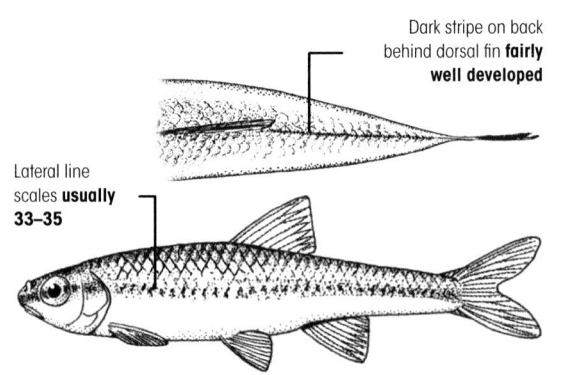

Dark stripe on back behind dorsal fin **fairly well developed**

Lateral line scales **usually 33–35**

CENTRAL STONEROLLER *Campostoma anomalum* (Rafinesque 1820)

IDENTIFICATION The Central Stoneroller has a hard cartilaginous ridge on the lower jaw of its subterminal mouth. The body is thick, barely compressed, and strongly arched at the nape. The dorsal-fin origin is over or slightly behind the pelvic-fin origin. The intestine is long (about 18 inches in a 5-inch individual) and coiled around the gas bladder. The body is tan to brown on the back and side with irregular dark brown to black blotches. Large males have a black band on orange dorsal and anal fins (see Remarks), white lips, a bright red eye, and a crescent-shaped row of 1–3 large tubercles on the inner edge of the nostril; numerous smaller tubercles cover the top of the head, nape, and body. The lateral line is complete with usually 46–55 lateral scales. There are 36–46 scales around the body at the dorsal-fin origin. The pharyngeal teeth are 0,4-4,0. To 6¾ in. (17 cm).

SIMILAR SPECIES The Largescale Stoneroller, *C. oligolepis*, lacks the tubercles on the inner edge of the nostril, usually has 43–47 lateral scales and 31–36 scales around the body at the dorsal-fin origin.

HABITAT The species is found primarily in rocky riffles, runs, and pools of headwaters, creeks, and small to medium rivers.

DISTRIBUTION IN ILLINOIS The Central Stoneroller is abundant throughout Illinois except in an area in the south-central part of the state where the streams are mostly low gradient and turbid and have clay or mud rather than gravel substrates. The Illinois range has not changed substantially, and the species may be as widespread as it was in previous eras. The Largescale Stoneroller, *C. oligolepis*, is found in Hardin and Pope Counties, in extreme southeastern Illinois, where Central Stonerollers had been misidentified and recorded by Smith (1979).

REMARKS Three subspecies have often been recognized but are in need of further study (Blum et al. 2008). *Campostoma anomalum pullum* is apparently the form found in Illinois and usually has 18–20 scales over the body from lateral line to lateral line at the dorsal-fin origin (including lateral-line scales); other subspecies have 15–17 scales. An early examination of nontuberculate males (Burr 1976) is in error regarding the taxonomic status of stoneroller populations in southeastern Illinois.

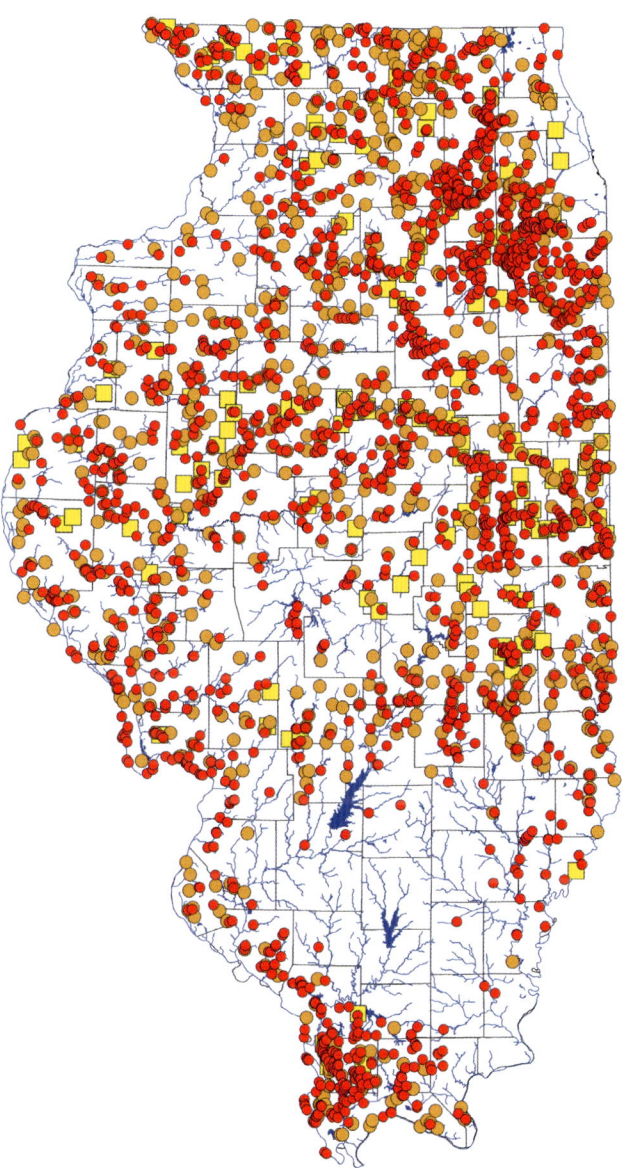

MINNOWS (Leuciscidae)

153

LARGESCALE STONEROLLER
Campostoma oligolepis Hubbs & Greene 1935

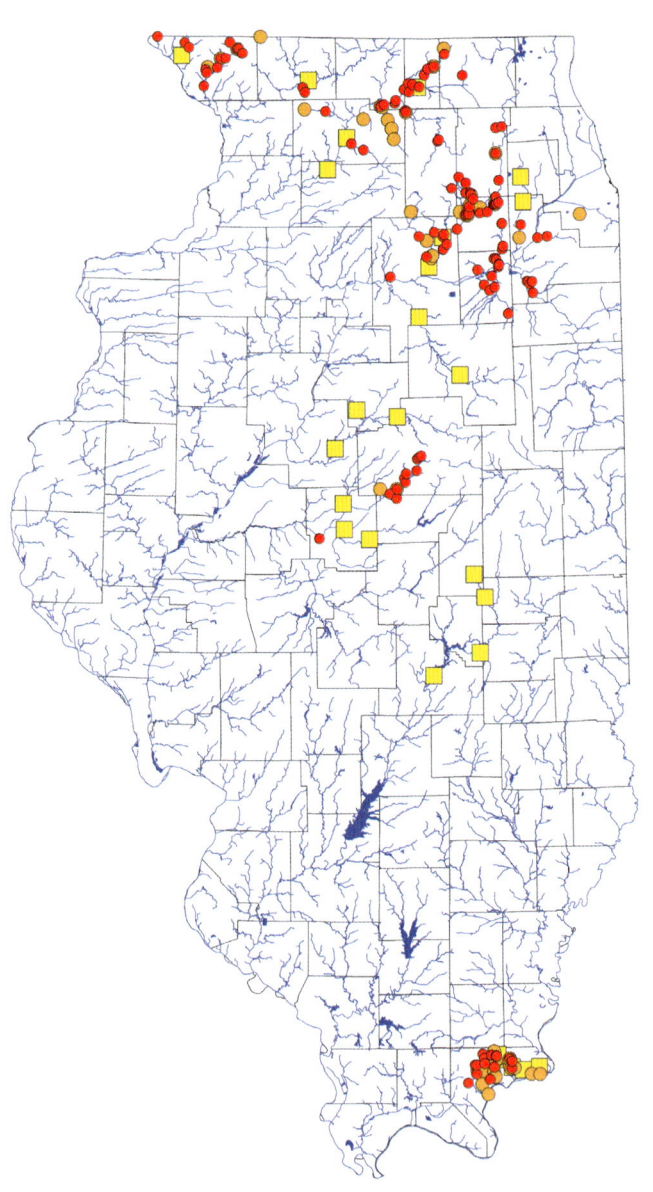

IDENTIFICATION The Largescale Stoneroller has a hard carti-laginous ridge on the lower jaw of its subterminal mouth. The body is thick, barely compressed, and slightly arched at the dorsal-fin origin. The dorsal-fin origin is over or slightly behind the pelvic-fin origin. The intestine is long (about 18 inches in a 5-inch individual) and coiled around the gas bladder. The body is tan to brown on the back and side with irregular dark brown to black blotches; the lower side and belly are cream to white. Large males have white lips, a bright red eye, no (or only a weak) black band on the orange anal fin, and no large tubercles on the inner edge of the nostril; numerous smaller tubercles cover the top of the head, nape, and body. The lateral line is complete with usually 43–47 lateral scales. There are 31–36 scales around the body at the dorsal-fin origin. The pharyngeal teeth are 0,4-4,0 or 1,4-4,1 (see Remarks). To 8½ in. (22 cm).

SIMILAR SPECIES The Central Stoneroller, *C. anomalum*, has a crescent-shaped row of 1–3 tubercles on the inner edge of the nostril. There are usually 46–55 lateral scales, and 36–46 scales around the body at the dorsal-fin origin.

HABITAT The Largescale Stoneroller occupies rocky riffles and runs of clear creeks and small to medium rivers. It occurs with the Central Stoneroller in many streams in the northern half of Illinois.

DISTRIBUTION IN ILLINOIS Smith (1979) noted that the range of the Largescale Stoneroller was greatly reduced from what it had been during the Forbes and Richardson era, with the species be-ing restricted to the northern one-third of the state and Kickapoo Creek in the Sangamon River basin. The distribution is essentially the same in the contemporary era as it was in the Smith era, with no records from the Kaskaskia, Mackinaw, and Vermilion River ba-sins where the species occurred during the Forbes and Richardson era. Although records from Big, Big Grande Pierre, and Lusk Creeks in Pope and Hardin Counties were treated as Central Stonerollers by Smith (1979), the Largescale Stoneroller is the only species of *Campostoma* in those streams.

REMARKS Largescale Stonerollers from extreme southeastern Illinois have the tubercle pattern as described here and have 1,4-4,1 pharyngeal teeth. The pharyngeal teeth are 0,4-4,0 in other Illinois populations.

SOUTHERN REDBELLY DACE *Chrosomus erythrogaster* (Rafinesque 1820)

IDENTIFICATION The Southern Redbelly Dace has a slender, fairly compressed body with the dorsal-fin origin above or slightly behind the pelvic-fin origin. The snout is moderately pointed, and the small, slightly subterminal mouth ends in front of the eye. The back and upper side are olive-brown with a dusky black stripe along the middle of the back, and black spots (sometimes absent) are on the upper side. There are 2 black stripes along the silver-yellow side of the body; the upper stripe is thin and usually broken into spots posteriorly; the lower stripe is wider, becoming thinner on the caudal peduncle. There is a black wedge-shaped spot at the base of the caudal fin. The breast and belly are white, yellow, or red. Large males are brilliantly colored, with yellow fins and a bright red breast and belly, lower part of the head, and dorsal-fin base. The lateral line is incomplete, and there are 67–95 lateral scales, 8 dorsal rays, 7–8 anal rays, and the pharyngeal teeth are 0,5-5,0. To 3½ in. (9.1 cm).

SIMILAR SPECIES No other Illinois minnow has 2 black stripes along the side of the body.

HABITAT The Southern Redbelly Dace occupies rocky, cool, usually groundwater-fed, shallow pools of headwaters and small creeks. Schools of Southern Redbelly Dace often are found among tree roots under overhanging banks.

DISTRIBUTION IN ILLINOIS Forbes and Richardson (1909) recorded the Southern Redbelly Dace from the cool streams in northern Illinois and in groundwater-fed streams draining into the Illinois River and in extreme southeastern and southwestern Illinois. Smith (1979) found a similar distribution but added records from the lower Illinois River basin and the Wabash River basin in east-central Illinois. He did not find the species in southern Illinois or in the small streams draining into Lake Michigan in Lake County where Forbes and Richardson (1909) had recorded them. Contemporary-era records document the continued presence of the species in extreme southeastern Illinois but not from southwestern Illinois or from the small streams draining into Lake Michigan in Lake County. Otherwise, except for a few records in central Illinois from tiny groundwater-fed streams, the range of this species has changed little since Smith (1979).

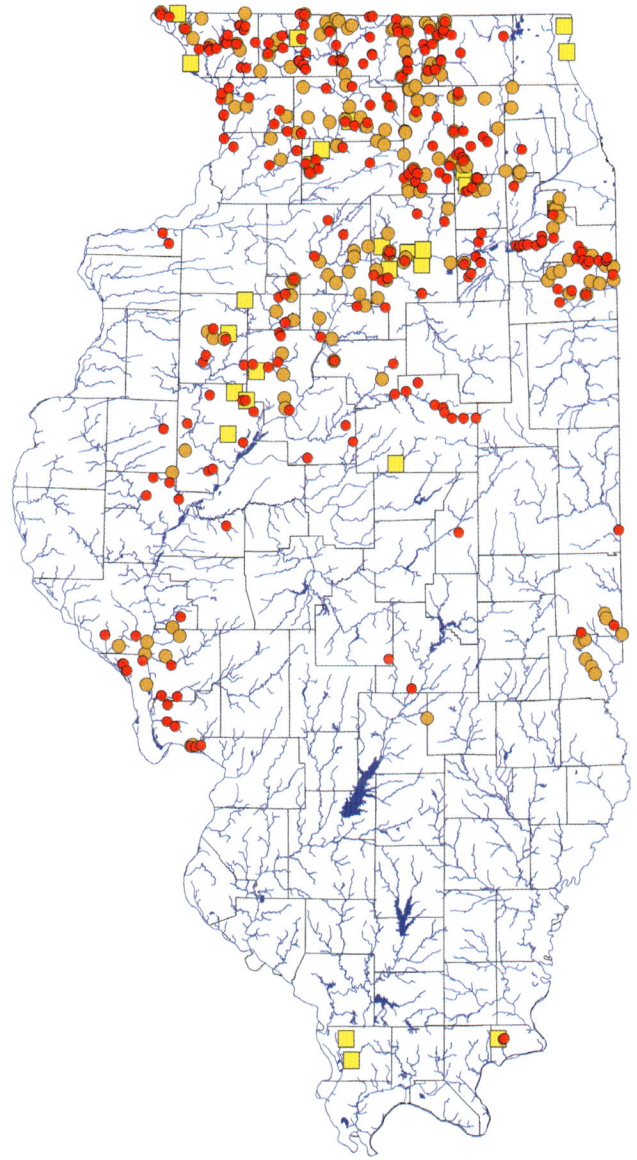

REDSIDE DACE   *Clinostomus elongatus* (Kirtland 1840)

IDENTIFICATION The Redside Dace has a compressed body, a large, oblique mouth, and a long, pointed snout. The dorsal-fin origin is behind the pelvic-fin origin. The back and upper side of the body are olive with numerous black specks or scattered dark blotches. The side is silver with a bright red streak (on adults) from the opercle to below the dorsal fin followed by a dusky to black stripe that extends to the caudal fin. Large males are steel blue above and have a yellow-gold stripe along the bright red side of the body. The lateral line is complete with 62–75 lateral scales. There are 8 dorsal rays and 9 anal rays, and the pharyngeal teeth are usually 2,5-4,2. To 4½ in. (11.4 cm).

SIMILAR SPECIES The Common Shiner, *Luxilus cornutus*, and the Striped Shiner, *L. chrysocephalus*, have much larger scales (usually 37–40 lateral scales) and obvious dark stripes on the back and upper side. Species of *Lythrurus* usually have 10–12 anal rays, 35–56 lateral scales, pharyngeal teeth are 2,4-4,2, and large males develop bright red or bright yellow fins.

HABITAT The Redside Dace occupies rocky and sandy pools of headwaters, creeks, and small rivers.

DISTRIBUTION IN ILLINOIS The Redside Dace was not known to occur in Illinois in earlier eras. It is known only from East Fork Raccoon Creek in the Pecatonica River basin in Winnebago County, where it was captured in 1998.

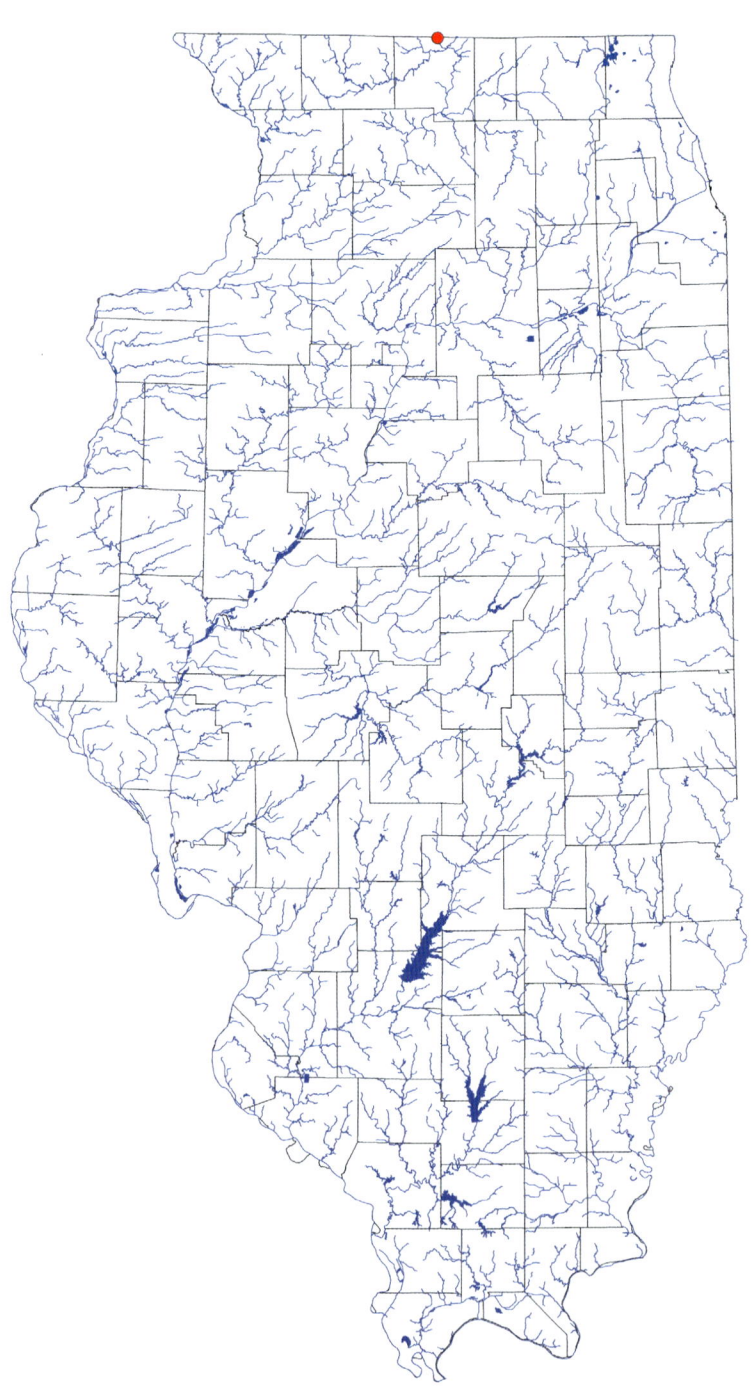

LAKE CHUB *Couesius plumbeus* (Agassiz 1850)

IDENTIFICATION The Lake Chub has a moderately compressed body, a large head that is somewhat flattened above and below, a large eye, a moderately pointed snout, and a small barbel at the corner of a large, slightly subterminal mouth. The dorsal-fin origin is over or slightly behind the pelvic-fin origin. The back and upper side are light brown to pale green, and the side is silver-gray with a dusky stripe, which is darkest on juveniles and large males. Black specks on the side and belly and a dusky caudal spot may be present. Large males have red at the corner of the mouth and at the pectoral fin and pelvic-fin origins. The lateral line is complete with 53–70 scales. There are 8 anal rays, and the pharyngeal teeth are 2,4-4,2. To 9 in. (23 cm).

SIMILAR SPECIES The Creek Chub, *Semotilus atromaculatus*, has a large, black spot at the front of the dorsal-fin base and a flaplike barbel in the groove above the upper lip; large males have pink on the lower half of the head and body and orange anal and paired fins. The Hornyhead Chub, *Nocomis biguttatus*, has a bright red or brassy spot behind the eye, 7 anal rays, 38–45 lateral scales, and 1,4-4,1 pharyngeal teeth.

HABITAT In Illinois the Lake Chub is found along the pebble and gravel-sand shoreline of Lake Michigan and in small tributaries of the lake. It rarely occurs in the cold, deep water of the lake.

DISTRIBUTION IN ILLINOIS As noted by Smith (1979), Forbes and Richardson (1909) did not report the Lake Chub from Illinois even though it had been recorded from the Illinois portion of Lake Michigan in the 1870s by Jordan (1878). The Lake Chub is restricted in Illinois to the shoreline of Lake Michigan and the lower Waukegan River and other small tributaries of the lake. Its contemporary distribution is the same as reported by Smith.

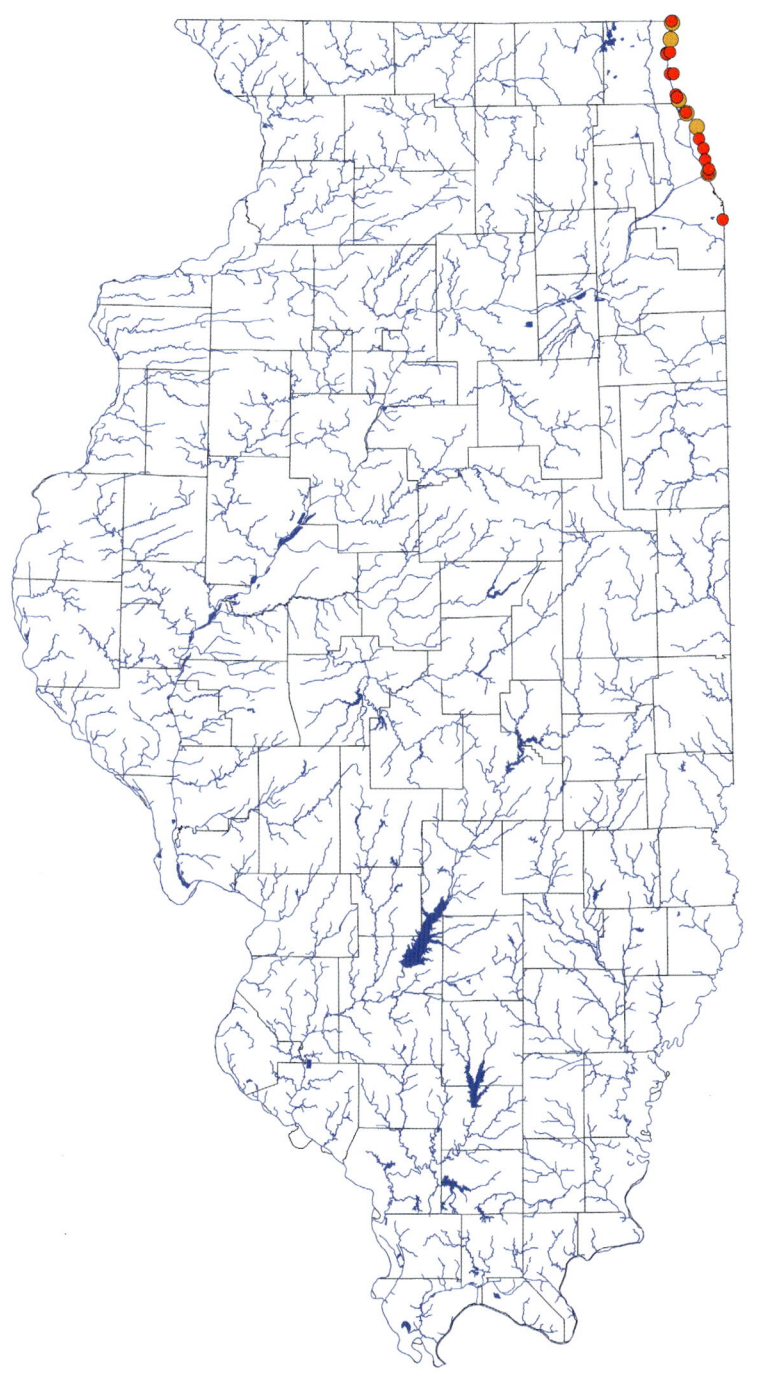

RED SHINER   *Cyprinella lutrensis* (Baird & Girard 1853)

IDENTIFICATION The Red Shiner has a deep compressed body, a terminal mouth, and a somewhat rounded snout. The dorsal-fin origin is slightly behind the pelvic-fin origin. The back and upper side are dusky olive to blue, and the scales on the upper side appear diamond shaped. There is usually a black stripe along the back and a diffuse dark stripe along the posterior half of the silver side. The dorsal fin is dusky and without a black blotch posteriorly. There is a dusky blue triangular bar behind the head (darkest on adults). Large males have red fins (except the dorsal), a blue back and side, and a dark blue bar before a pink bar behind the head. The lateral line is complete with 32–36 scales, and there are usually 26 scales around the body at the dorsal-fin origin, 9 anal rays, and 0,4-4,0 pharyngeal teeth. To 3½ in. (9 cm).

SIMILAR SPECIES The Spotfin Shiner, *C. spiloptera*, has a black blotch in the posterior half of the dorsal fin, usually 8 anal rays, 1,4-4,1 pharyngeal teeth, and yellow fins on large males. The Steelcolor Shiner, *C. whipplei*, has a black blotch on the posterior half of the dorsal fin, a more pointed snout, usually 37–38 scales along the lateral line, and 1,4-4,1 pharyngeal teeth. The Blacktail Shiner, *C. venusta*, has a large, black spot at the base of the caudal fin, a black blotch on the posterior half of the dorsal fin, usually 8 anal rays, 36–38 scales along the lateral line, and 1,4-4,1 pharyngeal teeth.

HABITAT The Red Shiner occurs in silty, sandy, and rocky pools and runs, sometimes riffles, of creeks and small to medium rivers. It is tolerant of siltation, low dissolved-oxygen levels, and high turbidity and often is the most commonly encountered minnow in polluted, channelized, or otherwise highly modified streams.

DISTRIBUTION IN ILLINOIS Forbes and Richardson found the Red Shiner to be widespread in west-central and southwestern Illinois and in a few localities in the Illinois River basin in east-central Illinois. By mid-20th century, the species had spread eastward and was widespread throughout Illinois except in the Wabash River basin and in northern and northeastern Illinois (Larimore & Smith 1963). Smith (1979) noted that the species had recently gained access to the Vermilion River system (Wabash) in Ford County, and contemporary-era records show that it occurs throughout much of the Vermilion River system and has gained access to other tributaries of the Wabash River and has moved up the Ohio River and into the Saline River basin. The Red Shiner has long been considered a pioneering species that seems to be aggressively expanding its range. However, it also is a colorful species that frequently is used as bait, which no doubt facilitates its range expansion. Some scattered records in the state may be the result of "bait bucket" releases.

REMARKS As the Red Shiner has moved eastward, it has hybridized with other species of *Cyprinella*, sometimes forming hybrid swarms (e.g., with the Spotfin Shiner, *C. spiloptera*, in central and western Illinois) that last for many years, and other times has competitively displaced related species (Page & Smith 1970).

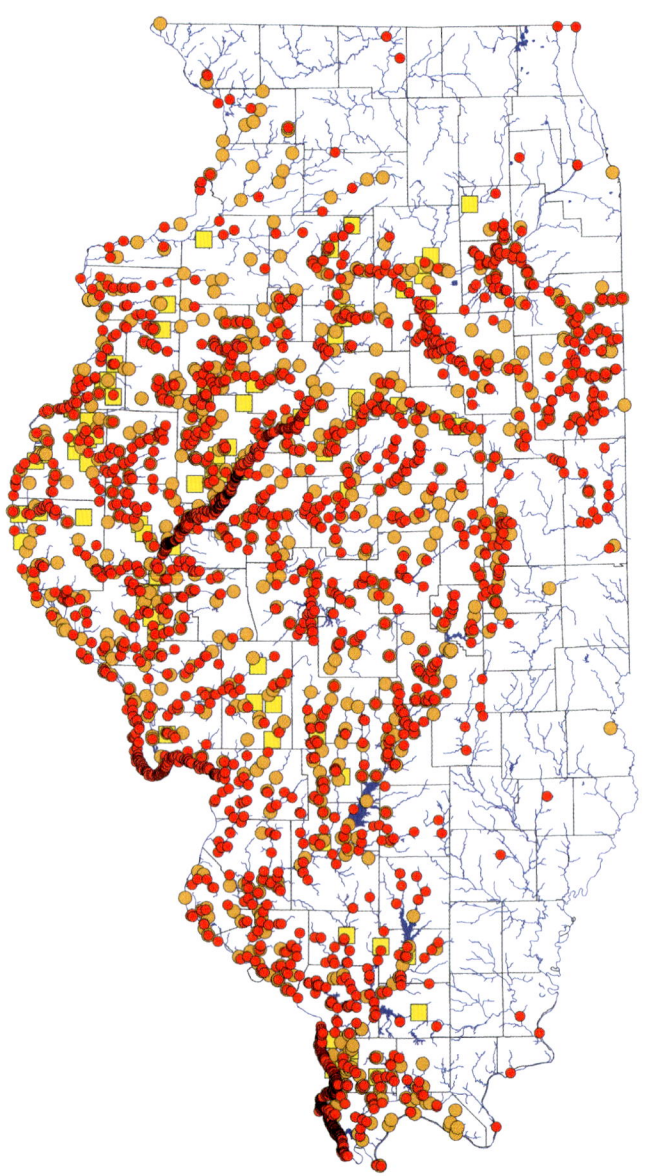

SPOTFIN SHINER *Cyprinella spiloptera* (Cope 1867)

IDENTIFICATION The Spotfin Shiner has a fairly deep and compressed body, a terminal mouth, and a pointed snout. The dorsal-fin origin is slightly behind the pelvic-fin origin. The back and upper side are dusky olive, and the scales on the upper side of the body appear diamond shaped. Often there is a dusky bar on the side behind the head and a diffuse dark stripe along the posterior half of the silver side. There is a black blotch on the posterior half of the dorsal fin and little or no black pigment on the membranes in the front half of the dorsal fin (except in large males). Large males have a metallic-blue back and side, yellow-white fins, and a dusky dorsal fin. The lateral line is complete with 35–39 scales, and there are usually 26 scales around body at the dorsal-fin origin, 8 anal rays, and 1,4-4,1 pharyngeal teeth. To 4¾ in. (12 cm).

SIMILAR SPECIES The Steelcolor Shiner, *C. whipplei*, usually has 9 anal rays and black specks on all membranes of the dorsal fin; large males have an enlarged dorsal fin. The Red Shiner, *C. lutrensis*, has red fins on large males, a deeper body, no black blotch on the posterior half of the dorsal fin, usually 9 anal rays, and 0,4-4,0 pharyngeal teeth. The Blacktail Shiner, *C. venusta*, has a large, black spot on the caudal-fin base and usually has 28–29 scales around the body at the dorsal-fin origin.

HABITAT The Spotfin Shiner occurs chiefly in sand and gravel runs and pools of creeks and small to large rivers. Occasionally it is found in lakes and reservoirs.

DISTRIBUTION IN ILLINOIS Forbes and Richardson (1909) found the Spotfin Shiner throughout the northern half of Illinois and in the Wabash and Kaskaskia River basins. By the mid-20th century, it had disappeared from much of central Illinois (Smith 1979), at least in part due to its replacement by the related Red Shiner, *C. lutrensis* (Page & Smith 1970). In the contemporary era, it remains common in northern and eastern Illinois, is much more widespread in the Mississippi and lower Illinois River basins than it was in previous eras, is in the Lake Michigan basin, and has re-established itself in areas of central Illinois where it was not found in mid-century.

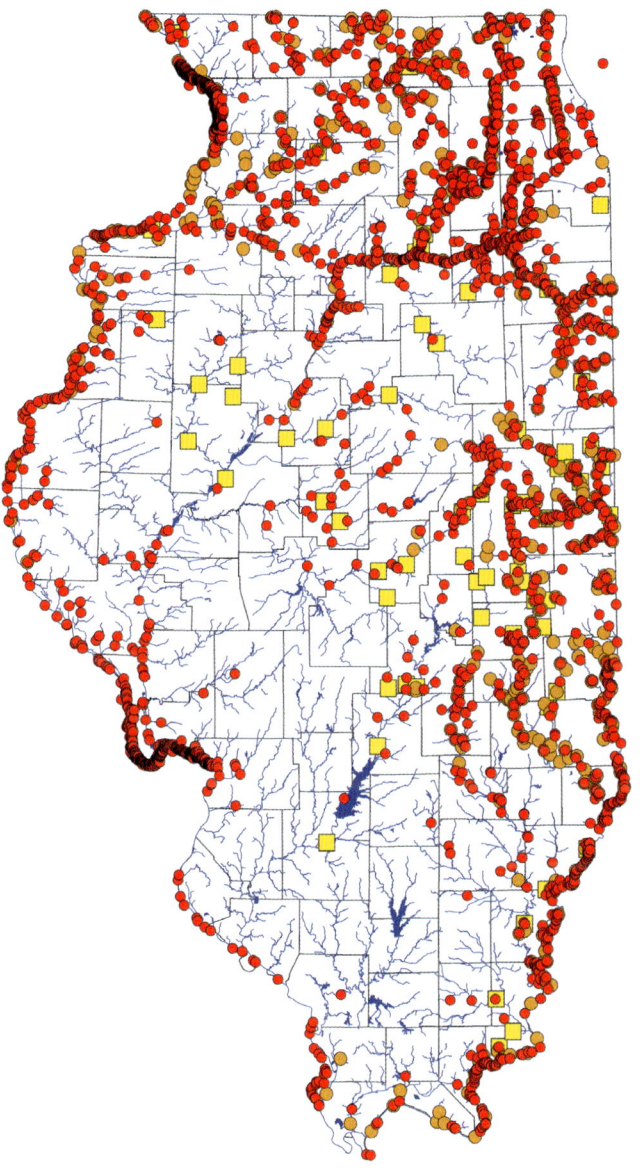

BLACKTAIL SHINER *Cyprinella venusta* Girard 1856

IDENTIFICATION The Blacktail Shiner has a fairly deep and compressed body, a terminal mouth, and a pointed snout. The dorsal-fin origin is slightly behind the pelvic-fin origin. The back and upper side are dusky olive, and the scales on the upper side of the body appear diamond shaped. There is a narrow black stripe along the back, a diffuse dark stripe along the posterior half of the silver side, and a large, black spot at the base of the caudal fin. The dorsal fin is dusky with a black blotch posteriorly (darkest in large males). Large males have a bluish back and side, yellow-white fins, and an overall dusky dorsal fin. The lateral line is complete with 36–38 scales, and there are usually 28–29 scales around the body at the dorsal-fin origin, 8 anal rays, and 1,4-4,1 pharyngeal teeth. To 7½ in. (19 cm).

SIMILAR SPECIES The Spotfin Shiner, *C. spiloptera*, has no large, black spot at the base of the caudal fin and usually has 26 scales around the body at the dorsal-fin origin. The Steelcolor Shiner, *C. whipplei*, has no large, black spot at the base of the caudal fin, usually 9 anal rays and 26 scales around the body at the dorsal-fin origin; large males have a red snout. The Red Shiner, *C. lutrensis*, has no large, black spot at the base of the caudal fin, no black blotch on the posterior half of the dorsal fin, bright red and bright blue on large males, and usually 9 anal rays, 32–36 scales along the lateral line, and 0,4-4,0 pharyngeal teeth.

HABITAT The Blacktail Shiner is found in sandy pools and runs, less often in rocky areas, of small to medium rivers.

DISTRIBUTION IN ILLINOIS Forbes and Richardson (1909) recorded the Blacktail Shiner from Clear Creek in extreme southwestern Illinois, as did Smith (1979), who also added records from the Mississippi and Ohio Rivers of southern Illinois. In the contemporary era, the Blacktail Shiner is more widespread with records in the Big Muddy River and tributaries of the Ohio River in southeastern Illinois.

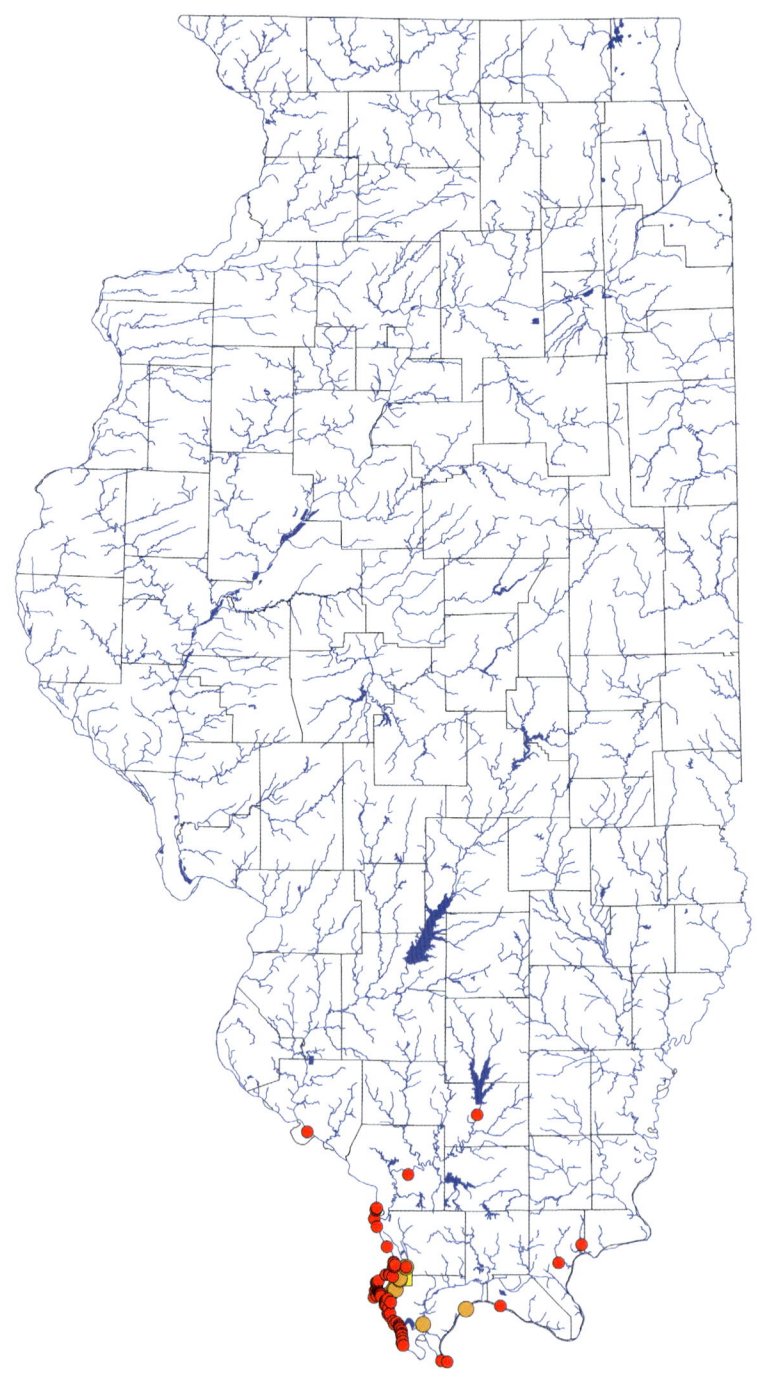

STEELCOLOR SHINER *Cyprinella whipplei* Girard 1856

IDENTIFICATION The Steelcolor Shiner has a fairly deep and compressed body, a terminal mouth, and a pointed snout. The dorsal-fin origin is slightly behind the pelvic-fin origin. The back and upper side are dusky olive, often with a blue sheen, and the scales on the upper side of the body appear diamond shaped. A dark stripe is along the back, and a diffuse dark stripe is along the posterior half of the silver side. A black blotch is on the posterior half of the dorsal fin, and black pigment is on the membranes in the front half of the dorsal fin. Large males have a metallic-blue back and side, a red snout, white-edged yellow fins, and an enlarged and rounded dusky dorsal fin. The lateral line is complete with 37–38 scales, and there are usually 26 scales around the body at the dorsal-fin origin, 9 anal rays, and 1,4-4,1 pharyngeal teeth. To 6¼ in. (16 cm).

SIMILAR SPECIES The Spotfin Shiner, *C. spiloptera*, usually has 8 anal rays and little or no black on the membranes on the front half of the dorsal fin; large males do not have an enlarged dorsal fin. The Red Shiner, *C. lutrensis*, lacks a black blotch on the posterior half of the dorsal fin, a blunter snout, 32–36 scales along the lateral line, and 0,4-4,0 pharyngeal teeth; the male has red fins. The Blacktail Shiner, *C. venusta*, has a large, black spot on the caudal-fin base, usually 8 anal rays, and 28–29 scales around the body at the dorsal-fin origin.

HABITAT The Steelcolor Shiner inhabits rocky and sandy runs and less often pools of creeks and small to medium rivers. It usually is found near riffles or "beside raceways, especially in relatively unmodified, tree-margined streams" (Smith 1979:127).

DISTRIBUTION IN ILLINOIS Forbes and Richardson (1909) found the Steelcolor Shiner in Illinois River tributaries in central, including west-central Illinois, and in the Wabash and Kaskaskia River basins. Smith found the species to be less widespread in the Illinois and Kaskaskia River basins but still common in the Wabash River basin. In the contemporary era, the species has reclaimed some of its former territory in the Illinois and Kaskaskia basins and occurs sporadically across southern Illinois.

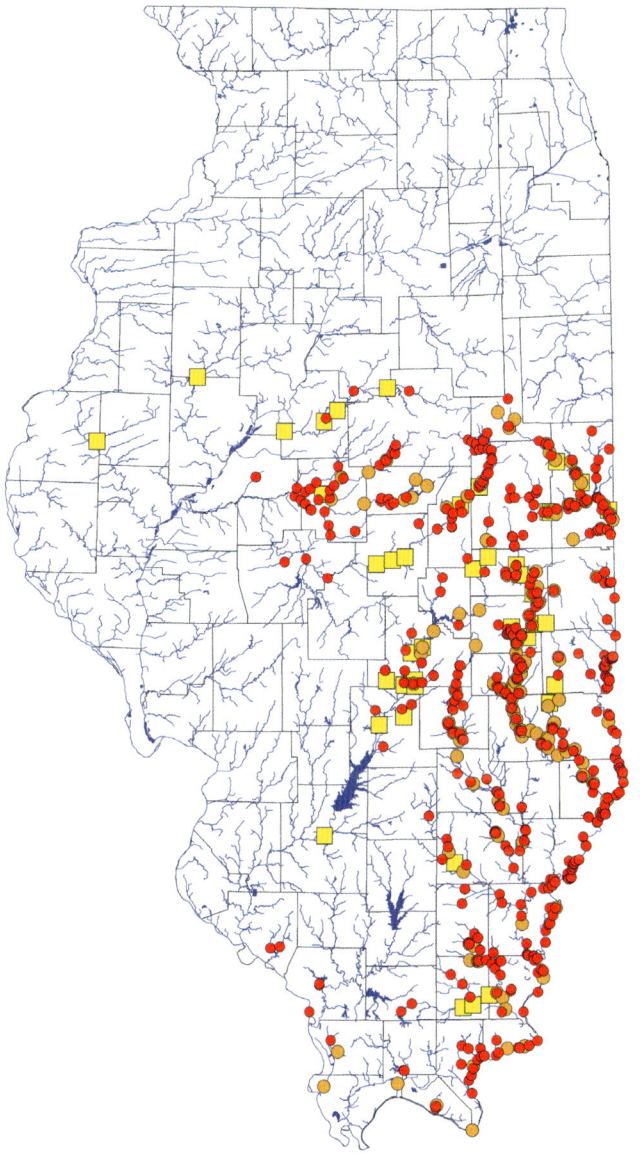

SILVERJAW MINNOW *Ericymba buccata* Cope 1865

IDENTIFICATION The Silverjaw Minnow has large, modified infraorbital and preoperculomandibular canal pores that appear as large, silver-white chambers on the cheek and on the flattened underside of the head. The body is slightly compressed and deepest at the nape. The mouth is subterminal, the eye is high on the head and directed upward, and the dorsal-fin origin is over the pelvic-fin origin. The back and upper side are light tan to olive-yellow with darkly outlined scales, and there is a dark streak along the back that is darkest in front of the dorsal fin. The side is transparent with a silver sheen and sometimes has a dusky stripe. The lateral line is complete with 31–37 scales. There are 8 anal rays and 1,4-4,1 pharyngeal teeth. To 3¾ in. (9.8 cm).

SIMILAR SPECIES No other Illinois minnow has silver-white chambers on the cheek and underside of the head.

HABITAT The Silverjaw Minnow is abundant in shallow sandy riffles and runs of creeks and small to medium rivers.

DISTRIBUTION IN ILLINOIS Both Forbes and Richardson (1909) and Smith (1979) found the Silverjaw Minnow to be widespread in east-central Illinois in the Iroquois, Wabash, and Kaskaskia River basins. The species appeared to be more widespread in the Smith era, having expanded its range in the upper Illinois River basin and perhaps in southeastern and southwestern Illinois. The contemporary range is similar to that of the Smith era except that it is now known from the Saline River basin in southern Illinois.

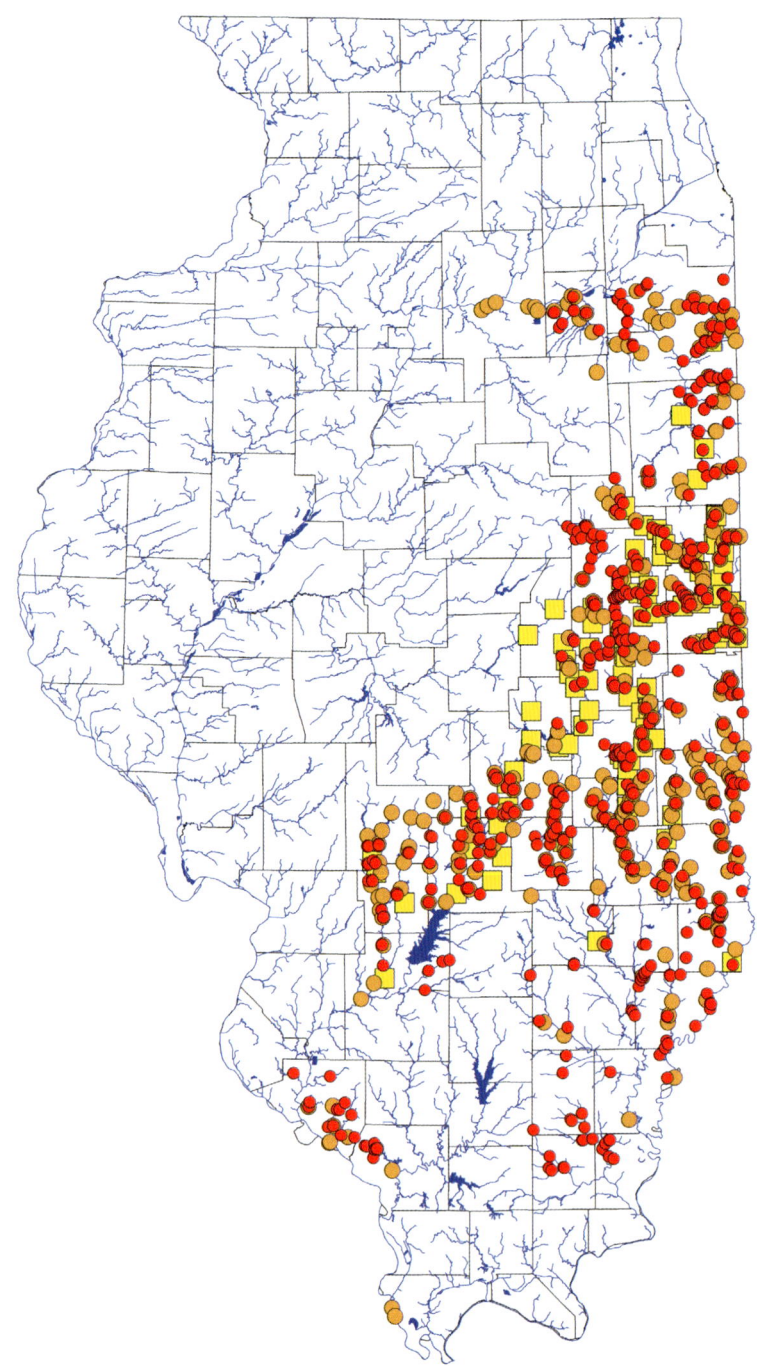

GRAVEL CHUB *Erimystax x-punctatus* (Hubbs & Crowe 1956)

IDENTIFICATION The Gravel Chub has a long, slender body, deepest at the nape and flattened below. The dorsal-fin origin is in front of the pelvic in origin. There is a small barbel at the corner of the subterminal mouth, a long, bulbous snout, a large eye, and large, horizontal pectoral fins. The back and upper side of the body are light olive with darkly outlined scales, the side is silver with a blue sheen, and the breast and belly are silver-white. There is a dusky stripe along the midline of the back, small, black Xs on the back and side of the body, and a dusky caudal spot. The lateral line is complete with 38–45 scales, and there are 7 anal rays and 0,4-4,0 pharyngeal teeth. To 4¼ in. (11 cm).

SIMILAR SPECIES The Shoal Chub, *Macrhybopsis hyostoma*, has distinct black spots rather than Xs on the back and side of the body, a somewhat elliptical and upwardly directed eye, and usually 8 anal rays.

HABITAT The Gravel Chub is found in moderate to swift gravel riffles and runs of small creeks and small to large rivers.

DISTRIBUTION IN ILLINOIS Forbes and Richardson (1909) found the Gravel Chub to be widespread but sporadic, with records from the Rock, Illinois, Kaskaskia, and Embarras River basins. Smith (1979) recorded the species from the Rock River basin and the Mississippi (1 record) and from the Vermilion (1 record) and Wabash Rivers, indicating that the species had disappeared from the Illinois, Kaskaskia, and Embarras River basins. Contemporary records indicate that the species has disappeared from most of its former range in Illinois but is now more widespread in the Rock River basin and also is found in the Vermilion River (Wabash Basin) in Vermilion County.

REMARKS Two subspecies of Gravel Chub occur in Illinois. *Erimystax x. trautmani*, in the Wabash River basin and eastward, usually has 12 scales around the caudal peduncle and is more slender. *Erimystax x. x-punctatus*, west of the Wabash River basin, usually has 16 scales around the caudal peduncle and a stouter body.

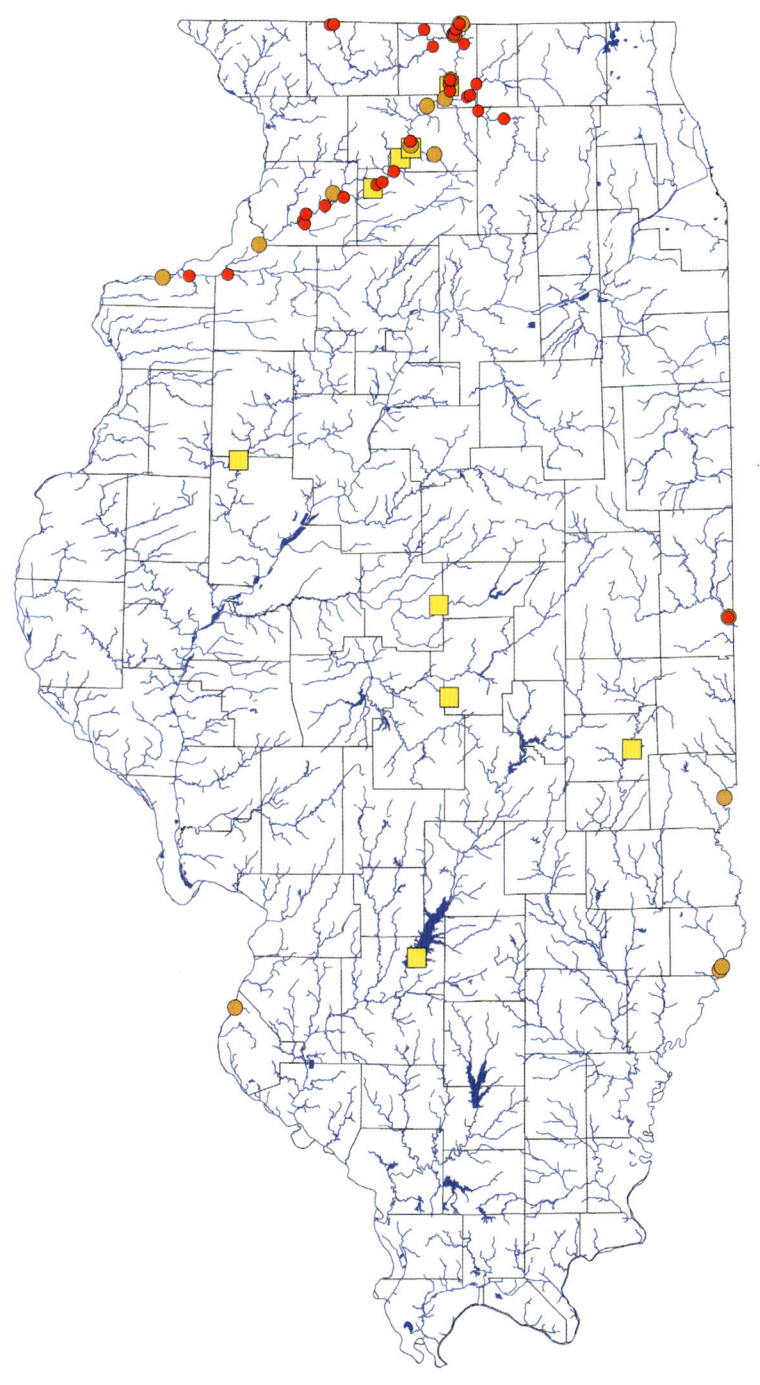

MINNOWS (Leuciscidae)

WESTERN SILVERY MINNOW   *Hybognathus argyritis* Girard 1856

IDENTIFICATION The Western Silvery Minnow has a stout body that is moderately compressed and deepest and widest in front of the dorsal fin. The snout is rounded, and the mouth is small and slightly oblique with the posterior edge of the mouth anterior to the eye. The dorsal fin is pointed with its origin in front of the pelvic-fin origin. The caudal peduncle is relatively deep. The eye is small (about one-fifth of head length). The intestine is long and coiled, and the peritoneum is black. The back and upper side of the body are light brown to yellow-olive with a silver sheen, and there is a wide dusky to yellow-green stripe along the back. The side of the body is dusky silver, and the underside is white. The posterior margin of the basioccipital process is straight or slightly concave. The lateral line is complete and straight with 34–41 scales. There are usually 8 anal rays, 15–16 pectoral rays, and 0,4-4,0 pharyngeal teeth. To 4¾ in. (12 cm).

SIMILAR SPECIES The Mississippi Silvery Minnow, *H. nuchalis*, is extremely similar but has a slightly larger eye, and the posterior margin of the basioccipital process is broad and distinctly concave. The Plains Minnow, *H. placitus*, is extremely similar but has a peglike posterior margin on the basioccipital process.

HABITAT The Western Silvery Minnow is found in pools and backwaters, usually over sand or mud, in small to large rivers. In Illinois it has been found only along the shore of the Mississippi River in shallow sandy and muddy areas with noticeable current.

DISTRIBUTION IN ILLINOIS The Western Silvery Minnow was not recognized as an inhabitant of Illinois by Forbes and Richardson (1909). Smith (1979) found it to be widespread in the Mississippi River mainstem below the mouth of the Missouri River. Although still found in this area, it is dramatically less frequently encountered, with only 2 contemporary-era records.

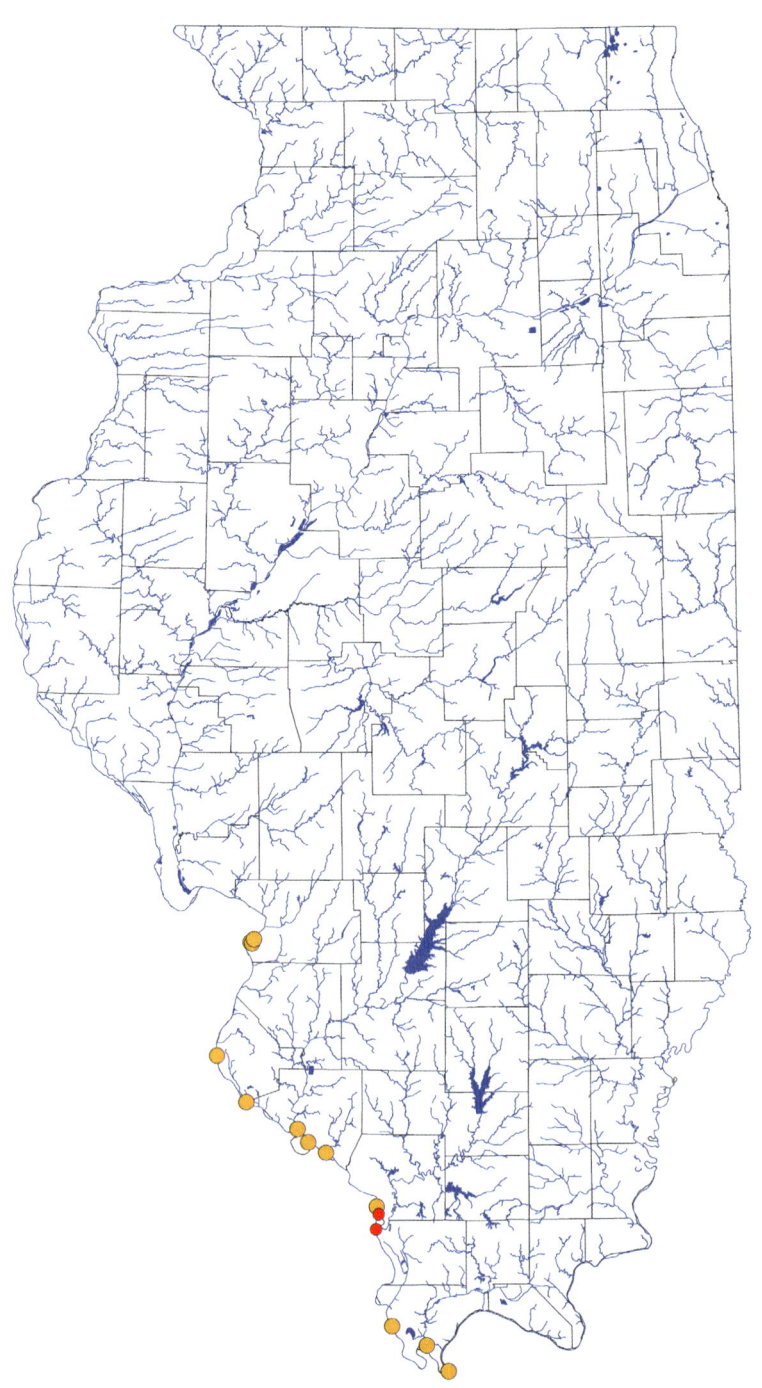

BRASSY MINNOW *Hybognathus hankinsoni* Hubbs 1929

IDENTIFICATION The Brassy Minnow has a stout body that is moderately compressed and deepest and widest in front of the dorsal fin. The snout is rounded, and the mouth is small and slightly subterminal with the posterior edge of the mouth anterior to the eye. The dorsal fin is rounded with its origin in front of the pelvic-fin origin. The caudal peduncle is relatively deep. The eye is small (about one-fourth of head length). The intestine is long and coiled, and the peritoneum is black. The back and upper side of the body are dusky olive, and there is a wide, dusky to yellow-green stripe along the back. Large individuals are brassy yellow with a wide, dusky black stripe along the side and often with thin dark stripes on the upper side parallel to the stripe along the back. The posterior margin of the basioccipital process is straight or barely concave. The lateral line is complete with 36–39 scales. There are usually 8 anal rays, 13–15 pectoral rays, and 0,4-4,0 pharyngeal teeth. To 3¾ in. (9.7 cm).

SIMILAR SPECIES The Mississippi Silvery Minnow, *H. nuchalis*, and Western Silvery Minnow, *H. argyritis*, have a pointed dorsal fin and lack the brassy yellow color and dark stripes on the upper side.

HABITAT The Brassy Minnow occupies gravel-, sand-, and mud-bottomed sluggish pools of clear creeks and small rivers.

DISTRIBUTION IN ILLINOIS The Brassy Minnow was not recognized as an inhabitant of Illinois by Forbes and Richardson (1909). Smith (1979) found it to be fairly widespread in extreme northern Illinois, including along the shoreline of Lake Michigan. Although still showing essentially the same distribution, it appears to be somewhat less widespread. There are no contemporary records for the shoreline of Lake Michigan.

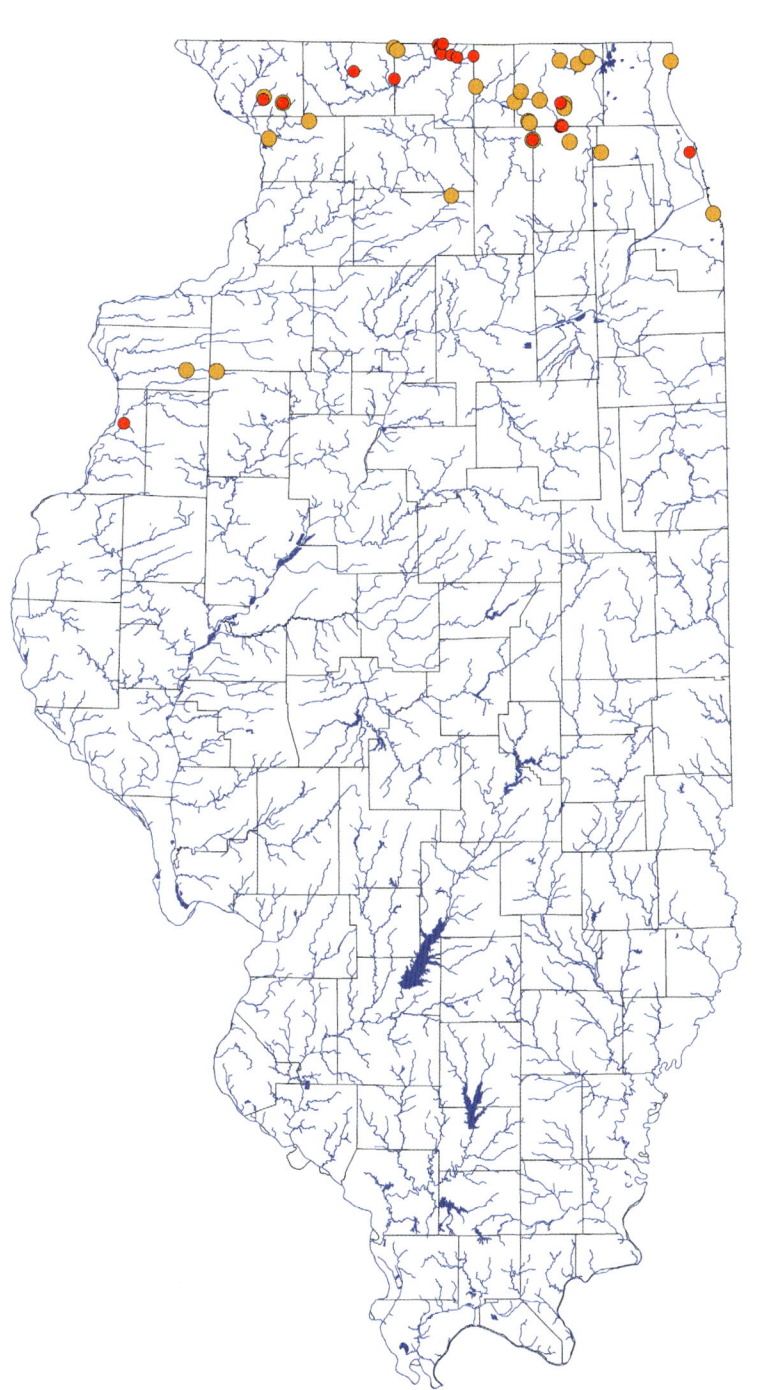

MINNOWS (Leuciscidae)

165

CYPRESS MINNOW *Hybognathus hayi* Jordan 1885

IDENTIFICATION The Cypress Minnow has a compressed body that is deepest and widest near the dorsal-fin origin. The snout is rounded, and the mouth is small and slightly subterminal with the posterior edge of the mouth anterior to the eye. The dorsal fin is pointed with its anterior rays extending past the posterior rays when pressed to the body. The dorsal-fin origin is in front of the pelvic-fin origin. The eye is moderately large (about one-third of head length). The intestine is long and coiled, and the peritoneum is black. The back and upper side of the body are dusky olive with a silver sheen, and there is a thin, dusky to yellow-green stripe along the back. The scales on the back and upper side are thinly outlined with black and appear diamond shaped. The side is silver and sometimes overlaid by a dusky stripe usually best developed on the caudal peduncle. The posterior margin of the basioccipital process is broad and straight to slightly concave. The lateral line is complete with 34–41 lateral scales. There are usually 8 anal rays, 14–16 pectoral rays, and 0,4-4,0 pharyngeal teeth. To 4½ in. (12 cm).

SIMILAR SPECIES Other species of *Hybognathus* lack scales that appear diamond shaped, have a smaller eye, and a more slender, less compressed body. Species of *Cyprinella* have scales that appear diamond shaped but have a less pointed dorsal fin with its anterior rays not extending past the posterior rays.

HABITAT The Cypress Minnow is found in swamps, oxbows, and backwaters and pools of sluggish streams, usually over mud and near detritus.

DISTRIBUTION IN ILLINOIS In the Forbes and Richardson era, the Cypress Minnow was recorded only from the lower Cache River basin in extreme southern Illinois. In the early 1940s the species was collected in large numbers at 3 localities in the Big Muddy River system. Other Smith-era collections were taken from the LaRue–Pine Hills Research Natural Area in Union County and Horseshoe Lake in Alexander County. Beginning in the mid-1980s, adults and juveniles have been found in a 5 mi. (8 km) reach of the middle Cache River and occasionally at Horseshoe Lake. Although Smith (1979) considered the species to be extirpated, it seems to be as frequently encountered in Illinois as it ever was, with the exception of the Big Muddy River basin, where it has not been recorded during the contemporary era.

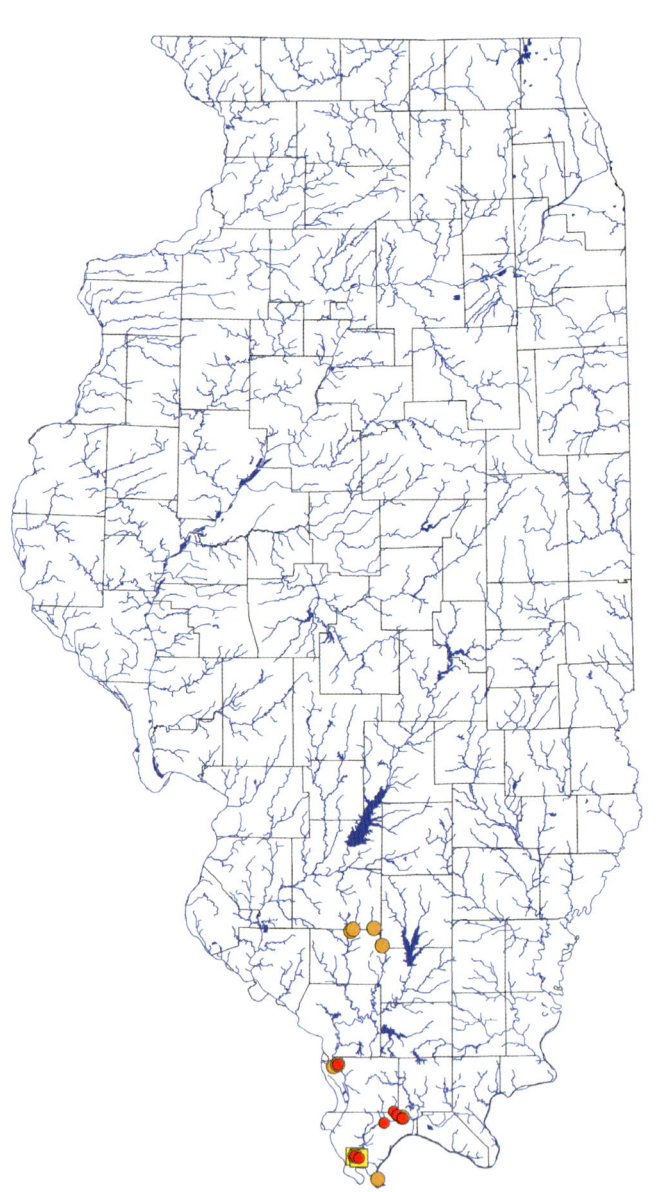

MISSISSIPPI SILVERY MINNOW
Hybognathus nuchalis Agassiz 1855

IDENTIFICATION The Mississippi Silvery Minnow has a stout body that is moderately compressed and deepest and widest in front of the dorsal fin. The snout is rounded, and the mouth is small and slightly oblique with the posterior edge of the mouth anterior to the eye. The dorsal fin is pointed with its origin in front of the pelvic-fin origin. The caudal peduncle is relatively deep. The eye is small (about one-fourth of head length). The intestine is long and coiled, and the peritoneum is black. The back and upper side of the body are light brown to yellow-olive with a silver sheen, and there is a wide, dusky to yellow-green stripe along the back. The side of the body is dusky silver, and the underside is white. The posterior margin of the basioccipital process is broad and distinctly concave. The lateral line is complete with 34–41 scales. There are usually 8 anal rays, 15–16 pectoral rays, and 0,4-4,0 pharyngeal teeth. To 7 in. (18 cm).

SIMILAR SPECIES The Western Silvery Minnow, *H. argyritis*, is extremely similar but has a slightly smaller eye, and the posterior margin of the basioccipital process is straight or slightly concave. The Plains Minnow, *H. placitus*, is extremely similar but has a slightly smaller eye and a peglike posterior margin on the basioccipital process.

HABITAT The Mississippi Silvery Minnow lives over sand and mud in pools and backwaters of sluggish creeks and small to large rivers.

DISTRIBUTION IN ILLINOIS Forbes and Richardson (1909) found the Mississippi Silvery Minnow to be nearly statewide in distribution but rare in northeastern Illinois and especially widespread in central, southern, and southeastern Illinois. Smith (1979) found the species to be common in the Illinois, Kaskaskia, and Wabash River basins but much less widespread elsewhere than it had been in the Forbes and Richardson era. Contemporary data show a dramatic shift in distribution with the species remaining widespread in the Wabash River basin and more widespread in the Mississippi River and its small tributaries than it had been in the Smith era but near extirpation in the Illinois and Kaskaskia River basins.

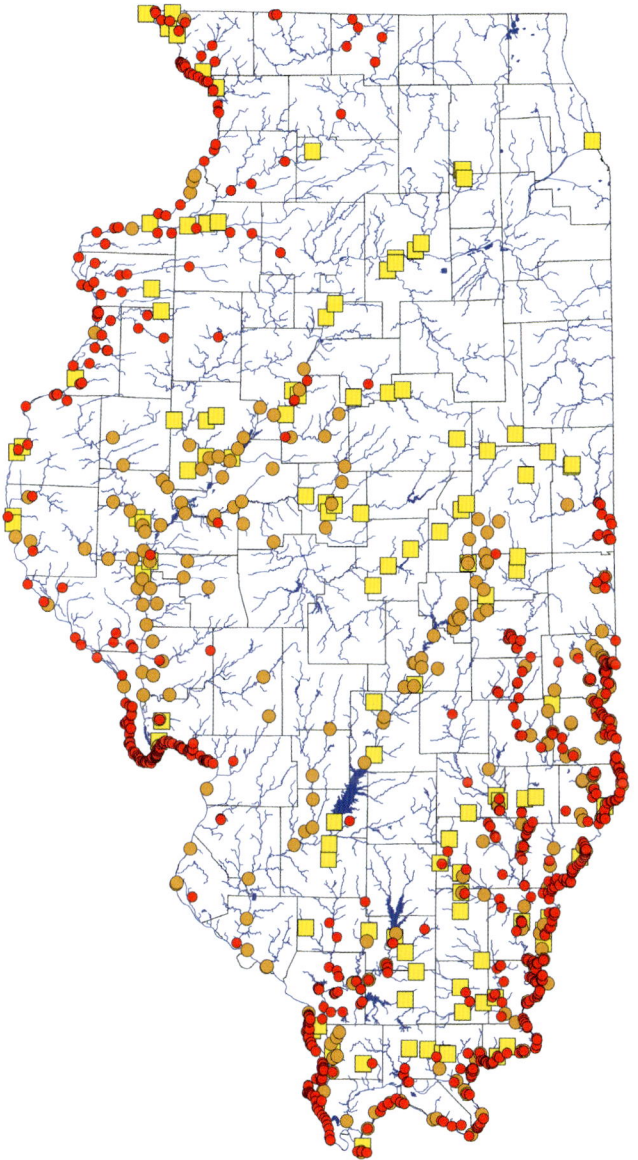

PLAINS MINNOW *Hybognathus placitus* Girard 1856

IDENTIFICATION The Plains Minnow has a somewhat cylindrical body, rounded in cross section and deepest and widest in front of the dorsal fin. The snout is rounded, the underside of the head is flattened, and the mouth is small and slightly oblique with its posterior edge anterior to the eye. The dorsal fin is pointed with its origin in front of the pelvic-fin origin. The caudal peduncle is relatively deep. The eye is small (about one-fifth of head length). The intestine is long and coiled, and the peritoneum is black. The back and upper side of the body are light brown to olive with a silver sheen, and there is a wide, dusky to yellow-green stripe along the back; the side of the body is dusky silver, and the underside is white. The posterior margin of the basioccipital process is peg-like. The lateral line is complete with 34–41 scales. There are usually 8 anal rays, 16–17 pectoral rays, and 0,4-4,0 pharyngeal teeth. To 5 in. (13 cm).

SIMILAR SPECIES The Mississippi Silvery Minnow, *H. nuchalis*, is extremely similar but has a slightly larger eye, and the posterior margin of the basioccipital process is broad and distinctly concave. The Western Silvery Minnow, *H. argyritis*, is extremely similar but the posterior margin of the basioccipital process is straight or slightly concave.

HABITAT The Plains Minnow is found in pools and backwaters, usually over sand or mud, in creeks and small to large rivers. In Illinois it has been found only along the shore of the Mississippi River in shallow, sandy areas with noticeable current.

DISTRIBUTION IN ILLINOIS The Plains Minnow was not recognized as an inhabitant of Illinois by Forbes and Richardson (1909). Smith (1979) found it to be widespread in the Mississippi River mainstem below the mouth of the Missouri River. Although still found in this area, it is dramatically less frequently encountered, with only 3 contemporary records, 2 of which are slightly north of the mouth of the Missouri River.

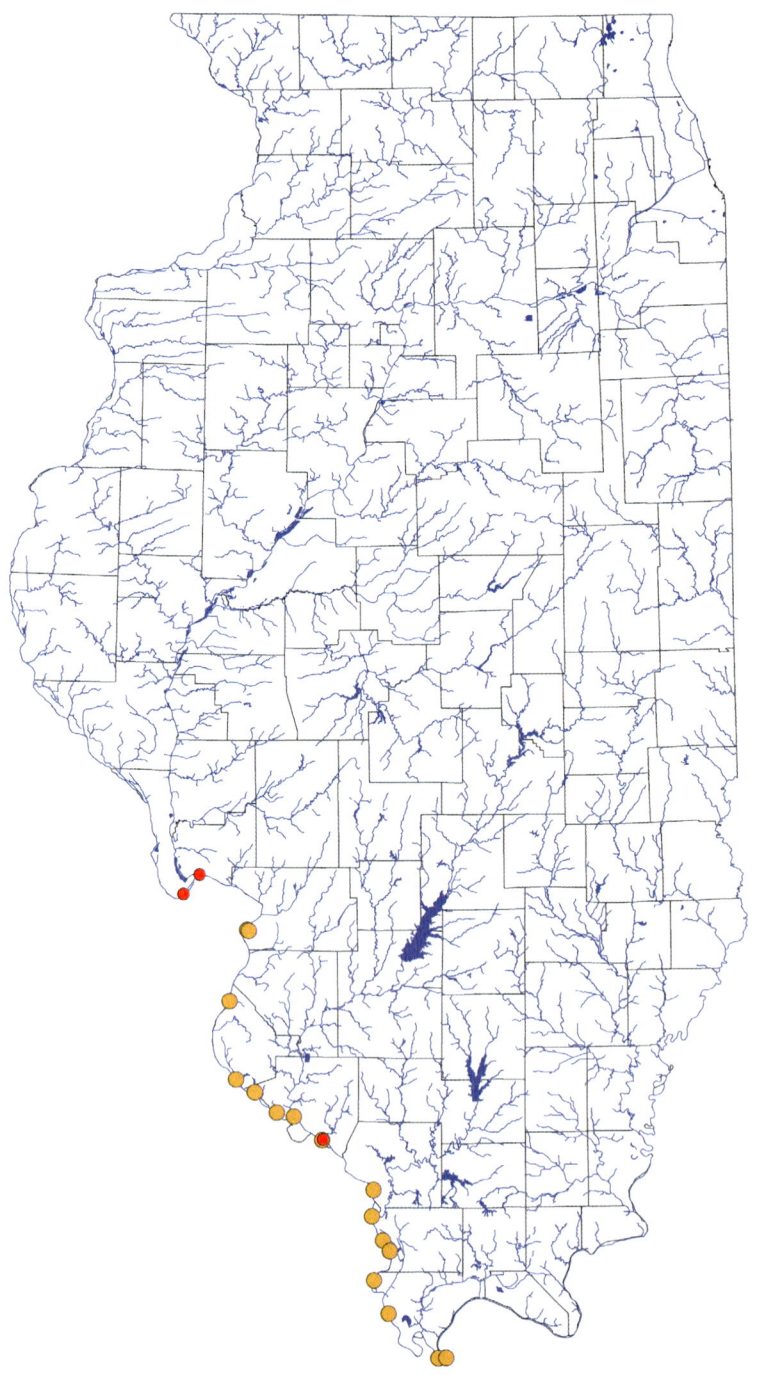

BIGEYE CHUB *Hybopsis amblops* (Rafinesque 1820)

IDENTIFICATION The Bigeye Chub is slender with a slightly compressed body, a rounded snout that projects well beyond the upper lip, and a small mouth with a small barbel in each corner. The large eye is about equal in length to the snout. The dorsal-fin origin is over or slightly behind the pelvic-fin origin. The back is light yellow-olive with a dark streak along the midline in front of the dorsal fin. The scales on the back and side are dark-edged, producing wavy lines. There is a black stripe along the silver side and around the snout, and often a yellow streak is above the black stripe, and a small, black spot is at the base of the caudal fin. Large males have many small tubercles on the head. The lateral line is complete with 33–38 lateral scales. There are 8 anal rays and 1,4-4,1 pharyngeal teeth. To 3½ in. (9 cm).

SIMILAR SPECIES The Pallid Shiner, *H. amnis*, has no barbel at the corner of the mouth, a more arched and compressed body, and the dorsal-fin origin is over or slightly in front of the pelvic-fin origin; males have fewer but larger tubercles on the head.

HABITAT The Bigeye Chub lives over sand and gravel in clean, clear creeks and small to medium rivers. It is most often found in flowing pools near riffles and emergent vegetation.

DISTRIBUTION IN ILLINOIS Misidentifications led to errors in the distributions reported by Smith (1979) for the Bigeye Chub and Pallid Shiner (Warren & Burr 1988). With errors corrected, the species has essentially the same contemporary distribution as it had in past eras, except it seems to have disappeared from the Kaskaskia River basin and Big Creek in Hardin County.

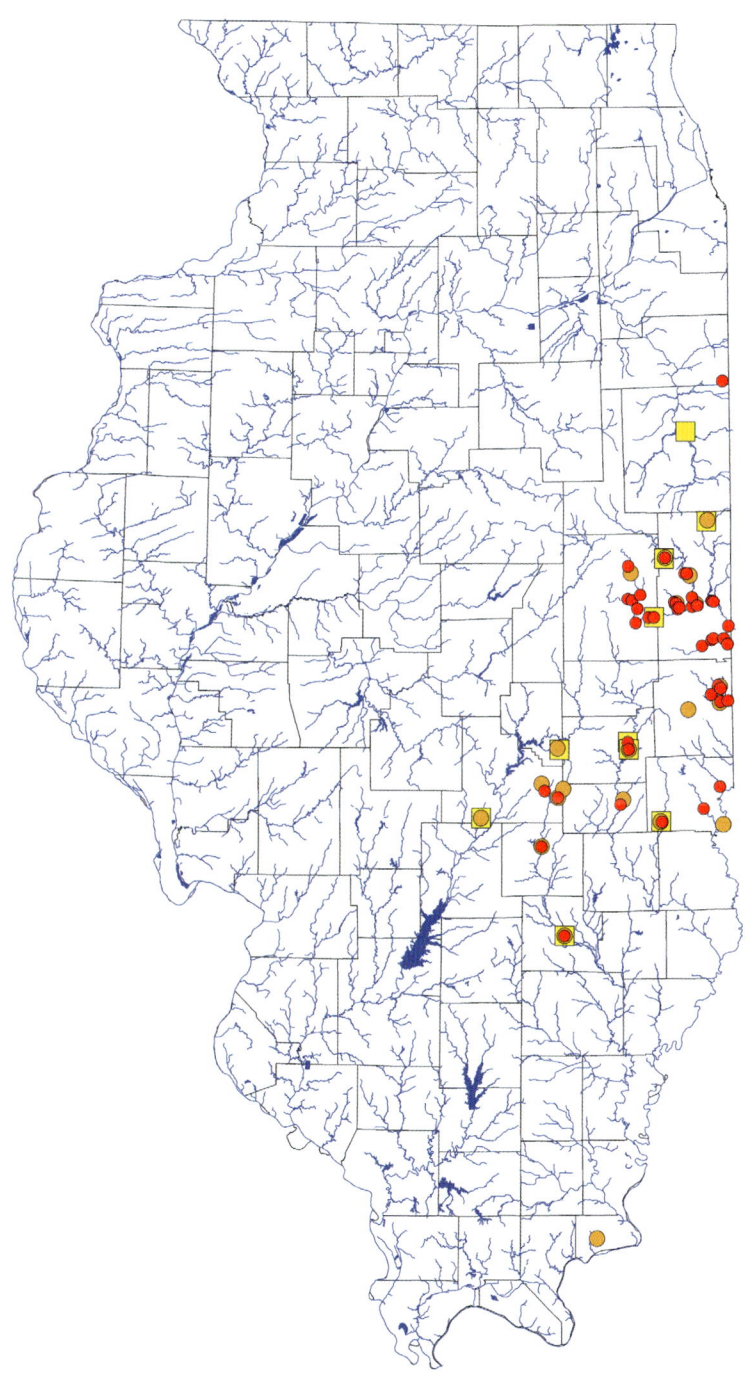

PALLID SHINER *Hybopsis amnis* (Hubbs & Greene 1951)

IDENTIFICATION The Pallid Shiner has a compressed body, a rounded snout that projects well beyond the upper lip, and a small mouth with no barbels. The back is arched at the dorsal-fin origin, which is over or slightly in front of the pelvic-fin origin. The large eye is about equal to the length of the snout. The back and upper side are straw yellow with dark-edged scales. There is a dusky to black stripe (absent in turbid water) along the silver side and around the snout. Large males have tubercles concentrated on the lower half of the head. The lateral line is complete with 33–38 scales. There are 8 anal rays and 1,4-4,1 pharyngeal teeth. To 3¼ in. (8.4 cm).

SIMILAR SPECIES The Bigeye Chub, *H. amblops*, has a less arched and compressed body, a barbel at the corner of the mouth, and the dorsal-fin origin over or slightly behind the pelvic-fin origin; males have more, smaller tubercles on the head. The Blacknose Shiner, *Notropis heterolepis*, has a less arched body, a smaller eye, black crescents in the black stripe along the side of the body, the dorsal-fin origin slightly behind the pelvic-fin origin, and 0,4-4,0 pharyngeal teeth.

HABITAT The Pallid Shiner is found in sandy and silty pools with moderate flow in medium to large rivers.

DISTRIBUTION IN ILLINOIS Misidentifications led to errors in the distributions reported by Smith (1979) for the Pallid Shiner and Bigeye Chub (Warren & Burr 1988). With errors corrected, it is apparent that the Pallid Shiner has suffered one of the greatest declines of any fish in Illinois. In the Forbes and Richardson era, it occurred sporadically throughout Illinois, with records in the Mississippi, upper Illinois, Kaskaskia, Big Muddy, and Saline River basins. In the Smith era, records were found for those basins, although they were highly dispersed. It had disappeared from the Saline River basin in the Ohio River basin but was found in the Cache River (also Ohio River basin). The species appears to be gone from all areas of the state except in the Mississippi River mainstem and tributaries in extreme northern Illinois, and in the lower Kankakee River and adjacent reaches of the Des Plaines and Illinois Rivers in northeastern Illinois. Ironically, the extant populations appear to be reasonably large.

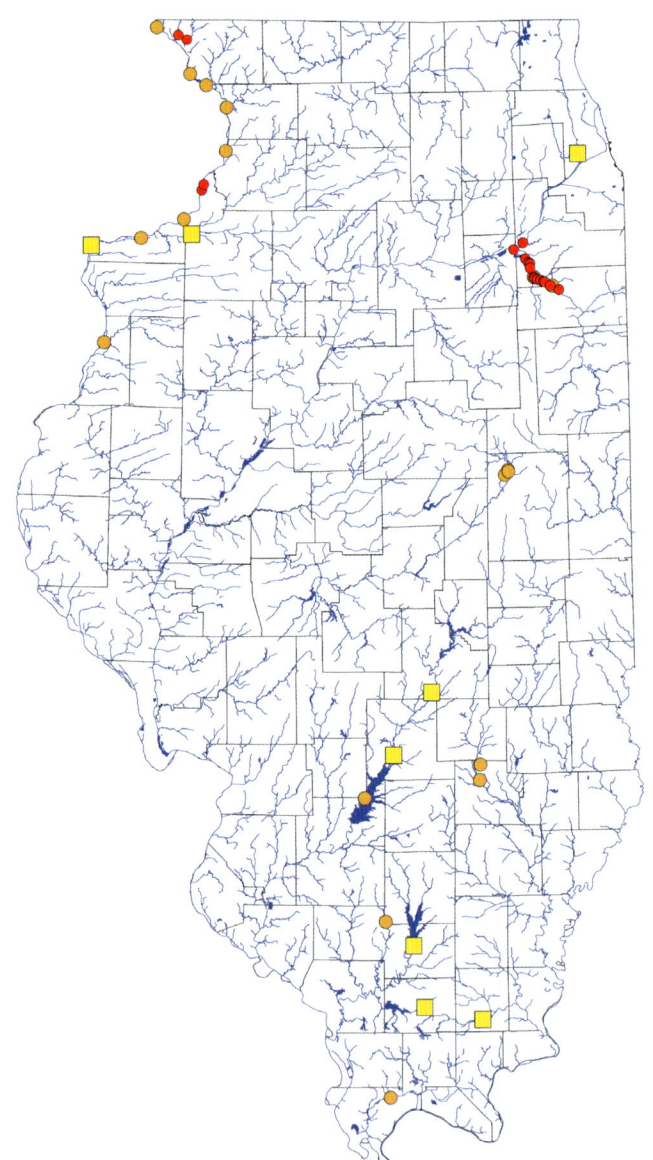

STRIPED SHINER *Luxilus chrysocephalus* Rafinesque 1820

IDENTIFICATION The Striped Shiner has a deep and strongly compressed body with large scales; scales on the front half of the side of the body are much deeper than wide. The mouth is large, terminal, and oblique. The dorsal-fin origin is over or slightly behind the pelvic-fin origin. The back and upper side of the body are light to dark olive, and there is a dark stripe along the middle of the back. When viewed from above, 3 dark stripes on one upper side of the body meet the 3 dark stripes from the other side of the body behind the dorsal fin meet to form large Vs. The side is silver-bronze with a dusky to black bar behind the gill cover; dark crescents are on the side of large individuals. Large males and sometimes females have pink or red on the body and fins. Large males have hooked tubercles on the snout during the breeding season. The lateral line is complete with 36–46 scales, and there are 12–19, usually 13–16, predorsal scales, usually 24–29 scales around the body at the dorsal-fin origin, 9 anal rays, and 2,4-4,2 pharyngeal teeth. To 7¼ in. (18 cm).

SIMILAR SPECIES The Common Shiner, *L. cornutus*, has 1 or 2 dark stripes (often faint) on the upper side that are parallel to the stripe along the back and do not form Vs, 16–30, usually 18–24, predorsal scales; and usually 30–35 scales around the body at the dorsal-fin origin.

HABITAT The Striped Shiner lives in pools and runs over gravel or mixed sand and gravel in creeks and small to medium rivers.

DISTRIBUTION IN ILLINOIS The Striped Shiner has essentially the same distribution as it had in previous surveys except that it has disappeared from several streams in west-central Illinois and from the Cache River basin in the Shawnee Hills, and it is now more widespread in the most southern reaches of the Sangamon River basin.

REMARKS The Striped Shiner and Common Shiner are closely related, extremely similar morphologically, and hybridize frequently where their ranges overlap (Gilbert 1961). Gilbert (1964) noted that the Striped Shiner is more tolerant of warmer and more turbid water and has supplanted the Common Shiner in agricultural areas. Smith (1979) was of the opinion that the Striped Shiner had moved north into the range of the Common Shiner. There is little evidence today that the presence of the Striped Shiner within the range of the Common Shiner has changed except in Lake and McHenry Counties where the Striped Shiner was absent but has appeared in recent years.

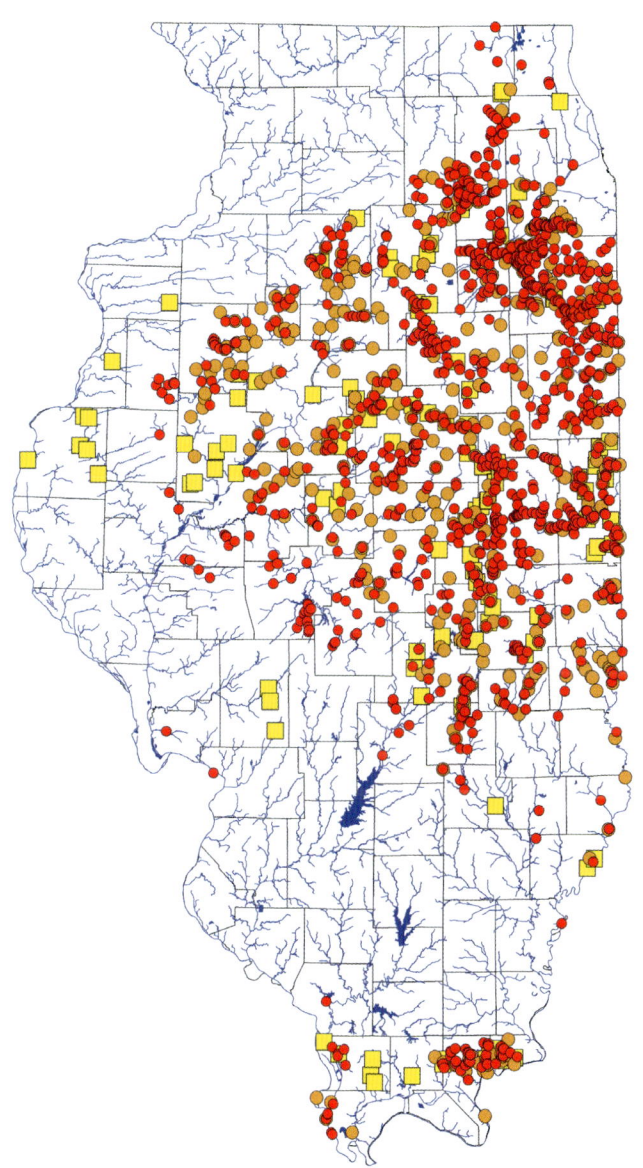

MINNOWS (Leuciscidae)

COMMON SHINER *Luxilus cornutus* (Mitchill 1817)

IDENTIFICATION The Common Shiner has a deep and strongly compressed body with large scales; scales on the front half of the side of the body are much deeper than wide. The mouth is large, terminal, and oblique. The dorsal-fin origin is over or slightly behind the pelvic-fin origin. The back and upper side of the body are light to dark olive, and there is a dark stripe along the middle of the back. One or 2 dark stripes on the upper side are parallel to the stripe along the back. The side is silver-bronze, there is a dusky to black bar behind the gill cover, and dark crescents are on the side of large individuals. Large males and sometimes females have pink or red on the body and fins. Large males have hooked tubercles on the snout during the breeding season. The lateral line is complete with 36–46 scales, and there are 16–30, usually 18–24, predorsal scales, usually 30–35 scales around the body at the dorsal-fin origin, 9 anal rays, and 2,4-4,2 pharyngeal teeth. To 7 in. (18 cm).

SIMILAR SPECIES The Striped Shiner, *L. chrysocephalus*, has 3 dark stripes on one side of the body that meet 3 dark stripes from the other side behind the dorsal fin to form Vs when viewed from above, and 12–19, usually 13–16, predorsal scales and usually 24–29 scales around the body at the dorsal-fin origin.

HABITAT The Common Shiner lives in rocky pools and runs in creeks and small to medium rivers.

DISTRIBUTION IN ILLINOIS The Common Shiner is restricted to the northern quarter of Illinois. It has essentially the same distribution as it had in previous eras, although it may be more widespread in the southwestern part of its range than previously.

REMARKS The Striped Shiner and Common Shiner are closely related, extremely similar morphologically, and hybridize frequently where their ranges overlap (Gilbert 1961). Gilbert (1964) noted that the Striped Shiner is more tolerant of warmer and more turbid water and has supplanted the Common Shiner in agricultural areas. Smith (1979) was of the opinion that the Striped Shiner had moved north into the range of the Common Shiner. There is little evidence today that the presence of the Striped Shiner within the range of the Common Shiner has changed except in Lake and McHenry Counties where the Striped Shiner was absent but has appeared in recent years.

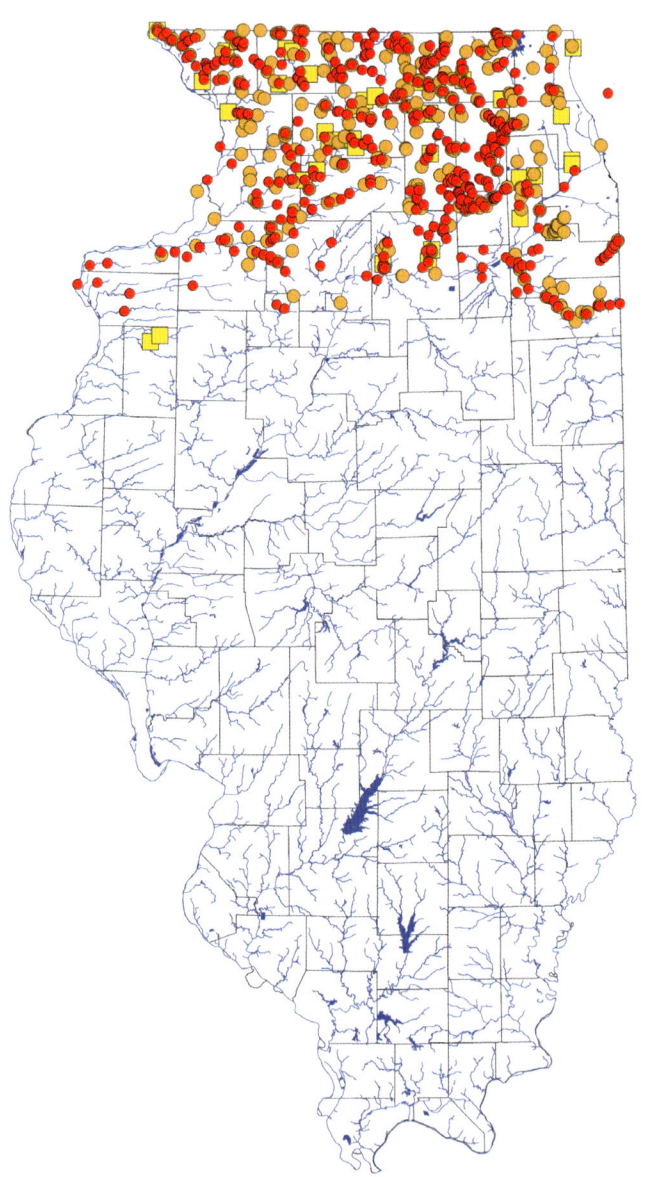

BLEEDING SHINER *Luxilus zonatus* (Putnam 1863)

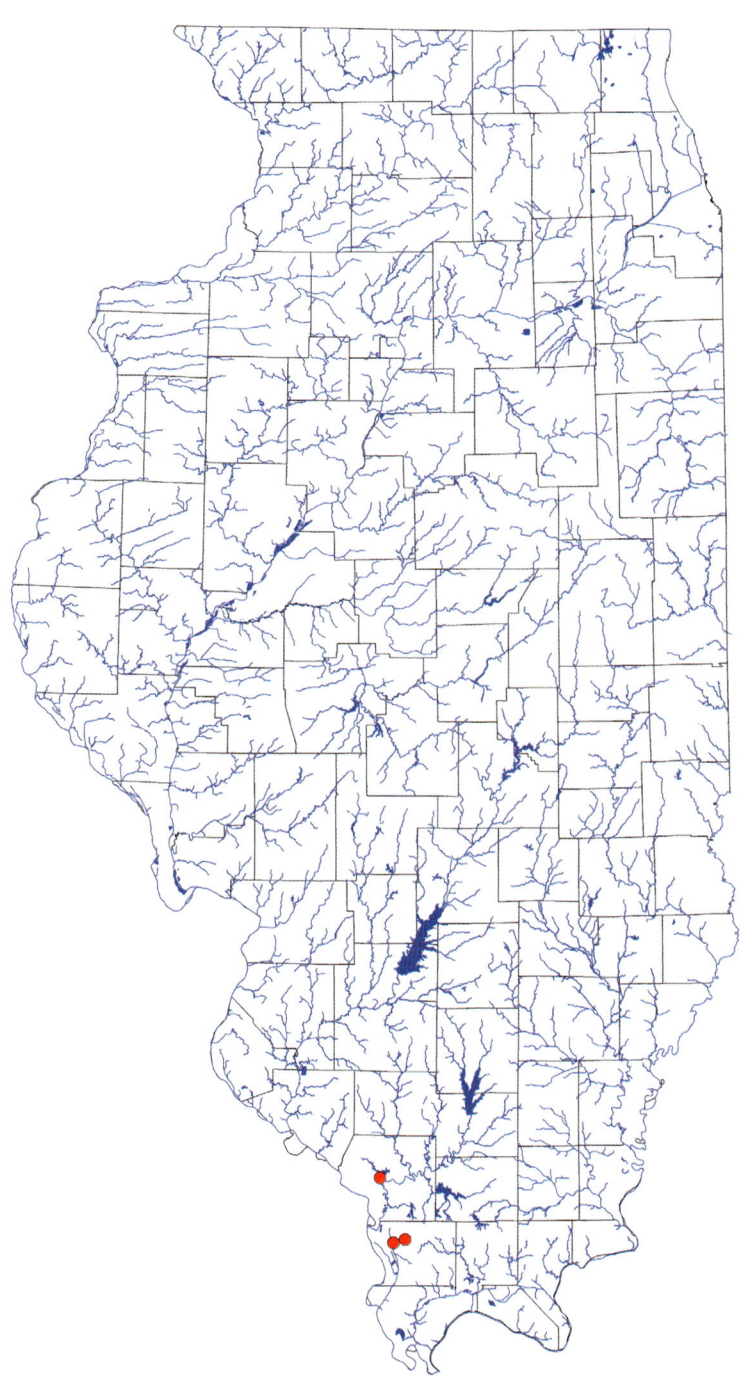

IDENTIFICATION The Bleeding Shiner has a deep and strongly compressed body with large scales; scales on the front half of the side of the body are much deeper than wide. The mouth is large, terminal, and oblique. The dorsal-fin origin is over or slightly behind the pelvic-fin origin. The back and upper side of the body are light to dark olive, and there is a dark stripe along the middle of the back. Three dark stripes on one upper side of the body meet the 3 dark stripes from the other side of the body behind the dorsal fin to form large Vs when viewed from above. The side is silver bronze, and there is a bold, black bar behind the gill cover and a black stripe along the side and around the snout. Large individuals have a red head and fins, brightest on the breeding males, which also have hooked tubercles on the snout. The lateral line is complete with 36–46 scales, and there are usually 26 scales around the body at the dorsal-fin origin, 9 anal rays, and 2,4-4,2 pharyngeal teeth. To 5 in. (13 cm).

SIMILAR SPECIES The Striped Shiner, *L. chrysocephalus*, and Common Shiner, *L. cornutus*, lack a black stripe along the side of the body and around the snout and lack bright red on the body and fins.

HABITAT The Bleeding Shiner inhabits rocky runs, riffles, and flowing pools of clear, fast creeks and small to medium rivers.

DISTRIBUTION IN ILLINOIS The Bleeding Shiner was first reported from Illinois by Hiland and Poly (2000) when a single individual was captured in July 1999 from Kincaid Creek in the Big Muddy River basin in Jackson County. Since 2010 it has been captured several times in the Clear Creek system in Union County, where it is established in Hutchins and Green Creeks.

REMARKS It is assumed that this species entered Illinois as a "bait bucket" transfer from the nearby Missouri Ozarks.

MINNOWS (Leuciscidae)

SCARLET SHINER *Lythrurus fasciolaris* (Gilbert 1891)

IDENTIFICATION The Scarlet Shiner has a deep and compressed body and a terminal, oblique mouth. The dorsal-fin origin is behind the pelvic-fin origin. Scales on the nape are extremely small. The back and upper side of the body are light olive, there is a dusky stripe along the back, and a black blotch is at the dorsal-fin origin. Large males are steel-blue with dusky bars over the back. The breeding male has a keeled nape, dark blue-gray bars on the back and upper side of the body, and bright red or bright orange on the fins and lower side. The lateral line is complete with 38–53 scales, and there are 9–12, usually 10, anal rays, and 2,4-4,2 pharyngeal teeth. To 3½ in. (8.6 cm).

SIMILAR SPECIES The Redfin Shiner, *L. umbratilis*, has herringbone lines on the upper side of large males and lacks dusky bars on the back. The Ribbon Shiner, *L. fumeus*, is more slender, lacks a dark blotch at the dorsal-fin origin, has yellow fins on large males, and usually 11–12 anal rays.

HABITAT The Scarlet Shiner is found in rocky and sandy runs and flowing pools of clear headwaters, creeks, and small rivers.

DISTRIBUTION IN ILLINOIS During the Forbes and Richardson era, 6 specimens of the Scarlet Shiner were taken in 1900 in a spring branch of Big Creek, Hardin County. These individuals were captured with the Redfin Shiner, *L. umbratilis*, the only species of *Lythrurus* found in the Big Creek system today. It seems likely that the Scarlet Shiner was competitively displaced by the Redfin Shiner, with humans serendipitously recording its final presence in Illinois.

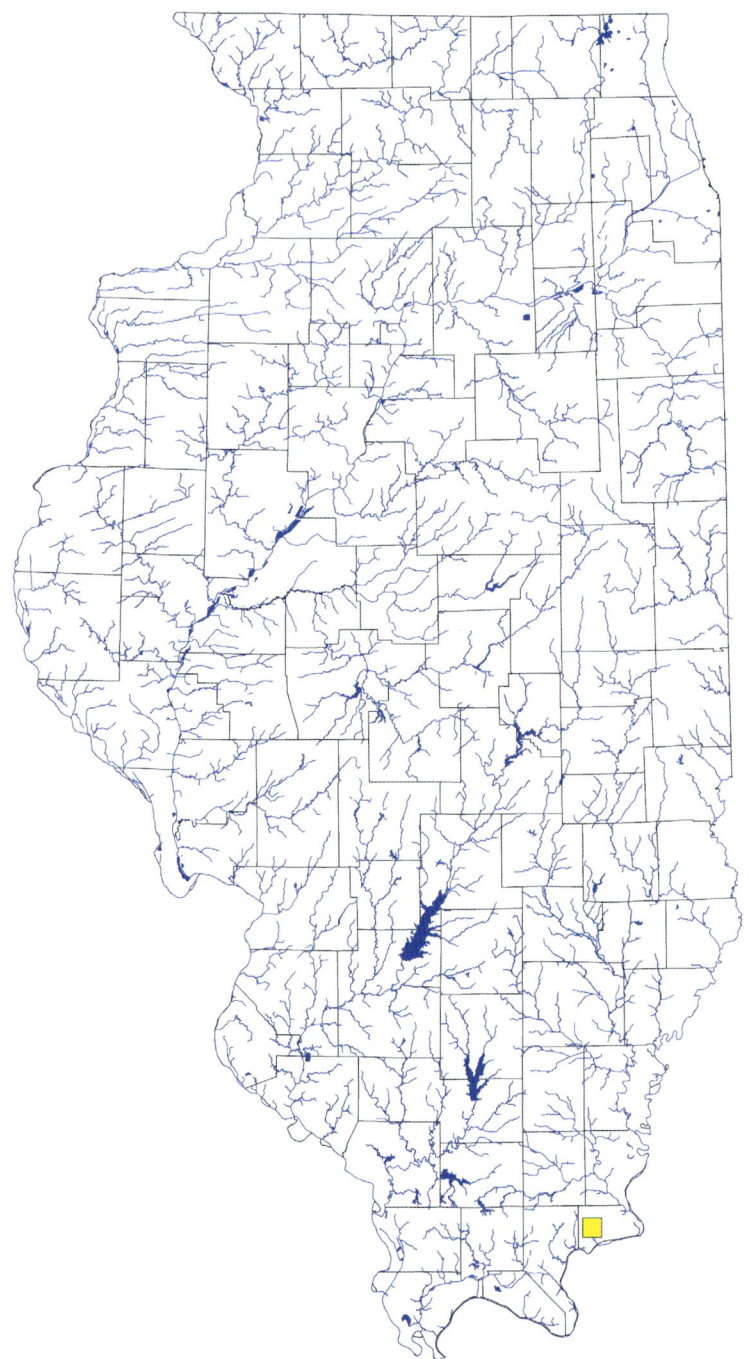

RIBBON SHINER *Lythrurus fumeus* (Evermann 1892)

IDENTIFICATION The Ribbon Shiner has a slender and compressed body and a terminal, oblique mouth. The dorsal-fin origin is behind the pelvic-fin origin. Scales on the nape are small but outlined in black, which makes them readily apparent. The back and upper side of the body are light olive, and there is a dusky stripe along the back. There is a silver-black stripe along the side (often faint but darkest at rear) and around the snout; the lips and chin are dusky. Large males have yellow eyes and fins. The lateral line is complete with 35–45 scales, and there are usually 11–12 anal rays and 2,4-4,2 pharyngeal teeth. To 2¾ in. (7 cm).

SIMILAR SPECIES The Redfin Shiner, *L. umbratilis*, has a deeper body, a dark blotch at the dorsal-fin origin, scales on the nape not outlined in black, herringbone lines on the upper side, and red fins on large males. The Scarlet Shiner, *L. fasciolaris*, usually has 9–10 anal rays and dusky bars over the back.

HABITAT The Ribbon Shiner is found in quiet, often turbid, mud- or sand-bottomed pools of headwaters, creeks, and small rivers.

DISTRIBUTION IN ILLINOIS Forbes and Richardson (1909) recorded the Ribbon Shiner from only the Kaskaskia, Skillet Fork, and Saline River basins. Smith (1979) found it to be much more widespread in the southern half of Illinois and attributed the increase to the species's tolerance for turbid water and silty substrates. The species has essentially the same distribution as in the Smith era, although it appears to be slightly less widespread in the northern part of its range and more widespread in lower Wabash River tributaries.

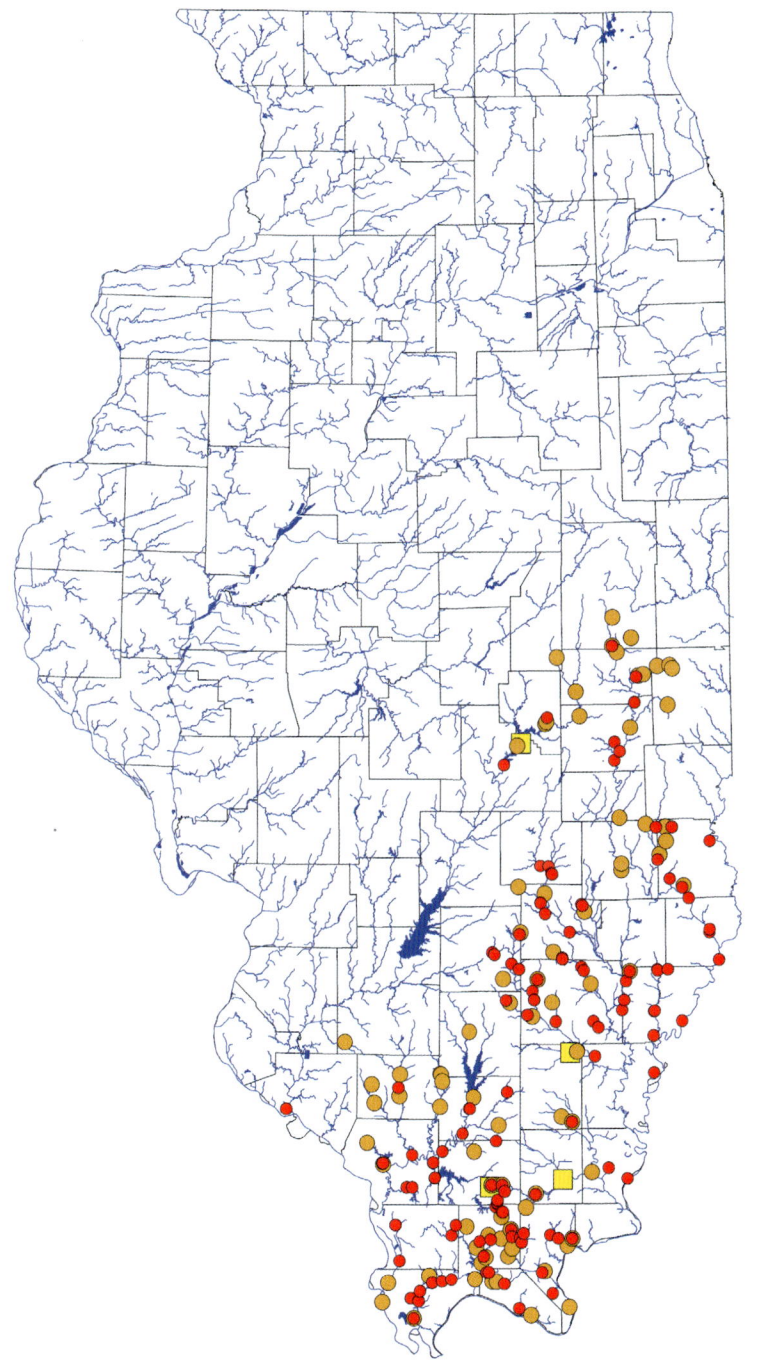

MINNOWS (Leuciscidae)

175

REDFIN SHINER *Lythrurus umbratilis* (Girard 1856)

IDENTIFICATION The Redfin Shiner has a deep and compressed body and a terminal, oblique mouth. The dorsal-fin origin is behind the pelvic-fin origin. Scales on the nape are extremely small. The back and upper side of the body are light olive, there is a dusky stripe along the back, and a dusky black blotch is at the dorsal-fin origin. There is a silver-black stripe along the side (often faint but darkest at rear). Large individuals have distinctive herringbone lines on the upper side. Large males have a light to bright blue head and body, black membranes on red fins, and often a large, dark blotch on the side. The lateral line is complete with 37–56 scales, and there are 10–11 anal rays and 2,4-4,2 pharyngeal teeth. To 3½ in. (8.6 cm).

SIMILAR SPECIES The Scarlet Shiner, *L. fasciolaris*, has dusky bars on the back of large males and lacks herringbone lines on the upper side of the body. The Ribbon Shiner, *L. fumeus*, is more slender, has scales on the nape outlined in black, and lacks the dark blotch at the dorsal-fin origin, herringbone lines, and red fins.

HABITAT The Redfin Shiner inhabits quiet to flowing mud- and sand-bottomed pools (often turbid) of headwaters, creeks, and small to medium rivers.

DISTRIBUTION IN ILLINOIS As in the earlier eras (Forbes & Richardson 1909, Smith 1979), the Redfin Shiner is statewide in distribution but much more widespread in eastern and southern Illinois than in western Illinois. Although still present in the Des Plaines and Fox Rivers, it is less frequently encountered in extreme northeastern Illinois than historically.

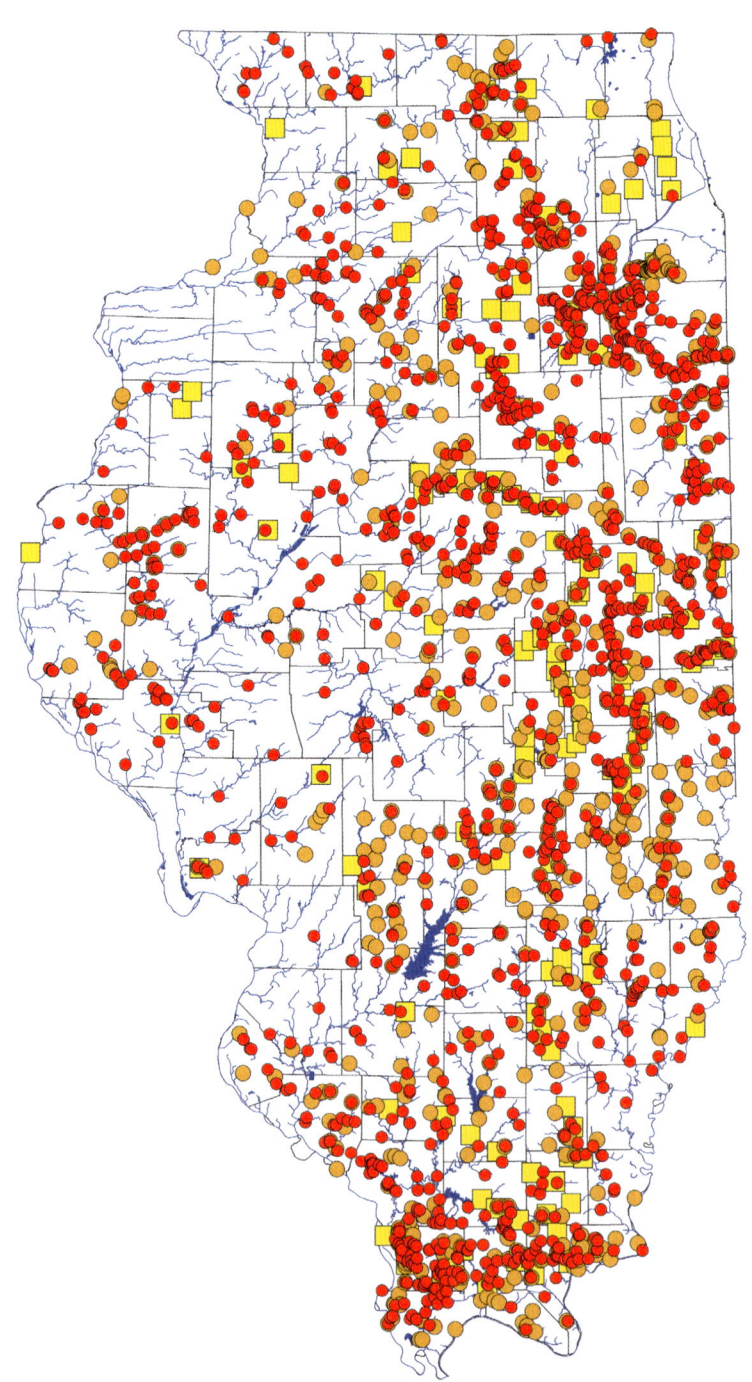

STURGEON CHUB   *Macrhybopsis gelida* (Girard 1856)

IDENTIFICATION The Sturgeon Chub has a slender body, deepest at the nape and strongly tapering to a narrow caudal peduncle. The eye is small; the snout is long and fleshy and projects well beyond the upper lip. A long barbel is in the corner of the subterminal mouth, and large, sensory papillae are on the underside of the head. The fins have straight or only slightly concave edges, and the first dorsal ray does not extend beyond the last ray when the fin is depressed. The dorsal-fin origin is over or slightly behind the pelvic-fin origin. The scales on the back and side of the body are keeled (with small, bony ridges). The back and side of the body are pallid to tan, often with brown specks. The side of the body has a silver sheen, and the underside is white. Fins are mostly transparent, but the lower lobe of the caudal fin is dusky black with a white edge. The lateral line is complete with 39–45 scales, and there are 8 anal rays and 1,4-4,1 pharyngeal teeth. To 3¼ in. (8.4 cm).

SIMILAR SPECIES The Sicklefin Chub, *M. meeki*, lacks keels on the scales and has large, sharply pointed, sickle-shaped fins with the pectoral fin reaching the pelvic-fin origin, and the first dorsal ray extending beyond the last ray when the fin is depressed. The Flathead Chub, *Platygobio gracilis*, lacks keels on the scales and has a broader head, the dorsal-fin origin over or in front of the pelvic-fin origin, the first dorsal ray extending beyond the last ray when the fin is depressed, no papillae on the underside of the head, and 2,4-4,2 pharyngeal teeth.

HABITAT The Sturgeon Chub lives over sand, silt, and small gravel in current in medium to large, turbid rivers.

DISTRIBUTION IN ILLINOIS The Sturgeon Chub was not known to occur in Illinois during the Forbes and Richardson era, but Smith (1979) reported it from the Mississippi River mainstem below the mouth of the Missouri River. Contemporary-era records are available only from the southernmost section of the Mississippi River in Illinois and suggest that the species may be more restricted now than previously.

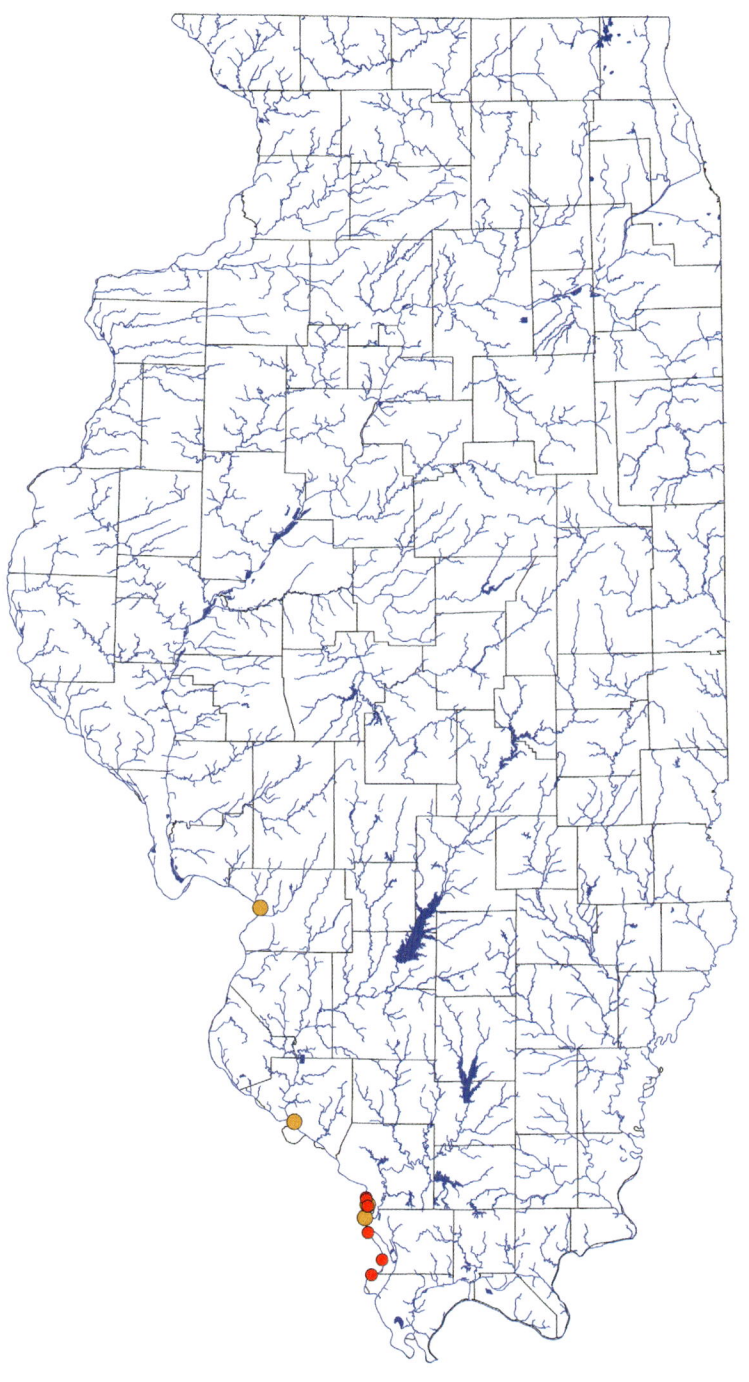

SHOAL CHUB *Macrhybopsis hyostoma* (Gilbert 1884)

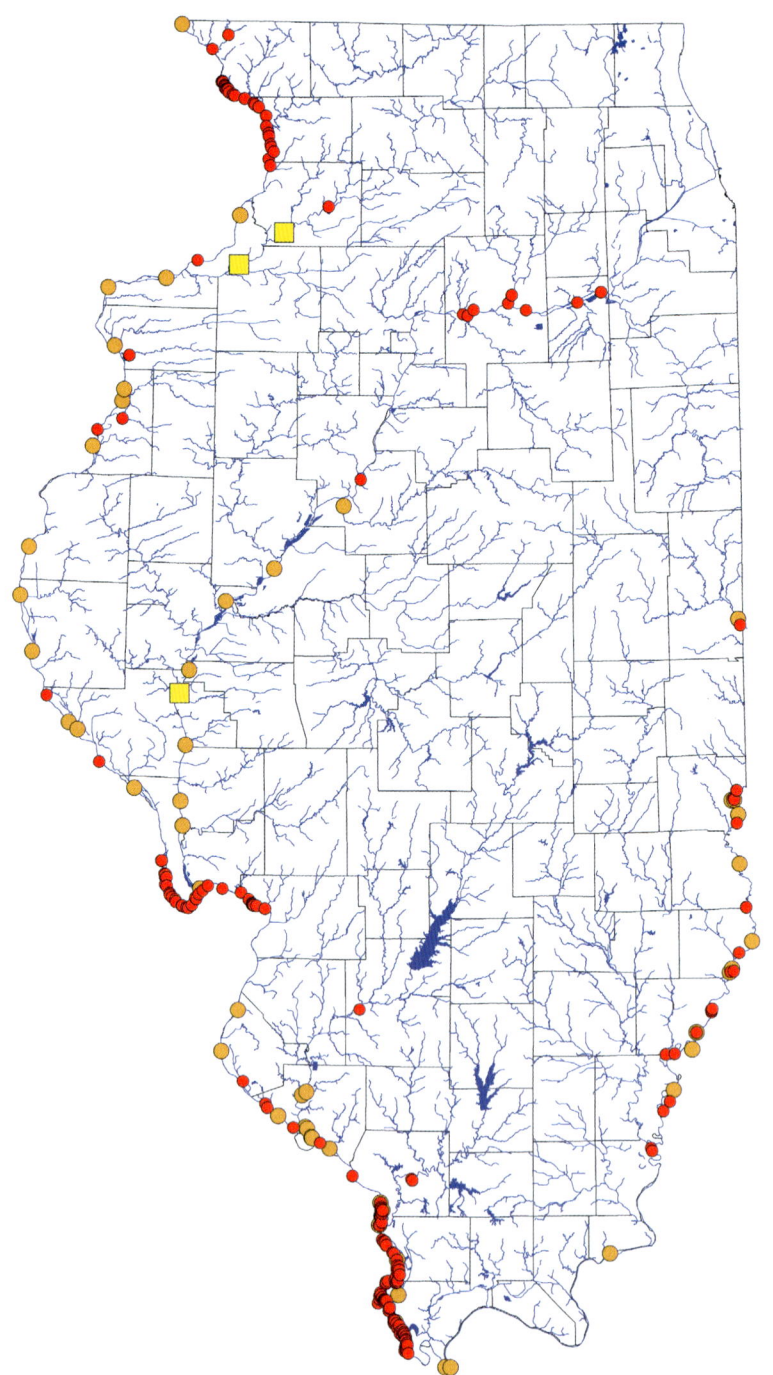

IDENTIFICATION The Shoal Chub has a slender body that is flattened below, deepest at the nape, and strongly tapering to a narrow caudal peduncle. The snout is long and fleshy and projects well beyond the upper lip. There is a long barbel in the corner of the subterminal mouth. The eye is somewhat elliptical and directed upward. The dorsal-fin origin is over or slightly in front of the pelvic-fin origin. The back and side of the body are a translucent light olive to gray with distinct black spots and dusky scale margins. The side of the body has a silver to iridescent-blue stripe (darkest on the caudal peduncle), and the underside is silver-white. The lateral line is complete with 32–43 scales, and there are usually 8 anal rays and 0,4-4,0 pharyngeal teeth. To 3 in. (7.6 cm).

SIMILAR SPECIES The Gravel Chub, *Erimystax x-punctatus*, has black Xs on the back and side of the body, a rounder and more laterally directed eye, and usually 7 anal rays.

HABITAT The Shoal Chub is found in current over sand and gravel in runs and riffles of medium to large rivers.

DISTRIBUTION IN ILLINOIS Forbes and Richardson (1909) recorded the Shoal Chub in the Rock and Illinois River basins, and Smith (1979) found it in the Mississippi, lower Illinois, Wabash, and Ohio Rivers and a couple of their large tributaries. Recent records show nearly the same distribution as that found by Smith, although the species now is present in the upper Illinois River, appears to be gone from the lower Illinois River, and is less widespread in the middle Mississippi River.

REMARKS Forbes and Richardson (1909) treated this species as *Hybopsis hyostomus*, and Smith (1979) treated it as *Hybopsis aestivalis*, the Speckled Chub.

SICKLEFIN CHUB *Macrhybopsis meeki* (Jordan & Evermann 1896)

IDENTIFICATION The Sicklefin Chub has a slender body, deepest at the nape and strongly tapering to a narrow caudal peduncle. The eye is small; the snout is long and fleshy and projects well beyond the upper lip. There is a long barbel in the corner of the subterminal mouth and small sensory papillae on the throat. The fins are large and sickle-shaped; the first dorsal ray extends beyond the last ray when the fin is depressed, and the tip of the pectoral fin reaches beyond the pelvic-fin origin. The dorsal-fin origin is over or slightly behind the pelvic-fin origin. The back and side of the body are light green with a silver sheen, often with brown and silver specks, and the underside is white. Fins are mostly transparent, but the lower lobe of the caudal fin is dusky black with a white edge. The lateral line is complete with 43–50 scales, and there are 8 anal rays and 1,4-4,1 pharyngeal teeth. To 4¼ in. (11 cm).

SIMILAR SPECIES The Sturgeon Chub, *M. gelida*, has keeled scales on the back and side of the body, straight-edged to slightly concave fins with the pectoral fin not reaching the pelvic-fin origin, and the first dorsal ray not extending beyond the last ray when the fin is depressed. The Flathead Chub, *Platygobio gracilis*, has a broader head, no papillae on the throat, the dorsal-fin origin over or in front of the pelvic-fin origin, the pectoral fin not reaching the pelvic-fin origin, and 2,4-4,2 pharyngeal teeth.

HABITAT The Sicklefin Chub lives over sand and small gravel in current in large, turbid rivers.

DISTRIBUTION IN ILLINOIS Like the Sturgeon Chub, the Sicklefin Chub was not known to occur in Illinois during the Forbes and Richardson era. Smith (1979) reported it from the Mississippi River mainstem below the mouth of the Missouri River, and contemporary-era records suggest that the species may be more restricted now than previously. These 2 species occupy similar habitats and have been taken recently using trawling nets. The Sicklefin Chub appears to be somewhat more widespread than the Sturgeon Chub.

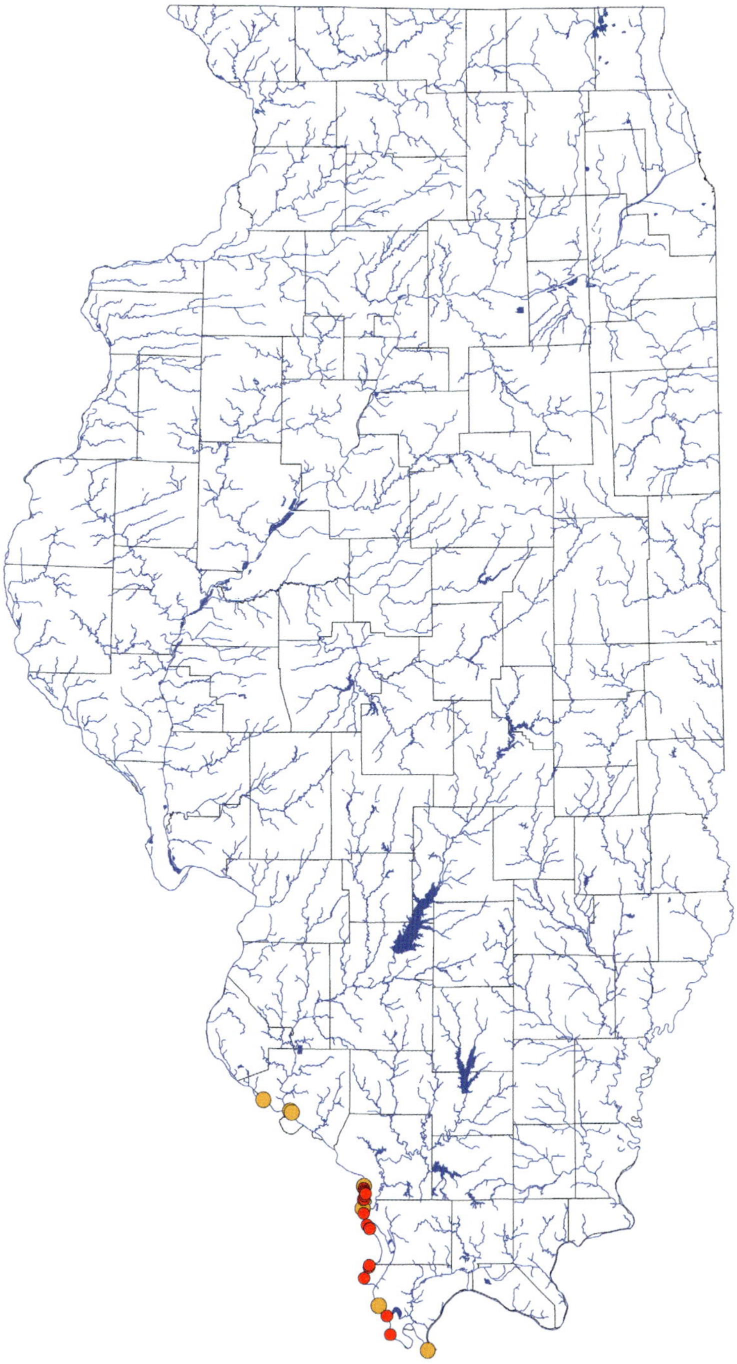

MINNOWS (Leuciscidae)

179

SILVER CHUB   *Macrhybopsis storeriana* (Kirtland 1845)

IDENTIFICATION The Silver Chub has a slender, fairly compressed body that is flattened below and deepest at the nape and that tapers to a narrow caudal peduncle. The large eye is on the upper half of the head. The snout is short and rounded, and there is a barbel in the corner of the subterminal mouth. The dorsal-fin origin is in front of the pelvic-fin origin. The back and upper side of the body are light olive, the side of the body is silver-white and often has a faint dusky stripe, and the underside is white. Fins are mostly transparent, but the lower lobe of the caudal fin is dusky black with a white edge. The lateral line is complete with 35–48 scales, and there are 8 anal rays and 1,4-4,1 pharyngeal teeth. To 9 in. (23 cm).

SIMILAR SPECIES The Bigeye Chub, *Hybopsis amblops*, has a bold black stripe along the side of the body and onto the snout and reaches only 3½ in. (9 cm). The Sturgeon Chub, *M. gelida*, and Sicklefin Chub, *M. meeki*, also have a white edge on the dusky lower lobe of the caudal fin, but they have a much smaller eye and longer snout.

HABITAT The Silver Chub is found in sand-, silt-, and gravel-bottomed pools and backwaters of small to large rivers.

DISTRIBUTION IN ILLINOIS The Silver Chub remains widespread in the largest rivers in Illinois and is found sporadically in smaller rivers. However, it seems to have nearly disappeared from the Vermilion (Wabash) and Saline River basins where Forbes and Richardson (1909) found it and from the Rock and Kaskaskia basins where Smith (1979) found it to be rather widespread. It also is gone from Lake Michigan where it was recorded by Jordan (1878).

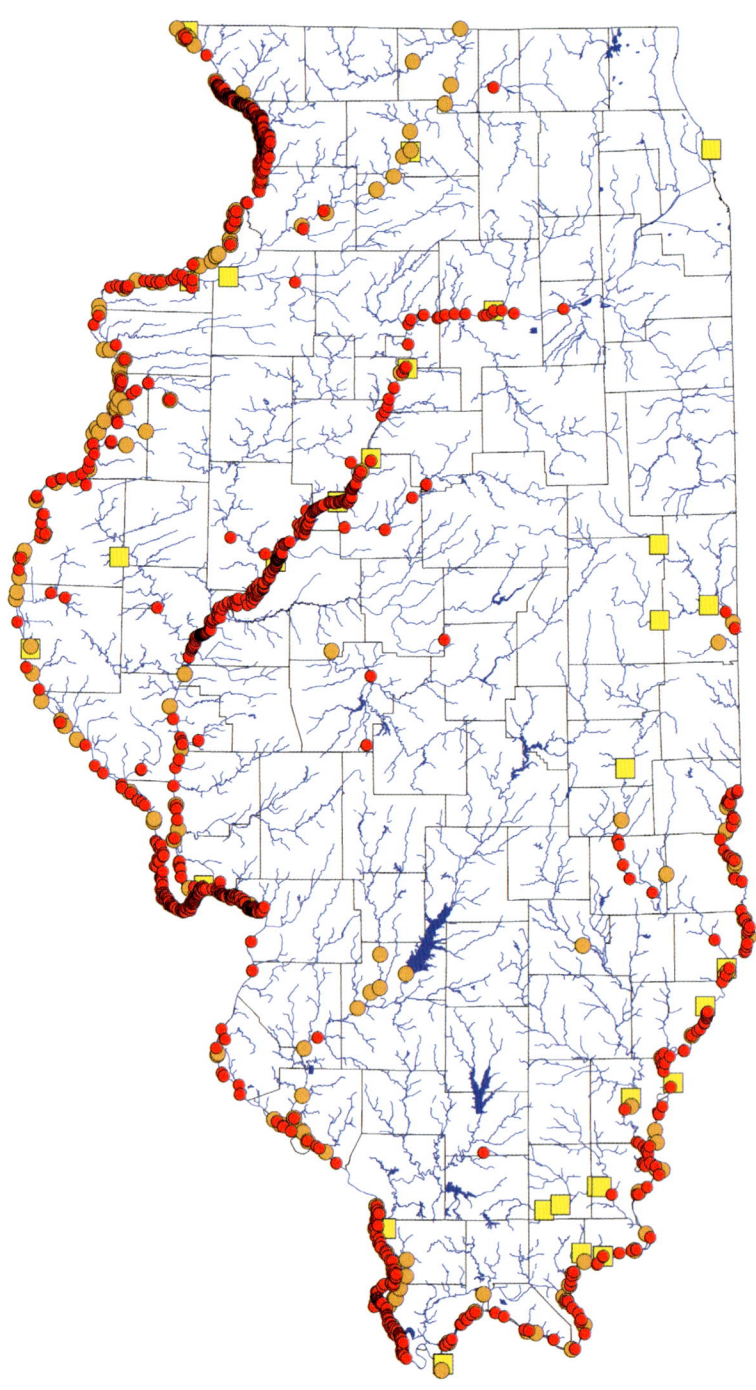

HORNYHEAD CHUB *Nocomis biguttatus* (Kirtland 1840)

IDENTIFICATION The Hornyhead Chub has a stout body, a large head, a rounded snout, and a barbel in the corner of the slightly subterminal mouth. The dorsal-fin origin is slightly in front of to slightly behind the pelvic-fin origin. The back and side of the body are dark olive to yellow-brown with an iridescent green sheen and dark-edged scales. There are an iridescent yellow stripe along the back and a yellow streak above a dusky black stripe along the side and around the snout. The underside is white to light yellow, and there is a black caudal spot. The stripe along the side and the caudal spot are darkest on juveniles. The caudal fin is red on juveniles and yellow on adults; other fins are yellow to orange. Males have a bright red spot behind the eye; females have a brassy spot. Large males have a pink breast and belly, pink-orange fins, and many large tubercles on top of the head. The lateral line is complete with 38–45 scales, and there are 7 anal rays and 1,4-4,1 pharyngeal teeth. To 10¼ in. (26 cm).

SIMILAR SPECIES The River Chub, *N. micropogon*, has a longer snout (about the same as the length of the head behind the eye), a smaller eye higher on the head, no red or brassy spot behind the eye, an olive to orange caudal fin, 0,4-4,0 pharyngeal teeth, no tubercles on top of the head, and a large hump on the head of the breeding male. The Creek Chub, *Semotilus atromaculatus*, has a large, black spot at the front of the dorsal-fin base, a small, flaplike barbel in the groove above the upper lip and near the corner of the mouth, 47–65 lateral scales, usually 8 anal rays, and 2,5-4,2 pharyngeal teeth.

HABITAT The Hornyhead Chub inhabits clear, rocky pools and runs of creeks and small to medium rivers; it rarely is found in sluggish water, silty substrates, and large rivers.

DISTRIBUTION IN ILLINOIS As in previous eras, the Hornyhead Chub is widespread throughout the northern half of Illinois. It appears to be more widespread in the lower Illinois River basin, including the Sangamon River, than in the Forbes and Richardson era but has disappeared from extreme southern Illinois where Forbes and Richardson recorded the species in Clear Creek, which has been well sampled over the past 60 years.

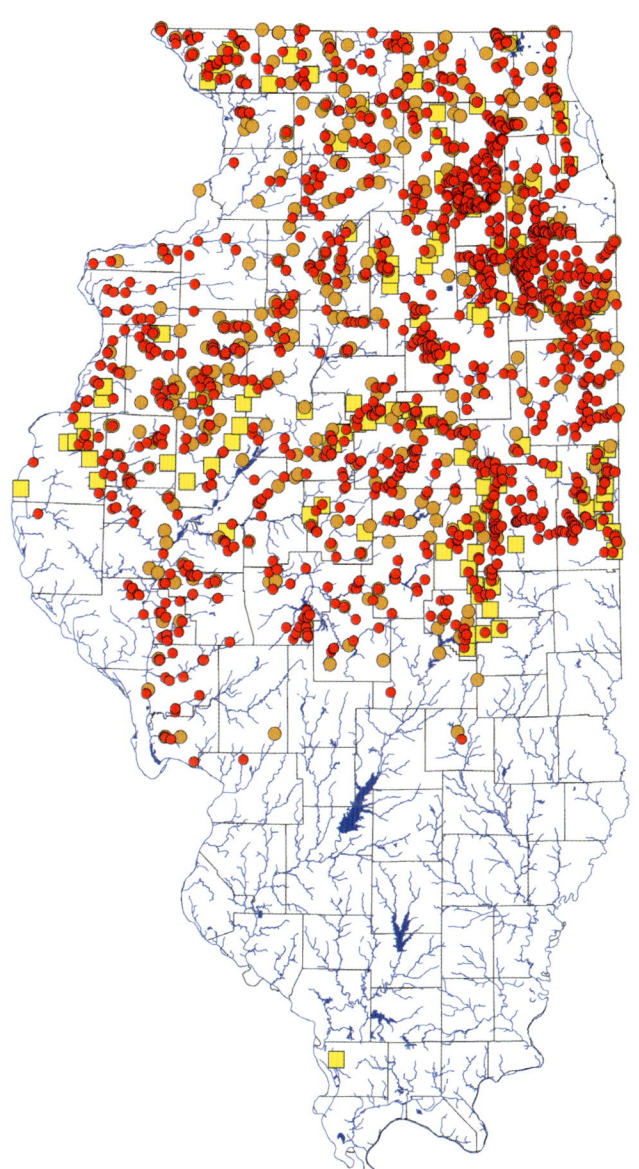

RIVER CHUB *Nocomis micropogon* (Cope 1865)

IDENTIFICATION The River Chub has a stout body and a barbel in the corner of the slightly subterminal mouth. The eye is high on the large head, and the snout is somewhat pronounced (about the same as the length of the head behind the eye). The dorsal-fin origin is slightly in front of to slightly behind the pelvic-fin origin. The back and side of the body are dark olive to yellow-brown with an iridescent green sheen and dark-edged scales. There are an iridescent yellow stripe along the back and a yellow streak above a dusky black stripe along the side and around the snout. The underside is white to light yellow, and there is a black caudal spot. The stripe along the side and the caudal spot are darkest on juveniles. The caudal fin is olive to orange; other fins are clear to yellow-pink. Large males have a bluish-pink head, body, and fins and during the breeding season have large tubercles on the snout and a large hump on top of the head. The lateral line is complete with 38–41 scales, and there are 7 anal rays and 0,4-4,0 pharyngeal teeth. To 12½ in. (32 cm).

SIMILAR SPECIES The Hornyhead Chub, *N. biguttatus*, has a shorter snout (less than the length of the head behind the eye), a larger eye lower on the head, a red or brassy spot behind the eye, a more distinct dark stripe along the side of the body, a red caudal fin on juveniles, 1,4-4,1 pharyngeal teeth, large tubercles on top of the head, and no hump on the head of the breeding male. The Creek Chub, *Semotilus atromaculatus*, has a large, black spot at the front of the dorsal-fin base, a small, flaplike barbel in the groove above the upper lip and near the corner of the mouth, 47–65 lateral scales, usually 8 anal rays, and 2,5-4,2 pharyngeal teeth.

HABITAT The River Chub inhabits clear, rocky runs and pools of creeks and small to medium rivers.

DISTRIBUTION IN ILLINOIS The River Chub was first reported from Illinois in the Wabash River basin by O'Donnell (1935). It is found only in the Wabash River basin with 2 Smith-era records from the Wabash River mainstem and 3 contemporary-era records from the Little Vermilion River system in Vermilion County.

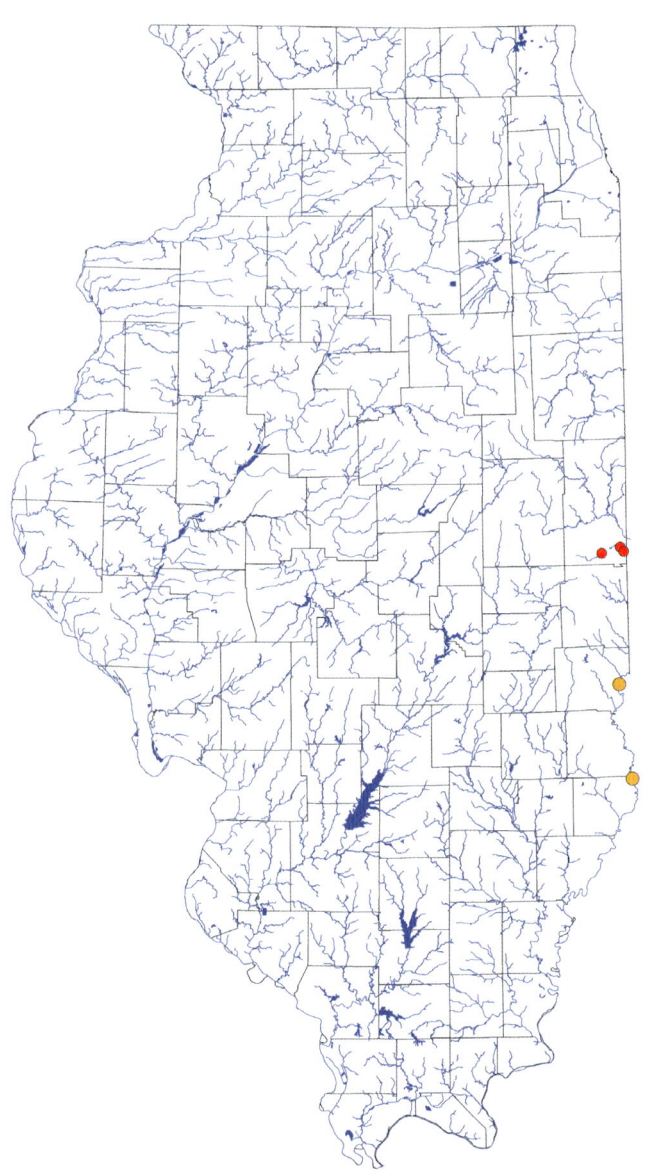

GOLDEN SHINER *Notemigonus crysoleucas* (Mitchill 1814)

IDENTIFICATION The Golden Shiner has an extremely compressed body, a pointed snout, a strongly decurved lateral line, and a scaleless keel along the belly from the pelvic fin to the anal fin. The mouth is small and slightly upturned. The dorsal-fin origin is far behind the pelvic-fin origin. The side of the body is silver with a greenish cast in clear and turbid waters and brassy gold in lakes and coffee-colored waterbodies. Fins are orange or red in large individuals. In clear water, there is a dusky stripe along the midside and herringbone lines on the upper side (more prominent on juveniles). The lateral line is complete with 44–54 scales, and there are 7–9 dorsal rays, usually 11–14 anal rays, and 0,5-5,0 pharyngeal teeth. To 12½ in. (32 cm).

SIMILAR SPECIES The Redfin Shiner, *Lythrurus umbratilis*, lacks the keel on the belly, has a less compressed body, and 2,4-4,2 pharyngeal teeth. Large males have a blue head and body.

HABITAT The Golden Shiner frequents vegetated ponds, lakes, swamps, backwaters, and pools of creeks and small to large rivers. It is most often found over mud.

DISTRIBUTION IN ILLINOIS The Golden Shiner is essentially statewide in occurrence but most widespread in northeastern, east-central, and southern Illinois, as it was in earlier eras. The Golden Shiner is frequently sold as a bait fish, and some records could be the result of "bait bucket" introductions.

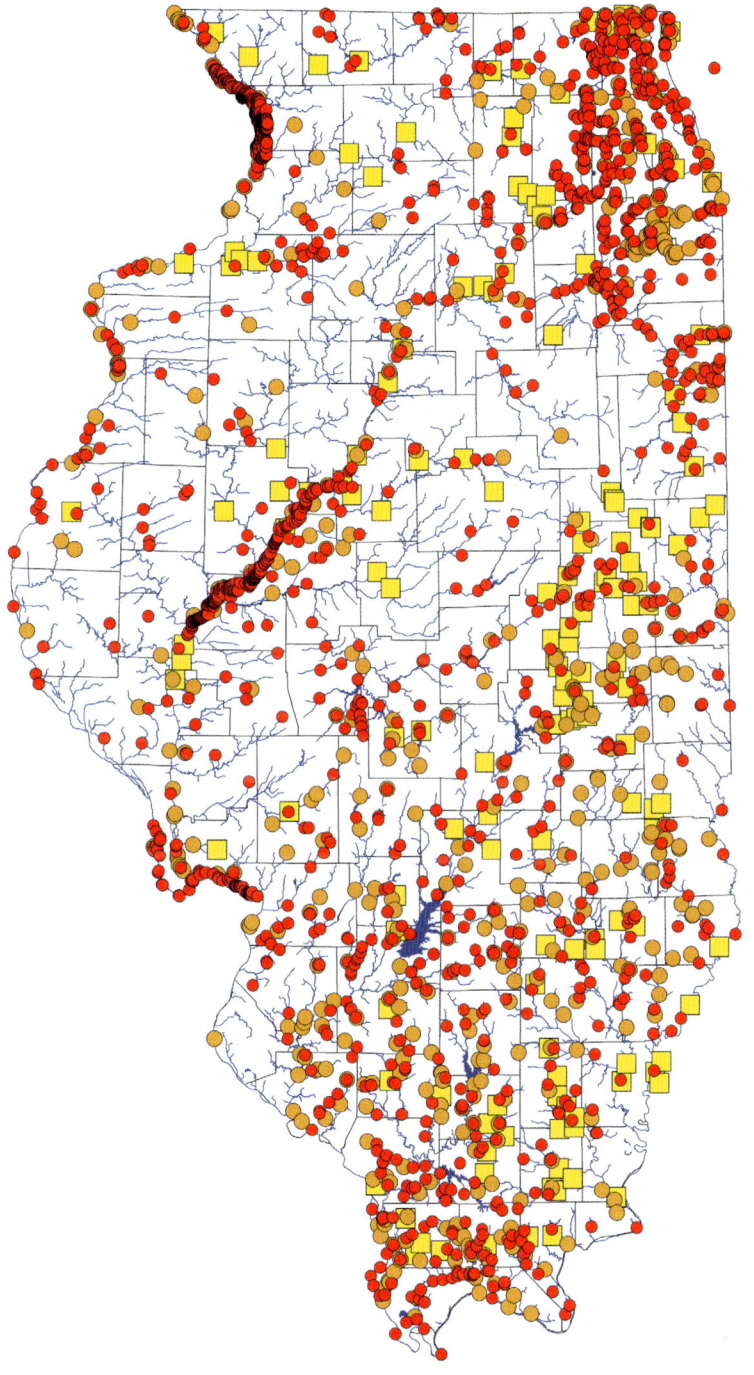

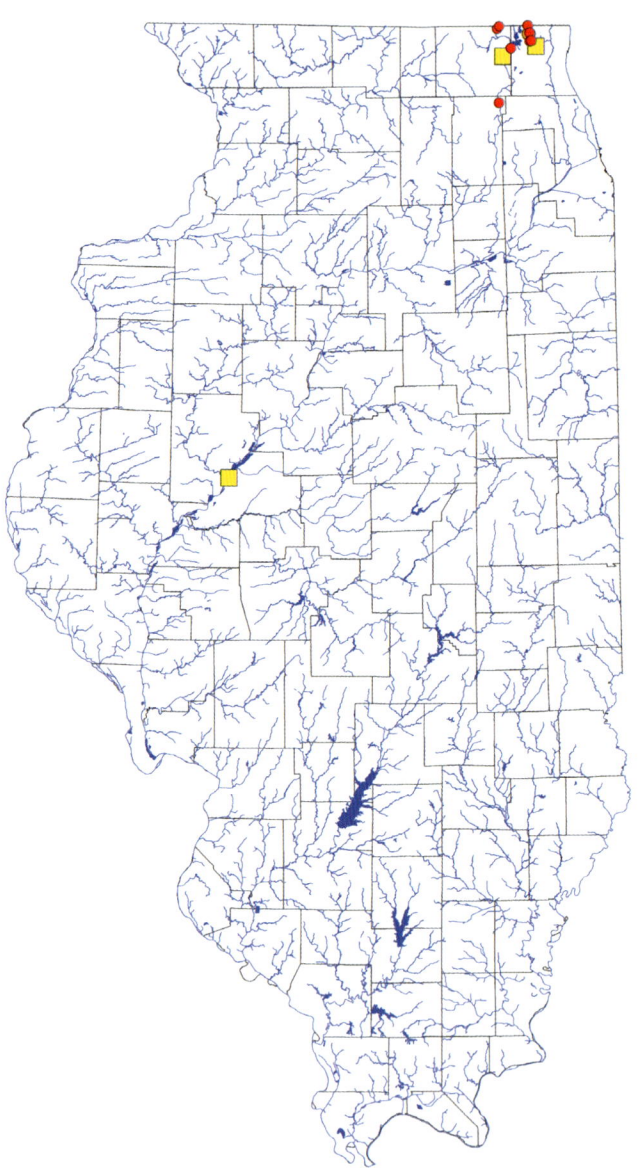

PUGNOSE SHINER *Notropis anogenus* Forbes 1885

IDENTIFICATION The Pugnose Shiner is a small minnow with a slightly compressed body—deepest at the nape and tapering to a narrow caudal peduncle—and with a small, sharply upturned mouth. The dorsal-fin origin is over the pelvic-fin origin. The back and upper side are light olive-green, and there are a thin dark line along the back and a black stripe along the side of the body. Anteriorly, the black stripe continues onto the side of the head, chin, lower lip, and side of the upper lip. Posteriorly, the stripe ends in a black wedge at the base of the caudal fin. Scales on the upper side are darkly outlined except along the upper surface of the black stripe where there is a clear to iridescent greenish-yellow stripe. The lower side is silver-white; the black peritoneum gives the belly a dusky black appearance. Large males have a yellow body and fins. The lateral line is usually complete with 34–38 scales. There are 8 dorsal rays, 8 (sometimes 7) anal rays, and 0,4-4,0 pharyngeal teeth. To 2¼ in. (5.8 cm).

SIMILAR SPECIES The Pugnose Minnow, *Opsopoeodus emiliae*, also has a sharply upturned mouth but has a dark crosshatched pattern on the upper side of the body, a silver-white belly, a clear window between 2 dark areas on the dorsal fin, 9 dorsal rays, and 0,5-5,0 pharyngeal teeth.

HABITAT The Pugnose Shiner is found over sand and mud in clear, vegetated lakes and pools of sluggish creeks and small to medium rivers.

DISTRIBUTION IN ILLINOIS The Pugnose Shiner was originally described from the Fox River in McHenry County, Illinois. Soon thereafter it was found in nearby glacial lakes in Lake County and in 1909 was found much farther south in an Illinois River floodplain lake in Mason County. Smith (1979) had additional records for the species in the glacial lakes in Lake County but noted the disappearance of the species from Mason County. There are several contemporary-era records from the glacial lakes in Lake and McHenry Counties in extreme northeastern Illinois and 1 record from the Fox River in Kane County. The species appears to be gone from the lower Illinois River basin.

EMERALD SHINER
Notropis atherinoides Rafinesque 1818

IDENTIFICATION The Emerald Shiner has a slender, compressed body, an oblique terminal mouth that extends posteriorly to the front of the eye, and a fairly pointed snout. The dorsal-fin origin is well behind the pelvic-fin origin (over the middle of the pelvic fin). The back and upper side of the body are light olive, and the side of the body is silvery with an emerald-green sheen and often a dusky stripe along the side—darkest on the posterior half. There is a narrow dusky stripe along the back. Fins are clear. The lateral line is complete with 35–40 scales. There are 8 pelvic rays, 10–12 anal rays, and 2,4-4,2 pharyngeal teeth. To 5 in. (13 cm).

SIMILAR SPECIES The Rosyface Shiner, *N. rubellus*, and Carmine Shiner, *N. percobromus*, have a sharper snout, a thin black streak above the dusky silver stripe on the side of the body, and red on the head and body. The Silver Shiner, *N. photogenis*, has black crescents between the nostrils, a sharper snout, the lower jaw projects beyond the upper jaw, and 9 pelvic rays. The Silverband Shiner, *N. shumardi*, has 9 pelvic rays, 9 anal rays, a less slender body, and a long and more pointed dorsal fin with the origin over or slightly behind the pelvic-fin origin.

HABITAT The Emerald Shiner occupies pools and runs of small to large rivers and lakes. It is found most often in large rivers over sand or mixed sand and gravel.

DISTRIBUTION IN ILLINOIS The Emerald Shiner is found statewide in Illinois, as in earlier surveys, in the largest rivers and their principal tributaries and along the shoreline of Lake Michigan.

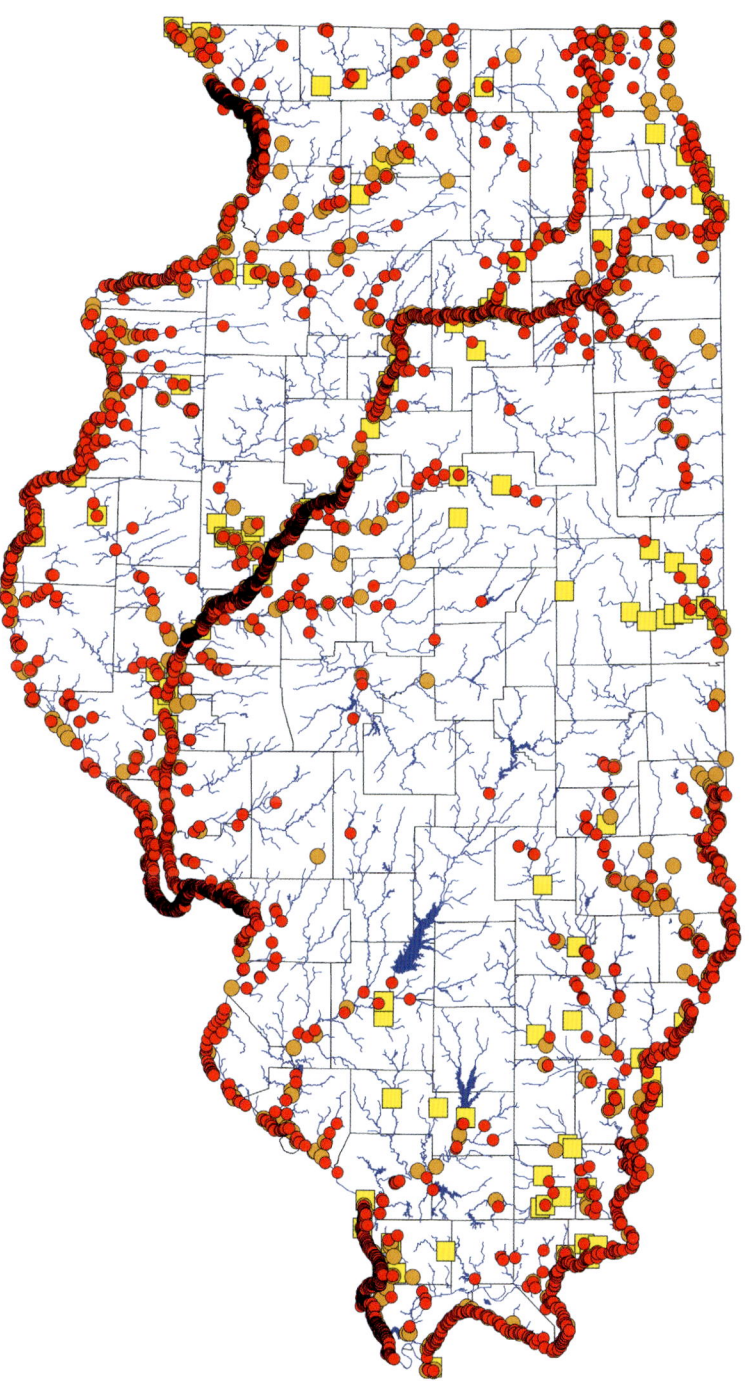

IDENTIFICATION The River Shiner has a slightly compressed body—deepest at the nape—and a rounded snout. The slightly subterminal mouth extends to beneath the anterior margin of the eye. The dorsal-fin origin is over or slightly behind the pelvic-fin origin. The back and upper side of the body are light olive to straw, and the scales on the upper side are faintly outlined. There is a uniformly dark stripe along the back that encircles the dorsal-fin base. The side of the body is silvery, often with a dusky stripe that is usually darkest on the posterior half. The underside is silver-white. The lateral line is complete with 34–41 (usually 35–36) scales, and there are 7 anal rays, and 2,4-4,2 (often 1,4-4,1) pharyngeal teeth. To 5¼ in. (13 cm).

SIMILAR SPECIES The Mississippi Silvery Minnow, *Hybognathus nuchalis*, and other species of *Hybognathus*, have a smaller mouth with the posterior edge of the mouth anterior to the eye, the dorsal-fin origin in front of the pelvic-fin origin, 8 anal rays, and a long, coiled gut. The Sand Shiner, *N. stramineus*, has a dusky stripe along the back expanded at the dorsal-fin origin but not encircling the dorsal fin, a punctate lateral line, and a smaller mouth.

HABITAT The River Shiner lives over sand and gravel in the main channels and flowing pools of medium to large rivers.

DISTRIBUTION IN ILLINOIS The River Shiner is found in the large rivers that border Illinois, in much of the Illinois River, and in a few tributaries of these rivers. It appears to have disappeared from the uppermost localities where it was found in previous eras in the Illinois River basin and from most of the Embarras River.

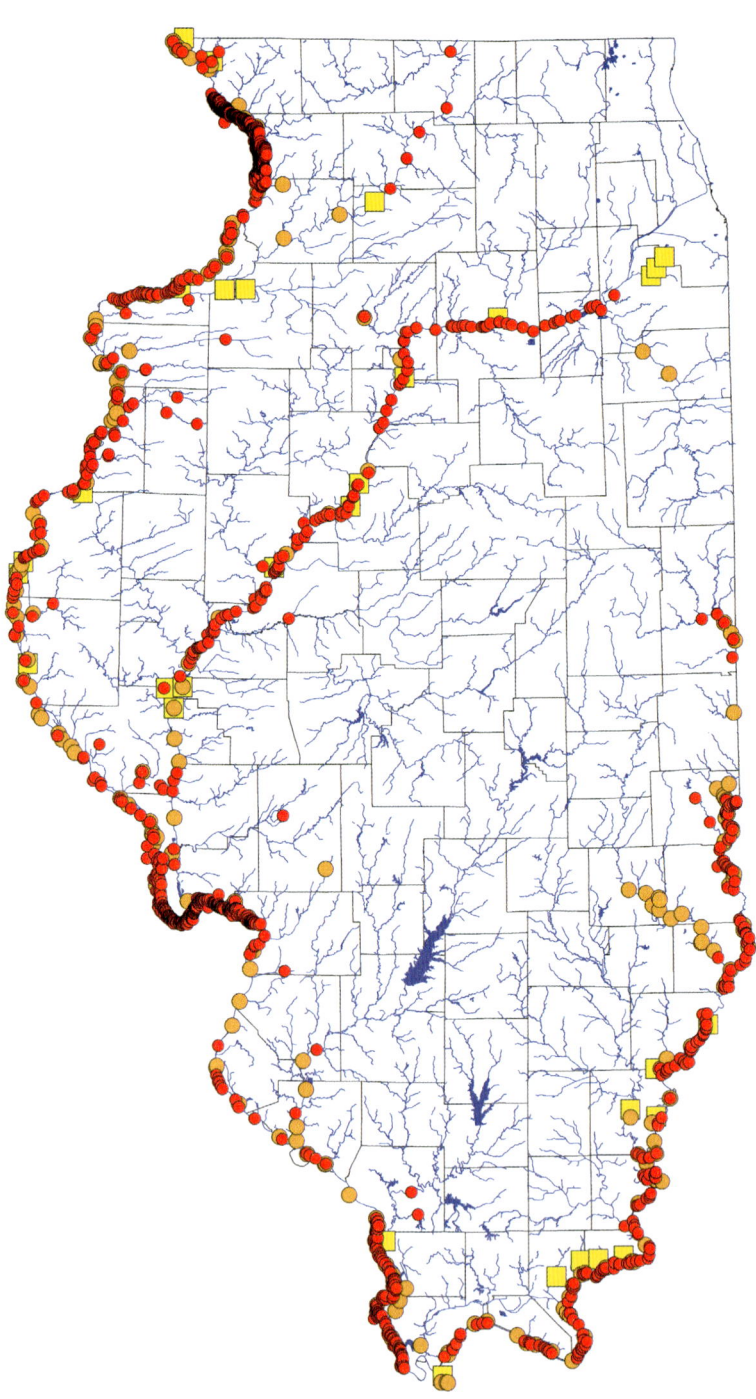

BIGEYE SHINER *Notropis boops* Gilbert 1884

IDENTIFICATION The Bigeye Shiner is slender and fairly compressed with the body deepest under the nape and tapering to a narrow caudal peduncle. The large eye is much wider than the length of the snout, and the posterior edge of the large, terminal mouth reaches below the eye. The lateral line is punctate, at least on the anterior half of the body. The dorsal-fin origin is over the pelvic-fin origin. The back and upper side are light olive-yellow with darkly outlined scales, and there is a thin dark stripe along the back. There is a clear stripe above a black stripe along the silver side of the body. The black stripe extends anteriorly around the snout and onto both lips and posteriorly to a small, black spot at the base of the caudal fin. The lower side is silver-white; the black peritoneum gives the belly a dusky black appearance. The lateral line is complete with 34–40 scales. There are 8 anal rays, and the pharyngeal teeth are 1,4-4,1. To 3½ in. (9 cm).

SIMILAR SPECIES The Blackchin Shiner, *N. heterodon*, has a smaller mouth reaching only to below the nostril, a smaller eye, and an incomplete lateral line. The Bigeye Chub, *Hybopsis amblops*, has a smaller eye about equal in length to the snout, and a smaller mouth reaching only to below the nostril and with a barbel in each corner.

HABITAT The Bigeye Shiner is found in clear, rocky pools of creeks and small to medium rivers where it is often taken near emergent vegetation near riffles and stream margins.

DISTRIBUTION IN ILLINOIS As noted by Smith (1979), the Bigeye Shiner is sporadic in Illinois but more widespread than during the Forbes and Richardson era. Most Smith-era records were from the Wabash River basin with only a few from the Illinois and Mississippi River basins. It is much less widely distributed in the contemporary era than previously, especially in the Wabash basin of eastern Illinois. It is frequently encountered in the Clear Creek system in southwestern Illinois and in the Little Vermilion River in Vermilion County.

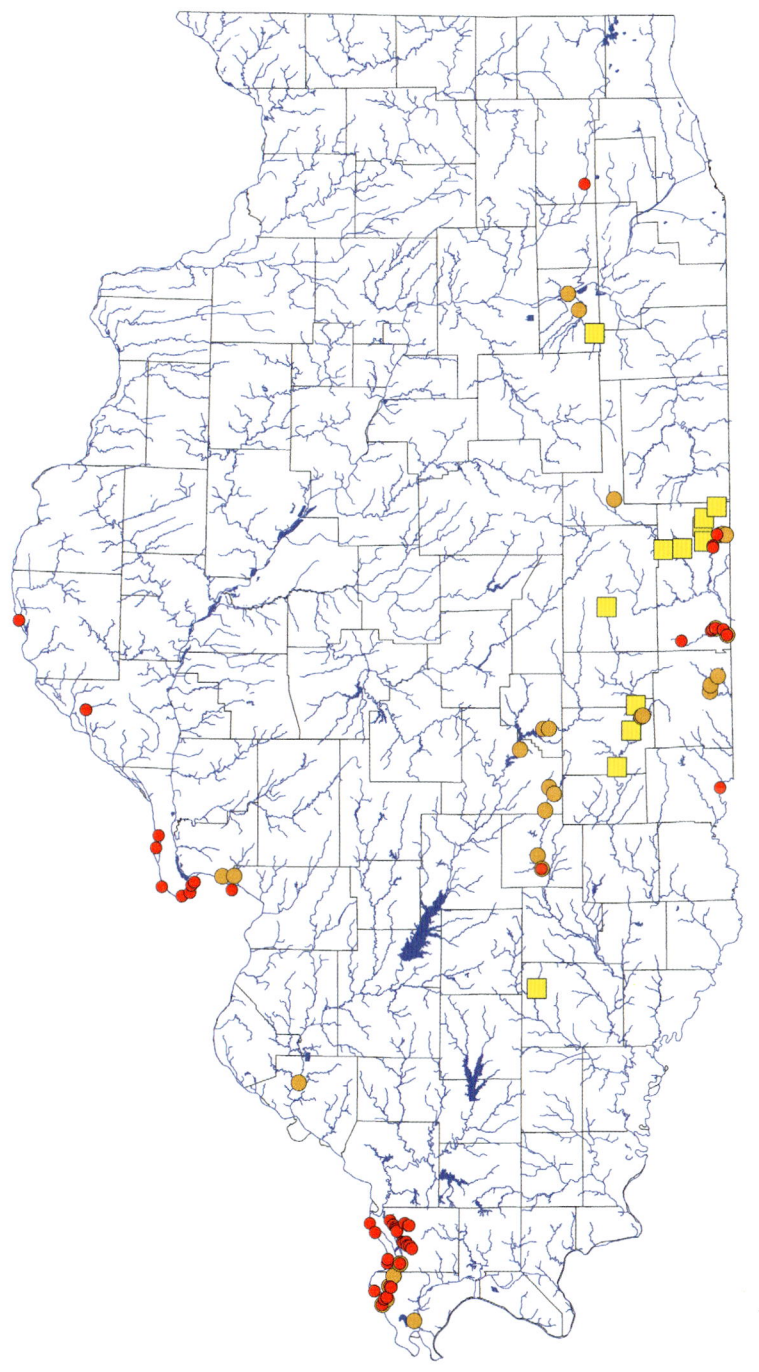

MINNOWS (Leuciscidae)

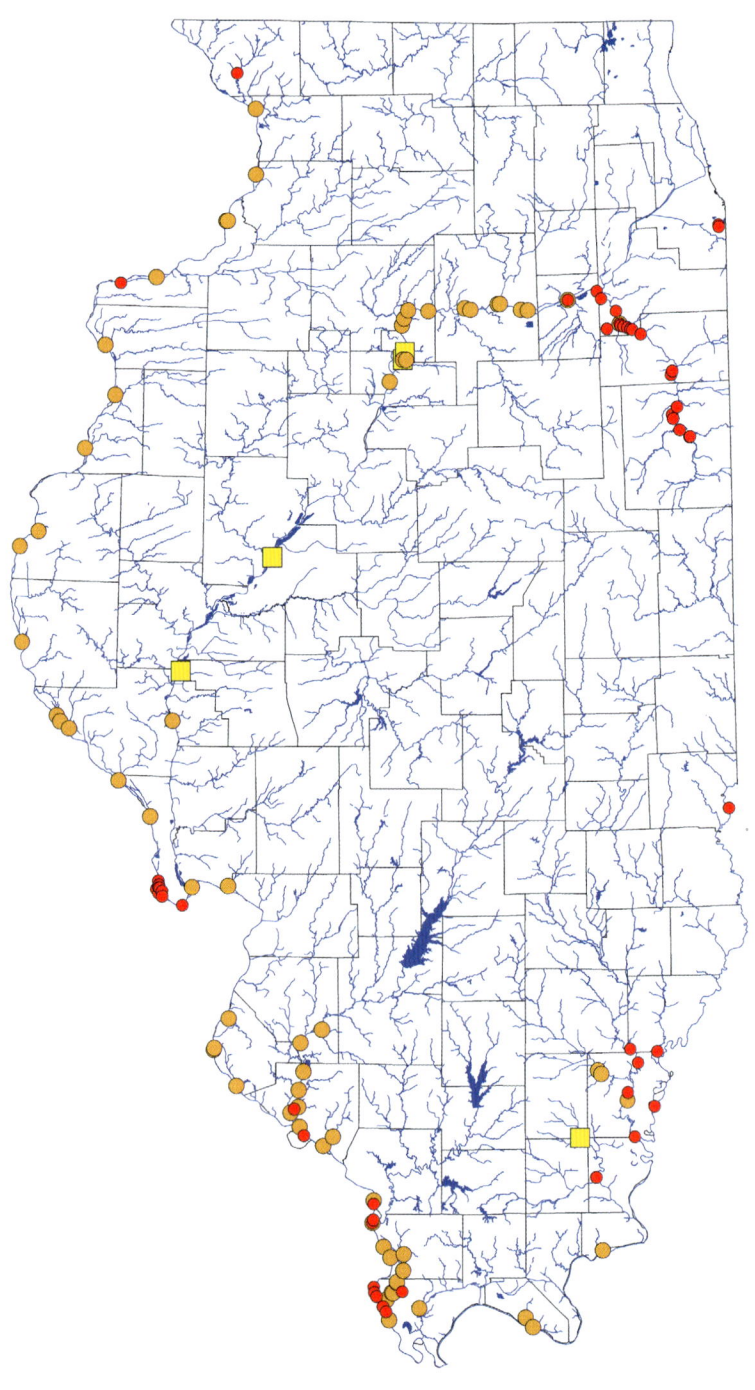

GHOST SHINER *Notropis buchanani* Meek 1896

IDENTIFICATION The Ghost Shiner has a compressed body, arched at the nape, deepest at the dorsal-fin origin, and strongly tapering to a narrow caudal peduncle. The fins are large and pointed, and the depressed pelvic fin reaches past the anal-fin origin. The lateral-line scales on the anterior half of the body are more deeper than wide. The eye is fairly large—about as long as the rounded snout, and the mouth is small and slightly subterminal. The dorsal-fin origin is over the pelvic-fin origin. There is usually no infraorbital canal (rarely, a short segment is present). Appropriately named, the Ghost Shiner is translucent and pale overall. In turbid water, individuals lack dark pigment; in clear water, the scales on the back are faintly outlined, and black specks may be present on the snout, along the lateral line, and along the underside of the caudal peduncle. The lateral line is complete with 30–35 scales, and there are 8 anal rays and 0,4-4,0 pharyngeal teeth. To 2½ in. (6.4 cm).

SIMILAR SPECIES The Mimic Shiner, *N. volucellus*, also has the lateral-line scales on the front half of the body more deeper than wide, but the depressed pelvic fin does not reach the anal-fin origin. The Mimic Shiner also lacks the strongly arched body, the scales on the back in front of the dorsal fin are wider than those on the upper side, and it has an infraorbital canal.

HABITAT The Ghost Shiner is most often found over sand or silt in pools and backwaters of small to large rivers.

DISTRIBUTION IN ILLINOIS Forbes and Richardson (1909) recorded the Ghost Shiner only from the Illinois and Saline River basins. Smith (1979) found it to be more widespread, especially in the Mississippi, upper Illinois, and Kaskaskia Rivers, but contemporary-era records indicate a significant decline for the species throughout most of its former range. It is frequently encountered only in the Kankakee River system and is sporadic elsewhere—mostly in southern Illinois.

IRONCOLOR SHINER   *Notropis chalybaeus* (Cope 1867)

IDENTIFICATION The Ironcolor Shiner's body is compressed, deepest and arched at the dorsal-fin origin and tapering to the caudal peduncle, which is much narrower than the rest of the body. The dorsal-fin origin is over the pelvic-fin origin. The snout is pointed, the eye is longer than the snout, and the mouth is small, terminal, and black on the inside. The back and upper side of the body are olive-yellow and have darkly outlined scales. A dusky stripe along the back is widest and darkest anterior to the dorsal fin. A black stripe on the side extends forward on the head and around the snout, covering both lips and the chin, and posteriorly to a black spot at the base of the caudal fin. Above the black stripe is a gold-orange streak. The lower side and belly are silver-white. Large males have a bright orange-gold stripe on the side of the body and light orange-gold fins. The lateral line is usually incomplete with 31–37 scales. There are 8 anal rays and 2,4-4,2 pharyngeal teeth. To 2½ in. (6.5 cm).

SIMILAR SPECIES The Weed Shiner, *N. texanus*, has black-edged scales below the lateral line as well as above, a less compressed and less arched body, a blunter snout, 7 anal rays, and no (or little) black inside the mouth. The Bluehead Shiner, *Pteronotropis hubbsi*, has the dorsal-fin origin behind the pelvic-fin origin, 9–10 dorsal and 9–10 anal rays, 0,4-4,0 pharyngeal teeth, and enlarged fins on large males.

HABITAT The Ironcolor Shiner lives in clear, vegetated, sand-bottomed pools and slow runs of creeks and small rivers. It is sometimes found in soft-bottomed swamps adjacent to the Kankakee River (Smith 1979).

DISTRIBUTION IN ILLINOIS Forbes and Richardson (1909) recorded a collection of the Ironcolor Shiner from the Des Plaines River in Cook County in 1901, but it has not been taken in that basin since then. Smith recorded the species from 2 disjunct areas characterized by sand-bottomed streams, the Iroquois-Kankakee River basin in Kankakee and Iroquois Counties and the Salt-Sangamon River basin in Mason and Tazewell Counties. The species remains frequently encountered in these 2 areas.

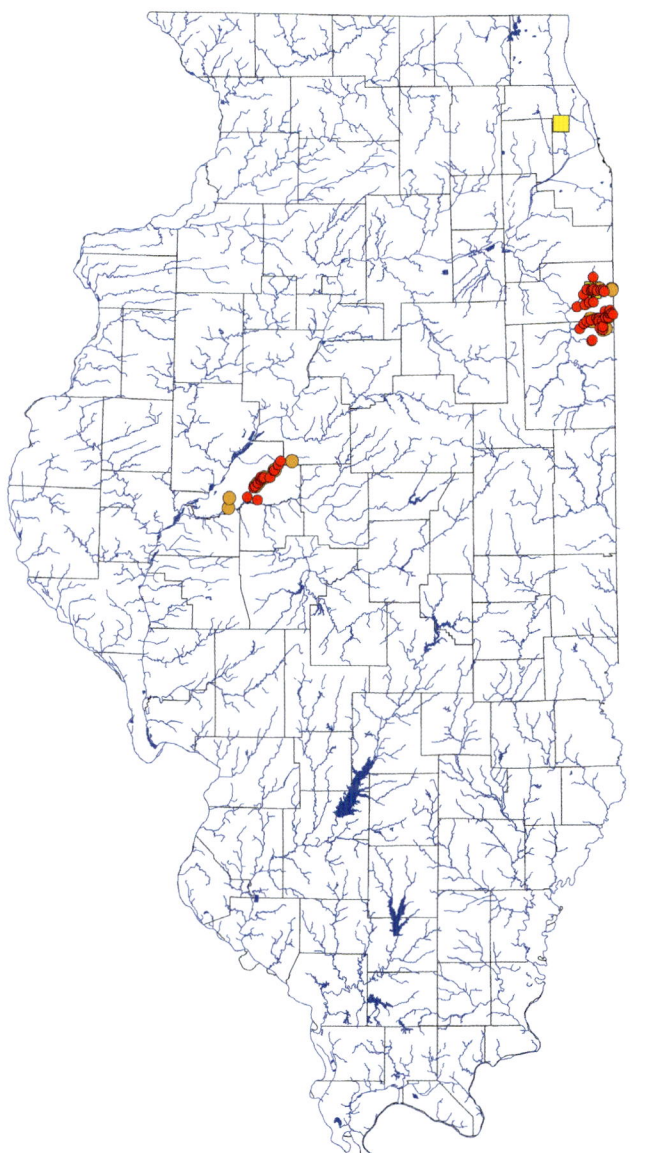

BIGMOUTH SHINER *Notropis dorsalis* (Agassiz 1854)

IDENTIFICATION The Bigmouth Shiner's body is compressed, deepest and arched at the dorsal-fin origin and tapering to the narrow caudal peduncle. The dorsal-fin origin is over to slightly in front of the pelvic-fin origin. The snout is long, the eye is directed slightly forward and upward, and the head is flattened on the underside. The mouth is large, reaching under the eye, and distinctly subterminal. The lateral line is punctate on the anterior half of the body. The back and upper side of the body are light tan to olive with faintly outlined scales, and there is a dark stripe along the back. The lower side and belly are silver-white. The lateral line is complete with 33–39 scales. There are 8 anal rays and 1,4-4,1 pharyngeal teeth. To 3¼ in. (8 cm).

SIMILAR SPECIES The Sand Shiner, *N. stramineus*, lacks an upwardly directed eye and flattened head and has a dark wedge at the dorsal-fin origin and 7 anal rays. The Silverjaw Minnow, *Ericymba buccata*, has large, silver-white chambers on the cheek and underside of the head.

HABITAT The Bigmouth Shiner occupies shallow sandy and mixed sand and silt runs and pools of headwaters, creeks, and small to medium rivers.

DISTRIBUTION IN ILLINOIS Forbes and Richardson (1909) found the Bigmouth Shiner to be widespread in the Mississippi and Illinois River basins. Smith (1979) noted that the species appeared to be more widespread than previously and documented its presence in the Sangamon and Kaskaskia River basins, where it had not been recorded by Forbes and Richardson. In the contemporary era the Bigmouth Shiner has essentially the same distribution as recorded by Smith.

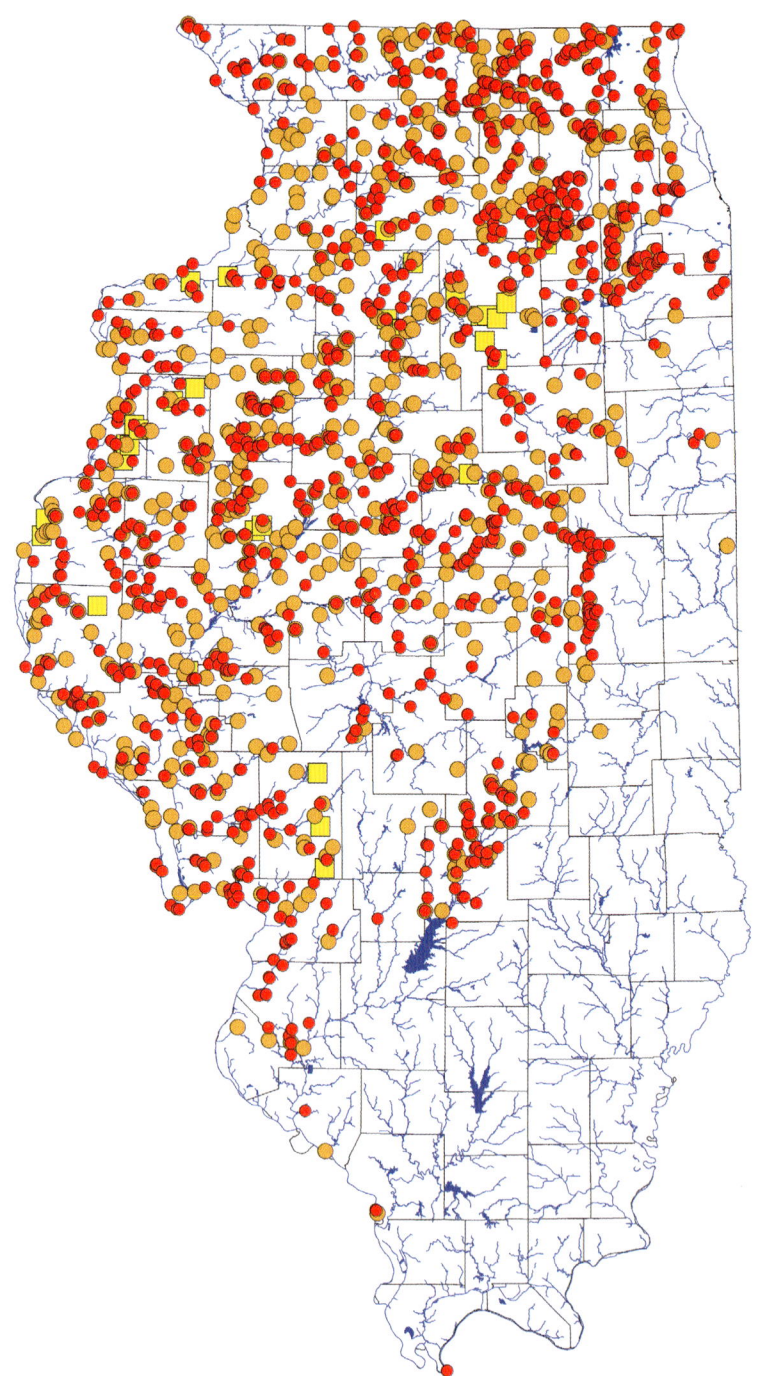

BLACKCHIN SHINER *Notropis heterodon* (Cope 1865)

IDENTIFICATION The Blackchin Shiner's body is fairly compressed, and the caudal peduncle is slender. The mouth is small with the posterior edge reaching below the nostril. The black stripe along the side of the body extends anteriorly around the short, pointed snout and onto the lips and chin. Concentration of the black pigment on the lateral-line pores creates a zigzag in the stripe, at least on the front half of the body. The back and upper side of the body are olive to pale yellow, and the scales are darkly outlined except for a pale stripe above the black stripe on the side. A dusky stripe along the back is darker and much wider in front of the dorsal fin; the lower side and belly are silver-white. The dorsal-fin origin is over to slightly in front of the pelvic-fin origin. The lateral line is usually incomplete with 34–38 scales. There are 8, often 7, anal rays and 1,4-4,1 pharyngeal teeth. To 2¾ in. (7.1 cm).

SIMILAR SPECIES The Bigeye Shiner, *N. boops*, has a larger eye, a larger mouth with the posterior edge reaching below the eye, a black peritoneum that gives the belly a dusky black appearance, and a complete lateral line. The Blacknose Shiner, *N. heterolepis*, is more slender and has no black on the chin; the dorsal-fin origin is behind the pelvic-fin origin, and there are 0,4-4,0 pharyngeal teeth.

HABITAT The Blackchin Shiner is usually found over sand, in clear, vegetated lakes and pools and slow runs of creeks and small rivers.

DISTRIBUTION IN ILLINOIS As in the earlier eras, the Blackchin Shiner is restricted to the glacial lakes and tributaries of the Fox and Des Plaines Rivers in northeastern Illinois. Although Smith (1979) had a record from the shore of Lake Michigan (Dead River), there are no contemporary-era records for the lake.

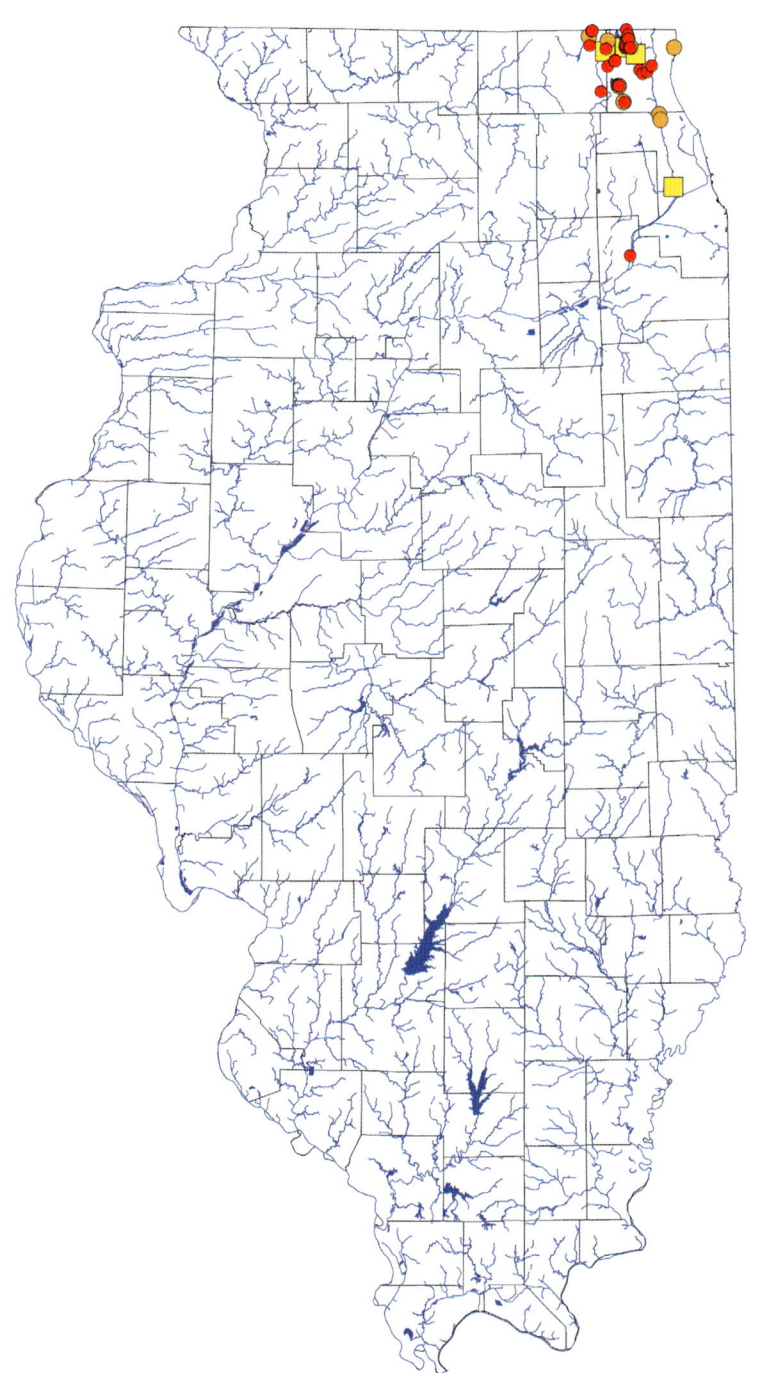

BLACKNOSE SHINER *Notropis heterolepis* Eigenmann & Eigenmann 1893

IDENTIFICATION The Blacknose Shiner's body is slender and slightly compressed, and the dorsal-fin origin is slightly behind the pelvic-fin origin. The snout is rounded and somewhat elongated, the eye is round, and the mouth is small and slightly subterminal. The black stripe along the side of the body extends around the snout and reaches onto the upper lip but not on the chin. There are bold black crescents within the stripe. The back and upper side are light olive to yellow-brown, and there is a faint streak in front of the dorsal fin. The scales are darkly outlined on the back and upper side except within a light stripe above the black stripe along the side. The lower side and belly are silver-white. The lateral line is incomplete with 32–39 lateral scales. There are 8 anal rays and 0,4-4,0 pharyngeal teeth. To 3¾ in. (9.8 cm).

SIMILAR SPECIES The Blackchin Shiner, *N. heterodon*, is less slender, has black on the chin as well as the lower lip; the dorsal-fin origin is over to slightly in front of the pelvic-fin origin; and there are 1,4-4,1 pharyngeal teeth. The Pallid Shiner, *Hybopsis amnis*, has a larger eye, a more arched body, and no black crescents in the black stripe along the side of the body; the dorsal-fin origin is over or in front of the pelvic-fin origin; and there are 1,4-4,1 pharyngeal teeth.

HABITAT The Blacknose Shiner lives over sand or mixed sand and mud in clear, vegetated lakes and pools of creeks and small rivers.

DISTRIBUTION IN ILLINOIS The Blacknose Shiner has shown a dramatic decrease in Illinois since the Forbes and Richardson era, when it was generally distributed in the northern half of the state and in a few localities in southern Illinois. The decrease has continued since the Smith era as the species has disappeared from the Kankakee River basin, most of the upper Rock River basin, and several localities in northeastern Illinois, including along Lake Michigan. The Blacknose Shiner persists only in the glacial lakes of Lake and McHenry Counties and in a few scattered localities in the Green-Rock and upper Illinois River basins.

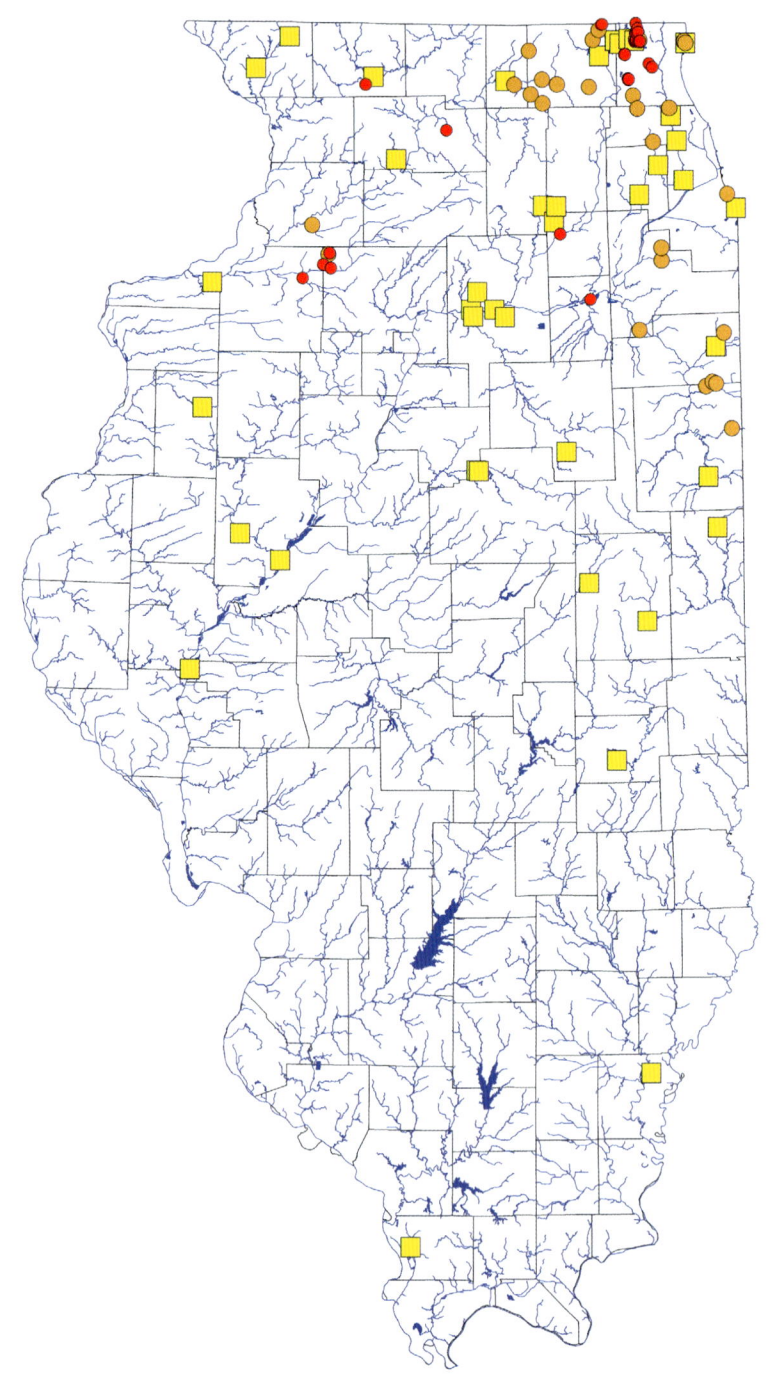

MINNOWS (Leuciscidae)

192

SPOTTAIL SHINER *Notropis hudsonius* (Clinton 1824)

IDENTIFICATION The Spottail Shiner has a large eye, a short, rounded snout, and a nearly horizontal, slightly subterminal mouth. There is a large, black caudal spot. The body is somewhat compressed, deepest at the dorsal-fin origin and with a narrow caudal peduncle. The dorsal-fin origin is over or slightly in front of the pelvic-fin origin. The back and upper side are olive to gray-brown with dark-edged scales that often form wavy lines, and there is a dusky stripe along the back. The side of the body is silvery with a punctate lateral line (often faint) on the anterior half and a dusky stripe along the posterior half. The lower side and belly are silver-white. The lateral line is incomplete with 36–42 lateral scales. There are 8 anal rays and 2,4-4,2 pharyngeal teeth. To 5¾ in. (15 cm).

SIMILAR SPECIES The Silver Chub, *Macrhybopsis storeriana*, has no black spot on the caudal-fin base and has a barbel at the corner of the mouth; the eye is high on the head. The Bigeye Chub, *H. amblops*, has a small, black spot on the caudal-fin base, a barbel at the corner of the mouth, and 1,4-4,1 pharyngeal teeth; the dorsal-fin origin is over or slightly behind the pelvic-fin origin. The Taillight Shiner, *Notropis maculatus*, has black spots above and below the caudal spot and 0,4-4,0 pharyngeal teeth.

HABITAT The Spottail Shiner lives in flowing, sandy and rocky pools and runs of medium to large rivers and along the sandy and rocky shores of lakes.

DISTRIBUTION IN ILLINOIS The Spottail Shiner is frequently encountered in the shallow waters of Lake Michigan and in the Fox, Rock, and Illinois Rivers. It also is widespread in the Mississippi River above the mouth of the Missouri River, less so in the lower Mississippi. Its distribution has changed little since the earlier eras.

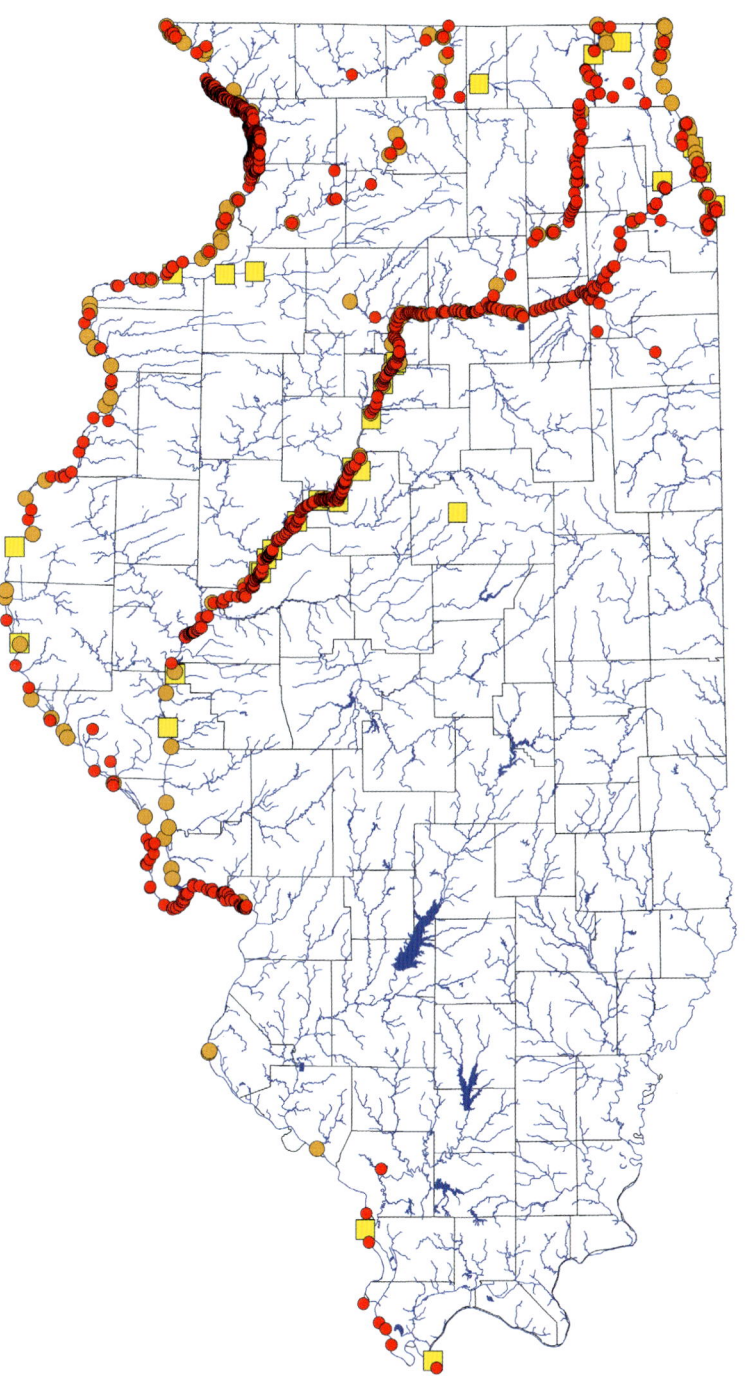

TAILLIGHT SHINER   *Notropis maculatus* (Hay 1881)

IDENTIFICATION   The Taillight Shiner is slender and compressed with a large and conspicuous black spot at the base of the caudal fin. There are black spots above and below the caudal spot, often with bright red between the spots, and a large, black blotch (darkest on males) along the front edge of the dorsal fin. It has a fairly long, rounded snout, a subterminal mouth, and large, pointed fins. The dorsal-fin origin is behind the pelvic-fin origin. The back and side are yellow-olive with a crosshatched pattern, and there is a thin dusky stripe along the back. There is usually a dusky or black stripe along the side of the body and around the snout; the snout is reddish. Large males have dusky black bands on the posterior edges of the fins. During the breeding season, the male has a bright red head and body and red-black edges on the fins. The lateral line is incomplete with 8–10 pores. There are 34–39 lateral scales, 8 anal rays, 8 dorsal rays, and 0,4-4,0 pharyngeal teeth. To 3 in. (7.6 cm).

SIMILAR SPECIES   The Pugnose Minnow, *Opsopoeodus emiliae*, lacks the bold black spot at the base of the caudal fin and has a nearly vertical mouth, 9 dorsal rays, and 0,5-5,0 or 0,5-4,0 pharyngeal teeth; large males lack red on the body and have bright white anal and pelvic fins. The Spottail Shiner, *Notropis hudsonius*, has 2,4-4,2 pharyngeal teeth and a silvery side without a dusky or black stripe; it lacks black spots above and below the caudal-fin base.

HABITAT   The Taillight Shiner lives near vegetation or woody debris in swamps, lakes, and backwaters and pools of small to large rivers. The 2 locations in Illinois where it has been found are wetlands with murky water over mud.

DISTRIBUTION IN ILLINOIS   The Taillight Shiner was discovered in a wetland and an adjacent small creek south of Unionville in Massac County, in 1987 and collected there again in 1988. More recently it was recorded at 1 locality in the Cache River basin, Pulaski County, in 2020. Although the species appears to be rare in Illinois, it is frequently encountered in oxbow lakes across the Ohio River in Kentucky (Burr & Warren 1986).

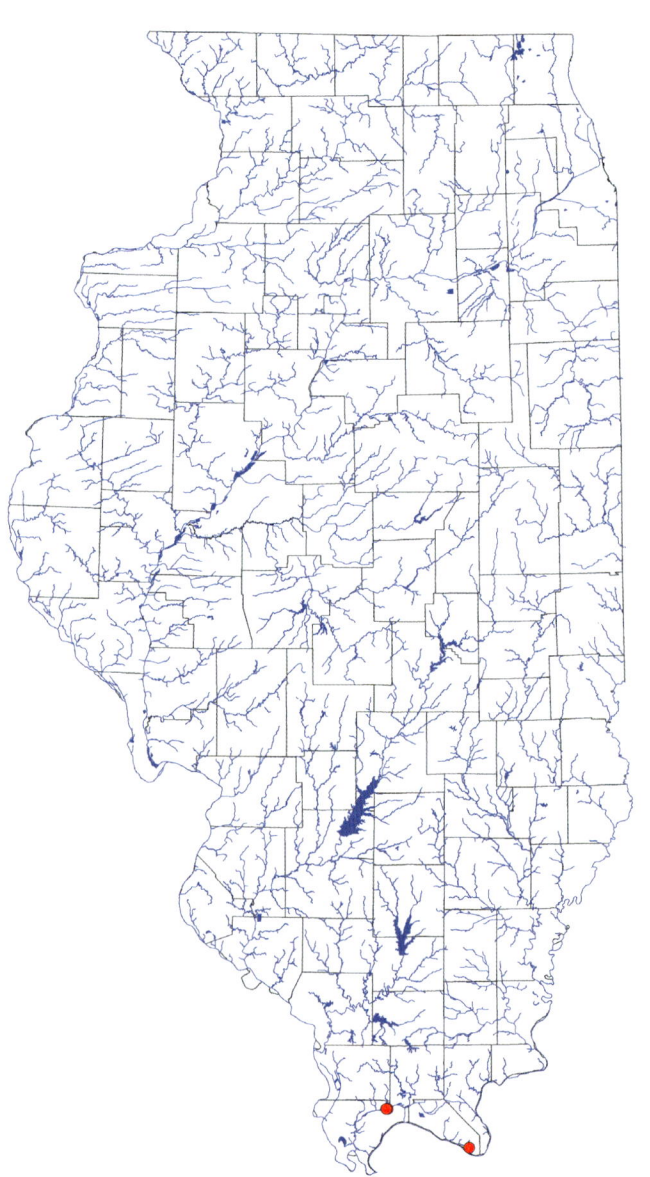

OZARK MINNOW *Notropis nubilus* (Forbes 1878)

IDENTIFICATION The Ozark Minnow has a slender and barely compressed body. The mouth is slightly subterminal, and the dorsal-fin origin is over the pelvic-fin origin. It is strongly bicolored, with the back and upper side of the body dark olive-brown and the lower side silver-white to yellow-orange. There is a thin, faint-yellow stripe above a silver-black stripe along the side of the body and around the snout. The lateral line is punctate, ending at a small, black spot on the caudal-fin base. There are gold spots on a dark stripe along the back. Large individuals are yellow-orange below and have yellow to orange fins (brighter on the male). The gut is long and coiled (at least twice the length of the body). The long gut and black peritoneum give the belly a dusky black appearance when viewed from below. There are 33–38 scales along the lateral line, 8 anal rays, and 0,4-4,0 pharyngeal teeth. To 3¾ in. (9.3 cm).

SIMILAR SPECIES Species of *Hybognathus* have a long, coiled gut but are not strongly bicolored, lack a well-defined dark stripe along the side of the body, and have a smaller eye. All other species of *Notropis* in Illinois lack a long, coiled gut.

HABITAT The Ozark Minnow is found in flowing, rocky pools and runs of creeks and small to medium rivers.

DISTRIBUTION IN ILLINOIS The Ozark Minnow is found predominantly in northern Illinois, where it appears to be about as widespread as in previous surveys, although 2 recent records from the Fox River system in DeKalb County are the first records in the Illinois River basin. As noted by Smith (1979), records in extreme southern Illinois may indicate that the species is dispersing from streams in the neighboring Missouri Ozarks, where it is widespread.

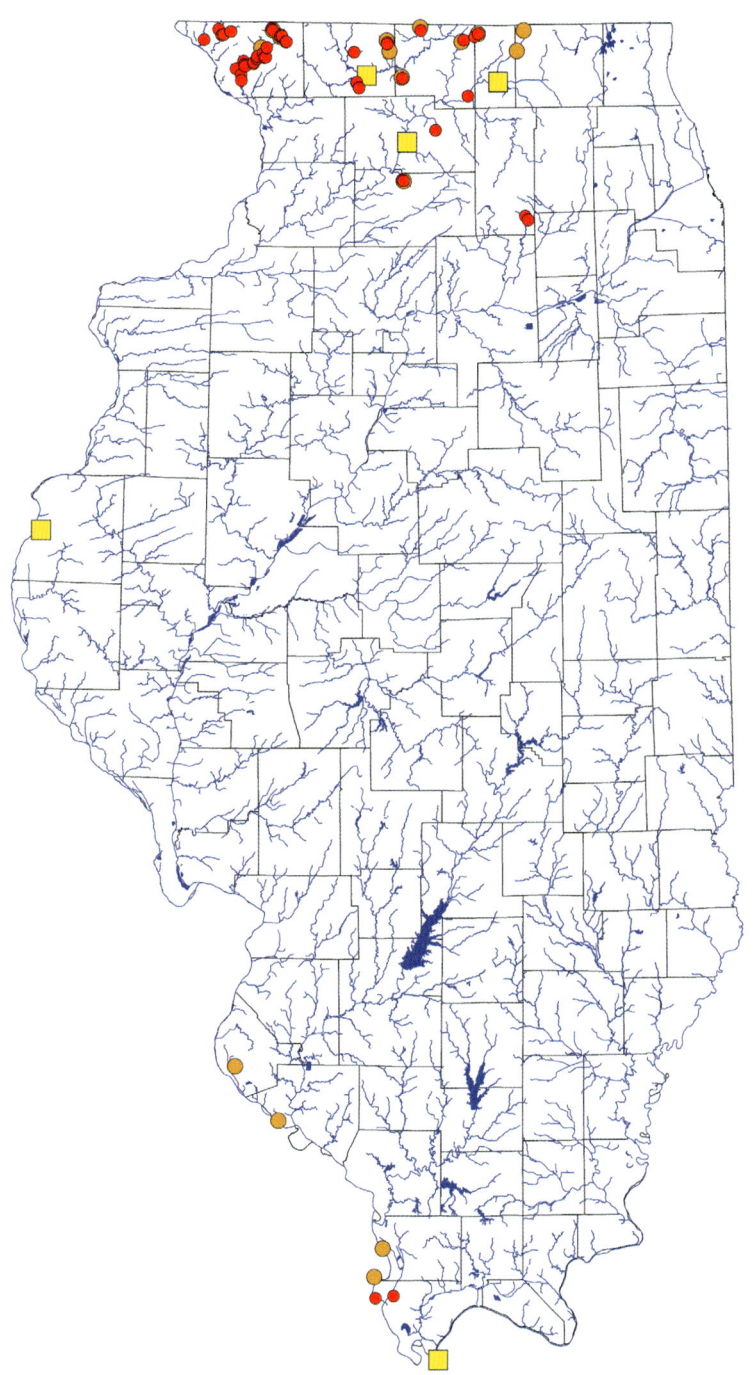

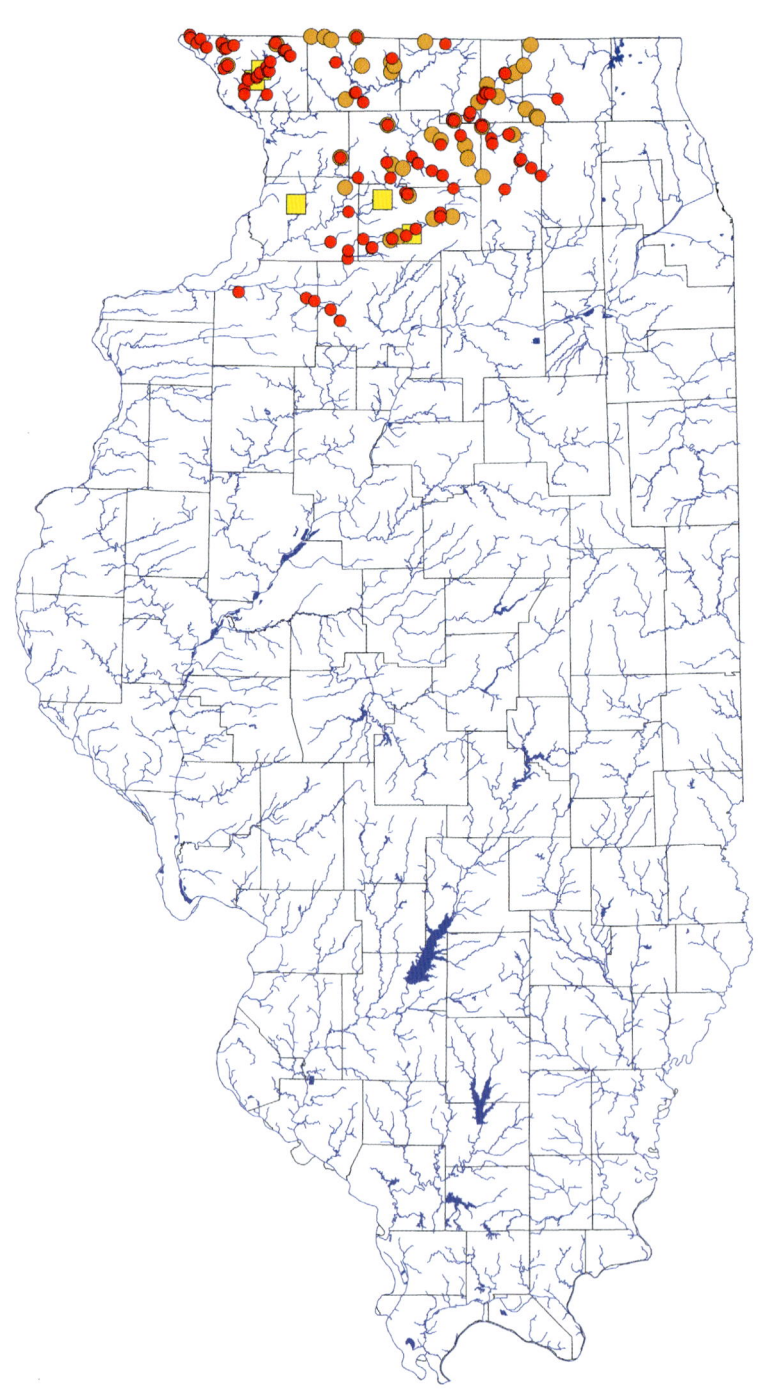

CARMINE SHINER   *Notropis percobromus* (Cope 1871)

IDENTIFICATION   The Carmine Shiner has a slender, compressed body and a sharply pointed snout that is longer than the eye diameter. The dorsal-fin origin is well behind the pelvic-fin origin (over the middle of the pelvic fin). The back and upper side are light olive with dusky margins on the scales. There are a narrow dusky stripe along the back, a faint-red spot at the base of the dorsal fin, a thin black streak immediately above a dusky silver stripe along the side, and a blue sheen overall to the side of the body. The lateral-line pores are outlined by black specks. Large males are pale blue with orange to bright red on the head, anterior half of the body, and fins. Large females often have some red on the head and body. The lateral line is complete with 36–45 scales. There are 9–11 anal rays and 2,4-4,2 or 1,4-4,1 pharyngeal teeth. To 3½ in. (9 cm).

SIMILAR SPECIES   The Rosyface Shiner, *N. rubellus*, is genetically distinct and allopatrically distributed but morphologically indistinguishable from the Carmine Shiner (Scott et al. 2018). The Emerald Shiner, *N. atherinoides*, has a shorter snout, no thin black streak above the dusky silver stripe on the side of the body, and no red on the head or body. The Silver Shiner, *N. photogenis*, has 2 black crescents between the nostrils.

HABITAT   The Carmine Shiner is found in rocky runs and flowing pools of creeks and small to medium rivers, usually in clear water.

DISTRIBUTION IN ILLINOIS   The Carmine Shiner and Rosyface Shiner, *N. rubellus*, were not distinguished in earlier eras, although the distributions of the 2 species in Illinois are allopatric. The Carmine Shiner has essentially the same distribution in the upper Mississippi River basin in northwestern Illinois as in previous eras.

SILVER SHINER *Notropis photogenis* (Cope 1865)

IDENTIFICATION The Silver Shiner has a slender, compressed body, a long snout, and a large, terminal mouth that reaches to below the front of the eye. There are 2 black crescents between the nostrils, and the lips are black anteriorly. The thickened tip of the lower jaw projects forward beyond the upper jaw, the eye is large, and the dorsal-fin origin is behind the pelvic-fin origin. The back and upper side of the body are light olive, and there are a black stripe along the back and a dusky silver stripe with a blue sheen along the side. The lower side and belly are silver-white. The lateral line is complete with 36–40 scales. There are 9 pelvic rays, 10–12 anal rays, and 2,4-4,2 pharyngeal teeth. To 5½ in. (14 cm).

SIMILAR SPECIES The Emerald Shiner, *N. atherinoides*, lacks black crescents between the nostrils and has a smaller eye, a lower jaw not projecting beyond the upper jaw, and 8 pelvic rays. The Rosyface Shiner, *N. rubellus*, and the Carmine Shiner, *N. percobromus*, have sharper snouts, a black streak above a silver stripe along the side, and red on the head and body of large males; both species lack black crescents between the nostrils. The Silverband Shiner, *N. shumardi*, lacks black crescents between the nostrils, has 9 anal rays, a less slender body, and a long and more pointed dorsal fin with the origin over or slightly behind the pelvic-fin origin.

HABITAT The Silver Shiner occupies rocky and mixed gravel-and-sand-bottomed runs and pools near riffles of small to large rivers.

DISTRIBUTION IN ILLINOIS The Silver Shiner is reported here for the first time in Illinois, although it is widespread in the Wabash River basin in neighboring Indiana. It was collected in the Wabash River at 2 localities in Clark County in 2011.

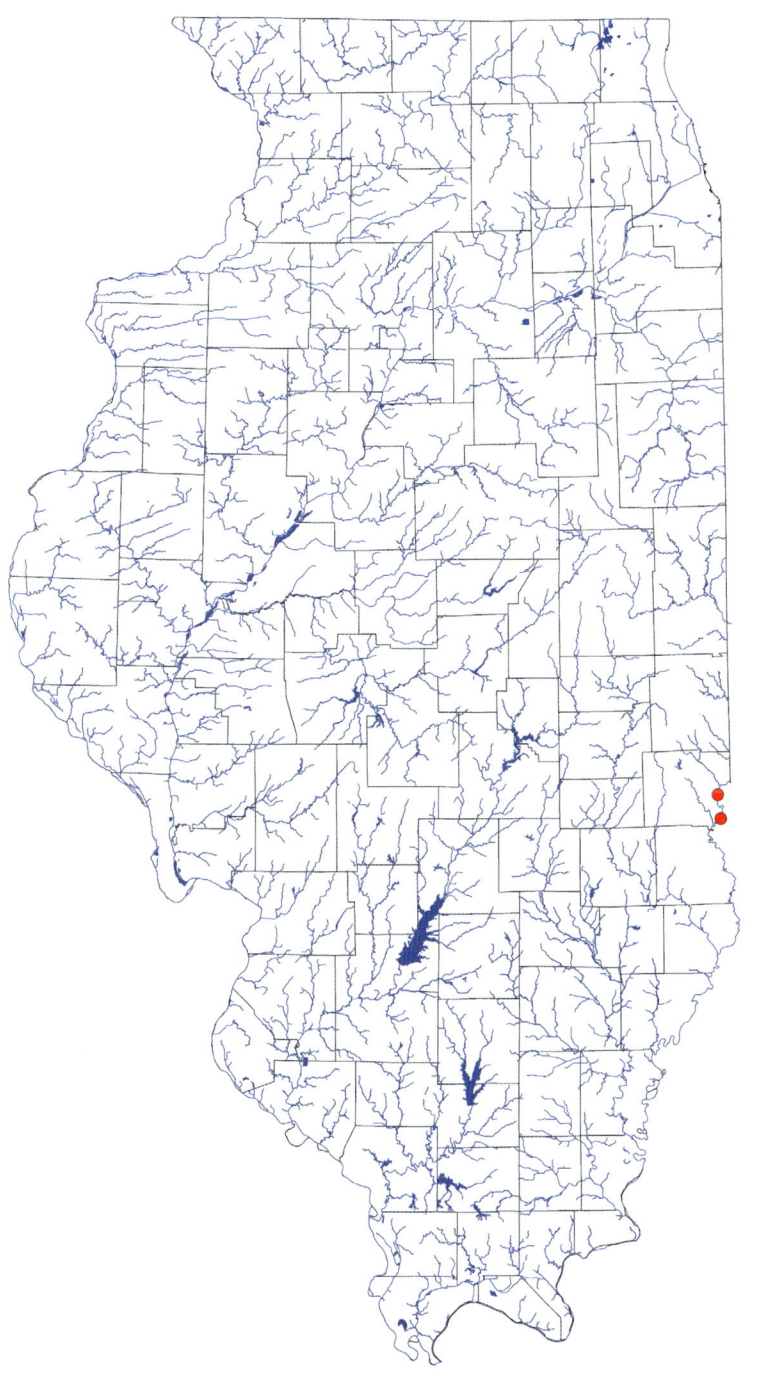

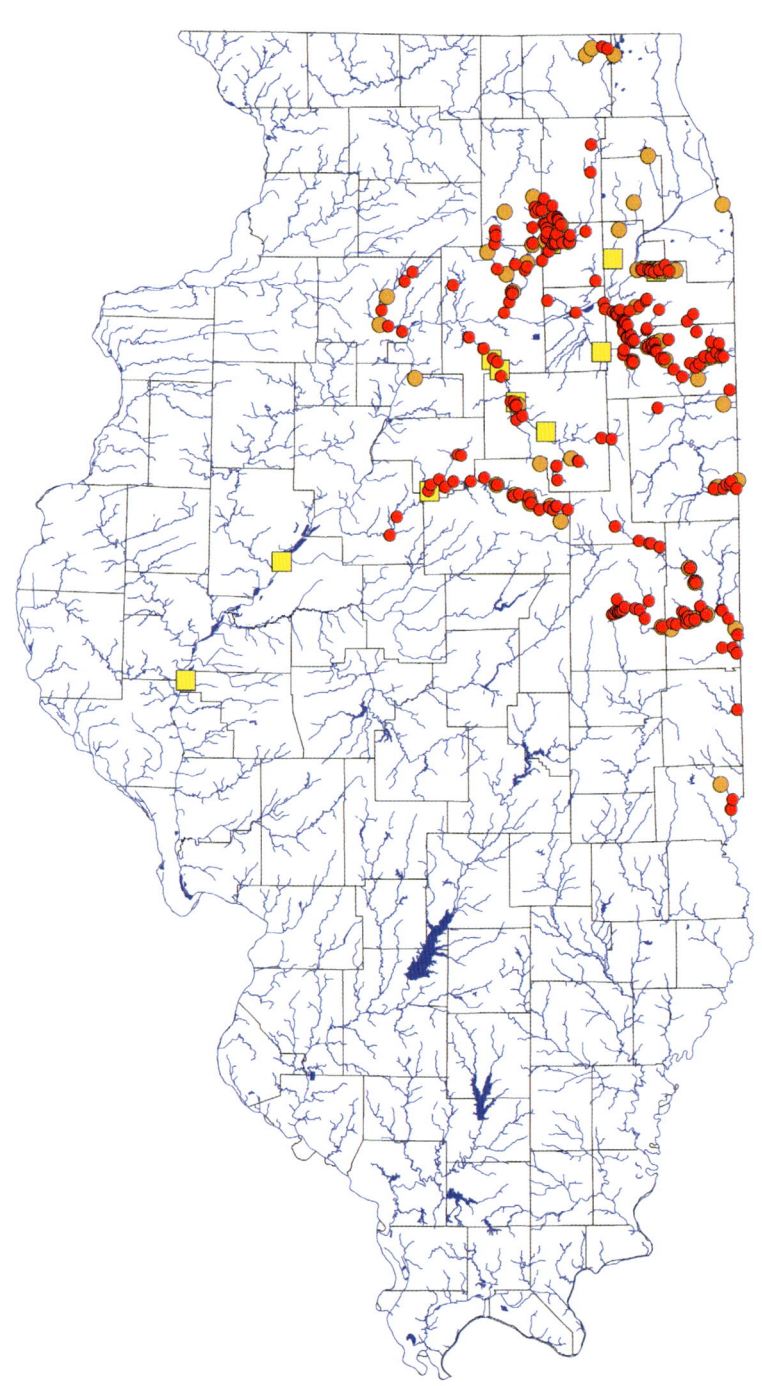

ROSYFACE SHINER   *Notropis rubellus* (Agassiz 1850)

IDENTIFICATION   The Rosyface Shiner has a slender, compressed body and a sharply pointed snout that is longer than the eye diameter. The dorsal-fin origin is well behind the pelvic-fin origin (over the middle of the pelvic fin). The back and upper side are light olive with dusky margins on the scales. There are a narrow dusky stripe along the back, a faint red spot at the base of the dorsal fin, a thin black streak immediately above a dusky silver stripe along the side, and a blue sheen overall to the side of the body. The lateral-line pores are outlined by black specks. Large males are pale blue with orange to bright red on the head, anterior half of the body, and fins. Large females often have some red on the head and body. The lateral line is complete with 36–45 scales. There are 9–11 anal rays and 2,4-4,2 or 1,4-4,1 pharyngeal teeth. To 3½ in. (9 cm).

SIMILAR SPECIES   The Carmine Shiner, *N. percobromus*, is genetically distinct and allopatrically distributed but morphologically indistinguishable from the Rosyface Shiner (Scott et al. 2018). The Emerald Shiner, *N. atherinoides*, has a shorter snout, no thin black streak above the dusky silver stripe on the side of the body, and no red on the head or body. The Silver Shiner, *N. photogenis*, has 2 black crescents between the nostrils.

HABITAT   The Rosyface Shiner is found in rocky runs and flowing pools, often near riffles, of small to medium rivers.

DISTRIBUTION IN ILLINOIS   Earlier surveys did not distinguish between the Rosyface Shiner and Carmine Shiner, although the distributions of the 2 species in Illinois are allopatric. The Rosyface Shiner has essentially the same distribution as in previous eras—in the upper Illinois and upper Wabash River basins, although it no longer is found in the lower Illinois River as it was in the Forbes and Richardson era or along Lake Michigan as in the Smith era.

SILVERBAND SHINER *Notropis shumardi* (Girard 1856)

IDENTIFICATION The Silverband Shiner has a tall, pointed dorsal fin with the front rays extending well beyond the posterior rays when the dorsal fin is depressed. The body is compressed with a relatively deep caudal peduncle, and the dorsal-fin origin is over to slightly behind the pelvic-fin origin and about midway between the tip of the snout and the caudal-fin base. The mouth is terminal and slightly upturned on a short, pointed snout. The back and upper side are light olive, and there is a dusky stripe along the back. The side of the body is silvery, often with a dusky stripe on the posterior half. The lateral line is complete with 33–39 scales. There are 9 pelvic rays (unusual in *Notropis*), 9 (sometimes 8) anal rays, and 2,4-4,2 pharyngeal teeth. To 4 in. (10 cm).

SIMILAR SPECIES The Emerald Shiner, *N. atherinoides*, has 10–12 anal rays, a more slender body, and a shorter dorsal fin with the origin well behind the pelvic-fin origin and closer to the caudal-fin base than to the tip of the snout.

HABITAT The Silverband Shiner lives over sand and gravel in flowing pools and main channels of large, often turbid rivers.

DISTRIBUTION IN ILLINOIS The Silverband Shiner was not recognized by Forbes and Richardson (1909). Smith (1979) found it present but not widespread in the Illinois, lower Mississippi, Wabash, and Ohio Rivers and in a few of the tributaries of these rivers. In the contemporary era it is more widespread in the Illinois and Wabash Rivers, where it occurs much farther north than previously. In contrast, the species is less widespread in the Ohio River, where it is known from only 2 recent records, and in the Big Muddy River, for which there are no contemporary records.

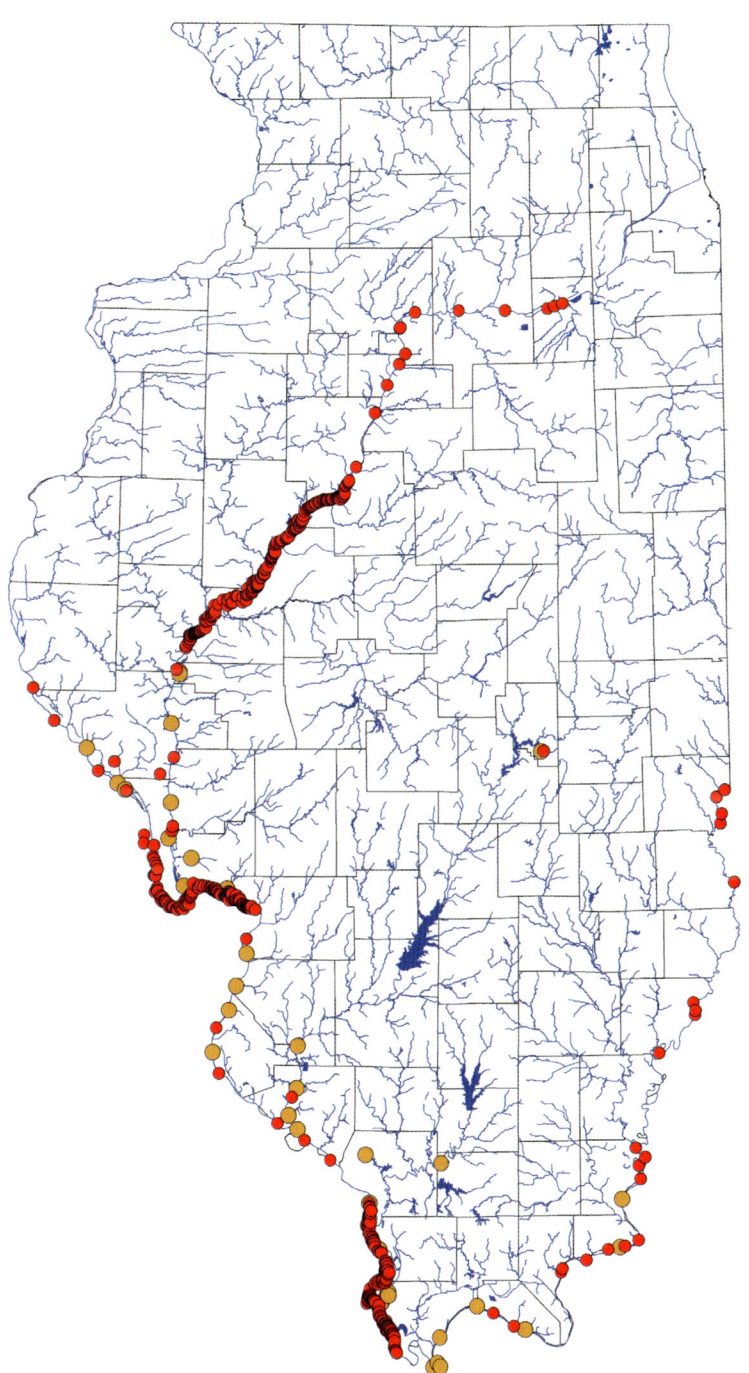

SAND SHINER   *Notropis stramineus* (Cope 1865)

IDENTIFICATION   The Sand Shiner has a punctate lateral line and a dusky black wedge at the dorsal-fin origin. The body is fairly slender and compressed, the snout is rounded, and the mouth is small and slightly subterminal. The dorsal-fin origin is over to slightly behind the pelvic-fin origin. There is a nipple at the front of the pupil of the fairly large eye. The back and upper side are light straw-yellow with darkly outlined scales, and there is a dusky stripe along the back. The side of the body is silvery, often with a dusky stripe—usually darker on the posterior half. There is a small, black wedge-shaped caudal spot. Fins are clear but often tinged with white on large individuals. The lateral line is complete with 31–38 lateral scales. There are 7 anal rays and 0,4-4,0 pharyngeal teeth. To 3¼ in. (8.1 cm).

SIMILAR SPECIES   The Mimic Shiner, *N. volucellus*, lacks the dusky stripe along the back and the wedge at the dorsal-fin origin, has 8 anal rays, scales along the back are wider than those on the upper side, and the lateral-line scales on the anterior half of the body are much deeper than wide. The Bigmouth Shiner, *N. dorsalis*, has an upwardly directed eye and 8 anal rays and lacks the dark wedge at the dorsal-fin origin. The River Shiner, *N. blennius*, lacks a dusky black wedge at the dorsal-fin origin and a punctate lateral line and has a larger mouth.

HABITAT   The Sand Shiner, appropriately named, is found along the sandy margins of lakes and sandy or mixed sand and gravel runs and pools of creeks and small to large rivers.

DISTRIBUTION IN ILLINOIS   As in previous eras, the Sand Shiner occurs nearly statewide but is much less widespread in the south-central part of the state than elsewhere and is absent from extreme southern Illinois except for the Clear Creek system. It is now in the Saline River system, where it previously was absent and is more widespread in the Skillet Fork–Little Wabash River system.

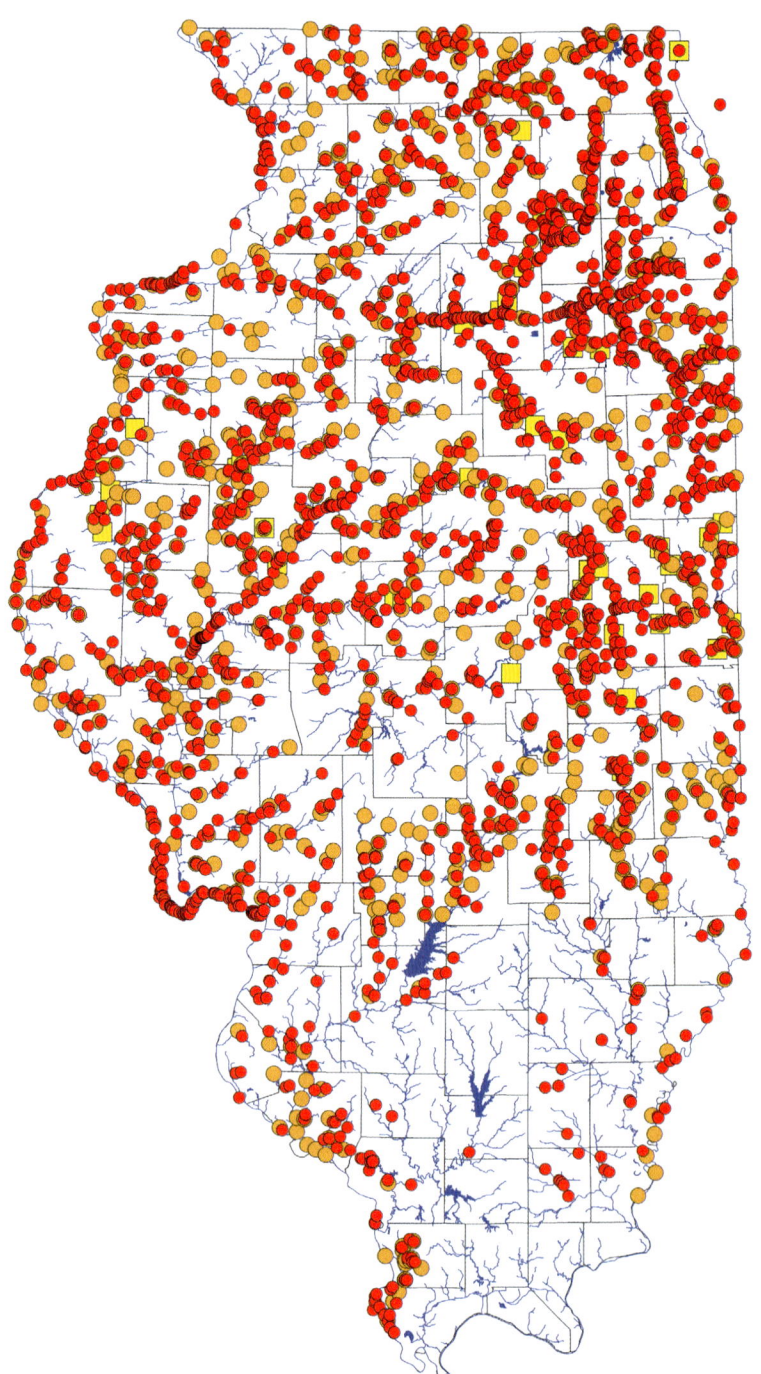

MINNOWS (Leuciscidae)

200

WEED SHINER *Notropis texanus* (Girard 1856)

IDENTIFICATION The Weed Shiner has a black stripe along the silvery side of the body. The stripe extends around the snout on both lips and posteriorly to a black spot at the base of the caudal fin; black streaks extend from the spot to the end of the caudal fin. There is a pale yellow stripe between the black stripe and black-edged scales on the light olive-yellow upper side. Scales just below the black stripe have black edges, creating a crosshatched pattern on the lower side. A dark stripe along the back is much wider in front of than behind the dorsal fin and is often expanded into a blotch at the front of the dorsal fin. The body is fairly compressed, and the dorsal-fin origin is in front of the pelvic-fin origin. The snout is fairly blunt, and the terminal mouth is small with the posterior edge reaching under the nostril. The lateral line is complete or nearly complete with 32–39 scales. There are usually 7 anal rays and 2,4-4,2 pharyngeal teeth. To 3½ in. (8.6 cm).

SIMILAR SPECIES The Blackchin Shiner, *N. heterodon*, and Bigeye Shiner, *N. boops*, have a pointed snout, the dorsal-fin origin over the pelvic-fin origin, usually 8 anal rays, and 1,4-4,1 pharyngeal teeth. The Ironcolor Shiner, *N. chalybaeus*, lacks black-edged scales below the lateral line and has a more arched body, a more pointed snout, 8 anal rays, and black inside the mouth.

HABITAT The Weed Shiner lives near vegetation in clear, sandy runs and pools of creeks and small to medium rivers.

DISTRIBUTION IN ILLINOIS The Weed Shiner is much less widespread than it was in previous eras. It remains widespread in the Kankakee River basin in Iroquois, Kankakee, and Will Counties and in the Green and Rock River basins in Bureau, Henry, and Whiteside Counties. Other populations in northern Illinois, the middle Illinois River, and the Wabash River have disappeared.

REMARKS Individuals in Illinois tend to be deeper bodied and more yellow than those from southern states, and the black spot at the base of the caudal fin is continuous with the black stripe along the side. Most significant, they lack the black pigment along the posterior rays of the anal fin.

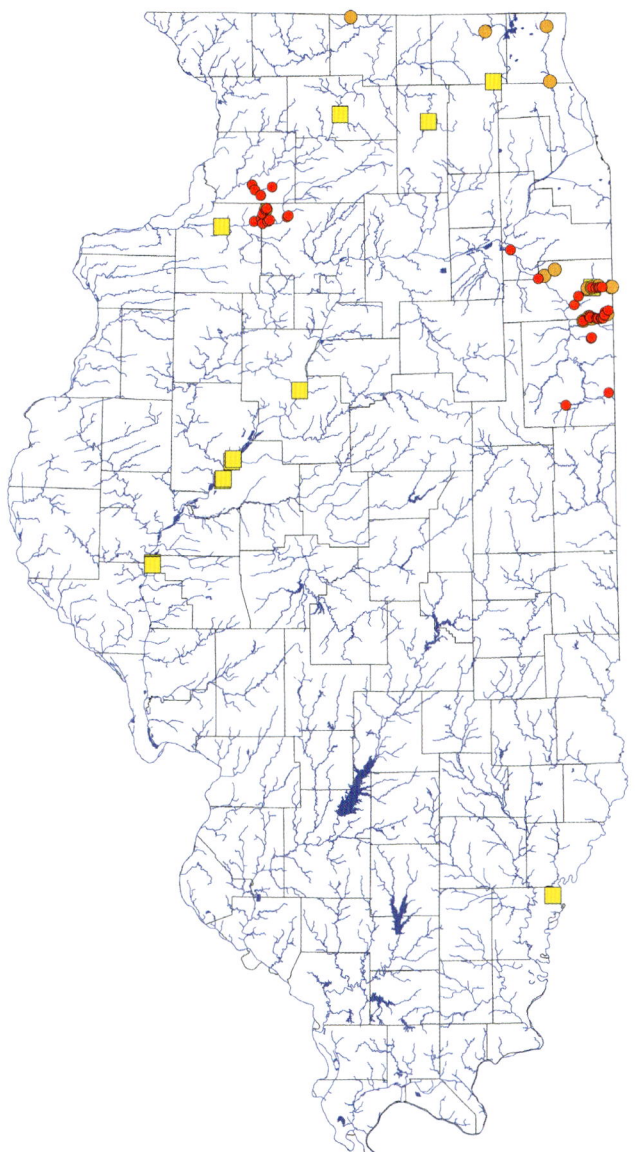

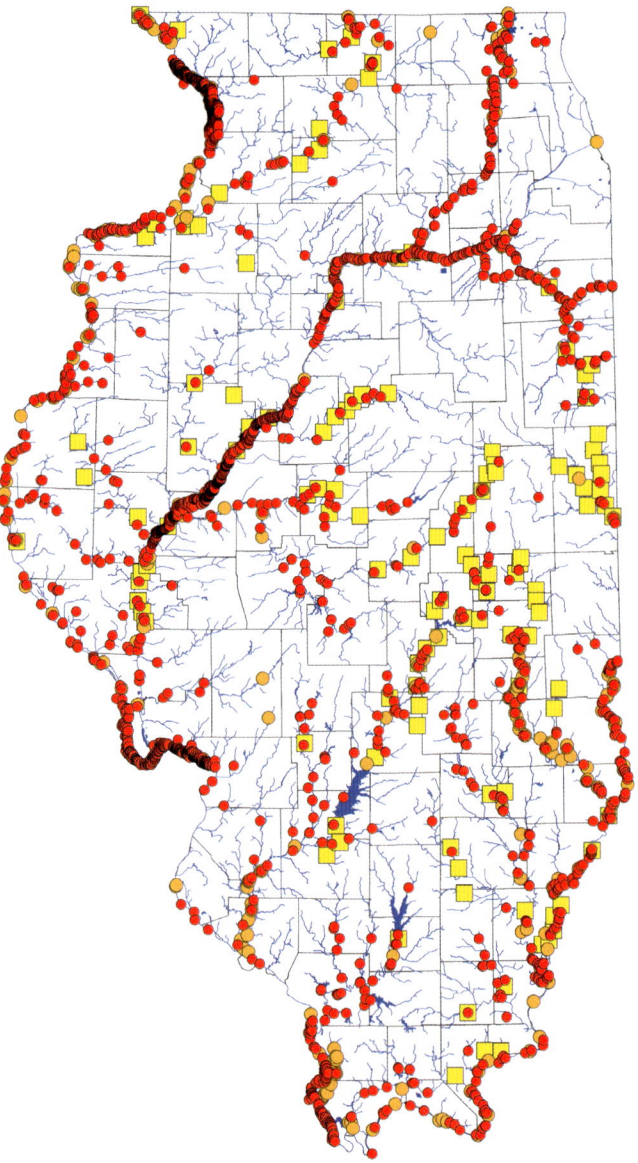

BULLHEAD MINNOW   *Pimephales vigilax* (Baird & Girard 1853)

IDENTIFICATION The Bullhead Minnow has a somewhat compressed body that is squarish in cross section anteriorly and more slender posteriorly. The top of the head and nape is flattened. The snout is blunt and overhangs the small, terminal mouth. The eye is large, on the upper half of the head, and directed somewhat upward. The dorsal-fin origin is over the pelvic-fin origin, and the second ray of the dorsal fin is short and stout. Scales on the nape are much smaller (usually more than 20 in a row from the head to the dorsal fin) than elsewhere on the body. The back and upper side of the body are light to dark olive-tan, with dark-edged scales forming a crosshatched pattern. There is a dusky to black stripe around the snout and along the silver-bluish-white side that ends in a conspicuous black spot at the base of the caudal fin. The intestine is short, and the peritoneum is silvery with black specks (belly appears white). The fins are clear except for a black blotch at the front of the dorsal fin about midway from the body. During the breeding season, large males are black with a silver bar along the posterior edge of the opercle and have a large, gray fleshy pad on the nape and 5–9 large tubercles in 1–2 rows on the snout. The lateral line is complete with 37–45 scales, and there are usually 7 anal rays and 0,4-4,0 pharyngeal teeth. To 3½ in. (8.9 cm).

SIMILAR SPECIES The Bluntnose Minnow, *P. notatus*, has a smaller head, the eye lower on the head and directed laterally, a more subterminal mouth, a dusky black belly, and no bluish sheen on the side of the body. The Fathead Minnow, *P. promelas*, has a deeper body with herringbone lines on the upper side and an incomplete lateral line.

HABITAT The Bullhead Minnow lives in pools and runs over sand, silt, or gravel in small to large rivers. It is most frequently encountered in medium and large rivers.

DISTRIBUTION IN ILLINOIS Forbes and Richardson (1909) found the Bullhead Minnow to be nearly statewide in distribution. Smith (1979) also found it to be statewide but absent from some of the rivers where it had been recorded previously. In the contemporary era, the Bullhead Minnow has returned to most of those rivers except, notably, the upper reaches of the Vermilion River basin in Vermilion County.

FLATHEAD CHUB *Platygobio gracilis* (Richardson 1836)

IDENTIFICATION The Flathead Chub has a slender body that is deepest and widest anteriorly and more strongly compressed at the narrow caudal peduncle. The head is broad and flat, the snout is pointed, and there is a small barbel in the corner of the large, subterminal mouth. The dorsal-fin origin is in front of the pelvic-fin origin. The dorsal and pectoral fins are large and sickle shaped; the first dorsal ray extends beyond the last ray when the fin is depressed. The back and upper side of the body are light dusky brown or olive, the side of the body is silver-white often with a faint, thin dusky stripe, and the underside is white. Fins are mostly transparent, but the lower lobe of the caudal fin is dusky black, and the pectoral fin is orange in large individuals. The lateral line is complete with 42–59 scales, and there are 8 anal rays and 2,4-4,2 pharyngeal teeth. To 12½ in. (32 cm).

SIMILAR SPECIES The Sicklefin Chub, *Macrhybopsis meeki*, and Sturgeon Chub, *M. gelida*, have a narrower head, small papillae on the throat or underside of the head, the dorsal-fin origin over or behind the pelvic-fin origin, and 1,4-4,1 pharyngeal teeth. On the Sicklefin Chub, the tip of the pectoral fin reaches beyond the pelvic-fin origin. On the Sturgeon Chub, the first dorsal ray does not extend beyond the last ray when the fin is depressed.

HABITAT The Flathead Chub lives in sandy runs of small to large rivers. In Illinois it is restricted to the sandy shoreline of the Mississippi River.

DISTRIBUTION IN ILLINOIS In the Forbes and Richardson era, the Flathead Chub was known in Illinois only from the Mississippi River at the mouth of the Ohio River. In contrast, Smith (1979) found it to be widespread in the Mississippi River below the mouth of the Missouri River. In the contemporary era, this species has nearly disappeared from Illinois, with only 2 recent records from the Mississippi River near Grand Tower in Jackson County.

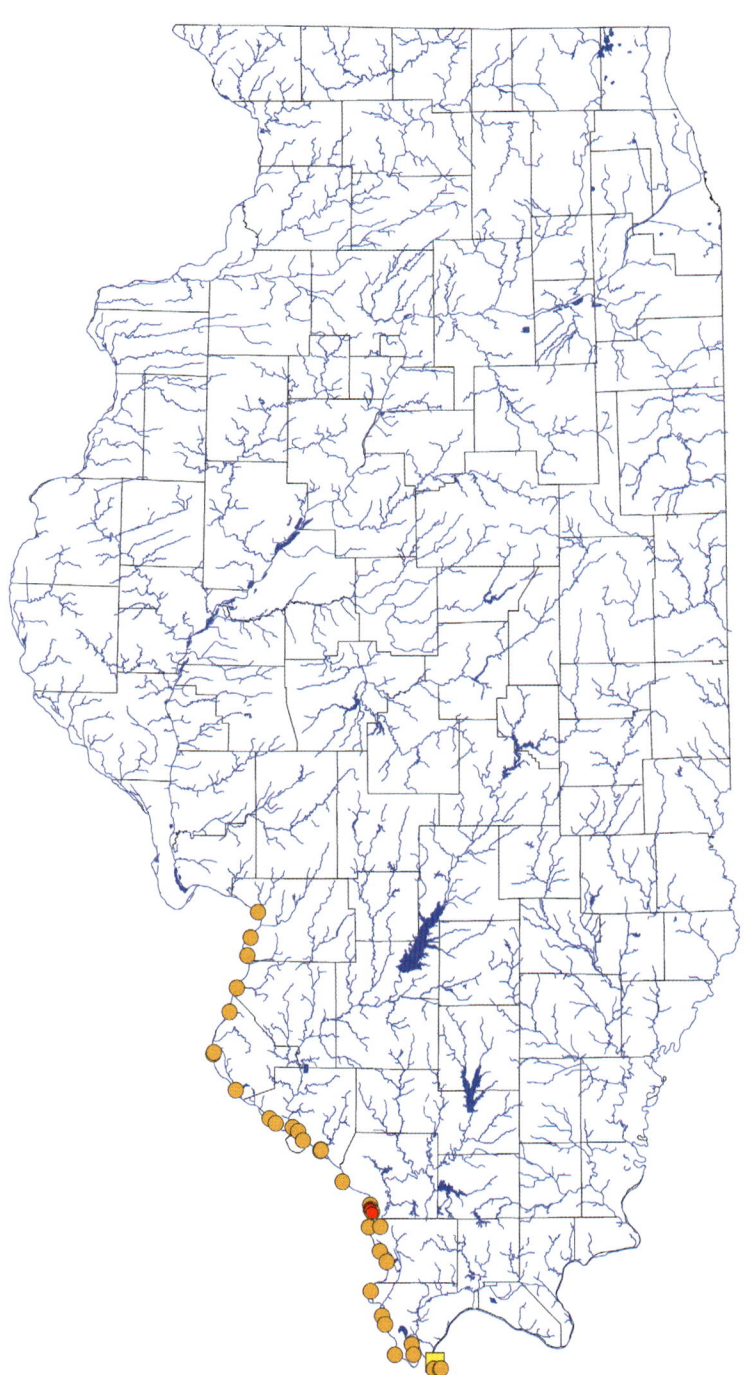

BLUEHEAD SHINER *Pteronotropis hubbsi* (Bailey & Robison 1978)

IDENTIFICATION The Bluehead Shiner has a deep compressed body, deepest near the dorsal-fin origin, tapering to a narrow caudal peduncle. The snout is short and blunt, and the mouth is terminal. The dorsal-fin origin is behind the pelvic-fin origin. The back and upper side of the body are dusky orange-brown with the scales outlined in black. There is a dusky black stripe along the back from the head to the dorsal fin, and a wide black stripe along the side of the body from the chin (but not the upper lip and snout) to the caudal fin, where it expands into a black spot. There is often a light orange stripe above the black stripe. The dorsal fin is dusky; other fins are clear to faint yellow or orange. Large males have greatly enlarged dorsal and anal fins and bright iridescent blue on the top of the head and on the dorsal, anal, caudal, and pelvic fins. The lateral line is incomplete with 2–9 pores. There are 34–38 lateral scales, 9–10 dorsal rays, 9–10 anal rays, and 0,4-4,0 pharyngeal teeth. To 2¼ in. (6 cm).

SIMILAR SPECIES The Ironcolor Shiner, *Notropis chalybaeus*, has the black stripe along the side extending onto the upper lip and snout, the dorsal-fin origin over the pelvic-fin origin, 8 dorsal and anal rays, and 2,4-4,2 pharyngeal teeth; it lacks blue color and enlarged fins.

HABITAT In Illinois the Bluehead Shiner is known only from a single oxbow lake (Wolf Lake) that is lined with woody debris and submergent and emergent vegetation. The water is tannin-stained, at least 6 ft. (1.8 m) deep, and the substrate is mud and organic matter. Elsewhere, the species is found in swamps, backwaters, oxbows, and sluggish pools of creeks and small rivers usually near vegetation over mud or sand.

DISTRIBUTION IN ILLINOIS The Bluehead Shiner was discovered in Wolf Lake in Union County in 1954. No Bluehead Shiners have been found in Wolf Lake since 1974, and an attempt to reestablish a population with several hundred adults from Caddo Lake, Texas, in the 1990s, was unsuccessful.

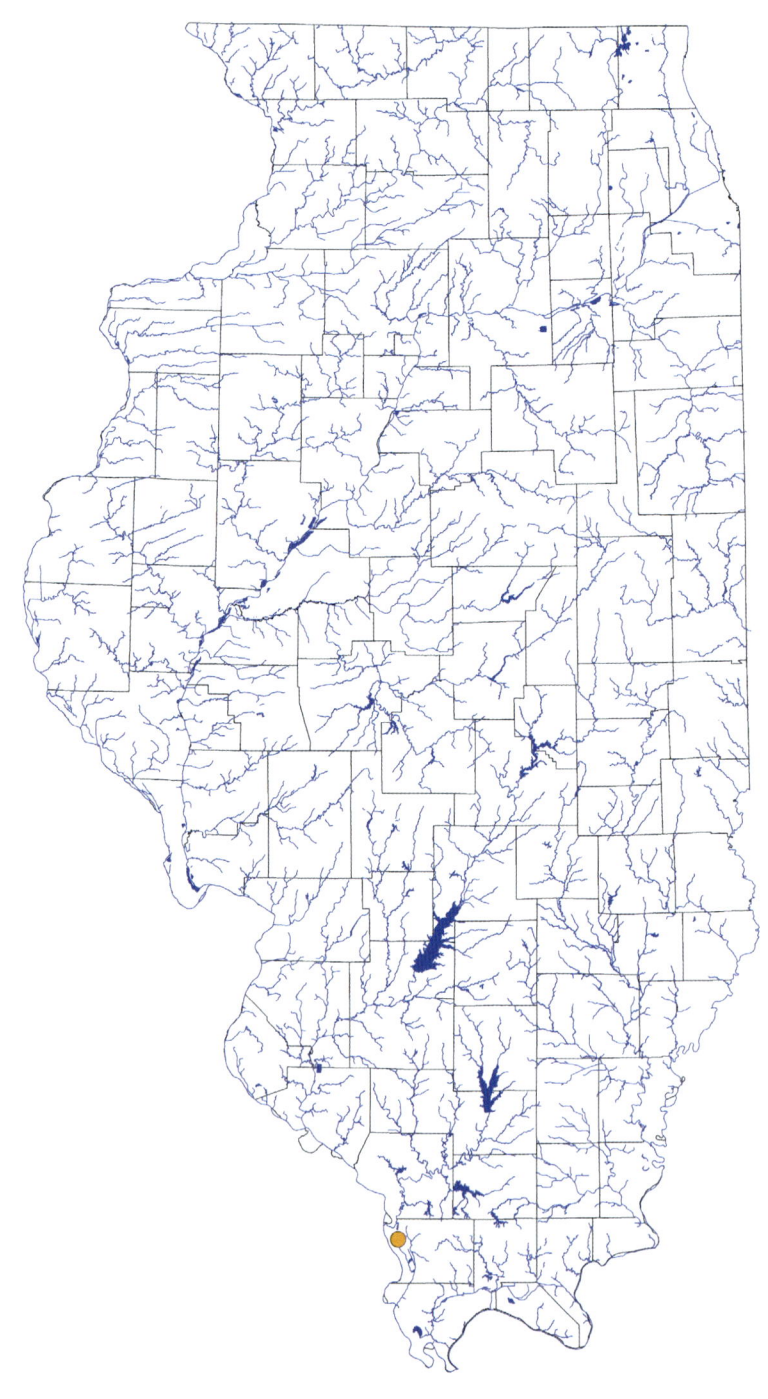

MINNOWS (Leuciscidae)

210

LONGNOSE DACE   *Rhinichthys cataractae* (Valenciennes 1842)

IDENTIFICATION The Longnose Dace has a slender body, flattened below, deepest at the nape but with a deep caudal peduncle. The eyes are high on the head; the snout is long and fleshy and projects well beyond the upper lip. There is a barbel in the corner of the mouth; no groove separates the upper lip from the snout. The dorsal-fin origin is behind the pelvic-fin origin. The back and side of the body are olive-brown to dark red-purple, with brown-black spots and mottling. The lower side of the body and belly are silver-white to yellow, strongly contrasting with the dark upper side to create a strongly bicolored appearance. There often are a dusky black stripe along the side (darkest on young) and a dusky black spot on the caudal-fin base. Fins are clear to dusky yellow-orange. Large males have bright red on the lips and along the bases of fins. The lateral line is complete with 61–75 lateral scales, and there usually are 8 anal rays and 2,4-4,2 pharyngeal teeth. To 6¼ in. (16 cm).

SIMILAR SPECIES The Western Blacknose Dace, *R. obtusus*, has a less projecting snout, eyes lower on the side of the head, and 7 anal rays.

HABITAT The Longnose Dace occurs in rocky riffles, runs, and shallow pools of creeks and small to medium rivers with fast current. In Illinois it is found on the rocky shores of Lake Michigan.

DISTRIBUTION IN ILLINOIS In the Forbes and Richardson era, the Longnose Dace was known to occur only along the shore of Lake Michigan and at 3 localities in extreme southern Illinois. Smith (1979) recorded the species along the shore of Lake Michigan and in extreme northwestern Illinois in fast-flowing streams in the Driftless Area; however, the species had disappeared from southern Illinois. In the contemporary era, it has the same distribution as during the Smith era.

REMARKS The records from the Forbes and Richardson era for Hardin, Union, and Alexander Counties in southern Illinois are far removed from the known range of the Longnose Dace, which does not occur in adjacent areas of Missouri, Kentucky, or Indiana and must have represented relictual populations of a much-wider distribution for the species.

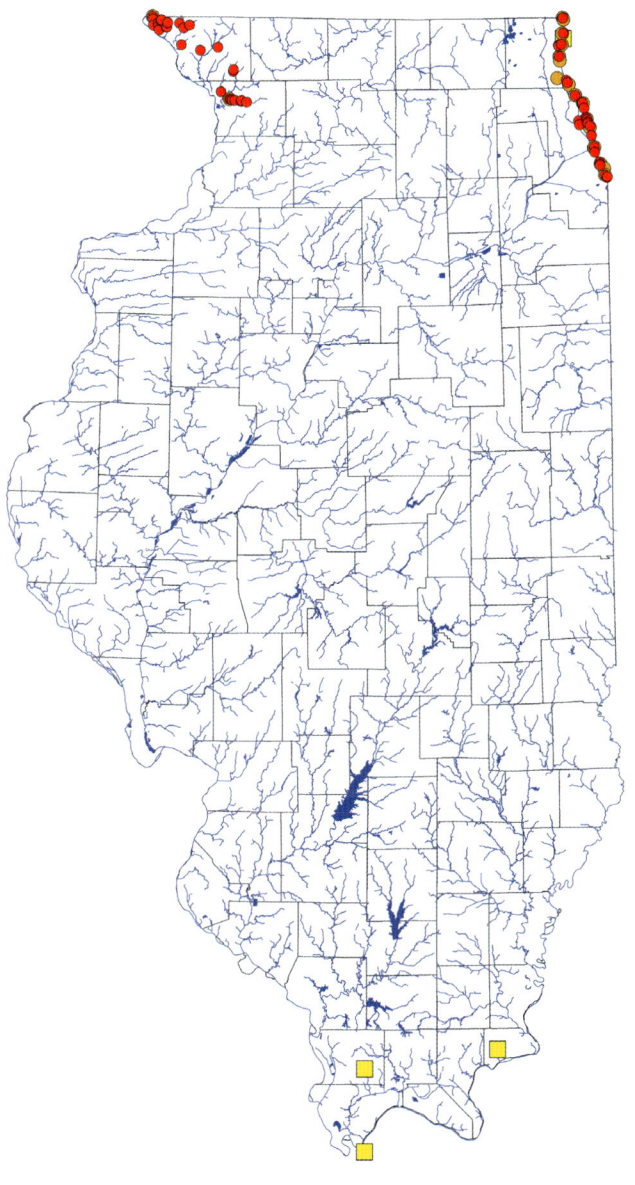

WESTERN BLACKNOSE DACE   *Rhinichthys obtusus* Agassiz 1854

IDENTIFICATION The Western Blacknose Dace has a slender body, flattened below, and deepest at the nape but also with a fairly deep caudal peduncle. There is a barbel in the corner of the mouth, and no groove separates the upper lip from the snout. The dorsal-fin origin is behind the pelvic-fin origin. The snout is long, fleshy, and projects slightly beyond the upper lip. The back and upper side of the body are light brown to gray, with brown-black spots and mottling, and a black spot followed by a silver spot at the base of the dorsal fin. The lower side of the body and belly are silver-white, strongly contrasting with the dark upper side to create a bicolored appearance. There is usually a black stripe along the side (continuous in juveniles, as blotches in adults) extending forward through the eye and onto the snout and posteriorly to a dusky black spot on the caudal-fin base. There is often a silver stripe above the black stripe. Fins are clear to dusky yellow. Large males have yellow-white pectoral and pelvic fins and a brick-red stripe below the black stripe on the side. The lateral line is complete with 53–70 lateral scales, and there are 7 anal rays and 2,4-4,2 pharyngeal teeth. To 4 in. (10 cm).

SIMILAR SPECIES The Longnose Dace, *R. cataractae*, has a longer, more projecting snout, eyes higher on the head, and usually 8 anal rays.

HABITAT The Western Blacknose Dace is found most frequently in rocky riffles, runs, and pools of fast-flowing headwaters, creeks, and small rivers.

DISTRIBUTION IN ILLINOIS In the Forbes and Richardson era, the Western Blacknose Dace was known to occur only in the Rock and Illinois River basins, along the shore of Lake Michigan, and in Union County in extreme southern Illinois. Smith (1979) found the species to be more widespread in the Illinois River basin, including in the Kankakee River system, and in tributaries of the Mississippi and Wabash Rivers. However, the species had disappeared from southern Illinois. In the contemporary era, it seems to be at least as widespread as during the Smith era and has been found in additional tributaries of the Mississippi River in northern Illinois and in tributaries of the lower Illinois River.

REMARKS Forbes and Richardson (1909) treated this species as *Rhinichthys atronasus*, and Smith (1979) treated it as *R. atratulus*, the Blacknose Dace. The records in southern Illinois from Forbes

and Richardson are far outside the known range of the Western Blacknose Dace, which does not include adjacent areas of Missouri or Kentucky, and must have represented relictual populations from a much-wider distribution of the species.

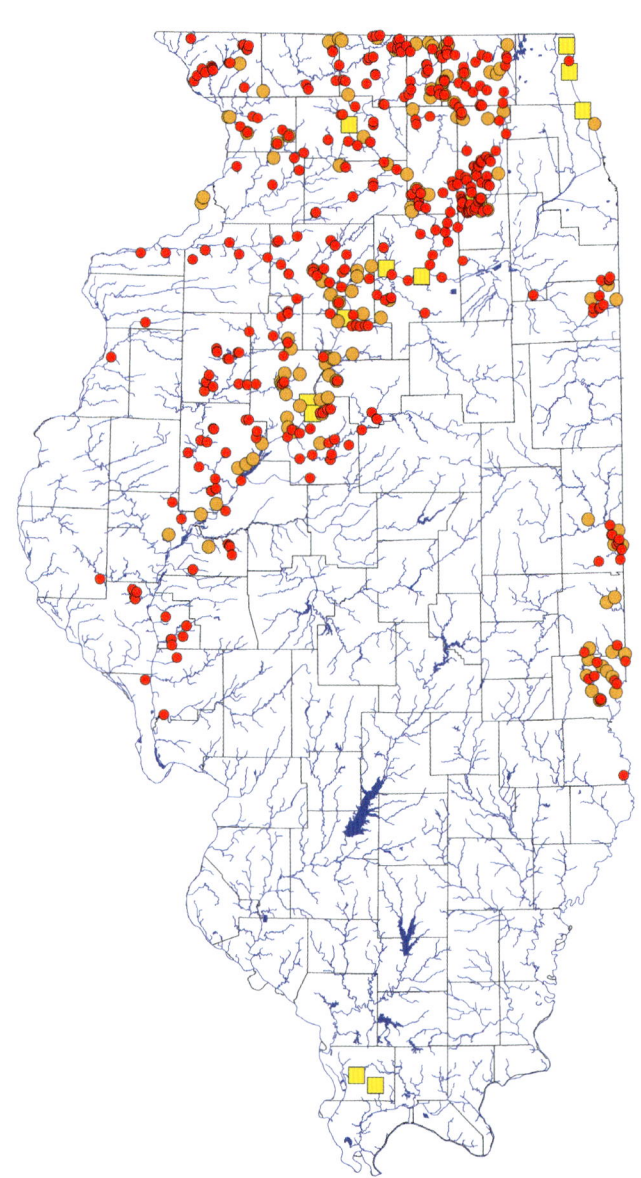

CREEK CHUB *Semotilus atromaculatus* (Mitchill 1818)

IDENTIFICATION The Creek Chub has a robust body, deep and barely compressed anteriorly, slender and compressed posteriorly; a large head that is somewhat flattened above; a moderately pointed snout; and a large, terminal mouth that reaches past the anterior margin of the eye. There is a small, flaplike barbel in the groove above the upper lip and near the corner of the mouth. The dorsal-fin origin is behind the pelvic-fin origin. The back and upper side of the body are gray-brown with a dark stripe along the back; juveniles have herringbone lines on the upper side. There is a large, black spot anteriorly at the base of the dorsal fin, a black bar along the posterior margin of the gill cover, and a dusky black stripe along the yellow-olive side of the body; the stripe extends anteriorly around the snout and onto the upper lip and ends posteriorly at a black caudal spot (often indistinct in large individuals). Large males have orange at the base of the dorsal fin, blue on the side of the head, pink on the lower half of the head and body, orange anal and paired fins, and 6–12 large tubercles on the head. The lateral line is complete with 47–65 lateral scales, and there are 8 dorsal rays, 8 anal rays, and 2,5-4,2 pharyngeal teeth. To 12 in. (30 cm).

SIMILAR SPECIES The Lake Chub, *Couesius plumbeus*, lacks a large, black spot at the front of the dorsal-fin base and has a barbel at the corner (not in the groove above the upper lip) of the mouth, and large males have red at the corner of the mouth and at the pectoral fin and pelvic-fin origins. The Hornyhead Chub, *Nocomis biguttatus*, and River Chub, *Nocomis micropogon*, lack a large, black spot at the front of the dorsal-fin base, have a barbel at the corner of the mouth, 36–45 lateral scales, usually 7 anal rays, and 0,4-4,0 or 1,4-4,1 pharyngeal teeth.

HABITAT The Creek Chub is most frequently encountered in rocky and sandy pools of headwaters, creeks, and small rivers, less so in medium and large rivers. In Illinois, it is most often found over mud or clay near undercut banks and tree roots.

DISTRIBUTION IN ILLINOIS The Creek Chub is statewide in distribution, as it was in earlier eras. Although Smith (1979) suggested that the Creek Chub was absent from the Lake Michigan basin, he overlooked several records from the basin. The Creek Chub was in Lake Michigan in the 2 previous eras as it is in the contemporary era.

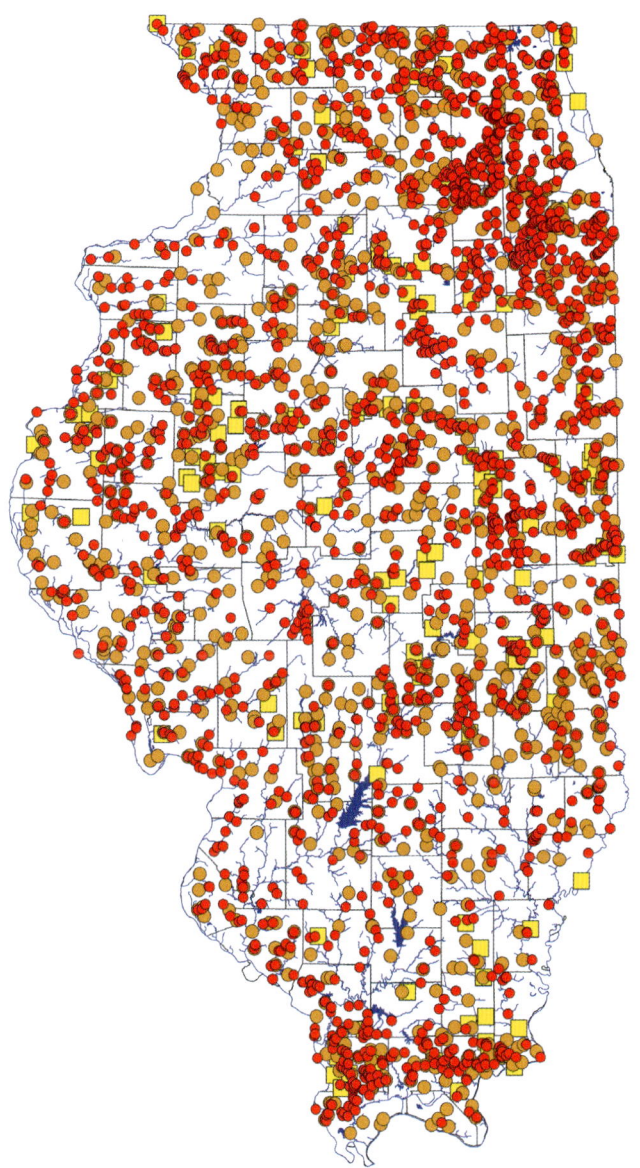

MINNOWS (Leuciscidae)

213

SPINED LOACHES

Family Cobitidae Swainson 1838

Spined loaches have a long, slender body, an erectile bifid spine under the eye (hidden in some species), a subterminal mouth with 3–6 pairs of barbels, tiny or no scales, and 1 row of pharyngeal teeth. Some species have an adipose fin. Males usually are smaller than females, and in many genera males have modified pectoral rays. Spined loaches are bottom-dwelling fishes native to the fresh waters of Eurasia and Morocco where they occupy a range of habitats from fast mountain streams to lowland streams and swamps. Some species occasionally bury in soft substrates with only the head exposed. At least 221 species are known, with the greatest diversity in Southeast Asia. Some species are popular as aquarium fishes, and 1, the Oriental Weatherfish, *Misgurnus anguillicaudatus*, has repeatedly been released and established in the United States, beginning in the 1930s.

ORIENTAL WEATHERFISH *Misgurnus anguillicaudatus* (Cantor 1842)

IDENTIFICATION The Oriental Weatherfish is long and slender with 10–12 barbels around the mouth, a small eye, a low adipose crest, and rounded dorsal and caudal fins. The body is gray-brown, dark above and light below with many dusky to black spots and blotches on the back and side of the body. The dorsal and caudal fins have small, black spots, and there is a bold black spot at the upper edge of the caudal-fin base. Scales are very small. There is a large, stout spine on the pectoral fin and a small, subcutaneous spine under the eye. To 10 in. (25 cm).

SIMILAR SPECIES No other fish In Illinois has a long, slender body with more than 8 barbels around the mouth.

HABITAT The Oriental Weatherfish lives in mud-bottomed pools and backwaters, often burying in the substrate. It is capable of swallowing air with oxygen, which is then absorbed in the intestine, and often occupies poorly oxygenated water where other fishes cannot survive.

DISTRIBUTION IN ILLINOIS The Oriental Weatherfish is native to eastern Asia but is common in the aquarium trade and has been released and is established in streams in several places in the United States, including northeastern Illinois, southern Florida, and the Pacific Northwest. It was first recorded in Illinois in 1987 in an artificial canal near Chicago. It is reproducing and is now widely distributed in the upper Illinois River basin as far south as Will County.

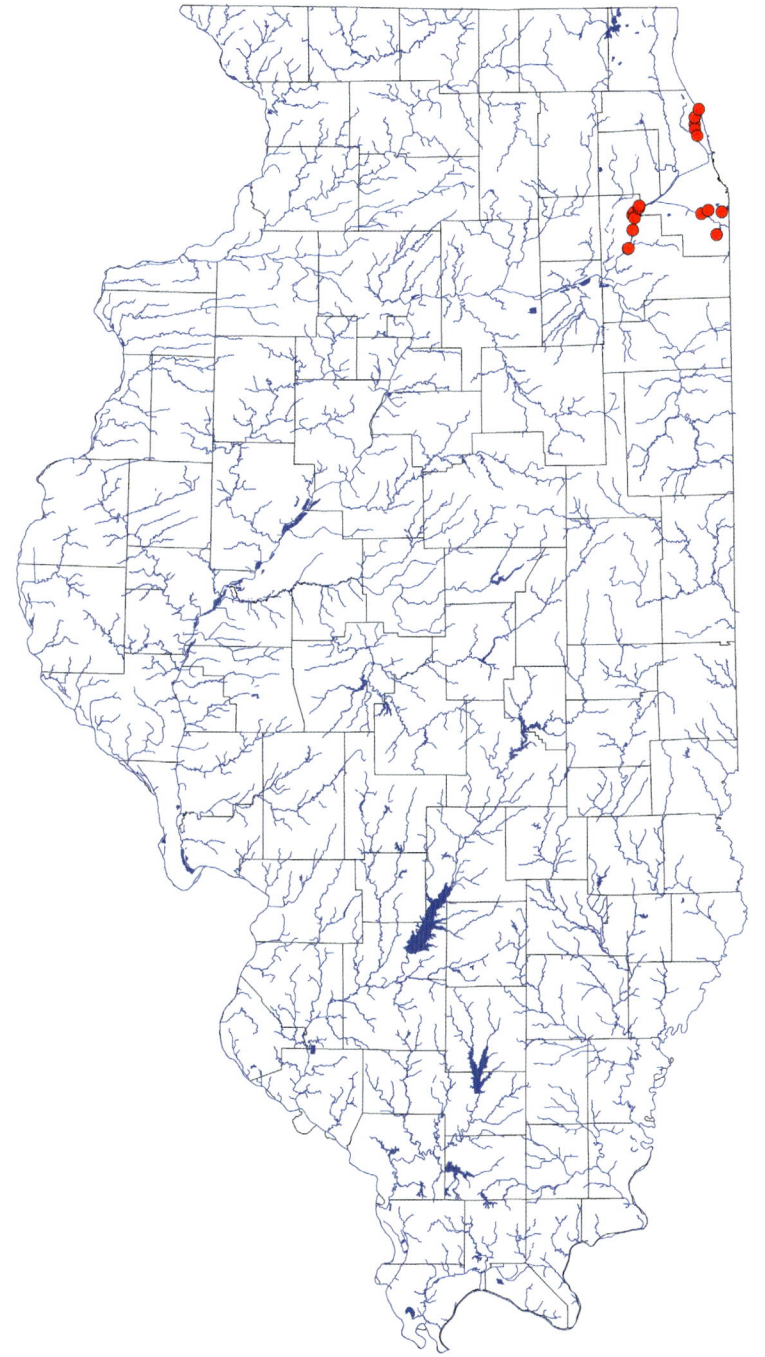

NORTH AMERICAN CATFISHES

Family Ictaluridae Gill 1861

North American catfishes have 8 barbels ("whiskers"), 4 on the chin, 2 on the snout, and 1 on each corner of the mouth, an adipose fin, stout spines in the dorsal and pectoral fins, and no scales. The family is endemic to North America, ranging from Canada to Guatemala and Belize. Larger species, especially the widely marketed Channel Catfish, *Ictalurus punctatus*, are of major commercial and angling value. The Flathead Catfish, *Pylodictis olivaris*, and Blue Catfish, *Ictalurus furcatus*, reach around 130 lb. (60 kg) and over 60 in. (150 cm) in length. The smaller species, referred to as madtoms (*Noturus*), have venom in cells surrounding fin spines and are capable of stinging. Of the 51 species in the family, 13 are native to Illinois, and 1, the White Catfish, *Ameiurus catus*, has been introduced.

KEY TO THE NORTH AMERICAN
CATFISHES (Ictaluridae)

1a Adipose fin a free lobe, widely separated from caudal fin **2**

1b Adipose fin a low, keel-like ridge, connected to caudal fin or with, at most, a slight notch between them **8**

Madtoms (*Noturus*)

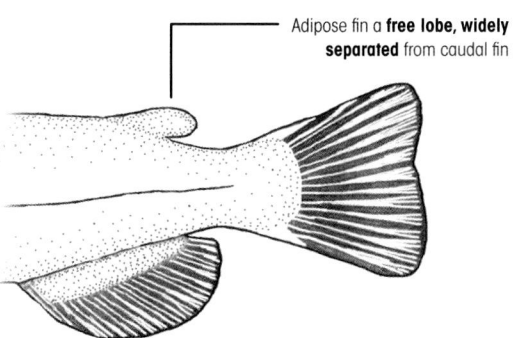

Adipose fin a **free lobe, widely separated** from caudal fin

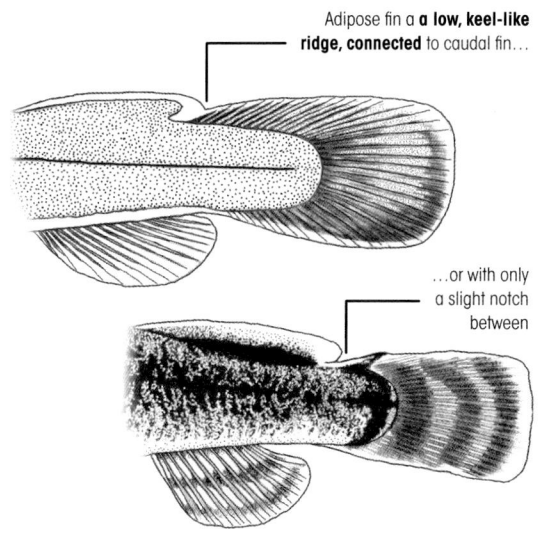

Adipose fin a **a low, keel-like ridge, connected** to caudal fin…

…or with only a slight notch between

2a Caudal fin moderately to deeply forked, center of fin distinctly shorter than upper and lower margins **3**

2b Caudal fin rounded, truncate, or only slightly emarginate (notched) **5**

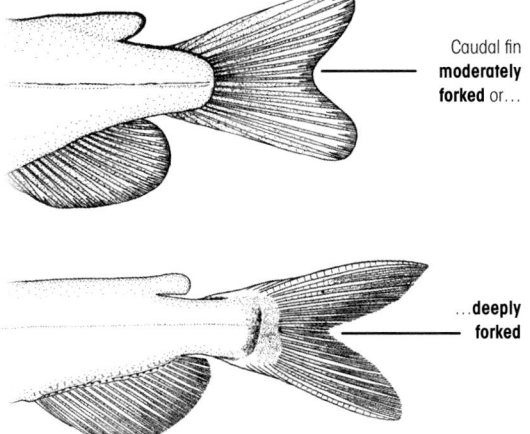

Caudal fin **moderately forked** or…

…**deeply forked**

Caudal fin **rounded** or…

…**slightly emarginate**

3a Caudal fin moderately forked, length of center of fin (A) going less than 2 times into length of upper lobe (B); anal fin rays usually 22–24; supraoccipital bone extending backwards to dorsal fin spine as a hard ridge but not reaching spine, creating soft region in front of dorsal fin. . . . . **Page 224**

3b Caudal fin deeply forked, length of center of fin (A) going more than 2 times into length of upper lobe (B); anal fin rays usually 24 or more; supraoccipital bone extending backwards and contacting dorsal fin spine, creating continuous hard ridge in front of dorsal fin **4**

White Catfish (*Ameiurus catus*)

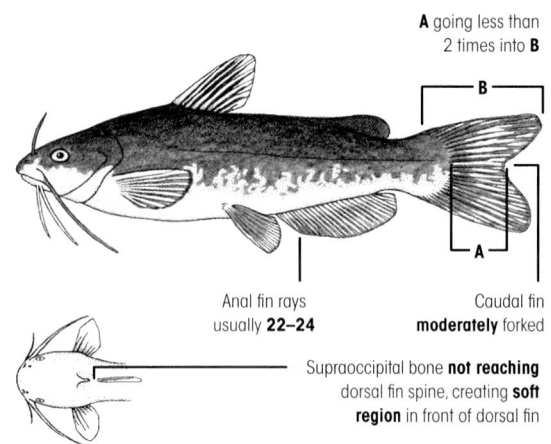

A going less than 2 times into **B**

B

A

Anal fin rays usually **22–24**

Caudal fin **moderately** forked

Supraoccipital bone **not reaching** dorsal fin spine, creating **soft region** in front of dorsal fin

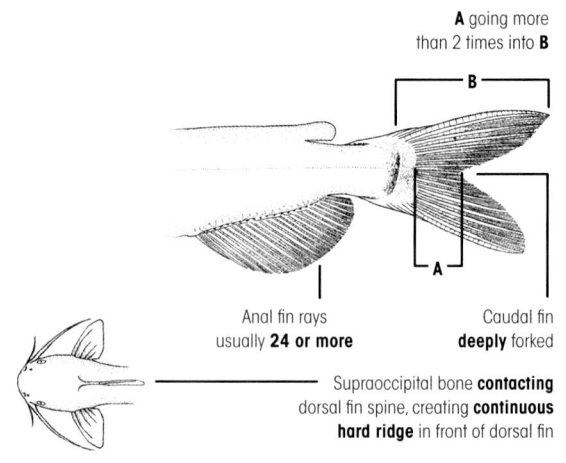

A going more than 2 times into **B**

B

A

Anal fin rays usually **24 or more**

Caudal fin **deeply** forked

Supraoccipital bone **contacting** dorsal fin spine, creating **continuous hard ridge** in front of dorsal fin

---

4a Outer margin of anal fin rounded; anal fin rays 24–29; body with scattered dark round spots, except in large adults and small young; eye positioned on upper half of head. . . . . **Page 229**

4b Outer margin of anal fin straight; anal fin rays usually 30–36; body without dark spots at all ages; eye positioned about equidistant between top and bottom of head . . . .**Page 228**

Channel Catfish (*Ictalurus punctatus*)

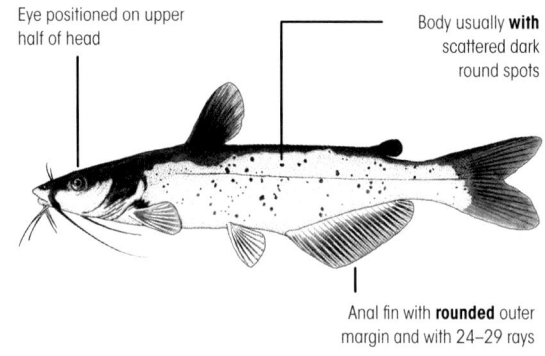

Eye positioned on upper half of head

Body usually **with** scattered dark round spots

Anal fin with **rounded** outer margin and with 24–29 rays

Blue Catfish (*Ictalurus furcatus*)

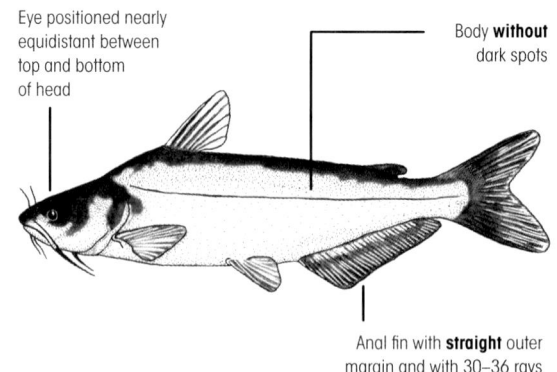

Eye positioned nearly equidistant between top and bottom of head

Body **without** dark spots

Anal fin with **straight** outer margin and with 30–36 rays

5a Lower jaw projecting beyond upper jaw except in small young; head wide and flat, distance from back of eye to rear margin of gill cover (A) greater than 1.5 times depth of head at eye (B); anal fin short, the length of its base (C) less than distance from back of eye to rear margin of gill cover (A); upper tip of caudal fin lighter in color than rest of fin, except in large adults; pad of teeth in upper jaw with a backward extension on each side; body mottled or marbled, except in large adults. . . . . **Page 237**

5b Lower jaw not projecting beyond upper jaw; head short and deep, distance from back of eye to rear margin of gill cover (A) less than 1.5 times depth of head at eye (B); anal fin longer, the length of its base (C) greater than the distance from back of eye to rear margin of gill cover (A); upper tip of caudal fin not lighter in color than rest of fin; pad of teeth in upper jaw without backward extensions; body may or may not be mottled or marbled **6**

Flathead Catfish (*Pylodictis olivaris*)

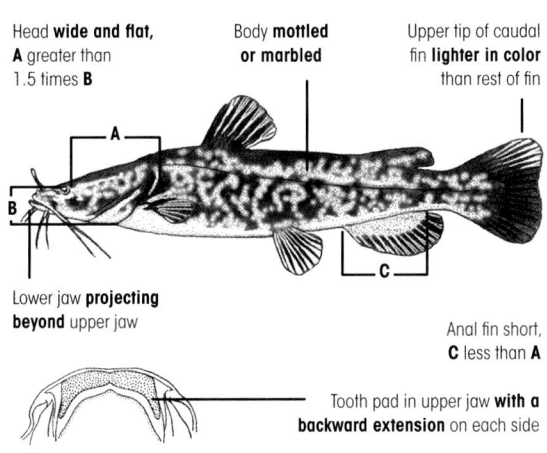

Head **wide and flat, A** greater than 1.5 times **B**

Body **mottled or marbled**

Upper tip of caudal fin **lighter in color** than rest of fin

Lower jaw **projecting beyond** upper jaw

Anal fin short, **C** less than **A**

Tooth pad in upper jaw **with a backward extension** on each side

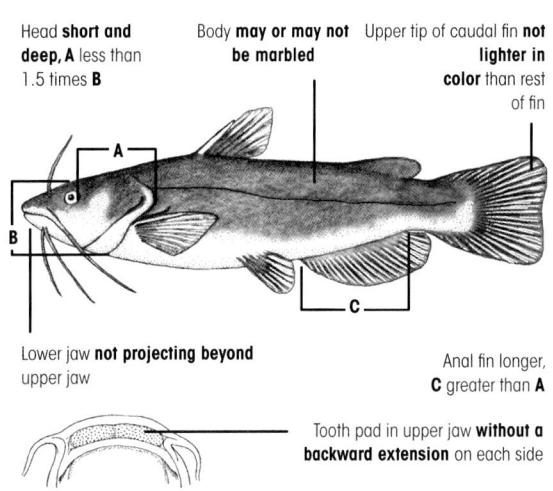

Head **short and deep, A** less than 1.5 times **B**

Body **may or may not be marbled**

Upper tip of caudal fin **not lighter in color** than rest of fin

Lower jaw **not projecting beyond** upper jaw

Anal fin longer, **C** greater than **A**

Tooth pad in upper jaw **without a backward extension** on each side

---

6a Chin barbels whitish or lighter in color than nasal barbels; anal fin rays (counting all elements) usually 24–27; rear margin of caudal fin usually rounded in adults, straighter in small young; basal third of anal fin usually lighter in color than rest of fin . . . . **Page 226**

6b Chin barbels dark, usually black or gray, similar in color to nasal barbels; anal fin rays usually fewer than 24; rear margin of caudal fin often emarginate (notched); basal third of anal fin may or may not be lighter in color than rest of fin **7**

Yellow Bullhead (*Ameiurus natalis*)

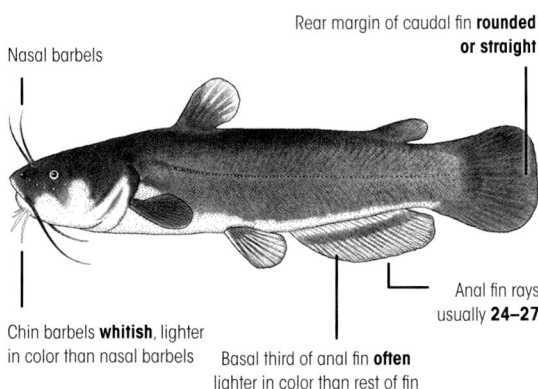

Nasal barbels

Rear margin of caudal fin **rounded or straight**

Chin barbels **whitish**, lighter in color than nasal barbels

Basal third of anal fin **often** lighter in color than rest of fin

Anal fin rays usually **24–27**

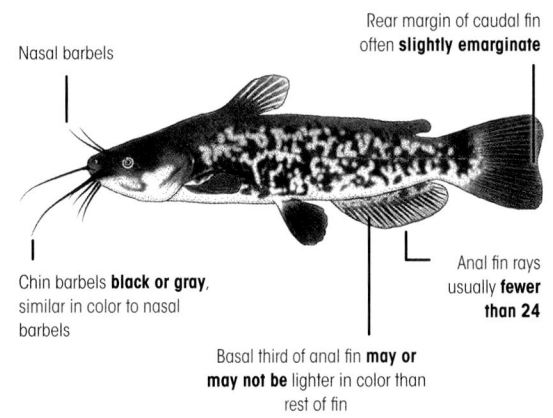

Nasal barbels

Rear margin of caudal fin often **slightly emarginate**

Chin barbels **black or gray**, similar in color to nasal barbels

Basal third of anal fin **may or may not be** lighter in color than rest of fin

Anal fin rays usually **fewer than 24**

7a Gill rakers on first arch usually 16–20; body uniformly dark; pectoral fin spine without obvious teeth on back margin. . **Page 225**

Black Bullhead (*Ameiurus melas*)

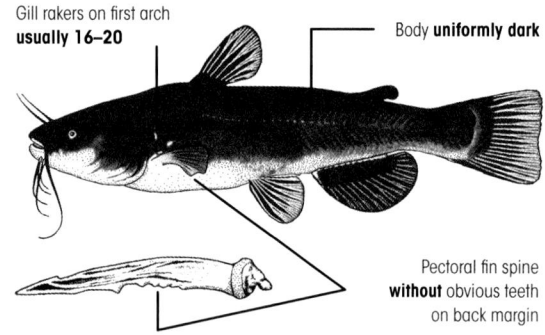

Gill rakers on first arch **usually 16–20**

Body **uniformly dark**

Pectoral fin spine **without** obvious teeth on back margin

7b Gill rakers on first arch usually 12–15; body usually mottled in our region, especially in large juveniles and adults; pectoral spine usually with obvious teeth on back margin . . . .**Page 227**

Brown Bullhead (*Ameiurus nebulosus*)

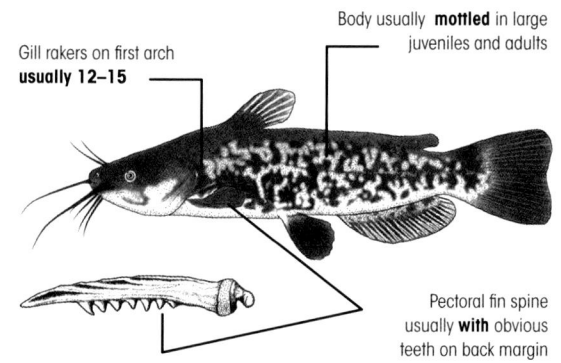

Gill rakers on first arch **usually 12–15**

Body usually **mottled** in large juveniles and adults

Pectoral fin spine usually **with** obvious teeth on back margin

8a Upper jaw not projecting beyond lower jaw, the two jaws about equal length. **9**

Upper jaw **not projecting beyond** lower jaw, the two jaws about equal length

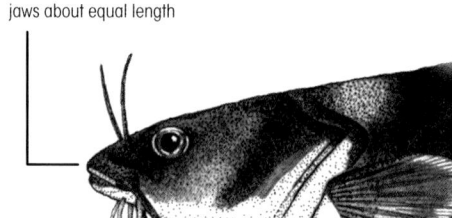

8b Upper jaw projecting well beyond lower jaw **10**

Upper jaw **projecting well beyond** lower jaw

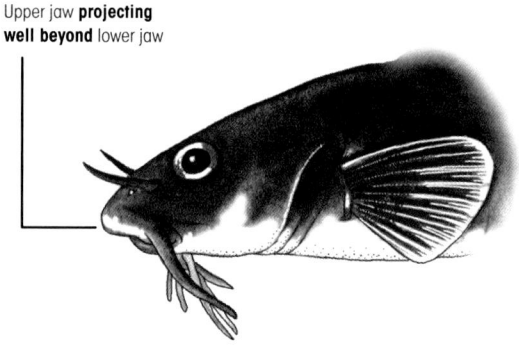

9a Anal, caudal, and dorsal fins dark-edged; body generally slender, its depth (A) going more than 4.5 times into standard length (B); pectoral fin spine with obvious teeth on back margin . . . . **Page 231**

Slender Madtom (*Noturus exilis*)

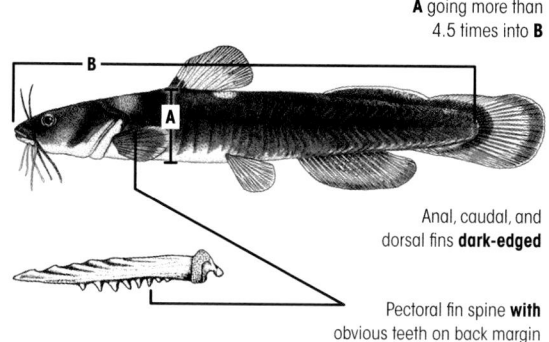

A going more than 4.5 times into **B**

Anal, caudal, and dorsal fins **dark-edged**

Pectoral fin spine **with** obvious teeth on back margin

9b Anal, caudal, and dorsal fins not dark-edged; body deeper, its depth (A) going less than 4.5 times into standard length (B); pectoral spine without obvious teeth on back margin . . . . **Page 233**

Tadpole Madtom (*Noturus gyrinus*)

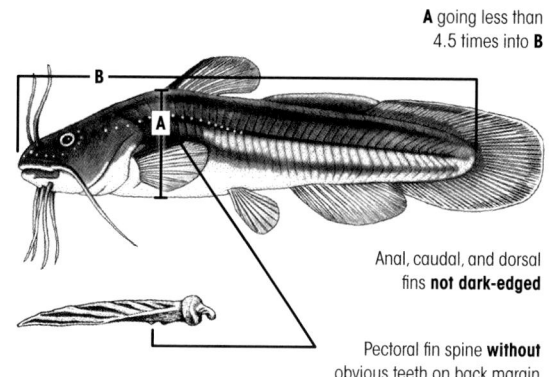

A going less than 4.5 times into **B**

Anal, caudal, and dorsal fins **not dark-edged**

Pectoral fin spine **without** obvious teeth on back margin

10a Body rather uniformly colored, without distinct blotches, saddles, or bars; pectoral fin spine without prominent teeth on back margin **11**

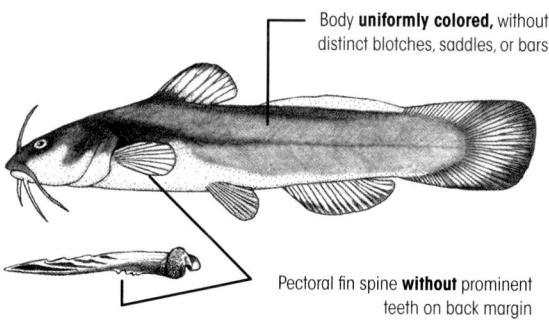

Body **uniformly colored,** without distinct blotches, saddles, or bars

Pectoral fin spine **without** prominent teeth on back margin

10b Body with some combination of distinct blotches or bars on side or saddles on back; pectoral fin spine with prominent saw-like teeth on back margin **12**

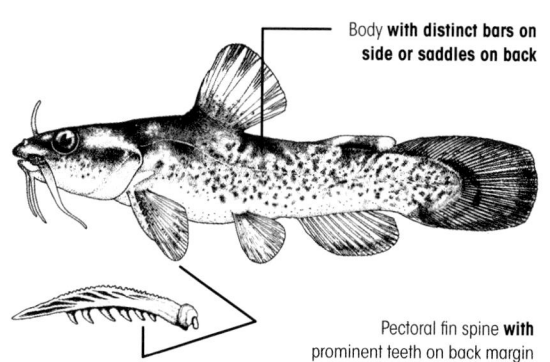

Body **with distinct bars on side or saddles on back**

Pectoral fin spine **with** prominent teeth on back margin

11a Pad of teeth in upper jaw with a backward extension on each side; upper tip of caudal fin lighter in color than center of fin; white blotch behind back edge of dorsal fin base; lower lip and chin without dark pigment . . . . **Page 232**

Stonecat (*Noturus flavus*)

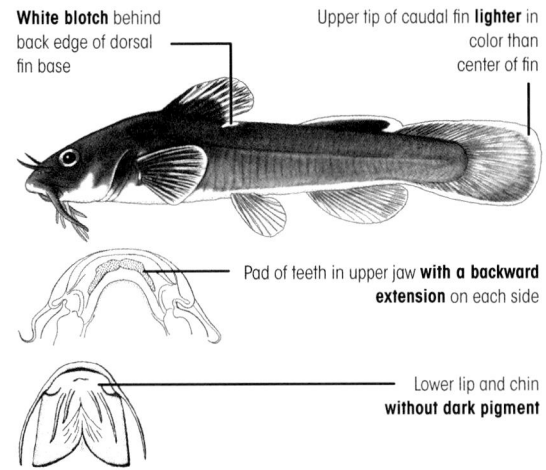

White blotch behind back edge of dorsal fin base

Upper tip of caudal fin **lighter** in color than center of fin

Pad of teeth in upper jaw **with a backward extension** on each side

Lower lip and chin **without dark pigment**

11b Pad of teeth in upper jaw without backward extensions; upper tip of caudal fin not lighter in color than center of fin; no light blotch behind back edge of dorsal fin base; lower lip and chin heavily speckled with dark pigment . . . . **Page 235**

Freckled Madtom (*Noturus nocturnus*)

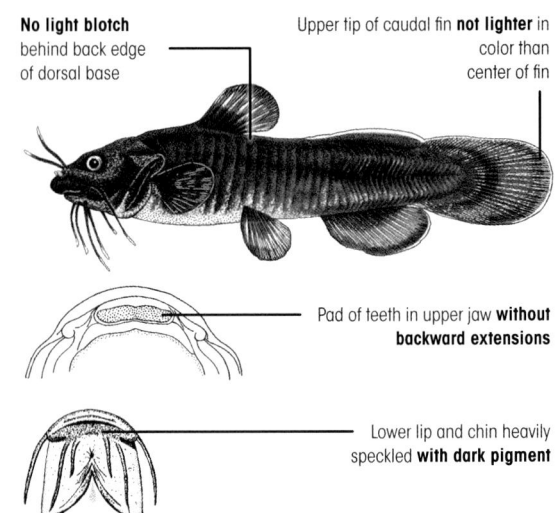

No light blotch behind back edge of dorsal base

Upper tip of caudal fin **not lighter** in color than center of fin

Pad of teeth in upper jaw **without backward extensions**

Lower lip and chin heavily speckled **with dark pigment**

12a Prominent dark blotch on upper edge of dorsal fin; rear margin of caudal fin rounded; dark bar or blotch at base of adipose fin extending upward to near fin margin; body usually mottled. . . . . **Page 234**

Brindled Madtom (*Noturus miurus*)

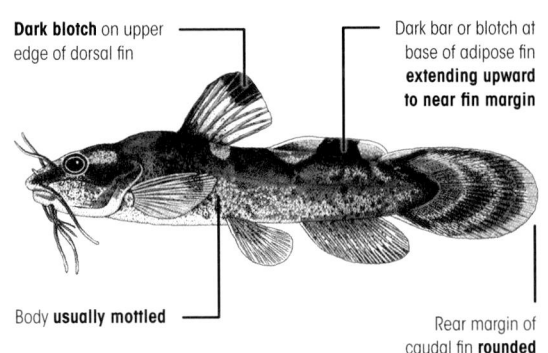

Dark blotch on upper edge of dorsal fin

Dark bar or blotch at base of adipose fin **extending upward to near fin margin**

Body **usually mottled**

Rear margin of caudal fin **rounded**

12b No dark blotch on upper edge of dorsal fin; rear margin of caudal fin straight, or slightly rounded on upper and lower edges; dark bar or blotch at base of adipose fin extending only into bottom half of fin; body not heavily mottled **13**

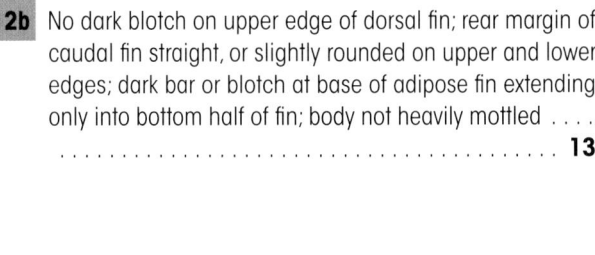

No dark blotch on upper edge of dorsal fin

Dark bar or blotch at base of adipose fin **extending only into bottom half of fin**

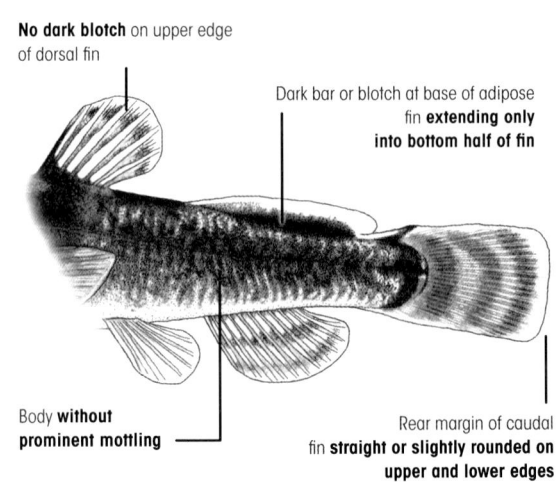

Body **without prominent mottling**

Rear margin of caudal fin **straight or slightly rounded on upper and lower edges**

13a Usually 10 sensory pores (i.e., preoperculomandibular pores) on lower jaw and edge of bone in front of gill cover (i.e., preopercle bone); no band or faint band near upper edge of dorsal fin; dark bar or blotch at base of adipose fin; bar at base of caudal fin usually dark and prominent; body more slender in large young and adults, depth of caudal peduncle (A) going 5 to 6.5 times into distance from front of dorsal fin to notch behind adipose fin (B); caudal fin rays usually 44–48 . . . . **Page 230**

13b Usually 11 sensory pores on lower jaw and edge of bone in front of gill cover; dark band near upper edge of dorsal fin; dark bar or blotch at base of adipose fin extending into upper half of fin; bar at base of caudal fin usually faint, blending into body color; body somewhat chunkier in large young and adults, depth of caudal peduncle (A) going 4 to 5 times into distance from front of dorsal fin to notch behind adipose fin (B); caudal fin rays usually 49–53 . . . . **Page 236**

Mountain Madtom (*Noturus eleutherus*)

No band or only a faint band near edge of dorsal fin

Body **more slender, A** going 5 to 6.5 times into **B** in large young and adults

Bar at base of caudal fin usually **dark and prominent**

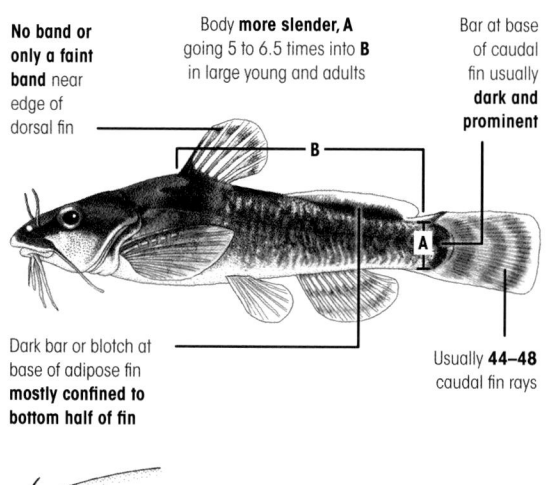

Dark bar or blotch at base of adipose fin **mostly confined to bottom half of fin**

Usually **44–48** caudal fin rays

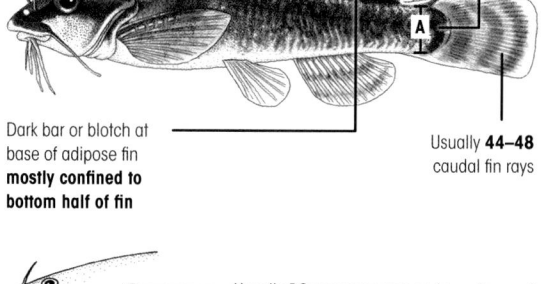

Usually **10** sensory pores on lower jaw and edge of preopercle bone

Northern Madtom (*Noturus stigmosus*)

Dark band near upper edge of dorsal fin

Body **chunkier, A** going 4 to 5 times into **B** in large young and adults

Bar at base of caudal fin usually **faint**, blending into body color

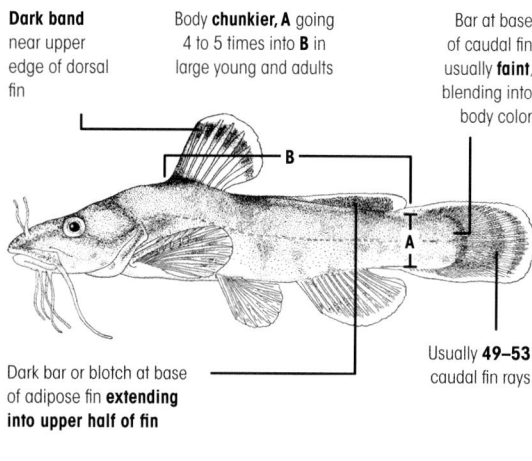

Dark bar or blotch at base of adipose fin **extending into upper half of fin**

Usually **49–53** caudal fin rays

Usually **11** sensory pores on lower jaw and edge of preopercle bone

WHITE CATFISH *Ameiurus catus* (Linnaeus 1758)

IDENTIFICATION The White Catfish is depressed anteriorly and strongly compressed posteriorly and has a moderately forked caudal fin. The adipose fin is small with a short base, and its posterior edge is free from the back and far from the caudal fin. The upper jaw projects beyond the lower jaw. The eye is relatively small and high on the side of the head. There are 11–15 moderately large, sawlike teeth on the posterior edge of the pectoral-fin spine. The anal fin is rounded in outline and has 22–25 rays. The back and side of the body are gray to blue-black, the adipose fin is dusky to black, and the breast and belly are white to light yellow. The chin barbels are white or yellow. There are 18–21 rakers on the first gill arch. To 24¼ in. (62 cm).

SIMILAR SPECIES The Channel Catfish, *I. punctatus* and Blue Catfish, *I. furcatus*, lack a dusky to black adipose fin and have a more deeply forked caudal fin. The Channel Catfish usually has scattered black spots on the side of the body. The Blue Catfish has a long, straight-edged anal fin.

HABITAT The White Catfish occupies lakes, impoundments, and sluggish mud-bottomed pools and backwaters of small to large rivers.

DISTRIBUTION IN ILLINOIS Native to Atlantic Ocean and Gulf of Mexico slope basins, the White Catfish was first captured in Illinois in 1965 (Smith 1979). Although it has been stocked at various localities in the state over 5 decades, it rarely if ever reproduces here. As in the Smith era, records are widely scattered in the western half of the state, and all may be the result of stocking.

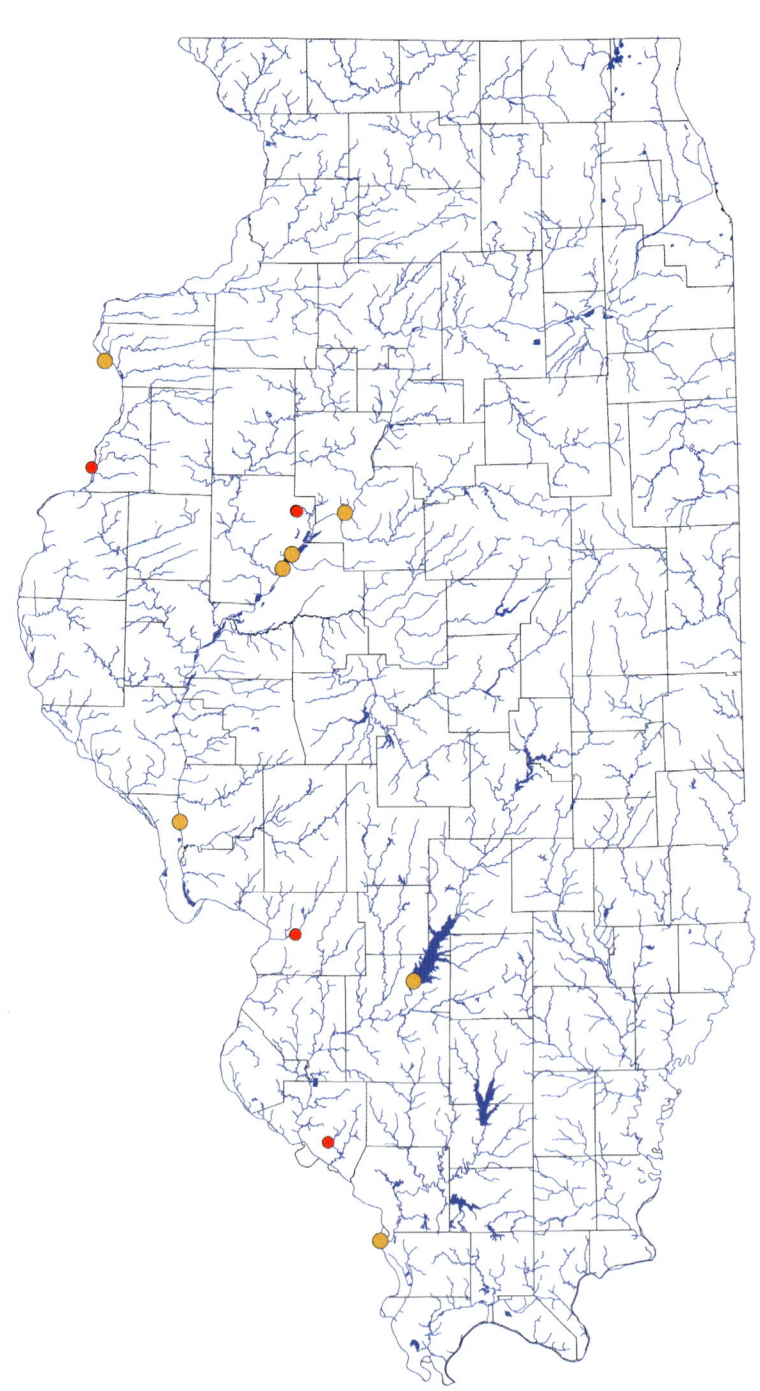

BROWN BULLHEAD *Ameiurus nebulosus* (Lesueur 1819)

IDENTIFICATION The Brown Bullhead is depressed anteriorly and strongly compressed posteriorly, and has a straight-edged or slightly notched caudal fin. The adipose fin is small, has a short base, and its posterior edge is free from the back and far from the caudal fin. The upper jaw projects slightly or is even with the lower jaw. The eye is relatively small and high on the side of the head. The anal fin is moderately long, rounded in outline, and has 21–24 rays; the anterior rays are distinctly longer than the posterior rays. There are 5–8 large, sawlike teeth on the rear of the pectoral-fin spine. The back and side of the body are dark olive or yellow-brown with dark brown or black mottling or spots. The breast and belly are yellow to white. The fins are dusky gray. The chin barbels are dusky to black. There are 11–15 rakers on the first gill arch. To 21 in. (50 cm).

SIMILAR SPECIES The Black Bullhead, *A. melas*, lacks brown or black mottling or spots on the body, no large, sawlike teeth on the posterior edge of the pectoral-fin spine, and has usually 15–21 anal rays, and 15–21 rakers on the first gill arch. The Yellow Bullhead, *A. natalis*, has white or yellow chin barbels, 24–27 anal rays, and no pronounced brown or black mottling or spots on the side of the body.

HABITAT The Brown Bullhead lives over soft substrates in lakes, ponds, and sluggish pools and backwaters of small to large rivers.

DISTRIBUTION IN ILLINOIS Smith (1979) found the Brown Bullhead to be restricted to the glacial lakes in northeastern Illinois, floodplain lakes along the Illinois River, and a few other localities including in southwestern Illinois. It had been much more generally distributed in the central part of the state during the Forbes and Richardson era. The distribution is nearly the same in the contemporary era as in the Smith era, although it is recorded for the first time in the Wabash River. A record of this species in Boone County (Smith 1979) was re-identified as a Black Bullhead, *A. melas*.

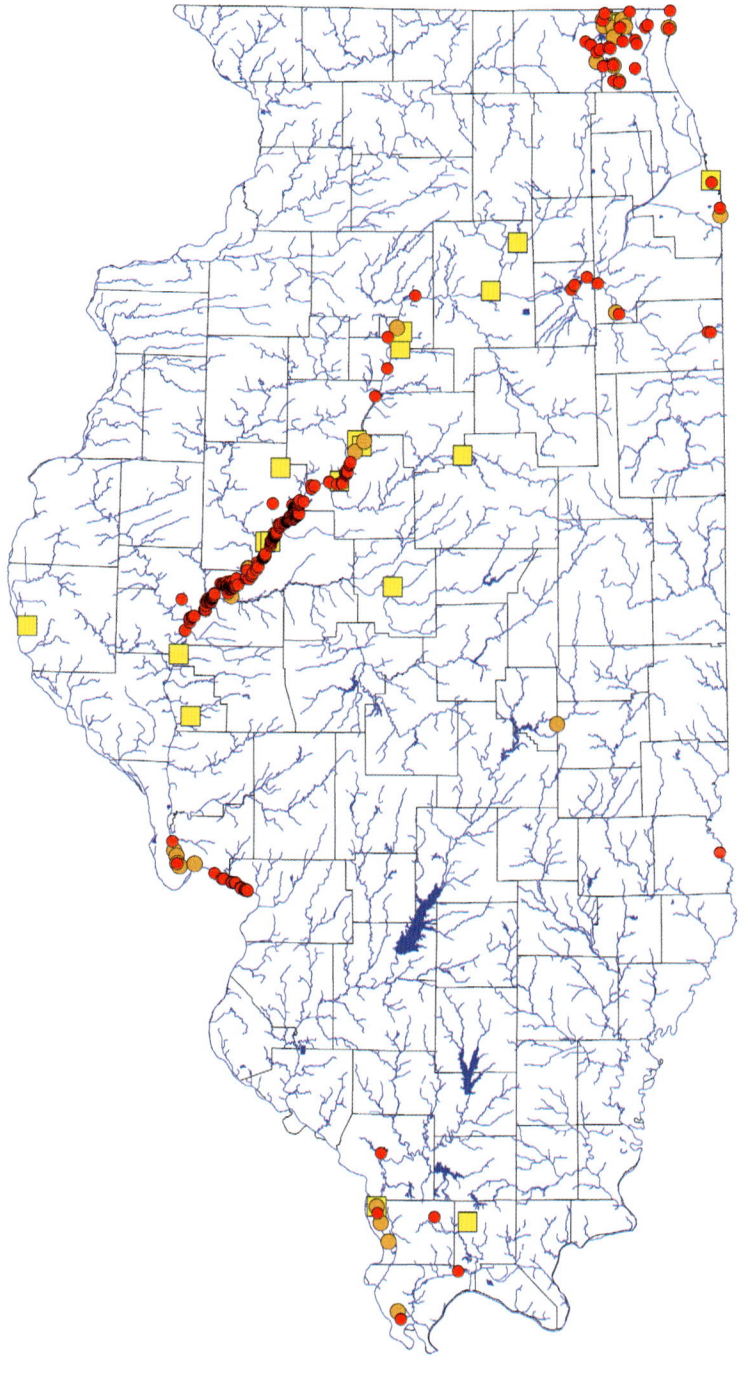

NORTH AMERICAN CATFISHES (Ictaluridae)

BLUE CATFISH *Ictalurus furcatus* (Valenciennes 1840)

IDENTIFICATION The Blue Catfish is depressed anteriorly and strongly compressed posteriorly and has a deeply forked caudal fin. The upper jaw projects beyond the lower jaw. The eye is fairly large and on the side of the head. The adipose fin is small, has a short base, and its posterior edge is free from the back and far from the caudal fin. The back and side of the body are silver in juveniles and pale blue to olive on adults. The largest individuals are blue-black on the back and upper side. There are no black spots on the body. The chin barbels are white to dusky, and the fins are silver to dusky black. The anal fin is straight edged and deeper anteriorly than posteriorly and has 30–35 rays. The gas bladder has separate chambers. To 65 in. (165 cm).

SIMILAR SPECIES The Channel Catfish, *I. punctatus*, typically has black spots on the back and side of the body, a rounded anal fin with 24–32 rays, and no chambers in the gas bladder. The White Catfish, *Ameiurus catus*, has a rounded anal fin with 22–25 rays.

HABITAT The Blue Catfish lives in lakes, impoundments, and the main channels and backwaters of medium to large rivers.

DISTRIBUTION IN ILLINOIS The Blue Catfish occurs in large rivers in the southern half of the state and extends up the Illinois River to Peoria. Forbes and Richardson (1909) had only 1 record for the species, in the Mississippi River at St. Louis, but Smith (1979) found the species to have essentially the same distribution as it has in the contemporary era, that is, in the Illinois, Mississippi, Wabash, Kaskaskia, and Big Muddy Rivers. Smith's enigmatic record in Jo Daviess County, far north of other records in Illinois, was based on a specimen collected in 1972.

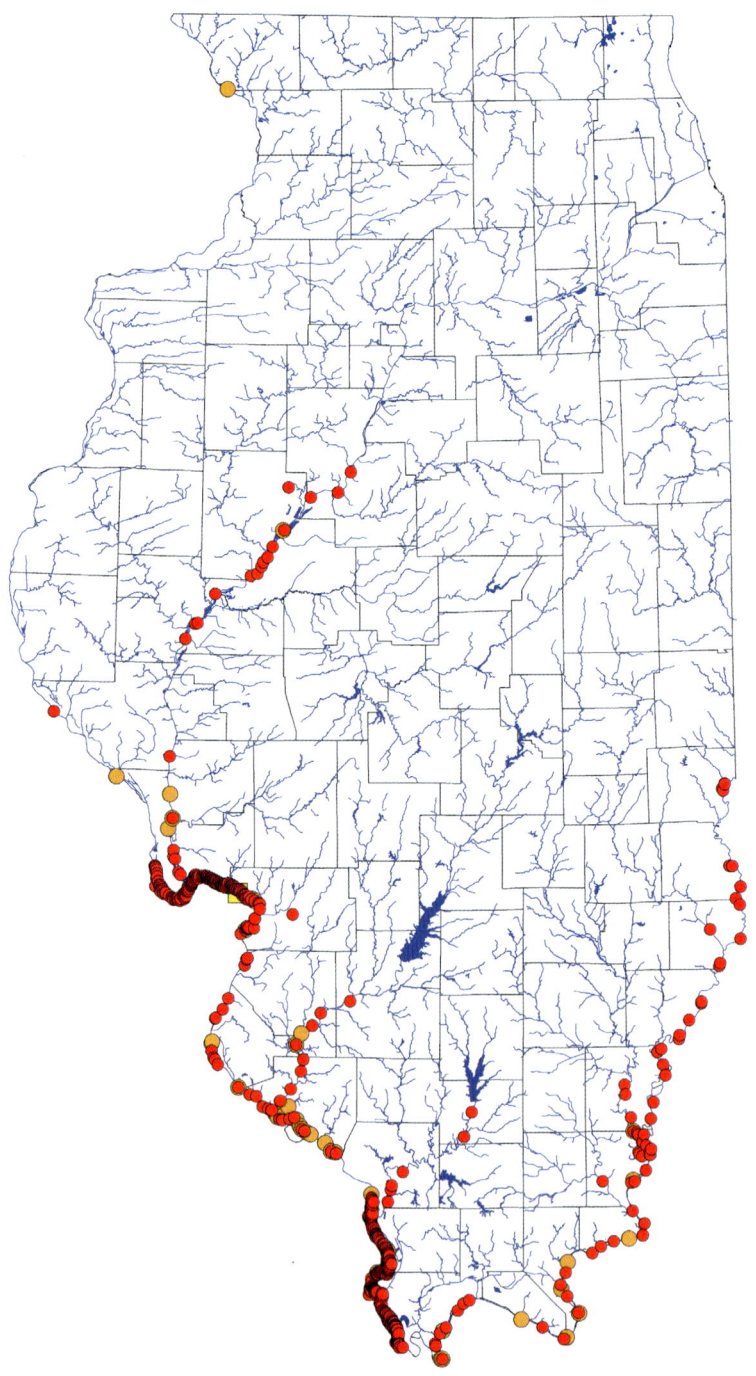

CHANExNEL CATFISH   *Ictalurus punctatus* (Rafinesque 1818)

IDENTIFICATION The Channel Catfish is depressed anteriorly and strongly compressed posteriorly and has a deeply forked caudal fin. The upper jaw projects beyond the lower jaw. The eye is large in juveniles, much smaller in adults, and high on the side of the head. The adipose fin is small and has a short base, and its posterior edge is free from the back and far from the caudal fin. Most individuals have scattered black spots on a silver back and side of the body; however, small juveniles are silver to gold, have no black spots, and have black-tipped fins, and the largest adults are blue-black on the back and upper side and have no black spots. The chin barbels are white to dusky, and the fins are similar in color to the adjacent body. The anal fin is rounded with 24–32 rays. The gas bladder is without constrictions and chambers. To 50 in. (127 cm).

SIMILAR SPECIES The Blue Catfish, *I. furcatus*, lacks black spots on the body and has a straight-edged anal fin with 30–35 rays and a gas bladder with separate chambers. The White Catfish, *Ameiurus catus*, has a moderately forked caudal fin and lacks black spots on the side of the body.

HABITAT Channel Catfish are frequently encountered in deep pools and runs over sand or rocks in small to large rivers, lakes, and impoundments. The species is rarely found in headwaters or upland streams.

DISTRIBUTION IN ILLINOIS Smith (1979) considered the Channel Catfish to be less widespread than during the Forbes and Richardson era but now, either through stocking or natural range expansion, the species is widespread in all basins including in northeastern Illinois, where it was infrequently encountered in earlier eras.

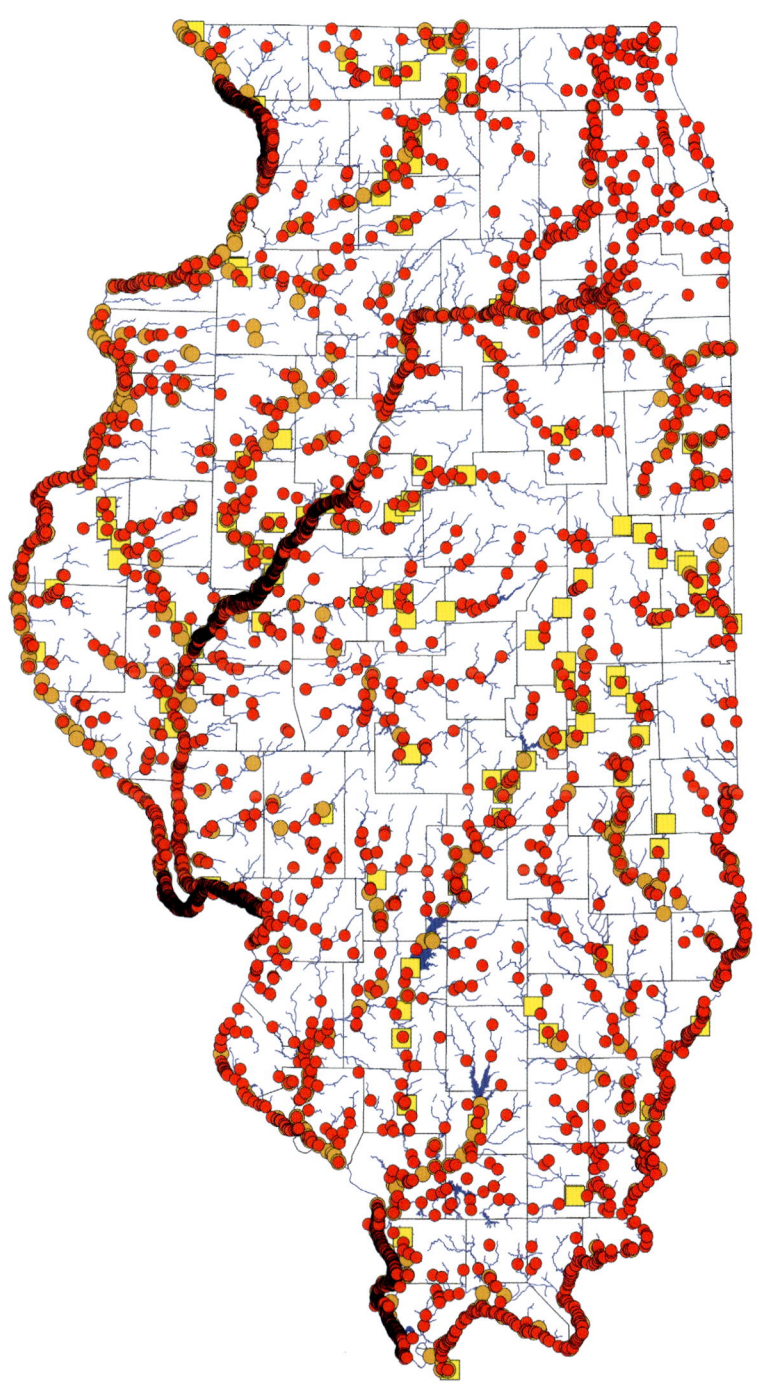

NORTH AMERICAN CATFISHES (Ictaluridae)

229

MOUNTAIN MADTOM *Noturus eleutherus* Jordan 1877

IDENTIFICATION The Mountain Madtom is depressed anteriorly and compressed posteriorly with a moderately deep caudal peduncle and a straight-edged caudal fin. The adipose fin is long, low, and barely separated from the caudal fin. The upper jaw projects beyond the lower jaw. The back and side of the body are gray-brown with dark brown speckling and up to 4 saddles—usually only the first saddle is well developed. The anterior edge of the first saddle is at the dorsal-fin origin. The adipose fin has dark pigment that is confined to the lower half of the fin. There are a dark brown bar at the base of the caudal fin and a dark brown band near the clear edge of the caudal fin. The breast and belly are white to pale yellow, usually without dark specks. Other fins have poorly defined dark bands. The pectoral-fin spine is long with 6–10 large, sawlike teeth on the posterior edge and large teeth on the anterior edge. There are 12–16 anal rays. To 5 in. (13 cm).

SIMILAR SPECIES The Northern Madtom, *N. stigmosus*, lacks a dark brown bar at the base of the caudal fin and has a dark brown crescent in the middle of the caudal fin, 2 large, light yellow spots in front of the dorsal fin, and a dark brown band on the adipose fin that continues into the upper half of the fin. The Brindled Madtom, *N. miurus*, has a large, black blotch on the dorsal fin, and the black band on the adipose fin extends to the edge of the fin.

HABITAT The Mountain Madtom occupies clean, rocky riffles and runs of small to large rivers and is often taken near vegetation or woody debris.

DISTRIBUTION IN ILLINOIS The Mountain Madtom is restricted to the Vermilion, Embarras, and Wabash Rivers. The range of this species has not changed appreciably since the Smith era. As Smith (1979) noted, this species is the most widespread madtom in the Wabash River, where Forbes and Richardson (1909) had their only record.

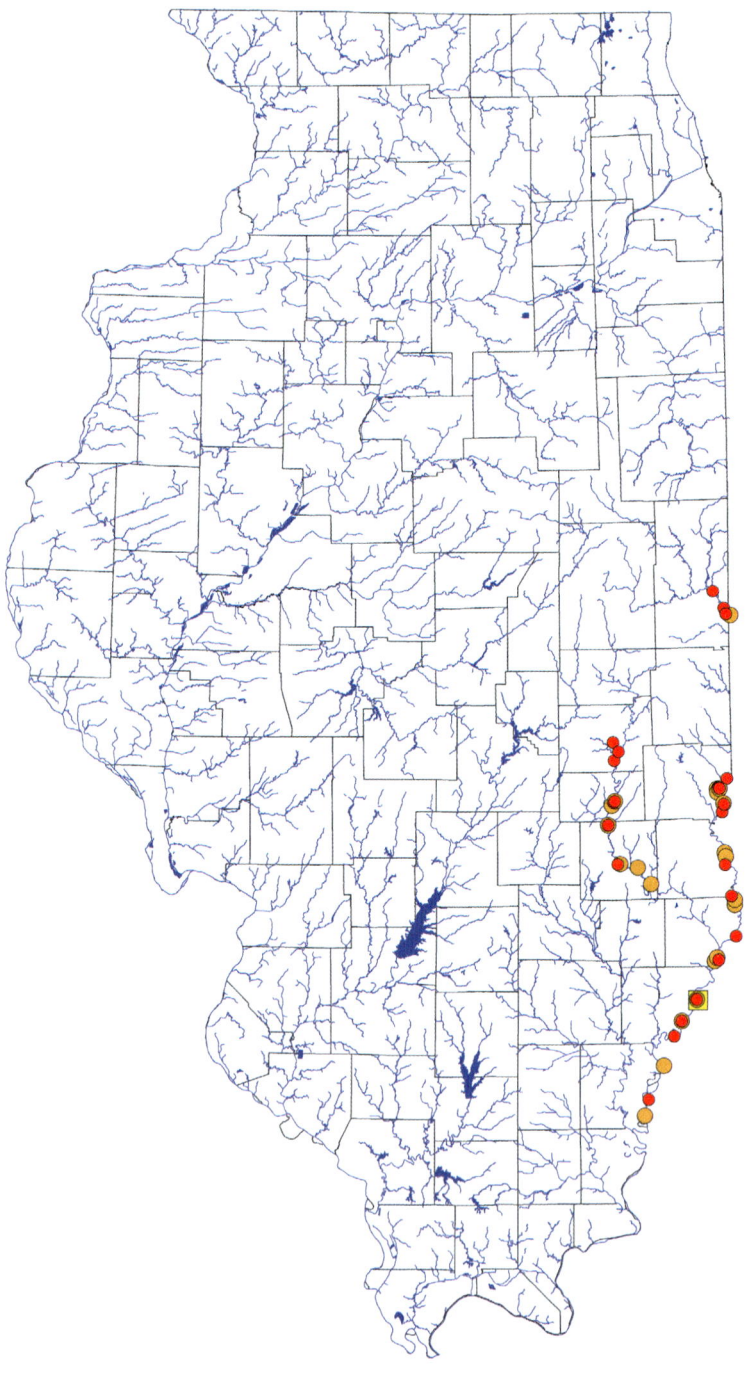

NORTH AMERICAN CATFISHES (Ictaluridae)

SLENDER MADTOM *Noturus exilis* Nelson 1876

IDENTIFICATION The Slender Madtom is slender, depressed anteriorly, and compressed posteriorly and has a straight-edged or slightly rounded caudal fin. The mouth is terminal with the upper and low jaws equal in length. The adipose fin is long, low, and joined to or slightly separated from the caudal fin. The back and side of the body are yellow-brown to gray-black, and most individuals have a large, pale yellow spot on the nape and a smaller spot at the posterior edge of the dorsal-fin base. The breast and belly are light yellow, and there is a dusky to black border on the dorsal, caudal, and anal fins. The pectoral fin has 6 strong sawlike teeth on the posterior edge. There are 17–22 anal rays. To 5¾ in. (15 cm).

SIMILAR SPECIES The Freckled Madtom, *N. nocturnus*, has the upper jaw projecting beyond the lower jaw, many tiny black dots on the body and fins, and 2–3 small, sawlike teeth on the posterior edge of the pectoral-fin spine. The Tadpole Madtom, *N. gyrinus*, is thicker-bodied and has herringbone lines emanating from a dark gray stripe along the side.

HABITAT The Slender Madtom lives in clear, rocky riffles, runs, and flowing pools of creeks and small rivers.

DISTRIBUTION IN ILLINOIS Forbes and Richardson (1909) and Smith (1979) found this species to be sporadically distributed throughout the state. Contemporary-era records show the species to be present in most of the same areas as in the earlier eras, except that it now seems to be gone from the upper Kaskaskia River basin, where Smith recorded it at several localities; it has been found in a few additional streams, including in the Saline River system. The Slender Madtom was originally described from the Mackinaw River in McLean County in 1876, but it has not been found in the Mackinaw River system since then.

REMARKS Slender Madtoms from cool clear streams are more slender and have more boldly edged fins than individuals from warm, turbid waters. Some individuals lack black fin borders.

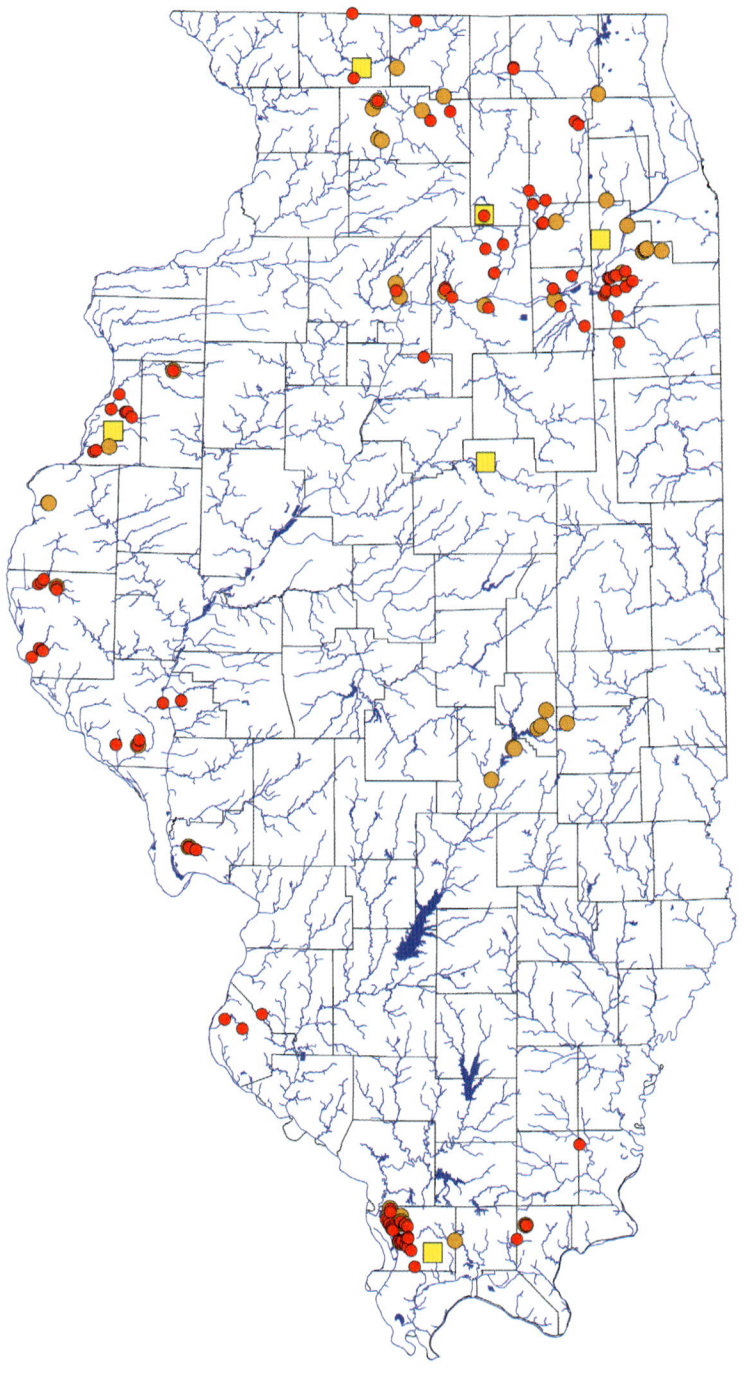

STONECAT *Noturus flavus* Rafinesque 1818

IDENTIFICATION The Stonecat is slender, depressed anteriorly, and compressed posteriorly and has a straight-edged or slightly rounded caudal fin. The upper jaw projects beyond the lower jaw. The premaxillary tooth patch (on the roof of the mouth) has a large, backward extension on each side. The adipose fin is long, low, and slightly separated from the caudal fin. The back and side of the body are pale yellow to slate-olive, and there is a yellow-white patch on the nape, a cream-white spot at the posterior edge of the dorsal-fin base, and a cream-white blotch on the upper edge of the dusky gray caudal fin. The pectoral, dorsal, and adipose fins are dark at the base and pale or white at the edge. The breast and belly are yellow-white. There are no or only a few weak teeth on the posterior edge of the pectoral-fin spine. There are usually 15–18 anal rays. To 10 in. (25 cm).

SIMILAR SPECIES All other species of *Noturus* lack a backward extension from each side of the premaxillary tooth patch and lack the cream-white blotch on the upper edge of the caudal fin.

HABITAT The Stonecat lives in rocky riffles and runs of creeks and small to large rivers.

DISTRIBUTION IN ILLINOIS Forbes and Richardson (1909) found the Stonecat to be distributed throughout the northern half of Illinois. Smith (1979) found it in the northern half of the state and in the Wabash River and at 1 locality near the confluence of the Kaskaskia and Mississippi Rivers in southern Illinois. In the contemporary era it is found in the same areas and is much more widespread in the lower Mississippi River.

REMARKS In the Mississippi River below the mouth of the Missouri River, Stonecats have eyes that are much smaller than in other populations in Illinois.

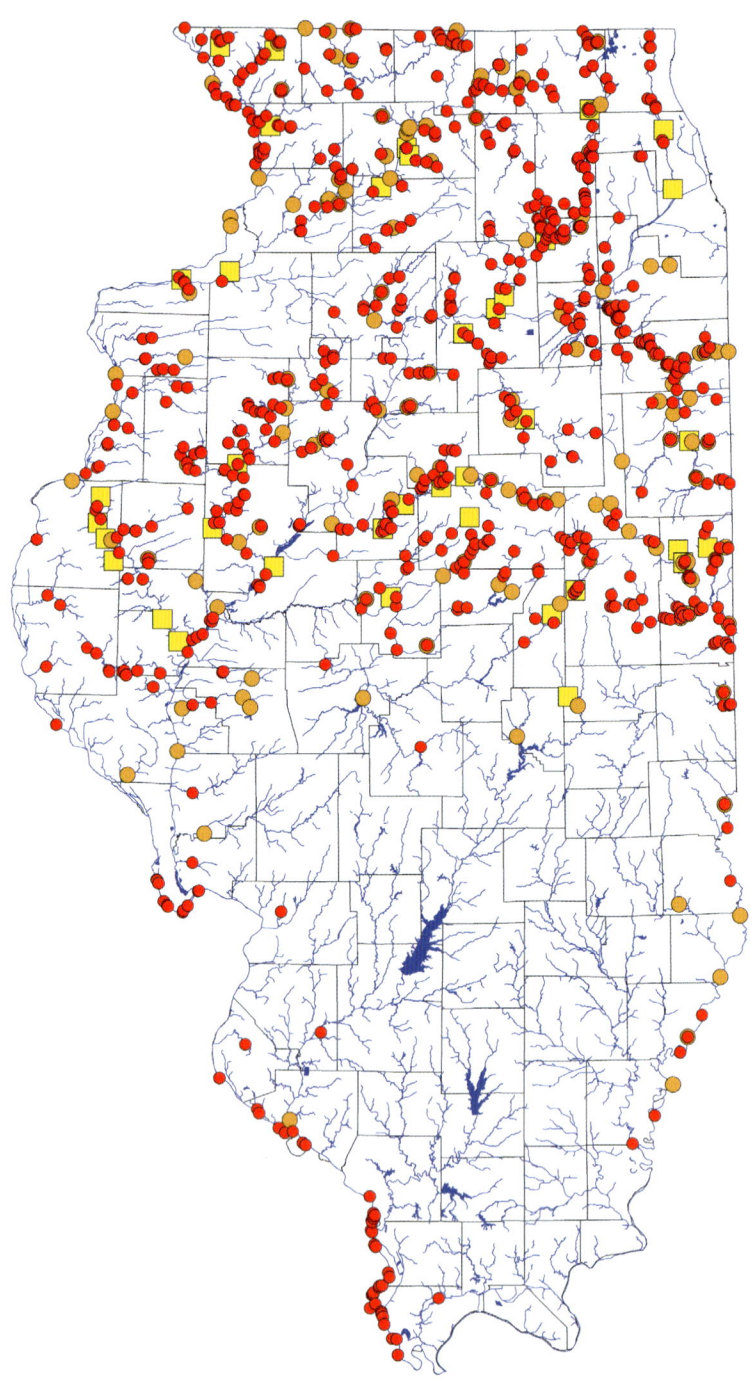

TADPOLE MADTOM   *Noturus gyrinus* (Mitchill 1817)

IDENTIFICATION The Tadpole Madtom is thick-bodied anteriorly and compressed posteriorly and has a rounded caudal fin. The mouth is terminal with the upper and lower jaws equal in length. The adipose fin is long, low, and joined to or slightly separated from the caudal fin. The back and side of the body are reddish-tan or gray, and there are herringbone lines emanating from a distinctive dark gray veinlike stripe along the side. The breast and belly are dusky gray, white, or yellow, and the fins are gray or brown. There are no teeth on the posterior edge of the pectoral-fin spine. There are 13–18 anal rays. To 5 in. (13 cm).

SIMILAR SPECIES The Slender Madtom, *N. exilis*, the only other madtom in Illinois with the upper and low jaws equal in length, lacks herringbone lines emanating from a dark gray stripe along the side and has black edges on the median fins and a more slender body. The Freckled Madtom, *N. nocturnus*, has the upper jaw projecting beyond the lower jaw and lacks herringbone lines and a dark gray veinlike stripe on the side of the body.

HABITAT The Tadpole Madtom occupies lakes, ponds, swamps, and rock- or woody-debris-bottomed pools and backwaters of lowland creeks and small to large rivers.

DISTRIBUTION IN ILLINOIS The Tadpole Madtom is statewide in occurrence but is somewhat less generally distributed in the western and northwestern parts of the state than elsewhere. Smith (1979) commented that the species appeared to be less common than it had been in the Forbes and Richardson era, and this may be true in the contemporary era as well. The species is much less widespread in the Vermilion River (Wabash) basin in Vermilion County, and there are no contemporary-era records from the Lake Michigan basin.

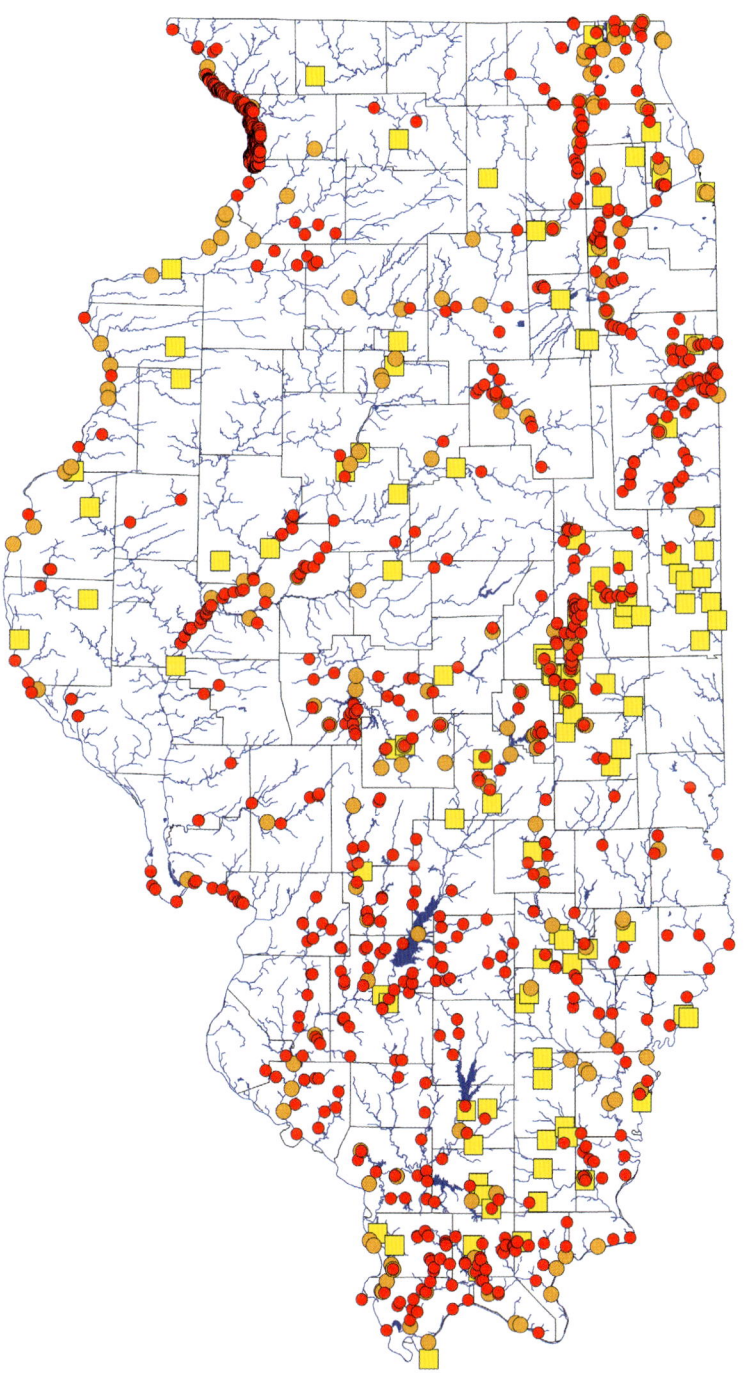

BRINDLED MADTOM *Noturus miurus* Jordan 1877

IDENTIFICATION The Brindled Madtom is depressed anteriorly and compressed posteriorly with a moderately deep caudal peduncle and a rounded caudal fin. The upper jaw projects beyond the lower jaw. The adipose fin is long, low, and joined to or barely separated from the caudal fin. The back and side of the body are light yellow or brown with dark brown mottling and 3–4 dark brown saddles; the saddle under the adipose fin continues as a dark band to the dorsal edge of the fin. There is a conspicuous black blotch on the outer third of the dorsal fin across the first 3–5 rays. The caudal fin has a brown-black border; other fins have brown mottling or poorly formed dark bands. The breast and belly are white. There are usually 5–9 large, sawlike teeth on the posterior edge of the pectoral-fin spine and large teeth on the anterior edge. There are 13–17 anal rays. To 5 in. (13 cm).

SIMILAR SPECIES The Mountain Madtom, *N. eleutherus*, and Northern Madtom, *N. stigmosus*, lack a large, black blotch on the dorsal fin, and the black band on the adipose fin does not extend to the edge of the fin.

HABITAT The Brindled Madtom lives over gravel and sand, often near woody debris, in riffles, runs, and flowing pools of creeks and small rivers.

DISTRIBUTION IN ILLINOIS Forbes and Richardson (1909) documented the presence of the Brindled Madtom throughout the Wabash River basin and in the upper Kaskaskia River. Records from the Smith era expanded the distribution to include tributaries of the Ohio River but also suggested that the species had disappeared from the Kaskaskia, Skillet Fork, and Saline River basins. Contemporary-era records indicate the species is widespread in the Vermilion (Wabash) basin and remains in or has reinvaded the Skillet Fork and Saline basins but is gone from the Kaskaskia River.

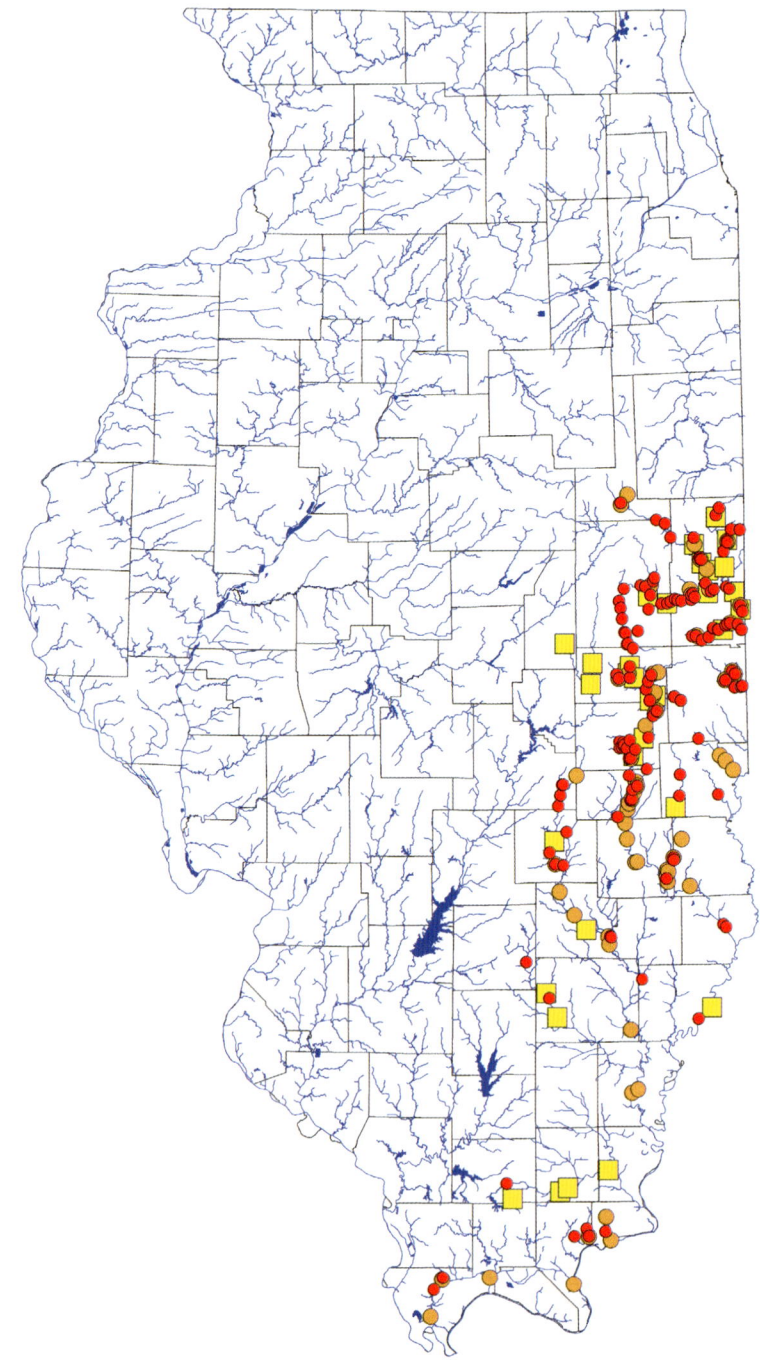

FRECKLED MADTOM   *Noturus nocturnus* Jordan & Gilbert 1886

IDENTIFICATION The Freckled Madtom is depressed anteriorly and compressed posteriorly and has a slightly rounded caudal fin. The upper jaw projects beyond the lower jaw. The adipose fin is long, low, and joined to or slightly separated from the caudal fin. The back and side of the body are light to dark brown or gray and peppered with many tiny black dots. The breast and belly are dusky white to yellow. The anal fin and sometimes the dorsal and caudal fins have a dusky black edge. There are 2–3 small, sawlike teeth on the posterior edge of the large pectoral-fin spine and 16–18 anal rays. To 6 in. (15 cm).

SIMILAR SPECIES The Tadpole Madtom, *N. gyrinus*, has the upper and lower jaws equal in length and herringbone lines and a dark gray veinlike stripe on the side of the body and lacks dusky black edges on the median fins. The Slender Madtom, *N. exilis*, has the upper and lower jaws equal in length, a large, pale yellow spot on the nape, a smaller spot at the posterior edge of the dorsal-fin base, and 6 large, sawlike teeth on the posterior edge of the pectoral-fin spine.

HABITAT The Freckled Madtom is found in creeks and small to large rivers in sand- and gravel-bottomed riffles and runs near woody debris and among tree roots along undercut banks.

DISTRIBUTION IN ILLINOIS The Freckled Madtom is more widely distributed in the contemporary era than it was in previous eras, especially in the Mississippi River and northern third of Illinois, where it occurs sporadically, and in tributaries of the Illinois and Wabash Rivers. However, it seems to be absent from the Saline and lower Des Plaines River basins where it occurred in the Forbes and Richardson era.

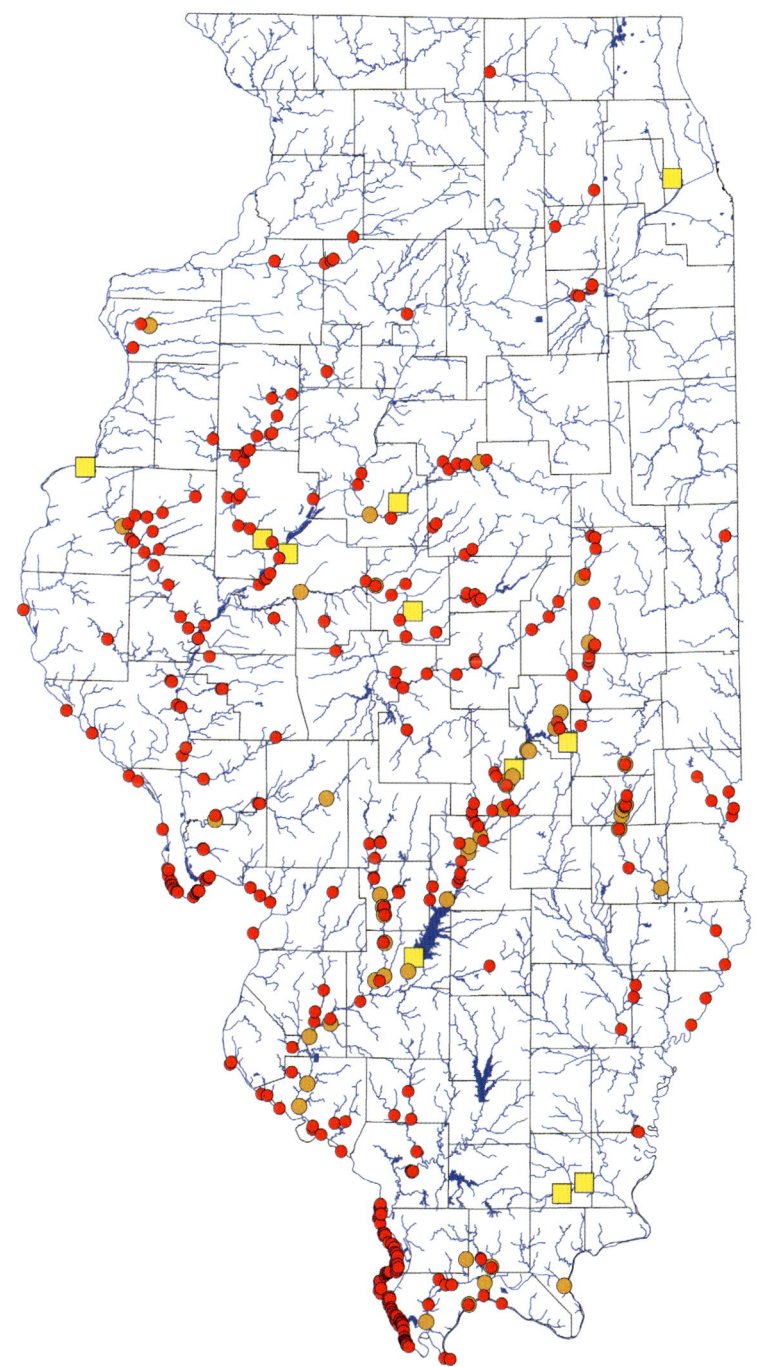

NORTH AMERICAN CATFISHES (Ictaluridae)

235

NORTHERN MADTOM *Noturus stigmosus* Taylor 1969

IDENTIFICATION The Northern Madtom is depressed anteriorly and compressed posteriorly with a moderately deep caudal peduncle and a straight-edged caudal fin. The upper jaw projects beyond the lower jaw. The adipose fin is long, low, and joined to or barely separated from the caudal fin. The back and side of the body are yellow-brown with dark brown speckling and up to 4 dark brown saddles. The anterior edge of the first saddle is irregular, usually enclosing 2 large, light yellow spots just in front of the dorsal fin. The adipose fin has a dark band that continues into the upper half of the adipose fin but not to the edge. There is a dark crescent-shaped dark brown band in the middle of the caudal fin that often extends forward across the upper and lower caudal rays to the caudal peduncle; there is another dark brown band near the clear edge of the caudal fin. Other fins have poorly defined dark bands. The breast and belly are white to pale yellow, usually without dark specks. There are 5–10 large, sawlike teeth on the rear edge of the large pectoral-fin spine, large teeth on the anterior edge, and 13–16 anal rays. To 5 in. (13 cm).

SIMILAR SPECIES The Mountain Madtom, *N. eleutherus*, lacks the 2 large, light yellow spots in front of the dorsal fin and lacks a dark crescent in the middle of the caudal fin; it has a dark brown bar at the base of the caudal fin and has a dark band on the adipose fin that is confined to the lower half of the fin. The Brindled Madtom, *N. miurus*, has a large, black blotch on the dorsal fin, and the black band on the adipose fin extends to the edge of the fin.

HABITAT The Northern Madtom lives in sandy and rocky riffles and runs in small to large, often swift, rivers. Most recent observations of the species were around or under dead mussel shells and rocks during extremely low water in the lower Ohio River.

DISTRIBUTION IN ILLINOIS The Northern Madtom was not recognized by Forbes and Richardson (1909) as distinct from the Mountain Madtom, *N. eleutherus*. In the Smith era, it was recorded from the Vermilion River in Vermilion County, the Wabash River in Wabash County, and the Ohio River in Pulaski County. There are no contemporary-era records from the Wabash River basin, but the species has been found at several localities in the Ohio River in Pulaski and Massac Counties.

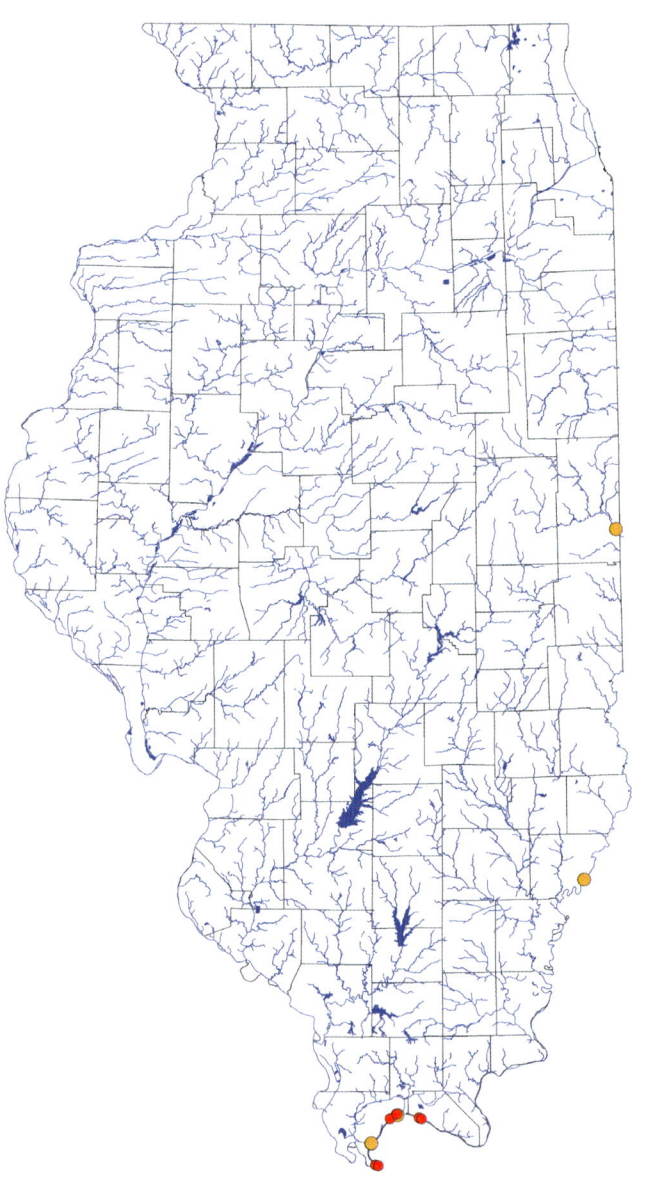

FLATHEAD CATFISH *Pylodictis olivaris* (Rafinesque 1818)

IDENTIFICATION The Flathead Catfish is depressed anteriorly and compressed posteriorly with a large, caudal fin. The head is wide and flat, and the lower jaw projects beyond the upper jaw (except in juveniles). The caudal fin is straight edged or slightly forked and has a distinctive white tip on the upper lobe (the white tip is lost in large adults). The eye is small and on top of the head. The adipose fin is small and has a short base, and its posterior edge is free from the back and far from the caudal fin. The back and side of the body are dusky yellow-brown to dark purple-brown with black or brown mottling. The breast and belly are dusky white. The fins are dusky black with dark mottling and specks. The premaxillary tooth patch (on the roof of the mouth) has a large, backward extension on each side. The anal fin is short and rounded in outline with 14–17 rays. To 61 in. (155 cm).

SIMILAR SPECIES Other catfishes in Illinois lack a lower jaw that projects beyond the upper jaw and a white tip on the upper lobe of the caudal fin.

HABITAT The Flathead Catfish is usually found in lakes, impoundments, and pools with logs and other woody debris in low- to moderate-gradient, small to large rivers. Juveniles are sometimes found in rocky and sandy runs and riffles.

DISTRIBUTION IN ILLINOIS The Flathead Catfish occurs statewide and appears to be more widespread in the contemporary era than in previous eras. Smith (1979) commented that the species was probably more common and widespread than available records indicated because large adults are difficult to collect and store in institutional collections. The same may be true today, but the species is found throughout the state, including in the northeastern region where there were only 2 records during the Smith era and none from the Forbes and Richardson era. Contemporary-era records include Lake Michigan.

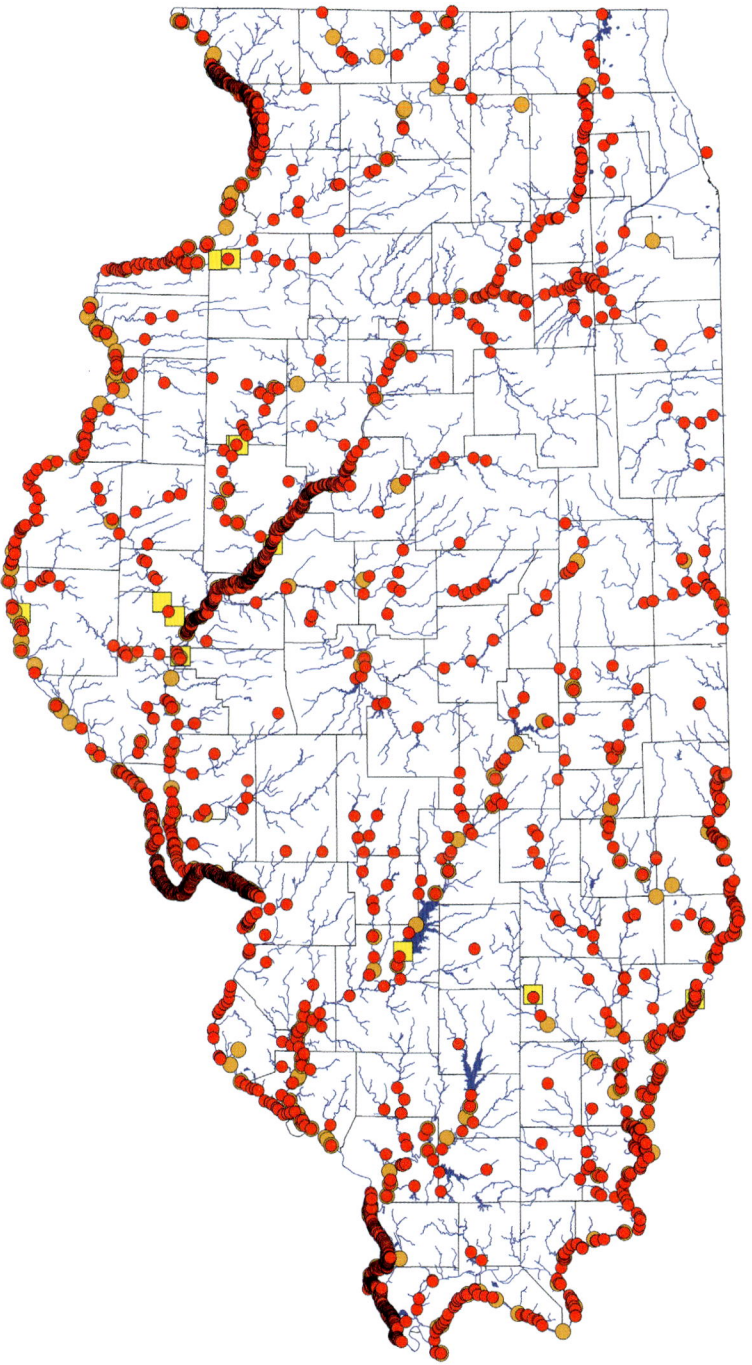

CENTRAL MUDMINNOW *Umbra limi* (Kirtland 1840)

IDENTIFICATION The Central Mudminnow is a small fish with an elongate body that is more-or-less cylindrical anteriorly and compressed posteriorly, a short snout, a terminal mouth, and a large, rounded caudal fin. The anal and dorsal fins are far back on the body. The back and side of the body are dark green to brown. There are up to 14 dark irregular bars on the side of the body and a black bar at the base of the caudal fin. The underside of the body is white to yellow. Fins are green to red-brown, often with dark blotches. During the breeding season, males have iridescent blue-green anal and pelvic fins. There are 30–37 lateral scales and 7–10 anal rays. To 6 in. (15 cm).

SIMILAR SPECIES Juvenile Bowfin, *Amia calva*, have a long dorsal fin and a bony gular plate on the throat.

HABITAT The Central Mudminnow occurs in swamps, sloughs, bogs, and other wetlands over mud and woody debris and in backwaters of creeks and small rivers. It is usually found near aquatic vegetation.

DISTRIBUTION IN ILLINOIS The Central Mudminnow is less widespread in some areas of the state than in earlier eras and more widespread in others. While it is more commonly encountered in the Kankakee River basin than in previous eras, it is much less widespread in northwestern Illinois and appears to have disappeared from the Wabash River basin and from the Big Muddy basin except in the southernmost reaches of the Big Muddy River.

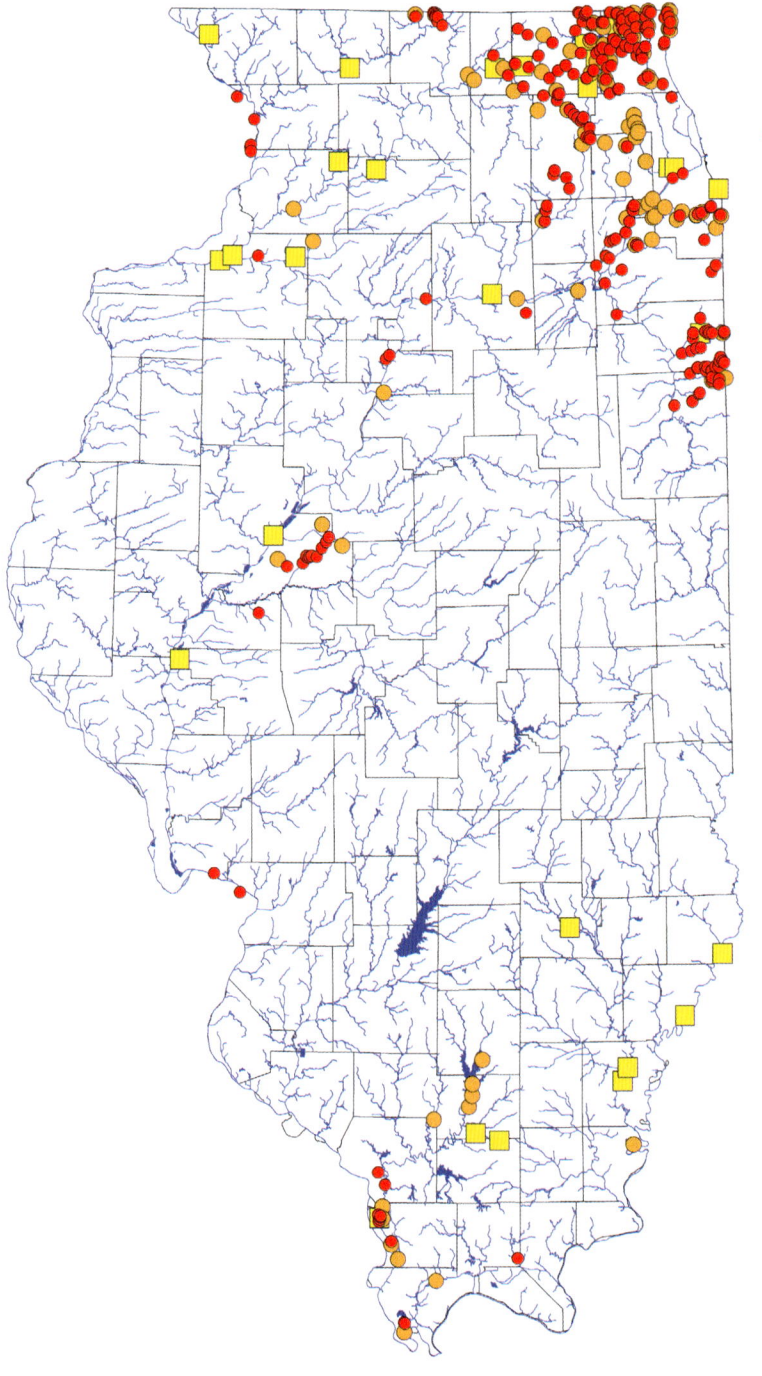

TROUTS, SALMONS, AND WHITEFISHES

Family Salmonidae Jarocki or Schinz 1822

Salmonidae includes the well-known species of trout and salmon and the less familiar ciscoes and whitefishes. All salmonids have 1 dorsal fin, an adipose fin, no spines in the fins, small scales (60 or more in the lateral line), well-developed pyloric caeca, an axillary process at the base of the pelvic fin, and a fusiform body shape. The 228 species of salmonids are native to cool and cold streams and lakes throughout Europe, northern Asia, the Atlas Mountains in extreme northern Africa, Taiwan, and North America as far south as northwestern Mexico.

Most salmonid species spend their entire lives in fresh water, but some living in coastal basins spawn in fresh water, migrate to the ocean as juveniles, and return as adults to fresh water to spawn. Because of the popularity of trout and salmon as sport and food fishes, they are commonly stocked outside their natural ranges. All of the trout and salmon in Illinois are non-native, except for the Lake Trout, *Salvelinus namaycush*.

Four species of *Coregonus* are extant in Illinois and are discussed. Five others, *C. alpenae*, *C. johannae*, *C. kiyi*, *C. reighardi*, and *C. zenithicus*, have been recorded from Lake Michigan (O'Donnell 1935, Smith 1979, Eshenroder et al. 2016) and may have occurred in Illinois but were never documented with vouchers. If they were present, which is debatable given that ciscoes (*Coregonus*) are among the most difficult North American freshwater fishes to identify, they seem to be no longer present.

KEY TO THE TROUTS, SALMON, AND WHITEFISHES (Salmonidae)

1a Mouth large, upper jaw extending behind center of eye; scales tiny, difficult to count, 20–27 scale rows above lateral line, more than 110 along lateral line; teeth in lower jaw (dentary) well developed **6**

Teeth in lower jaw **well developed**

Scales **tiny, 20–27** scale rows above lateral line

More than 110 scales along lateral line

Mouth **large**, upper jaw **extending** behind center of eye

1b Mouth small, upper jaw not extending beyond center of eye; scales large, 6–11 scale rows above lateral line, less than 100 along lateral line; teeth in lower jaw absent or poorly developed **2**

Teeth in lower jaw **absent or poorly developed**

Scales **large, 6–11** scale rows above lateral line

Less than 100 scales along lateral line

Mouth small, upper jaw **not extending** behind center of eye

2a Single flap between nostrils; gill rakers on first arch fewer than 20; body rounded in cross section . . . . . . **Page 258**

Round Whitefish (*Prosopium cylindraceum*)

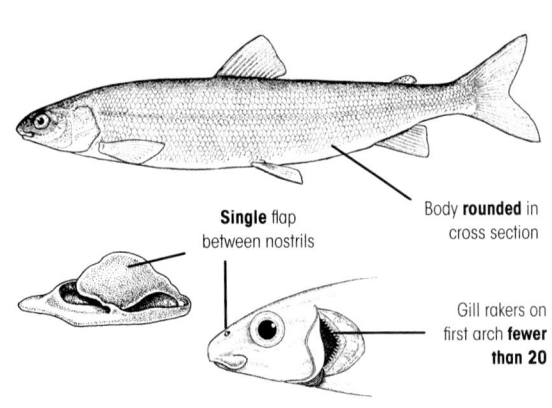

Single flap between nostrils

Body **rounded** in cross section

Gill rakers on first arch **fewer than 20**

2b Two flaps between nostrils; gill rakers on first arch usually more than 22; body usually laterally compressed in cross section (except Cisco) **3**

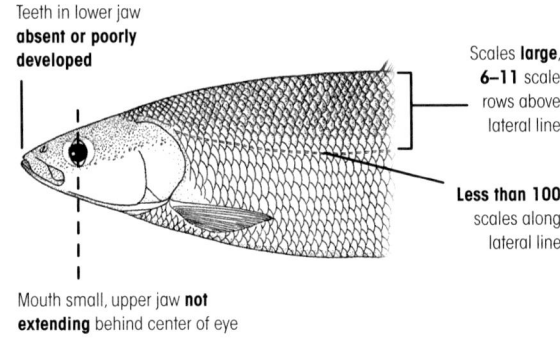

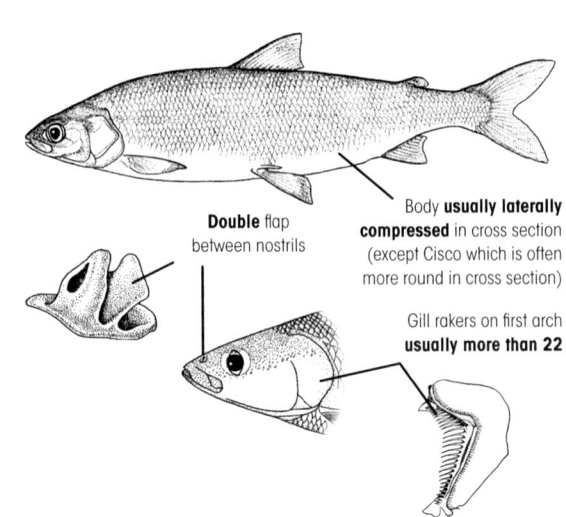

Double flap between nostrils

Body **usually laterally compressed** in cross section (except Cisco which is often more round in cross section)

Gill rakers on first arch **usually more than 22**

3a Premaxillaries (bones at front of upper jaw) pointing backward, giving front of snout a rounded profile; length of upper jaw (A) usually going 3 or more times into length of head (B); gill rakers on first arch usually 22–33 . . . . **Page 252**

3b Premaxillaries pointing forward, giving front of snout a pointed profile; length of upper jaw (A) usually going 3 or fewer times into length of head (B); gill rakers on first arch usually more than 37 . . . . **4**

Lake Whitefish (*Coregonus clupeaformis*)

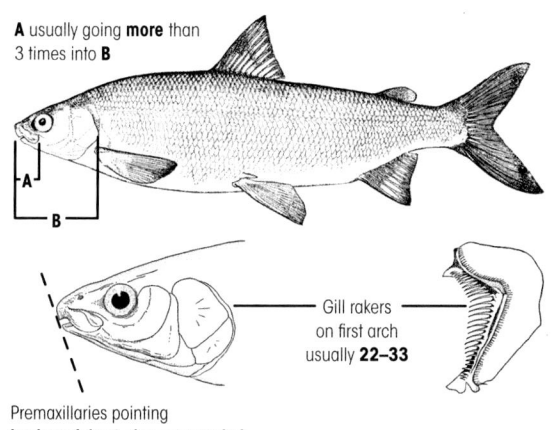

A usually going **more** than 3 times into **B**

Gill rakers on first arch usually **22–33**

Premaxillaries pointing **backward**, front of snout **rounded**

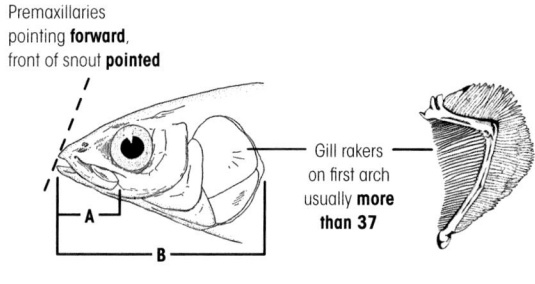

Premaxillaries pointing **forward**, front of snout **pointed**

Gill rakers on first arch usually **more than 37**

A usually going **less** than 3 times into **B**

4a Tip of lower jaw projects slightly beyond upper jaw; gill rakers on first arch usually 41–46; body deepest at middle . . . . **Page 253**

4b Lower jaw about equal to or shorter than upper jaw; gill rakers on first arch usually 46–50; body either elongate or deeper forward than at middle **5**

Bloater (*Coregonus hoyi*)

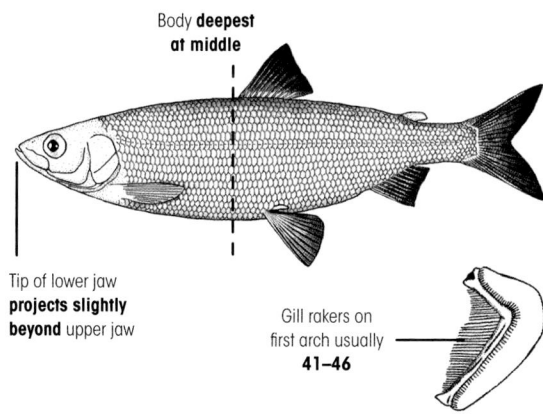

Body **deepest at middle**

Tip of lower jaw **projects slightly beyond** upper jaw

Gill rakers on first arch usually **41–46**

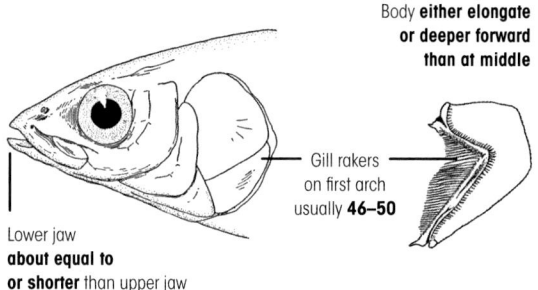

Body **either elongate or deeper forward than at middle**

Gill rakers on first arch usually **46–50**

Lower jaw **about equal to or shorter** than upper jaw

5a Body elongate, almost round in cross section; deepest at middle; lower jaw frail, weakly developed (with bone). . . .
. . . .**Page 251**

Cisco (*Coregonus artedi*)

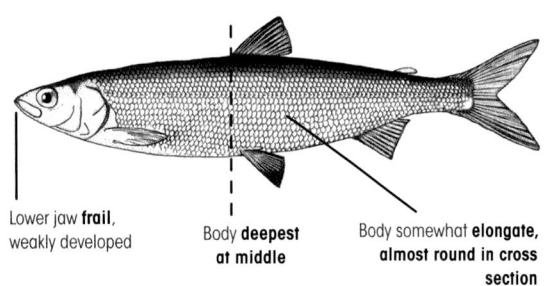

Lower jaw **frail**, weakly developed

Body **deepest at middle**

Body somewhat **elongate, almost round in cross section**

5b Body somewhat deeper and thicker; body deepest in front of middle; lower jaw stout. . . . .**Page 254**

Blackfin Cisco (*Coregonus nigripinnis*)

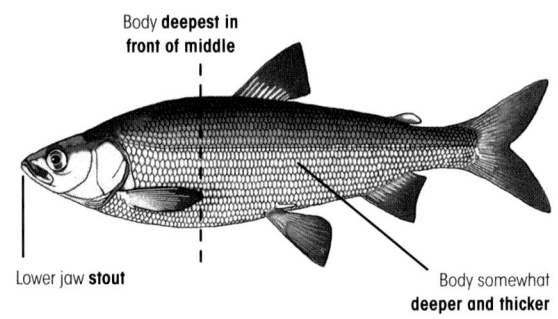

Body **deepest in front of middle**

Lower jaw **stout**

Body somewhat **deeper and thicker**

6a Anal fin with 8–12 principal rays; lower mouth whitish in adults. **7**

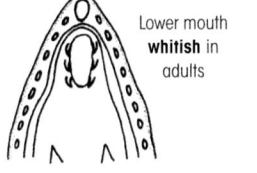

Lower mouth **whitish** in adults

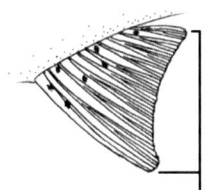

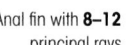

Anal fin with **8–12** principal rays

6b Anal fin with 13–19 principal rays; lower mouth blackish in adults. **10**

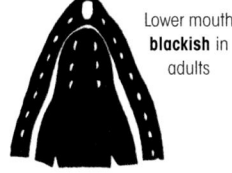

Lower mouth **blackish** in adults

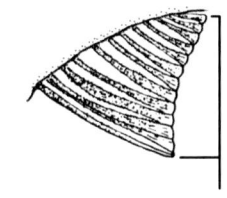

Anal fin with **13–19** principal rays

7a Body with dark brown or black spots on a light background; teeth on head and shaft of vomer, forming a strip down the center of the roof of the mouth; scales in lateral series fewer than 140. . . . .**8**

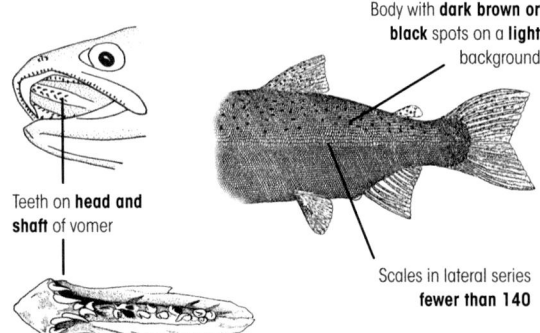

Body with **dark brown or black** spots on a **light** background

Teeth on **head and shaft** of vomer

Scales in lateral series **fewer than 140**

7b Body with light (red, pink, orange, yellow, or gray) spots on a dark background; teeth on head of vomer only, in patch at the front of the mouth; scales in lateral series more than 190 . . . .**9**

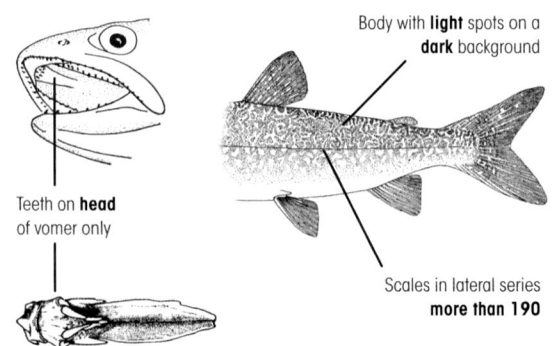

Body with **light** spots on a **dark** background

Teeth on **head** of vomer only

Scales in lateral series **more than 190**

8a Caudal fin with regular rows of black spots; side (in life) without orange or red spots, but with a pink or reddish longitudinal stripe; adipose fin often with a black edge. . .
. . . .**Page 256**

Rainbow Trout (*Oncorhynchus mykiss*)

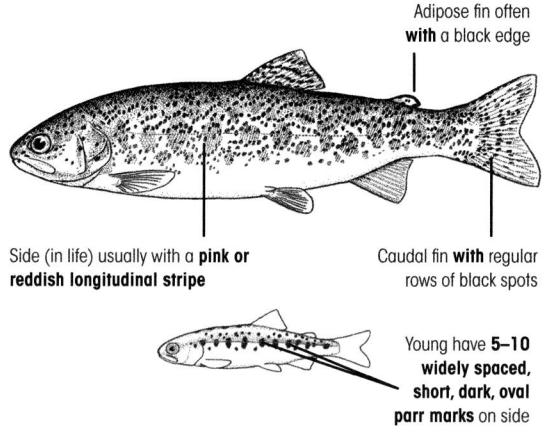

Adipose fin often **with** a black edge

Side (in life) usually with a **pink or reddish longitudinal stripe**

Caudal fin **with** regular rows of black spots

Young have **5–10 widely spaced, short, dark, oval parr marks** on side

8b Caudal fin usually unspotted, never with regular rows of black spots; gill covers with many dark spots; side (in life) sometimes with orange or reddish spots, but without a pink or reddish longitudinal stripe; adipose fin usually with red margin and no black edge . . . . **Page 259**

Brown Trout (*Salmo trutta*)

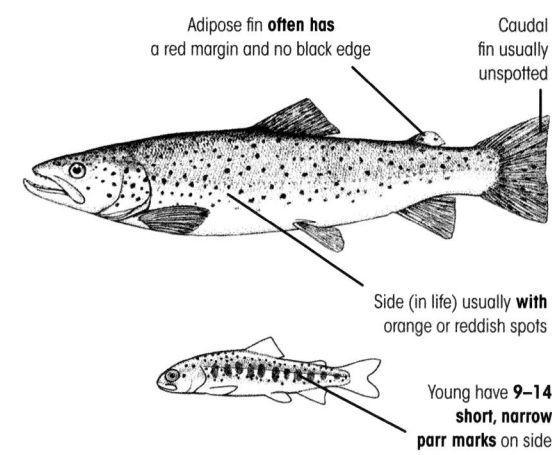

Adipose fin **often has** a red margin and no black edge

Caudal fin usually unspotted

Side (in life) usually **with** orange or reddish spots

Young have **9–14 short, narrow parr marks** on side

9a Caudal fin deeply forked; dorsal and caudal fins, body, and head covered with small, often bean-shaped, cream or yellow spots; side (in life) never with orange or red spots on dark green or gray background; lower fins without black stripe; young have 7–12 narrow, often interrupted parr marks on side. . . . . **Page 261**

Lake Trout (*Salvelinus namaycush*)

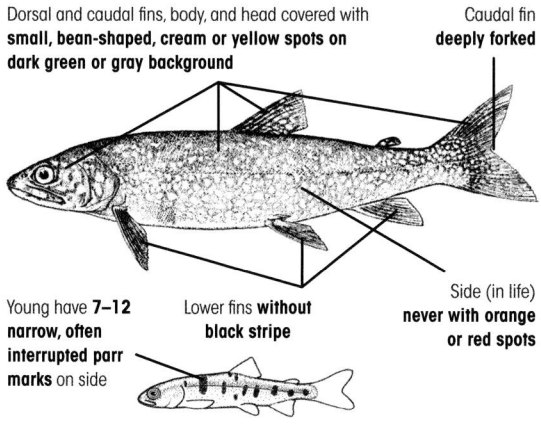

Dorsal and caudal fins, body, and head covered with **small, bean-shaped, cream or yellow spots on dark green or gray background**

Caudal fin **deeply forked**

Young have **7–12 narrow, often interrupted parr marks** on side

Lower fins **without black stripe**

Side (in life) **never with orange or red spots**

9b Caudal fin slightly forked or truncate (straight-edged); dorsal fin and upper body with light green or cream wavy lines or blotches; side (in life) has pink or red spots surrounded by blue halos; lower fins with black stripe near leading edge; young have 8–10 regularly arranged parr marks on side . . . .**Page 260**

Brook Trout (*Salvelinus fontinalis*)

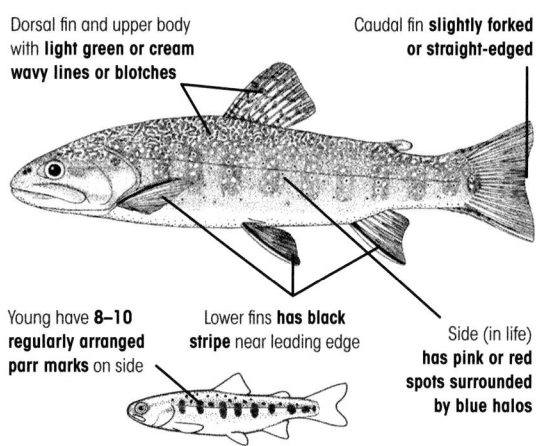

Dorsal fin and upper body with **light green or cream wavy lines or blotches**

Caudal fin **slightly forked or straight-edged**

Young have **8–10 regularly arranged parr marks** on side

Lower fins **has black stripe** near leading edge

Side (in life) **has pink or red spots surrounded by blue halos**

10a Flesh along base of teeth in lower jaw black; caudal fin with small back spots on both lobes; anal fin rays usually 14–19; young have 6–12 large parr marks on side . . . . . .
. . . .**Page 257**

10b Flesh along base of teeth in lower jaw lightly pigmented; caudal fin without small black spots or with spots on upper lobe only; anal fin rays usually 11–15; young have 8–12 narrow parr marks on side . . . .**Page 255**

Chinook Salmon (*Oncorhynchus tshawytscha*)

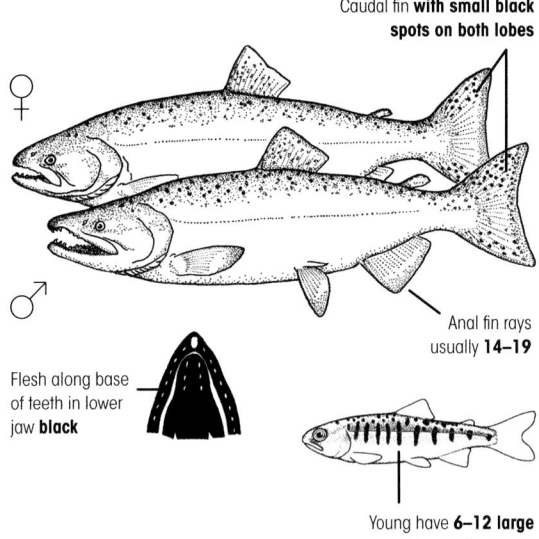

Caudal fin **with small black spots on both lobes**

Anal fin rays usually **14–19**

Flesh along base of teeth in lower jaw **black**

Young have **6–12 large** parr marks on side

Coho Salmon (*Oncorhynchus kisutch*)

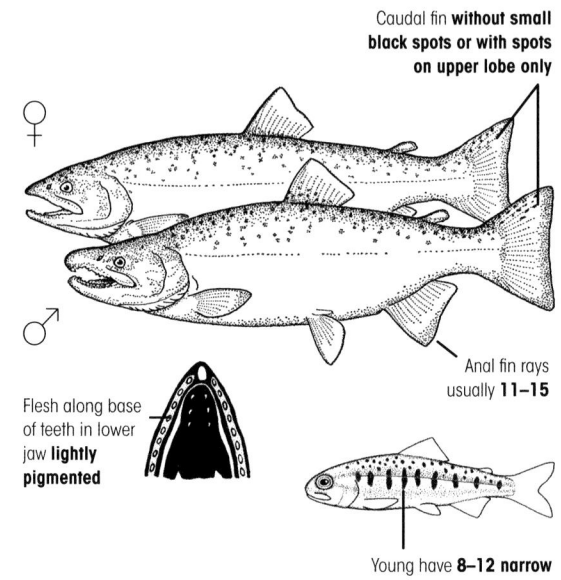

Caudal fin **without small black spots or with spots on upper lobe only**

Anal fin rays usually **11–15**

Flesh along base of teeth in lower jaw **lightly pigmented**

Young have **8–12 narrow** parr marks on side

CISCO *Coregonus artedi* Lesueur 1818

IDENTIFICATION The Cisco's body is long and slightly compressed to round in cross section and deepest at the middle of the body. The lower jaw projects farther forward than the upper jaw, which reaches beneath the front of the pupil. A symphyseal knob is present at the tip of the lower jaw. There are 2 small flaps of skin between the nostrils, a fairly broad snout, a small, subterminal to terminal mouth, and a forked caudal fin. The species is dusky blue to green on the back and silver-green to silver-blue on the side of the body. Individuals over about 6 in. (15 cm), often have dusky or black-tipped pelvic fins. No parr marks are present on juveniles. There are 36–64, usually 40–50, long, slender rakers on the first gill arch. To 22½ in. (57 cm).

SIMILAR SPECIES The Bloater, *C. hoyi*, has a more pointed snout, a more projecting lower jaw, a dusky black upper lip, and black edges on the dorsal and caudal fins. The Blackfin Cisco, *C. nigripinnis*, has the body deepest at the nape and no symphyseal knob at the tip of the lower jaw. The Lake Whitefish, *C. clupeaformis*, has a more subterminal mouth, a pronounced hump behind the head in the adult, 24–33 rakers on the first gill arch, and no dusky black pigment on the fins.

HABITAT The Cisco lives in open water of large rivers and lakes.

DISTRIBUTION IN ILLINOIS Forbes and Richardson (1909) recorded the Cisco in Lake Michigan, where it now appears to be much less frequently encountered than in previous eras, and in the Illinois River. The latter was presumably based on a strays from Lake Michigan.

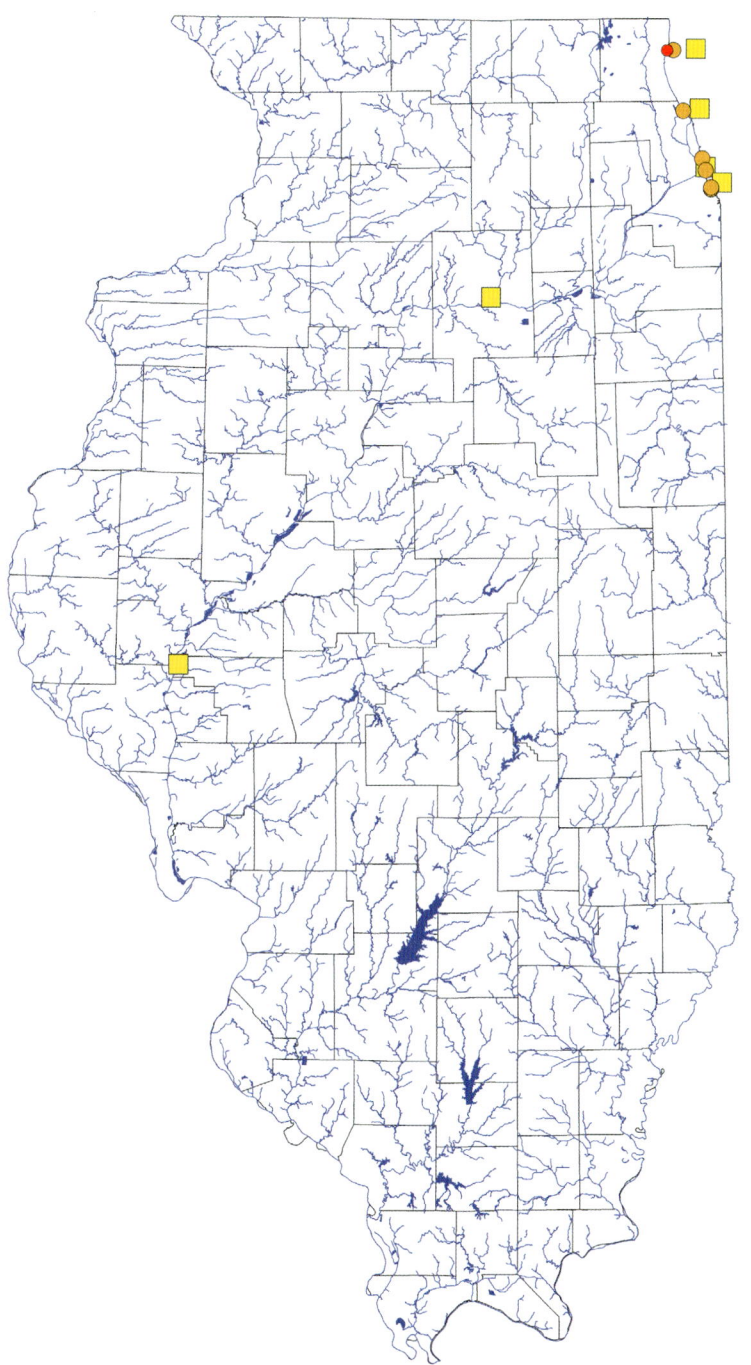

LAKE WHITEFISH *Coregonus clupeaformis* (Mitchill 1818)

IDENTIFICATION The Lake Whitefish's body is compressed to round in cross section, and there is a pronounced hump behind the head in the adult. There are 2 small flaps of skin between the nostrils, a fairly broad snout, a small, subterminal mouth, and a forked caudal fin. The species is blue-brown on the back and silver-green to silver-blue on the side of the body. No dark spots or other marks are present on the adult, and no parr marks are present on juveniles. Fins lack black pigment. There are 24–33 long, slender rakers on the first gill arch. To 31 in. (80 cm).

SIMILAR SPECIES The Cisco, *C. artedi*, Bloater, *C. hoyi*, and Blackfin Cisco, *C. nigripinnis*, have a more terminal mouth, no pronounced hump behind the head, usually more than 35 rakers on the first gill arch, and dusky black lower fins.

HABITAT The Lake Whitefish lives in open water of large rivers and lakes.

DISTRIBUTION IN ILLINOIS Forbes and Richardson (1909) noted that the Lake Whitefish had been the most important food fish in Lake Michigan but was becoming less common in the Illinois portion of the lake. The species had nearly disappeared by the time of Smith (1979), but in the contemporary era is frequently encountered in the Illinois portion of Lake Michigan.

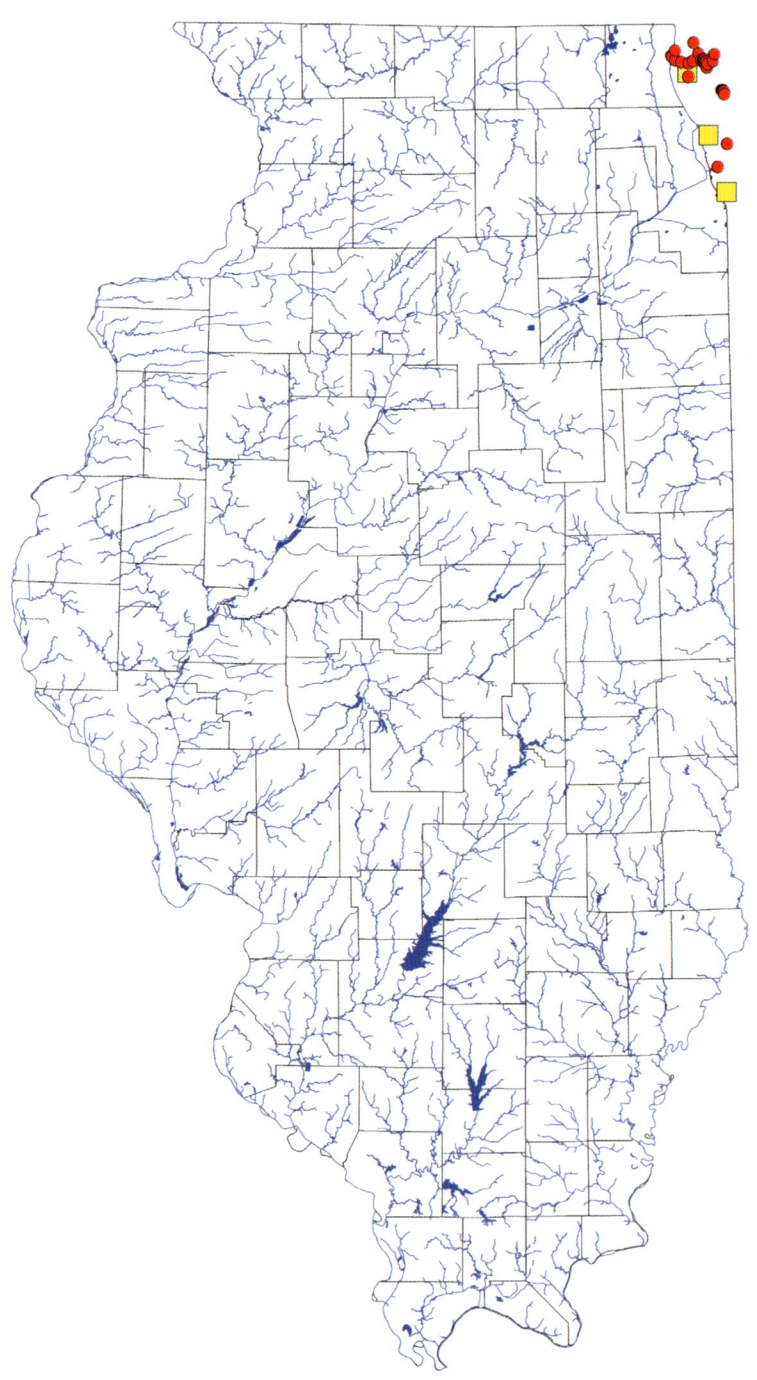

BLOATER *Coregonus hoyi* (Milner 1874)

IDENTIFICATION The Bloater's body is elongate and slightly compressed to round in cross section, and deepest at the middle of the body. A symphyseal knob is present at the tip of the lower jaw. There are 2 small flaps of skin between the nostrils, a small, terminal mouth, and a forked caudal fin. The snout is pointed, and the darkly pigmented lower jaw projects farther forward than the upper jaw. The species is dusky brown dorsally and pinkish-silver on the side of the body; it has a dusky black upper lip and often black edges on the dorsal and caudal fins. No parr marks are present on juveniles. There are usually 40–47 long, slender rakers on the first gill arch. To 14½ in. (37 cm).

SIMILAR SPECIES The Cisco, *C. artedi*, has a less pointed snout, a less projecting lower jaw, and no black on the upper lip or on the dorsal and caudal fins. The Blackfin Cisco, *C. nigripinnis*, has the body deepest at the nape and no symphyseal knob at the tip of the lower jaw. The Lake Whitefish, *C. clupeaformis*, has a more subterminal mouth, a pronounced hump behind the head in the adult, 24–33 rakers on the first gill arch, and no dusky black pigment on the fins.

HABITAT The Bloater lives in the deep, open areas of large lakes.

DISTRIBUTION IN ILLINOIS Forbes and Richardson (1909) had no records of the Bloater from Illinois, but Smith-era and contemporary-era records show the species to be the most widespread species of *Coregonus* in Lake Michigan.

REMARKS When this fish is brought up from deep water, the gas bladder expands and gives this fish a bloated appearance and its common name.

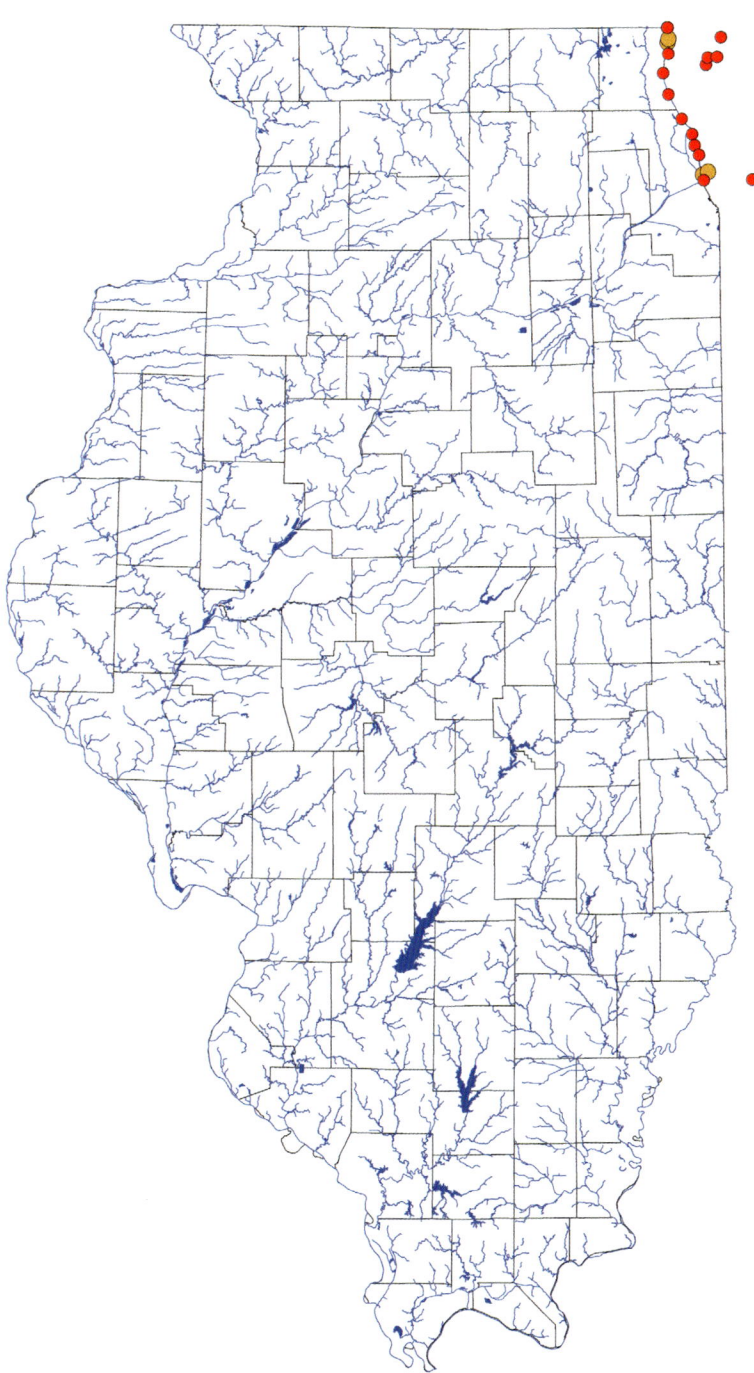

TROUTS, SALMONS, AND WHITEFISHES (Salmonidae)

253

BLACKFIN CISCO *Coregonus nigripinnis* (Milner 1874)

IDENTIFICATION The Blackfin Cisco's body is long and slightly compressed to round in cross section and usually deepest at the nape. The lower jaw does not project farther forward than the upper jaw, which reaches beneath the front of the pupil. No symphyseal knob is present at the tip of the lower jaw. There are 2 small flaps of skin between the nostrils, a fairly broad snout, a small, subterminal to terminal mouth, and a forked caudal fin. The species is dusky blue-green to blue-black on the back, silver on the side of the body, has dusky black fins (darkest, sometimes blue-black, on large individuals), and a dusky black upper lip. No parr marks are present on juveniles. There are 46–50 long, slender rakers on the first gill arch. To 15¼ in. (39 cm).

SIMILAR SPECIES The Cisco, *C. artedi*, and the Bloater, *C. hoyi*, are deepest at the middle of the body and have a symphyseal knob at the tip of the lower jaw. The Lake Whitefish, *C. clupeaformis*, has a more subterminal mouth, a pronounced hump behind the head in the adult, 24–33 rakers on the first gill arch, and no dusky black pigment on the fins.

HABITAT The Blackfin Cisco lives in the deep areas of Lakes Michigan, Huron, and Nipigon.

DISTRIBUTION IN ILLINOIS The Blackfin Cisco was not recognized by Forbes and Richardson (1909) as distinct from the Cisco. Only 1 record from the Smith era and 1 from the contemporary era document the species in the Illinois portion of Lake Michigan.

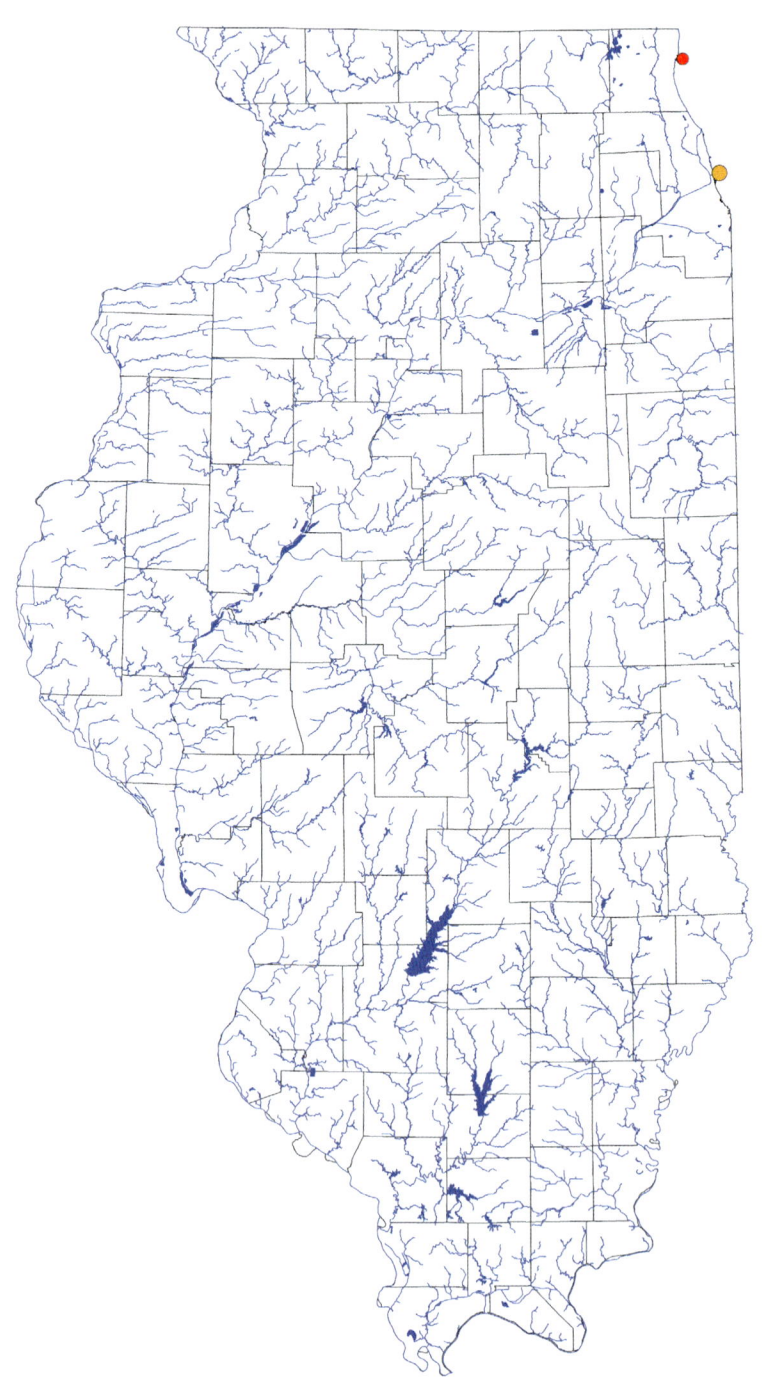

COHO SALMON *Oncorhynchus kisutch* (Walbaum 1792)

IDENTIFICATION The Coho Salmon is a large fish with a streamlined, compressed body, a large mouth extending past the eye, and straight-edged to slightly forked caudal fin. It is steel blue to dull brown on the back and side of the body and silver to white on the underside. Black spots are scattered on the back and upper lobe of the caudal fin. Gums are white at the base of the teeth. Large males have a dusky green-brown back and head, red side, and hooked upper jaw. The female has a bronze to pink-red side. Juveniles have 8–12, narrow, dark parr marks along the side of the body. Teeth are present on the front (head) and on the shaft of the vomer. Scales along the lateral line are as large or larger than scales in adjacent rows. There are 120–170 lateral scales and 13–19 anal rays. To 38½ in. (98 cm).

SIMILAR SPECIES The Chinook Salmon, *O. tshawytscha*, has black spots on both lobes of the caudal fin, and the gums are black at the base of the teeth. The Rainbow Trout, *O. mykiss*, has a pink to red stripe on the midside, small, irregular black spots on the back and most fins, radiating rows of black spots on the caudal fin, and 8–12 anal rays.

HABITAT The Coho Salmon lives in open waters of lakes and cool, clear, gravel-bottomed rivers.

DISTRIBUTION IN ILLINOIS The Coho Salmon is native to the basins of the Arctic and Pacific Oceans of the United States, Canada, and Northeast Asia. The Coho was stocked in Lake Michigan beginning in 1967 (Smith 1979). All contemporary-era records from Illinois are from the shallower waters of Lake Michigan.

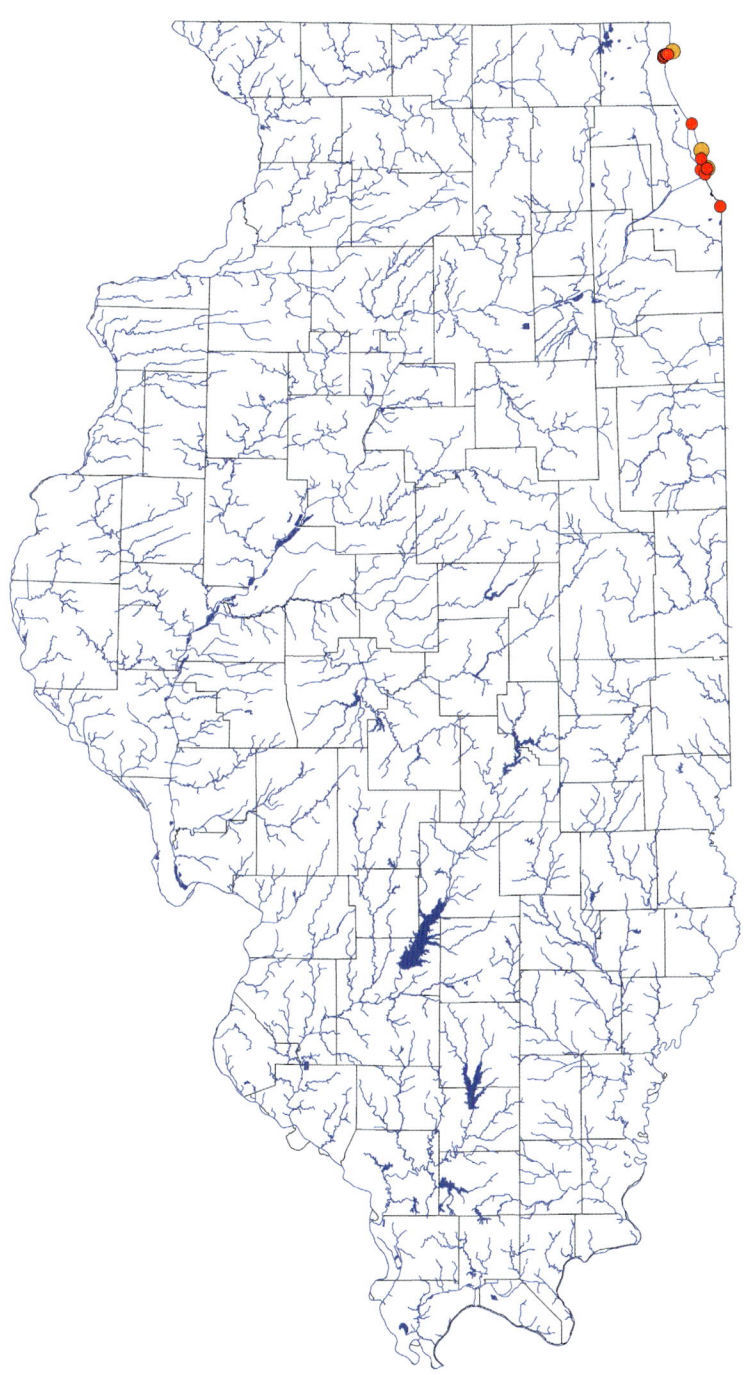

TROUTS, SALMONS, AND WHITEFISHES (Salmonidae)

RAINBOW TROUT *Oncorhynchus mykiss* (Walbaum 1792)

IDENTIFICATION The Rainbow Trout has an elongate, compressed body, a large mouth extending to the rear edge or just past the eye, and a slightly forked caudal fin. It is steel blue to yellow-brown on the back and silver to pale yellow-green on the underside. There is a pink to red stripe on the midside, small, irregular black spots on the back and most fins, and radiating rows of black spots on the caudal fin. The adipose fin usually has a black edge. Individuals in lakes tend to be much lighter in color and more silvery than those in streams. The upper jaw reaches beneath the eye in juveniles and females and well past the eye in large males. Juveniles have 5–10 widely spaced, dark, oval parr marks. Teeth are present on the front (head) and on the shaft of the vomer. Scales along the lateral line are as large or larger than scales in adjacent rows. There are 120–170 lateral scales and 8–12 anal rays. To 45 in. (114 cm).

SIMILAR SPECIES The Brook Trout, *Salvelinus fontinalis*, and Lake Trout, *S. namaycush*, have light-colored spots on a dark body, a white edge on the pelvic and anal fins, and teeth on the front (head) but not on the shaft of the vomer; the 150–200 scales along the lateral line are smaller than surrounding scales. The Coho Salmon, *O. kisutch*, and Chinook Salmon, *O. tshawytscha*, have no pink-red stripe on the midside, no radiating rows of black spots on the caudal fin, and 13–19 anal rays.

HABITAT The Rainbow Trout lives in lakes and in clear, cold, gravel-bottomed headwaters, creeks, and small to medium rivers.

DISTRIBUTION IN ILLINOIS The Rainbow Trout is native to the Pacific Ocean basin of the United States and Canada. It has been transplanted to many areas of the world, including in the United States outside its native range. Forbes and Richardson (1909) did not mention the species as having been stocked in Illinois, but Smith (1979) noted that stocking had occurred in Illinois prior to 1900. Contemporary-era occurrence records in the state are a reflection of widespread stocking, not of reproduction.

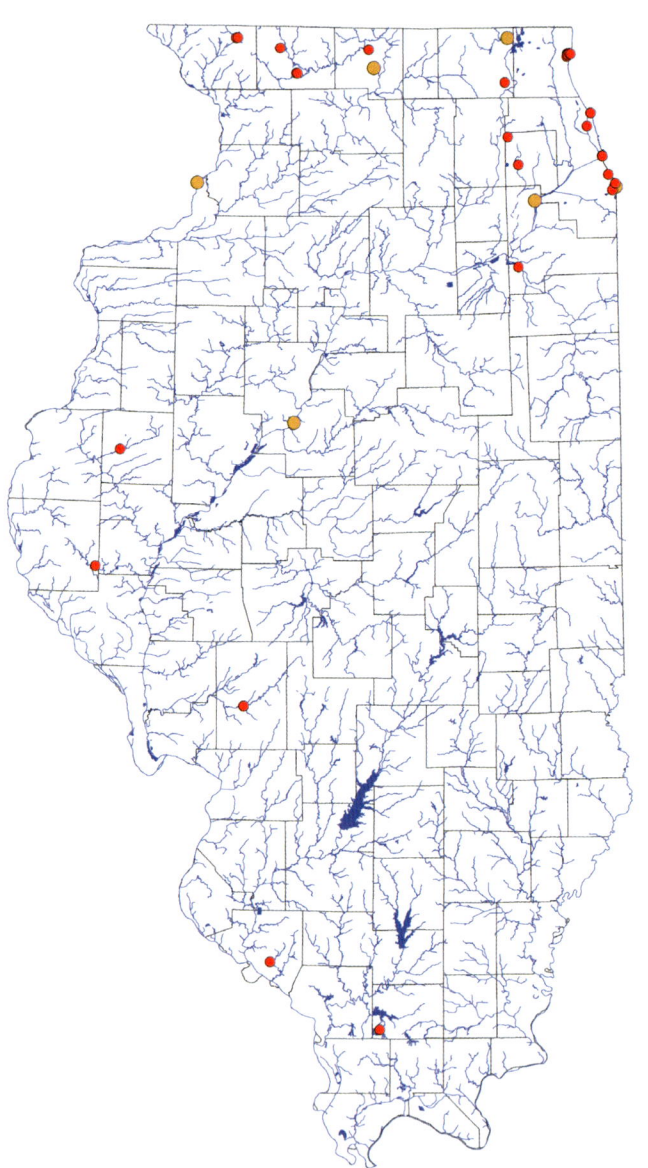

CHINOOK SALMON *Oncorhynchus tshawytscha* (Walbaum 1792)

IDENTIFICATION The Chinook Salmon is a large fish with a streamlined, compressed body, a large mouth extending past the eye, and a forked caudal fin. It is steel blue to dull yellow-brown on the back and upper side and silver to white on the lower side. Irregular, black spots are scattered on the back, the dorsal and adipose fins, and both lobes of the caudal fin. The gums are black at the base of the teeth. Large males are dark olive-brown to purple with dull-red blotches on the side. Juveniles have 6–12 narrow, dark parr marks along the side of the body. Teeth are present on the front (head) and on the shaft of the vomer. Scales along the lateral line are as large or larger than scales in adjacent rows. There are 120–170 lateral scales and 13–19 anal rays. To 58 in. (147 cm).

SIMILAR SPECIES The Coho Salmon, *O. kisutch*, has no black spots on the lower lobe of the caudal fin, and the gums are white at the base of the teeth. The Rainbow Trout, *O. mykiss*, has a pink to red stripe on the midside, radiating rows of black spots on the caudal fin, and 8–12 anal rays.

HABITAT The Chinook Salmon lives in open waters of lakes and cool, clear, gravel-bottomed rivers.

DISTRIBUTION IN ILLINOIS The Chinook Salmon is native to the Pacific Ocean basin of the United States and Canada. The Chinook was stocked in Lake Michigan beginning in 1967 (Smith 1979). All contemporary-era records from Illinois are from the shallower waters of Lake Michigan, except for 1 record from the middle Illinois River.

REMARKS The Chinook Salmon is the largest salmon in North America. Salmon over 30 lb. (14 kg) are almost always this species.

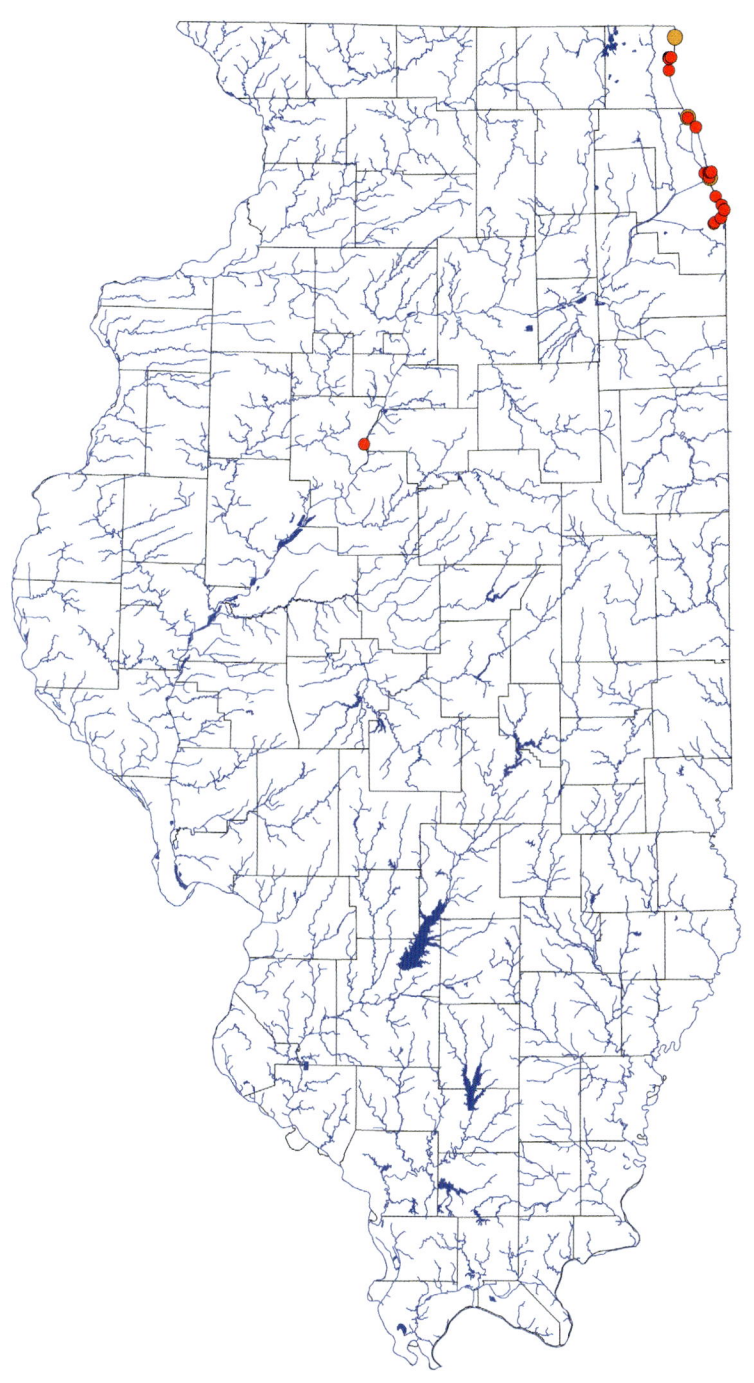

TROUTS, SALMONS, AND WHITEFISHES (Salmonidae)

257

ROUND WHITEFISH   *Prosopium cylindraceum* (Pennant 1784)

IDENTIFICATION The Round Whitefish has a small mouth that is subterminal, a round body that is about as wide as deep, and a forked caudal fin. There is 1 flap of skin between the nostrils, and the moderately pointed snout appears pinched when viewed from above. The Round Whitefish has dark-edged scales on the light brown to green-bronze back and upper side. The lower side of the body is silver-white, the dorsal and caudal fins are dusky gray, and the anal and paired fins are pale yellow to amber. Small individuals are silver with dark parr marks on the side of the body and 2 or more rows of black spots that sometimes merge with a row of black spots along the back. There are 76–89 lateral scales and 14–21 short, stout rakers on the first gill arch. To 22 in. (56 cm).

SIMILAR SPECIES Species of *Coregonus* are more compressed, have 2 small flaps of skin between the nostrils, usually no distinct black spots on the body, and long, slender rakers on the first gill arch. Young *Coregonus* lack parr marks.

HABITAT The Round Whitefish lives in clear, cool, rocky streams and relatively shallow areas of lakes.

DISTRIBUTION IN ILLINOIS Smith (1979) noted that the Round Whitefish had always been rare in Lake Michigan, but neither he nor Forbes and Richardson (1909) cited vouchered records from the Illinois portion of the lake. Contemporary-era records show it to be present and perhaps more widespread in the lake than in previous eras.

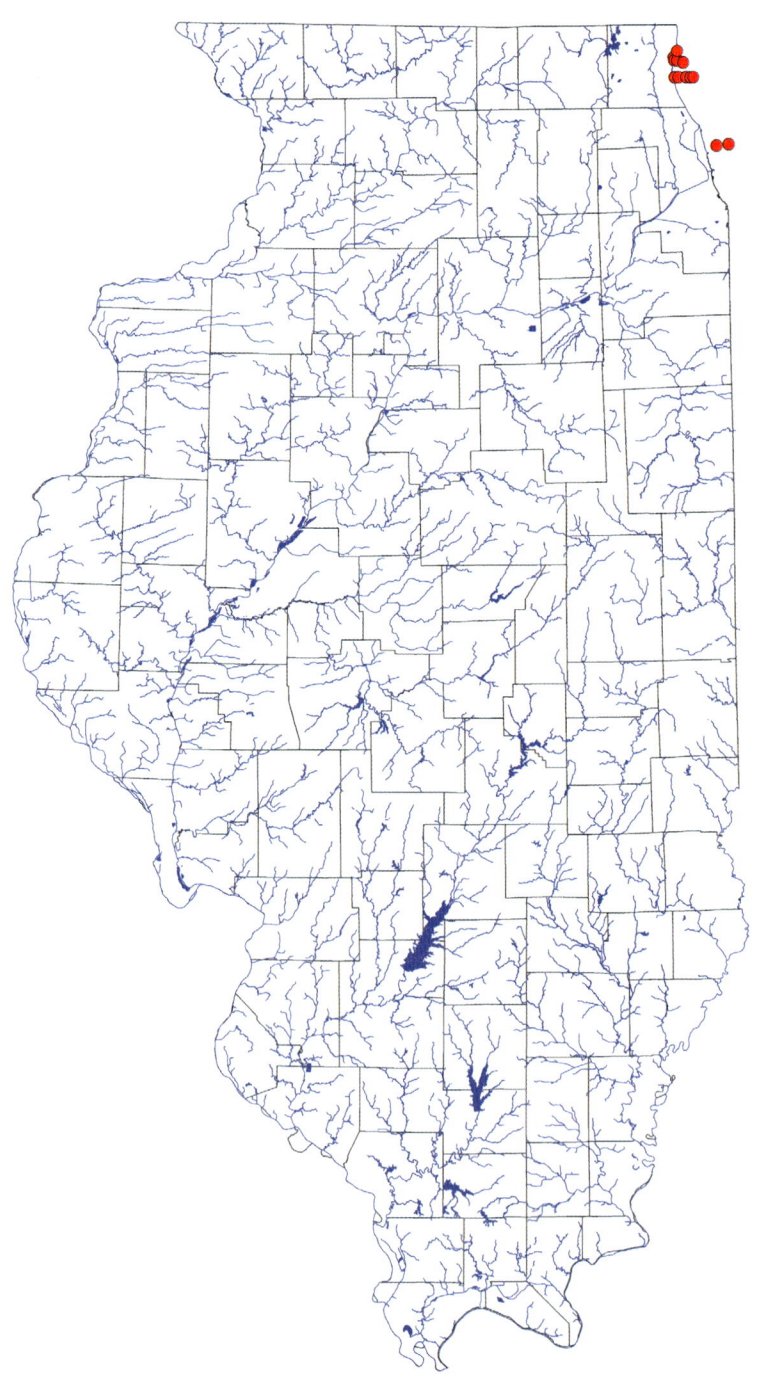

BROWN TROUT *Salmo trutta* Linnaeus 1758

IDENTIFICATION The Brown Trout has an elongate, compressed body, a large mouth extending past the eye, and a shallowly forked to straight-edged caudal fin. It has red and black spots on the olive-green to light brown head, back, side of the body, and dorsal, adipose, and caudal fins. The spots often are surrounded by pale halos. Large, black spots are present on the gill cover. The lower side of the body is white to yellow. The adipose fin is often red-orange on large individuals. The upper jaw reaches beneath the center of the eye in small individuals and well beyond the eye in larger fish. Large males develop a hooked lower jaw and become red on the lower side during the breeding season. Juveniles have 9–14 short, narrow parr marks along the side of the body and often have small, red spots along the lateral line. There are 120–130 lateral scales. To 40½ in. (103 cm).

SIMILAR SPECIES The Brook Trout, *Salvelinus fontinalis*, and Lake Trout, *Salvelinus namaycush*, have light-colored spots on a dark body and a white edge on the pelvic and anal fins.

HABITAT Brown Trout live in cool, fast-flowing streams and cold lakes. They are most often found in fast-flowing parts of streams.

DISTRIBUTION IN ILLINOIS The Brown Trout, native to Europe, northern Africa, and western Asia, was introduced to North America in 1883 (Mather 1889, Courtenay et al. 1984) and has been widely stocked throughout Canada and the United States. O'Donnell (1935) reported Brown Trout from northern Illinois, and Smith (1979) noted its presence along the Lake Michigan shoreline (although he cited no vouchers) as well as in streams in extreme northern Illinois. Contemporary-era records document the same distribution in the state due to continued stocking.

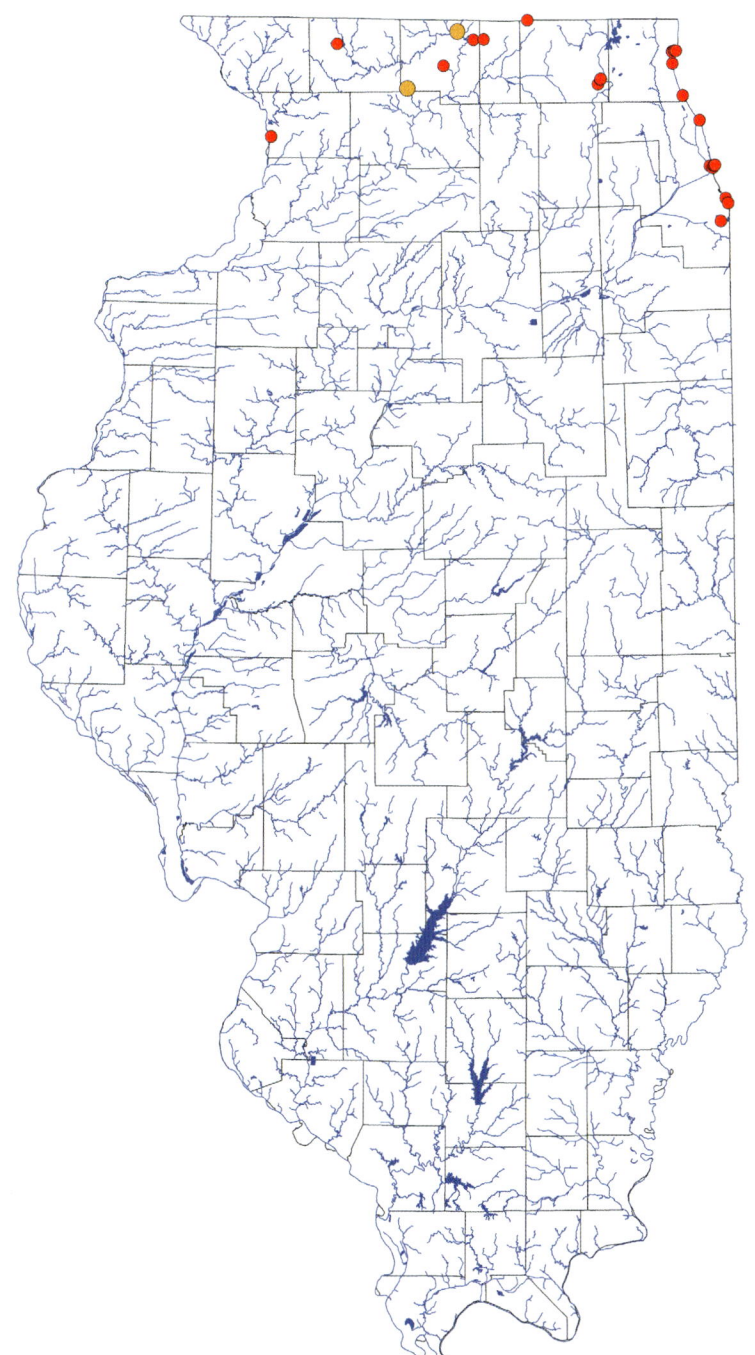

TROUTS, SALMONS, AND WHITEFISHES (Salmonidae)

259

BROOK TROUT *Salvelinus fontinalis* (Mitchill 1814)

IDENTIFICATION The Brook Trout has an elongate, slightly compressed body, a large mouth extending past the eye, and a shallowly forked to straight-edged caudal fin. The back and dorsal fin are olive to light brown with light green or cream-colored wavy lines or blotches. The side of the body has blue halos around pink or red spots. The pectoral, pelvic, and anal fins are red, and each has a black line behind a narrow, white leading edge. Large males become bright orange or bright red on the lower side of the body during the breeding season. Juveniles have 8–10 regularly arranged dark parr marks on the side of the body. Scales along the lateral line are smaller than surrounding scales. There are 23–55 pyloric caeca. Teeth are present on the front (head) but not on the shaft of the vomer. To 28 in. (70 cm).

SIMILAR SPECIES The Lake Trout, *S. namaycush*, has cream-colored spots on a green body, a deeply forked caudal fin, and 90–210 pyloric caeca. The Brown Trout, *Salmo trutta*, has black spots on the head and body and no white edge on the pelvic and anal fins.

HABITAT Brook Trout live in cold lakes and in clear, cool creeks and small to medium rivers.

DISTRIBUTION IN ILLINOIS Although Forbes and Richardson (1909) had no records for Illinois, Smith (1979) opined that the Brook Trout was native to Lake Michigan. The species is not known to reproduce in Illinois, and all records from streams—and probably those from Lake Michigan—are the result of stocking in Illinois.

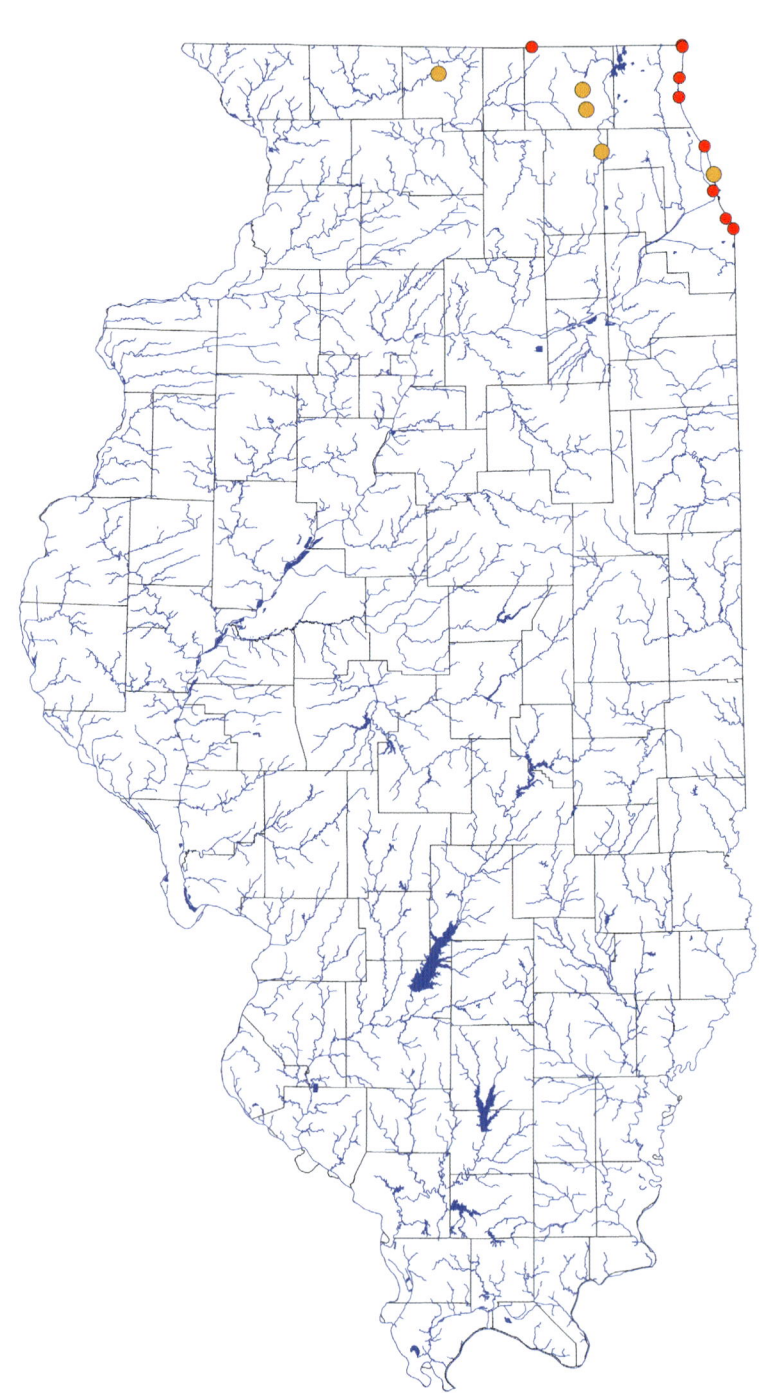

LAKE TROUT *Salvelinus namaycush* (Walbaum 1792)

IDENTIFICATION The Lake Trout is a large fish with an elongate, slightly compressed body, a large mouth extending past the eye, and a forked caudal fin. The head, body, and fins are green with many small, bean-shaped cream-yellow spots. The pectoral, pelvic, and anal fins are reddish-orange with a narrow, white leading edge. Individuals in large lakes may be silver overall. Large males develop a dark-green stripe along the side of the body. Juveniles have 7–12 narrow, often interrupted, dark parr marks. Scales along the lateral line number 150–200 and are smaller than surrounding scales. There are 90–210 pyloric caeca. Teeth are present on the front (head) but not on the shaft of the vomer. To 49½ in. (126 cm).

SIMILAR SPECIES The Brook Trout, *S. fontinalis*, has blue halos around pink or red spots on the side of the body, a shallowly forked caudal fin, and 23–55 pyloric caeca. The Brown Trout, *Salmo trutta*, has black spots on the head and body, no white edge on the pelvic and anal fins, and a shallowly forked caudal fin. The Coho Salmon, *Oncorhynchus kisutch*, and Chinook Salmon, *O. tshawytscha*, have black spots on the body and fins.

HABITAT The Lake Trout lives in the deep areas of lakes in the southern part of its range, including Illinois. In more-northern regions, it is found in clear, cool rivers and in shallow as well as deep areas of lakes.

DISTRIBUTION IN ILLINOIS The Lake Trout is native to Lake Michigan. It was severely impacted by the introduction of the Sea Lamprey in the 1940s and 1950s (Smith 1979) but is now widespread in the lake as the result of stocking and natural reproduction.

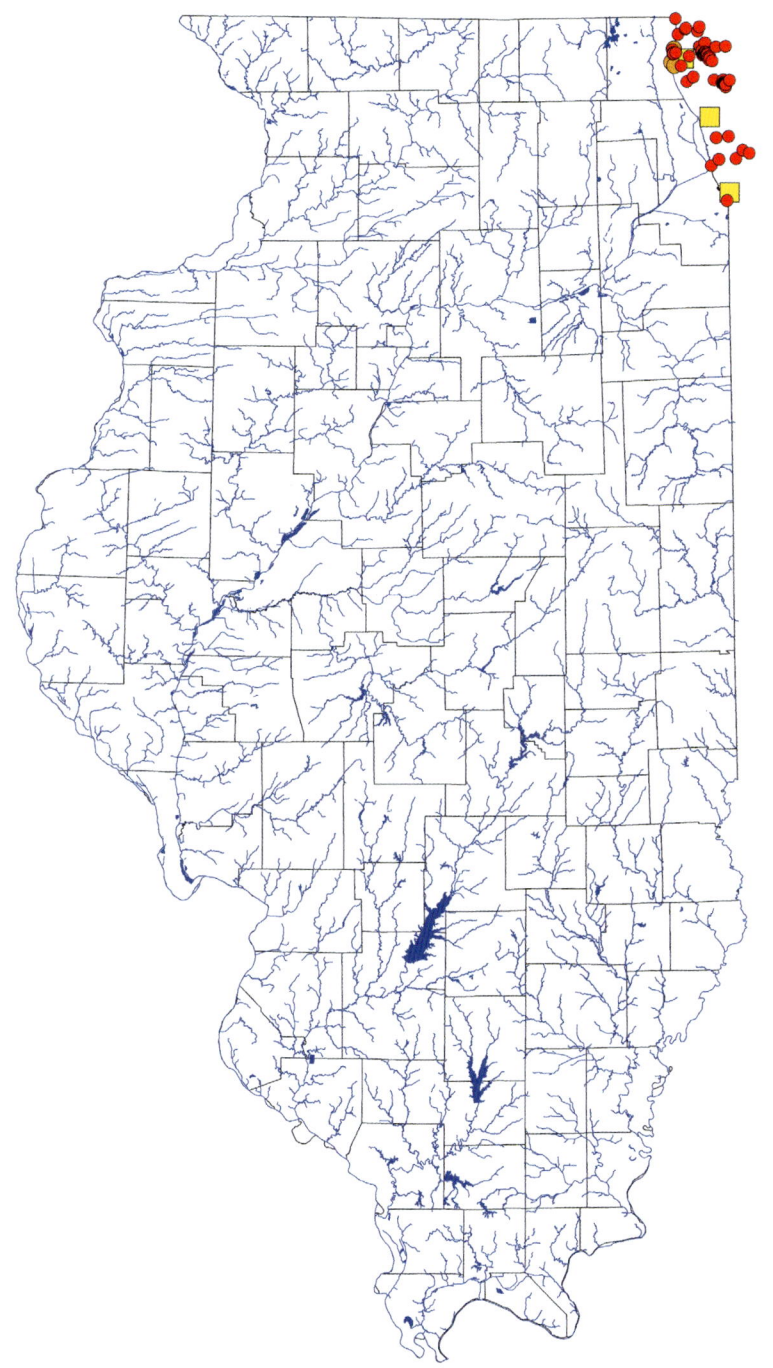

TROUTS, SALMONS, AND WHITEFISHES (Salmonidae)

SMELTS

Family Osmeridae Regan 1913

Smelts, of which there are 14 species, live in cold and temperate coastal waters, both marine and fresh, in the Northern Hemisphere. They are small, slender, silvery fishes with a large mouth, a protruding lower jaw, teeth on the jaws, an incomplete lateral line, and abdominal pelvic fins. They have 1 dorsal fin and an adipose fin. There are no spines in the fins and no axillary process on the pelvic fins. Smelt feed mainly on crustaceans. Most species are anadromous and spawn in spring over gravel or sand in streams or on gravel shores of lakes. Smelts are known to have a strong cucumber odor when captured and taken out of water. A single species occurs in Illinois.

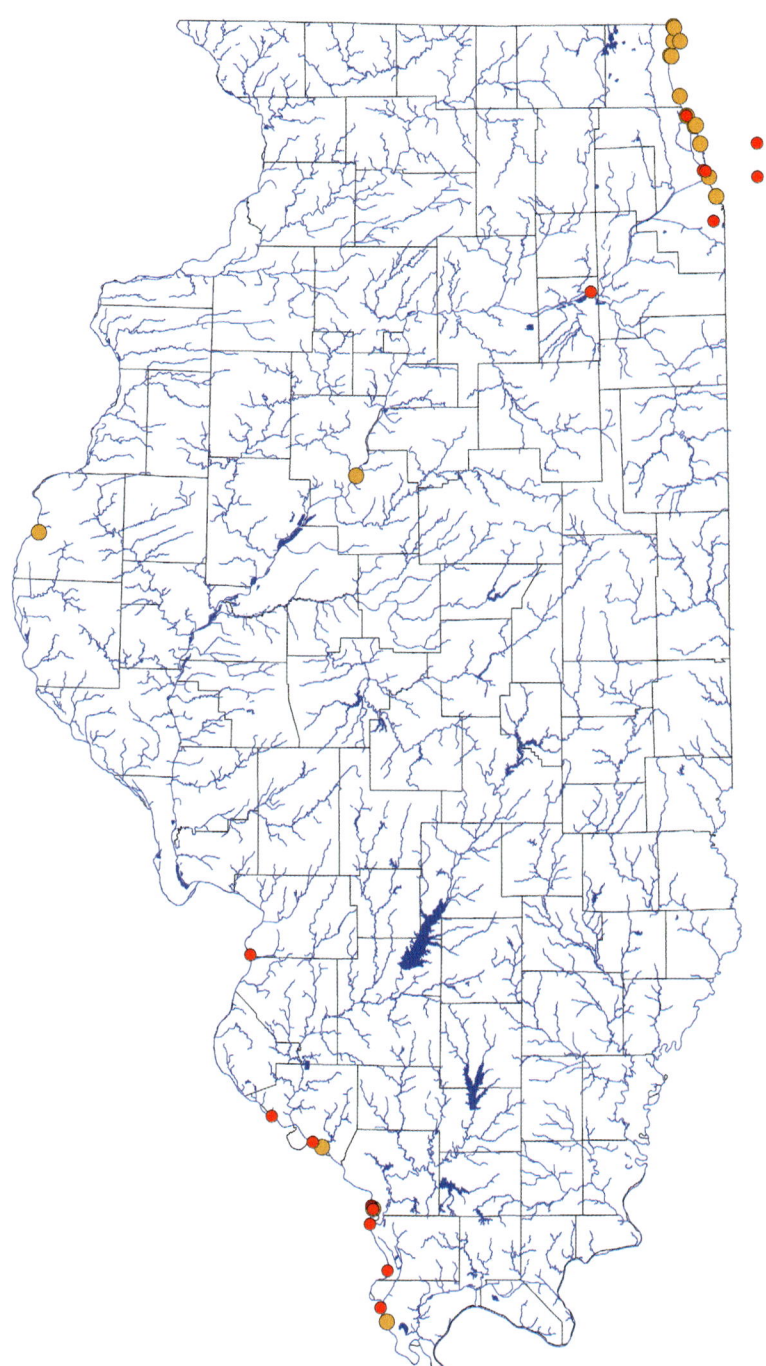

RAINBOW SMELT *Osmerus mordax* (Mitchill 1814)

IDENTIFICATION The Rainbow Smelt is a compressed, slender fish with 1 dorsal fin, an adipose fin, and a large mouth that reaches to the middle of the eye. There are teeth on the jaws, including 2 large, canine teeth, and large teeth on the tongue. The dorsal-fin origin is directly above or in front of the pelvic-fin origin. The pelvic fins are on the abdomen, the anal fin is concave, and the caudal fin is forked. The back is light olive with an iridescent sheen, the side of the body is silver-white, sometimes with a blue or green iridescence, and the lower side and belly are white. Often there is a broad, silver stripe along the side of the body and small, black specks on the upper side. The lateral line is incomplete. There are 62–72 lateral scales and 11–16 anal rays. To 13 in. (33 cm).

SIMILAR SPECIES Species of *Coregonus* have an axillary process at the base of pelvic fin and lack conspicuous jaw teeth.

HABITAT The Rainbow Smelt inhabits clear, cool lakes and medium to large rivers. It often is found in schools in open water.

DISTRIBUTION IN ILLINOIS The Rainbow Smelt is essentially a marine species found along the coasts of northern North America. It was stocked in the Great Lakes as forage for game fishes as early as 1906 and was first found in Lake Michigan in 1923 (Becker 1983). Smith (1979) considered it to be common in the Illinois portion of Lake Michigan and noted that it occasionally entered the Illinois River. In the contemporary era, the species is in Lake Michigan, the Illinois River, and the lower Mississippi River. Individuals in the Mississippi River may come from Lake Michigan via the Illinois River (Burr & Mayden 1980), or they may come from the Missouri River as the result of a separate introduction in the Missouri River in North Dakota in 1971 (Mayden et al. 1987).

TROUT-PERCHES

Family Percopsidae Agassiz 1850

Percopsidae is a small family endemic to North America with 1 genus and 2 species. Both have small, somewhat transparent bodies, a large, unscaled head, ctenoid scales on the body, 1 dorsal fin, an adipose fin, subthoracic pelvic fins, and spines as well as soft rays in the dorsal, anal, and pelvic fins. One species, the Sand Roller, *Percopsis transmontana*, occurs in the Columbia River basin, while the other species, the Trout-perch, *P. omiscomaycus* occurs from Alaska, across Canada, into the midwestern and eastern United States, including in Illinois.

TROUT-PERCH *Percopsis omiscomaycus* (Walbaum 1792)

IDENTIFICATION The Trout-perch has an elongate body that is deep and arched anteriorly, large head, subterminal mouth, large eye that is directed upward, adipose fin, long and thin caudal peduncle, and forked caudal fin. Preoperculomandibular canal pores appear as large, silver-white chambers on the cheek and flattened underside of the head. There are weak spines in the dorsal, anal, and pelvic fins. The head and body are somewhat translucent, and the back and side are yellow-olive to straw-colored with 2 rows of 7–12 dusky brown-black spots. There are silver flecks on the head and upper side and often a silver stripe along the midside. Fins are clear to slightly dusky, and the breast and belly are silver-white. To 7¾ in. (20 cm).

SIMILAR SPECIES No other fish in Illinois has a similar appearance. The combination of an adipose fin and large, silver-white chambers on the cheek and underside of the head readily separate this species from all others.

HABITAT The Trout-perch lives over sand in lakes and in pools and backwaters of large rivers.

DISTRIBUTION IN ILLINOIS In earlier eras the Trout-perch was widespread in the Illinois River and Lake Michigan, and Smith (1979) recorded it in the Mississippi River. Contemporary-era records include those for the shoreline of Lake Michigan, 2 localities in the Illinois River, and 1 locality in the Mississippi River in southern Illinois. The species was not recorded in the state between 1999 and 2011 but has been recorded in Will and Grundy Counties 7 times between 2012 and 2018. The species-distribution map contains records through 2018.

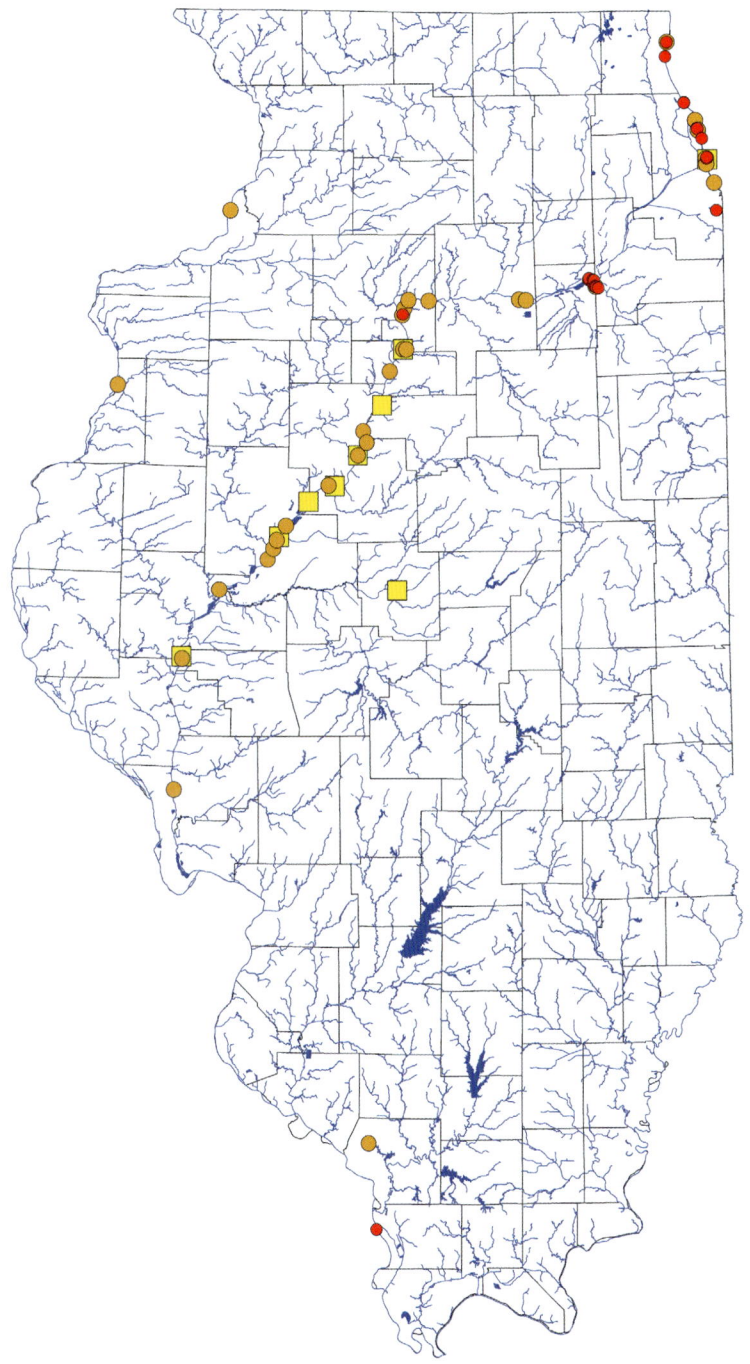

PIRATE PERCHES

Family Aphredoderidae Bonaparte 1845

The family Aphredoderidae contains a single distinctive species endemic to the eastern United States. The Pirate Perch has a large head and mouth, ctenoid scales, 1 dorsal fin, subthoracic pelvic fins, and spines as well as soft rays in the dorsal, anal, and pelvic fins. The anal and urogenital openings are located near the anal fin in the juvenile, but migrate forward as the fish matures and are located on the throat in the adult. The Pirate Perch is found in lower-elevation streams and is nocturnal and rarely seen in open water.

PIRATE PERCH

Aphredoderus sayanus (Gilliams 1824)

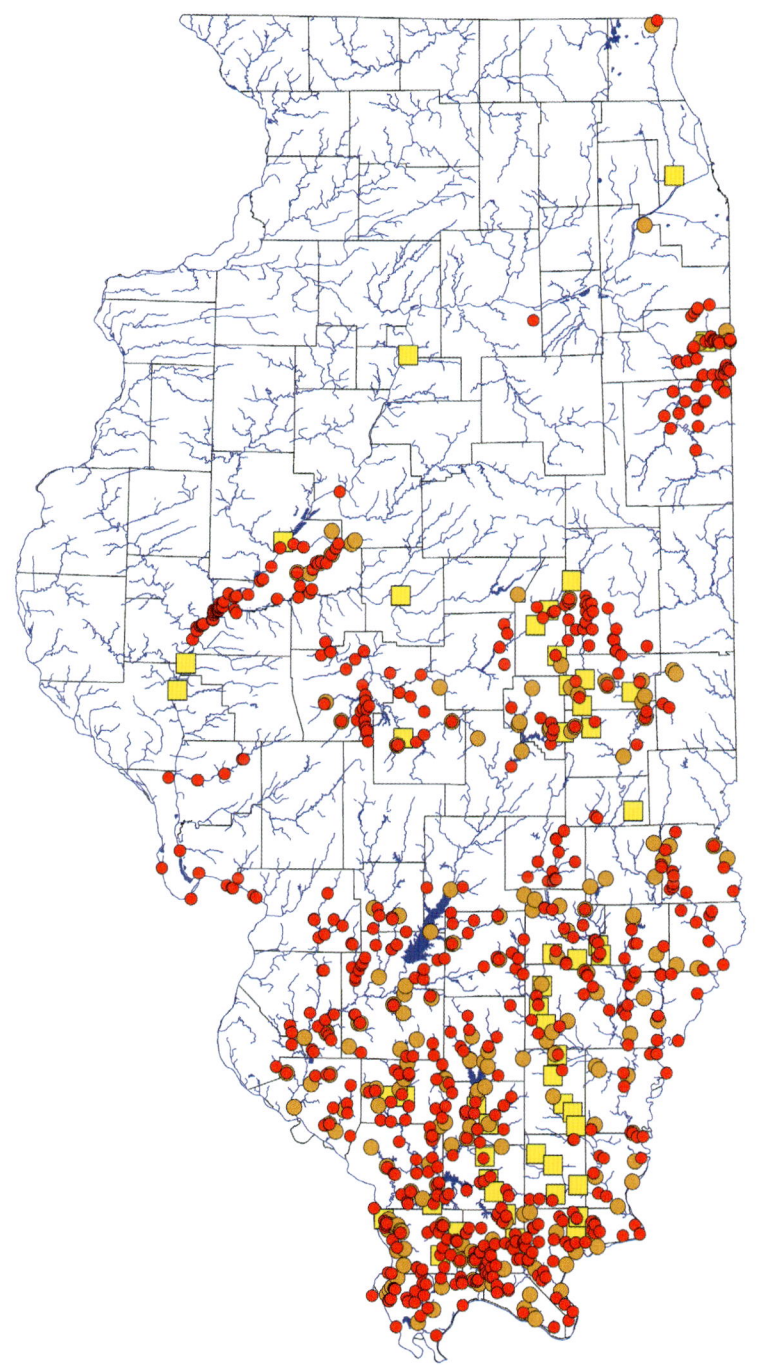

IDENTIFICATION The Pirate Perch has a short, deep, somewhat compressed body, a large head with a large, slightly upturned mouth, subthoracic pelvic fins, ctenoid scales, and a slightly forked caudal fin. The anal and urogenital openings are on the belly in juveniles but migrate to the throat in the adult. The lateral line is absent or very short. The back and side of the body are gray-black to iridescent purple and often speckled with small, black spots. The breast and belly are light yellow to white. There are a black teardrop, a black bar on the caudal-fin base, and dusky to black fins. To 5½ in. (14 cm).

SIMILAR SPECIES The Shawnee Hills Cavefish, *Forbesichthys papilliferus*, is the only other species in Illinois with the anus and genital pore in the jugular position, but it lacks pelvic fins. Smaller individuals superficially resemble the Banded Pygmy Sunfish, *Elassoma zonatum*, which has a smaller head and mouth, 2 dorsal fins (broadly joined), the first with spines and the second with rays, anal and urogenital openings near the anal fin, and a rounded caudal fin and is strongly mottled or with bright iridescent colors.

HABITAT The Pirate Perch lives in swamps, sloughs, lakes, backwaters, and slow-flowing creeks. It is usually found near woody debris or roots along shorelines.

DISTRIBUTION IN ILLINOIS The Pirate Perch has essentially the same distribution as in previous eras. It is absent from the upper Mississippi River basin in northwestern Illinois and is heavily clustered in large areas of the state.

REMARKS The subspecies *A. s. gibbosus*, the form in Illinois, is distinguished from *A. s. sayanus* by having usually 2 (vs. 3) anal spines, 3 (vs. 4) dorsal spines, 12 (vs. 11) pectoral rays, and more than 45 (vs. < 42) lateral-line scales and lacks a dark stripe along the side of the body (Boltz & Stauffer 1993).

PIRATE PERCHES (Aphredoderidae)

CUSKFISHES

Family Lotidae Bonaparte 1835

Cuskfishes have a wide head with a single chin barbel, a long, slender body, 1 or 2 dorsal fins, 1 anal fin, no barbels on the snout, no spines in the fins, thoracic or jugular pelvic fins, and a round caudal fin. Most of the 22 members of this family are marine. The lone Illinois species, Burbot, *Lota lota*, occurs in cool to cold fresh waters of northern North America and northern Eurasia. Small cuskfishes feed on invertebrates while large cuskfishes feed on fishes.

BURBOT *Lota lota* (Linnaeus 1758)

IDENTIFICATION The Burbot is an elongate, bottom-dwelling fish with a single, long barbel at the tip of the chin. The head is wide, and the long, slender body is strongly compressed posteriorly. The back and side of the body are light brown to yellow-brown with dark brown to black mottling. The underside is light brown, and the second dorsal and anal fins often have a black edge. The first dorsal fin is short with 8–16 rays, and the second dorsal fin is much longer with 60–80 rays. The small pelvic fin is located anterior to the pectoral fin, the anal fin is long, and the caudal fin is rounded. Scales on the body are small and embedded. To 33 in. (84 cm).

SIMILAR SPECIES No other fish in Illinois has a single, long barbel at the tip of the chin.

HABITAT The Burbot lives in lakes and rocky runs, pools, and backwaters of large rivers.

DISTRIBUTION IN ILLINOIS The Burbot was reported by Forbes and Richardson (1909) from Lake Michigan and the Mississippi and Illinois Rivers. In the Smith era, the species was found at a few localities in these same waterbodies and in the Wabash River, although Smith (1979) opined that the Burbot was less frequently encountered overall than previously, especially in Lake Michigan where it once was common. Smith also discussed a record from the Big Muddy River reported by Lewis (1955) but did not include it on his map, presumably because there was no voucher specimen. In the contemporary era, the Burbot appears to be restricted to Lake Michigan and the Illinois River, for which there is only 1 recent record.

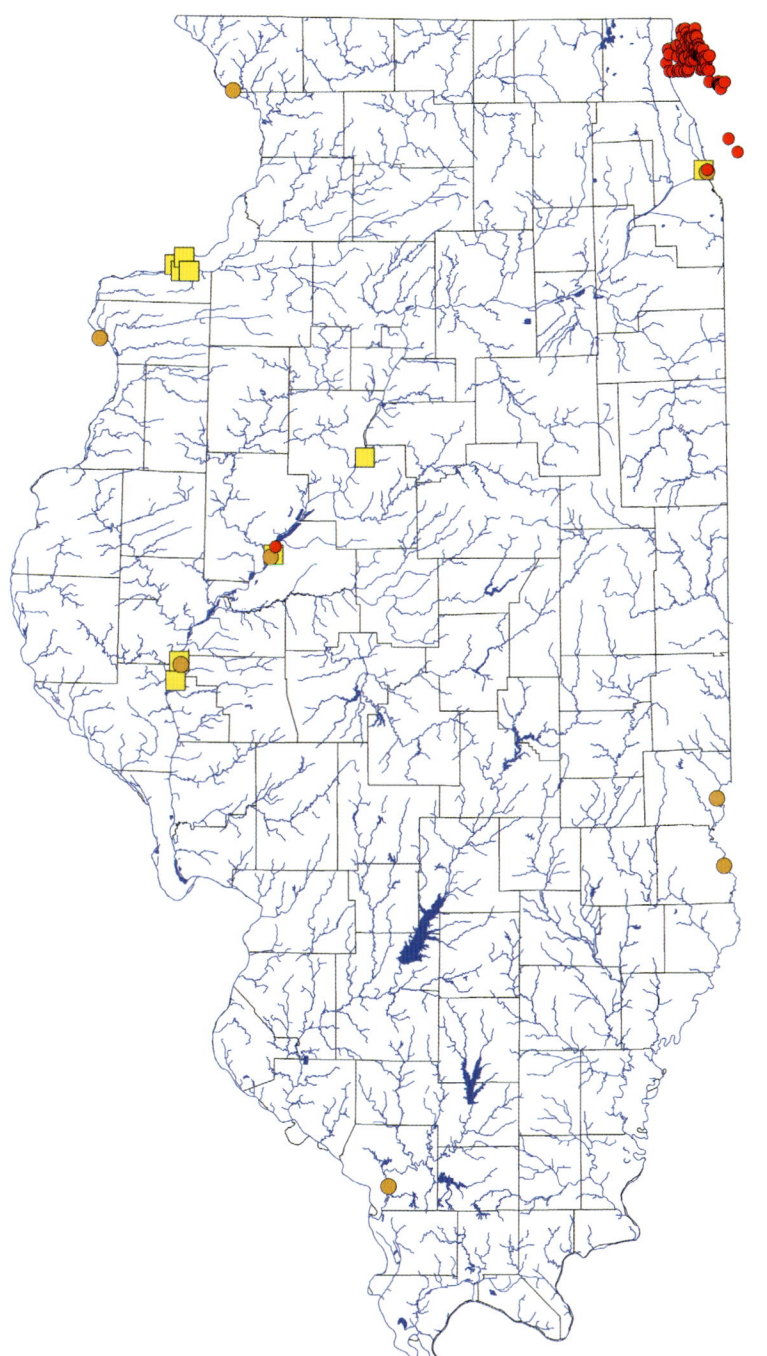

GOBIES

Family Gobiidae Cuvier 1816

Gobies comprise 1 of the largest families of fishes, with about 1,900 species, most of which live in tropical coastal marine or brackish environments. Relatively few live in fresh water. In most species (and in all freshwater species), the pelvic fins are positioned directly below the pectoral fins and are fused to form a suction disc with an anterior transverse membrane. The disc is used to attach to objects and maintain position on the substrate. There is no lateral line, 5 branchiostegal rays, and 2 well-separated dorsal fins—the first with spines and the second fin with 0–1 spine and 9–25 rays. Gobies are mostly under 4 in. (10 cm) in length, and many are brightly colored.

Two freshwater gobies native to Europe are established in the Great Lakes, presumably as a result of the discharge of ballast water from ships. The Round Goby, *Neogobius melanostomus*, is found throughout the Great Lakes, including Illinois. The Freshwater Tubenose Goby, *Proterorhinus semilunaris* (Heckel 1837), has been recorded from Lake St. Clair, Michigan, and from Lake Erie where it is known to be reproducing, but it has not been recorded from Illinois.

ROUND GOBY *Neogobius melanostomus* (Pallas 1814)

IDENTIFICATION The Round Goby has a large head with short, tubular nostrils and eyes high on the head. The body is wide anteriorly and compressed posteriorly. The species is gray to brown with blue, dark brown, or black specks and mottling on the back and side of the body and light gray to white on the lower side and belly. There is a small, dusky black spot at the base of the pectoral fin, a large, black spot at the rear of the first dorsal fin, and often an interrupted black stripe along the side of the body. During the breeding season, the male is dark brown or black overall. The first dorsal fin is short, the second dorsal and anal fins are much longer. The caudal fin is rounded. There are 45–54 lateral scales. To 12 in. (30 cm).

SIMILAR SPECIES The Freshwater Tubenose Goby, *Proterorhinus semilunaris,* has long barbel-like nostrils overhanging the upper lip and no black spot on the rear of the first dorsal fin. It is native to eastern Europe but established in the Great Lakes. It has not yet been recorded from Illinois.

HABITAT The Round Goby lives along rocky lake shores and rocky areas of large rivers.

DISTRIBUTION IN ILLINOIS The Round Goby is native to the Azov, Black, and Caspian Seas in Eurasia. It was first recorded in the Great Lakes in 1990 and was first recorded in Illinois in Lake Michigan in 1994. It is now established and is widespread in the Illinois River, upper Illinois River basin, and Lake Michigan tributaries in the Chicago metropolitan area, and has been recorded in the Mississippi River downstream of the Illinois River.

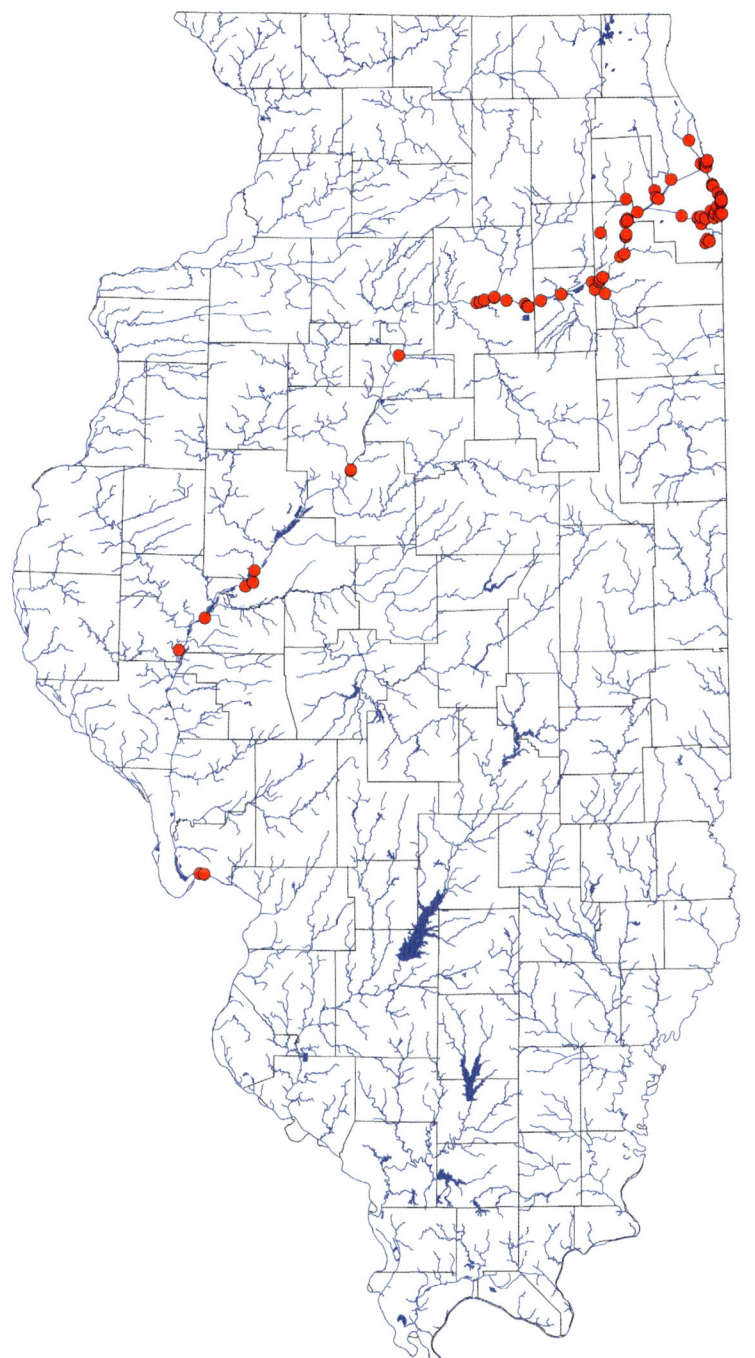

NEW WORLD SILVERSIDES

Family Atherinopsidae Fitzinger 1873

Atherinopsids are small, strongly compressed, silvery to nearly translucent fishes with a flattened head and have a large eye, a terminal mouth, a long snout, protractile premaxillae, no lateral line, a long, sickle-shaped anal fin, abdominal pelvic fins, and 2 widely separated dorsal fins. The first dorsal fin is small and has spines. Members of the family often swim in large schools near the surface of the water. Some species leap out of the water and glide through air for short distances when spawning or disturbed. The family is restricted to the Western Hemisphere. About 110 species are known from North and South America. Two species occur in Illinois.

KEY TO THE NEW WORLD
SILVERSIDES (Atherinopsidae)

1a Jaws forming a beak; snout length (A) much greater than diameter of eye (B); anal fin rays usually 22–26; scales small, usually 74–94 in lateral series; about 40 predorsal scales on dorsal midline . . . . **Page 276**

Brook Silverside (*Labidesthes sicculus*)

A much greater than **B**

┌ **A** ┬ **B** ┐

About **40** predorsal scales

Small scales, usually **74–94** in lateral series

Anal fin rays usually **22–26**

1b Jaws not forming a beak; snout length (A) equal to or less than diameter of eye (B); anal fin rays usually 15–20; scales larger, usually 38–46 in lateral series; about 20 predorsal scales on dorsal midline . . . . **Page 277**

Inland Silverside (*Menidia beryllina*)

A equal to or less than **B**

┌ **A** ┬ **B** ┐

About **20** predorsal scales

Scales **larger**, usually **38–46** in lateral series

Anal fin rays usually **15–20**

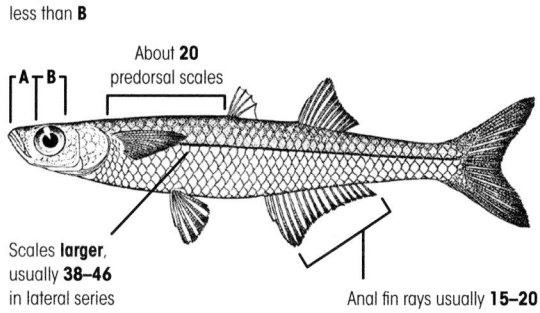

BROOK SILVERSIDE *Labidesthes sicculus* (Cope 1865)

IDENTIFICATION The Brook Silverside is a slender and compressed fish with a flat head, slightly upturned mouth, a long, beak-like snout that is about 1½ times the eye diameter, 2 widely separated dorsal fins, and a long, sickle-shaped anal fin. The back and upper side are silvery, translucent, and light straw-colored to pale or iridescent green with darkly outlined scales. The lower side is silver-white, and there is a bright silver stripe over a dusky black stripe along the side of the body. The origin of the first dorsal fin is above or slightly in front of the origin of the anal fin. There are 74–87 lateral scales and 22–25 anal rays. To 5 in. (13 cm).

SIMILAR SPECIES The Inland Silverside, *Menidia beryllina*, lacks a long, beaklike snout (snout length equal to or less than the diameter of the eye), has the origin of the first dorsal fin in front of the origin of the anal fin, 34–40 lateral scales, and 14–21 anal rays.

HABITAT The Brook Silverside lives near the surface, often in schools, of lakes, ponds, and quiet pools of creeks and small to large rivers.

DISTRIBUTION IN ILLINOIS The Brook Silverside occurs statewide in Illinois, and its distribution is similar to that of the Smith era. It is missing from several large creeks and small rivers where habitats appear to be suitable, especially in western Illinois. By the mid-20th century, the Brook Silverside had disappeared from the Spoon River and the Vermilion River (Illinois basin; Smith 1979) where Forbes and Richardson (1909) had recorded the species; it remains absent from those basins in the contemporary era. Smith also failed to record it in the Mackinaw River, where Forbes and Richardson had found it, but there is 1 contemporary-era record for the Mackinaw River. Forbes and Richardson did not record the species from the Kaskaskia River basin, but it has been recorded from a large portion of that basin beginning in the mid-20th century.

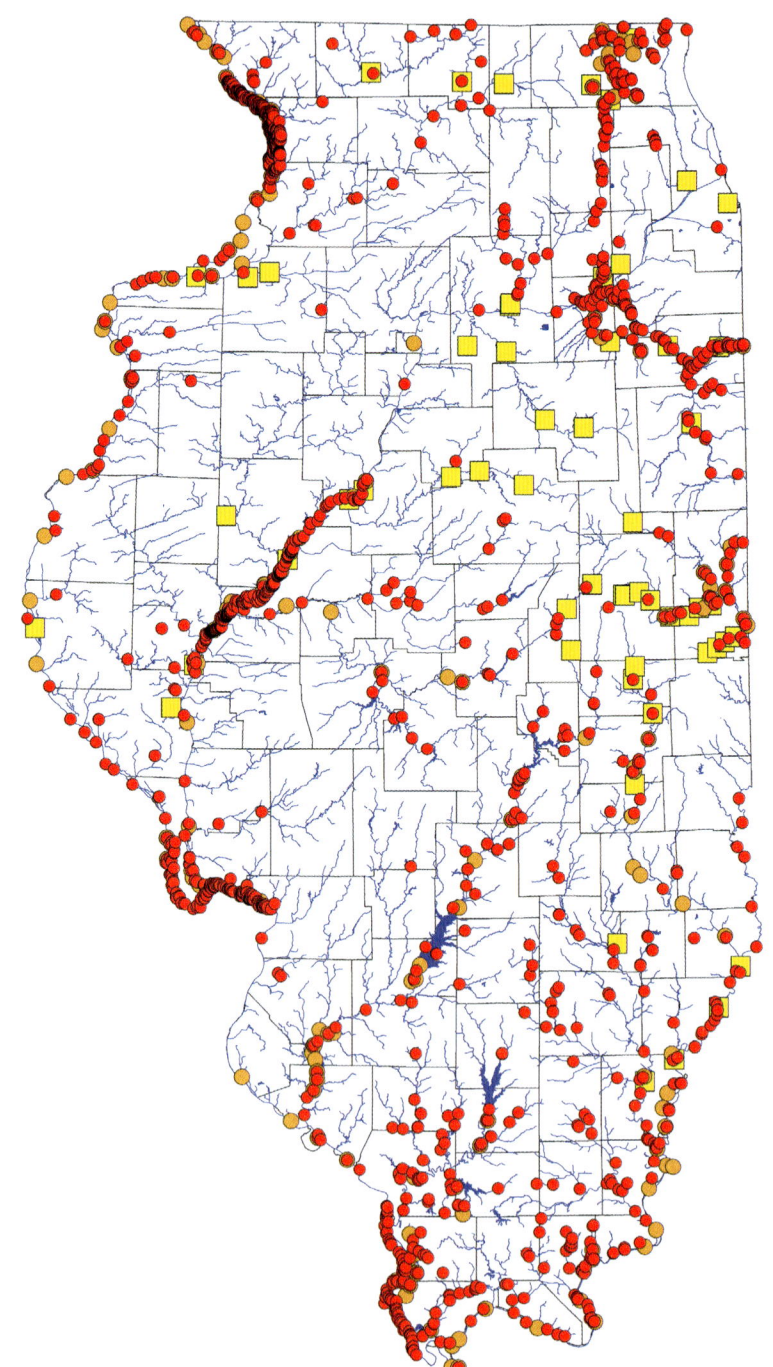

INLAND SILVERSIDE   *Menidia beryllina* (Cope 1867)

IDENTIFICATION The Inland Silverside is slender and compressed and has a flat head, slightly upturned mouth, a pointed snout about equal in length to the diameter of the eye, 2 widely separated dorsal fins, and a long, sickle-shaped anal fin. The back and upper side are silvery, translucent, and light straw-colored to pale or iridescent green. Scales on the back are darkly outlined. Along the side of the body is a clear stripe above a silver-black stripe that extends from the head onto the caudal fin. The lower side is silver-white. The origin of the first dorsal fin is far in front of the origin of the anal fin. There are 34–40 lateral scales and 14–21 anal rays. To 5 in. (15 cm).

SIMILAR SPECIES The Brook Silverside, *Labidesthes sicculus*, has a longer beak-like snout (length about 1½ times the eye diameter), the origin of the first dorsal fin above the origin of the anal fin, 74–87 lateral scales, and 22–25 anal rays.

HABITAT The Inland Silverside occurs near the surface of lakes, ponds, and quiet pools of small to large rivers.

DISTRIBUTION IN ILLINOIS The Inland Silverside was first reported from Illinois in 1978, and beginning in 1980 the species was stocked in several impoundments in southern Illinois as forage for sportfish species (Burr et al. 1988). The species is now widespread in the lower Mississippi, Kaskaskia, Big Muddy, Ohio, and Wabash Rivers.

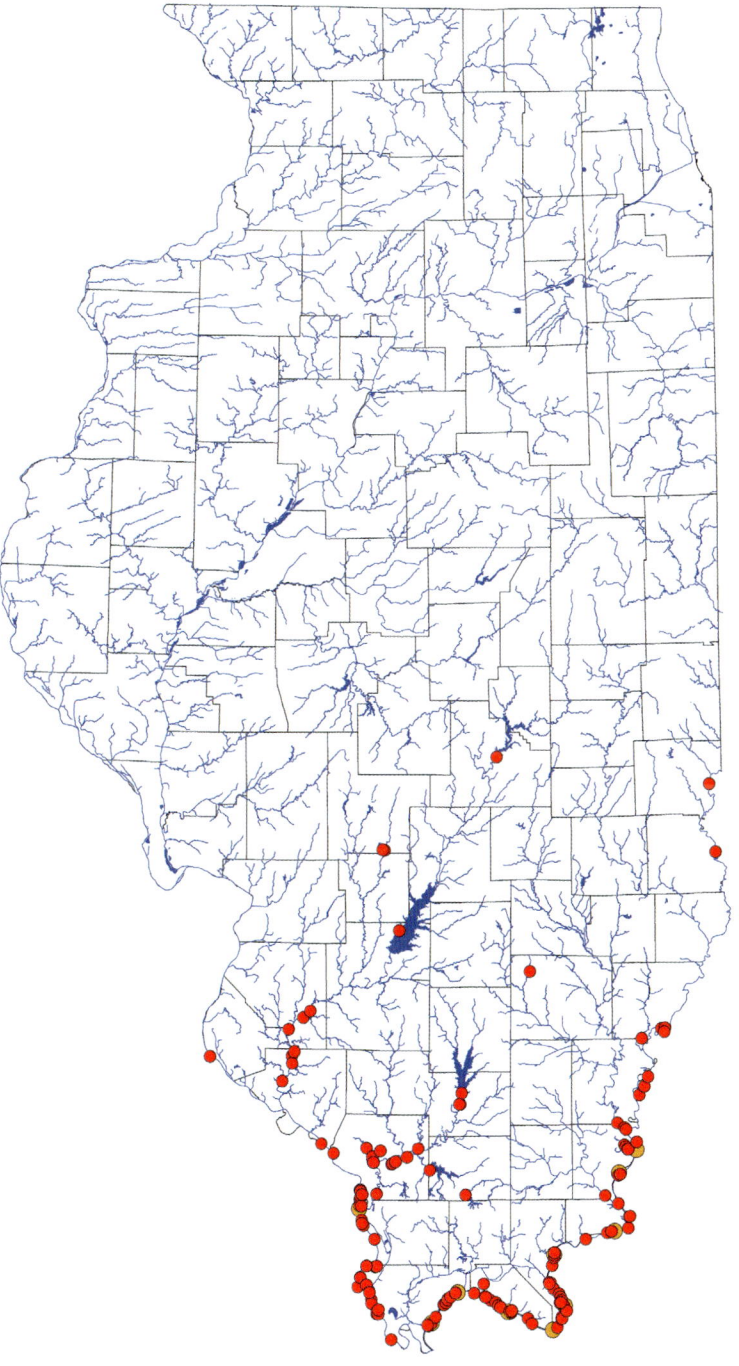

NEW WORLD SILVERSIDES (Atherinopsidae)

3a Blue-black vertical bar or teardrop beneath eye; side of females with prominent horizontal lines, males with faint horizontal lines and prominent narrow vertical bars . . . . .
. . . .**Page 283**

Starhead Topminnow (*Fundulus dispar*)

Side of females with **prominent horizontal lines**

Side of males with **faint horizontal lines** and **prominent narrow vertical bars**

Blue-black vertical bar or teardrop beneath eye

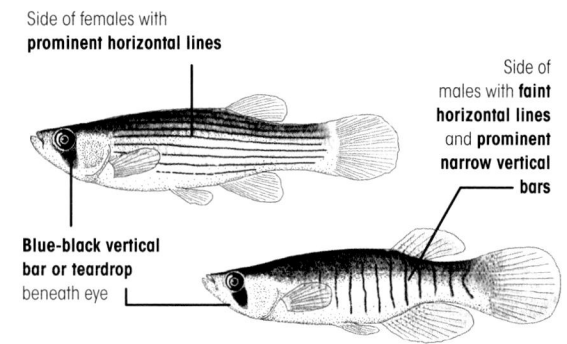

3b No blue-black vertical bar or teardrop beneath eye; side of both sexes with a single broad, black stripe extending from tip of snout to base of caudal fin **4**

Side with a **single broad, black stripe** extending from snout to base of caudal fin

No blue-black bar or teardrop beneath eye

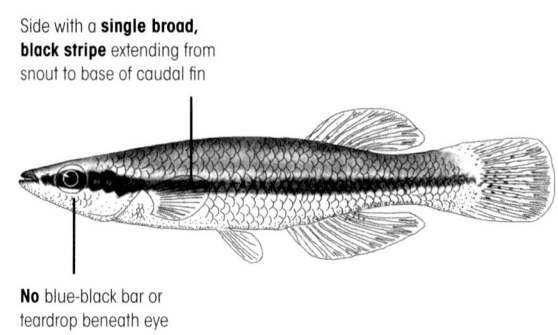

4a Upper side with few to many black spots that are regular in outline and as dark as stripe along midside . . .**Page 285**

Blackspotted Topminnow (*Fundulus olivaceus*)

Upper side **with** few to many black spots that are regular in outline and **as dark** as stripe along midside

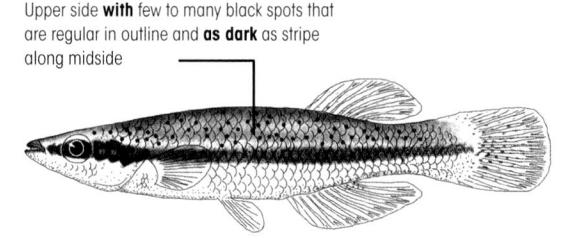

4b Upper side without black spots or with spots that are ir-regular in outline and not as dark as stripe along midside
. . . .**Page 284**

Blackstripe Topminnow (*Fundulus notatus*)

Upper side **without** black spots or with spots irregular in outline and **not as dark** as stripe along midside

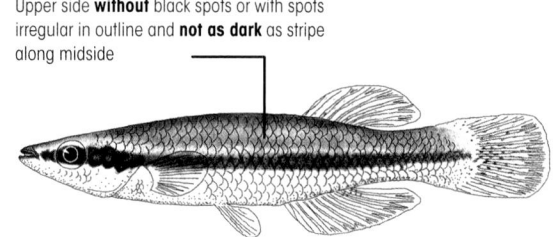

NORTHERN STUDFISH *Fundulus catenatus* (Storer 1846)

IDENTIFICATION The Northern Studfish has a flattened head and nape, an upturned mouth, and a large eye. The dorsal fin is located far back on the body with its origin over or slightly in front of the origin of the anal fin. The back and upper side of the body are light yellow-brown with a short, gold stripe in front of the dorsal fin, and the side of the body is silver-blue with horizontal rows of small, brown (female and young) or red-brown spots (male). The adult has rows of small, brown or red spots on the dorsal and caudal fins. During the breeding season, the male has a bright blue side, red spots on the head and fins, usually yellow paired fins, an orange edge and black submarginal band on the caudal fin, and tubercles on the side of the head, body, and the dorsal, anal, and paired fins. To 7 in. (18 cm).

SIMILAR SPECIES The Banded Killifish, *F. diaphanus*, lacks horizontal rows of small, brown or red-brown spots on the side of the body. The Starhead Topminnow, *F. dispar*, has a dark blue or black teardrop, and males have 3–13 dark vertical bars on the side of the body.

HABITAT The Northern Studfish lives in the margins, pools, and backwaters of creeks and small to medium rivers. It usually is found in shallow water over sand or gravel.

DISTRIBUTION IN ILLINOIS Smith (1979) reported the Northern Studfish from a single specimen collected in the Mississippi River. In the contemporary era it has been recorded from 2 additional localities in the Mississippi River and 1 locality in the lower Kaskaskia River basin. Each of these records is based on a single specimen, and it seems likely that they are stragglers from Missouri rivers where the species is widespread. There is no evidence that the Northern Studfish is reproducing in Illinois.

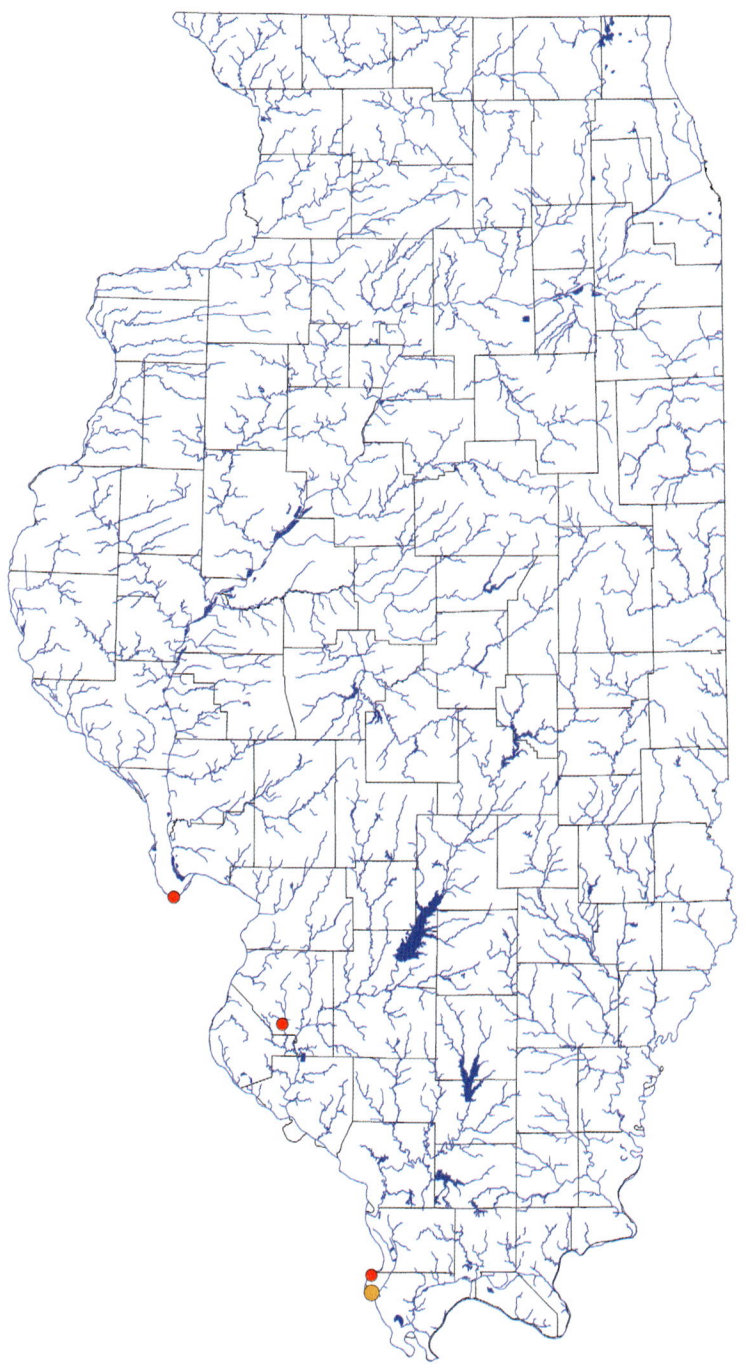

TOPMINNOWS (Fundulidae)

BANDED KILLIFISH *Fundulus diaphanus* (Lesueur 1817)

IDENTIFICATION The Banded Killifish is a small fish with a flattened head and body, an upturned mouth, and a large eye. The dorsal fin is located far back on the body with its origin slightly in front of the origin of the anal fin. The body is wide anteriorly and tapers to a slender caudal peduncle. The back and upper side of the body are dark olive to tan with many small, brown spots. There is a brown stripe along the back. The side of the body is silver with 10–20 green-brown bars; the bars are more numerous and wider on males. The breast and belly are white to yellow, and fins are clear to dusky olive-yellow. During the breeding season males have wide green bars along the side of the body and a yellow breast and fins. The origin of the dorsal fin is in front of the origin of the anal fin. To 5 in. (13 cm).

SIMILAR SPECIES The Starhead Topminnow, *F. dispar*, has a large, dark blue or black teardrop. The Northern Studfish, *F. catenatus*, has horizontal rows of small, brown or red-brown spots and no dark vertical bars on the side of the body.

HABITAT The Banded Killifish occurs in shallow, quiet margins of lakes, ponds, creeks, and rivers, usually over sand or mud. It is most often found near vegetation and frequently in schools just below the surface of the water.

DISTRIBUTION IN ILLINOIS Forbes and Richardson (1909) found the Banded Killifish in the Rock and upper Illinois River basins, Lake Michigan tributaries, Wolf and Powderhorn Lakes (Cook County), and a small area in the Salt Creek basin in McLean County. Smith (1979) noted the disappearance of the species in McLean County and Rock River basin as well as the continued presence in the glacial lakes in extreme northeastern Illinois. In the contemporary era the species has been recorded from several localities in the middle Illinois River basin and in the Rock and Mississippi Rivers. Records through 2019 are illustrated on the species-distribution map.

REMARKS The rapid range expansion in the contemporary era appears to be the result of invasion of the eastern subspecies, *F. d. diaphanus*, and its hybridization with the native western subspecies, *F. d. menona*. Willink et al. (2018, 2019) suggested hybridization between the subspecies may be extensive.

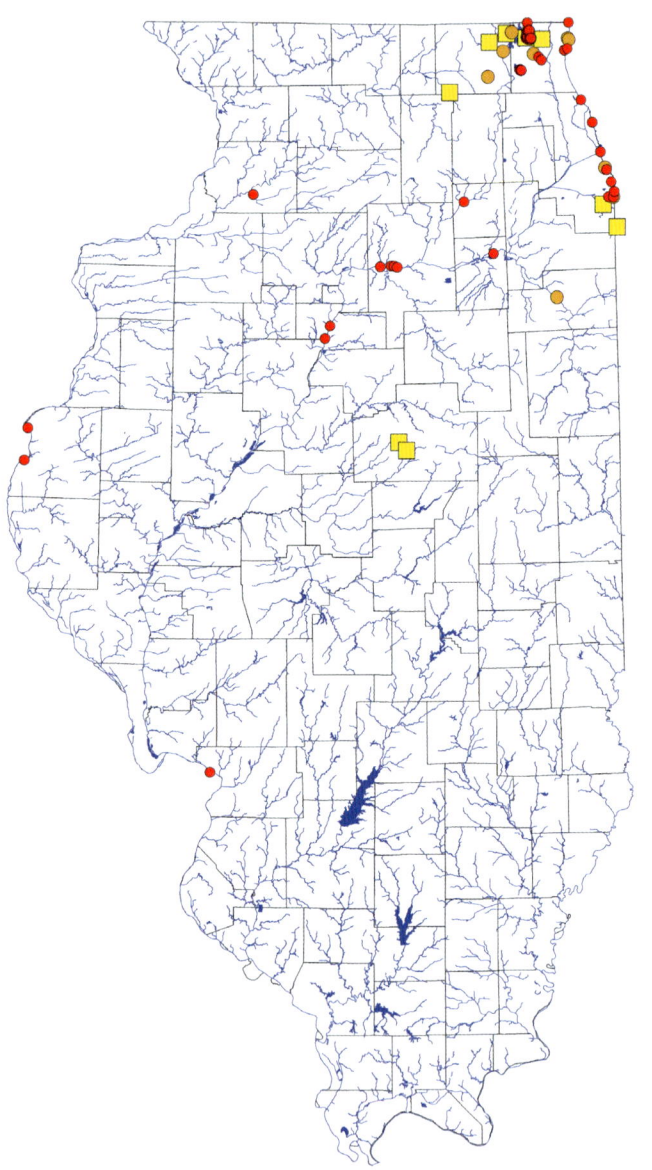

STARHEAD TOPMINNOW *Fundulus dispar* (Agassiz 1854)

IDENTIFICATION The Starhead Topminnow has a flattened head and nape, an upturned mouth, and a large eye. The body is deep anteriorly and tapers to a narrow caudal peduncle. The dorsal fin is located far back on the body with the origin of the fin over or slightly in front of the origin of the anal fin. The back and upper side of the body are olive to tan with a large, iridescent gold spot on the top of the head and a smaller, gold spot at the origin of the dorsal fin. The side of the body is yellow, with small, red, blue, or green flecks; the underside is white to yellow. There is a large, dark blue or black teardrop. Females have 6–8 thin, dark horizontal stripes formed by small, red to brown spots; males have 3–13 dark vertical bars. To 3 in. (7.8 cm).

SIMILAR SPECIES The Northern Studfish, *F. catenatus*, and Banded Killifish, *F. diaphanus*, lack the large teardrop. The Northern Studfish also lacks dark bars along the side of the body.

HABITAT The Starhead Topminnow occurs in vegetated standing waterbodies, especially floodplain lakes, and quiet pools and backwaters of small to medium rivers. Starhead Topminnows are known to jump from the water onto land or large leaves of water lilies and other plants to escape predators, then return to the water.

DISTRIBUTION IN ILLINOIS Smith (1979) noted that the Starhead Topminnow had essentially the same distribution as it had during the Forbes and Richardson era, except that it had disappeared from the Wabash River basin where floodplain lakes had been drained or polluted by oil mining. In the contemporary era the species remains widespread in areas of northeastern Illinois and is found sporadically in the Rock, Illinois, and Mississippi River basins. It has not been reported from the Wabash River basin in over 100 years.

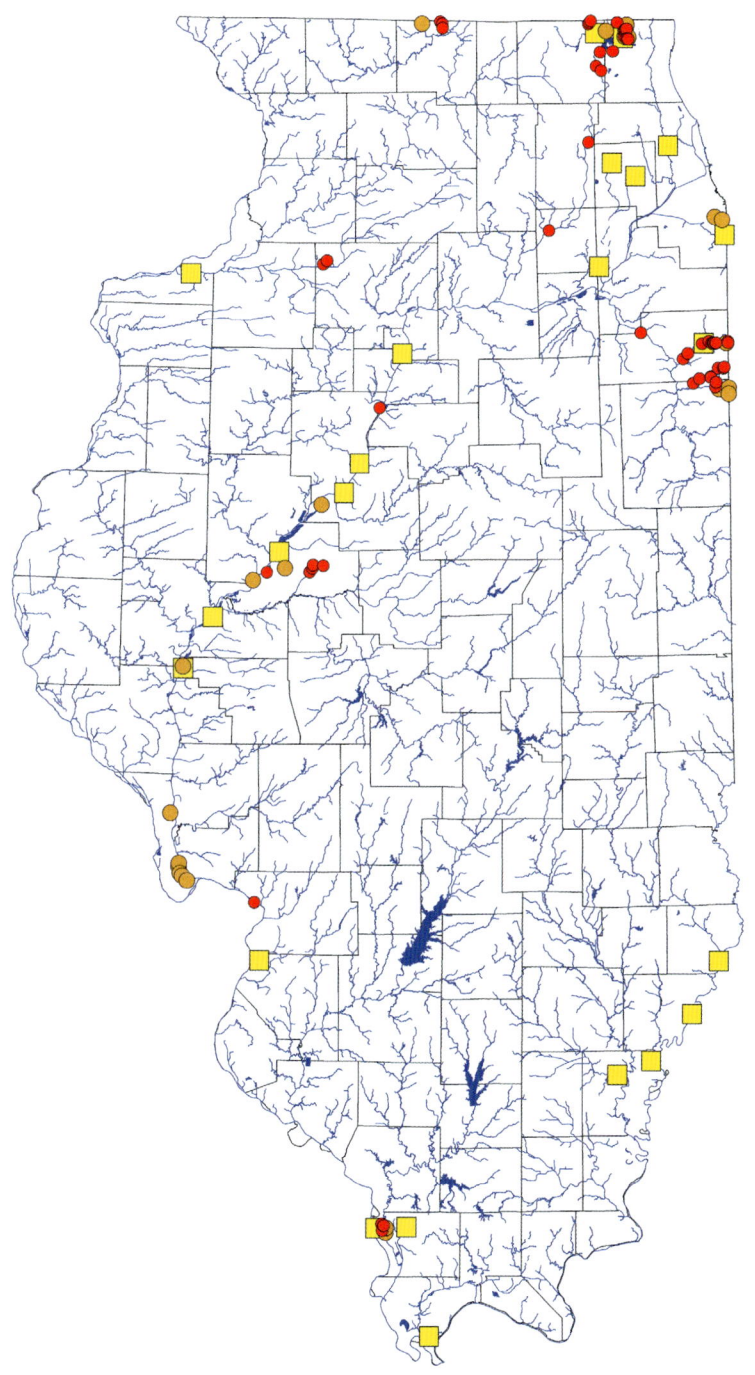

BLACKSTRIPE TOPMINNOW *Fundulus notatus* (Rafinesque 1820)

IDENTIFICATION The Blackstripe Topminnow has a flattened head and nape, an upturned mouth, and a large eye. The dorsal fin is located far back on the body with the origin of the fin behind the origin of the anal fin. The back and upper side of the body are olive-tan with a few dusky to dark (rarely, black) spots, and the lower side is white. There is a silver-white spot on top of the head, and a wide, blue-black stripe along the side of the body and around the snout. The edges of the stripe are smooth in females, jagged in males. Fins are clear to yellow, and the dorsal, caudal, and anal fins have many small, black spots. The male has larger dorsal and anal fins and is deeper bodied. To 3 in. (7.4 cm).

SIMILAR SPECIES The Blackspotted Topminnow, *F. olivaceus*, has numerous jet-black spots on the upper side of the body.

HABITAT The Blackstripe Topminnow lives at the surface of the water in ponds, lakes, and the margins of creeks and small to large rivers. It is often found near vegetation.

DISTRIBUTION IN ILLINOIS As in previous eras, the Blackstripe Topminnow occurs statewide with the exception of the Spoon River basin and some tributaries of the Mississippi River in western and northwestern Illinois. Smith (1979) stated that the Blackstripe Topminnow was replaced in the higher-gradient streams in the Shawnee Hills by the Blackspotted Topminnow, *F. olivaceus*. Contemporary-era records for the Blackstripe Topminnow in the Shawnee Hills, the report of hybridization between *F. notatus* and *F. olivaceus* in southern Illinois (Duvernell et al. 2007), and the overall similarity between the 2 species suggest that species limits have been hard to define in the Shawnee Hills. It is possible that the Blackstripe Topminnow was always there but not recognized.

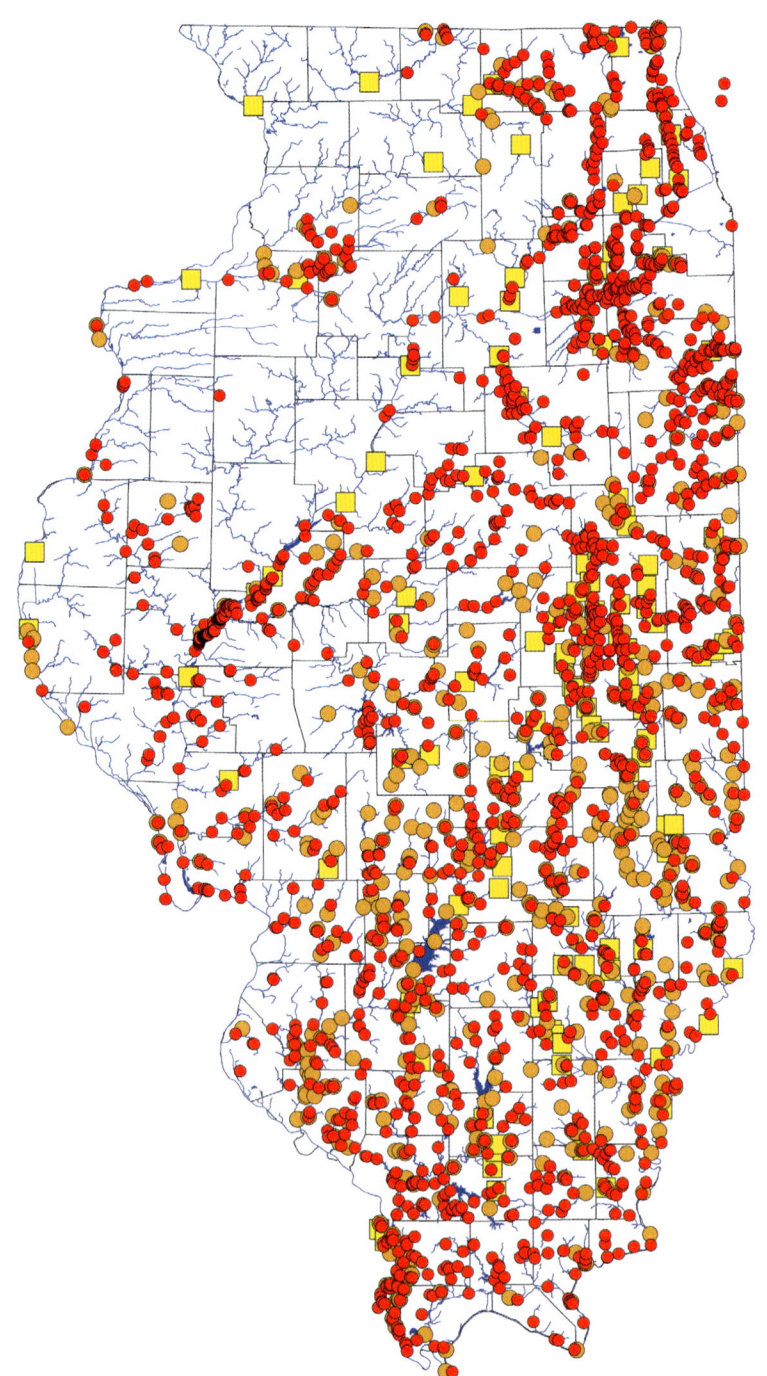

TOPMINNOWS (Fundulidae)

284

BLACKSPOTTED TOPMINNOW *Fundulus olivaceus* (Storer 1845)

IDENTIFICATION The Blackspotted Topminnow has a flattened head and nape, an upturned mouth, and a large eye. The dorsal fin is located far back on the body with the origin of the fin behind the origin of the anal fin. The back and upper side of the body are olive-tan with many jet-black spots, and the lower side is white. There is a silver-white spot on top of the head, and a wide blue-black stripe along the side of the body, around the snout, and onto the caudal fin. The edges of the stripe are smooth in females and jagged in males. Fins are clear to yellow, and the dorsal, caudal, and anal fins have many small, black spots. The male has larger dorsal and anal fins and is deeper bodied. To 3¾ in. (9.7 cm).

SIMILAR SPECIES The Blackstripe Topminnow, *F. notatus*, has dusky black, not jet-black spots, on the upper side of the body.

HABITAT The Blackspotted Topminnow is found near the surface of quiet to flowing water, usually near the margins of clear, sandy to gravelly headwaters, creeks, and small rivers. Less often it is found in ponds, lakes, and backwaters of streams.

DISTRIBUTION IN ILLINOIS The Blackspotted Topminnow occurs in every basin in the southern one-fourth of Illinois. Its range has remained largely unchanged since earlier eras.

REMARKS See discussion of distribution in Illinois for Blackstripe Topminnow, *F. notatus*.

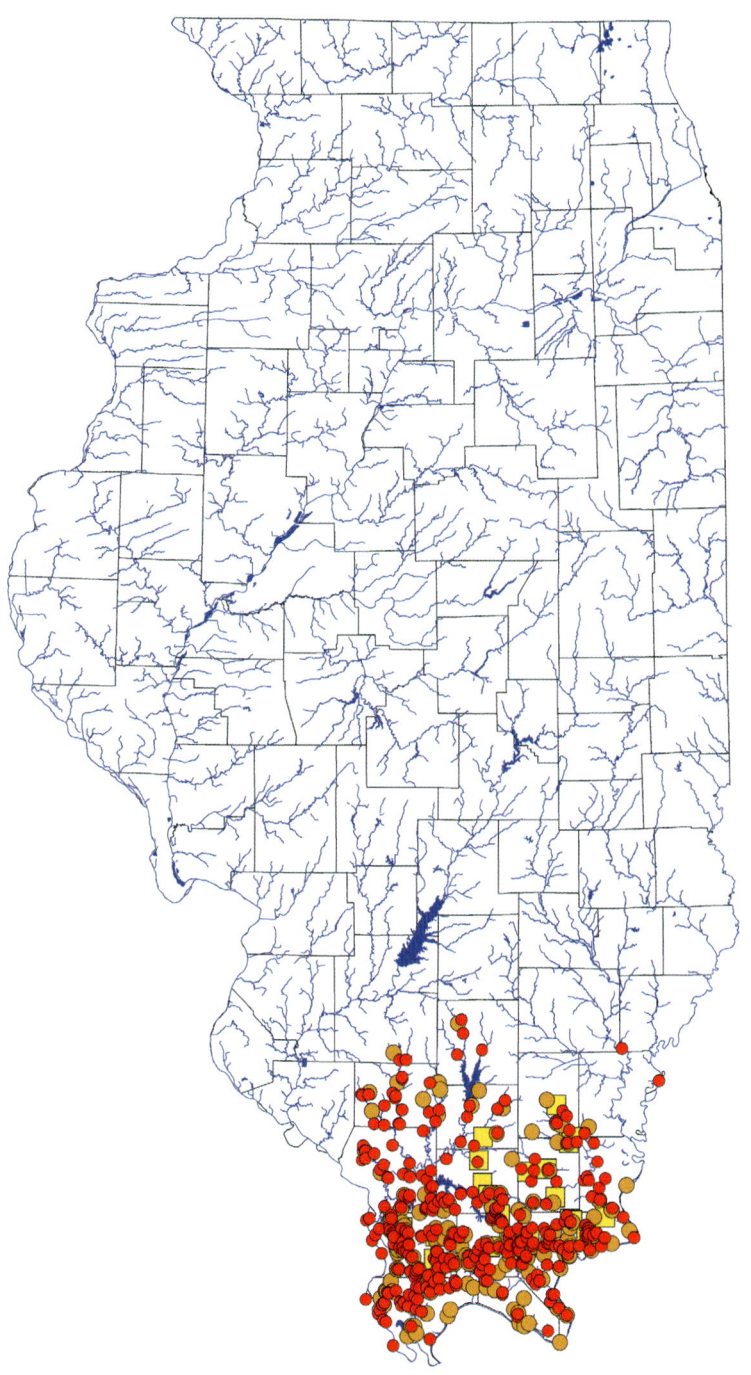

TOPMINNOWS (Fundulidae)

285

LIVEBEARERS

Family Poeciliidae Bonaparte 1831

Livebearers are small fishes that resemble Topminnows (Fundulidae) in having a flattened head, an upturned mouth, 1 dorsal fin, and no lateral line. They differ from Topminnows in that males have the anterior rays of the anal fin modified and elongated into an intromittent organ called a gonopodium used for internal fertilization, and females give birth to live young. There are about 274 species of Livebearers known from this North and South American family, only 1 of which occurs in Illinois. Members of this family swim near the surface of the water and occur in both fresh and brackish water. Many popular aquarium fishes are in this family, including guppies, mollies, platys, and swordtails.

WESTERN MOSQUITOFISH *Gambusia affinis* (Baird & Girard 1853)

IDENTIFICATION The Western Mosquitofish is a small fish with a flattened head and back and an upturned mouth. The origin of the dorsal fin is well behind the origin of the anal fin. The back and side of the body are light gray to yellow-brown with black margins on the scales creating a crosshatched appearance. There is a large, dusky to black teardrop, and a thin dark stripe is along the back from the head to the dorsal fin. The anterior side of the body has a blue iridescence; the underside is white. Fins are mostly clear; the dorsal and caudal fins have 1–3 rows of black spots. Males have the anal fin modified and elongated into an intromittent organ, the gonopodium, and the upper 4–6 pectoral rays of the adult male are thickened and curved upward. Females are much larger than males and usually pregnant as indicated by a large black spot near the urogenital opening. To 2½ in. (6.5 cm).

SIMILAR SPECIES The Starhead Topminnow, *Fundulus dispar*, has a dark blue or black tear drop like the Western Mosquitofish but lacks darkly outlined scales and has horizontal stripes (female) or vertical bars (male) on the side of the body.

HABITAT The Western Mosquitofish is found in standing to slow-flowing water in ponds, lakes, and swamps and in backwaters and quiet pools of headwaters and creeks.

DISTRIBUTION IN ILLINOIS The Western Mosquitofish was somewhat more widespread in the Smith era than in the Forbes and Richardson era, and in the contemporary era is much more widespread than it was in the Smith era. This is especially true in the Illinois River basin and in major tributaries of the Kaskaskia and Wabash Rivers. This may be due in part to its transport and release for mosquito control. It remains most common in the southern one-third of the state.

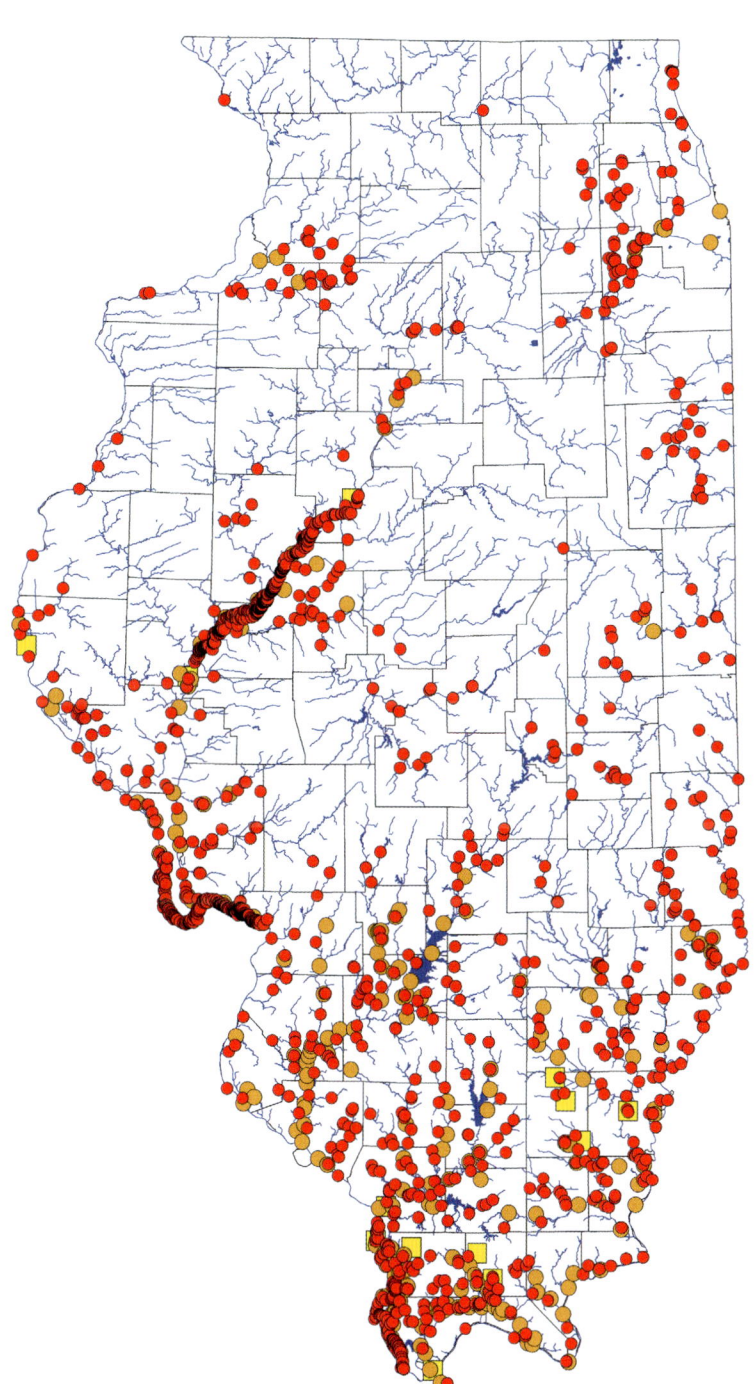

LIVEBEARERS (Poeciliidae)

287

MULLETS

Family Mugilidae Jarocki 1822

Mullets are moderately elongated, silver-gray fishes with a more or less cylindrical head and forebody, a flattened back, and 2 widely separated dorsal fins, the first with 4 spines, and the second with 1 spine and 7–10 rays. The pectoral fins are high on the body, and the lateral line is absent or barely discernible. Mullets live in open water, often swim in schools, and can be seen leaping from the water. Most species use long, sievelike gill rakers to feed on plankton in open water and on detritus extracted from sediment. They have a gizzardlike stomach and a very long intestine. About 77 species are recognized worldwide, mostly in coastal marine waters. Some species travel far inland, and 1 species is a recent arrival in Illinois.

STRIPED MULLET *Mugil cephalus* Linnaeus 1758

IDENTIFICATION The Striped Mullet has an elongated body that is rounded anteriorly and compressed posteriorly, a flattened head and back, and 2 widely separated dorsal fins, the first with 4 spines, and the second with 1 spine and 7–10 rays. The snout is short and blunt, the mouth is small and terminal, and there is an adipose eyelid on the large eye. The pectoral fins are high on the body. The back and side of the body are silver-white with a blue-green cast and dusky stripes formed of rows of small, black spots. Fins are dusky gray. Scales are large, with 38–42 along the side of the body. To 36 in. (91 cm).

HABITAT The Striped Mullet is a marine species that spawns in salt water and commonly enters creeks and rivers along the coast. Some individuals swim far up the Mississippi River to Illinois, where they are found in open water of large rivers.

DISTRIBUTION IN ILLINOIS The Striped Mullet was first found in Illinois in the Mississippi River at Cairo in 1988, although it had been recorded for the Mississippi River at New Madrid, Missouri, north of Cairo, in 1983 (Pflieger 1997). More recently it has been found in the lower Mississippi River almost to the mouth of the Illinois River and in the Ohio River.

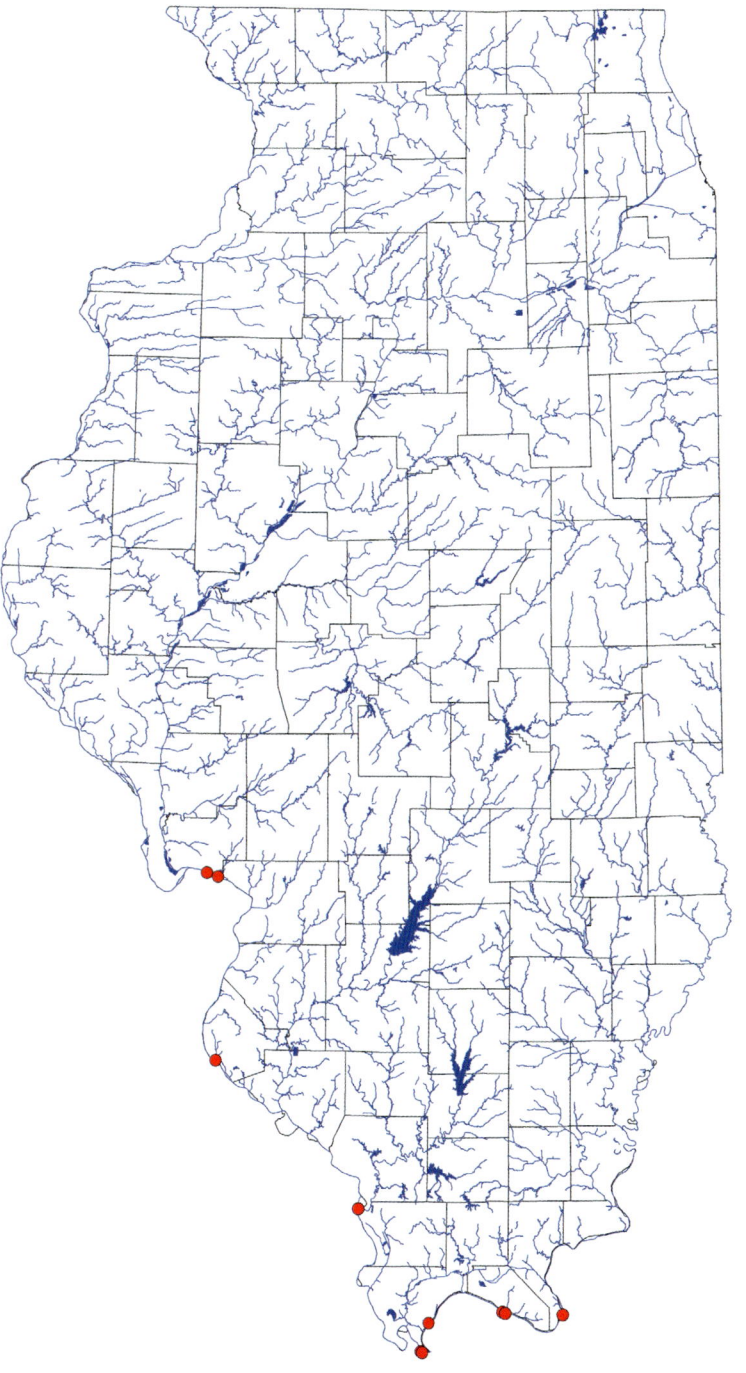

MULLETS (Mugilidae)

289

SUNFISHES

Family Centrarchidae Bleeker 1859

Sunfishes are compressed, some extremely so, and have 2 dorsal fins, the first with spines, and the second with rays. The dorsal fins are typically so broadly joined that they appear to be 1 fin. Sunfishes have 3–8 anal spines, pelvic fins on the thorax, and unlike white basses (Moronidae) have no sharp spine near the posterior edge of the gill cover. Species of *Lepomis* and most other genera in the family are deep bodied; those of *Micropterus* (basses) are more slender. The family includes some of the most popular freshwater sport fishes, and although they occur naturally only in North America, they have been introduced to many other parts of the world, and to parts of North America where they are not native. Seventeen of the 37 species in the family are found in Illinois.

KEY TO THE SUNFISHES (Centrarchidae)

1a Anal fin spines usually 3, rarely 4 . . . .**2**

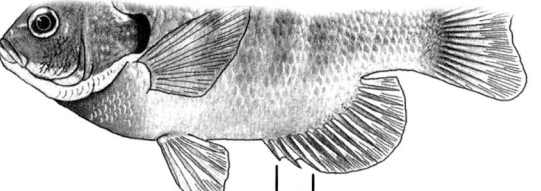

Anal fin spines
usually 3, rarely 4

1b Anal fin spines usually 5 to 8 . . . . **14**

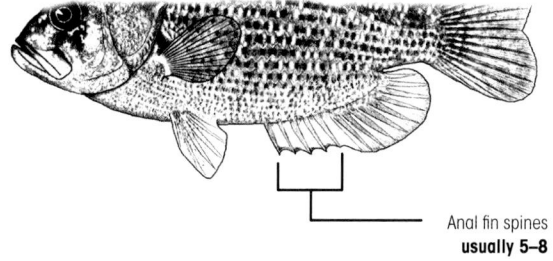

Anal fin spines
usually 5–8

2a Body more slender, its depth (A) going usually 3 to 5 times into standard length (B), except in largest females; scales small, with 53 or more in lateral line; mouth larger, upper jaw extending to or behind middle of eye; dorsal fins nearly separate. **3**

Black basses (*Micropterus*)

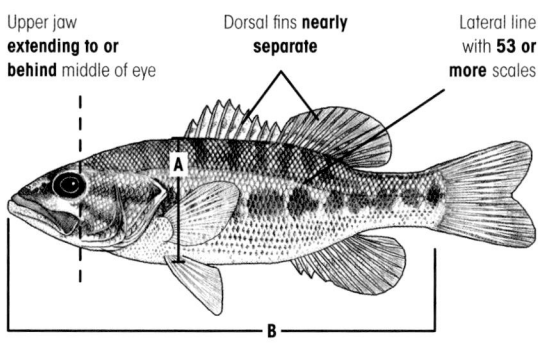

Upper jaw **extending to or behind** middle of eye

Dorsal fins **nearly separate**

Lateral line with **53 or more** scales

A going 3 to 5 times into **B** except in largest females

2b Body deeper, its depth (A) going usually 2 to 2.8 times into standard length (B); scales larger, with 52 or fewer in lateral line; mouth smaller, upper jaw not extending behind middle of eye except in adults of two species (e.g., Green Sunfish, Warmouth); dorsal fins continuous **5**

Sunfishes (*Lepomis*)

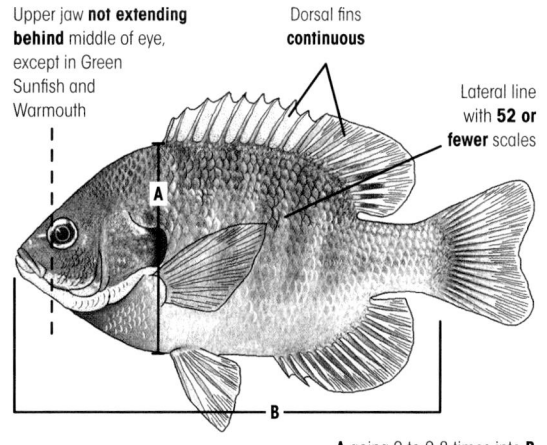

Upper jaw **not extending behind** middle of eye, except in Green Sunfish and Warmouth

Dorsal fins **continuous**

Lateral line with **52 or fewer** scales

A going 2 to 2.8 times into **B**

SUNFISHES (Centrarchidae)

291

3a Spinous dorsal and soft dorsal fins nearly separate; margin of spinous dorsal strongly convex, the length of spine near notch (A) less than half the length of longest spine (B); mouth large, upper jaw extending far behind eye in fish more than 6 inches (150+ mm TL) in length; caudal fin of young 2-colored, the rear edge darker than base . . . .
. . . .**Page 315**

3b Spinous dorsal and soft dorsal fins well connected; margin of spinous dorsal gently rounded to nearly straight, the length of shortest spine near notch (A) more than half the length of longest spine (B); mouth smaller, upper jaw not extending much behind back of eye; caudal fin of young 3-colored, with a prominent, dark vertical bar separating yellow or orange base from white fringe on rear edge . . .
. . . .**4**

Largemouth Bass (*Micropterus salmoides*)

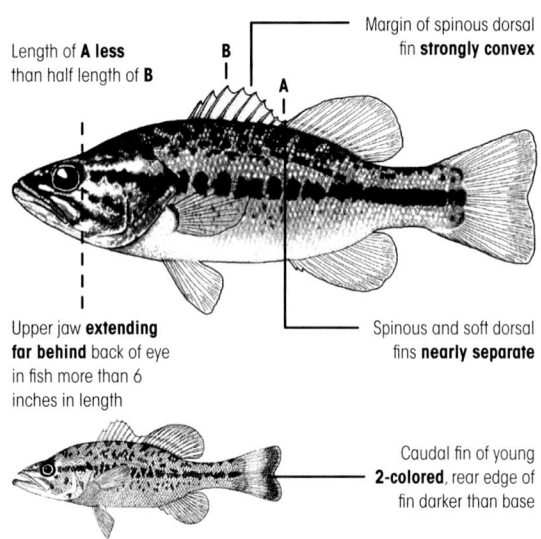

Length of **A less** than half length of **B**

Margin of spinous dorsal fin **strongly convex**

Upper jaw **extending far behind** back of eye in fish more than 6 inches in length

Spinous and soft dorsal fins **nearly separate**

Caudal fin of young **2-colored**, rear edge of fin darker than base

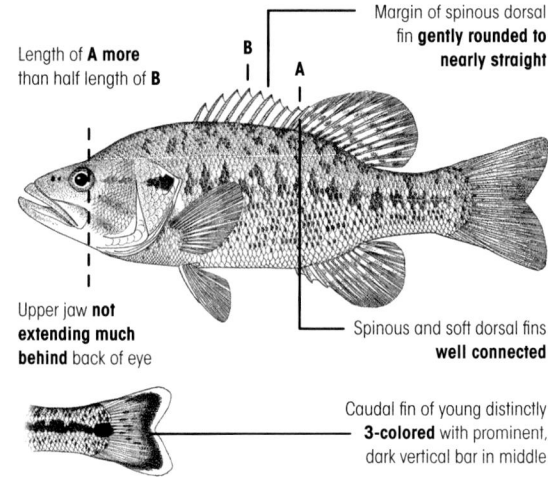

Length of **A more** than half length of **B**

Margin of spinous dorsal fin **gently rounded to nearly straight**

Upper jaw **not extending much behind** back of eye

Spinous and soft dorsal fins **well connected**

Caudal fin of young distinctly **3-colored** with prominent, dark vertical bar in middle

4a Side with a black horizontal stripe; lower side with a series of dark horizontal streaks or rows of dark spots in adults; juvenile with a prominent black spot at base of caudal fin; scales larger, usually 60–68 in lateral line, 8–9 above lateral line, and 23–27 around narrowest part of caudal peduncle . . . .**Page 314**

4b Side plain or with a series of separate vertical bars; lower side without dark horizontal streaks or rows of dark spots; juvenile without a prominent black spot at base of caudal fin; scales smaller, usually 71–77 in lateral line, 12–13 above lateral line, and 29–31 around narrowest part of caudal peduncle . . . .**Page 313**

Spotted Bass (*Micropterus punctulatus*)

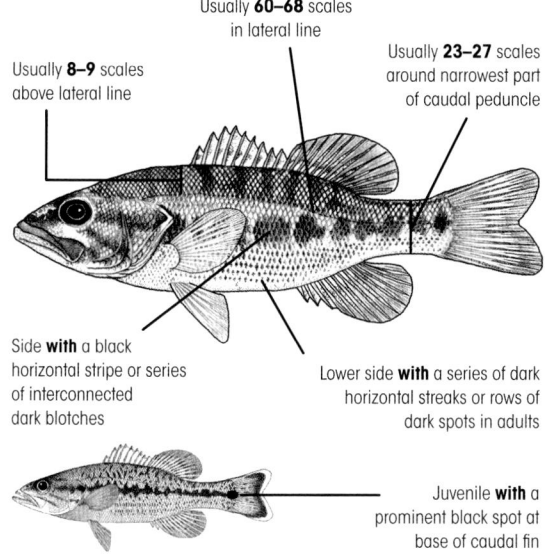

Usually **60–68** scales in lateral line

Usually **8–9** scales above lateral line

Usually **23–27** scales around narrowest part of caudal peduncle

Side **with** a black horizontal stripe or series of interconnected dark blotches

Lower side **with** a series of dark horizontal streaks or rows of dark spots in adults

Juvenile **with** a prominent black spot at base of caudal fin

Smallmouth Bass (*Micropterus dolomieu*)

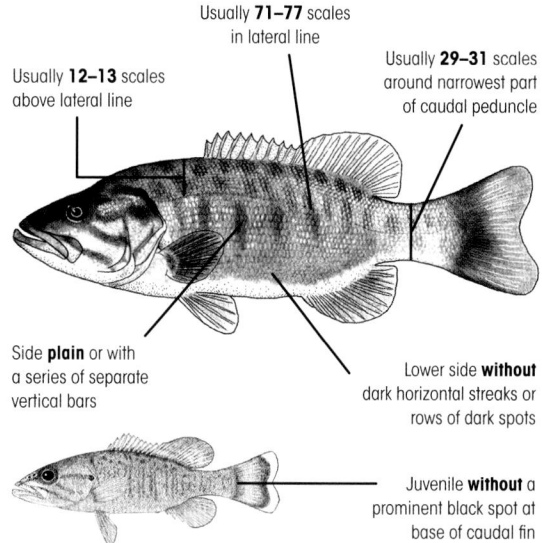

Usually **71–77** scales in lateral line

Usually **12–13** scales above lateral line

Usually **29–31** scales around narrowest part of caudal peduncle

Side **plain** or with a series of separate vertical bars

Lower side **without** dark horizontal streaks or rows of dark spots

Juvenile **without** a prominent black spot at base of caudal fin

5a Tongue with a patch of teeth on floor of mouth; several distinct dark lines radiating back from eye; mouth large, upper jaw extending to or behind middle of eye . . . .
. . . .**Page 305**

Warmouth (*Lepomis gulosus*)

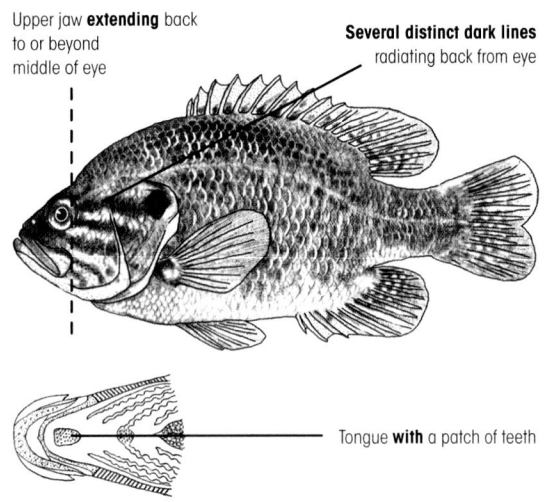

Upper jaw **extending** back to or beyond middle of eye

Several distinct dark lines radiating back from eye

Tongue **with** a patch of teeth

5b Tongue without teeth; no distinct dark lines radiating back from eye; mouth smaller, upper jaw not extending behind middle of eye except in adult *Lepomis cyanellus* **6**

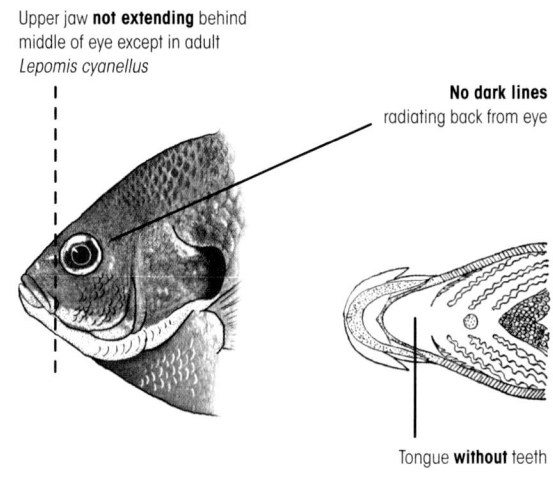

Upper jaw **not extending** behind middle of eye except in adult *Lepomis cyanellus*

No dark lines radiating back from eye

Tongue **without** teeth

6a Pectoral fin shorter and more rounded, usually not extending beyond front of eye when bent forward towards eye; mouth larger, upper jaw extending to beneath pupil of eye except in small young. **7**

Upper jaw **extending** to beneath pupil of eye, except in young

Pectoral fin **shorter and more rounded,** usually **not extending** beyond front of eye when bent toward eye

6b Pectoral fin longer and more pointed, usually extending far beyond front of eye when bent forward toward eye; mouth smaller, upper jaw not extending to beneath pupil of eye, except in young **12**

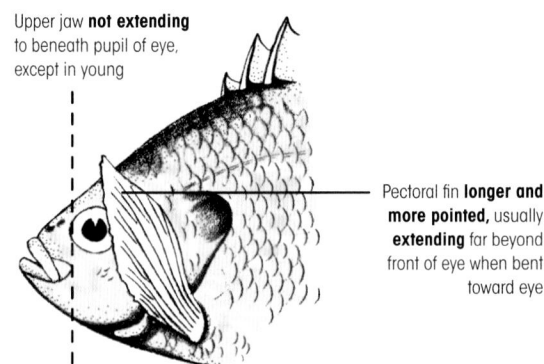

Upper jaw **not extending** to beneath pupil of eye, except in young

Pectoral fin **longer and more pointed,** usually **extending** far beyond front of eye when bent toward eye

7a Rear margin of gill cover (lying within base of, but not including membranous ear flap) stiff; membranous ear flap not greatly elongated **8**

7b Rear margin of gill cover thin and flexible; membranous ear flap elongated in adults, especially males **10**

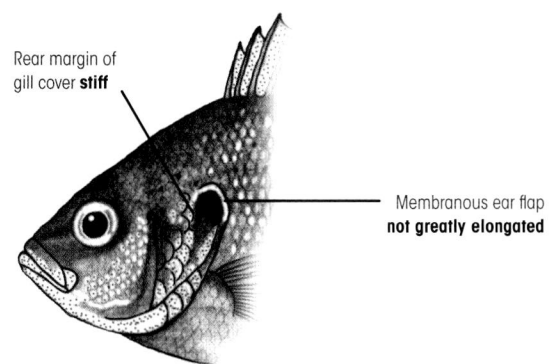

Rear margin of gill cover **stiff**

Membranous ear flap **not greatly elongated**

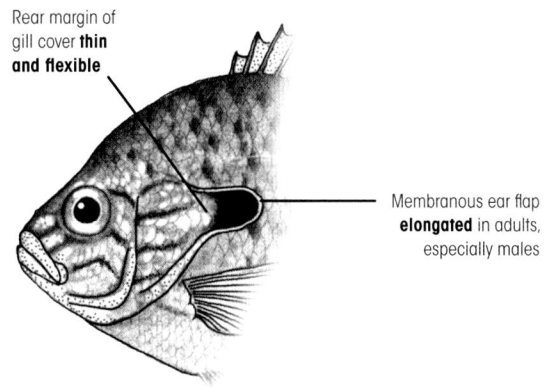

Rear margin of gill cover **thin and flexible**

Membranous ear flap **elongated** in adults, especially males

8a Body more slender, its depth (A) usually less than distance from tip of snout to front of dorsal fin (B); snout longer, its length (C) going less than 2 times into distance from back of eye to rear margin of ear flap (D); lateral line scales usually 43 or more; soft dorsal and anal fins with black blotches near bases of last few rays in adult; soft dorsal, caudal, and anal fins with yellow or orange margins in life . . . .**Page 303**

Green Sunfish (*Lepomis cyanellus*)

8b Body deeper, its depth (A) greater than distance from tip of snout to front of dorsal fin (B); snout shorter, its length (C) going about 2 times into distance from back of eye to rear margin of ear flap (D); lateral line scales usually 41 or fewer; soft anal fin without a black blotch; soft dorsal, caudal, and anal fins with white or plain margins **9**

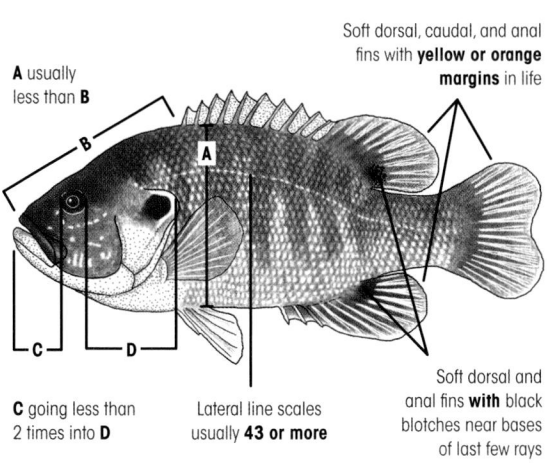

A usually less than **B**

Soft dorsal, caudal, and anal fins with **yellow or orange margins** in life

C going less than 2 times into **D**

Lateral line scales usually **43 or more**

Soft dorsal and anal fins **with** black blotches near bases of last few rays

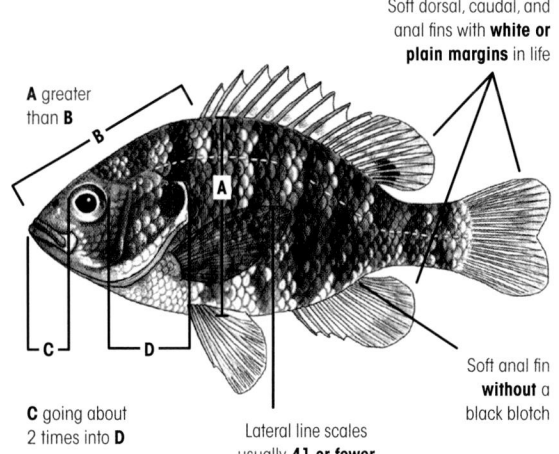

Soft dorsal, caudal, and anal fins with **white or plain margins** in life

A greater than **B**

C going about 2 times into **D**

Lateral line scales usually **41 or fewer**

Soft anal fin **without** a black blotch

9a Lateral line incomplete and interrupted; scales in lateral series usually 32–35; gill rakers on first arch longer, their length 5–6 times their width at base; soft dorsal fin with ink-black spot near base of last few rays in young, fading in adults; side with dark vertical bars or mottled . . . .
. . . .**Page 312**

9b Lateral line complete and uninterrupted; scales in lateral series usually 35 or more; gill rakers on first arch shorter, 2–3 times their width at base; soft dorsal fin without a black spot; sides with scattered small dark specks (young) or with rows of light spots (these are often bright orange-red in life). . . . .**Page 311**

Bantam Sunfish (*Lepomis symmetricus*)

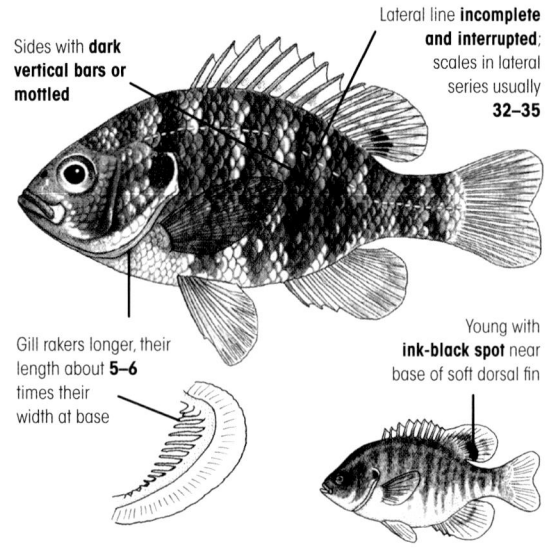

Sides with **dark vertical bars or mottled**

Lateral line **incomplete and interrupted**; scales in lateral series usually **32–35**

Gill rakers longer, their length about **5–6** times their width at base

Young with **ink-black spot** near base of soft dorsal fin

Spotted Sunfish (*Lepomis punctatus*)

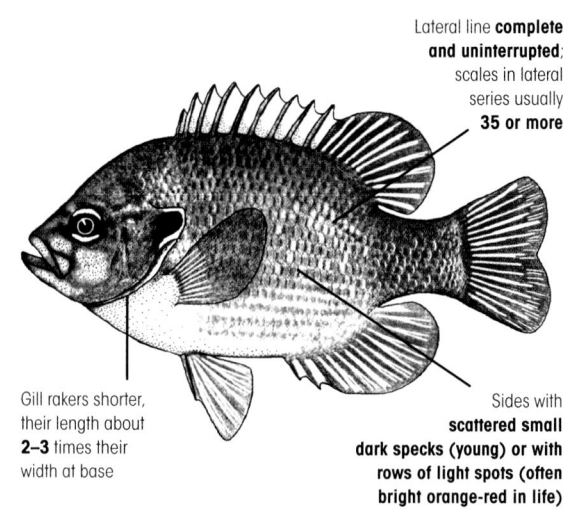

Lateral line **complete and uninterrupted**; scales in lateral series usually **35 or more**

Gill rakers shorter, their length about **2–3** times their width at base

Sides with **scattered small dark specks (young) or with rows of light spots (often bright orange-red in life)**

10a Ear flap black with broad white margin; sides of male with numerous scattered orange or red spots, that of female with scattered brown spots; gill rakers on first arch long, their length about 4–5 times their width at base; pectoral fin rays usually 15 . . . .**Page 306**

Orangespotted Sunfish (*Lepomis humilis*)

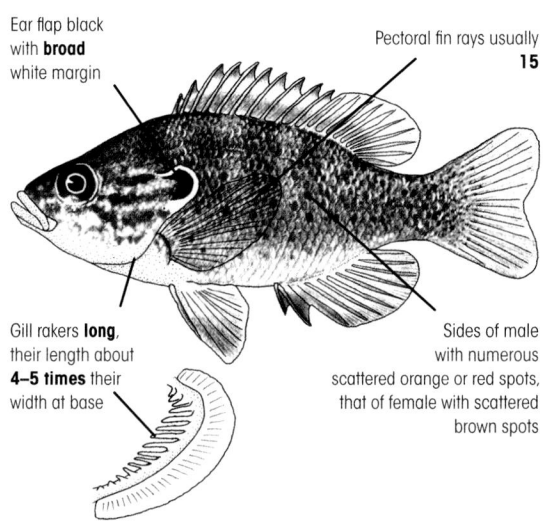

Ear flap black with **broad** white margin

Pectoral fin rays usually **15**

Gill rakers **long**, their length about **4–5 times** their width at base

Sides of male with numerous scattered orange or red spots, that of female with scattered brown spots

10b Ear flap black with narrow white or red margin; sides of male brilliantly colored with blues and reds, that of female without brown spots; gill rakers on first arch shorter, their length about 2 times their width at base; pectoral fin rays usually 13 or 14. . . . .**11**

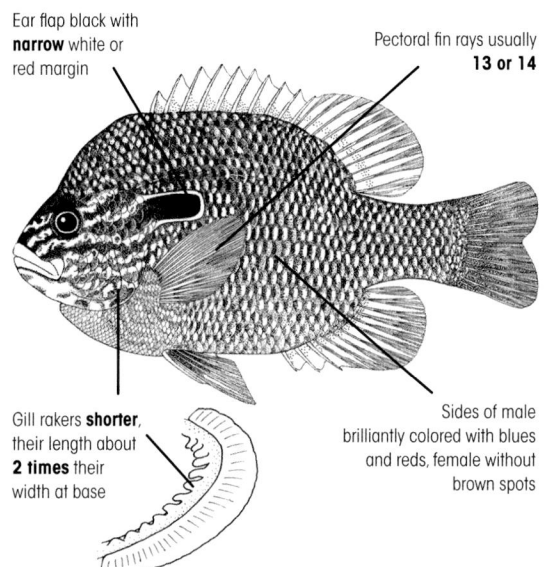

Ear flap black with **narrow** white or red margin

Pectoral fin rays usually **13 or 14**

Gill rakers **shorter**, their length about **2 times** their width at base

Sides of male brilliantly colored with blues and reds, female without brown spots

11a Ear flap black with narrow white margin; sides of male brilliantly colored with blues and reds, that of female without brown spots; pectoral fin rays usually 13 or 14. . . . .
. . . .**Page 308**

Longear Sunfish (*Lepomis megalotis*)

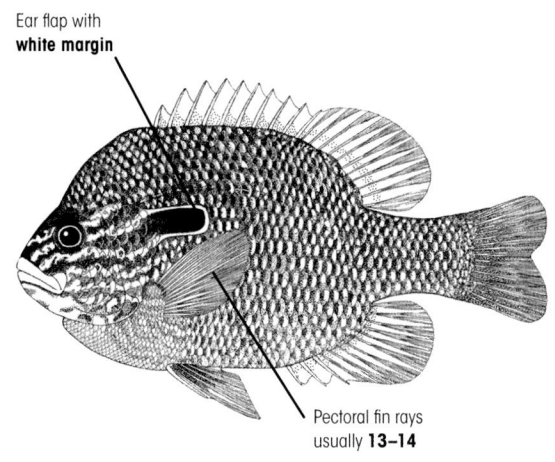

Ear flap with **white margin**

Pectoral fin rays usually **13–14**

11b Ear flap with narrow white and red margin; sides of male brilliantly colored with blues and reds, that of females without brown spots; ear flap usually angled upwards; pectoral fin rays usually 12 . . . .**Page 310**

Northern Sunfish (*Lepomis peltastes*)

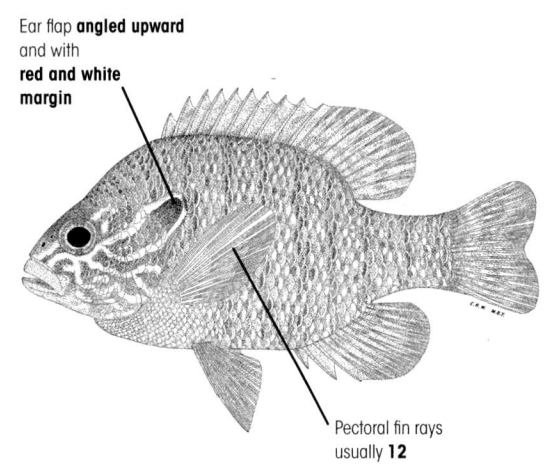

Ear flap **angled upward** and with **red and white margin**

Pectoral fin rays usually **12**

12a Soft dorsal fin with a distinct black blotch near base of last few rays; ear flap dark to its margin, without a light-colored border and without a red or orange spot; gill rakers on first arch long and thin, their length 4 or more times their width at base . . . .**Page 307**

12b Soft dorsal fin without a blotch near base of last few rays; ear flap not dark to its margin, with a light-colored border, and with a prominent red or orange spot in life; gill rakers on first arch short and thick, their length about 2 times their width at base **13**

Bluegill (*Lepomis macrochirus*)

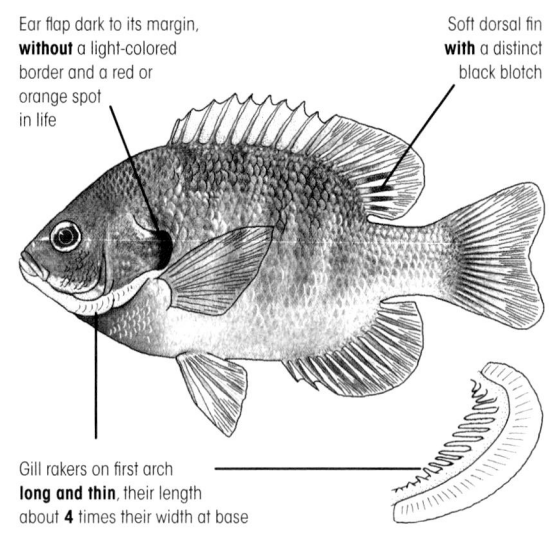

Ear flap dark to its margin, **without** a light-colored border and a red or orange spot in life

Soft dorsal fin **with** a distinct black blotch

Gill rakers on first arch **long and thin**, their length about **4** times their width at base

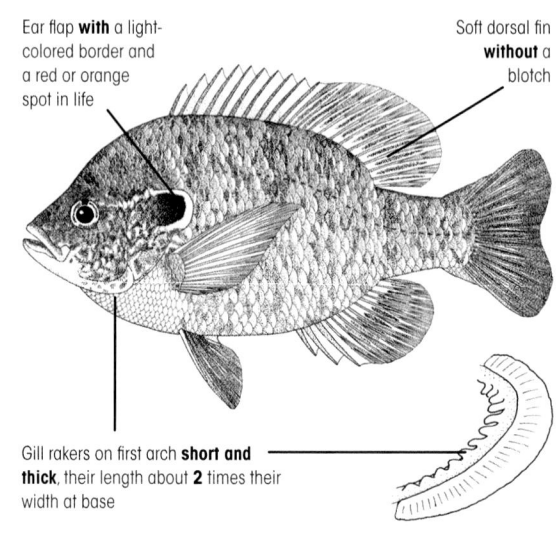

Ear flap **with** a light-colored border and a red or orange spot in life

Soft dorsal fin **without** a blotch

Gill rakers on first arch **short and thick**, their length about **2** times their width at base

---

13a Soft dorsal and anal fins mottled or with distinct spots; cheek with wavy lines that are bluish in life; sides often with pale spots encircled by dusky marks . . . . .**Page 304**

13b Soft dorsal and anal fins not mottled or spotted; cheek without wavy lines; sides lack distinct pale spots . . . .**Page 309**

Pumpkinseed (*Lepomis gibbosus*)

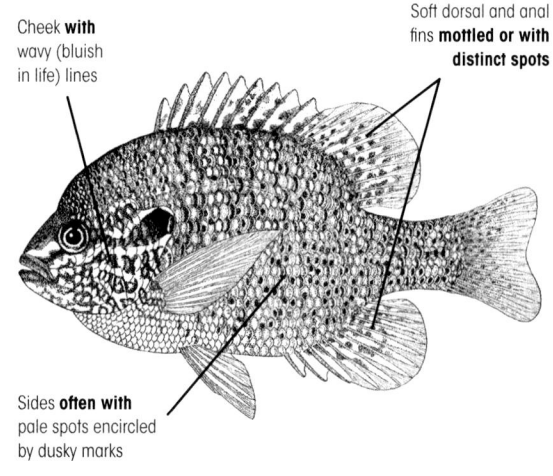

Cheek **with** wavy (bluish in life) lines

Soft dorsal and anal fins **mottled or with distinct spots**

Sides **often with** pale spots encircled by dusky marks

Redear Sunfish (*Lepomis microlophus*)

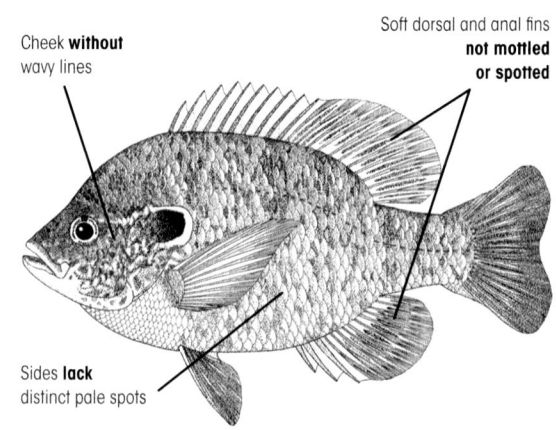

Cheek **without** wavy lines

Soft dorsal and anal fins **not mottled or spotted**

Sides **lack** distinct pale spots

14a Dorsal fin spines usually 11–13 . . . . **15**

Dorsal fin spines usually **11–13**

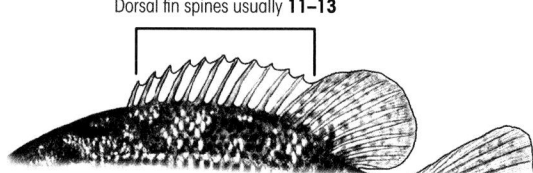

14b Dorsal fin spines usually 6–8 . . . . **16**

Dorsal fin spines usually **6–8**

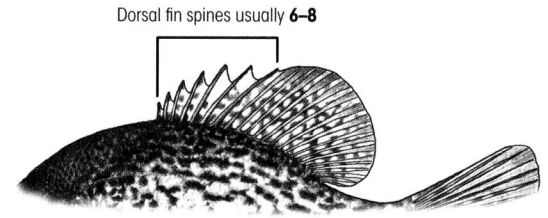

15a Anal fin much smaller than dorsal fin, with 5 or 6 spines and 10 or 11 rays; mouth large, upper jaw extending to below middle of eye; dark bar below eye (if present) extending backward . . . . **Page 301**

Rock Bass (*Ambloplites rupestris*)

Dark bar below eye (if present) **extending backward**

Mouth **large,** upper jaw extending **to below middle** of eye

Anal fin much smaller than dorsal fin, with **5 or 6** spines and **10 or 11** rays

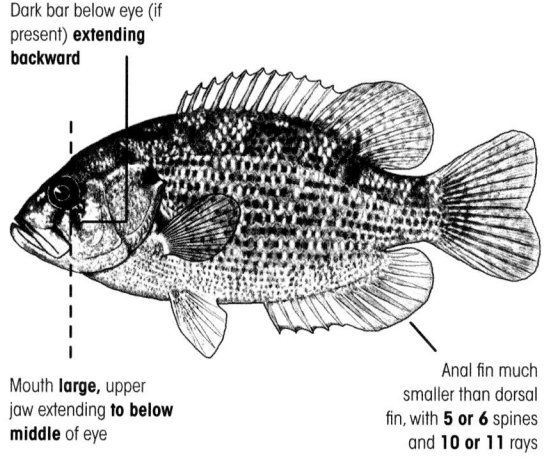

15b Anal fin nearly as large as dorsal fin, with 7 or 8 spines and 13–15 rays; mouth smaller, upper jaw extending to near front of eye; dark bar (tear drop) below eye prominent and vertically aligned; young with dark blotch on posterior base of dorsal fin . . . . **Page 302**

Flier (*Centrarchus macropterus*)

Dark bar below eye **prominent and vertically aligned**

Mouth **smaller,** upper jaw extending **to nearly front** of eye

Anal fin nearly as large as dorsal fin, with **7 or 8** spines and **13–15** rays

Young with dark blotch on posterior base of dorsal fin

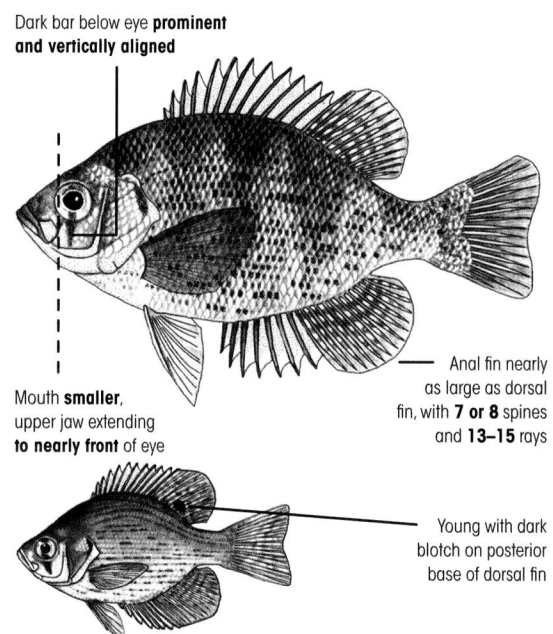

16a Dorsal fin spines usually 7 or 8; length of dorsal fin base (A) about equal to distance from front of dorsal fin to above eye or cheek (B); dark marking on sides consisting of irregularly arranged speckles and blotches. . **Page 317**

Black Crappie (*Pomoxis nigromaculatus*)

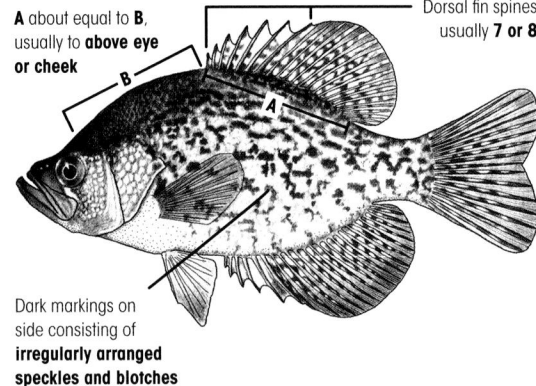

A about equal to **B**, usually to **above eye or cheek**

Dorsal fin spines usually **7 or 8**

Dark markings on side consisting of **irregularly arranged speckles and blotches**

16b Dorsal fin spines usually 6; length of dorsal fin base (A) much less than distance from front of dorsal fin to above eye or cheek (B); dark markings on sides consisting of regularly arranged vertical bars . . . .**Page 316**

White Crappie (*Pomoxis annularis*)

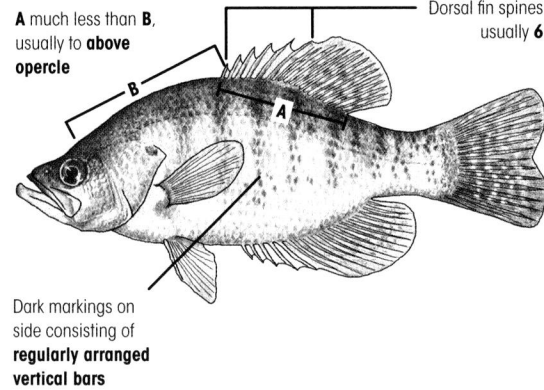

A much less than **B**, usually to **above opercle**

Dorsal fin spines usually **6**

Dark markings on side consisting of **regularly arranged vertical bars**

ROCK BASS

Ambloplites rupestris (Rafinesque 1817)

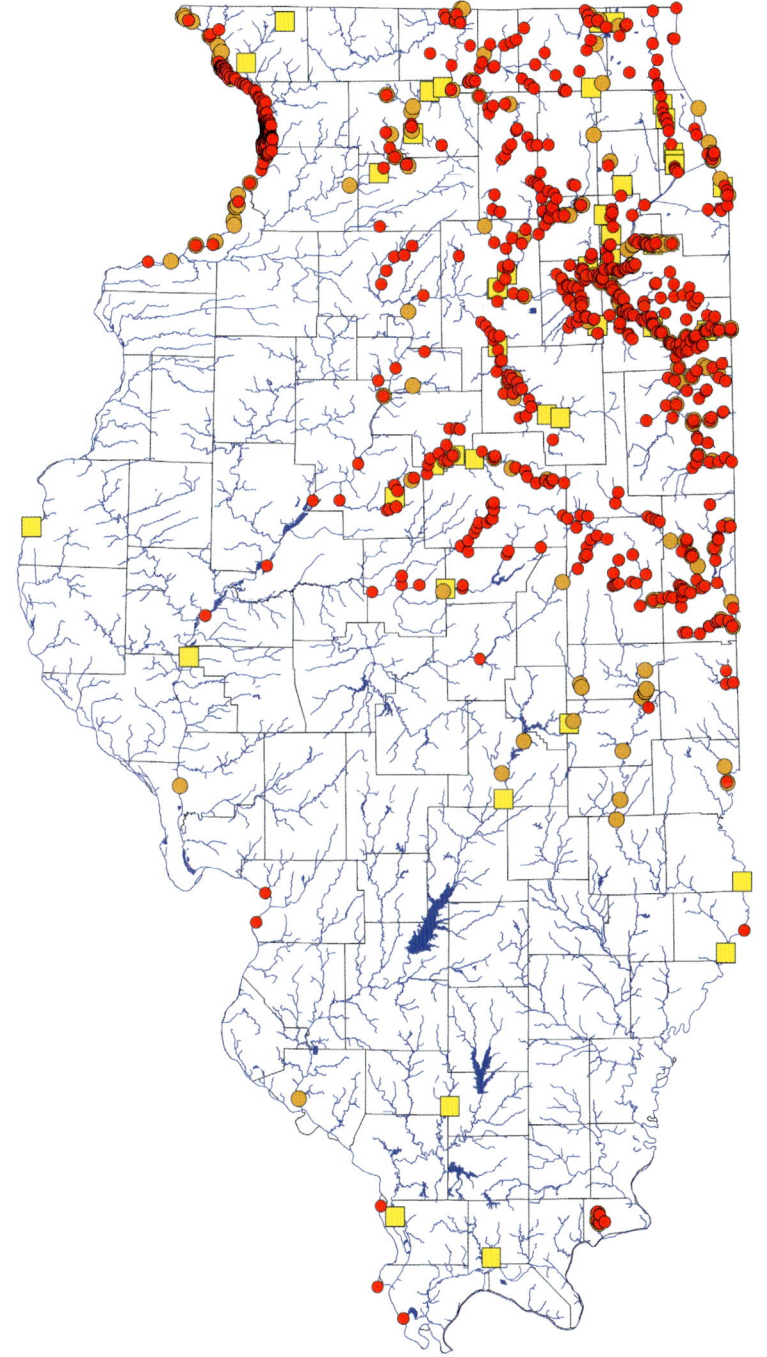

IDENTIFICATION Large Rock Bass are thick bodied. The mouth is large, extending under the pupil of the eye, and the pectoral fin is short and rounded. The Rock Bass has a dusky to black teardrop and distinctive rows of brown-black spots along the side of the body; the largest and darkest spots are below the lateral line. Juveniles have a much more compressed body with brown marbling on a gray side. Adults and young have a large, red eye and are light green on the back and upper side with brassy yellow flecks. There are 5 wide, dark saddles over the back and down to the midside; the breast and belly are white to bronze. The dorsal, caudal, and anal fins have distinctive black margins and are covered with dusky brown spots. There are 36–47 lateral scales, usually 11–13 dorsal spines, 5–7 anal spines, and 10–11 anal rays. To 17 in. (43 cm).

SIMILAR SPECIES The Warmouth, *Lepomis gulosus*, has dark lines radiating from the eye and only 3 anal spines.

HABITAT The Rock Bass is usually found along vegetated banks, in pools of clear creeks and small to medium rivers, and along rocky vegetated shores of lakes.

DISTRIBUTION IN ILLINOIS Forbes and Richardson (1909) found the Rock Bass to occur statewide but frequently encountered only in extreme northern Illinois and the upper Illinois River basin. Smith (1979) found nearly the same distribution but with the species somewhat more widespread in east-central Illinois. In the contemporary era, it remains widespread in northern Illinois and the upper Illinois River basin, but it is gone from the Kaskaskia River, where it was found in earlier eras, and from the Big Muddy and Cache River basins and is nearly gone from the Embarras River. A reproducing population occurs in Big Creek (a tributary to the Ohio River), Hardin County.

SUNFISHES (Centrarchidae)

301

FLIER

Centrarchus macropterus
(Lacepède 1801)

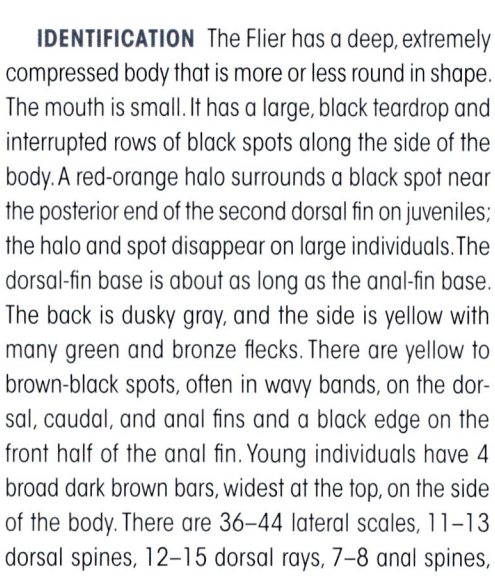

IDENTIFICATION The Flier has a deep, extremely compressed body that is more or less round in shape. The mouth is small. It has a large, black teardrop and interrupted rows of black spots along the side of the body. A red-orange halo surrounds a black spot near the posterior end of the second dorsal fin on juveniles; the halo and spot disappear on large individuals. The dorsal-fin base is about as long as the anal-fin base. The back is dusky gray, and the side is yellow with many green and bronze flecks. There are yellow to brown-black spots, often in wavy bands, on the dorsal, caudal, and anal fins and a black edge on the front half of the anal fin. Young individuals have 4 broad dark brown bars, widest at the top, on the side of the body. There are 36–44 lateral scales, 11–13 dorsal spines, 12–15 dorsal rays, 7–8 anal spines, and 13–17 anal rays. To 7½ in. (19 cm).

SIMILAR SPECIES The White Crappie, *Pomoxis annularis*, and Black Crappie, *P. nigromaculatus*, lack the bold black teardrop and rows of black spots on the side of the body and have 6–8 dorsal spines.

HABITAT The Flier occupies swamps, vegetated lakes, ponds, and sloughs and backwaters and pools of creeks and small rivers. It usually is found over mud.

DISTRIBUTION IN ILLINOIS The Flier is sporadically distributed in the southern one-fourth of Illinois, including the lower Kaskaskia River basin. Contemporary-era records indicate that it is present in tributaries of the lower Wabash River, where it was recorded by Forbes and Richardson (1909) but not by Smith (1979).

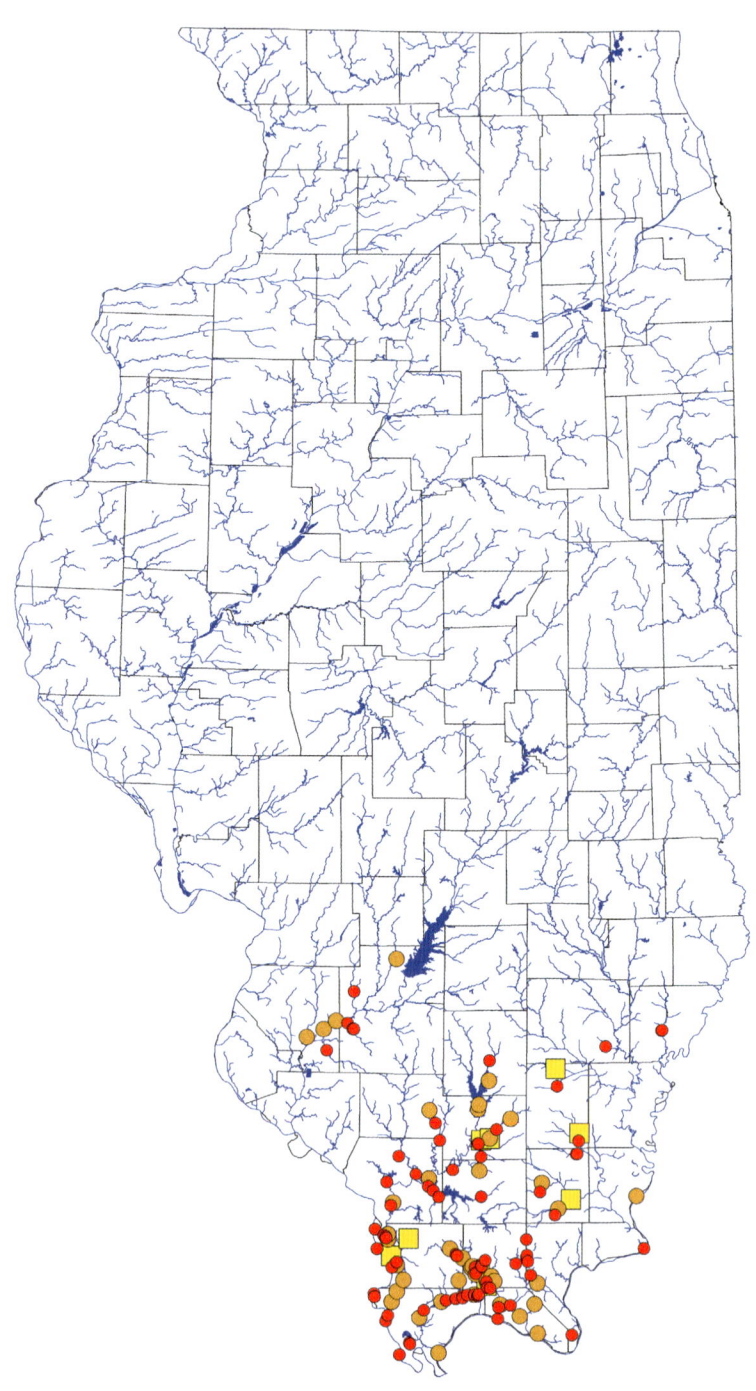

GREEN SUNFISH

Lepomis cyanellus Rafinesque 1819

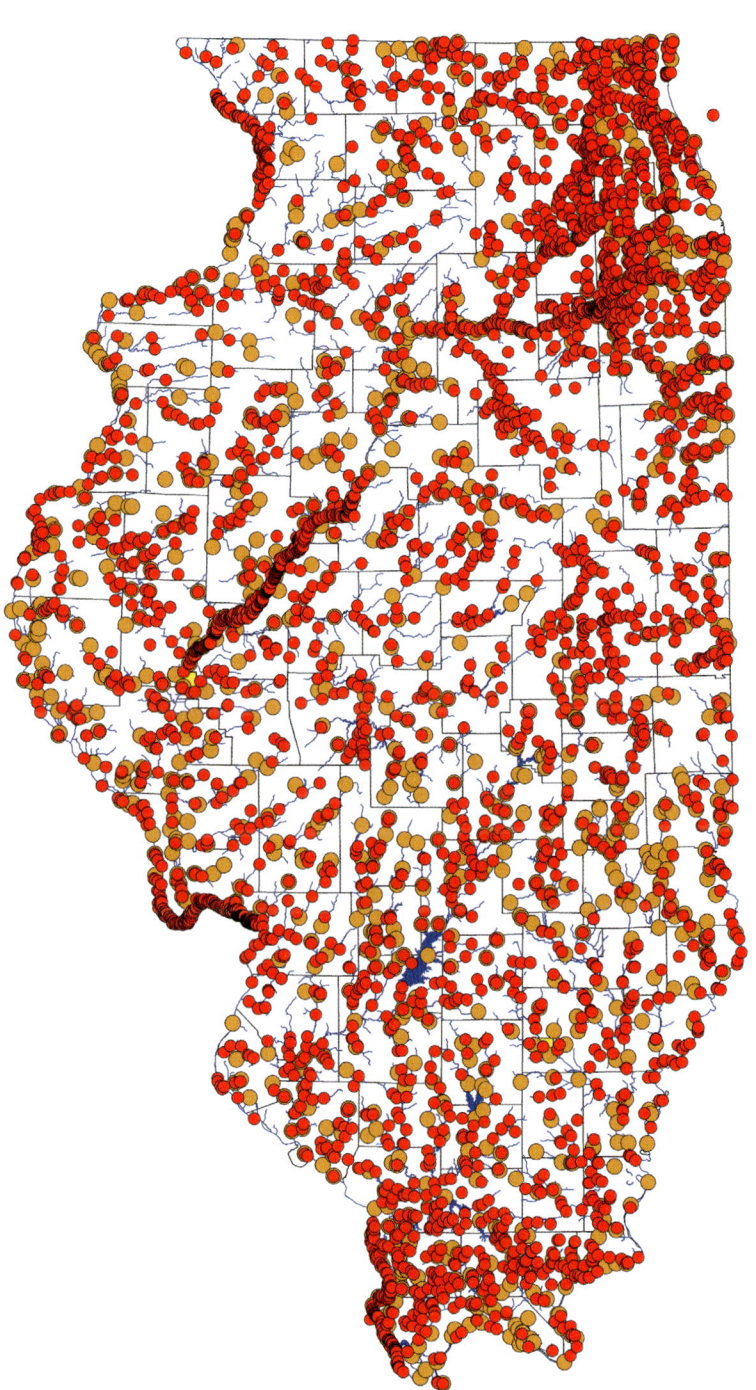

IDENTIFICATION The Green Sunfish has a thick body relative to other sunfishes. The mouth is large with the upper jaw extending beneath the pupil of the eye. There are no teeth on the tongue, and the pectoral fin is short and rounded—usually not reaching past the front of the eye when bent forward. Adults have a large, black spot at the posterior end of the second dorsal- and anal-fin bases. The back and side of the body are blue-green, often with yellow–metallic green flecks and dusky brown-black bars. There are green wavy lines on the cheek and opercle, a white to yellow edge (sometimes red on juveniles) on a short, black ear flap, and a white to yellow breast and belly. The second dorsal, caudal, and anal fins have white, yellow, or orange margins. There are 41–53 lateral scales, usually 13–14 pectoral rays, 3 anal spines, and 9 anal rays. Rakers on the first gill arch are long and thin. To 12 in. (31 cm).

SIMILAR SPECIES Other sunfishes lack yellow-orange margins on the median fins and a black spot near the posterior end of the anal-fin base (except the Bluegill, *L. macrochirus*) and have a smaller mouth (except the Warmouth, *L. gulosus*). The Warmouth has dark lines radiating from the eye, dark brown mottling on the side of the body, and teeth on the tongue.

HABITAT The Green Sunfish inhabits lakes and ponds and quiet pools and backwaters of sluggish creeks and rivers. It is often found near vegetation or woody debris.

DISTRIBUTION IN ILLINOIS The Green Sunfish is found throughout Illinois, including in Lake Michigan. It has been one of the most widespread species in Illinois at least since the Forbes and Richardson era (1909).

SUNFISHES (Centrarchidae)

303

PUMPKINSEED

Lepomis gibbosus (Linnaeus 1758)

IDENTIFICATION The Pumpkinseed has a deep, compressed body. The long, pointed pectoral fin usually extends far past the eye when bent forward. The mouth is small with the upper jaw not extending under the pupil of the eye. It is light to dark olive-green with many yellow-orange spots and gold flecks on the side, a bright red-orange spot on the white edge of the short, black ear flap, wavy blue lines on the cheek and opercle, and dark brown wavy lines and sometimes orange spots on the second dorsal, caudal, and anal fins. Juveniles and some adults have dusky chainlike bars on the side of the body. There are 35–47 lateral scales, usually 12–13 pectoral rays, 3 anal spines, and 10 anal rays. Rakers on the first gill arch are short and thick. To 16 in. (40 cm).

SIMILAR SPECIES The Redear Sunfish, *L. microlophus*, has a bright red margin (not a distinct spot) on a black ear flap and lacks wavy blue lines on the cheek and opercle. The Longear Sunfish, *L. megalotis*, lacks wavy lines or orange spots on the second dorsal fin and has a short, rounded pectoral fin.

HABITAT The Pumpkinseed inhabits vegetated lakes and ponds and quiet pools of creeks and small to large rivers.

DISTRIBUTION IN ILLINOIS The Pumpkinseed is generally distributed in northeastern Illinois and is widespread in the upper Mississippi River mainstem and along much of the Illinois River. Records for the southern one-third of Illinois are almost certainly the result of stocking. The range of this species has not changed substantially since the Forbes and Richardson or Smith eras.

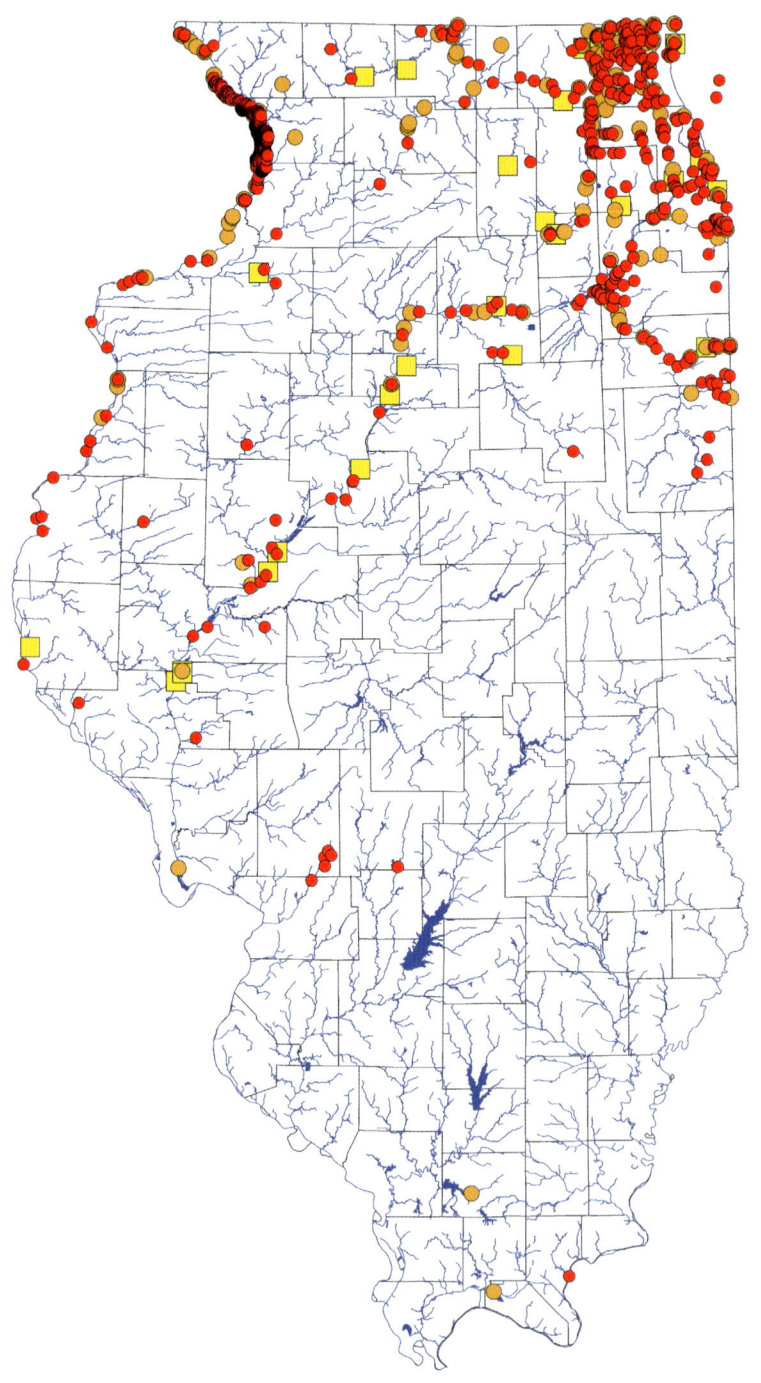

WARMOUTH

Lepomis gulosus (Cuvier 1829)

IDENTIFICATION The Warmouth has a deep, thick body. The mouth is large with the upper jaw extending under or beyond the pupil of the eye, and there is a patch of teeth on the tongue and a short, rounded pectoral fin that does not reach past the eye when bent forward. There are distinctive dark red-brown lines radiating posteriorly from the back of the red eye. The back and side of the body are olive-brown with dark brown mottling, an overall purple sheen, and often 6–11 chainlike dark brown bars. The lower side and belly are cream to bright yellow with dark brown spots. There are wavy dark brown bands on the fins. Large individuals have a red spot on the yellow edge of the short ear flap. During the breeding season, males have black pelvic fins and a bright red-orange spot at the base of the second dorsal fin. There are 36–44 lateral scales, usually 14 pectoral rays, 3 anal spines, and 9–10 anal rays. Rakers on the first gill arch are long and thin. To 12 in. (31 cm).

SIMILAR SPECIES The Green Sunfish, *L. cyanellus*, lacks the dark lines radiating from the eye and teeth on the tongue, and has a large, black spot at the rear of the second dorsal and anal fins and yellow or orange margins on the median fins. The Rock Bass, *Ambloplites rupestris*, has 6 anal spines and no dark lines radiating from the eye.

HABITAT The Warmouth inhabits vegetated lakes, ponds, swamps, and quiet-water areas of creeks and rivers, where it is often found over mud or amongst debris.

DISTRIBUTION IN ILLINOIS As Smith (1979) noted, the Warmouth occurs statewide but is much more generally distributed in the southern one-third of Illinois. It appears to be at least as widespread in the contemporary era as in earlier eras.

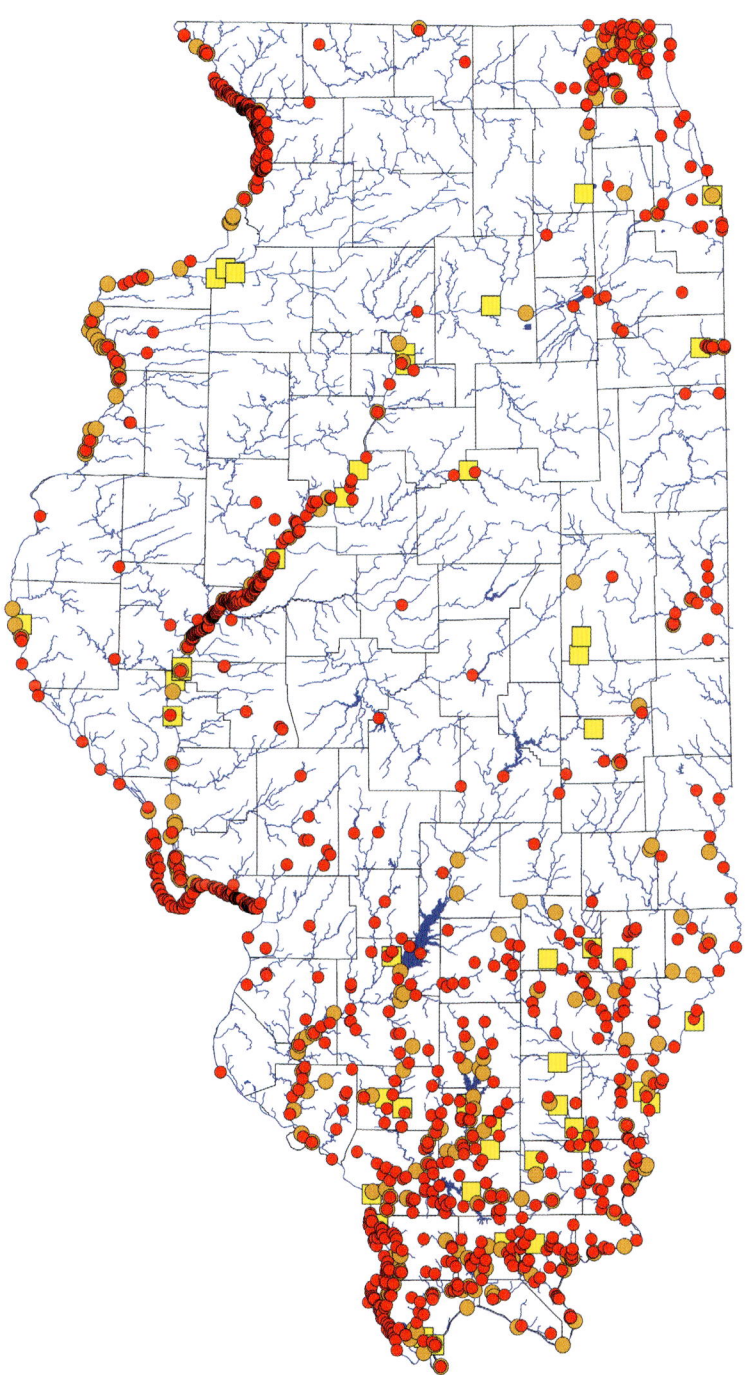

SUNFISHES (Centrarchidae)

305

ORANGESPOTTED SUNFISH

Lepomis humilis (Girard 1858)

IDENTIFICATION The Orangespotted Sunfish has a deep and strongly compressed body. The pectoral fin is short and rounded, not reaching past the eye when bent forward. The snout is fairly long, and the mouth is large with the upper jaw reaching under the pupil of the eye. There are greatly elongated pores along the edge of the preopercle and large, sensory pits between the eyes. There are many discrete bright orange (on large males) or red-brown (females) spots on the light olive to silver-blue side of the body. The lower side and belly are white to orange. Wavy lines on the cheek and opercle are orange on males and red-brown on females. The long, mostly black ear flap has a wide white margin. Juveniles have chainlike dusky bars on the side. During the breeding season, males are brilliantly colored, with a red eye, bright red-orange spots on the side, a red or orange breast and belly, and red edges on the anal and dorsal fins. There are 32–41 lateral scales, usually 14 pectoral rays, 3 anal spines, and 9 anal rays. Rakers on the first gill arch are moderately long and thin. To 6 in. (15 cm).

SIMILAR SPECIES Other Illinois sunfishes lack the wide white margin on a mostly black ear flap and the elongated pores on the preopercle. Other sunfishes with orange spots on the side have a much-darker body.

HABITAT The Orangespotted Sunfish inhabits quiet pools of creeks and small to large, often turbid, rivers. It is also known from ponds and large lakes usually near brush.

DISTRIBUTION IN ILLINOIS Smith (1979) found the Orangespotted Sunfish to be statewide in occurrence but more widespread in northern Illinois and less widespread in central Illinois than it had been during the Forbes and Richardson era. This pattern holds in the contemporary era, although the species seems to be more widespread in northeastern Illinois than in the earlier eras.

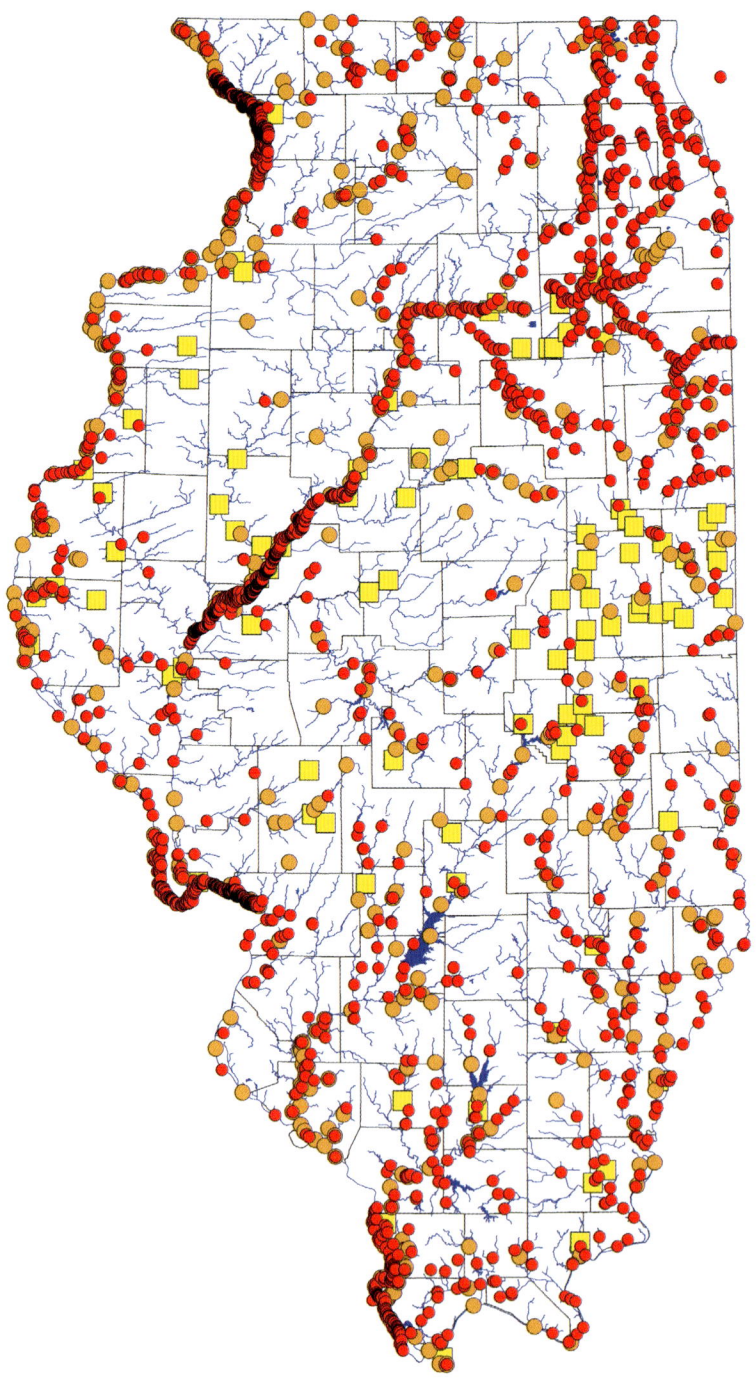

BLUEGILL

Lepomis macrochirus Rafinesque 1819

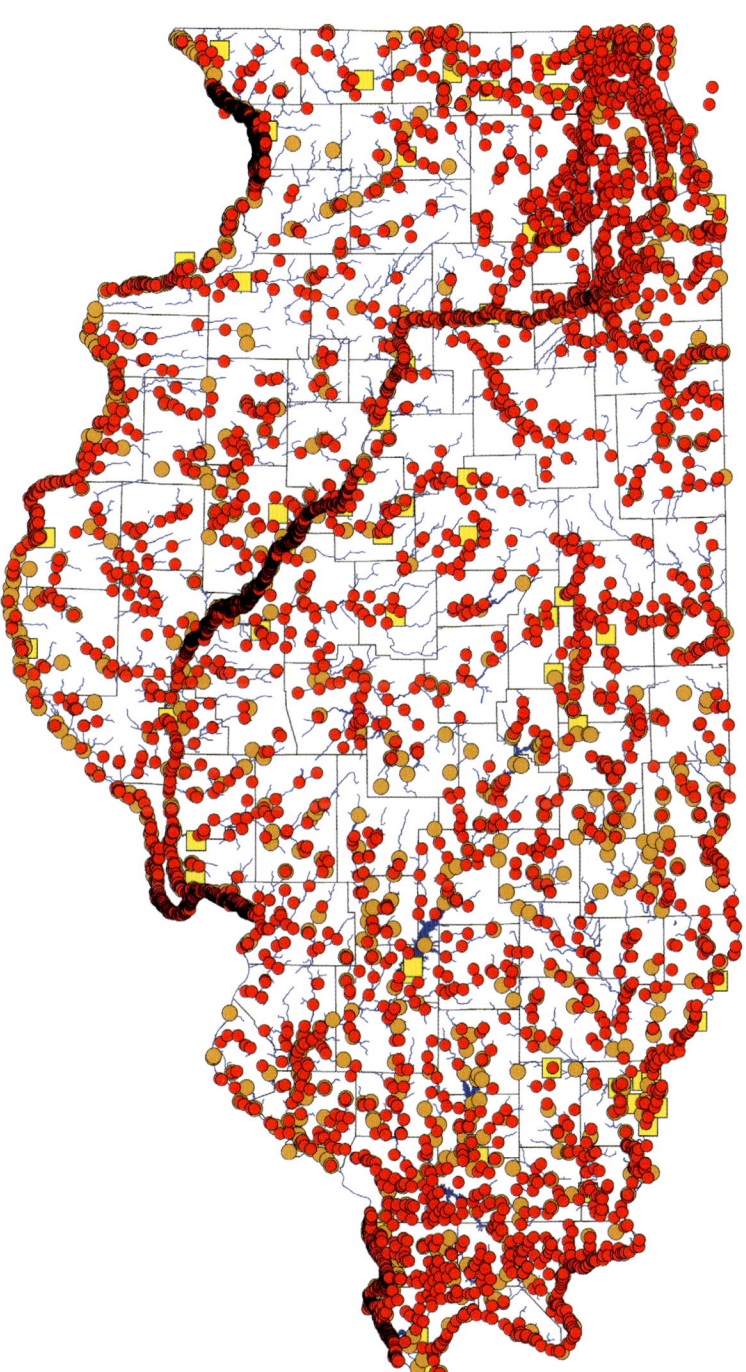

IDENTIFICATION The Bluegill has a deep and compressed body. The pectoral fin is long and pointed, usually extending far past the eye when bent forward. The mouth is relatively small with the upper jaw not extending under the pupil of the eye. There are dark bars on a light olive to silver-blue body. The bars are thin and chainlike on juveniles, dark blue-black on adults. There is a large, black spot near the posterior end of the dorsal fin (faint on young) and often a dusky spot near the posterior end of the anal fin. The ear flap is black to the edge and fairly long in adults. The fins are clear to dusky. Large males have a blue sheen overall, 2 blue streaks from the chin to the edge of the gill cover, and are white to yellow below. During the breeding season, males have a blue head and back, bright red-orange breast and belly, and black pelvic fins. There are 38–48 lateral scales, usually 13 pectoral rays, 3 anal spines, and 11 anal rays. Rakers on the first gill arch are long and thin. To 16¼ in. (41 cm).

SIMILAR SPECIES The Redear Sunfish, *L. micro-lophus*, lacks a large, dark spot in the second dorsal fin and has a bright red margin on the black ear flap and short, thick gill rakers.

HABITAT The Bluegill inhabits vegetated lakes, ponds, swamps, and pools of creeks and small to large rivers.

DISTRIBUTION IN ILLINOIS The Bluegill is statewide in occurrence, as it was in previous eras. It has been stocked for decades in farm ponds and impoundments.

LONGEAR SUNFISH

Lepomis megalotis (Rafinesque 1820)

IDENTIFICATION The Longear Sunfish has a deep and compressed body. The pectoral fin is short and rounded, not reaching past the eye when bent forward. The mouth is relatively large with the upper jaw extending to the pupil of the eye. There is a long, black ear flap. The flap is usually bordered by a white margin and proportionally longer in larger individuals. There are wavy blue-white lines on the cheek and opercle. Juveniles have an olive back and side of the body speckled with yellow flecks and often with chainlike bars. Adults are dark red above, bright orange below, marbled and spotted with blue, and sometimes have dusky bars on the side. Fins are clear to orange and blue. During the breeding season, males are brilliant orange and blue, and have a red eye, orange to red median fins, and blue-black pelvic fins. There are 33–46 (usually 39 or more) lateral scales, usually 13–14 pectoral rays, 3 anal spines, and 9–10 anal rays. Rakers on the first gill arch are very short and thick. To 9½ in. (24 cm).

SIMILAR SPECIES The Northern Sunfish, *L. peltastes*, also has short, rounded pectoral fins and wavy blue lines on the cheek and opercle but reaches a maximum size of 5 in. (13 cm), has a red margin on an upwardly directed ear flap, usually 12 (vs. 13–14) pectoral rays, and 40 or fewer (vs. 39 or more) lateral scales; large males of the 2 species are easy to distinguish, but females and juveniles are not. The Pumpkinseed, *L. gibbosus*, also has wavy blue lines on the cheek and opercle but also a bright red-orange spot on the white edge of the black earflap, and the Redear Sunfish, *L. microlophus*, has a bright red margin on the black earflap; both have a long, pointed pectoral fin that usually extends far past the eye when bent forward.

HABITAT The Longear Sunfish inhabits rocky and sandy pools of headwaters, creeks, and small to medium rivers and usually is found near vegetation or woody debris.

DISTRIBUTION IN ILLINOIS As in earlier eras, the Longear Sunfish is widespread in eastern and southern Illinois and is frequently encountered in the upper Illinois River basin. It is less widespread now in the Fox and Des Plaines Rivers and is gone from the Rock River and Lake Michigan. However, as discussed in the account for the Northern Sunfish, *L. peltastes*, most records of Northern and Longear Sunfishes cannot be distinguished, and all records for both species have been plotted on the map for *L. megalotis*. The decrease in distribution in northern Illinois appears to be that of the Northern Sunfish.

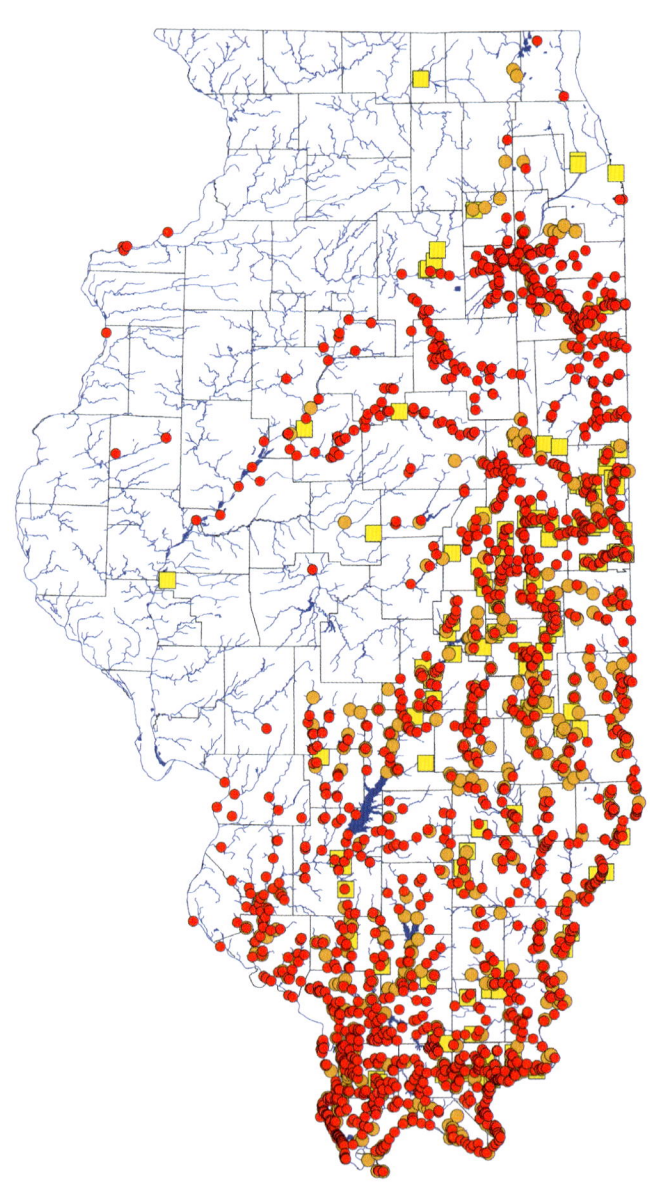

REDEAR SUNFISH

Lepomis microlophus (Günther 1859)

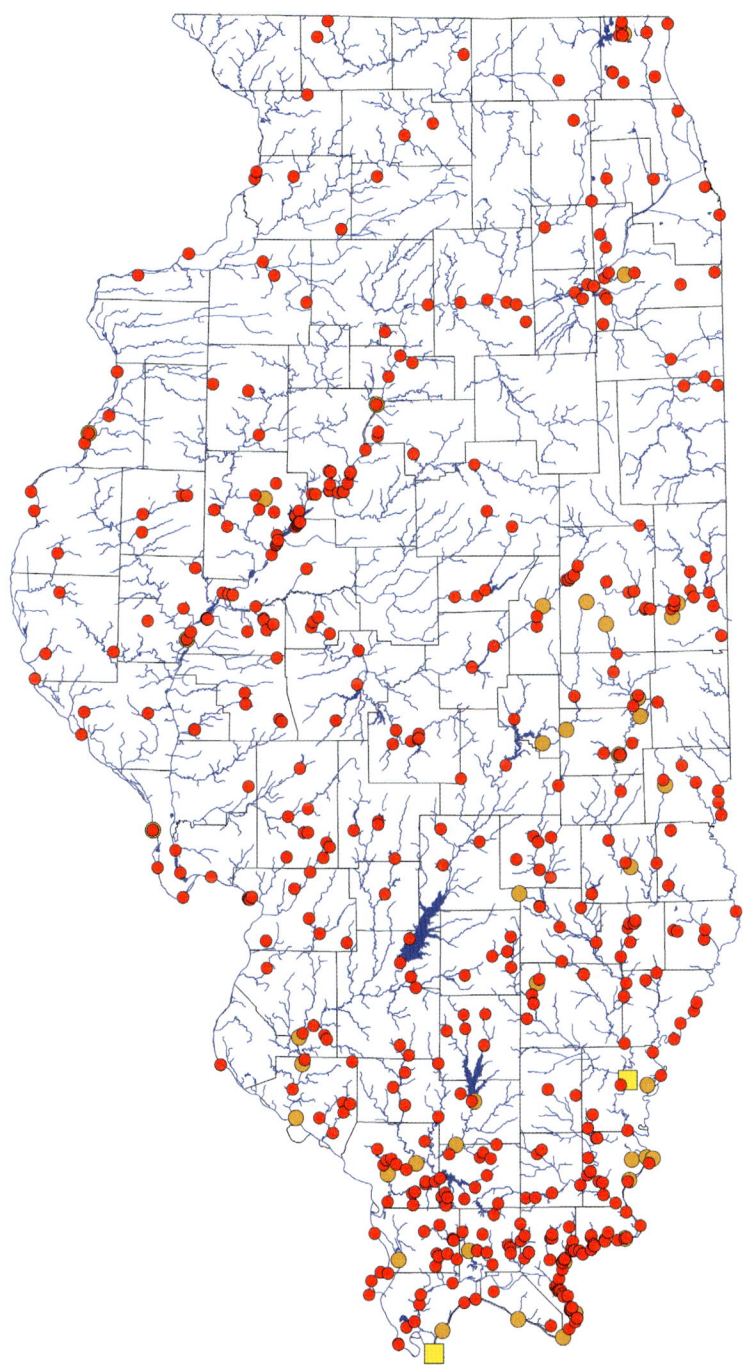

IDENTIFICATION The Redear Sunfish has a deep, compressed body. The snout is pointed, and the mouth is small with the upper jaw not extending under the pupil of the eye. The pectoral fin is long, and pointed and extends far past the eye when bent forward. The short, black ear flap is flexible to its margin, which is bright red. There are dark gray bars on a light olive to silver-green body. The second dorsal fin often has dusky black wavy lines; other fins are clear to dusky. Large males are brassy gold with dusky black fins. There are 34–39 lateral scales, usually 13–14 pectoral rays, 3 anal spines, and 10 anal rays. Rakers on the first gill arch are short and thick. To 10 in. (25 cm).

SIMILAR SPECIES The Pumpkinseed, *L. gibbosus*, has a bright red spot on the white margin of the earflap and wavy blue lines on the cheek and opercle. The Bluegill, *L. macrochirus*, lacks bright red on the ear flap and has long, thin gill rakers and a large, black spot at the posterior end of the second dorsal fin. The Longear Sunfish, *L. megalotis*, and Northern Sunfish, *L. peltastes*, have short, rounded pectoral fins, wavy blue lines on the cheek and opercle, and a long, black ear flap.

HABITAT The Redear Sunfish inhabits swamps, lakes, ponds, and pools of small to medium rivers.

DISTRIBUTION IN ILLINOIS Smith (1979) described the Redear Sunfish as being native to the southern one-third of Illinois and noted that it had become more widespread, including 1 record in Lake County in far northern Illinois, as a result of stocking. In the contemporary era the species is distributed throughout Illinois, in large part due to continued stocking in farm ponds and impoundments. It remains most common in southern Illinois.

SUNFISHES (Centrarchidae)

309

NORTHERN SUNFISH

Lepomis peltastes Cope 1870

IDENTIFICATION The Northern Sunfish has a deep and compressed body. The pectoral fin is short and rounded, not reaching past the eye when bent forward. The mouth is relatively large with the upper jaw extending to under the pupil of the eye. Wavy blue-white lines are on the cheek and opercle, and a long, black ear flap that is angled upward is bordered by a white margin in females and small males and a red margin in large males (breeding males have a large, red spot). Juveniles have an olive back, and the side of the body is speckled with yellow flecks and often with chainlike bars. Adults are dark red on the back and upper side of the body, orange below, marbled and spotted with blue, and sometimes have dusky bars on the side. Fins are clear to orange and blue. During the breeding season, males are brilliant orange and blue and have a red eye, a large, red spot on the ear flap, orange to red median fins, and blue-black pelvic fins. There are 33–46 (usually 40 or fewer) lateral scales, usually 12 pectoral rays, 3 anal spines, and 9–10 anal rays. Rakers on the first gill arch are very short and thick. To 5 in. (13 cm).

SIMILAR SPECIES The Pumpkinseed, *L. gibbosus*, and the Redear Sunfish, *L. microlophus*, have a long, pointed pectoral fin that usually extends far past the eye when bent forward. See Remarks in this entry and Similar Species for the Longear Sunfish.

HABITAT The Northern Sunfish inhabits rocky and sandy pools of headwaters, creeks, and small to medium rivers and usually is found near vegetation or woody debris.

DISTRIBUTION IN ILLINOIS Smith (1979) noted the presence of the "northern longear sunfish" in Kankakee, Will, Iroquois, and Grundy Counties. It occurs throughout the upper Illinois River basin but is less widespread now in the Fox and Des Plaines Rivers and is gone from the Rock River and Lake Michigan. See Remarks.

REMARKS Smith (1979) treated the "northern longear sunfish" as a subspecies, *Lepomis megalotis peltastes*. Bailey et al. (2004) treated it as *Lepomis peltastes* in their atlas of Michigan fishes, and it appears to be reproductively isolated and distinguished from the Longear Sunfish in Illinois by reaching a maximum size of 5 in. (13 cm) versus 9½ in. (24 cm) and having a red margin on an upwardly directed ear flap versus an all-black ear flap, usually 12 (vs. 13–14) pectoral rays, and usually 40 or fewer (vs. 40 or more) lateral scales. Large males of the 2 species are easy to distinguish, but females and juveniles are not, and the ranges of the 2 species in the Illinois River basin are uncertain. Records for both species are shown on the map for *L. megalotis*. The 2 species occur together in the Kankakee River basin; *L. peltastes* is probably the only species in the Fox and Des Plaines Rivers, and only *L. megalotis* occurs in the Wabash, Ohio, and Mississippi River basins, excluding the Illinois River basin.

SPOTTED SUNFISH

Lepomis punctatus (Valenciennes 1831)

IDENTIFICATION The Spotted Sunfish has a deep and compressed body. The pectoral fin is short and rounded, usually not reaching past the eye when bent forward. The mouth is relatively large with the upper jaw extending to the pupil of the eye. There are small, black spots on the side of the head and rows of spots of red-orange on the male and yellow-brown on the female along the side of the body. The back and side of the body are dark olive; the lower side and belly are white, yellow, or red-orange. There is an iridescent blue margin on the lower part of the eye and a white to yellow edge on the short, black ear flap. Fins are clear to dusky. The lateral line is complete with 35–41 scales. There are usually 13–14 pectoral rays, 3 anal spines, and 10 anal rays. Rakers on the first gill arch are moderately long and thin. To 8 in. (20 cm).

SIMILAR SPECIES The Bantam Sunfish, *L. symmetricus*, lacks black specks on the head and rows of red or yellow-brown spots on the side of the body and has a large, black spot at the posterior end of the second dorsal fin and an interrupted lateral line. The Longear Sunfish, *L. megalotis*, and the Northern Sunfish, *L. peltastes*, lack black spots on the head and have wavy blue lines on the cheek and opercle and short rakers on the first gill arch.

HABITAT This species is usually found in heavily vegetated ponds, lakes, pools of creeks and small to medium rivers, and swamps. It is often captured over mud or sand.

DISTRIBUTION IN ILLINOIS Forbes and Richardson (1909) found the Spotted Sunfish in bottomland lakes and wetlands along the Illinois and Wabash Rivers and at 1 locality each on the Mississippi and Iroquois Rivers. In the Smith era, the continued presence of the species along the Illinois and Wabash Rivers was documented, and records from the Ohio River, the LaRue–Pine Hills Research Natural Area in Union County, and tributaries of the Sangamon River were added. In the contemporary era, the Spotted Sunfish occurs in bottomland lakes and wetlands in southern Illinois along the Wabash, Ohio, and Mississippi Rivers and in tributaries of the Sangamon River in central Illinois. It appears to be gone from the bottomland lakes and wetlands along the Illinois River even though in 2010–2011 Spotted Sunfish were removed from Fish Creek in Mason County, cultured, and translocated into several lakes in the middle Illinois River basin in an attempt to reestablish the species.

REMARKS The subspecies *L. p. miniatus*, the form in Illinois, is distinguished from *L. p. punctatus* by having a pale to bright red-orange patch on the side just above the ear flap, rows of red or yellow spots on the side, no black specks on the body, and usually 35–41 (vs. 38–44) lateral scales (Page & Burr 2011).

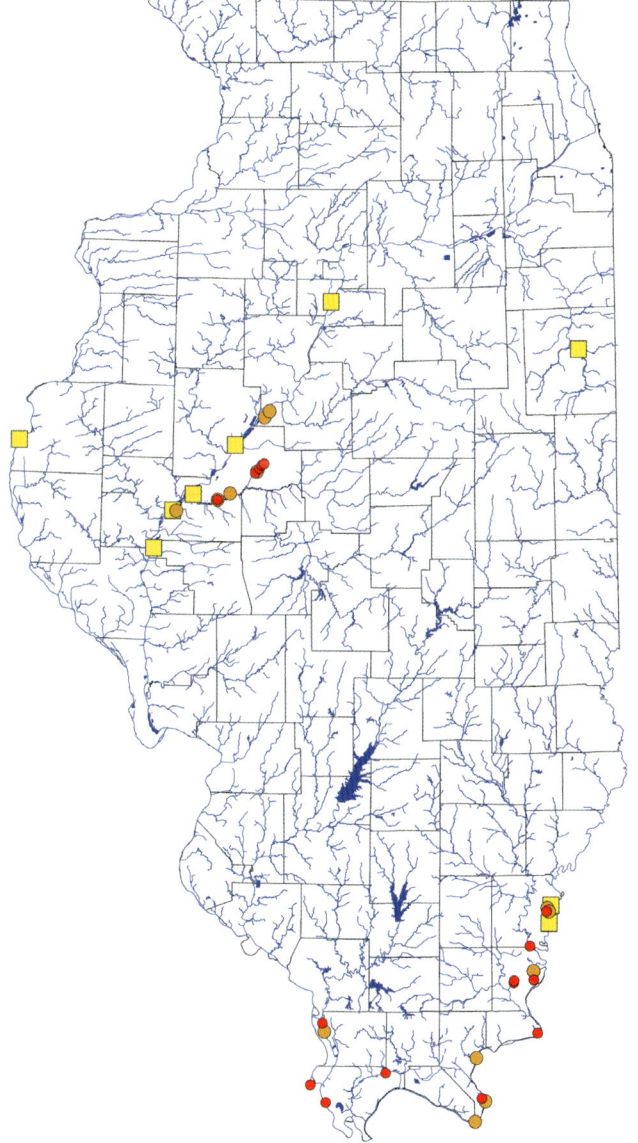

BANTAM SUNFISH

Lepomis symmetricus Forbes 1883

IDENTIFICATION The Bantam Sunfish has a somewhat chubby body (less compressed than other sunfishes). The lateral line is usually incomplete or interrupted. The pectoral fin is short and rounded and does not reach past the eye when bent forward. The mouth is fairly large with the upper jaw extending under the pupil of the eye. It has a light brown to dusky green back and side, yellow flecks and scattered small, dark brown spots on the side of adults and chainlike bars on the side (more prominent on juveniles). There is a red-brown halo around a bold black spot at the posterior edge of the dorsal fin on juveniles; the spot diminishes with growth and is absent in large adults. The ear flap is black with a narrow white edge. The dorsal and anal fins are red on young individuals and clear to dusky on adults. There are 30–40 lateral scales, usually 12–13 pectoral rays, 3 anal spines, and 10 anal rays. Rakers on the first gill arch are long and thin. To 3½ in. (9 cm).

SIMILAR SPECIES Other sunfishes in Illinois, except the Green Sunfish, *L. cyanellus,* and the Bluegill, *L. macrochirus,* lack a prominent black spot at the posterior edge of the dorsal fin. The Green Sunfish has a larger mouth, yellow or orange margins on the median fins, and 41–53 lateral scales. The Bluegill is more compressed and has a long pectoral fin, a dark edge on the ear flap, and 38–48 lateral scales.

HABITAT The Bantam Sunfish is found in swamps and heavily vegetated ponds, lakes, and sloughs, usually over a mud substrate.

DISTRIBUTION IN ILLINOIS The Bantam Sunfish was originally described by Forbes in 1883 from the Illinois River at Pekin, where it has not been taken since the 1880s. Forbes and Richardson (1909) also had records from the lower Wabash River basin, where the species also has not been seen since the 1880s, and from the LaRue–Pine Hills Research Natural Area in Union County. The species was found at only 2 locations in the Mississippi River floodplain in northern Union County during the Smith era. In the contemporary era it continues to exist in the Mississippi River floodplain in northern Union County, Horseshoe Lake in Alexander County, and the Clear Creek system in Union and Alexander Counties. Records from the Clear Creek system were made during the 1993 flood.

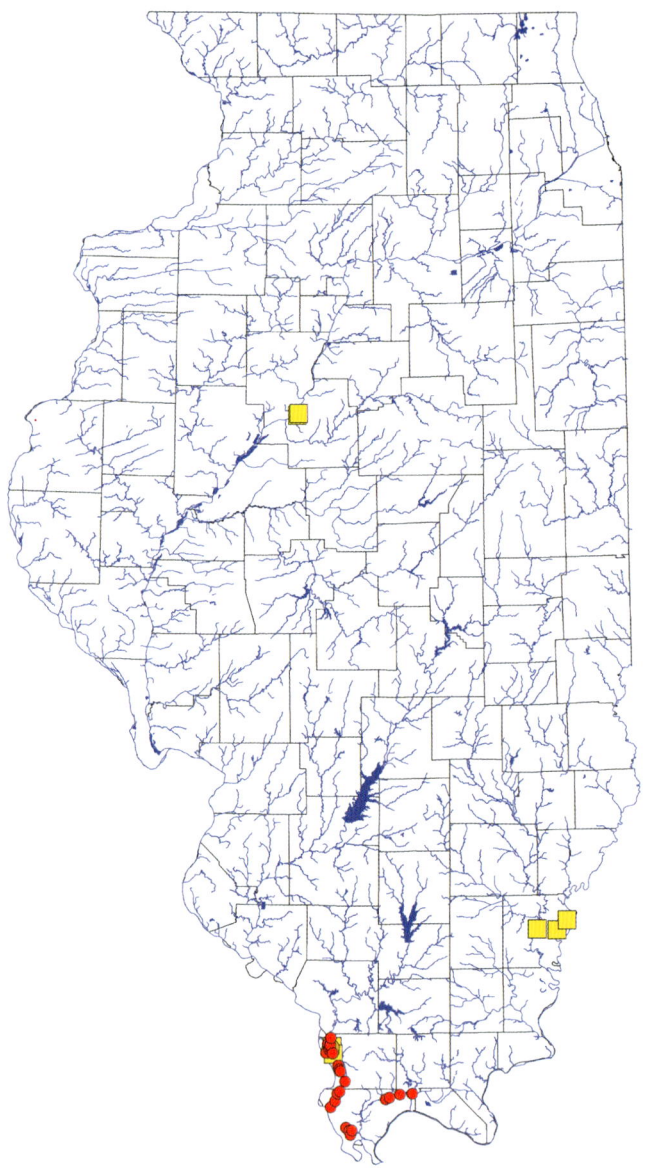

SMALLMOUTH BASS

Micropterus dolomieu Lacepède 1802

IDENTIFICATION The Smallmouth Bass has an elongate body, deepest under the first dorsal fin, and a large mouth with the upper jaw extending nearly to the posterior edge of the eye. The first dorsal fin is of nearly uniform height throughout, and the first and second dorsal fins are broadly joined at the base. The body usually has 8–16 dark brown bars and bronze specks on a yellow-brown to olive-green side; the underside is yellow-white. Young have a 3-colored (yellow, black, white edge) caudal fin. Large males are green-brown to bronze with black mottling on the back and bars on the side. The iris of the eye is often red in both juveniles and adults, and dark brown lines radiate posteriorly from the back of the eye. There is usually no patch of teeth on the tongue. There are usually 69–77 lateral scales, usually 10 dorsal spines, 3 anal spines, and 29–32 scales around the caudal peduncle. To 27¼ in. (69 cm).

SIMILAR SPECIES The Spotted Bass, *M. punctulatus*, has rows of black spots along the lower side and a black stripe (usually no bars) along the midside. The Largemouth Bass, *M. salmoides*, has a black stripe (no bars) along the side, a larger mouth, the first and second dorsal fins nearly separate, and usually 58–73 lateral scales.

HABITAT The Smallmouth Bass is found in clear, gravel-bottom runs and flowing pools of small to large rivers. Some are also found along the windswept, shallow, rocky areas of Lake Michigan.

DISTRIBUTION IN ILLINOIS In the contemporary era the Smallmouth Bass is widespread in the northern half of Illinois and sporadic in the southern half, as it was in earlier eras, but appears to be more widespread in the mainstems of the Wabash, Ohio, and Mississippi Rivers. Its recent disappearance from the Embarras River is enigmatic.

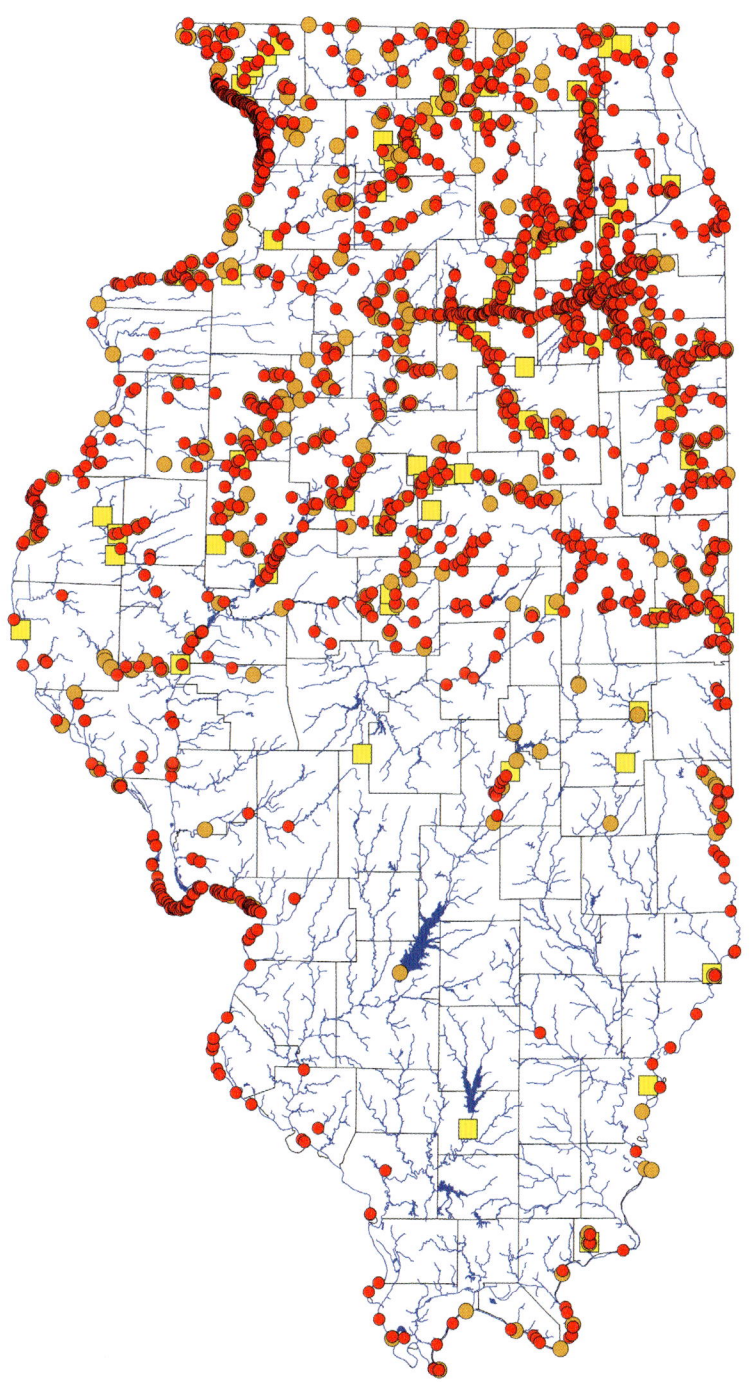

SUNFISHES (Centrarchidae)

313

SPOTTED BASS

Micropterus punctulatus
(Rafinesque 1819)

IDENTIFICATION The Spotted Bass has an elongate body and a large mouth, with the upper jaw extending to under the posterior half of the eye. The first dorsal fin is of nearly uniform height throughout, and the first and second dorsal fins are broadly joined at the base. Adults have rows of small, black spots on the lower side, a black stripe (or series of partly joined blotches) along the midside, and a black caudal spot (darkest on young). The body is light gold-green above with dark olive mottling and yellow-white below. The young have a 3-colored (yellow, black, white edge) caudal fin. The iris of the eye is often red in both juveniles and adults. There is a patch of teeth on the tongue. There are usually 61–73 lateral scales, 21–28 scales around the caudal peduncle, 9–10 dorsal spines, 12–13 dorsal rays, 3 anal spines, and 9–11 anal rays. To 24 in. (61 cm).

SIMILAR SPECIES The Largemouth Bass, *M. salmoides*, lacks rows of black spots on the lower side, has a deep notch between the dorsal fins, usually no teeth on the tongue, and a 2-colored (white, black edge) caudal fin on young. The Smallmouth Bass, *M. dolomieu*, lacks rows of black spots along the lower side and a black stripe along the midside and has no teeth on the tongue.

HABITAT The Spotted Bass occupies clear, gravel-bottomed runs and flowing pools of creeks and small to medium rivers. There are a few recent records from impoundments (e.g., Rend Lake) that are probably the result of purposeful introduction.

DISTRIBUTION IN ILLINOIS The Spotted Bass is native to the Wabash and Ohio River basins of Illinois, as documented by Forbes and Richardson (1909). It probably also is native to the Clear Creek system in Union and Alexander Counties (Smith 1979). Contemporary-era records of the Spotted Bass in the Mississippi River and a few of its tributaries and at 3 localities in northeastern Illinois are likely the result of stocking in Illinois and in adjacent Missouri (Pflieger 1997).

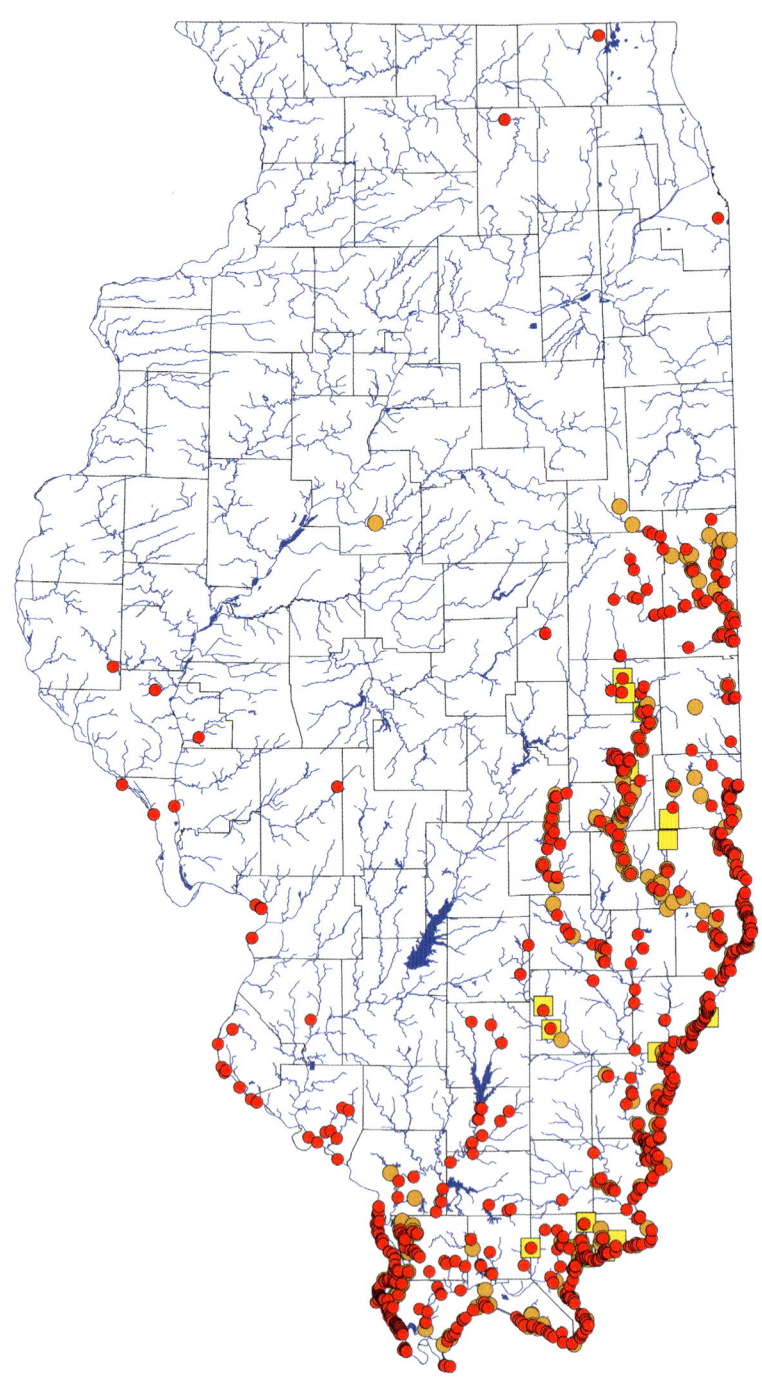

LARGEMOUTH BASS

Micropterus salmoides (Lacepède 1802)

IDENTIFICATION The Largemouth Bass has an elongate body and large mouth with the upper jaw extending past the eye in adults. The first dorsal fin is high at the middle and low at rear, with the first (spinous) and second (soft) dorsal fins barely joined at the base. The body is silver to brassy green (brown in dark water) above with dark olive mottling, a broad, black stripe (often broken into a series of blotches) along the side, and scattered black specks on the lower side; the underside is white. The iris of the eye is brown to red, and the fins are clear to yellow-olive to dusky black. Dark brown lines radiate from the snout and the back of the eye to the edge of the opercle. There is usually no patch of teeth on the tongue. There are usually 58–73 lateral scales, 9–11 dorsal spines, 3 anal spines, and 8 rakers on the first gill arch. To 38 in. (97 cm).

SIMILAR SPECIES The other 2 species of *Micropterus* in Illinois have more confluent dorsal fins and the upper jaw extending to or barely past the eye, juveniles have a 3-colored (yellow, black, white edge) caudal fin. The Spotted Bass, *M. punctulatus*, has rows of small, black spots on the lower side and a patch of teeth on the tongue. The Smallmouth Bass, *M. dolomieu*, usually has 8–16 dark brown bars and bronze specks on a yellow-brown to olive-green side.

HABITAT The Largemouth Bass is widespread in lakes, ponds, swamps, impoundments, and backwaters and pools of creeks of small to large rivers.

DISTRIBUTION IN ILLINOIS The Largemouth Bass is statewide in occurrence as it was in previous surveys and found in almost every water body in Illinois, including the open waters of Lake Michigan. The species has been stocked as a sport fish for decades.

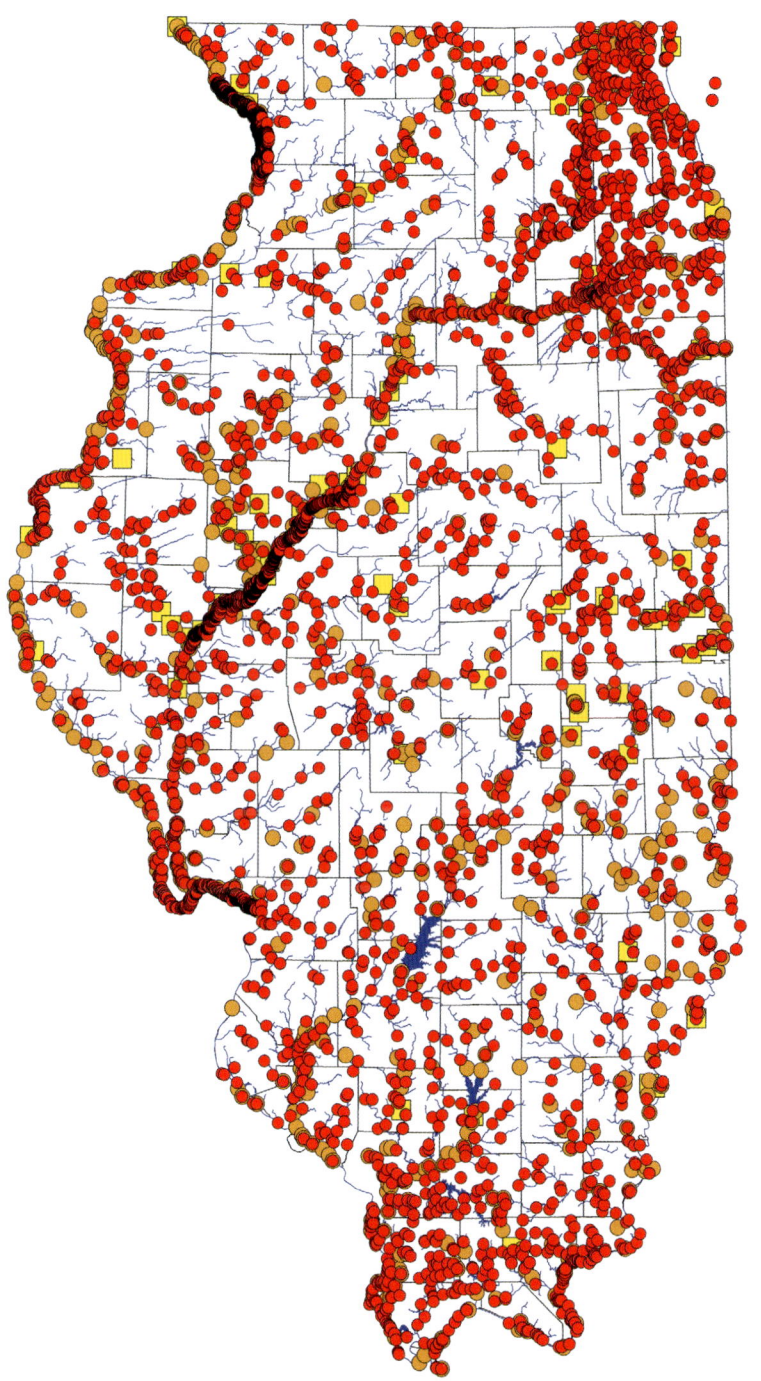

SUNFISHES (Centrarchidae)

315

WHITE CRAPPIE
Pomoxis annularis Rafinesque 1818

IDENTIFICATION The White Crappie has a deep, extremely compressed body with a very long predorsal region that is arched but with a sharp dip over the eye. The dorsal-fin base is shorter than the distance from the eye to the dorsal-fin origin and about as long as the anal-fin base. The mouth is large, and the upper jaw extends under the eye. The back and upper side are gray-green, and the side is silver with 6–9 dusky chainlike bars (widest at the top), black blotches, and green flecks. The underside is white. There are wavy black bands and spots on the dorsal, caudal, and anal fins. There are 6 dorsal spines, the first much shorter than the last, 14–15 dorsal rays, 6 anal spines, and 17–19 anal rays. To 21 in. (53 cm).

SIMILAR SPECIES The Black Crappie, *P. nigromaculatus,* has 7–8 dorsal spines and the dorsal-fin base is about the same length as the distance from the eye to the dorsal-fin origin.

HABITAT The White Crappie is found in sand- and mud-bottomed pools and backwaters of creeks and small to large rivers. It also is frequently encountered in lakes, ponds, and impoundments, often in turbid water.

DISTRIBUTION IN ILLINOIS The White Crappie is statewide in occurrence but somewhat more widespread in southern Illinois, as it was in the previous eras.

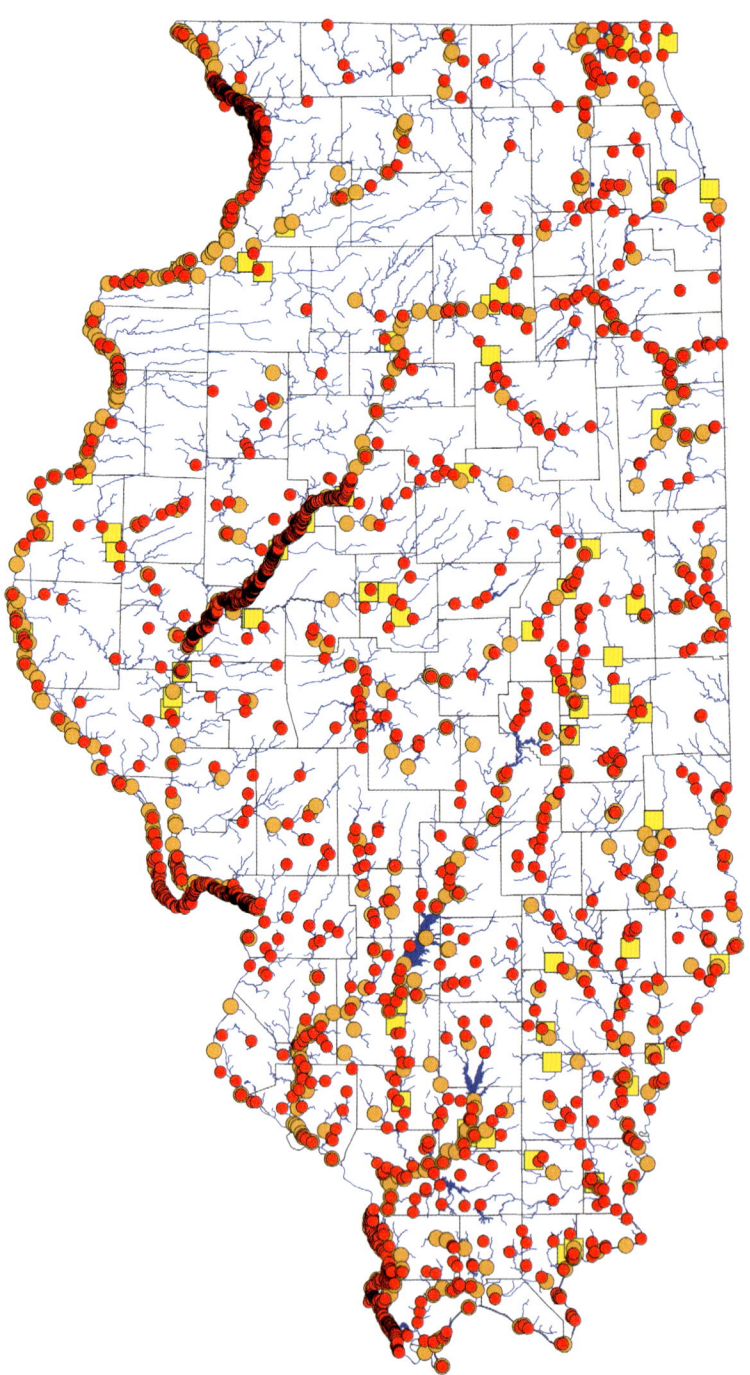

BLACK CRAPPIE
Pomoxis nigromaculatus (Lesueur 1829)

IDENTIFICATION The Black Crappie has a deep and extremely compressed body with a long predorsal region that is arched with a sharp dip over the eye. The dorsal-fin base is about as long as the distance from the eye to the dorsal-fin origin and about as long as the anal-fin base. The mouth is large with the upper jaw extending under the eye. The back and upper side are gray-green, and the side is silver-yellow with wavy black lines, black blotches, and green flecks. The underside is white. The dorsal, caudal, and anal fins have wavy black bands and spots. There are 7–8 dorsal spines, the first much shorter than the last, 15–16 dorsal rays, 6 anal spines, and 17–19 anal rays. To 19¼ in. (49 cm).

SIMILAR SPECIES The White Crappie, *P. annularis*, has 6 dorsal spines and usually dark bars on the side of the body, and the dorsal-fin base is shorter than the distance from the eye to the dorsal-fin origin.

HABITAT The Black Crappie inhabits lakes, ponds, sloughs, impoundments, and backwaters and pools of streams. It is most frequently encountered in medium to large rivers and usually is near vegetation over mud or sand in clear water.

DISTRIBUTION IN ILLINOIS The Black Crappie is statewide in occurrence as it was in the previous eras, but today is somewhat more widespread in northeastern and southern Illinois.

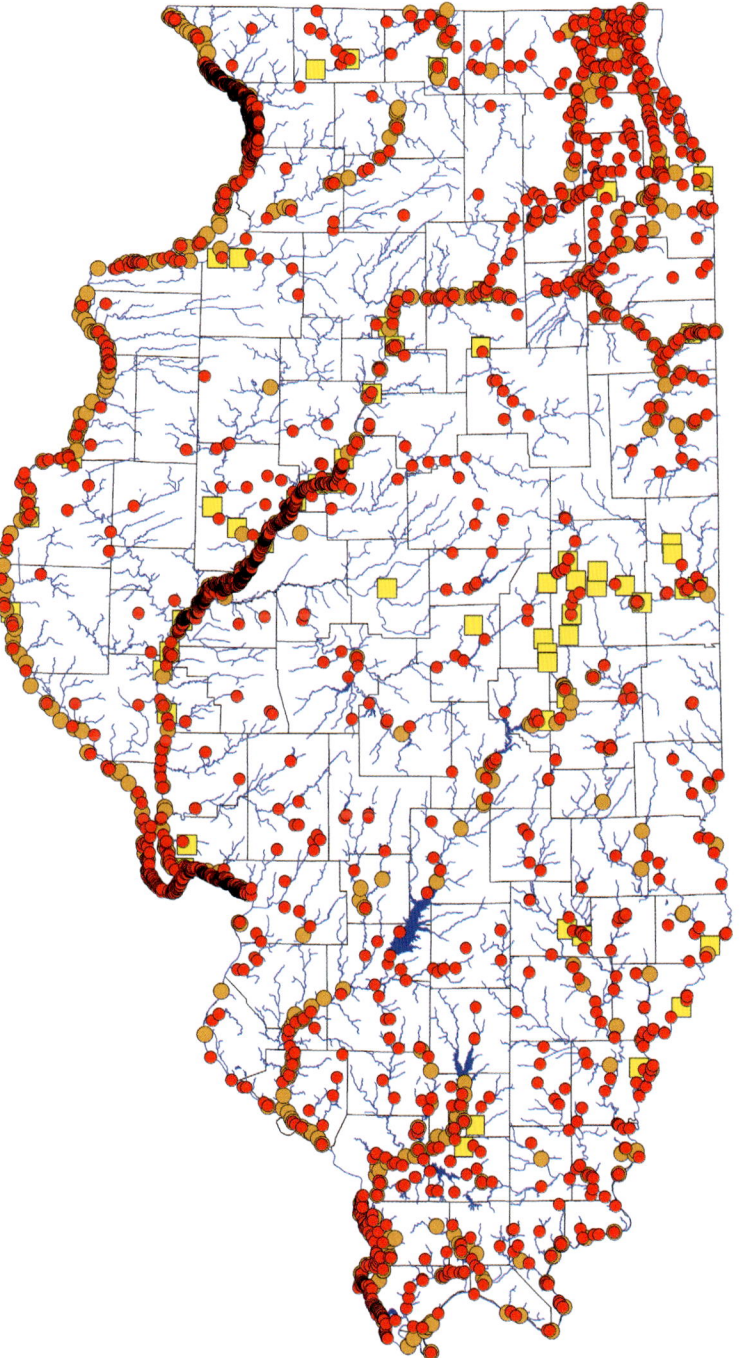

SUNFISHES (Centrarchidae)

317

PYGMY SUNFISHES

Family Elassomatidae Jordan 1877

Pygmy sunfishes, all in the genus *Elassoma*, are small, deep-bodied fishes endemic to the southeastern United States north to southern Illinois. They lack a lateral line, have an upturned mouth with a protruding lower jaw, a rounded caudal fin, 3–5 dorsal spines, and usually 3 anal spines. They have 2 dorsal fins, the first with spines and the second with rays. The dorsal fins are broadly joined and appear to be 1 fin. Breeding males are darkly pigmented overall and have brilliant iridescent colors, which makes them popular as aquarium pets. They occur almost exclusively in vegetated, swampy habitats or stream pools. Seven species are known, 1 of which occurs in Illinois.

BANDED PYGMY SUNFISH

Elassoma zonatum Jordan 1877

IDENTIFICATION The Banded Pygmy Sunfish has a small, deep body, a slightly superior mouth with a protruding lower jaw, a large eye, a short and deep caudal peduncle, and a rounded caudal fin. The back and side of the body are tan to dark brown, and the underside is cream to white. There are many black spots and mottling on the head and body and 7–12 blue-black bars on the side. There are often 1–2 large, black spots on the upper side of the body near the dorsal-fin origin and usually a black bar extending from the rear of the eye. Breeding males are black with iridescent blue or bronze horizontal bars on the side, green-gold flecks on the body, and a gold to iridescent bar under the eye. There are 4–5 dorsal spines, 9–10 dorsal rays, 5–6 anal rays, and 28–45 lateral scales. To 1¾ in. (4.7 cm).

SIMILAR SPECIES Juvenile Pirate Perch, *Aphredoderus sayanus*, have a larger head and mouth and lack mottling and iridescent coloration.

HABITAT The Banded Pygmy Sunfish occurs in heavily vegetated swamps, ditches, sloughs, and the margins of low-gradient creeks and small rivers.

DISTRIBUTION IN ILLINOIS As in the Smith era, the Banded Pygmy Sunfish is restricted to bottomland lakes and wetlands in the Mississippi and Ohio River basins in extreme southern Illinois, including the lower Big Muddy River and most of the Cache River system. Forbes and Richardson (1909) had records for this species from the Wabash River floodplain as far north as Lawrence County, but no recent records exist for the Wabash River basin in Illinois.

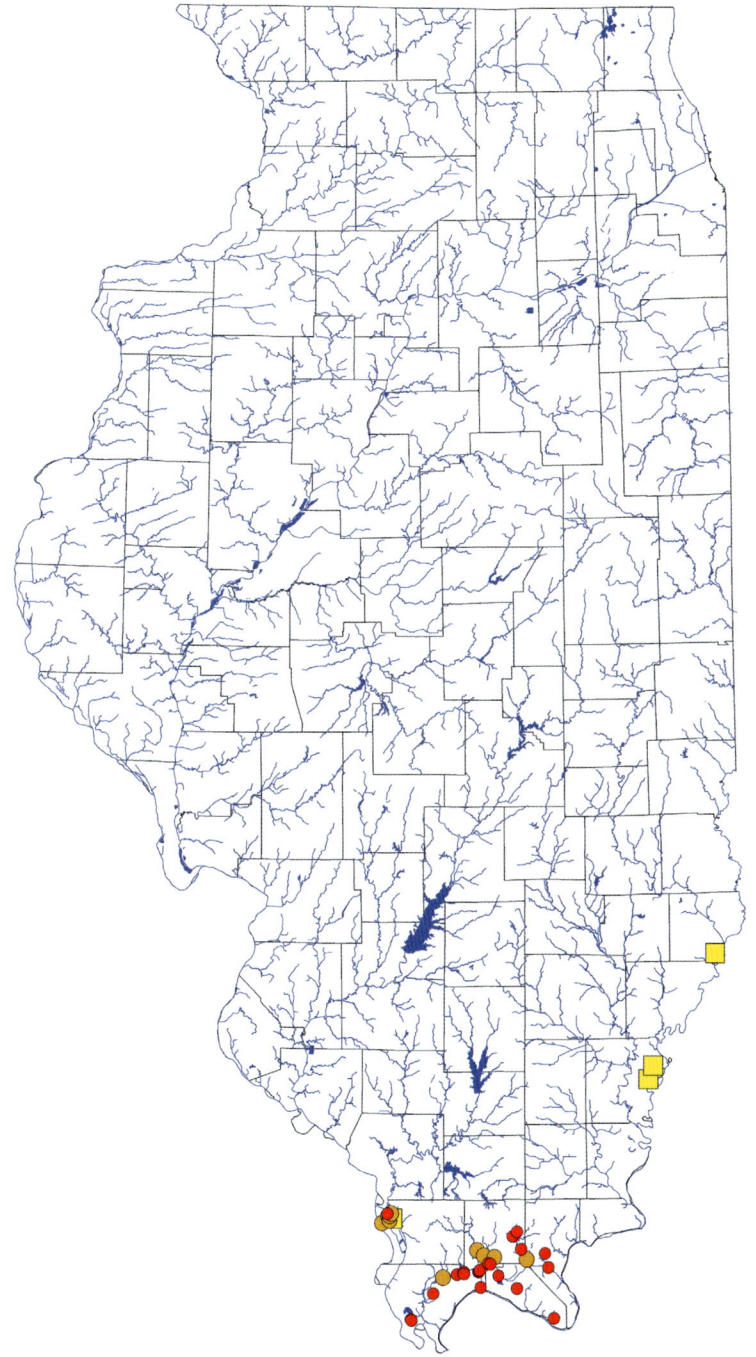

WHITE BASSES

Family Moronidae Jordan & Evermann 1896

White basses are compressed, deep-bodied fishes with 2 dorsal fins, the first with 7–10 spines and the second with 1 spine and 11–14 rays, a large, posteriorly directed spine on the gill cover, a small gill (pseudobranch) on the underside of the gill cover, and a strongly saw-toothed pre-opercle. They have a large mouth, ctenoid scales, thoracic pelvic fins, a complete lateral line, and 3 anal spines. They are silvery to brassy yellow, and 3 of the 6 species in the family have dark stripes on the side of the body. They are found in North America, Europe, and northern Africa. Two species are native to Illinois, and 2 species have entered Illinois as a result of human activities.

KEY TO THE WHITE BASSES (Moronidae)

1a Second spine of anal fin about same length as third spine and longer than anal fin base, extending beyond base of last anal fin ray when depressed; spinous dorsal and soft dorsal fins slightly connected by a membrane; no tooth patch on middle of tongue **2**

1b Second spine of anal fin distinctly shorter than third spine and anal fin base, not reaching to base of last anal fin ray when depressed; spinous dorsal and soft dorsal fins separate, or if bases in contact, not connected by a membrane; distinct patch of teeth present on middle of tongue **3**

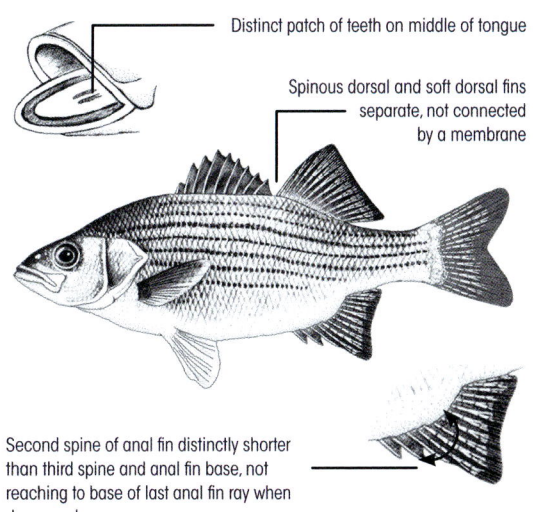

No tooth patch on middle of tongue

Spinous dorsal and soft dorsal fins slightly connected by a membrane

Second spine of anal fin about same length as third spine and longer than anal fin base, extending beyond base of last anal fin ray when depressed

Distinct patch of teeth on middle of tongue

Spinous dorsal and soft dorsal fins separate, not connected by a membrane

Second spine of anal fin distinctly shorter than third spine and anal fin base, not reaching to base of last anal fin ray when depressed

2a Distinct dark stripes along side usually sharply broken and offset above front of anal fin; scales in lateral series usually 50–55 . . . .**Page 325**

Yellow Bass (*Morone mississippiensis*)

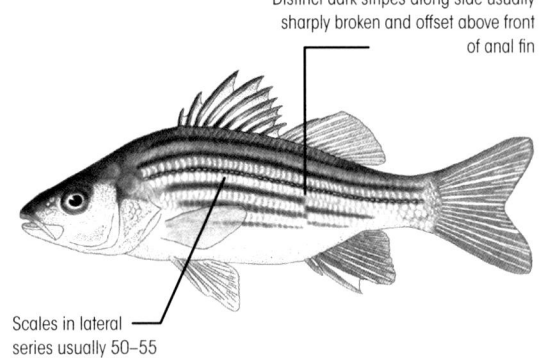

Distinct dark stripes along side usually sharply broken and offset above front of anal fin

Scales in lateral series usually 50–55

2b No dark stripes along side of adults, only faint stripes along side of juveniles; scales in lateral series usually 46–49 . . . .**Page 323**

White Perch (*Morone americana*)

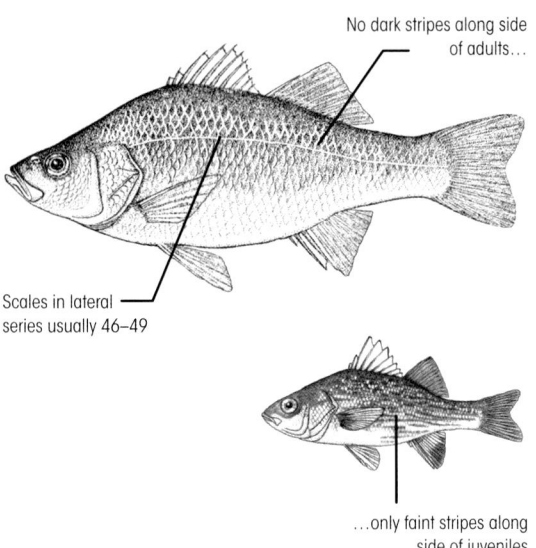

No dark stripes along side of adults…

Scales in lateral series usually 46–49

…only faint stripes along side of juveniles

3a Body deeper, its depth (A) usually going less than 3 times into standard length (B), the dorsal profile well arched; scales in lateral series usually 52–58; one patch of teeth or two narrowly divided patches of teeth on middle of tongue; juveniles never with "parr marks" . . . . .**Page 324**

White Bass (*Morone chrysops*)

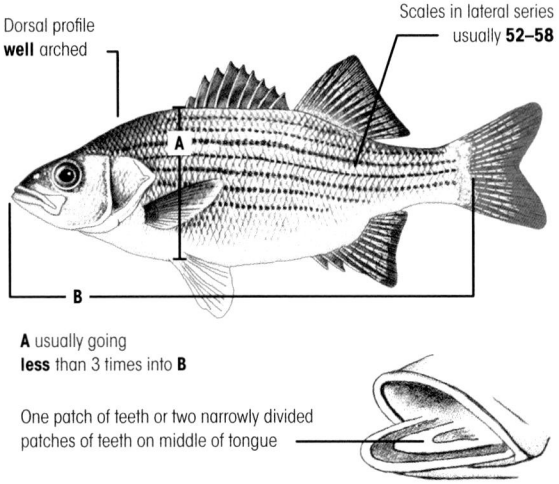

Dorsal profile **well** arched

Scales in lateral series usually **52–58**

A usually going **less** than 3 times into **B**

One patch of teeth or two narrowly divided patches of teeth on middle of tongue

3b Body more slender, its depth (A) usually going more than 3 times into standard length (B), the dorsal profile weakly arched; scales in lateral series usually 58–65; two patches of teeth on middle of tongue; juveniles (up to 150 mm total length) with narrow vertical "parr marks" along side . . . .**Page 326**

Striped Bass (*Morone saxatilis*)

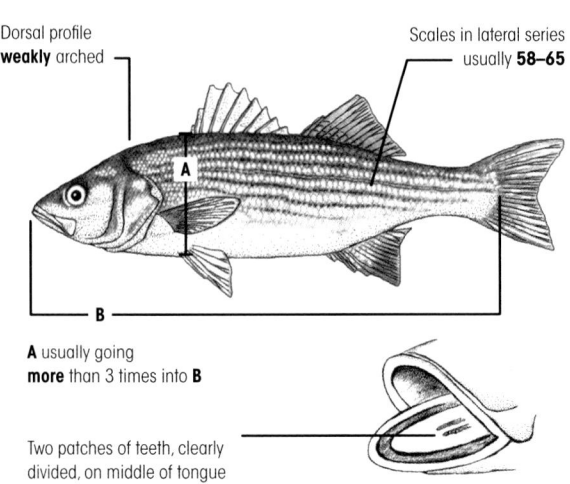

Dorsal profile **weakly** arched

Scales in lateral series usually **58–65**

A usually going **more** than 3 times into **B**

Two patches of teeth, clearly divided, on middle of tongue

WHITE PERCH

Morone americana (Gmelin 1789)

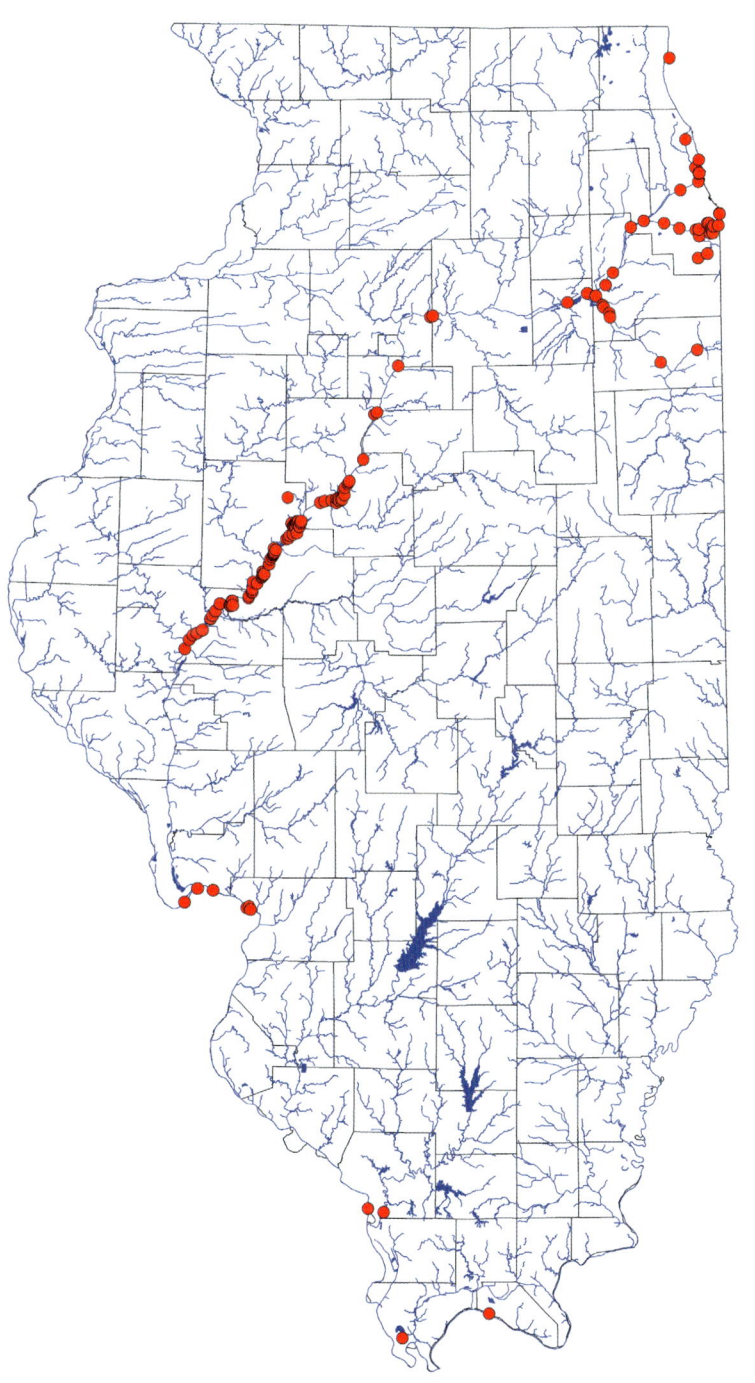

IDENTIFICATION The White Perch has a deep, compressed body that is deepest at the origin of the first dorsal fin and tapers to a narrow caudal peduncle and forked caudal fin. There are 2 dorsal fins, the first with usually 9–10 stout spines and the second with 1 spine and 10–14 rays. The mouth is large, pelvic fins are thoracic, and there are a large, posteriorly directed spine on the gill cover and a saw-toothed preopercle. The back and upper side are silver-green to brassy yellow, and the lower side and belly are white. The fins are clear to dusky. Juveniles have interrupted dark lines and bars on the side of the body; adults lack dark stripes and bars. Large adults have a silver-blue cast on the head and nape. The second anal spine is about as long as the third anal spine. There are no teeth on the tongue, 3 anal spines, and usually 9–10 anal rays. To 22¾ in. (58 cm).

SIMILAR SPECIES The White Bass, *M. chrysops*, and Striped Bass, *M. saxatilis*, have dark stripes on the side of the body, usually 11–13 anal rays, 1–2 patches of teeth on the rear of the tongue, and the second anal spine is distinctly shorter than the third spine. The Yellow Bass, *M. mississippiensis*, is yellow and has the stripes broken and offset on the lower side of the body.

HABITAT The White Perch lives in pools and other quiet-water areas of medium to large rivers. It is most often found over mud.

DISTRIBUTION IN ILLINOIS The White Perch was first recorded in Illinois in 1988 in Lake Michigan. It had gained access to the Great Lakes from Atlantic Slope basins, where it was native, through the Erie and Welland Canals (Savitz et al. 1989). It is now frequently encountered in the Illinois River and some of its upper tributaries and has reached the lower Mississippi and Ohio Rivers.

WHITE BASSES (Moronidae)

WHITE BASS

Morone chrysops (Rafinesque 1820)

IDENTIFICATION The White Bass has a deep, compressed body that is deepest between the dorsal fins and tapers to a narrow caudal peduncle and forked caudal fin. There are 2 dorsal fins, the first with usually 9 stout spines and the second with 1 spine and 11–14 rays. The mouth is large, pelvic fins are thoracic, and there are a large, posteriorly directed spine on the gill cover and a saw-toothed pre-opercle. The back and upper side of the body are silver-white to light blue-gray, and there are 4–7 dark gray-brown stripes on the side of the body. The lower side and belly are white. The eye is yellow. The dorsal, anal, and caudal fins are clear to gray, paired fins are clear to white. The second anal spine is distinctly shorter than the third anal spine. There are 1–2 patches of teeth on the rear of the tongue and 11–13 anal rays. To 17¾ in. (45 cm).

SIMILAR SPECIES The Striped Bass, *M. saxatilis*, is more slender and less strongly arched and reaches a much-larger size—to 6½ ft. (2 m). The White Perch, *M. americana*, is deepest at the origin of the first dorsal fin, lacks dark stripes on the side of the body, has no teeth on the tongue, and usually 9–10 anal rays, and the second anal spine is about as long as the third anal spine. The Yellow Bass, *M. mississippiensis*, is yellow, has the stripes broken and offset on the lower side of the body, no teeth on the tongue, and usually 9 anal rays, and the second anal spine is about as long as the third anal spine.

HABITAT The White Bass lives in open water and near woody debris or vegetation in lakes, ponds, and pools of creeks and small to large rivers. It is most frequently encountered in large rivers.

DISTRIBUTION IN ILLINOIS Smith (1979) noted that the White Bass is most common in large rivers and lakes but absent in most smaller streams and that it had disappeared from the Lake Michigan basin where Forbes and Richardson (1909) had recorded it. In the contemporary era, it is still frequently encountered in large rivers and much more widespread in smaller streams. It also is present again in the Lake Michigan basin.

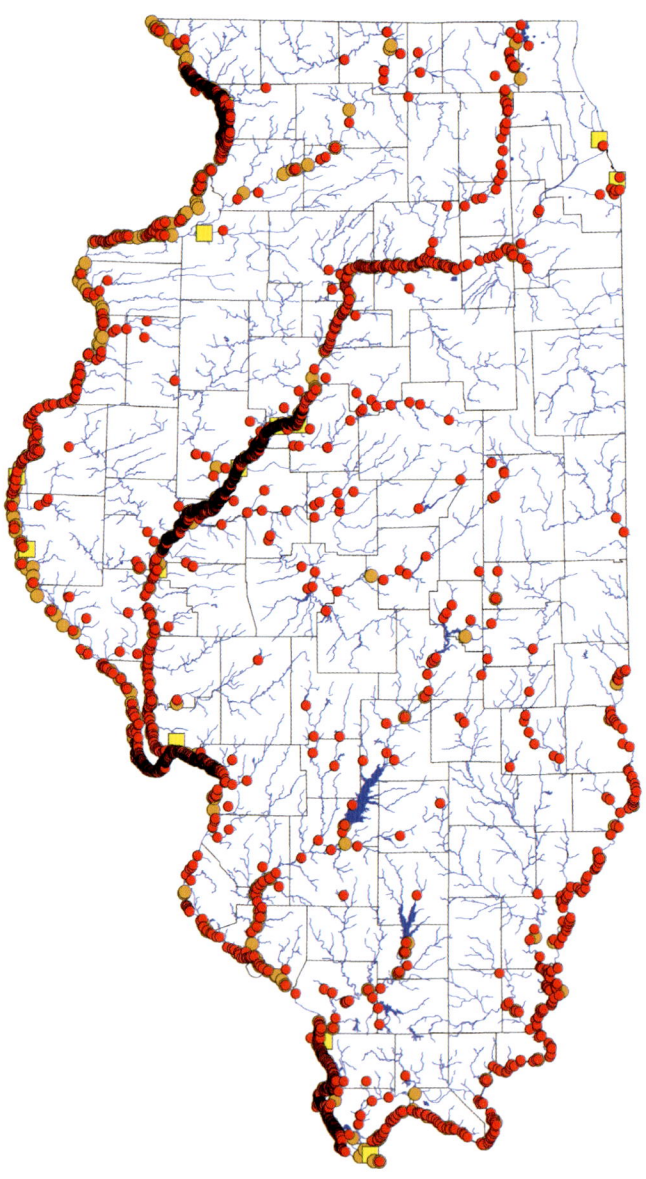

YELLOW BASS

Morone mississippiensis
Jordan & Eigenmann 1887

IDENTIFICATION The Yellow Bass has a deep, compressed body that is deepest under the first dorsal fin and tapers to a narrow caudal peduncle and forked caudal fin. There are 2 dorsal fins, the first with usually 9 stout spines and the second with 1 spine and 11–12 rays. The mouth is large, pelvic fins are thoracic, and there are a large, posteriorly directed spine on the gill cover and a saw-toothed preopercle. The back and upper side of the body are olive to silver-yellow, and the lower side and belly are white. There are 5–7 black stripes on the side of the body; the lower stripes are interrupted and offset. The fins are clear to dusky white. The second anal spine is about as long as the third anal spine. There are no teeth on the tongue and usually 9 anal rays. To 18 in. (46 cm).

SIMILAR SPECIES Other white basses, *Morone*, are silver-white and lack interrupted stripes on the side of the body.

HABITAT The Yellow Bass occupies lakes, ponds, and pools and backwaters of small to large rivers. It is most frequently encountered in medium to large rivers and usually is found in open water or near woody debris.

DISTRIBUTION IN ILLINOIS The Yellow Bass is much more widespread in the contemporary era than it was in previous eras. Forbes and Richardson (1909) found the species only in the northern two-thirds of Illinois, and Smith (1979) described the distribution as general in the lower Illinois River, upper Mississippi River, and the glacial lakes of northeastern Illinois. Today it occurs statewide.

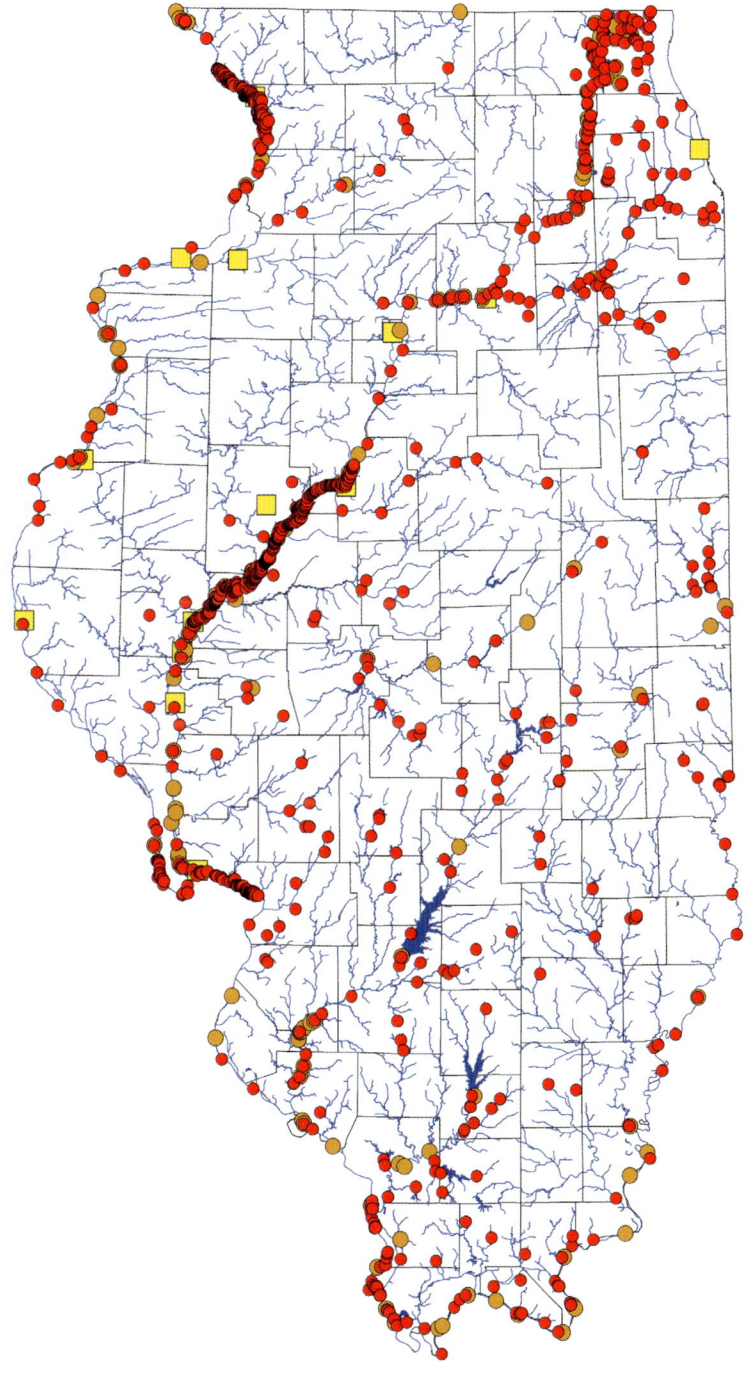

STRIPED BASS   *Morone saxatilis* (Walbaum 1792)

IDENTIFICATION The Striped Bass has a smoothly arched body that is deepest between the dorsal fins and tapers to a narrow caudal peduncle and forked caudal fin. There are 2 dorsal fins, the first with usually 8–11 stout spines and the second with 1 spine and 10–14 rays. The mouth is large, pelvic fins are thoracic, and there are a large, posteriorly directed spine on the gill cover and a saw-toothed preopercle. The back and upper side of the body are light olive to silver-blue or gray, often with a brassy yellow sheen. The fins are clear to dark gray. Large individuals have 6–9 dark gray stripes on the side of the body, white pelvic fins, and often white on the anal fin. Juveniles lack the dark stripes and often have dusky bars on the side of the body. The second anal spine is distinctly shorter than the third anal spine. There are 1–2 patches of teeth on the rear of the tongue and 9–13 (usually 11) anal rays. To 6½ ft. (2 m).

SIMILAR SPECIES The White Bass, *M. chrysops*, has a deeper body that is more strongly arched and reaches only 17¾ in. (45 cm). The White Perch, *M. americana*, lacks dark stripes on the side of the body, has no teeth on tongue, and the second anal spine about as long as the third anal spine. The Yellow Bass, *M. mississippiensis*, is yellow, has the stripes broken and offset on the lower side of the body, no teeth on the tongue, and usually 9 anal rays, and the second anal spine is about as long as the third anal spine.

HABITAT The Striped Bass occupies runs and flowing pools of medium to large rivers, lakes, and impoundments.

DISTRIBUTION IN ILLINOIS The Striped Bass first appeared in Illinois in 1974, presumably having entered the state from large impoundments in western Kentucky where it had been stocked (Smith 1979). In the contemporary era, it is stocked in Illinois by state agencies and is found as far north as extreme northern Illinois.

REMARKS Hybrids with the White Bass, *M. chrysops*, are frequently encountered where "Stripers" have been introduced.

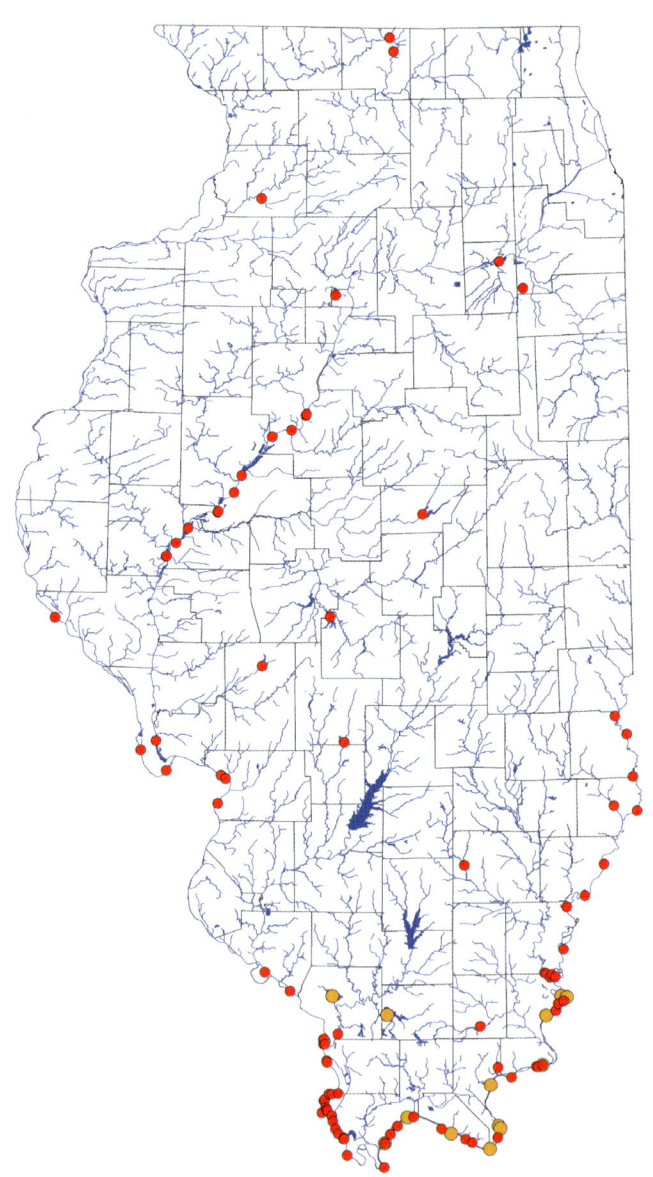

DRUMS AND CROAKERS

Family Sciaenidae Cuvier 1829

Drums and croakers have 2 dorsal fins; the first is short with 6–13 spines, and the second is considerably longer and has 1 spine and 20–35 rays. The lateralis system on the head consists of large, cavernous canals and pores, and the lateral line extends to the end of the caudal fin. There are 2 spines on the posterior edge of the opercle, thoracic pelvic fins and 1–2 anal spines. In most species the body is notably arched under the first dorsal fin and flattened below. Some species have a single barbel or a patch of small barbels on the chin. The names *drum* and *croaker* refer to the ability of these fishes to use the gas bladder as a resonating chamber. Most of the 210 species in the family occupy coastal areas of tropical and temperate oceans, although a few species live in fresh water—most notably in South America. One species is restricted to fresh waters of North America and found in Illinois.

FRESHWATER DRUM

Aplodinotus grunniens Rafinesque 1819

IDENTIFICATION The body of the Freshwater Drum is deep, highly arched near the origin of the first dorsal fin and flat below. The snout is rounded, and the mouth is subterminal. There are 2 dorsal fins—the first is relatively short and has usually 10 spines, and the second is longer with 29–32 rays. The back and side are silver-gray, the underside is white, and fins are clear to dusky gray. The outer pelvic ray is elongated and often bright white or yellow. The lateralis system on the head consists of large, cavernous canals and pores, and the lateral line extends to the end of the rounded caudal fin. To 35 in. (89 cm).

SIMILAR SPECIES No other fish in Illinois has a deep, highly arched body and 2 dorsal fins—the first short and with spines and the second longer with 29–32 rays.

HABITAT The Freshwater Drum lives on the rocky and sandy bottoms of lakes, creeks, and small to large rivers. It is usually found in large rivers.

DISTRIBUTION IN ILLINOIS The Freshwater Drum has the greatest latitudinal distribution of any North American freshwater fish, ranging from the Hudson Bay basin to Guatemala. It occurs statewide in Illinois except for its notable absence in the upper Iroquois and upper Vermilion River (Wabash) basins in east-central Illinois. Earlier eras had essentially the same distribution.

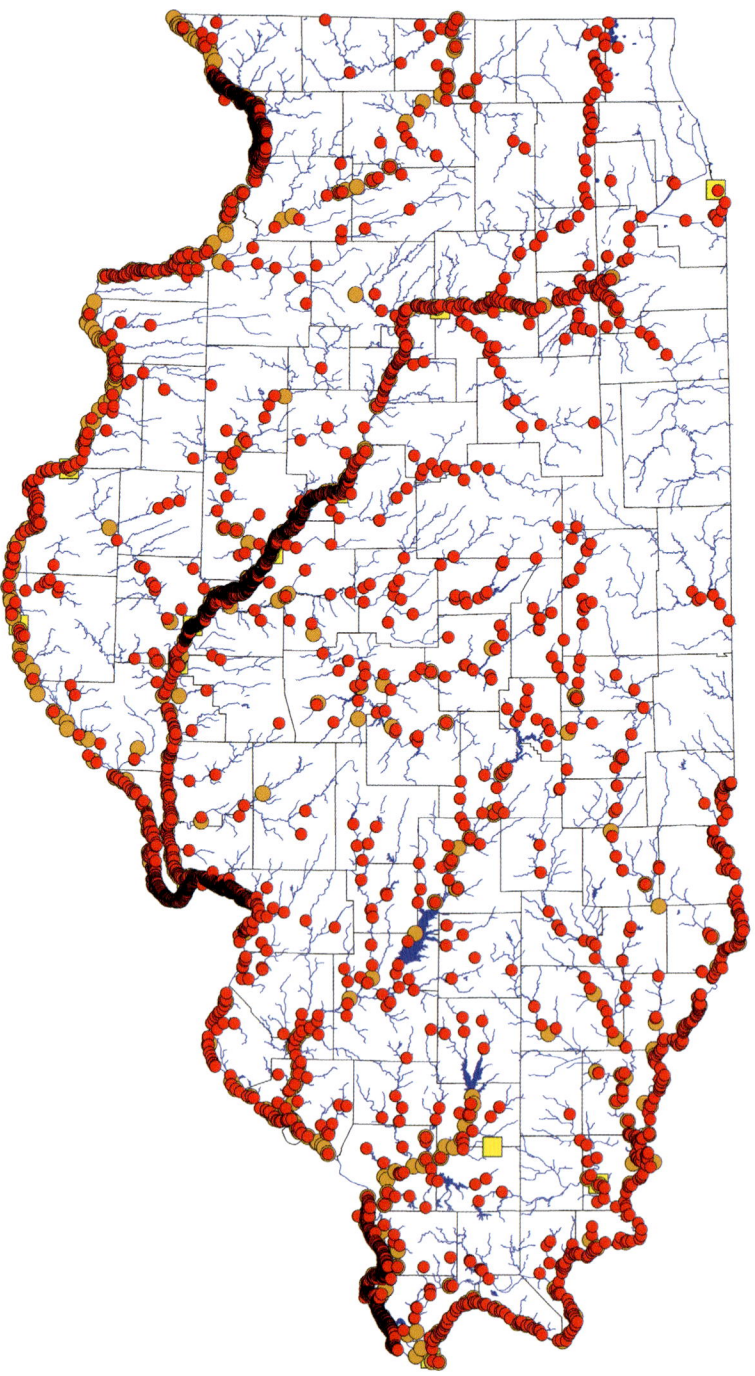

DARTERS AND PERCHES

Percidae Rafinesque 1815

Percidae is the second-most diverse family (after Leuciscidae) of freshwater fishes in North America and in Illinois. Percids have 2 dorsal fins that are separate or slightly joined (broadly joined in some Eurasian species); the first fin has spines, and the second has flexible branched rays. The pelvic fins are located on the thorax and have 1 spine and 5 rays.

Of the 220 species of percids native to North America, 31 are found in Illinois.[2] All but 3 of these species, Walleye, Sauger, and Yellow Perch, are darters. Darters are small—most are less than 4 in. (10 cm). Most have lost the gas bladder and dart about on bottoms of streams and lakes feeding on small invertebrates. Some darters are drab, but most are colorful, especially as breeding males. Characteristics useful in identifying darters include color patterns, shape and completeness of the lateral line and head canals, connection of branchiostegal membranes, presence or absence of a premaxillary frenum, and number of anal spines.

---

[2] As this book was going to press, the Tippecanoe Darter (*Etheostoma tippecanoe*) and Streamline Chub (*Erimystax dissimilis*) were recorded in the Vermilion River, Vermilion County, from Danville downstream to its confluence with the Wabash River. The Tippecanoe Darter is fairly widespread in the Wabash River basin in Indiana.

1a Rear margin of preopercle (bone just ahead of gill cover) moderately or strongly serrate (saw-toothed); branchiostegal rays (slender bones in membrane along lower margin of gill cover) usually 7, rarely 8; adult body size larger than 7 inches (17.8 cm). **2**

1b Rear margin of preopercle smooth (weakly serrate in Dusky Darter, see couplet 13a); branchiostegal rays usually 6, occasionally 5; size smaller, adults rarely exceed 7 inches (17.8 cm) **4**

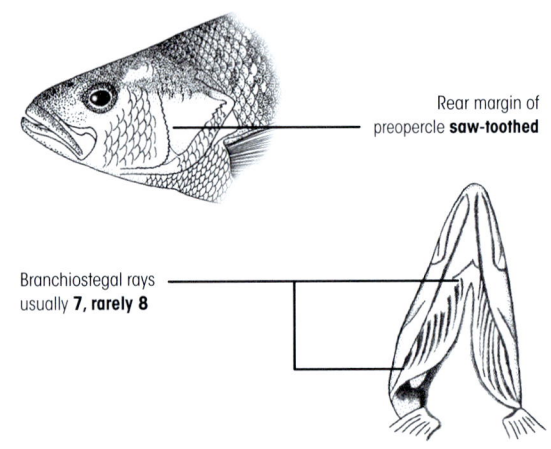

Rear margin of preopercle **saw-toothed**

Branchiostegal rays usually **7, rarely 8**

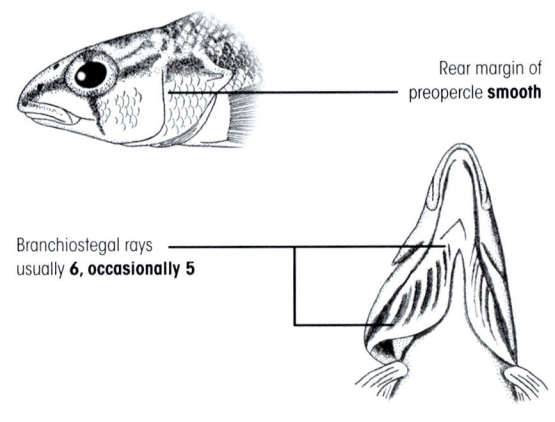

Rear margin of preopercle **smooth**

Branchiostegal rays usually **6, occasionally 5**

2a Jaws and roof of mouth without large, prominent canine teeth; anal fin with 6–9 soft rays; back and sides crossed by 5–9 dark vertical bars . . . . **Page 365**

2b Jaws and roof of mouth with large, prominent canine teeth; anal fin with 11–14 soft rays; back and sides without bars or crossed by several oblique saddles or blotches **3**

Yellow Perch (*Perca flavescens*)

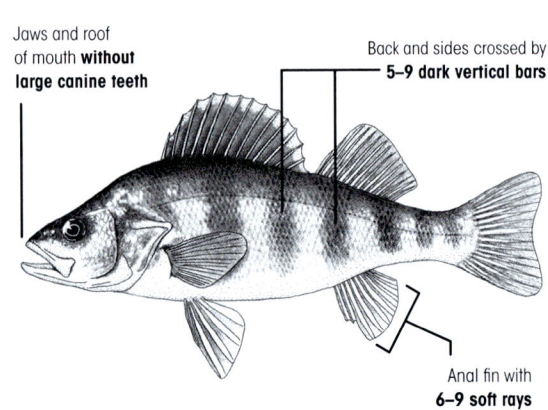

Jaws and roof of mouth **without large canine teeth**

Back and sides crossed by **5–9 dark vertical bars**

Anal fin with **6–9 soft rays**

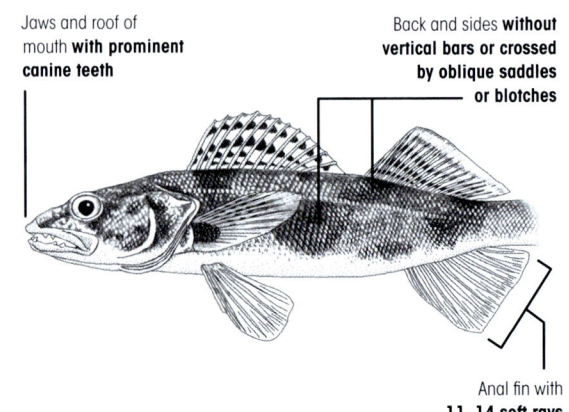

Jaws and roof of mouth **with prominent canine teeth**

Back and sides **without vertical bars or crossed by oblique saddles or blotches**

Anal fin with **11–14 soft rays**

3a Membrane of first dorsal fin with dark streaks or blotches, last 1 or 2 membranes with black blotch near base; back and sides without conspicuous dark blotches; pyloric caecae (fingerlike blind sacs between stomach and intestine) 3 . . . . **Page 375**

3b Membrane of first dorsal fin with distinct half-moon-shaped spots, last 1 or 2 membranes without black blotch near base; back with 4 dark saddles and side with 2 midlateral blotches conspicuous against paler background; pyloric caecae 2–6 . . . . **Page 374**

Walleye (*Sander vitreum*)

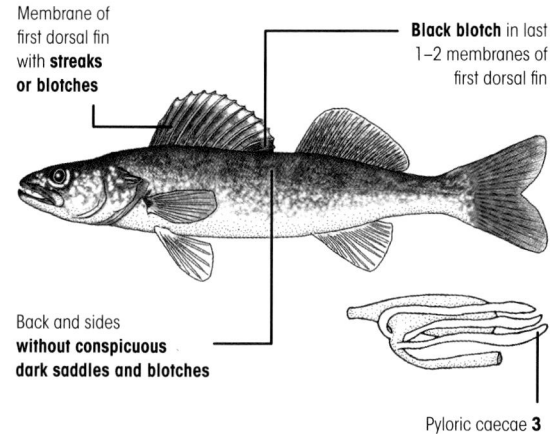

Membrane of first dorsal fin with **streaks or blotches**

Black blotch in last 1–2 membranes of first dorsal fin

Back and sides **without conspicuous dark saddles and blotches**

Pyloric caecae **3**

Sauger (*Sander canadense*)

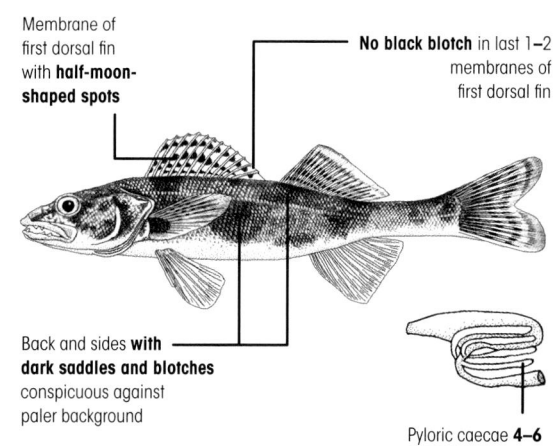

Membrane of first dorsal fin with **half-moon-shaped spots**

No black blotch in last 1–2 membranes of first dorsal fin

Back and sides **with dark saddles and blotches** conspicuous against paler background

Pyloric caecae **4–6**

4a Body slender, its depth (A) usually going 7 times or more into standard length (B); anal fin with 1 spine; midline of belly without scales; flesh translucent when alive **5**

4b Body deeper, its depth (A) going less than 7 times into standard length (B); anal fin usually with 2 spines, occasionally 1; midline of belly usually scaled; flesh opaque when alive **7**

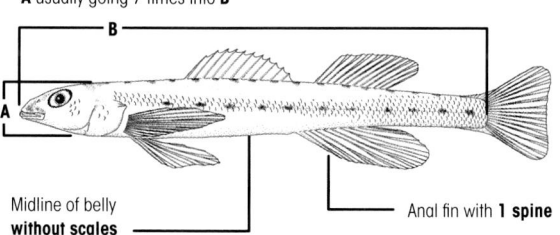

A usually going 7 times into **B**

B

A

Midline of belly **without scales**

Anal fin with **1 spine**

A usually going less than 7 times into **B**

B

A

Midline of belly **with scales**

Anal fin with **1–2 spines**

5a Caudal fin deeply forked; back with 3–4 dark saddles that extend obliquely forward onto upper sides; anal fin with 12–14 soft rays; second dorsal fin with 12–16 rays; lateral line scales more than 80 . . . .**Page 347**

Crystal Darter (*Crystallaria asprella*)

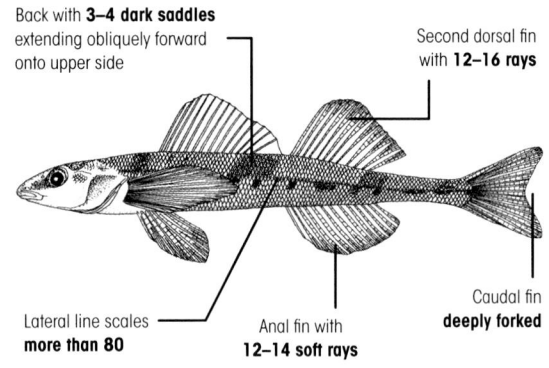

Back with **3–4 dark saddles** extending obliquely forward onto upper side

Second dorsal fin with **12–16 rays**

Lateral line scales **more than 80**

Anal fin with **12–14 soft rays**

Caudal fin **deeply forked**

5b Caudal fin emarginate (only shallowly forked); back with series of dark spots that do not extend onto sides; anal fin with 7–11 soft rays; second dorsal fin with 8–12 rays; lateral line scales usually fewer than 80 . . . .**6**

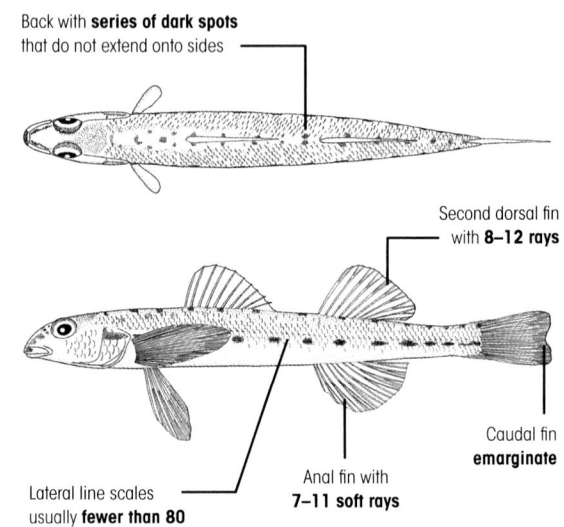

Back with **series of dark spots** that do not extend onto sides

Second dorsal fin with **8–12 rays**

Caudal fin **emarginate**

Anal fin with **7–11 soft rays**

Lateral line scales usually **fewer than 80**

6a Midside of body with row of oval or round blotches; spine near rear margin of gill cover short and less prominent, its length about equal to its width at base; transverse scale rows (counted downward at 45° angle from origin of second dorsal fin to base of anal fin) usually 8–16 . . . . . .
. . . .**Page 346**

Eastern Sand Darter (*Ammocrypta pellucida*)

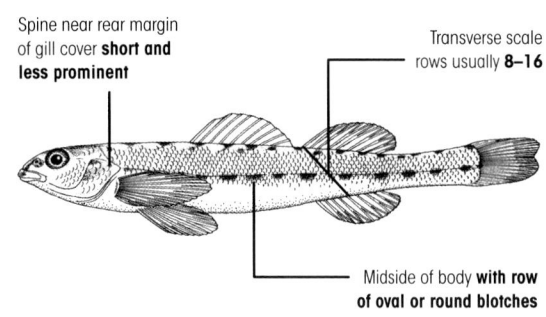

Spine near rear margin of gill cover **short and less prominent**

Transverse scale rows usually **8–16**

Midside of body **with row of oval or round blotches**

6b Midside of body without row of blotches (although a line of melanophores [small black specks] may be present); spine near rear margin of gill cover long and prominent; its length much greater than its width at base; transverse scale rows usually 3–6 . . . .**Page 345**

Western Sand Darter (*Ammocrypta clara*)

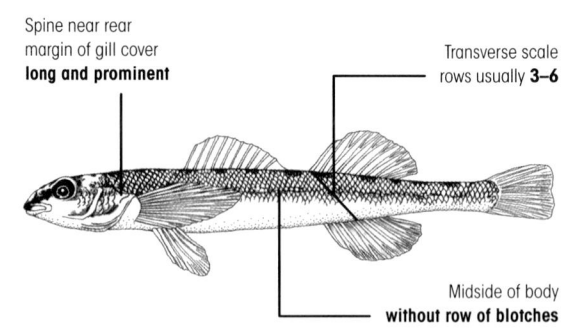

Spine near rear margin of gill cover **long and prominent**

Transverse scale rows usually **3–6**

Midside of body **without row of blotches**

7a One or more enlarged and modified scales between pelvic fins; midline of belly usually without scales or with a series of enlarged and modified scales (with scales of normal size and shape in some females); lateral line always complete, extending to base of caudal fin; anal fin often nearly as large or larger than second dorsal fin. **8**

Percina

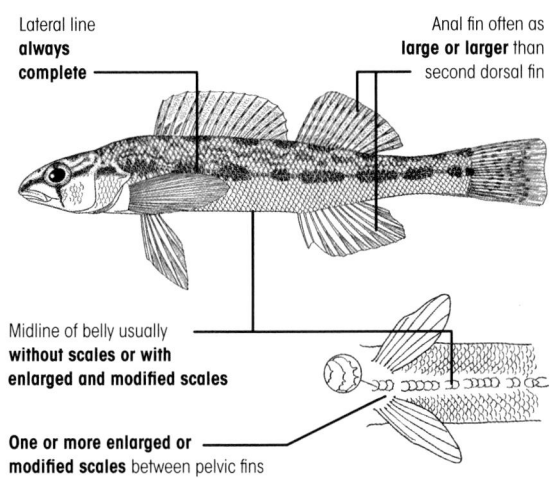

Lateral line **always complete**

Anal fin often as **large or larger** than second dorsal fin

Midline of belly usually **without scales or with enlarged and modified scales**

One or more enlarged or modified scales between pelvic fins

7b No enlarged and modified scales between pelvic fins; midline of belly with at least a few scales of normal size and shape; lateral line incomplete in several species; anal fin usually smaller than second dorsal fin. **15**

Etheostoma

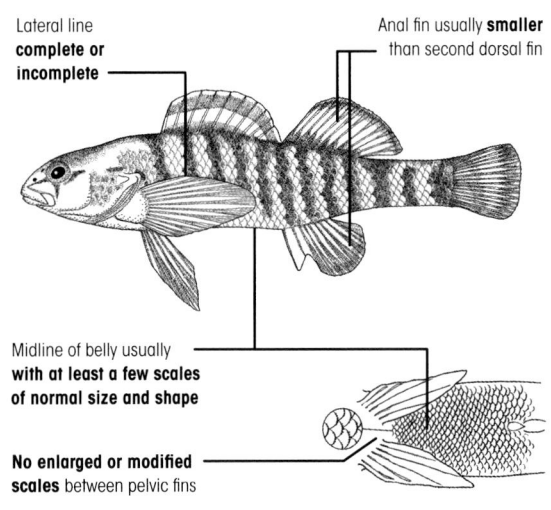

Lateral line **complete or incomplete**

Anal fin usually **smaller** than second dorsal fin

Midline of belly usually **with at least a few scales of normal size and shape**

No enlarged or modified scales between pelvic fins

8a Mouth overhung by distinctly conical snout, with a whitish pad at the tip; total dorsal fin elements 30 or more; color pattern consisting of 15–20 narrow, vertical brownish bars that are continuous across the back. . . . .**Page 366**

Logperch (*Percina caprodes*)

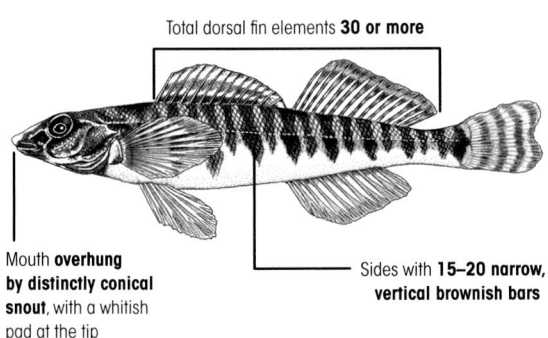

Total dorsal fin elements **30 or more**

Mouth **overhung by distinctly conical snout**, with a whitish pad at the tip

Sides with **15–20 narrow, vertical brownish bars**

8b Mouth pointed to blunt but not overhung by conical snout; total dorsal fin elements 29 or fewer; color pattern consisting of a series of blackish blotches, or of fewer than 12 vertical bars **9**

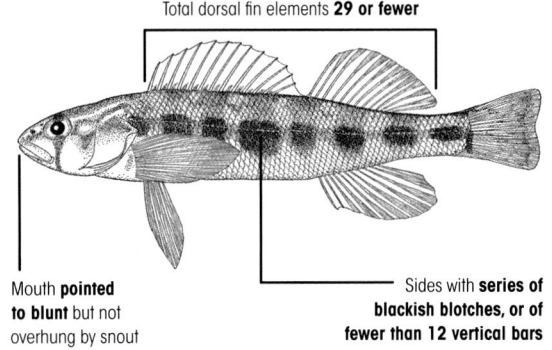

Total dorsal fin elements **29 or fewer**

Mouth **pointed to blunt** but not overhung by snout

Sides with **series of blackish blotches, or of fewer than 12 vertical bars**

9a Distance from front of upper lip to junction of gill covers (A) equal to or greater than distance from junction of gill covers to back of pelvic fin base (B); back and sides with 11–13 vague greenish or brownish blotches; membrane of first dorsal fin of male with a submarginal row of orange spots; snout distinctly elongated. . . . .**Page 370**

9b Distance from front of upper lip to junction of gill cover (A) much less than distance from junction of gill cover to back of pelvic fin base (B); back and sides with usually 5–10 blackish blotches; membrane of first dorsal fin of male without a submarginal row of spots; snout more rounded, not distinctly elongate **10**

Slenderhead Darter (*Percina phoxocephala*)

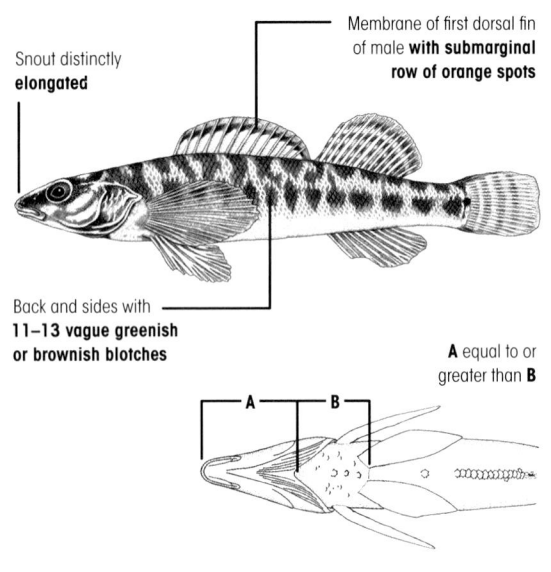

Snout distinctly **elongated**

Membrane of first dorsal fin of male **with submarginal row of orange spots**

Back and sides with **11–13 vague greenish or brownish blotches**

A equal to or greater than **B**

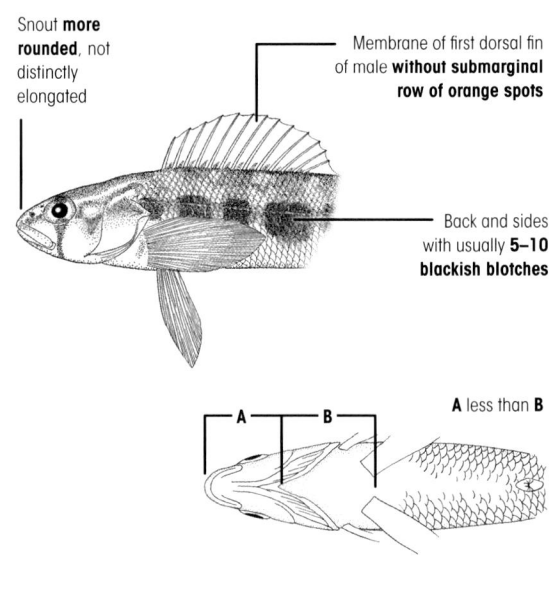

Snout **more rounded**, not distinctly elongated

Membrane of first dorsal fin of male **without submarginal row of orange spots**

Back and sides with usually **5–10 blackish blotches**

A less than **B**

10a Upper lip separated from snout at midline by a continuous deep groove, only rarely crossed by a narrow bridge of skin (frenum). **11**

10b Upper lip joined to snout at midline by a bridge of skin (frenum). **13**

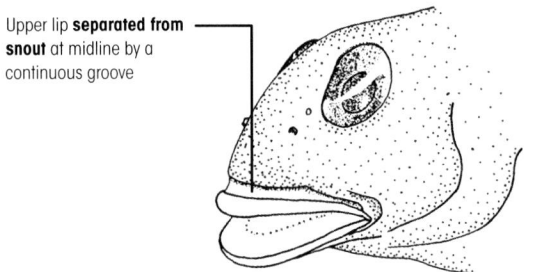

Upper lip **separated from snout** at midline by a continuous groove

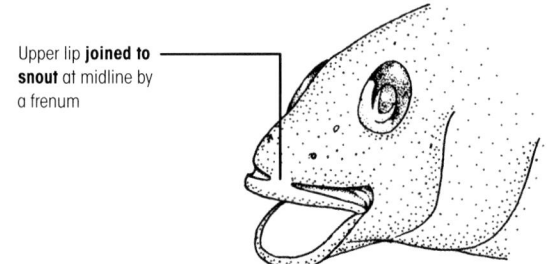

Upper lip **joined to snout** at midline by a frenum

11a Anal fin with 7–8 soft rays; anal fin of male similar in size to that of female. . . . .**Page 367**

11b Anal fin with usually 10–12 soft rays; anal fin of adult male greatly enlarged, extending nearly to caudal fin **12**

Channel Darter (*Percina copelandi*)

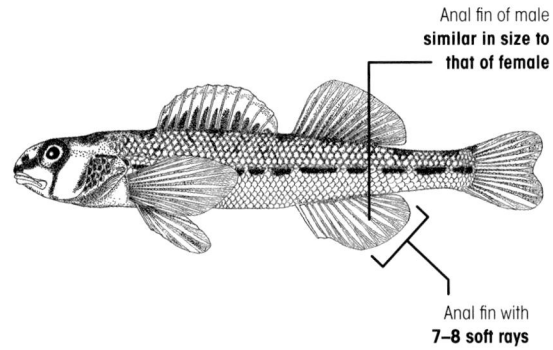

Anal fin of male
**similar in size to
that of female**

Anal fin with
7–8 soft rays

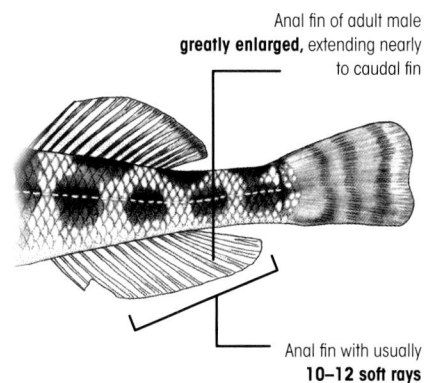

Anal fin of adult male
greatly enlarged, extending nearly
to caudal fin

Anal fin with usually
10–12 soft rays

12a First dorsal fin with a distinct black blotch at base of first membrane and another at bases of last 3 membranes; cheek fully scaled. . . . .**Page 372**

12b First dorsal fin without distinct black blotches; cheek without scales or with a few scattered scales . . . . .**Page 373**

River Darter (*Percina shumardi*)

Stargazing Darter (*Percina uranidea*)

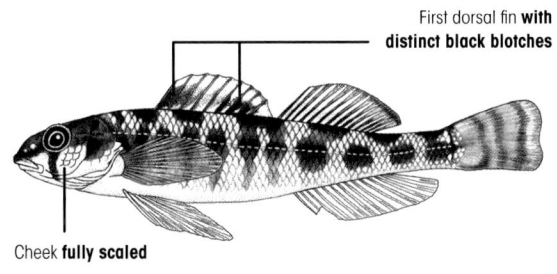

First dorsal fin **with
distinct black blotches**

Cheek **fully scaled**

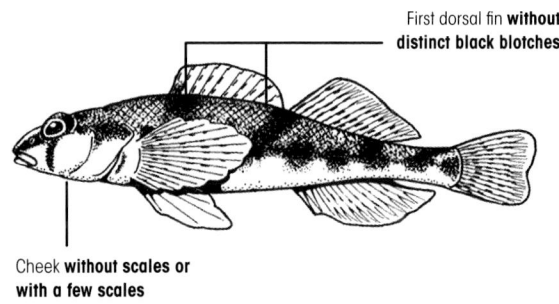

First dorsal fin **without
distinct black blotches**

Cheek **without scales or
with a few scales**

13a Gill covers broadly connected by membrane across throat; base of caudal fin with a vertical row of 3 black spots (lower 2 usually connected); teardrop short or absent . . . . **Page 371**

13b Gill covers not connected by membrane across throat; base of caudal fin without spots or with a single black spot; teardrop present. **14**

Dusky Darter (*Percina sciera*)

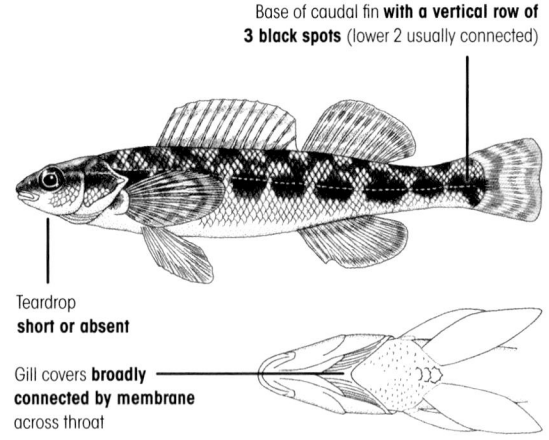

Base of caudal fin **with a vertical row of 3 black spots** (lower 2 usually connected)

Teardrop **short or absent**

Gill covers **broadly connected by membrane** across throat

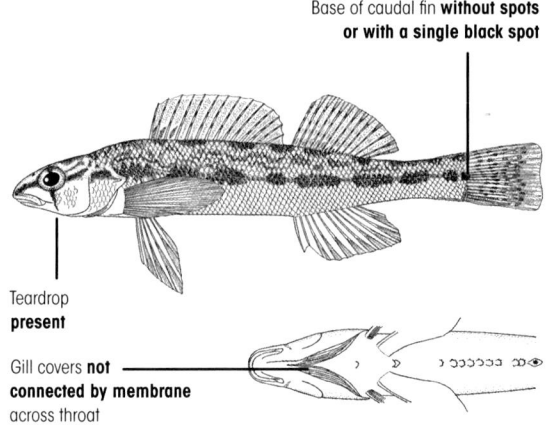

Base of caudal fin **without spots or with a single black spot**

Teardrop **present**

Gill covers **not connected by membrane** across throat

14a Dark blotches on side confined to midside and somewhat connected along lateral line; first dorsal fin with dark blotches at base of first 2–3 membranes; base of caudal fin with 1 discrete black spot; cheek fully scaled; adult male without bright colors on body and fins . . . **Page 369**

Blackside Darter (*Percina maculata*)

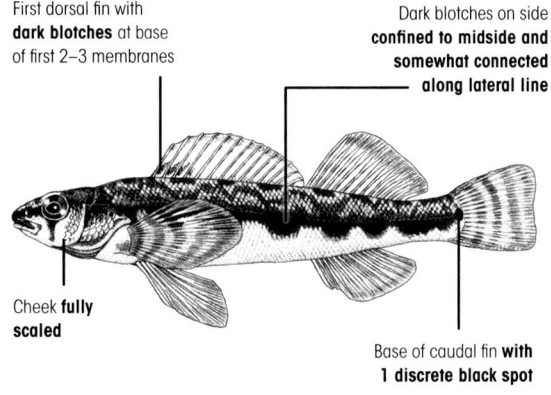

First dorsal fin with **dark blotches** at base of first 2–3 membranes

Dark blotches on side **confined to midside and somewhat connected along lateral line**

Cheek **fully scaled**

Base of caudal fin **with 1 discrete black spot**

14b Dark blotches on side vertically elongated and continuous with crossbars on back; first dorsal fin uniformly pigmented at bases of membranes; base of caudal fin without a discrete black spot; cheek without scales; adult male with bright orange and blue colors on body and fins . . . . **Page 368**

Gilt Darter (*Percina evides*)

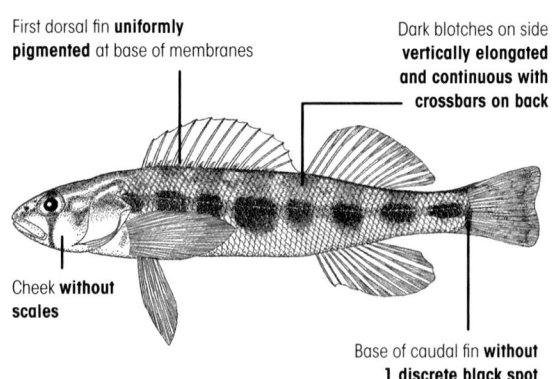

First dorsal fin **uniformly pigmented** at base of membranes

Dark blotches on side **vertically elongated and continuous with crossbars on back**

Cheek **without scales**

Base of caudal fin **without 1 discrete black spot**

15a Upper lip completely separated from snout by continuous deep groove. **16**

15b Upper lip joined to snout by a bridge of skin (frenum). . **19**

Upper lip **completely separated from snout** by continuous deep groove —

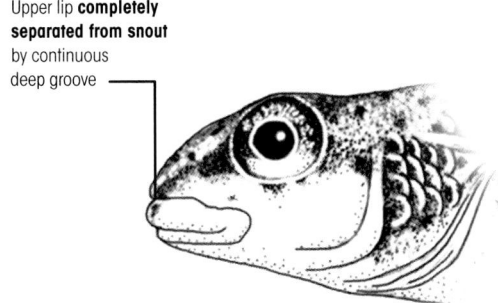

Upper lip **joined to snout** by a bridge of skin (frenum) —

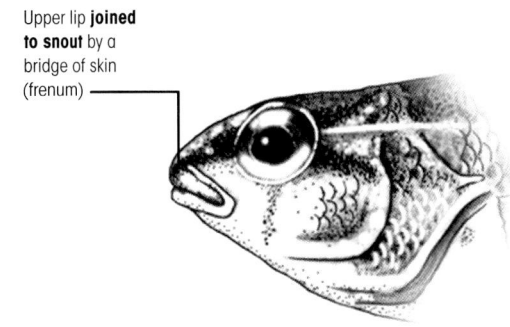

16a Anal fin with 2 thick and stiff spines; gill covers broadly connected by membrane across throat; sides of body with prominent W- or U-shaped markings, bars, or large irregular blotches. **17**

16b Anal fin with 1 thin and flexible spine; gill covers narrowly connected by membrane across throat; sides of body with many small flecks or W- and X-shaped markings **18**

Sides of body with **prominent W- or U-shaped markings, bars, or large irregular blotches** —

Anal fin with **2 thick and stiff spines** —

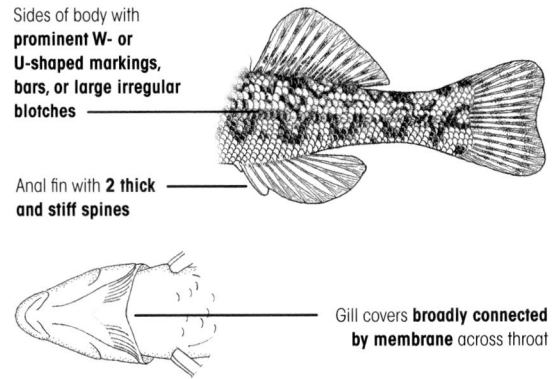

Gill covers **broadly connected by membrane** across throat

Sides of body with **many small flecks or W- and X-shaped markings** —

Anal fin with **1 thin and flexible spine** —

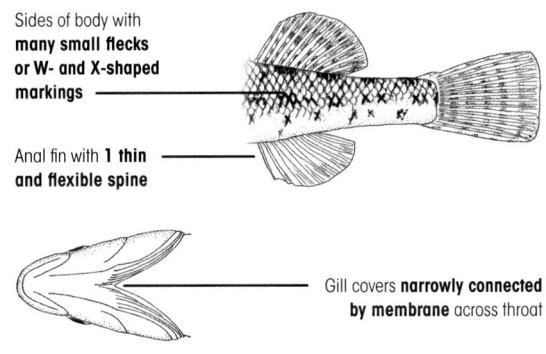

Gill covers **narrowly connected by membrane** across throat

17a Mouth overhung by blunt, rounded snout; first dorsal fin with usually 10 spines; pelvic fin with discrete brown markings; pectoral fin extremely long, extending backward beyond tip of pelvic fin . . . . **Page 357**

Harlequin Darter (*Etheostoma histrio*) in part

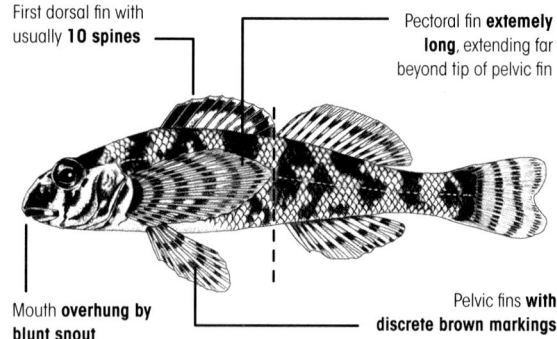

First dorsal fin with usually **10 spines**

Pectoral fin **extemely long**, extending far beyond tip of pelvic fin

Mouth **overhung by blunt snout**

Pelvic fins **with discrete brown markings**

17b Mouth not overhung by snout; first dorsal fin with usually 12–14 spines; pelvic fins without discrete markings; pectoral fin shorter, extending backward to or only slightly beyond tip of pelvic fin . . . . **Page 349**

Greenside Darter (*Etheostoma blennioides*)

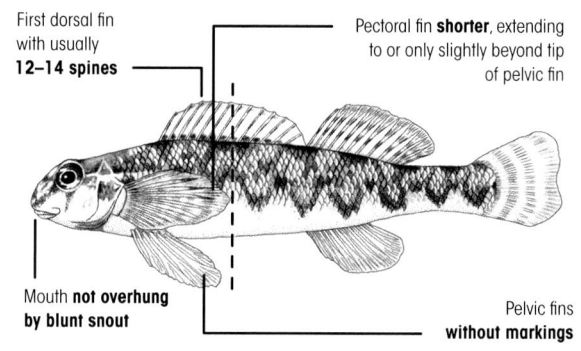

First dorsal fin with usually **12–14 spines**

Pectoral fin **shorter**, extending to or only slightly beyond tip of pelvic fin

Mouth **not overhung by blunt snout**

Pelvic fins **without markings**

18a Dark streak extending forward from front of eye not meeting its opposite on tip of snout; lateral line extending to near base of caudal fin; dorsal fins only slightly separated . . . . **Page 360**

Johnny Darter (*Etheostoma nigrum*)

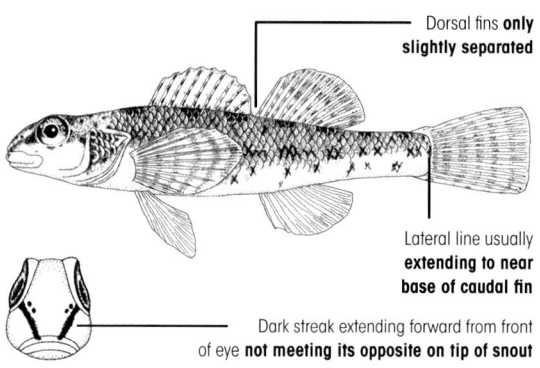

Dorsal fins **only slightly separated**

Lateral line usually **extending to near base of caudal fin**

Dark streak extending forward from front of eye **not meeting its opposite on tip of snout**

18b Dark streak extending forward from front of eye meeting its opposite on tip of snout; lateral line usually ending beneath second dorsal fin; dorsal fins widely separated . . . . **Page 302**

Bluntnose Darter (*Etheostoma chlorosoma*)

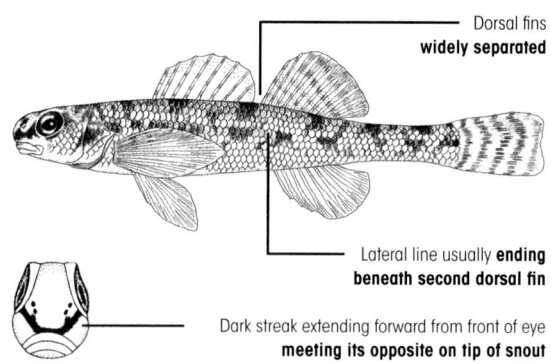

Dorsal fins **widely separated**

Lateral line usually **ending beneath second dorsal fin**

Dark streak extending forward from front of eye **meeting its opposite on tip of snout**

19a Lateral line always with more than 10 pored scales; scales in lateral series more than 40; body length commonly more than 2 inches (4.8 cm) **20**

19b Lateral line absent, or with 9 or fewer pored scales; scales in lateral series 38 or fewer; body length seldom more than 1.8 inches (4.4 cm) **31**

Lateral line always with **more than 10 pored scales**

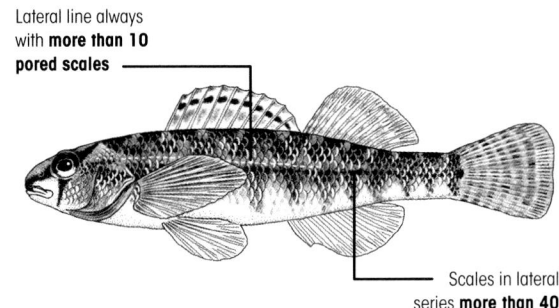

Scales in lateral series **more than 40**

Lateral line **absent, or with 9 or fewer pored scales**

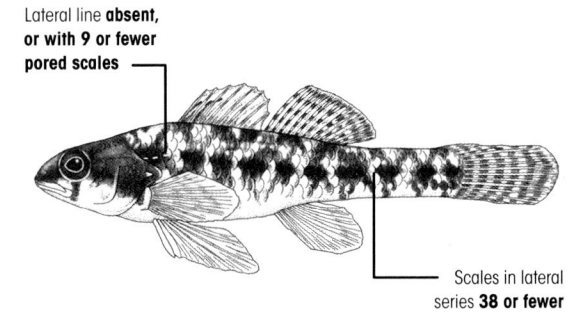

Scales in lateral series **38 or fewer**

20a First dorsal fin usually less than half the height of the second dorsal fin; caudal fin conspicuously crossed with 6 or more dark bands; dark humeral spot above base of pectoral fin. **21**

20b First dorsal fin nearly equal in height to second dorsal fin; caudal fin plain or indistinctly marbled, or crossed with usually fewer than 6 dark bands; dark humeral spot present or absent **24**

First dorsal fin usually **less than half the height** of the second dorsal fin

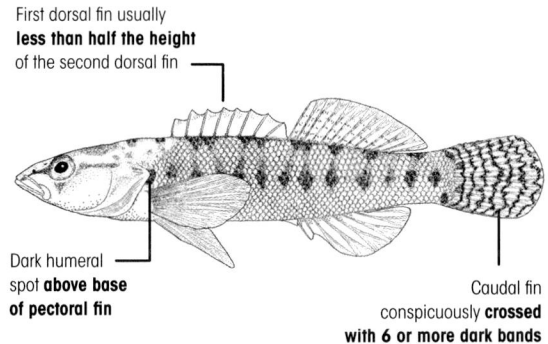

Dark humeral spot **above base of pectoral fin**

Caudal fin conspicuously **crossed with 6 or more dark bands**

First dorsal fin nearly **equal in height** to second dorsal fin

Caudal fin indistinctly **marbled, or crossed with usually fewer than 6 dark bands**

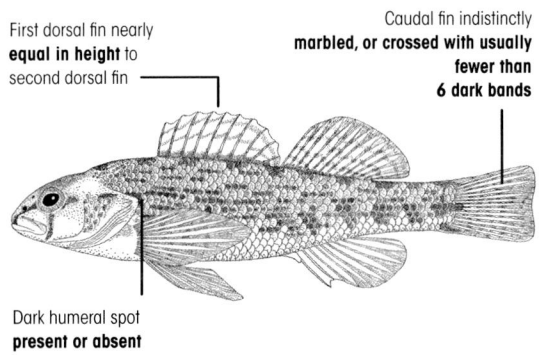

Dark humeral spot **present or absent**

21a Gill cover, breast, and nape without scales; base of caudal fin without vertically aligned dark spots **22**

21b Gill cover, breast, and nape scaled; base of caudal fin with usually 3 vertically aligned dark spots **23**

Nape **without scales**

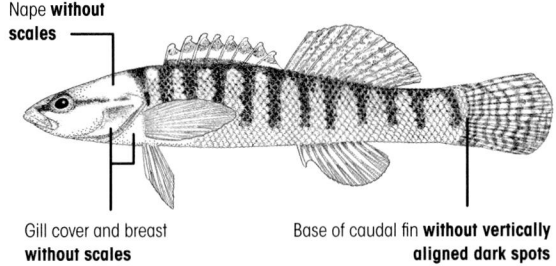

Gill cover and breast **without scales**

Base of caudal fin **without vertically aligned dark spots**

Nape **scaled**

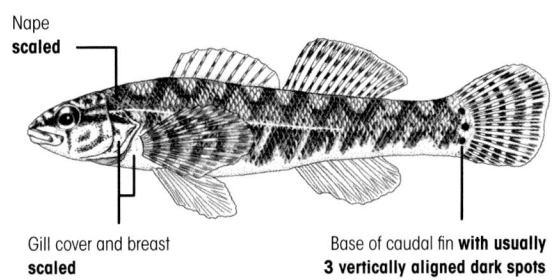

Gill cover and breast **scaled**

Base of caudal fin **with usually 3 vertically aligned dark spots**

22a Second dorsal fin rays usually 13–14; scale rows around caudal peduncle usually 24–30; first dorsal fin with dark base and golden-yellow at its top edge ......**Page 355**

Fantail Darter (*Etheostoma flabellare*)

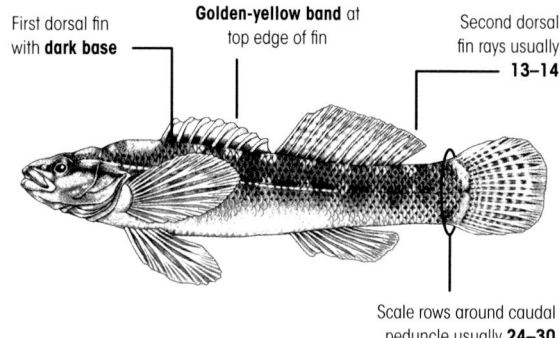

First dorsal fin with **dark base**

Golden-yellow band at top edge of fin

Second dorsal fin rays usually **13–14**

Scale rows around caudal peduncle usually **24–30**

22b Second dorsal fin rays usually 11–12; scale rows around caudal peduncle usually 17–22; first dorsal fin with clear base and separate dark and light orange bands at its top edge...............................**Page 358**

Stripetail Darter (*Etheostoma kennicotti*)

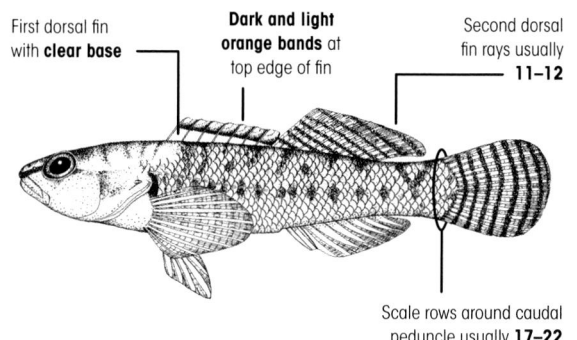

First dorsal fin with **clear base**

Dark and light orange bands at top edge of fin

Second dorsal fin rays usually **11–12**

Scale rows around caudal peduncle usually **17–22**

23a Adult (breeding) males with third branch of second dorsal fin rays greatly elongated; margin of second dorsal fin black, without white knobs; known only from Cache R. drainage...........................**Page 353**

Fringed Darter (*Etheostoma crossopterum*)

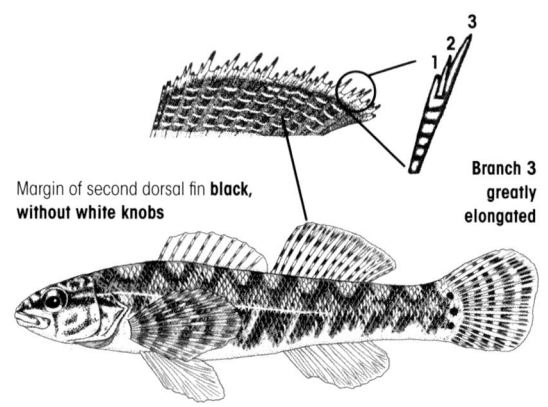

Margin of second dorsal fin **black, without white knobs**

Branch 3 greatly elongated

23b Adult (breeding) males with second and third branches of second dorsal fin rays equal in length; margin of second dorsal fin with white knobs ....**Page 363**

Spottail Darter (*Etheostoma squamiceps*)

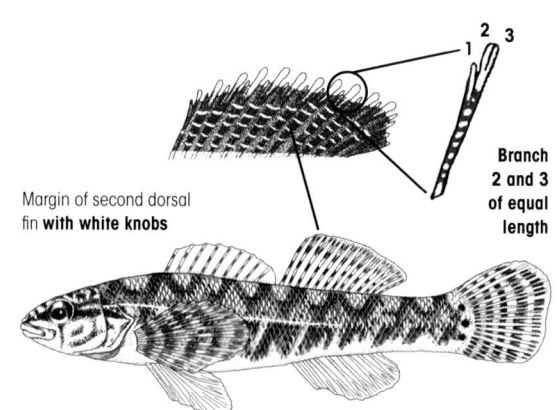

Margin of second dorsal fin **with white knobs**

Branch 2 and 3 of equal length

24a Forward part of lateral line arched upward, with 3 scale rows between lateral line and base of first dorsal fin. . . . . **Page 356**

24b Forward part of lateral line usually not arched upward (variable in *E. exile*), with 4 or more scale rows between lateral line and base of first dorsal fin **25**

Slough Darter (*Etheostoma gracile*)

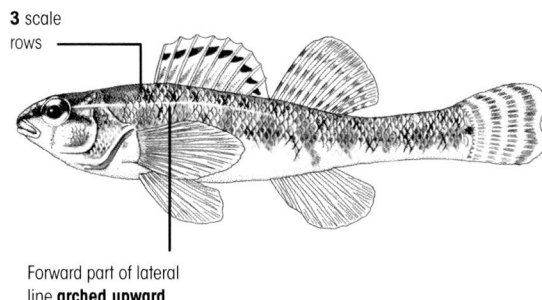

3 scale
rows

Forward part of lateral
line **arched upward**

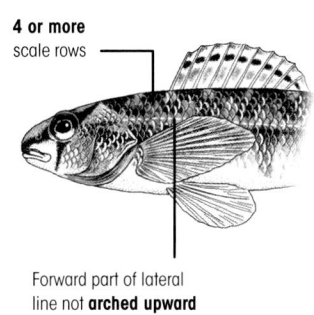

4 or more
scale rows

Forward part of lateral
line not **arched upward**

25a Gill covers broadly connected by membrane across throat; distance from membrane notch to tip of lower lip (A) greater than distance from notch to front of pelvic fin base (B) **26**

25b Gill covers separate or narrowly connected by membrane across throat; distance from membrane notch to tip of lower lip (A) less than distance from notch to front of pelvic fin base (B) **27**

A greater than **B**

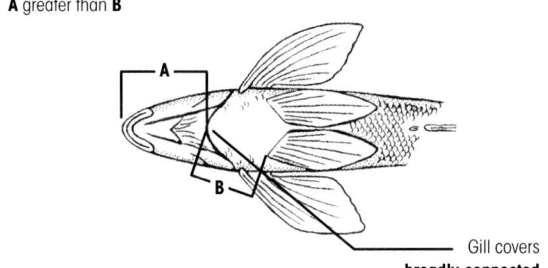

Gill covers
broadly connected

B less than **A**

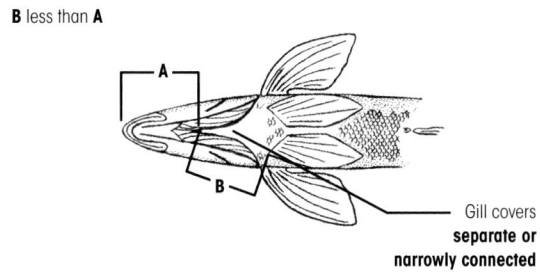

Gill covers
**separate or
narrowly connected**

26a Cheek without scales; first dorsal fin with usually 9–10 spines; pectoral fin longer, extending backward far beyond tip of pelvic fin; underside of head, breast, and pelvic fins with widely spaced, coarse dark spots; first dorsal fin with dark blotches in front and back . . . .**Page 357**

Harlequin Darter (*Etheostoma histrio*) in part

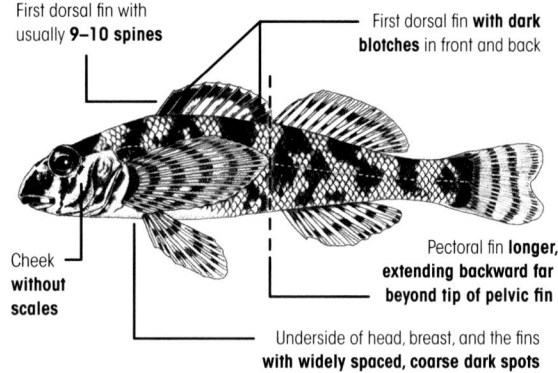

First dorsal fin with usually **9–10 spines**

First dorsal fin **with dark blotches** in front and back

Cheek **without scales**

Pectoral fin **longer, extending backward far beyond tip of pelvic fin**

Underside of head, breast, and the fins **with widely spaced, coarse dark spots**

26b Cheek entirely scaled; first dorsal fin with usually 11–12 spines; pectoral fin shorter, extending backward to or only slightly beyond tip of pelvic fin; underside of head, breast, and pelvic fins without coarse black dots; first dorsal fin without dark blotches in front and back . . . . . .**Page 364**

Banded Darter *(Etheostoma zonale)*

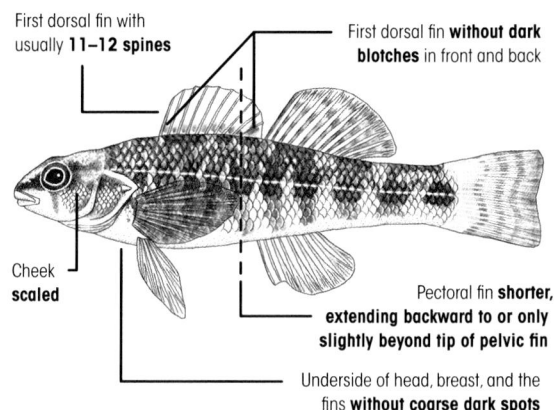

First dorsal fin with usually **11–12 spines**

First dorsal fin **without dark blotches** in front and back

Cheek **scaled**

Pectoral fin **shorter, extending backward to or only slightly beyond tip of pelvic fin**

Underside of head, breast, and the fins **without coarse dark spots**

27a Lateral line complete, extending to base of caudal fin; second dorsal and caudal fins bordered by broad whitish band with darkened edge; sides with numerous thin horizontal dark lines and scattered red dots; first dorsal fin with usually 11–12 spines . . . . **Page 351**

Bluebreast Darter (*Etheostoma camurum*)

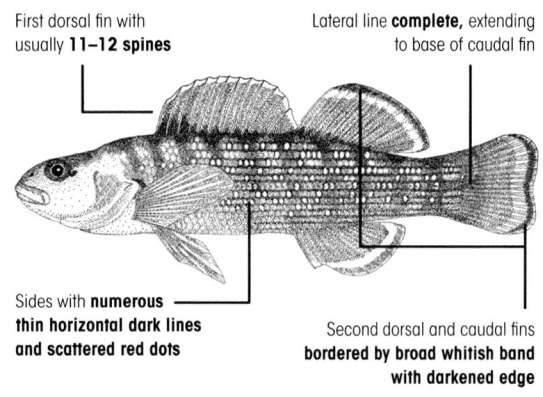

First dorsal fin with usually **11–12 spines**

Lateral line **complete,** extending to base of caudal fin

Sides with **numerous thin horizontal dark lines and scattered red dots**

Second dorsal and caudal fins **bordered by broad whitish band with darkened edge**

27b Lateral line incomplete, not extending behind second dorsal fin; second dorsal and caudal fins not bordered by broad whitish band with darkened edge; sides with vertical bands or blotches; first dorsal fin with usually 8–11 spines **28**

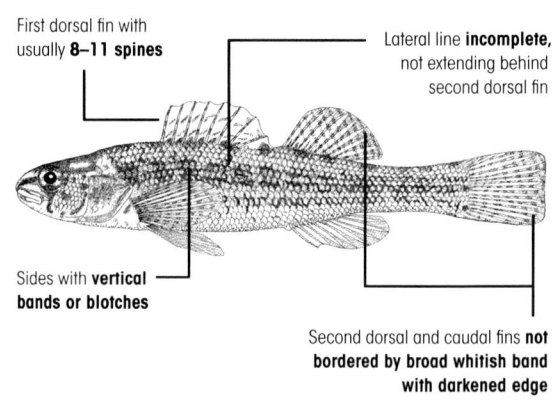

First dorsal fin with usually **8–11 spines**

Lateral line **incomplete,** not extending behind second dorsal fin

Sides with **vertical bands or blotches**

Second dorsal and caudal fins **not bordered by broad whitish band with darkened edge**

28a Lateral line short, not extending to front of second dorsal fin; and often arched upward; lateral series scales usually 52–60; eye diameter (A) greater than snout length (B)... ....**Page 354**

28b Lateral line longer extending to middle of second dorsal fin; and nearly straight; lateral series scales usually 41–51; eye diameter (A) less than snout length (B) **29**

Iowa Darter (*Etheostoma exile*)

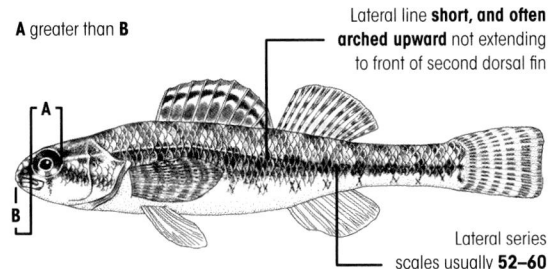

A greater than **B**

Lateral line **short, and often arched upward** not extending to front of second dorsal fin

Lateral series scales usually **52–60**

A less than **B**

Lateral line **longer and nearly straight**, extending to middle of second dorsal fin

Lateral series scales usually **41–51**

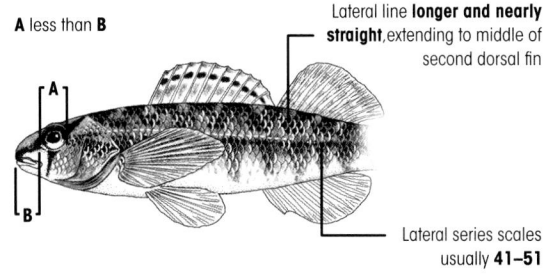

29a Cheek fully scaled; anal fin with usually 8–9 soft rays; vertical band near base of caudal fin more prominent than other lateral bands.....................**Page 348**

29b Cheek without scales or with a few small scales near eye; anal fin with usually 6–7 soft rays; vertical band or blotches near base of caudal fin no more prominent than other lateral bands or blotches..................**30**

Mud Darter (*Etheostoma asprigene*)

Vertical band near base of caudal fin **more prominent** than other lateral bands

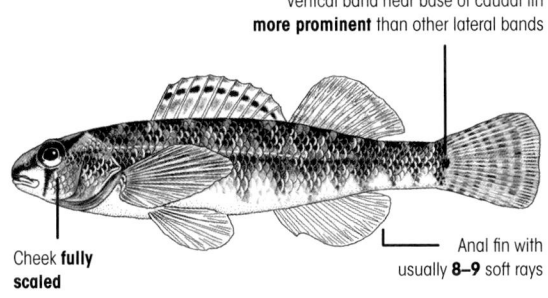

Cheek **fully scaled**

Anal fin with usually **8–9** soft rays

Vertical band or blotch near base of caudal fin **no more prominent** than other lateral bands

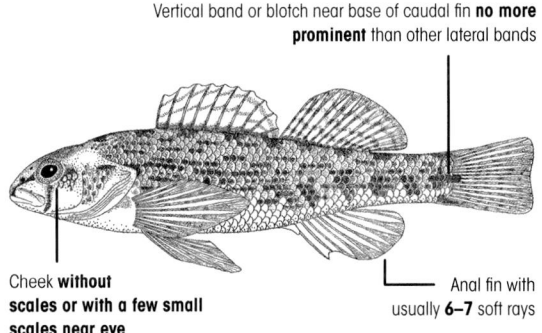

Cheek **without scales or with a few small scales near eye**

Anal fin with usually **6–7** soft rays

30a Pectoral fin rays usually 13–15; sensory canal (infraorbital canal) beneath eye complete; adult male with red or orange spot in blue-green anal fin; side without horizontal rows of dark spots ....**Page 350**

Rainbow Darter (*Etheostoma caeruleum*)

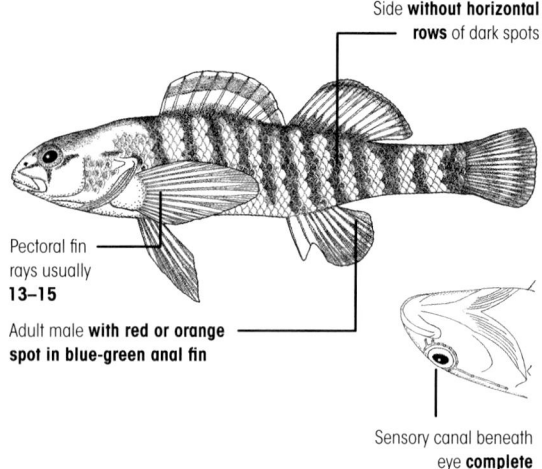

Side **without horizontal rows** of dark spots

Pectoral fin rays usually **13–15**

Adult male **with red or orange spot in blue-green anal fin**

Sensory canal beneath eye **complete**

30a Pectoral fin rays usually 11–12; sensory canal beneath eye interrupted; adult male without red or orange spot in blue-green anal fin; side with some short horizontal rows of dark spots...........................**Page 362**

Orangethroat Darter (*Etheostoma spectabile*)

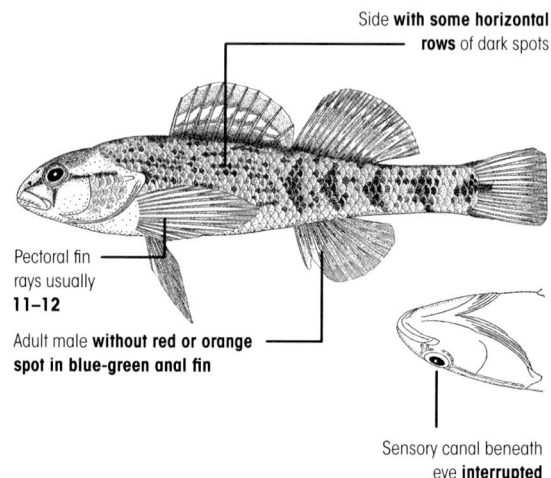

Side **with some horizontal rows** of dark spots

Pectoral fin rays usually **11–12**

Adult male **without red or orange spot in blue-green anal fin**

Sensory canal beneath eye **interrupted**

31a Cheek fully scaled; lateral series scales usually 35 or 36; forward part of lateral line with usually 2–4 pored scales; total counts of first dorsal fin spines and second dorsal fin rays usually 18–20.....................**Page 361**

Cypress Darter (*Etheostoma proeliare*)

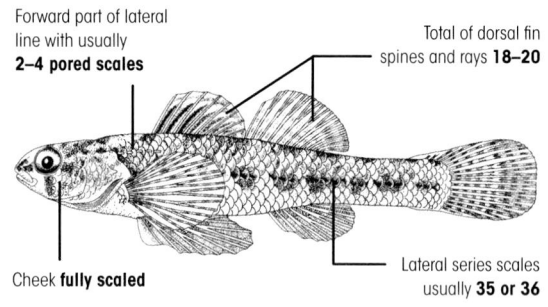

Forward part of lateral line with usually **2–4 pored scales**

Total of dorsal fin spines and rays **18–20**

Cheek **fully scaled**

Lateral series scales usually **35 or 36**

31b Cheek without scales; lateral series scales usually 32–34; forward part of lateral line with usually 0–1 pored scales; total counts of first dorsal fin spines and second dorsal fin rays usually 14–17.....................**Page 359**

Least Darter (*Etheostoma microperca*)

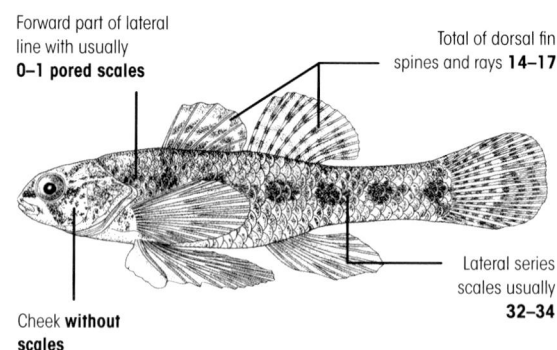

Forward part of lateral line with usually **0–1 pored scales**

Total of dorsal fin spines and rays **14–17**

Cheek **without scales**

Lateral series scales usually **32–34**

WESTERN SAND DARTER *Ammocrypta clara* Jordan & Meek 1885

IDENTIFICATION The Western Sand Darter has a long and slender, transparent body with 10–15 iridescent pale green blotches along the back and 10–20 horizontally elongated pale green blotches along the side of the body. The snout is pointed and lacks a premaxillary frenum, the opercle has a spine, and the caudal fin is shallowly forked. The lateral line is complete with 63–84 pored scales and slants downward posteriorly. There are 8–11 anal rays and 1 anal spine, and the head canals are uninterrupted. Scutes are absent on the breast and belly. To 2¾ in. (7.1 cm).

SIMILAR SPECIES The Eastern Sand Darter, *A. pellucida*, has no spine on the opercle and has darker blotches on the back and along the side of the body. The Crystal Darter, *Crystallaria asprella*, has a less transparent body, 4 brown saddles on the back, 12–16 anal rays, and a deeply forked caudal fin.

HABITAT The Western Sand Darter lives in flowing, moderately deep water (usually > 0.5 m) over sand in medium to large rivers. It buries in the substrate with only the snout and eyes protruding.

DISTRIBUTION IN ILLINOIS Smith (1979) noted that the Western Sand Darter was rare and less widespread than it had been in the Forbes and Richardson era. In the contemporary era it is found in the same general areas in the Mississippi River basin as it was in the Smith era, although it also has been discovered in the Kankakee River basin and the lower Mississippi River. It remains much less widespread than it was in the early 20th century.

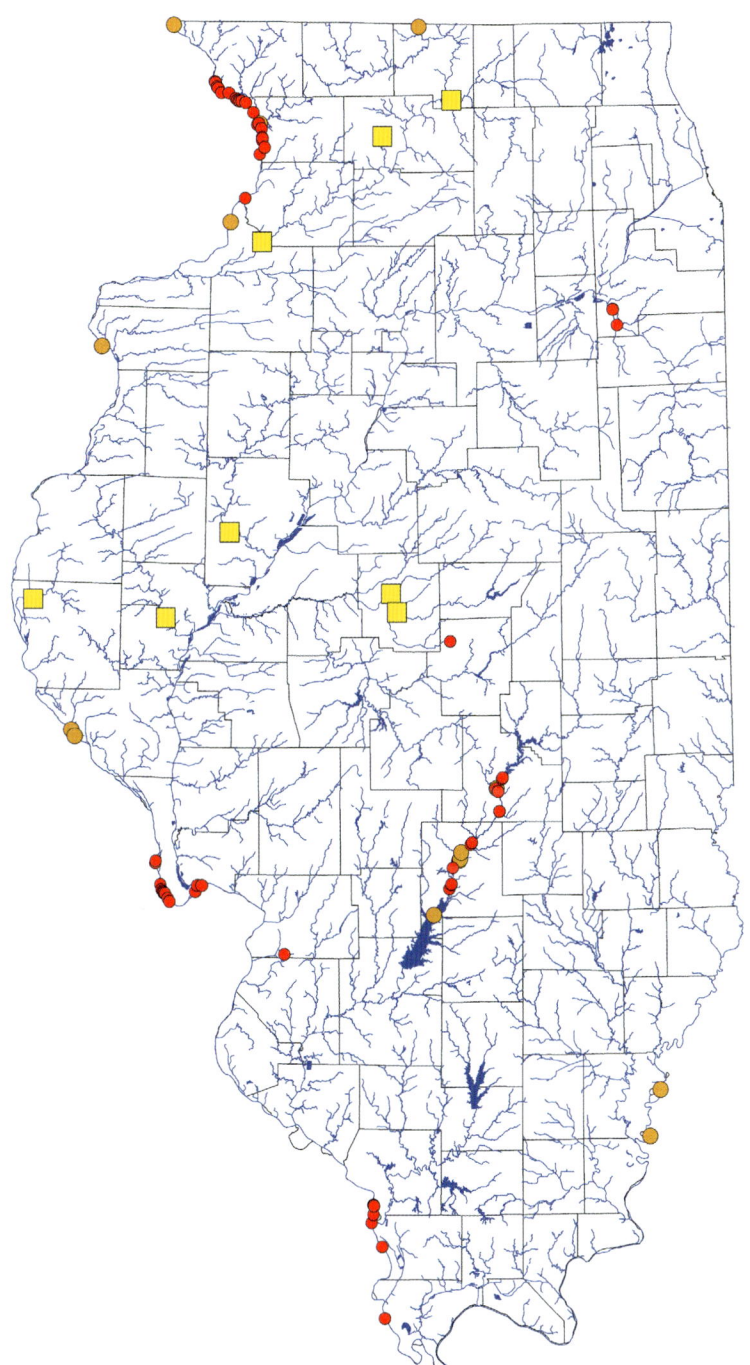

EASTERN SAND DARTER *Ammocrypta pellucida* (Putnam 1863)

IDENTIFICATION The Eastern Sand Darter has a long and slender, transparent body with 12–17 iridescent green blotches along the back and 10–20 horizontally elongated dark green blotches along the side of the body. Some black pigment usually is present on the pelvic fin of large males. The snout is pointed and lacks a premaxillary frenum, the opercle lacks a spine, and the caudal fin is shallowly forked. The lateral line is complete with 62–84 pored scales. There are 7–11 anal rays and 1 anal spine, and the head canals are uninterrupted. Scutes are absent on the breast and belly. To 3¼ in. (8.4 cm).

SIMILAR SPECIES The Western Sand Darter, *A. clara*, has a spine on the opercle and lighter green blotches on the back and along the side of the body. The Crystal Darter, *Crystallaria asprella*, has a less transparent body, 4 brown saddles on the back, 12–16 anal rays, and a deeply forked caudal fin.

HABITAT The Eastern Sand Darter lives in flowing, shallow to moderately deep water (usually > 0.5 m) over sand in small to medium rivers. It buries in the substrate with only the snout and eyes protruding.

DISTRIBUTION IN ILLINOIS Smith (1979) noted that the Eastern Sand Darter was less widely distributed than it had been in the Forbes and Richardson era and was commonly encountered only in the Embarras and Vermilion River basins in the Wabash River basin. In the contemporary era it is about as widespread as it was in the Smith era and remains common in the Embarras River. However, it appears to be less widely distributed in the Vermilion River (Wabash) basin than it was in the Smith era and is extirpated from the Little Wabash River.

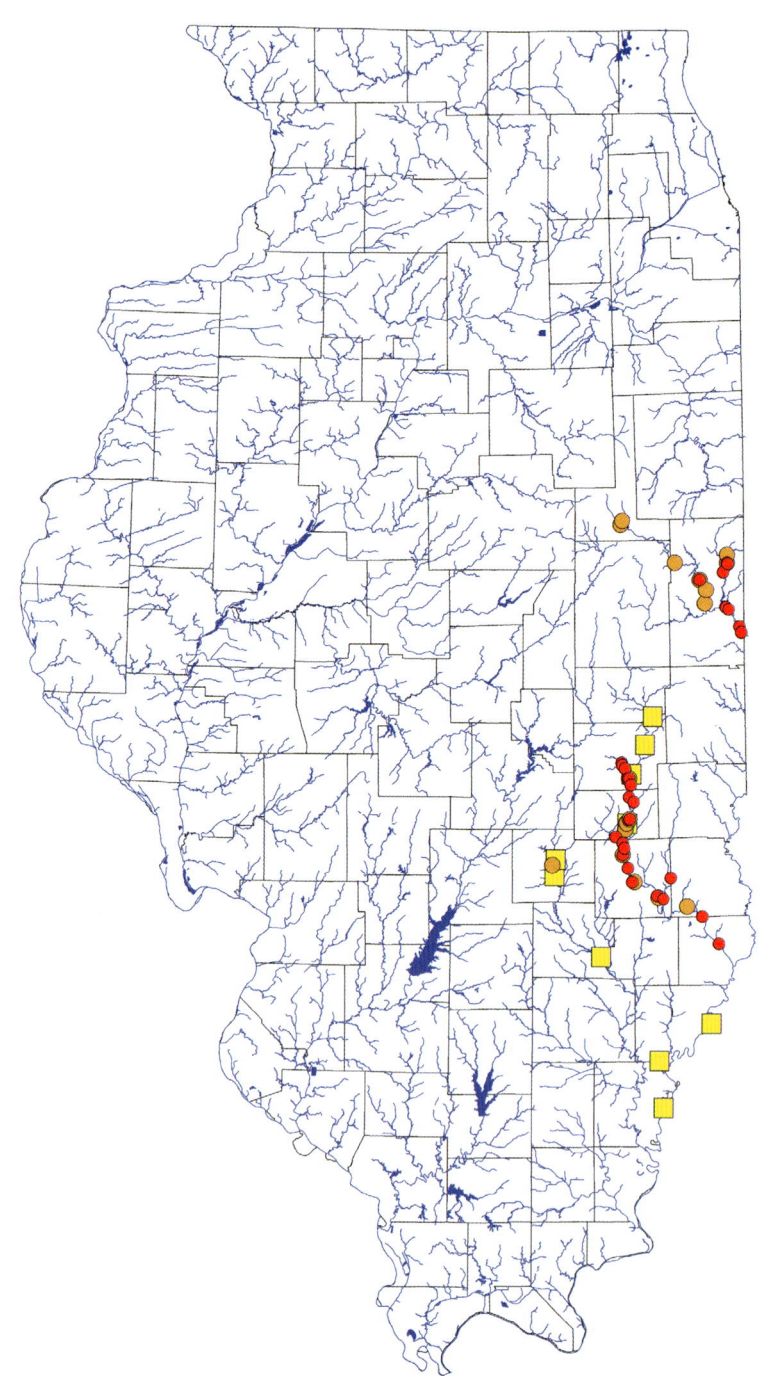

CRYSTAL DARTER   *Crystallaria asprella* (Jordan 1878)

IDENTIFICATION The Crystal Darter has a long, slender, transparent-yellow or white body with pale brown mottling and 4 dark brown saddles on the back and upper side of the body. The first 3 saddles are large; the fourth is on the caudal peduncle and much smaller. Dark brown, elongated blotches are present along the side. A black stripe continues around the snout from eye to eye. Dorsal and anal fins are large; the caudal fin is large and deeply forked. The lateral line is complete with 77–96 pored scales. There are 12–16 anal rays and 1 anal spine, and the head canals are uninterrupted. Scutes are absent on the breast and belly. To 6¼ in. (16 cm).

SIMILAR SPECIES The Eastern Sand Darter, *Ammocrypta pellucida*, and Western Sand Darter, *A. clara*, have more transparent bodies, no dark saddles on the back, 7–11 anal rays, and a much more shallowly forked caudal fin.

HABITAT The Crystal Darter lives in flowing, moderately deep water (usually > 0.5 m) over mixed sand and gravel in small to large rivers. It buries in the substrate when disturbed.

DISTRIBUTION IN ILLINOIS The Crystal Darter was more widely distributed in the Forbes and Richardson era than it is in the contemporary era, with the Forbes and Richardson records in the Mississippi, Rock, Little Wabash, and Wabash Rivers. Smith (1979) reported no records from the mid-20th century and considered the species to be extirpated from Illinois. Today it is nearly confined to the mainstem of the Mississippi River.

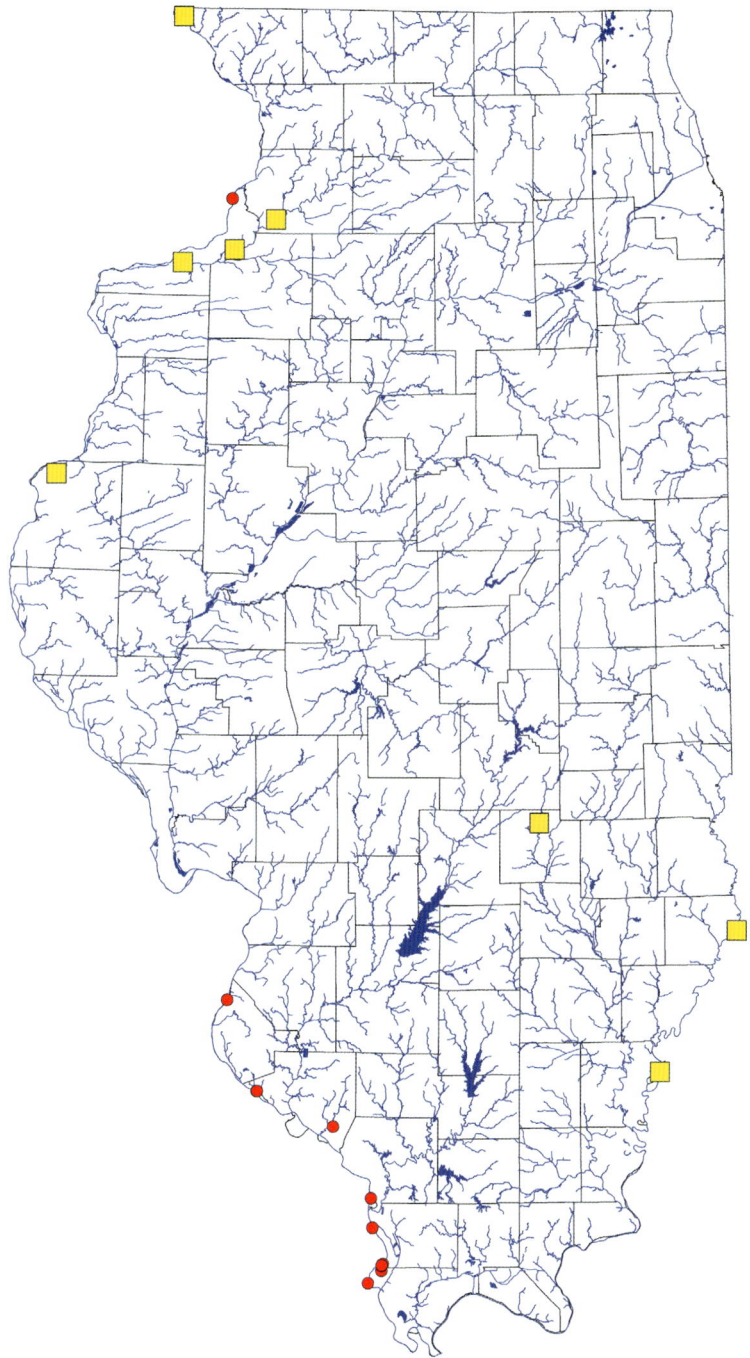

MUD DARTER *Etheostoma asprigene* (Forbes 1878)

IDENTIFICATION The Mud Darter is olive-green to light brown on the back and upper side of the body with 6–10 dark saddles and is white to pale orange on the lower side and belly. Dark bars are present on the side of the body; posterior bars are darkest. The bars are blue or green separated by dull orange on the male, dark brown separated by yellow-white on the female. The first dorsal fin has a middle red band, is blue-black on the edge and base, and has a large, black blotch posteriorly. There is a large, dusky teardrop. The body is deepest under the middle of the first dorsal fin. There are 2 anal spines, a wide premaxillary frenum, a fully scaled cheek, an uninterrupted infraorbital canal, and narrowly joined branchiostegal membranes. The lateral line is incomplete. There are 44–54 (usually 48–51) lateral scales. Scutes are absent on the breast and belly. To 2¾ in. (7.1 cm).

SIMILAR SPECIES The Rainbow Darter, *E. cae-ruleum*, and Orangethroat Darter, *E. spectabile*, lack a large blotch at the rear of the first dorsal fin. The Rainbow Darter has no scales on the cheek and has red in the anal fin. The Orangethroat Darter is deepest at the nape or front of the first dorsal fin.

HABITAT The Mud Darter usually is found near woody debris or large stones in shallow riffles in small to large rivers. It also lives among living or dead vegetation in lowland lakes.

DISTRIBUTION IN ILLINOIS Forbes and Richardson (1909) found the Mud Darter to be statewide in distribution and noted its occasional abundance. Smith (1979) noted that it was no longer found in many of the streams where it previously had been recorded. In the contemporary era, it appears to be as widespread as it was in the Smith era.

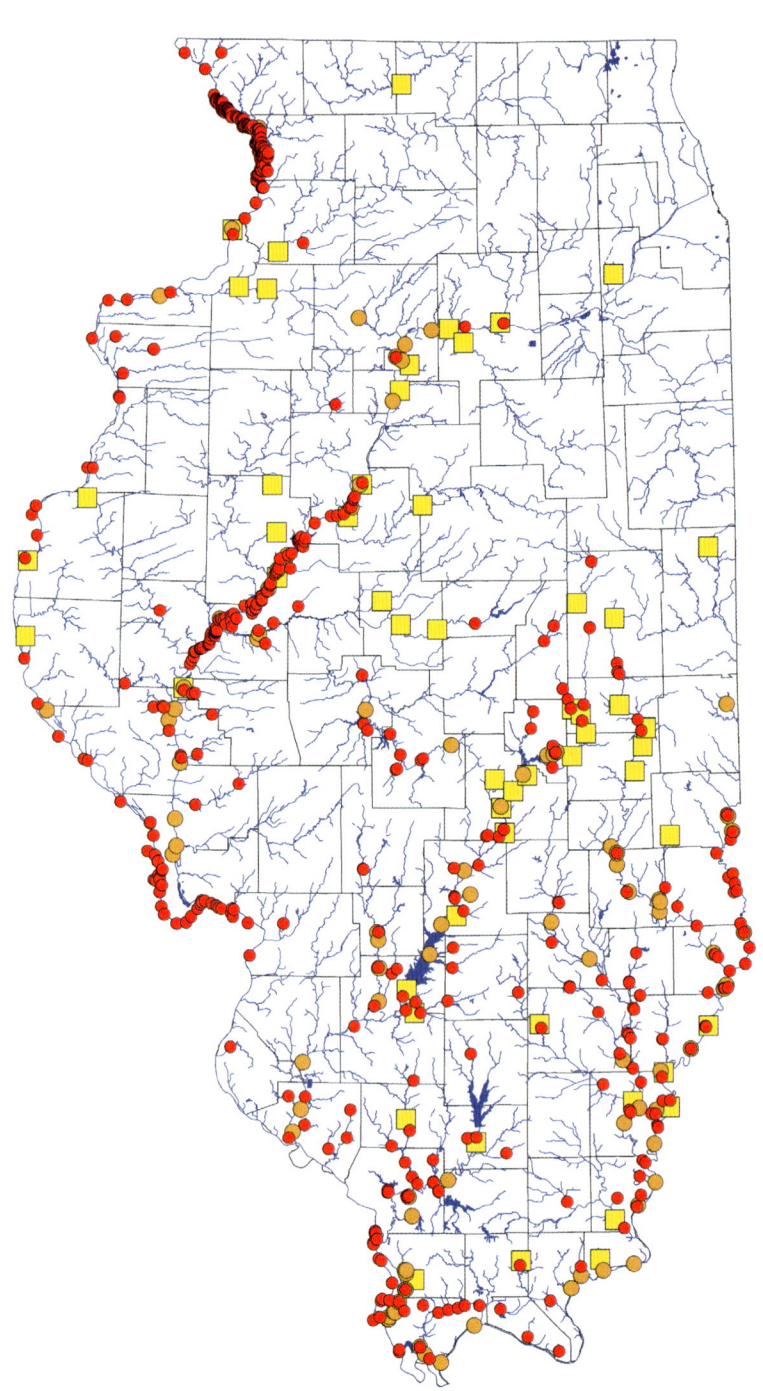

DARTERS AND PERCHES (Percidae)

348

GREENSIDE DARTER *Etheostoma blennioides* Rafinesque 1819

IDENTIFICATION The Greenside Darter has 6–7 dark green saddles on the light yellow-green back and upper side of the body, small, brown to dark red spots on the upper side, and 5–8 green Us, Ws, or bars along the side. The dorsal fins are green on the male, clear to dusky on the female, often with red at the base. A dusky teardrop usually is present. During the breeding season the male has bright green fins and bars on the side of the body. The snout is extremely blunt. There are 2 anal spines, no premaxillary frenum, an uninterrupted infraorbital canal, and broadly joined branchiostegal membranes. The lateral line is complete with 50–86 pored scales. Scutes are absent on the breast and belly. To 6¾ in. (17 cm).

SIMILAR SPECIES The Banded Darter, *E. zonale*, has a more pointed snout and 9 or more green bars on the side of the body that join those from the other side across the belly. The Harlequin Darter, *E. histrio*, lacks green Us, Ws, or bars along the side of the body and has 2 large, dark brown to dark green spots on the caudal-fin base.

HABITAT The Greenside Darter lives among large stones and gravel in riffles of creeks and small to medium rivers.

DISTRIBUTION IN ILLINOIS The Greenside Darter is found only in tributaries of the Wabash River in east-central Illinois. The distribution in the contemporary era is essentially the same as in earlier eras.

REMARKS The subspecies *E. b. pholidotum*, the form in Illinois, is distinguished by having low scale counts, including usually 55–65 lateral scales; other subspecies usually have 63 or more lateral scales (Piller et al. 2008).

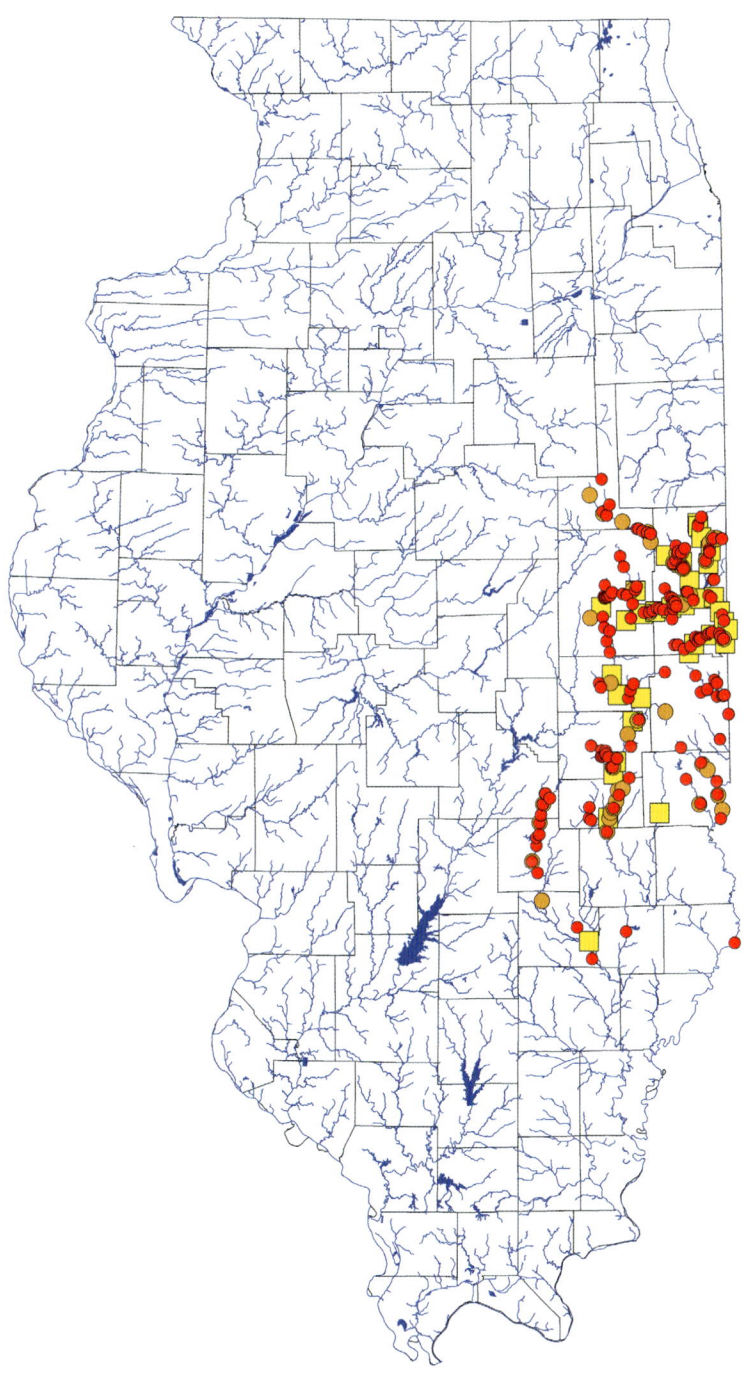

DARTERS AND PERCHES (Percidae)

349

RAINBOW DARTER *Etheostoma caeruleum* Storer 1845

IDENTIFICATION The Rainbow Darter is light brown to gray on the back with 6–10 dark saddles, 2 or 3 of which are much darker than the others, and yellow, green, or red on the lower side and belly. Dark blue bars on the side of the body are separated by red on the male; dark brown bars on the female are separated by yellow-white. The dorsal, caudal, and anal fins are red or orange with a blue edge (often faint on the female). The body is deepest under the middle of the first dorsal fin. Large males have a red pectoral fin, blue pelvic fin, orange branchiostegal membranes, and a blue cheek. There are 2 anal spines, usually 13 pectoral rays, a wide premaxillary frenum, no scales on the cheek, an uninterrupted infraorbital canal, and narrowly joined branchiostegal membranes. The lateral line is incomplete. There are 36–57 (usually 41–50) lateral scales. Scutes are absent on the breast and belly. To 3 in. (7.7 cm).

SIMILAR SPECIES The Orangethroat Darter, *E. spectabile*, is deepest at the nape or at the front of first dorsal fin, has no red in the anal fin, and 11–12 pectoral rays, and the infraorbital canal is interrupted. The Mud Darter, *E. asprigene*, has a fully scaled cheek, a large, black blotch at the rear of the first dorsal fin, and no red in the anal fin.

HABITAT The Rainbow Darter lives in fast, gravel and rocky riffles of creeks and small to medium rivers.

DISTRIBUTION IN ILLINOIS The Rainbow Darter has an unusual distribution in Illinois, being found in the Rock River basin, the upper Illinois River basin, the upper Wabash River basin, and in upland areas in extreme southern Illinois. Most of the records from the Forbes and Richardson era are missing because the Rainbow and Orangethroat Darters were not distinguished, and most specimens from that era that once were at INHS have been discarded. The species occurs in the contemporary era where it did in the Smith era except for its disappearance in the northwestern part of its former range. The absence from streams it previously occupied in the Rock and Illinois River basins is difficult to explain. It remains abundant elsewhere.

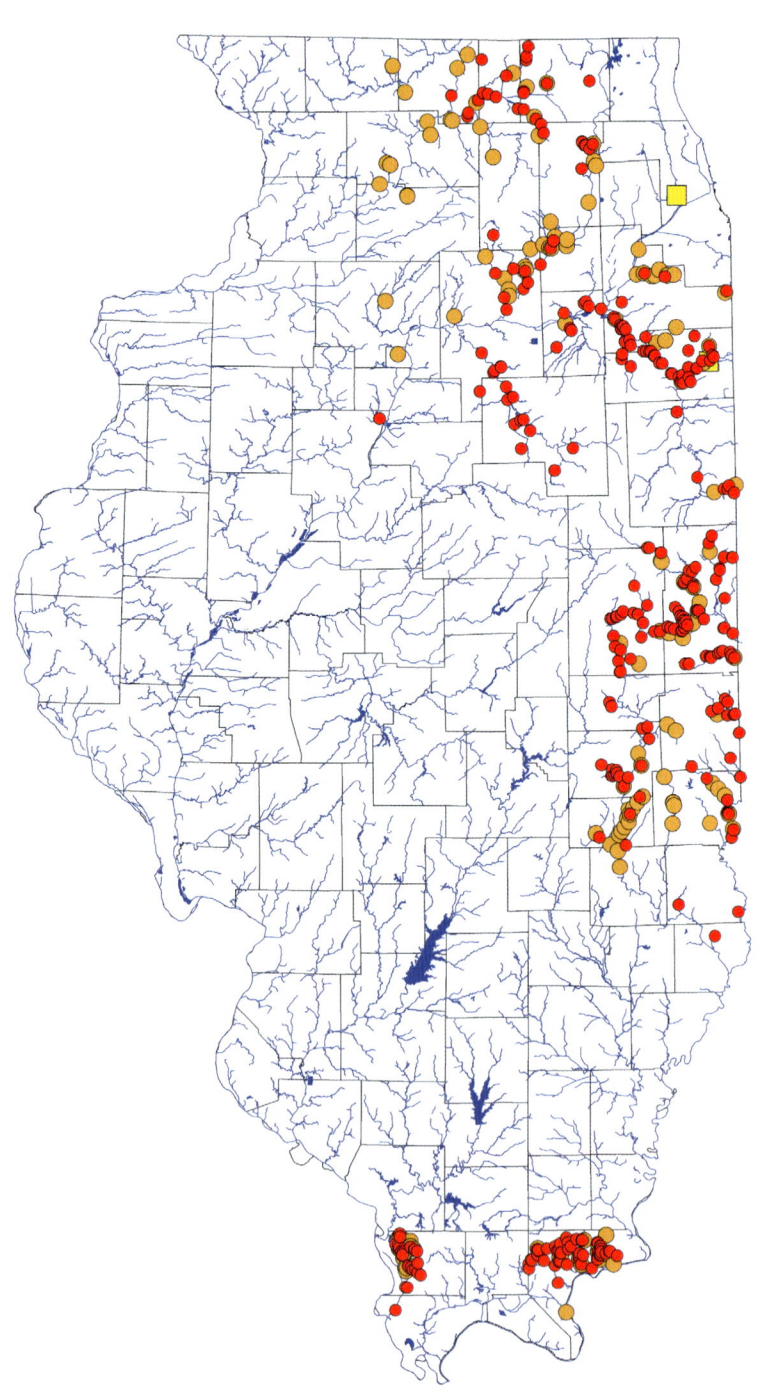

BLUEBREAST DARTER *Etheostoma camurum* (Cope 1870)

IDENTIFICATION The Bluebreast Darter is olive-green to dark gray on the back and upper side, has thin alternating dark and light stripes on the side of the body, and is light green to white on the lower side and belly. The second dorsal, caudal, and anal fins have a black edge. Males have bright red spots on the side of the body, red fins, and a blue breast. Females have brown spots on the side of the body, brown fins, and a white to light blue breast. The snout is moderately blunt. There are 2 anal spines, a wide premaxillary frenum, an uninterrupted infraorbital canal, and narrowly joined branchiostegal membranes. The lateral line is complete with 47–70 pored scales. Scutes are absent on the breast and belly. To 3¼ in. (8.4 cm).

SIMILAR SPECIES The Rainbow Darter, *E. caeruleum*, and Orangethroat Darter, *E. spectabile*, have alternating dark and light bars but no discrete red or brown spots on the side of the body.

HABITAT The Bluebreast Darter lives among large stones in fast, rocky riffles of small to medium rivers.

DISTRIBUTION IN ILLINOIS The Bluebreast Darter is confined in Illinois to the Vermilion River (Wabash) basin in Vermilion and Champaign Counties. Forbes and Richardson (1909) were not aware of the species's occurrence in the state, and Smith (1979) recorded it only from the Middle Fork of the Vermilion River and referenced a record from the Salt Fork. The species is more abundant in the contemporary era, almost certainly because of Illinois's efforts at protecting and improving the Vermilion River (Wabash) basin.

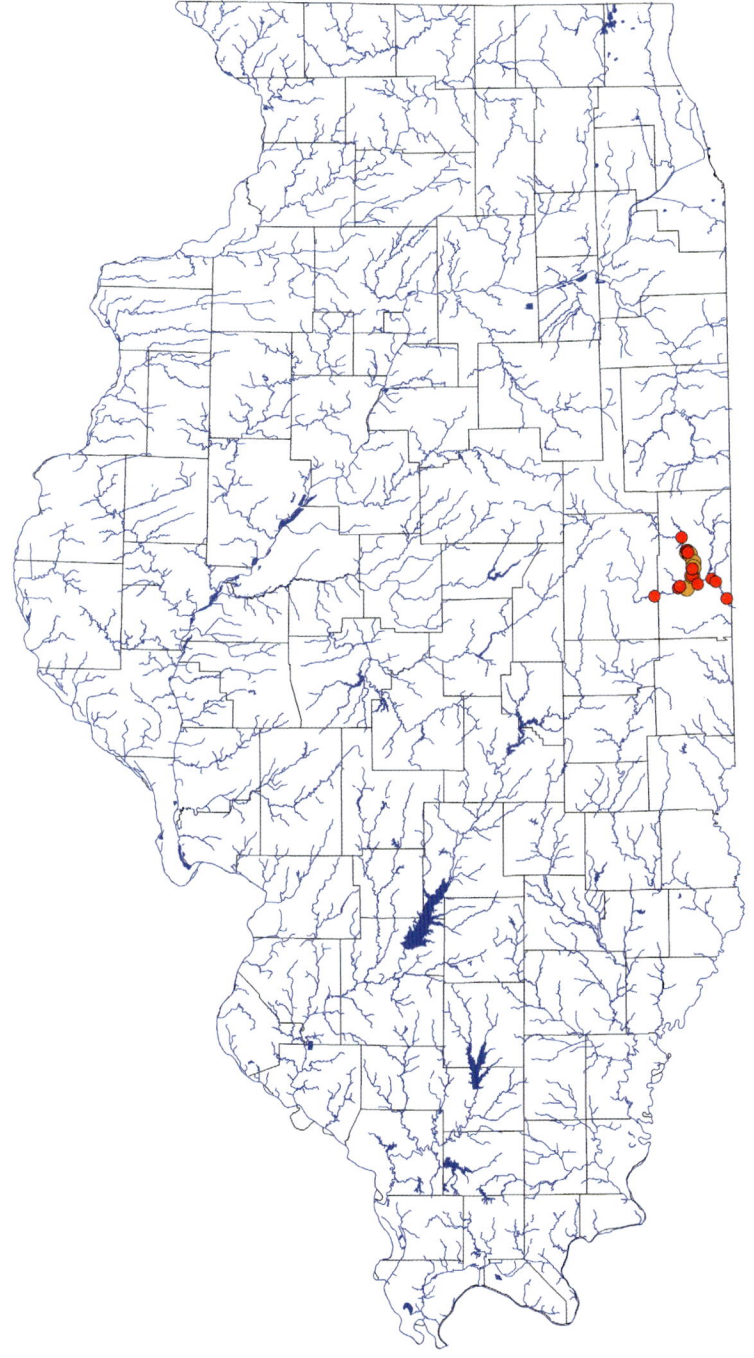

DARTERS AND PERCHES (Percidae)

351

BLUNTNOSE DARTER *Etheostoma chlorosoma* (Hay 1881)

IDENTIFICATION The Bluntnose Darter has a slender, semi-translucent light yellow body with 6 dusky brown-black saddles and brown Ws and Xs on the back and upper side. There is a black bridle around an extremely blunt snout, a series of horizontally elongate dusky black blotches along the midside, and often a black teardrop. During the breeding season the male has a black band (darkest anteriorly) and a dusky edge on the first dorsal fin. There is 1 anal spine, no premaxillary frenum, an interrupted infraorbital canal, and narrowly joined branchiostegal membranes. The lateral line is incomplete. There are 49–60 lateral scales. Scutes are absent on the breast and belly. To 2¼ in. (6 cm).

SIMILAR SPECIES The Johnny Darter, *E. nigrum*, lacks a black bridle around the snout and horizontal black blotches along the side of the body and has a more pointed snout.

HABITAT The Bluntnose Darter lives in creeks and small to medium rivers, lakes, and swamps. It is most often found in sand- or mud-bottomed pools and backwaters of streams or among living or dead vegetation in standing waterbodies.

DISTRIBUTION IN ILLINOIS Smith (1979) noted the substantial reduction in the range of the Bluntnose Darter that had occurred since the Forbes and Richardson era, and the decrease in the range of the species is even more evident in the contemporary era. It occurs in 2 extremely isolated areas in northern Illinois and in the southern one-third of the state. In the Forbes and Richardson era, the species was nearly statewide in distribution.

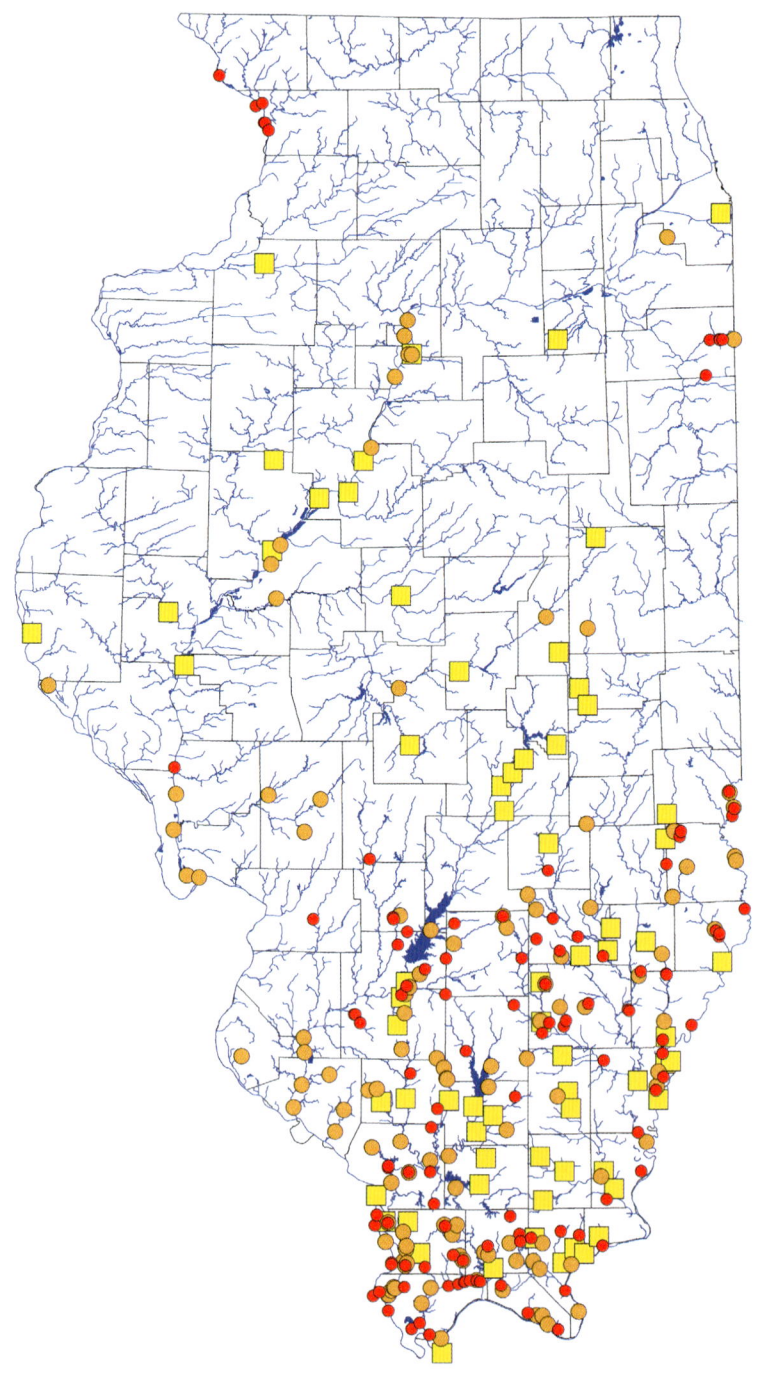

FRINGED DARTER

Etheostoma crossopterum Braasch & Mayden 1985

IDENTIFICATION The Fringed Darter has a light gray-brown back and side of the body with dark brown mottling. The lower side and belly are lighter brown or white. There is a black teardrop and 3 vertically aligned black spots on the caudal-fin base. The second dorsal and caudal fins have dark brown bands. The first dorsal fin of a large male has a dull orange band. During the breeding season the male has a black head, body, and fins (except the pectoral fin), and during spawning the male develops wide black and white bars on the side of the body. The second dorsal fin of the breeding male has a dusky white edge (fin rays often with small, black tips), and the third branch of each ray is much longer than the second branch. Branches of each ray are non-adnate. The snout is moderately blunt. There are 2 anal spines, a premaxillary frenum, an interrupted infraorbital canal, and narrowly joined branchiostegal membranes. The lateral line is incomplete. There are 45–62 lateral scales. Scutes are absent on the breast and belly. To 4 in. (10 cm).

SIMILAR SPECIES The Spottail Darter, *E. squamiceps*, is extremely similar, but the second dorsal fin of the breeding male has small, white knobs on the tips of the rays, and the second and third branches of each ray are adnate and equal in length.

HABITAT The Fringed Darter lives in shallow, rocky pools and riffles of headwaters, creeks, and small rivers.

DISTRIBUTION IN ILLINOIS Populations of the Fringed Darter in Illinois were unknown to Forbes and Richardson (1909) and treated by Smith (1979) as records for the Spottail Darter, *E. squamiceps*. The species remains widespread in the western part of the Cache River basin but may have disappeared from the only known locality for the species in Johnson County.

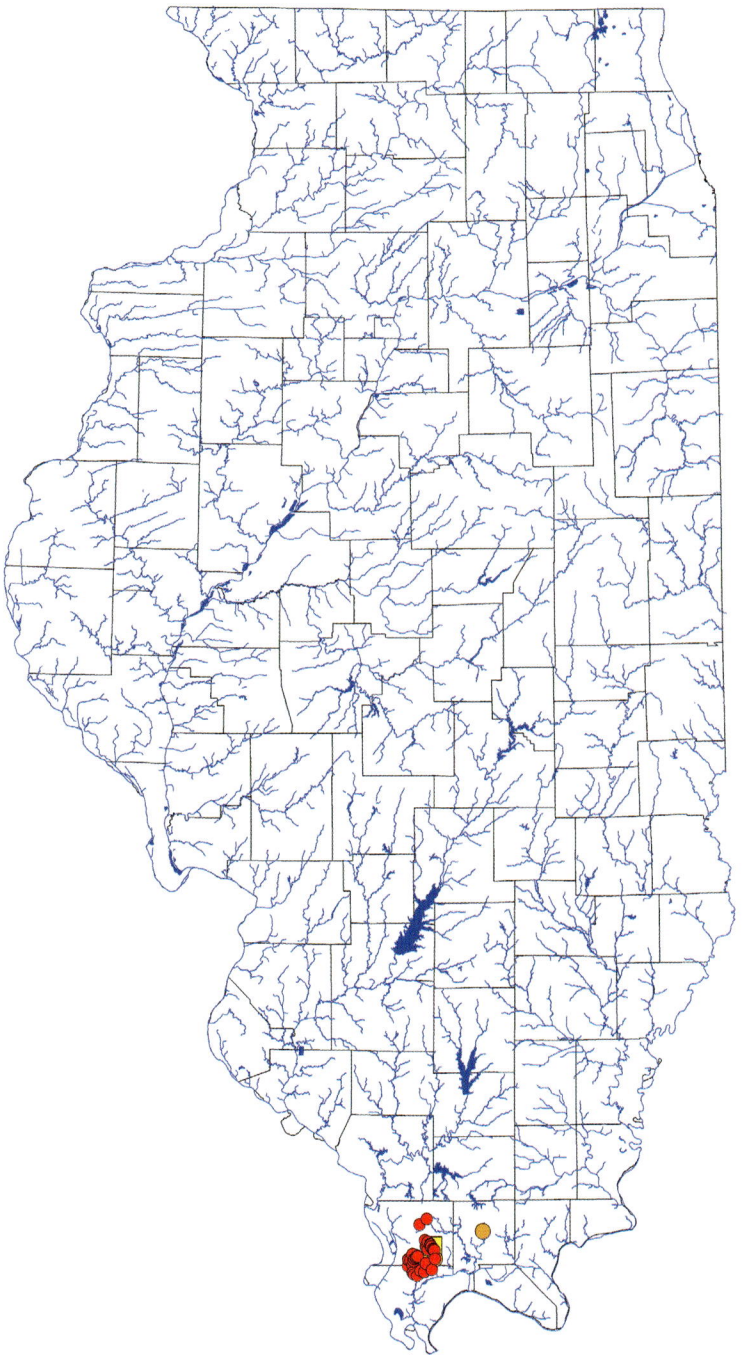

IOWA DARTER   *Etheostoma exile* (Girard 1859)

IDENTIFICATION   The Iowa Darter is slender with a long, narrow caudal peduncle. The back and upper side are tan with dark brown mottling. There are a black teardrop and usually short, dark bars on the side of the body. The first dorsal fin has a blue edge and base and a red submarginal band. During the breeding season the male has a bright orange breast and belly and alternating blue and red-orange bars on the side of the body. There are 2 anal spines, a premaxillary frenum, an uninterrupted infraorbital canal with 8 pores, and narrowly joined branchiostegal membranes. The lateral line is incomplete with 19–34 pores and often is arched anteriorly. There are 45–69 lateral scales. Scutes are absent on the breast and belly. To 2¾ in. (7.2 cm).

SIMILAR SPECIES   The Least Darter, *E. microperca*, is green and deeper bodied and has 0–3 lateral-line pores, 30–36 lateral scales, 2–3 infraorbital canal pores, and pelvic-fin flaps on the male.

HABITAT   The Iowa Darter lives in lakes and quiet pools of headwaters, creeks, and small to medium rivers. It is most often found among vegetation.

DISTRIBUTION IN ILLINOIS   The Iowa Darter suffered a substantial reduction in range between the Forbes and Richardson and Smith eras, having gone from a general distribution in the northern one-fourth of the state to being found only in glacial lakes and streams in northeastern Illinois. It seems not to have continued to decline since the Smith era, with contemporary-era records showing the same distribution. The presence of the species in old limestone quarries in Vermilion County suggests that the species at one time occurred even farther south than the Forbes and Richardson records indicated. Alternatively, the species could have been stocked, presumably accidently along with some sport fishes in the quarries sometime in the early 20th century.

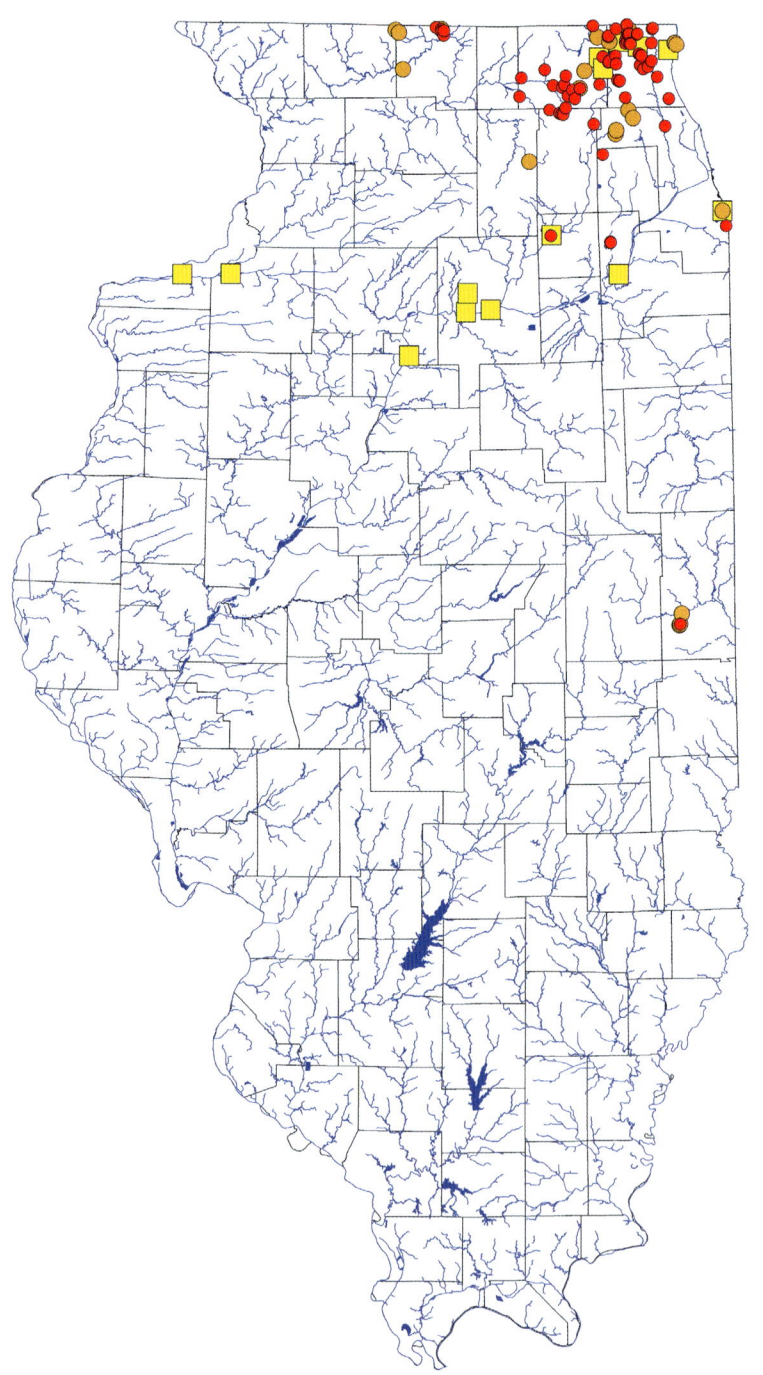

FANTAIL DARTER   *Etheostoma flabellare* Rafinesque 1819

IDENTIFICATION The Fantail Darter is brown-olive on the back and upper side and has thin black stripes on the side of the body and a dusky white to light brown lower side and belly. The first dorsal fin has gold knobs (faint on juveniles) on the tips of the spines. The second dorsal and caudal fins have black bands. The teardrop is faint or absent. During the breeding season, the male has a black head and large, gold knobs on the tips of the spines of the first dorsal fin and bold black bands on the caudal fin. The snout is pointed, and the lower jaw protrudes far forward. There are 2 anal spines, a premaxillary frenum, and a widely interrupted infraorbital canal with 4 pores anteriorly and 2 pores posteriorly. The lateral line is incomplete, and the branchiostegal membranes are broadly joined. There are 38–60 (usually 45–55) lateral scales. Scutes are absent on the breast and belly. To 3¼ in. (8.4 cm).

SIMILAR SPECIES The Stripetail Darter, *E. kennicotti*, has a less protruding lower jaw, black blotches, narrowly joined branchiostegal membranes, 1 posterior infraorbital canal pore, and no stripes on the side of the body.

HABITAT The Fantail Darter lives in rocky riffles of creeks and small to medium rivers.

DISTRIBUTION IN ILLINOIS As in previous eras, the Fantail Darter is widespread in the northern half of Illinois and found in a few rocky, clear-water streams in extreme southwestern Illinois. The range of the species seems to have changed little, although it might be slightly more widespread in the contemporary era in western Illinois.

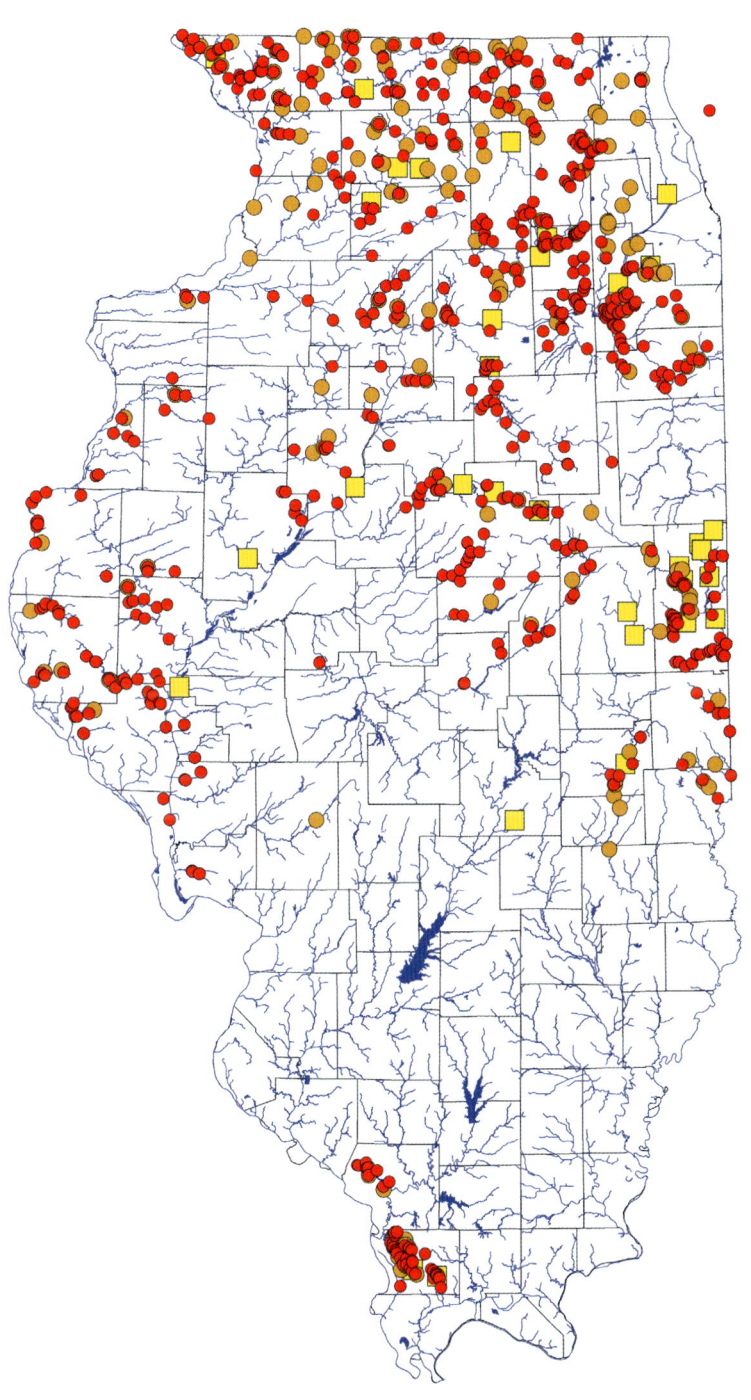

DARTERS AND PERCHES (Percidae)

355

SLOUGH DARTER *Etheostoma gracile* (Girard 1859)

IDENTIFICATION The Slough Darter has dark green saddles and wavy lines on a light yellow back and upper side of the body. There are green bars on the side of the male, green squares or mottling on the side of the female. The lower side and belly are yellow to white. There is a faint teardrop. The first dorsal fin has a blue-gray edge and base and a red submarginal band. During the breeding season the male has bright emerald-green bars on the yellow-brown side of the body and a bright red band in the first dorsal fin. The lateral line is incomplete with 13–27 pores and strongly arched near the front. There are 2 anal spines, a wide premaxillary frenum, an uninterrupted infraorbital canal (usually with 8 pores), and narrowly joined branchiostegal membranes. There are 40–55 lateral scales. Scutes are absent on the breast and belly. To 2¼ in. (6 cm).

SIMILAR SPECIES The Cypress Darter, *E. proeliare*, has brown or black dashes and no green bars or squares on the side of the body, 0–9 lateral-line pores, and 4 infraorbital canal pores.

HABITAT The Slough Darter lives among vegetation or woody debris in swamps, lakes, and backwaters or sluggish pools of headwaters, creeks, and small to medium rivers. It usually is found over sand or mud.

DISTRIBUTION IN ILLINOIS The Slough Darter remains, as it has since the Forbes and Richardson era, widespread in the southern one-third of the state. A record in Champaign County from 1928 (not recorded by Smith) indicates that the species at one time occurred farther north in Illinois than it does in the contemporary era, and absence of records in some of the most northern areas previously occupied (e.g., in Christian, Clark, and Cumberland Counties) suggests that the range of the species may be declining. Single records from the Smith and contemporary eras indicate infrequent occurrence in the lower Illinois River.

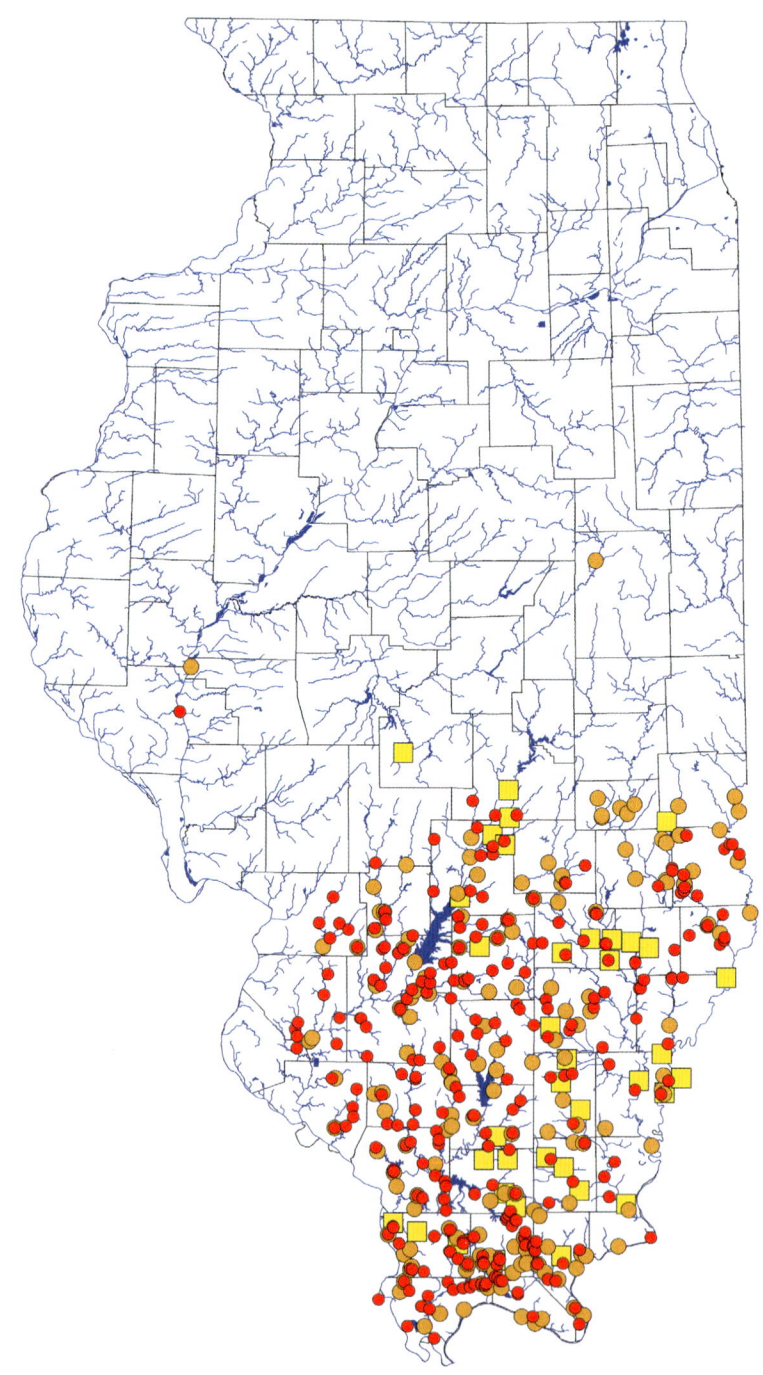

HARLEQUIN DARTER *Etheostoma histrio* Jordan & Gilbert 1887

IDENTIFICATION The Harlequin Darter is yellow-brown to green overall with 6–7 dark brown saddles, 7–11 dark brown blotches along the side of the body, and a black teardrop. There are many dark brown or black specks on the yellow belly and underside of the head and 2 large, dark brown to green spots (sometimes fused) on the caudal-fin base. The first dorsal fin is clear with a red-brown submarginal band. During the breeding season the male is emerald-green with brown and black mottling on the back and side of the body and dusky black dorsal and anal fins. There are 2 anal spines, a narrow or no premaxillary frenum, a relatively blunt snout, an uninterrupted infraorbital canal, and broadly joined branchiostegal membranes. The lateral line is complete with 45–58 pored scales. Scutes are absent on the breast and belly. To 3 in. (7.7 cm).

SIMILAR SPECIES The Greenside Darter, *E. blennioides*, lacks the large, dark brown to green spots on the caudal-fin base, has 5–8 green Us, Ws, or bars along the side of the body. The Banded Darter, *E. zonale*, has green bars on the side of the body and lacks the large, dark brown to green spots on the caudal-fin base.

HABITAT The Harlequin Darter lives near woody debris and large stones in sandy and gravelly runs of small to large rivers.

DISTRIBUTION IN ILLINOIS The Harlequin Darter was recorded by Forbes and Richardson (1909) from only 1 locality, in the Wabash River, and during the Smith era only from the Wabash River and the Embarras River in Cumberland and Jasper Counties. Contemporary-era records indicate that the species is much less frequently encountered in the Embarras River, where the only record is from 1983 in Coles County. It is more widespread in the Wabash River than previously and was recorded in the Ohio River at Pope County in 2019. The increased occurrence in the Wabash River, a large river that can sometimes be difficult to sample, is probably the result of targeted surveys.

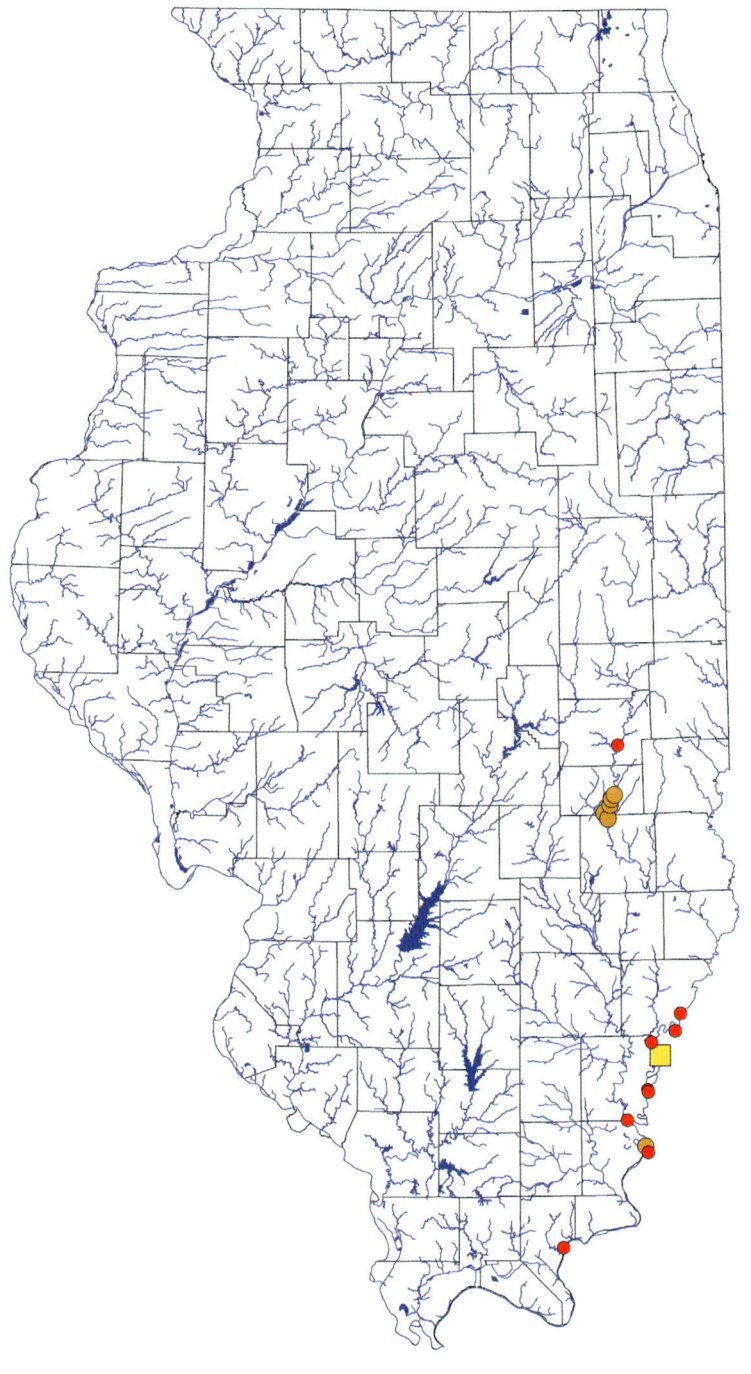

DARTERS AND PERCHES (Percidae)

357

STRIPETAIL DARTER *Etheostoma kennicotti* (Putnam 1863)

IDENTIFICATION The Stripetail Darter has a tan to yellow-brown back and side of the body, 6–7 dark brown saddles, dark brown blotches on the upper side, and larger dark brown blotches along the side of the body. The first dorsal fin has yellow-gold knobs (faint on juveniles) on the tips of the spines. The second dorsal, caudal, and pectoral fins have bold black bands. The teardrop is faint or absent. During the breeding season the male is yellow-brown with large, gold-orange knobs on the tips of the spines of the first dorsal fin. The snout is moderately pointed. There are 2 anal spines, a premaxillary frenum, and a widely interrupted infraorbital canal, with 4 pores anteriorly and 1 pore posteriorly. The branchiostegal membranes are narrowly joined. The lateral line is incomplete. There are 38–53 lateral scales. Scutes are absent on the breast and belly. To 3¼ in. (8.3 cm).

SIMILAR SPECIES The Fantail Darter, *E. flabellare*, has a pointed snout and protruding lower jaw, broadly joined branchiostegal membranes, black stripes on the side of the body, and 2 posterior infraorbital canal pores.

HABITAT The Stripetail Darter lives in shallow, rocky pools and riffles of headwaters, creeks, and small rivers.

DISTRIBUTION IN ILLINOIS Smith (1979) noted that although Forbes and Richardson (1909) had recorded only 1 locality for the species, the Stripetail Darter was abundant in the Ohio River basin of southern Illinois. The absence of records from the Cache River basin in Union County from the contemporary era is ironic given that the species was described from there in 1863.

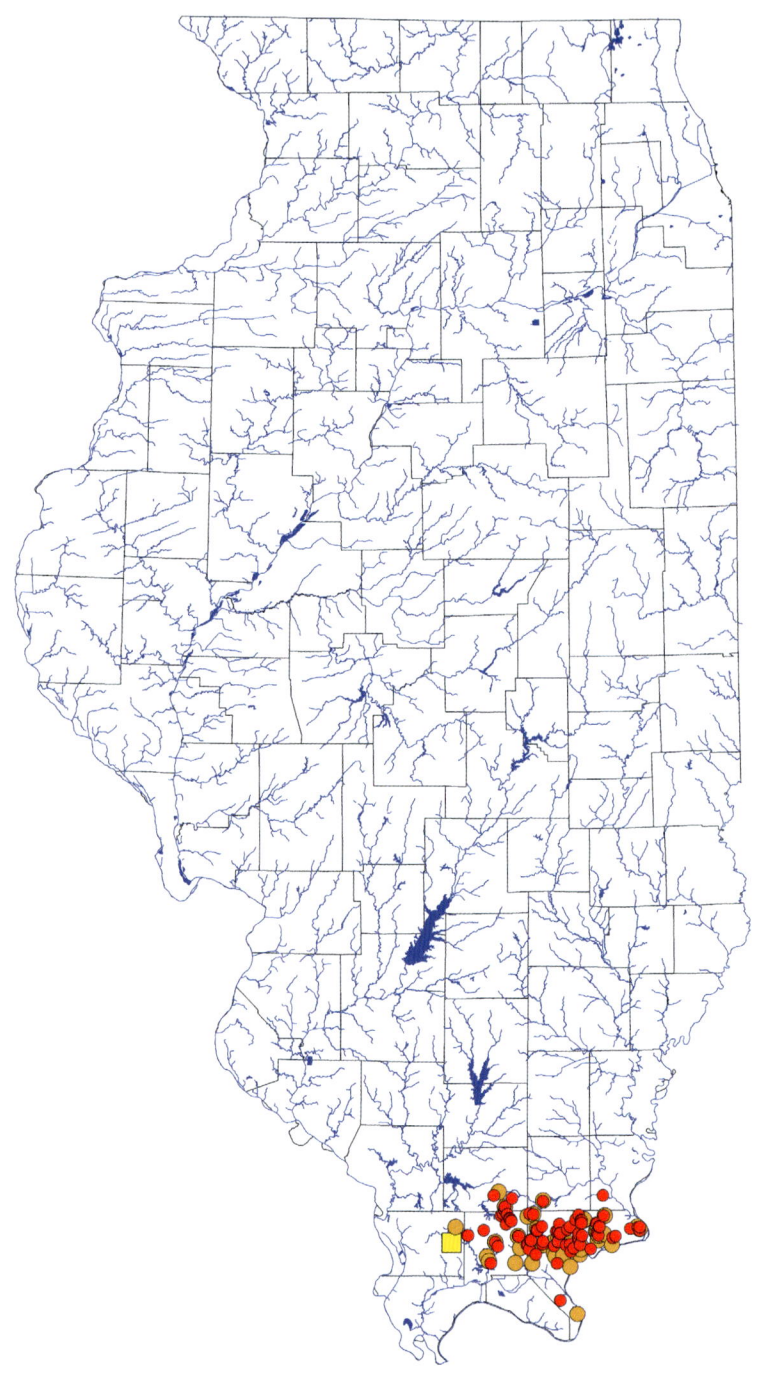

LEAST DARTER   *Etheostoma microperca* Jordan & Gilbert 1888

IDENTIFICATION The Least Darter has an extremely short lateral line with 0–3 pores. The body is deep and strongly compressed, and the snout is blunt. The back is olive-green with dark green or brown saddles, and the side of the body has rows of dark green or brown spots and blotches. The breast and belly are white to yellow, and there is a large, black teardrop. The first dorsal fin has a middle red band and a black edge and base. During the breeding season, the male has orange or red anal and pelvic fins and a large, lateral flap on the pelvic fins. There are 2 anal spines, a wide premaxillary frenum, 2–3 infraorbital canal pores, and narrowly joined branchiostegal membranes. There are 30–36 lateral scales. Scutes are absent on the breast and belly. To 1¾ in. (4.4 cm).

SIMILAR SPECIES The Cypress Darter, *E. proeliare*, and Iowa Darter, *E. exile*, are more slender and lack dark green on the body. The Cypress Darter has 4 infraorbital canal pores; the Iowa Darter has 8.

HABITAT The Least Darter is found over sand or mud in heavily vegetated lakes, headwaters, creeks, and small rivers.

DISTRIBUTION IN ILLINOIS The Least Darter is restricted to the Rock River and the Illinois River basin in the northeastern part of the state. It appears to be about as widespread as in earlier eras. Although Smith (1979) had 1 record from the shore of Lake Michigan, it has not been found there in more than 50 years.

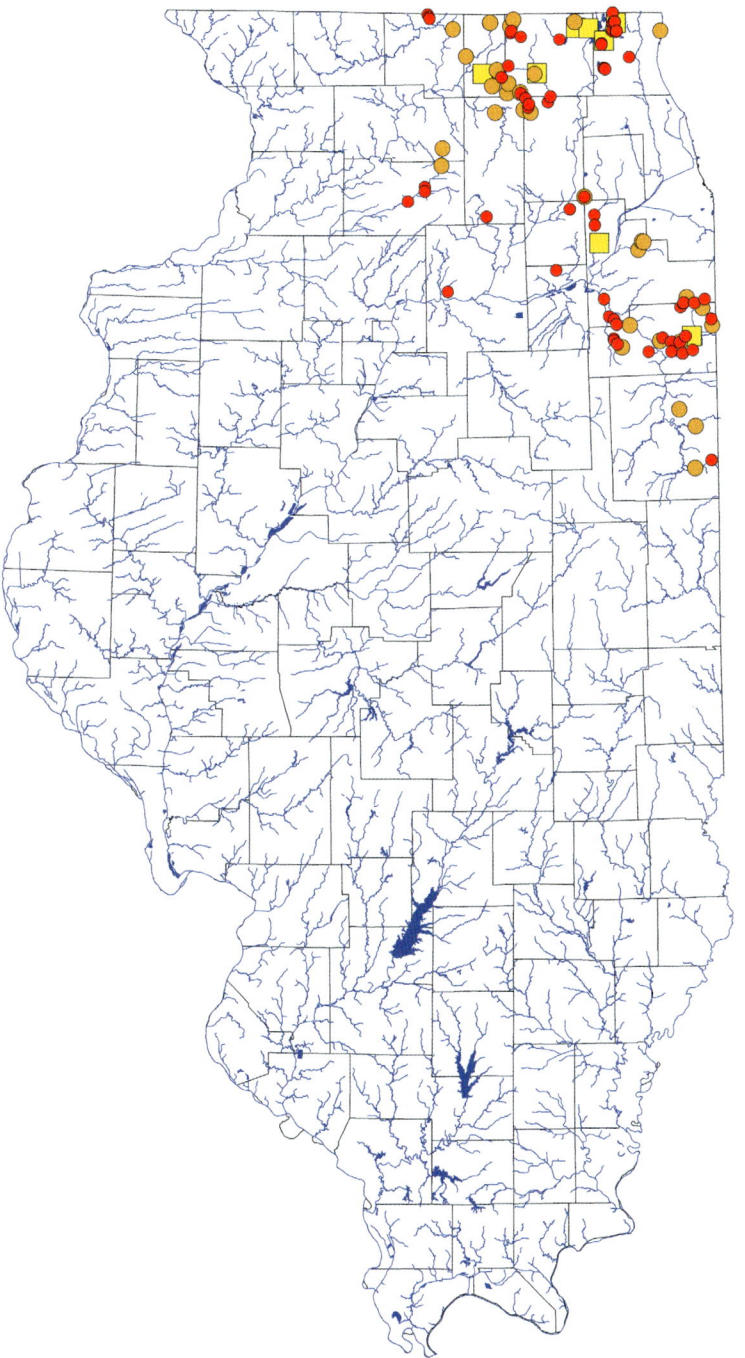

DARTERS AND PERCHES (Percidae)

JOHNNY DARTER *Etheostoma nigrum* Rafinesque 1820

IDENTIFICATION The Johnny Darter has a light yellow-brown body with 6 dark brown saddles on the back, wavy brown lines on the upper side of the body, and dark brown Ws and Xs along the side. A black bar extends from the eye onto the upper lip. During the breeding season the male has a black head, black on the anal and pelvic fins, and a black spot at the front of the first dorsal fin and may develop white knobs on the tips of the pelvic spine and rays. The snout is moderately blunt, and the body becomes increasingly slender to the caudal peduncle. There is 1 anal spine, no premaxillary frenum, an interrupted infraorbital canal, and narrowly joined branchiostegal membranes. The lateral line is complete with 35–56 pored scales. Scutes are absent on the breast and belly. To 2¾ in. (7.2 cm).

SIMILAR SPECIES The Bluntnose Darter, *E. chlorosoma*, has horizontal black blotches along the side of the body and a black bridle extending around an extremely blunt snout. The Channel Darter, *Percina copelandi*, has black (not brown) marks on the body and scutes on the breast and belly.

HABITAT The Johnny Darter lives in quiet to slow-flowing pools of headwaters, creeks, and small to medium rivers. It usually is found in shallow water over sand or mud, less often over a rocky substrate. It also lives on sandy shores of lakes.

DISTRIBUTION IN ILLINOIS The Johnny Darter has one of the most interesting distributions in the state in that it is extremely widespread except in a few areas that seem to provide suitable habitat. The species is extremely rare in most of the lower Illinois River basin and the immediate area to the south drained by the Mississippi River. Earlier eras showed a similar paucity of records in these areas, and although Smith (1979) suggested that the absence in this area may be related to the preponderance of low-gradient streams, the species is frequently encountered elsewhere in such environments. Overall, the species shows the same distribution it did in earlier eras.

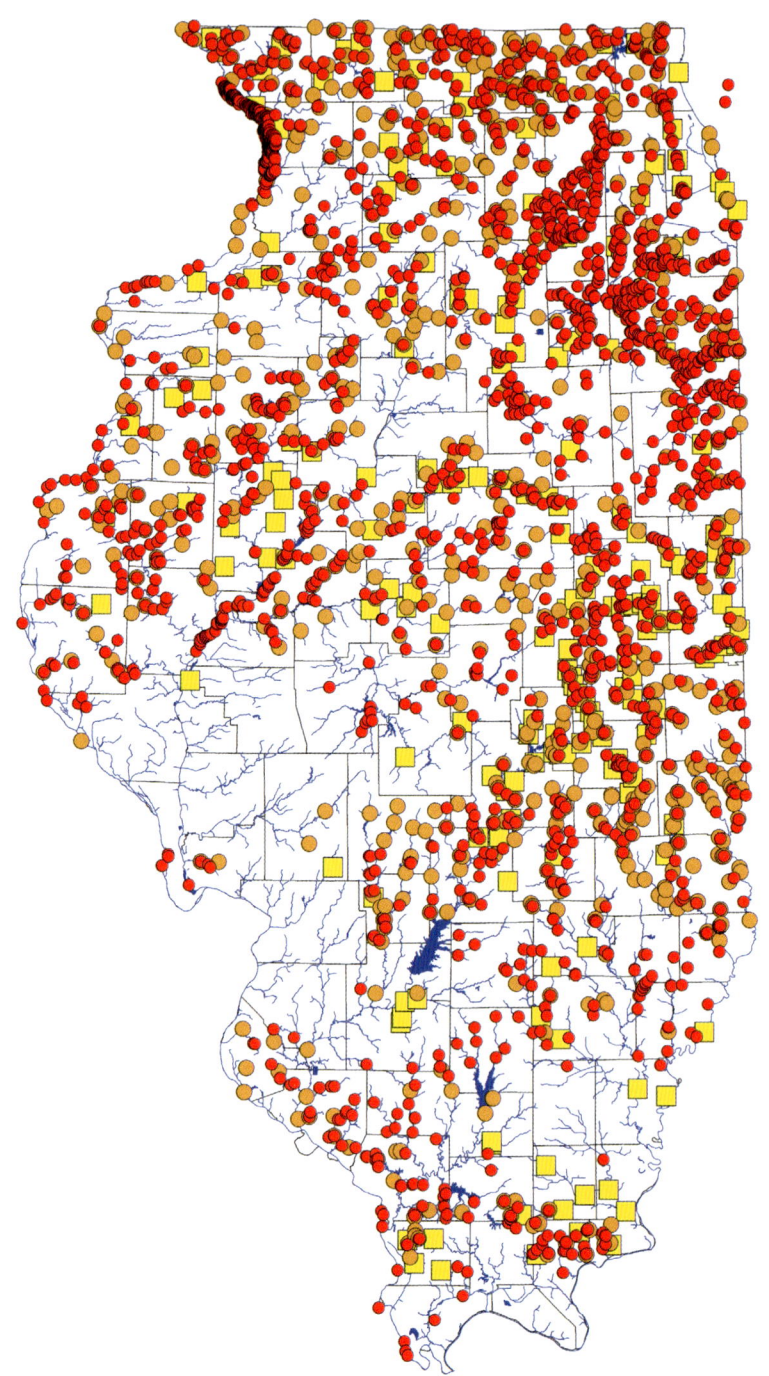

CYPRESS DARTER   *Etheostoma proeliare* (Hay 1881)

IDENTIFICATION The Cypress Darter has a very short and strongly arched lateral line with 0–9 pores. The back and upper side are olive-green with 6–9 dark brown saddles. There are black or brown dashes along the side of the body, many brown spots on the upper and lower sides of the body, and a faint teardrop. The first dorsal fin of the male often has a black edge and base, a black spot anteriorly, and a red submarginal band. During the breeding season the male has black anal and pelvic fins and a large, lateral flap on the pelvic fin. There are 2 anal spines, a wide premaxillary frenum, 4 infraorbital canal pores, and narrowly joined branchiostegal membranes. There are 34–38 lateral scales. Scutes are absent on the breast and belly. To 2 in. (4.8 cm).

SIMILAR SPECIES The Least Darter, *E. microperca*, is dark green and deeper bodied and has 2–3 infraorbital canal pores, orange or red anal and pelvic fins on the male, and a large, black teardrop. The Slough Darter, *E. gracile*, has green bars or squares on the side of the body, 13–27 lateral-line pores, and usually 8 infraorbital canal pores.

HABITAT The Cypress Darter lives among vegetation or woody debris in swamps, lakes, and backwaters or sluggish pools of headwaters, creeks, and small rivers. It usually is found over mud.

DISTRIBUTION IN ILLINOIS Forbes and Richardson (1909) recorded the Cypress Darter from the Little Wabash River basin in Wayne County, far north of any other records for the species. Other records are from 3 areas in southern Illinois where the distribution from the Smith and contemporary eras are remarkably similar. The species has been found only in a few localities but persists in those places.

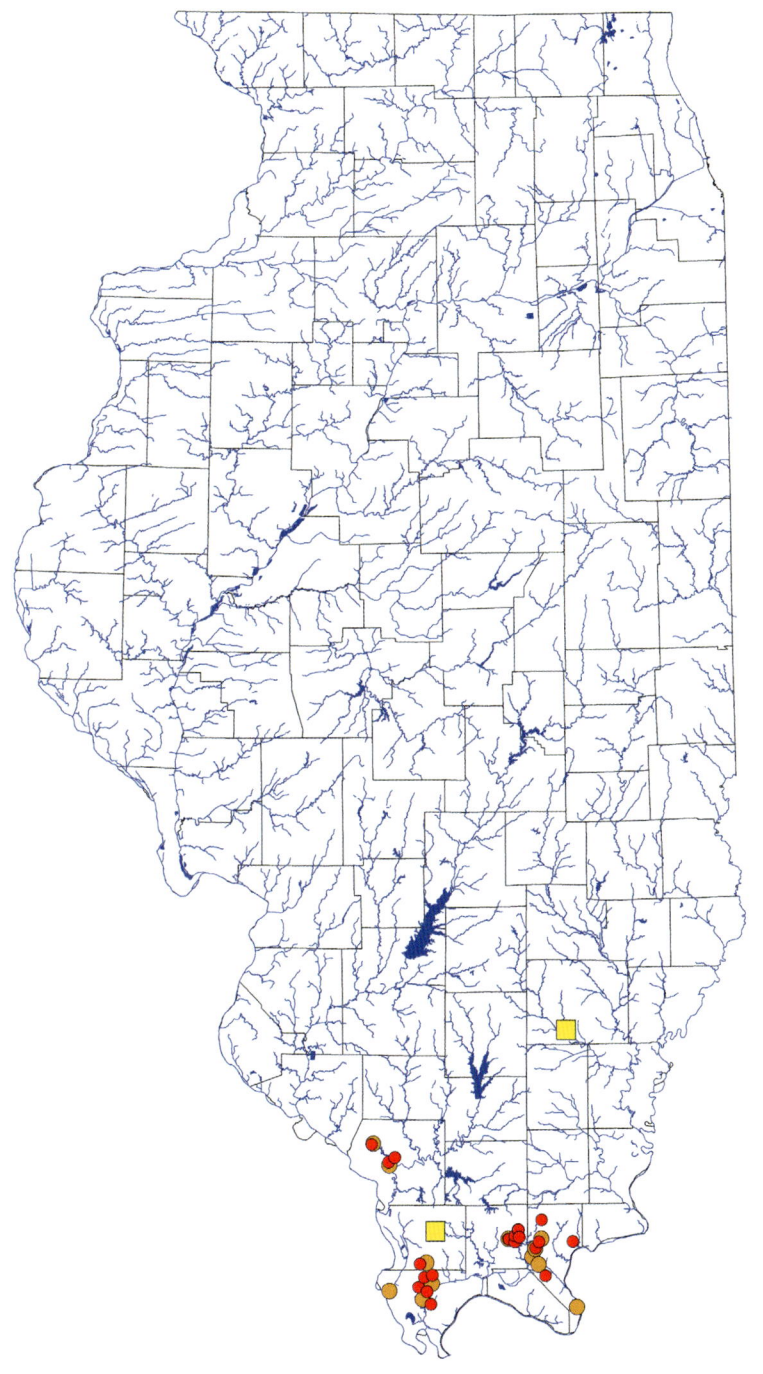

361

ORANGETHROAT DARTER *Etheostoma spectabile* (Agassiz 1854)

IDENTIFICATION The Orangethroat Darter has an arched body that is deepest at the nape or under the anterior portion of the first dorsal fin. The back and upper side are olive-green to light brown with 7–10 dark saddles. The side of the body usually has thin, dark brown stripes and 6–9 dark blue (on the male) or brown (on the female) bars separated by orange interspaces on the male and yellow-white interspaces on the female. The breast and belly are white to orange, and there is a thin black teardrop. The dorsal and caudal fins are clear with an orange band and a blue edge (brightest on males). During the breeding season the male has a blue or orange breast, blue anal fin, blue or black pelvic fin, an orange breast and branchiostegal membranes, and 2 orange spots on the caudal-fin base. There are 2 anal spines, a wide premaxillary frenum, a partly scaled cheek, usually 11–12 pectoral rays, an interrupted infraorbital canal with 3 posterior infraorbital pores, and narrowly joined branchiostegal membranes. The lateral line is incomplete with 17–35 pores. There are 38–55 (usually 42–50) lateral scales. Scutes are absent on the breast and belly. To 2¾ in. (7.2 cm).

SIMILAR SPECIES The Rainbow Darter, *E. caeruleum*, is deepest under the middle of the first dorsal fin and has red or orange on the anal fin, an uninterrupted infraorbital canal, and usually 13 pectoral rays. The Mud Darter, *E. asprigene*, is deepest under the middle of the first dorsal fin and has an uninterrupted infraorbital canal and a large, black blotch at the rear of the first dorsal fin.

HABITAT The Orangethroat Darter lives in gravel riffles and pools in headwaters, creeks, and small rivers.

DISTRIBUTION IN ILLINOIS The habitat of the Orangethroat Darter, a species that is found in all of the major basins in Illinois, is such that it seemingly could be found in virtually every small stream. However, it is inexplicably absent from large areas of the state. Records from the Forbes and Richardson era did not distinguish Rainbow and Orangethroat Darters, and most specimens from that era at the Illinois Natural History Survey (INHS) have been discarded. The distribution of the species in the contemporary era is very similar to that during the Smith era, although there are additional records in a few areas where it was not previously found.

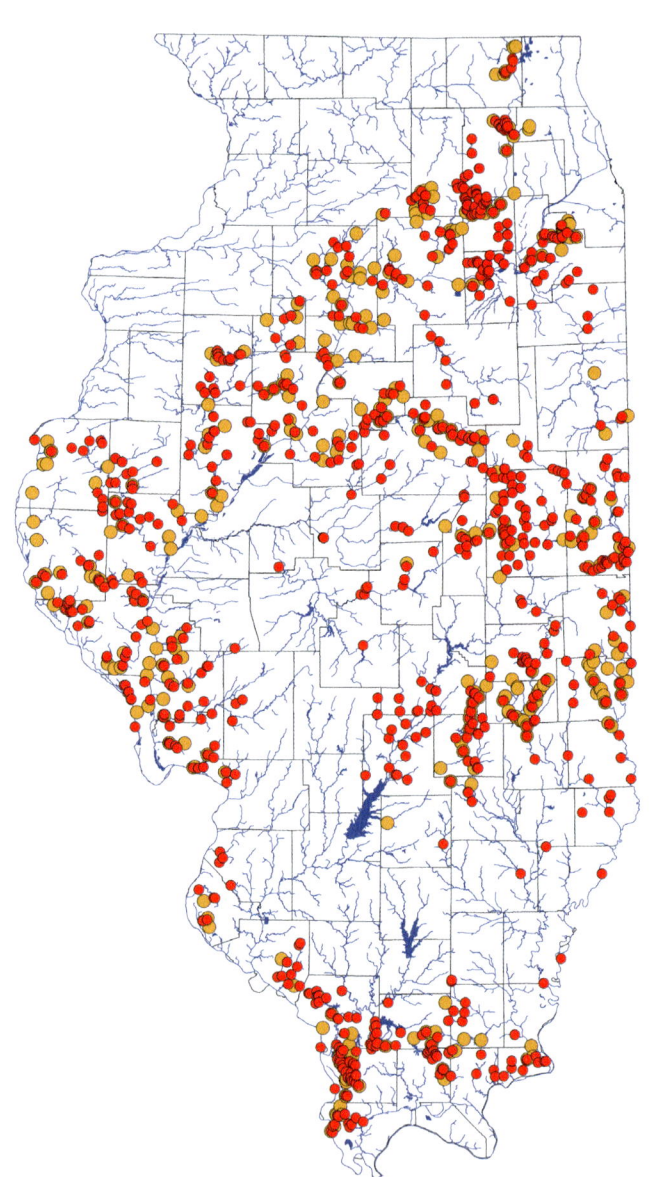

SPOTTAIL DARTER
Etheostoma squamiceps Jordan 1877

IDENTIFICATION The Spottail Darter has a light gray-brown back and side of the body with dark brown mottling. The lower side and belly are lighter brown or white. There are a black teardrop and 3 vertically aligned black spots on the caudal-fin base. The second dorsal and caudal fins have dark brown bands. During the breeding season the male has a black head, body, and fins (except the pectoral fin), and during spawning the male develops wide black and white bars on the side of the body. The second dorsal fin of the breeding male has small, white knobs on the tips of the rays, and the second and third branches of each ray are adnate and equal in length. The snout is moderately blunt. There are 2 anal spines, a premaxillary frenum, an interrupted infraorbital canal, and narrowly joined branchiostegal membranes. The lateral line is incomplete, there are 38–60 lateral scales. Scutes are absent on the breast and belly. To 3½ in. (8.8 cm).

SIMILAR SPECIES The Fringed Darter, *E. crossopterum*, is extremely similar but the second dorsal fin of the breeding male has a dusky white edge (fin rays often with small, black tips), and the third branch of each ray is much longer than the second branch. Branches of each ray are non-adnate.

HABITAT The Spottail Darter lives in shallow rocky pools and riffles of headwaters, creeks, and small rivers.

DISTRIBUTION IN ILLINOIS A record in White County from the Forbes and Richardson era indicates that the species at one time occurred farther north in Illinois than it does in the contemporary era. The species has the same distribution in southern Illinois in the contemporary era as it did in the Smith era except for records in the lower Saline River basin, where the species was not previously reported. The Saline basin was badly polluted in the past as a result of coal mining in the region, and the species may have recently invaded. A record in Forbes and Richardson (1909) in Shelby County was based on a misidentified Mud Darter, *E. asprigene*, and records recorded by Smith (1979) from the Cache River basin were based on the Fringed Darter, *E. crossopterum*.

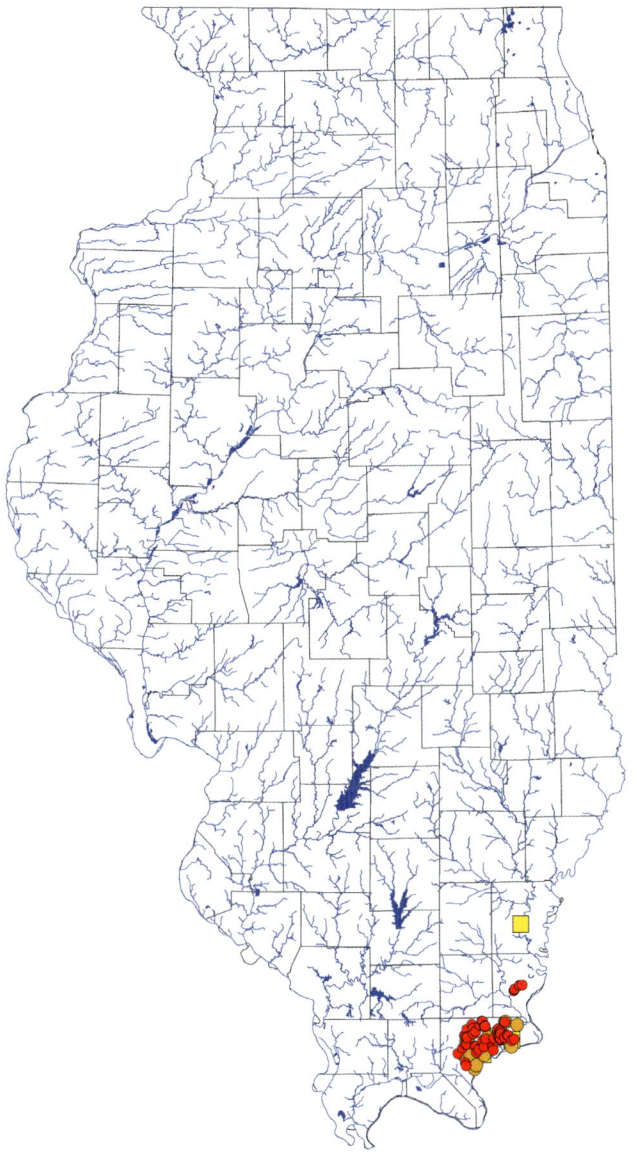

DARTERS AND PERCHES (Percidae)

363

BANDED DARTER *Etheostoma zonale* (Cope 1868)

IDENTIFICATION The Banded Darter is yellow-green with 6 dark green saddles and 9–13 large, dark green bars on the side of the body extending onto the belly and underside of the caudal peduncle to join those of the other side. There are 2 large, yellow spots on the caudal-fin base and a narrow black teardrop that often is broken into 2 black spots. The first dorsal fin has a green edge and red near the base. During the breeding season the bars on the male are bright green. There are 5 (rarely, 6) branchiostegal rays, 2 anal spines, a wide premaxillary frenum, a moderately pointed snout, an uninterrupted infraorbital canal, and broadly joined branchiostegal membranes. The lateral line is complete with 36–63 (usually 45 or more) pored scales. Scutes are absent on the breast and belly. To 3 in. (7.8 cm).

SIMILAR SPECIES Other darters in Illinois have 6 branchiostegal rays. The Greenside Darter, *E. blennioides*, has a blunter snout, fewer than 9 green bars on the side that rarely—except on large males—join those on the other side of the body. The Harlequin Darter, *E. histrio*, lacks bars on the side of the body and has 2 large, dark brown to green spots on the caudal-fin base.

HABITAT The Banded Darter lives among large stones and gravel in riffles of creeks and small to medium rivers.

DISTRIBUTION IN ILLINOIS The Banded Darter occurs in the northern half of Illinois. Almost all records are from the Illinois and Rock River basins, although the Galena River basin and Mississippi River each have 1 record, and the Vermilion River drainage (Wabash) has 2 records. The distribution of the species is similar in the contemporary era to what it was in earlier eras, except that the species is more widespread in Salt Creek, a tributary of the Sangamon River, and perhaps has disappeared from the upper Sangamon River mainstem. The 2 records, 1 from Smith (1979) and 1 contemporary, in the Vermilion River basin in Champaign and Vermilion Counties, are likely the result of introduction given the species's absence elsewhere in the Wabash River basin.

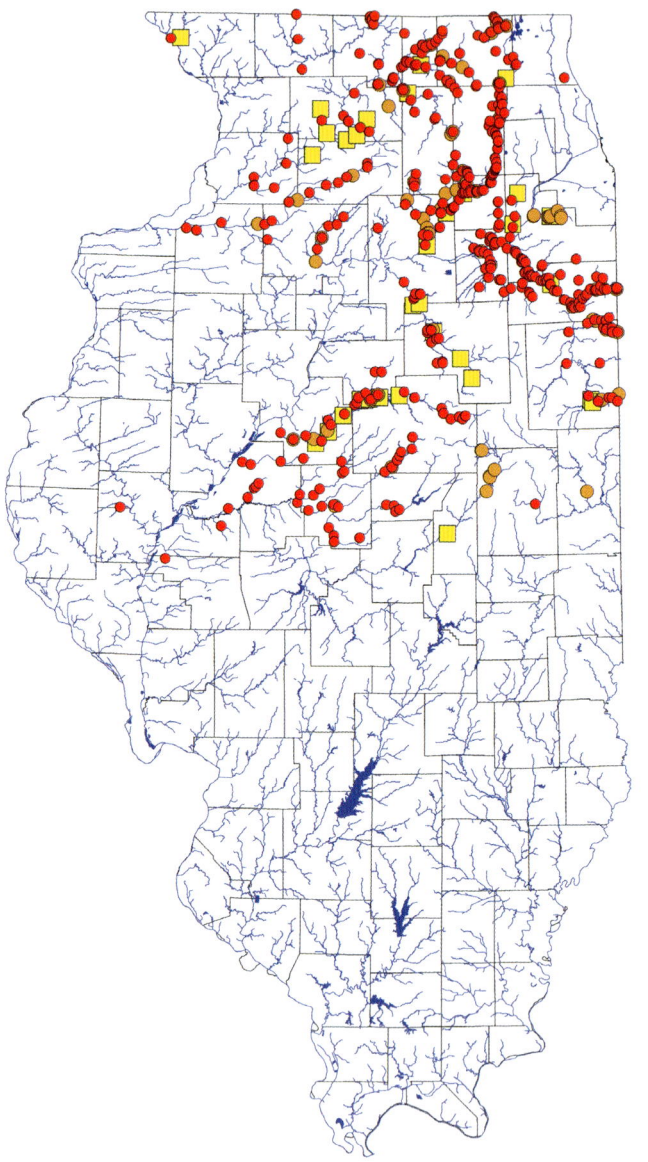

YELLOW PERCH *Perca flavescens* (Mitchill 1814)

IDENTIFICATION The Yellow Perch has a body that is relatively deep and compressed anteriorly and tapers to a slender caudal peduncle and a slightly forked caudal fin. The large mouth extends to the middle of the eye; there are no large, canine teeth. The back is light brown with 6–9 brown saddles that extend onto the yellow side of the body as triangular bars. There is a black blotch at the rear and often a black blotch at the front of the dusky first dorsal fin. The anal and paired fins are yellow to red. The lateral line is complete with 52–61 lateral scales, 12–14 dorsal rays, and 6–8 anal rays. To 16 in. (40 cm).

SIMILAR SPECIES The Sauger, *Sander canadense*, and the Walleye, *S. vitreum*, are more slender, lack the brown triangular bars on the side of the body, and have large, canine teeth, 17 or more dorsal rays, 11 or more anal rays, and more than 77 lateral scales.

HABITAT The Yellow Perch lives in lakes, ponds, and pools of creeks and small to large rivers. It is found most often in clear water near vegetation.

DISTRIBUTION IN ILLINOIS The Yellow Perch is generally distributed in the glacial lakes and streams in northeastern Illinois, along the shore of Lake Michigan, and in the Mississippi River above Quincy in Adams County. Elsewhere in the state it is less frequently encountered, and most contemporary-era records probably reflect stocking. It is decidedly less frequently found in the Illinois River than it was in the Forbes and Richardson and Smith eras.

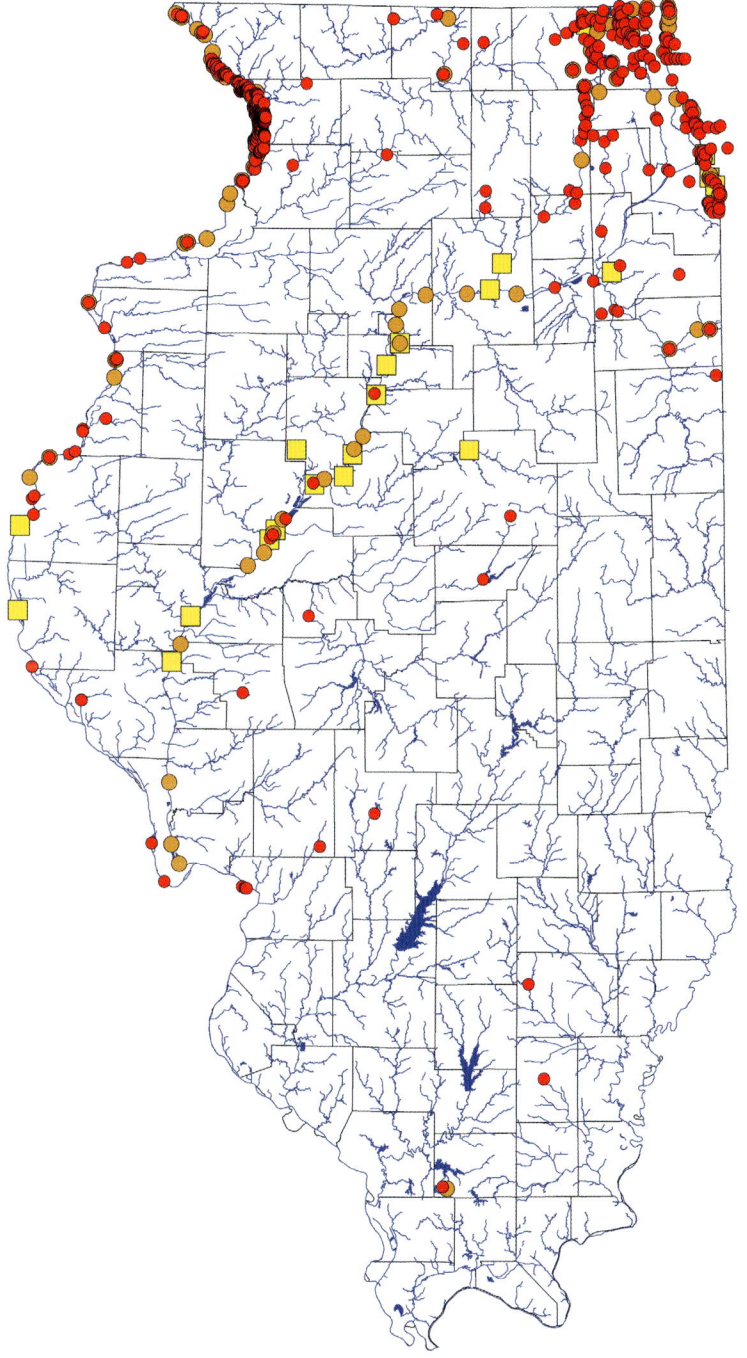

LOGPERCH *Percina caprodes* (Rafinesque 1818)

IDENTIFICATION The Logperch has many alternating long and short, dark brown bars along the yellow-brown side. The bars extend over the back to join those on the other side of the body. The lower side is yellow to white. There are a distinct black spot on the caudal-fin base and a dusky teardrop. The bulbous snout extends well beyond the upper jaw, there is a wide flat area between the eyes, and the body becomes increasingly slender to the caudal peduncle. The lateral line is complete with 67–100 pored scales. The head canals are uninterrupted. There are 2 anal spines. Scutes are present on the breast and on the male in a row along the midline of the belly. To 7¼ in. (18 cm).

SIMILAR SPECIES The Slenderhead Darter, *P. phoxocephala*, has round brown-black blotches along the midside rather than alternating long and short bars and lacks a wide flat area between the eyes.

HABITAT The Logperch is found most often over gravel and sand in medium to large rivers but can be found almost anywhere from small, fast-flowing rocky creeks to margins of lakes. As the common name suggests, individuals often are found near logs and brush.

DISTRIBUTION IN ILLINOIS The Logperch is statewide in distribution as it was in earlier eras, although it no longer is found in some of the smaller rivers in central Illinois, where it previously occurred. It also appears to be gone from the Illinois portion of the Lake Michigan shoreline.

REMARKS Two subspecies of *P. caprodes* and intergrades with a third, *P. c. fulvitaenia*, which has an orange band in the first dorsal fin and is found in adjacent Missouri, occur in Illinois (Morris & Page 1981). *Percina c. semifasciata*, with an unscaled nape, is in northern Illinois and intergrades with *P. c. fulvitaenia* in the Illinois and Kaskaskia Rivers and western Illinois. *Percina c. caprodes*, with a scaled nape and no orange band in the dorsal fin, is in the Wabash and Ohio Rivers.

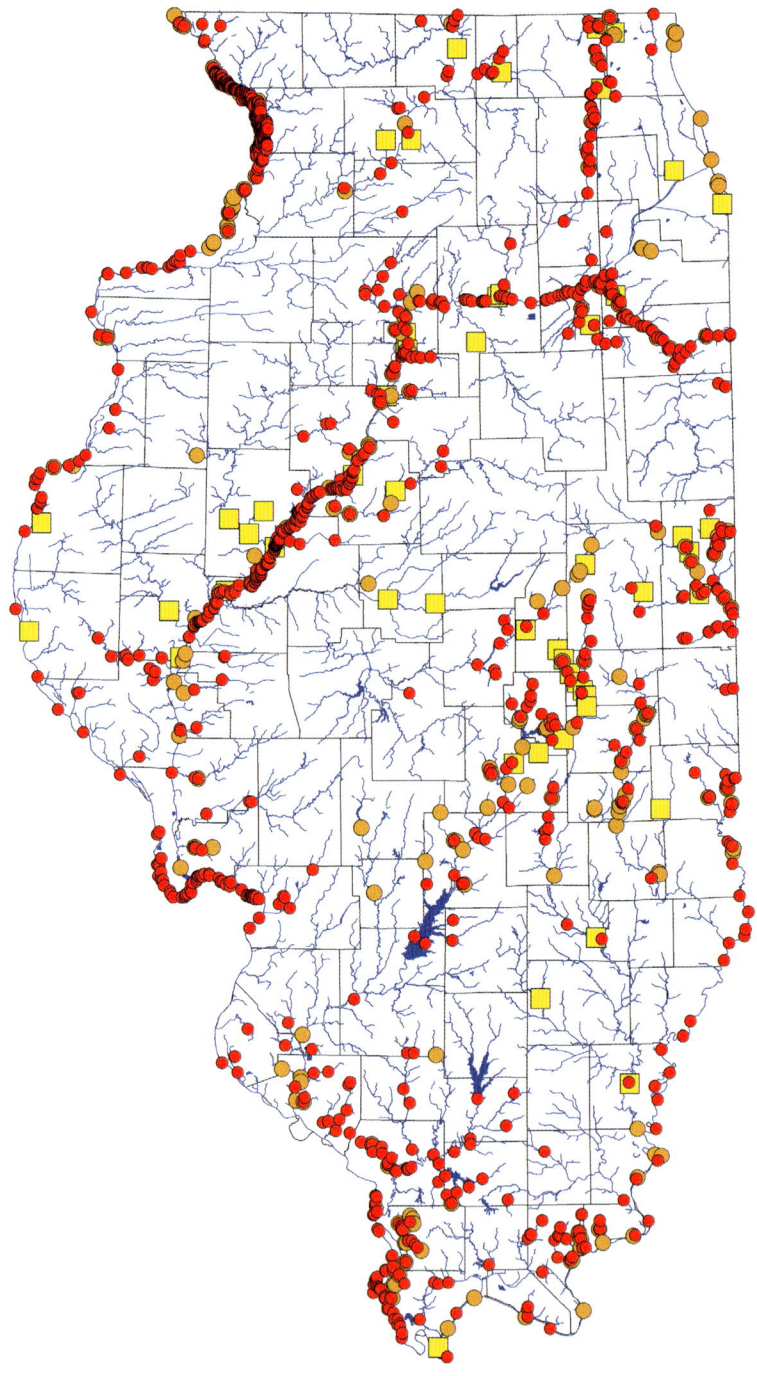

CHANNEL DARTER *Percina copelandi* (Jordan 1877)

IDENTIFICATION The Channel Darter has black Ws and Xs on the light olive to yellow back and upper side of the body and a medial black spot at the base of the caudal fin. There are 8–10 horizontally oblong to square black blotches along the midside of the body, and a dusky black teardrop (often reduced to a spot). Males have a black edge on the first dorsal fin and during the breeding season are black on the underside of the head and body and have black pelvic fins; tubercles may develop on the anal- and pelvic-fin rays. The body is slender, the snout is blunt, and there is no (or a very narrow) premaxillary frenum. The lateral line is complete with 42–67 pored scales, the head canals are uninterrupted, and there are 2 anal spines. Scutes are present on the breast and on the male in a row along the midline of the belly. To 2½ in. (6.2 cm).

SIMILAR SPECIES The Johnny Darter, *Etheostoma nigrum*, has brown (not black) marks on the body, no black edge on the first dorsal fin, and no scutes on the breast or belly. The Blackside Darter, *P. maculata*, and Dusky Darter, *P. sciera*, have larger, rounder black blotches along the side of the body and a wide premaxillary frenum. The River Darter, *P. shumardi*, has a large, black spot at the rear of the first dorsal fin and lacks oblong or square blotches along the side of the body.

HABITAT The Channel Darter lives in rocky and sandy, flowing pools and along the margins of riffles of small to large rivers.

DISTRIBUTION IN ILLINOIS Although Smith (1979) did not report the Channel Darter from Illinois, it had been collected along with the Gilt Darter, *P. evides*, in the Wabash River at New Harmony, Indiana, which he did report. Both the Channel and Gilt Darters were collected in the 1800s at Terre Haute, Vincennes, and New Harmony, Indiana, and certainly occurred on the Illinois side of the river as well. The last reported occurrence of the Channel Darter in the Wabash River was in 1958.

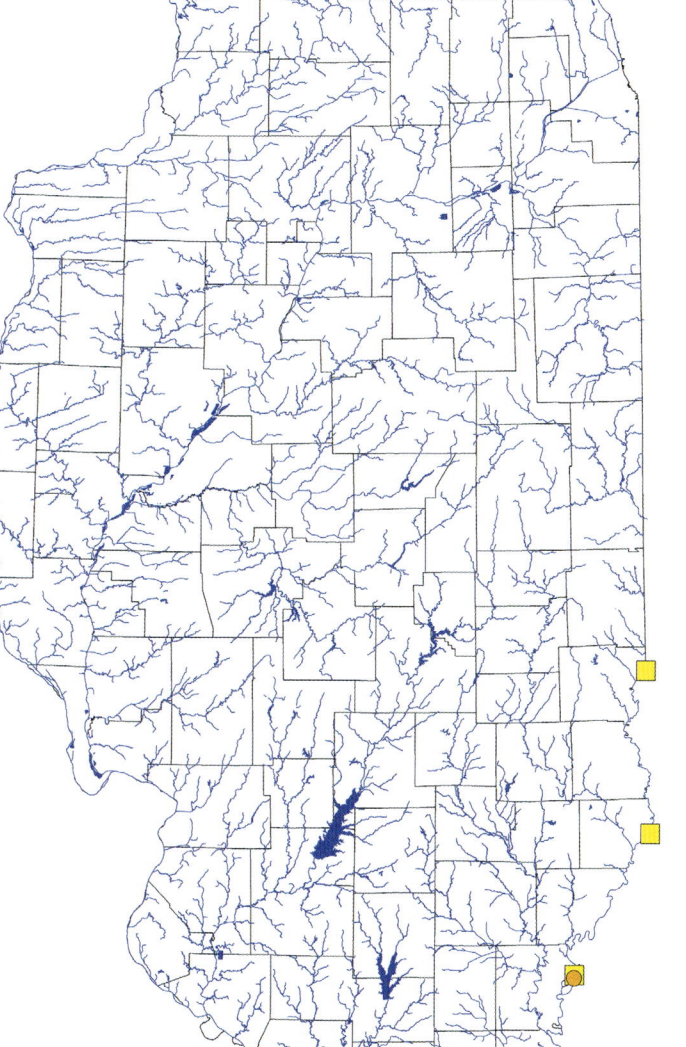

GILT DARTER *Percina evides* (Jordan & Copeland 1877)

IDENTIFICATION The Gilt Darter has 6–8 dark green saddles on the light olive-green back that extend ventrally to merge with dark green blotches along the midside of the body. The underside of the head, breast, and belly are yellow to orange. The first dorsal fin often is orange or amber; other fins are clear. There is a black teardrop and 2 large, white or pale yellow spots at the base of the caudal fin. During the breeding season the male has blue-green bars over the back and a bright orange breast and belly. The body is slender, deepest under the first dorsal fin, and the snout is moderately pointed. The lateral line is complete with 51–77 pored scales, and the head canals are uninterrupted. There are 2 anal spines. Scutes are present on the breast and, on the male in row along the midline of the belly. To 3¾ in. (9.6 cm).

SIMILAR SPECIES The Blackside Darter, *P. maculata*, Dusky Darter, *P. sciera*, and River Darter, *P. shumardi*, lack broad bands that extend from the back to the midside and large, white or yellow spots on the caudal-fin base. The River Darter also has a large, black spot at the rear of the first dorsal fin. The Stargazing Darter, *P. uranidea*, lacks large, white or yellow spots on the caudal-fin base and has a narrow or no premaxillary frenum.

HABITAT The Gilt Darter lives in rocky riffles of small to large rivers.

DISTRIBUTION IN ILLINOIS The Gilt Darter was collected at 3 localities in the Forbes and Richardson era, 1 in the lower Rock River and 2 in the Wabash River, and again in the lower Rock River in 1932 (Smith 1979), although the exact location of this Smith-era record could not be determined. No contemporary-era collections from Illinois exist, and the species appears to be extirpated.

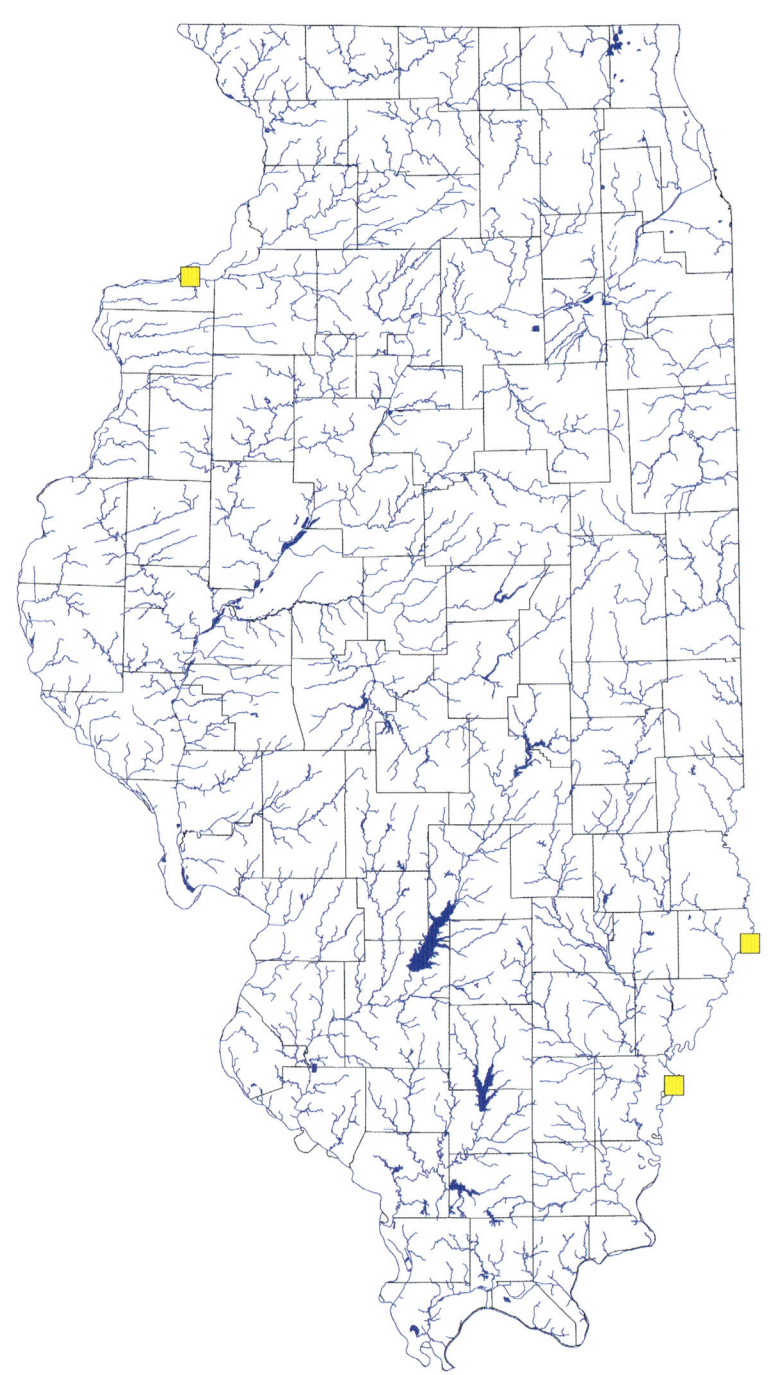

BLACKSIDE DARTER   *Percina maculata* (Girard 1859)

IDENTIFICATION The Blackside Darter has a distinct black spot at the base of the caudal fin and a bold black teardrop. The back and upper side of the body are olive-green to tan with black mottling and 8–9 dark brown to black saddles. Along the side are 6–9 large, oval, black blotches. The first dorsal fin is clear to dusky overall with a small, black blotch anteriorly. The second dorsal and caudal fins have dusky bands; other fins are clear. The body is slender, is deepest under the first dorsal fin, and tapers to a narrow caudal peduncle. The snout is pointed. The lateral line is complete with 53–81 pored scales, and the head canals are uninterrupted. There are 2 anal spines. Scutes are present on the breast and on the male in a row along the midline of the belly. To 4¼ in. (11 cm).

SIMILAR SPECIES The Dusky Darter, *P. sciera*, has a vertical row of 3 diffuse, dark brown to black spots on the caudal-fin base but no bold medial black spot. The River Darter, *P. shumardi*, has a large, black spot at the rear of the first dorsal fin, black bars on the side of the body, and a narrow or no premaxillary frenum.

HABITAT The Blackside Darter lives in gravelly and sandy, flowing pools of headwaters, creeks, and small to medium rivers.

DISTRIBUTION IN ILLINOIS The Blackside Darter is essentially statewide, except for an absence from a large area of southwestern Illinois from Randolph County north to Sangamon County. The species was absent from this area in the earlier eras as well, suggesting a natural explanation. The species is contemporarily unrecorded from a large area in the northwestern part of the state, where it was present during the Forbes and Richardson era. A few contemporary records suggest that this species may be reestablishing in this area.

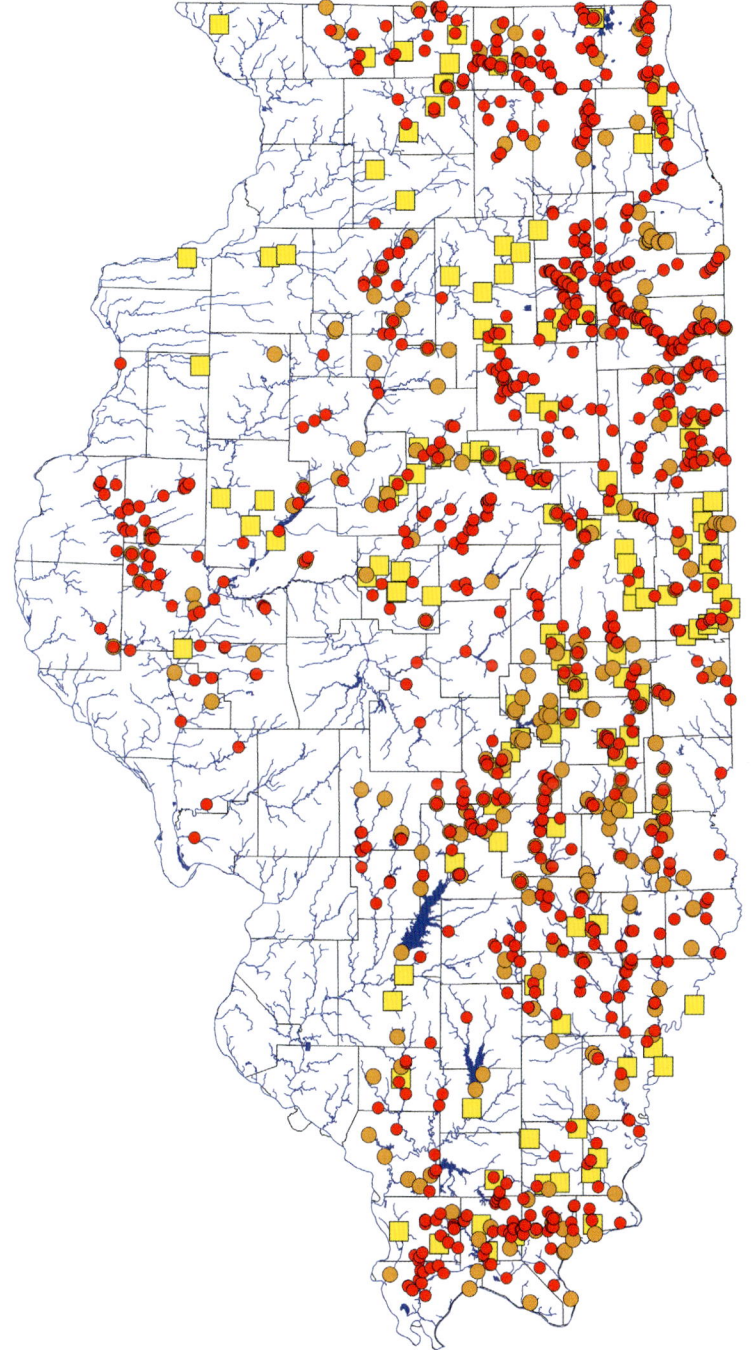

SLENDERHEAD DARTER *Percina phoxocephala* (Nelson 1876)

IDENTIFICATION The Slenderhead Darter has a back and upper side of the body that is yellow-brown with dark brown blotches or wavy lines. There are 10–16 round brown-black blotches along the midside, a bright orange band on the first dorsal fin, and a small, distinct, black spot on the caudal-fin base. The lower side is white to yellow, and often there is a black teardrop. The body is slender, and the snout is long and pointed. The lateral line is complete with 58–80 pored scales, and the head canals are uninterrupted. There are 2 anal spines. Scutes are present on the breast and on the male in row along the midline of the belly. To 3¾ in. (9.6 cm).

SIMILAR SPECIES The Logperch, *P. caprodes*, has alternating long and short bars on the back and side of the body and a wide, flat area between the eyes.

HABITAT The Slenderhead Darter lives in gravel runs and riffles of creeks and small to medium rivers.

DISTRIBUTION IN ILLINOIS The distribution of the Slenderhead Darter is essentially the same in the contemporary era as in earlier eras. It occurs statewide except in lowland areas in southern Illinois and in Lake Michigan and the upper Illinois River and its tributaries in extreme northeastern Illinois. Given the preference of this species for gravelly streams, its absence from lowland areas in southern Illinois is understandable. It is absent from the entire Big Muddy River and Cache River basins and is nearly absent from the Saline River and Skillet Fork basins. However, its absence in streams in northeastern Illinois is more likely explained by the fact that this is at the edge of the range for this species. The Slenderhead Darter is absent from all of the Great Lakes basin except for Lake Winnebago tributaries in Wisconsin (Becker 1983).

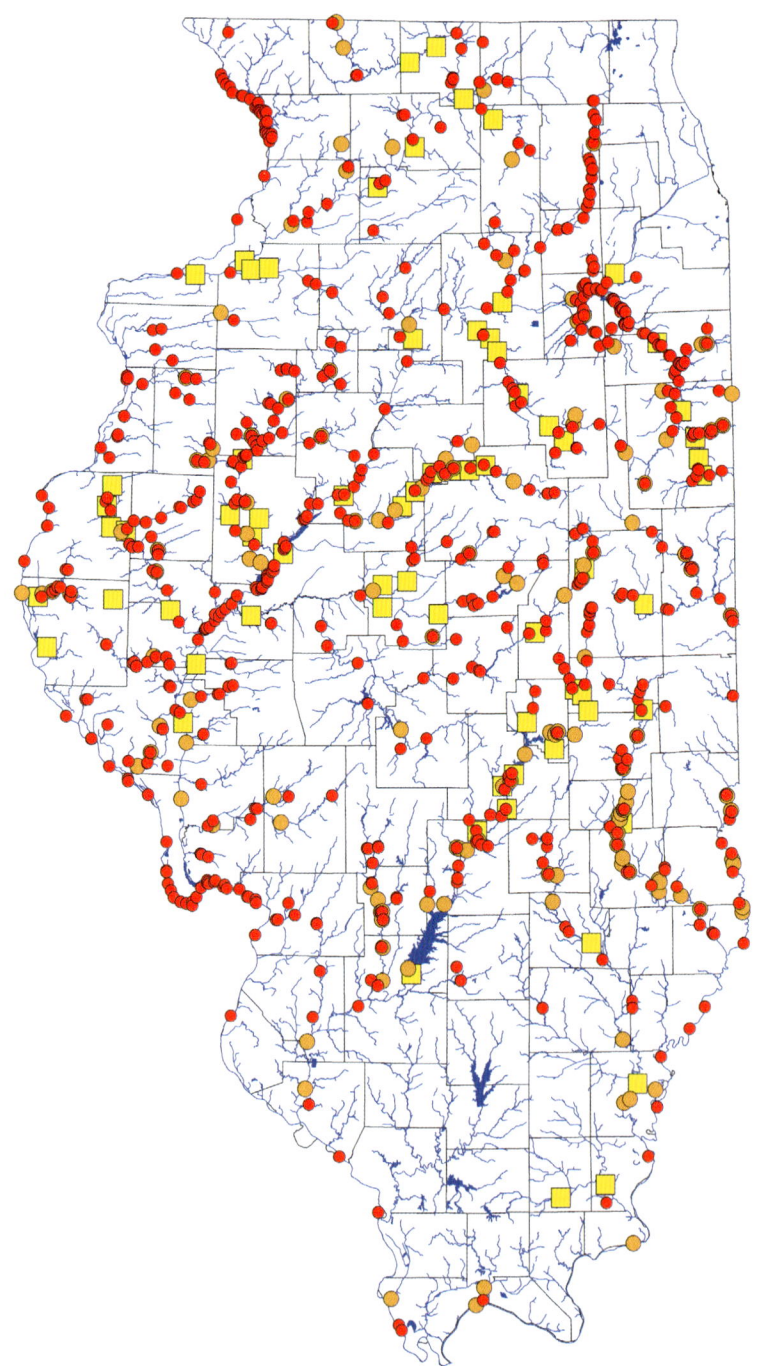

DUSKY DARTER *Percina sciera* (Swain 1883)

IDENTIFICATION The Dusky Darter has an olive to dusky black back and upper side of the body with 8–9 dark brown saddles and dark brown mottling on the upper side. There are 8–12 oval black blotches along the side of the body and a vertical row of 3 dark brown spots on the caudal-fin base; the lower 2 spots often are fused. Fins are mostly without black spots although often there is a dusky spot posteriorly on the first dorsal fin. There is usually no teardrop. The body is slender, is deepest under the first dorsal fin, and tapers to a narrow caudal peduncle. The snout is pointed. The lateral line is complete with 56–78 pored scales, the head canals are uninterrupted, and there are 2 anal spines. Scutes are present on the breast and on the male in a row along the midline of the belly. The preopercle usually has 5–16 serrae on the posterior edge. To 5 in. (13 cm).

SIMILAR SPECIES The Blackside Darter, *P. maculata*, has a distinct, medial black spot at the base of the caudal fin, usually has a black teardrop, and has no serrae on the edge of the preopercle. The River Darter, *P. shumardi*, has a large, black spot at the rear of the first dorsal fin, black bars on the side of the body, and a narrow or absent premaxillary frenum.

HABITAT The Dusky Darter lives in gravel runs of creeks and small to medium rivers. It often is near brush and sometimes in riffles.

DISTRIBUTION IN ILLINOIS The Dusky Darter is more widely distributed in the contemporary era than in earlier eras. Forbes and Richardson (1909) recorded only 1 locality for the species (in the Embarras River in Coles County), and multiple records for this species in the Smith era were available only for the Vermilion (Wabash), Embarras, and Little Wabash River basins. Contemporary-era records show the species to be more widely distributed in these basins, more frequently encountered in Ohio River tributaries, and present in the lower Mississippi River.

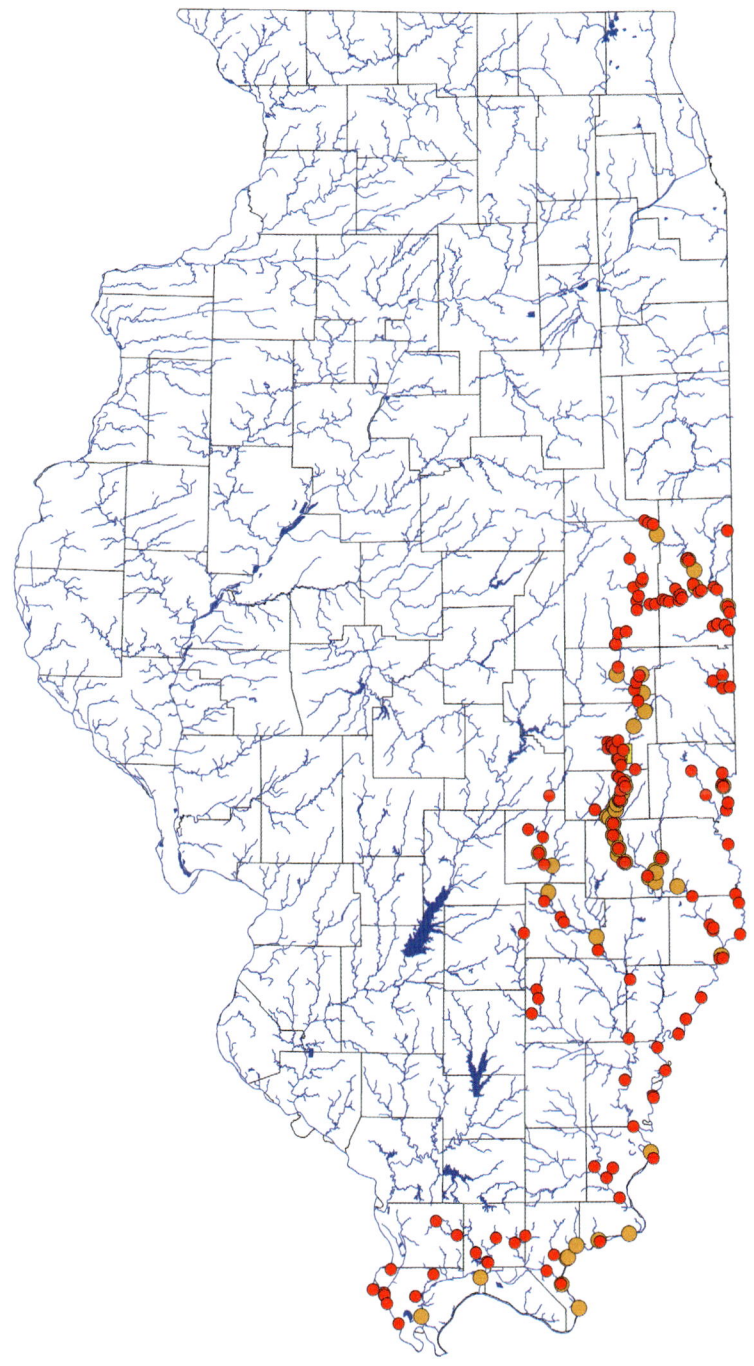

RIVER DARTER *Percina shumardi* (Girard 1859)

IDENTIFICATION The River Darter is dusky olive-green on the back and upper side and has 8–15 dusky to black bars along the side of the body, a small, black spot at the base of the caudal peduncle, and a black teardrop. There is a small, black spot at the front of the first dorsal fin, and a large, black spot is at the rear. The body is slender and tapers to a narrow caudal peduncle. The snout is moderately pointed, and the premaxillary frenum is absent or narrow. Large males have a long anal fin that extends to the caudal fin. The lateral line is complete with 46–62 pored scales, there are 2 anal spines, and the head canals are uninterrupted. Scutes are present on the breast and on the male in a row along the midline of the belly. To 3 in. (7.8 cm).

SIMILAR SPECIES The Blackside Darter, *P. maculata*, and Dusky Darter, *P. sciera*, lack a large, black spot at the rear of the first dorsal fin and have rounder black blotches along the side of the body and a wide premaxillary frenum. The Channel Darter, *P. copelandi*, lacks a large, black spot at the rear of the first dorsal fin and has horizontally oriented blotches along the side of the body. The Stargazing Darter, *P. uranidea*, has a red-brown back with 4 dark brown saddles and larger blotches on the side of the body and lacks a large, black spot at the rear of the first dorsal fin.

HABITAT The River Darter occupies rocky riffles and runs of medium to large rivers. Rarely it is found in creeks and small rivers.

DISTRIBUTION IN ILLINOIS The River Darter has a distribution in the contemporary era that is similar to that in the earlier eras, although it appears to be nearly gone from the Illinois and Kaskaskia Rivers, with a single recent record for each. It was widely distributed in these rivers in the Forbes and Richardson era.

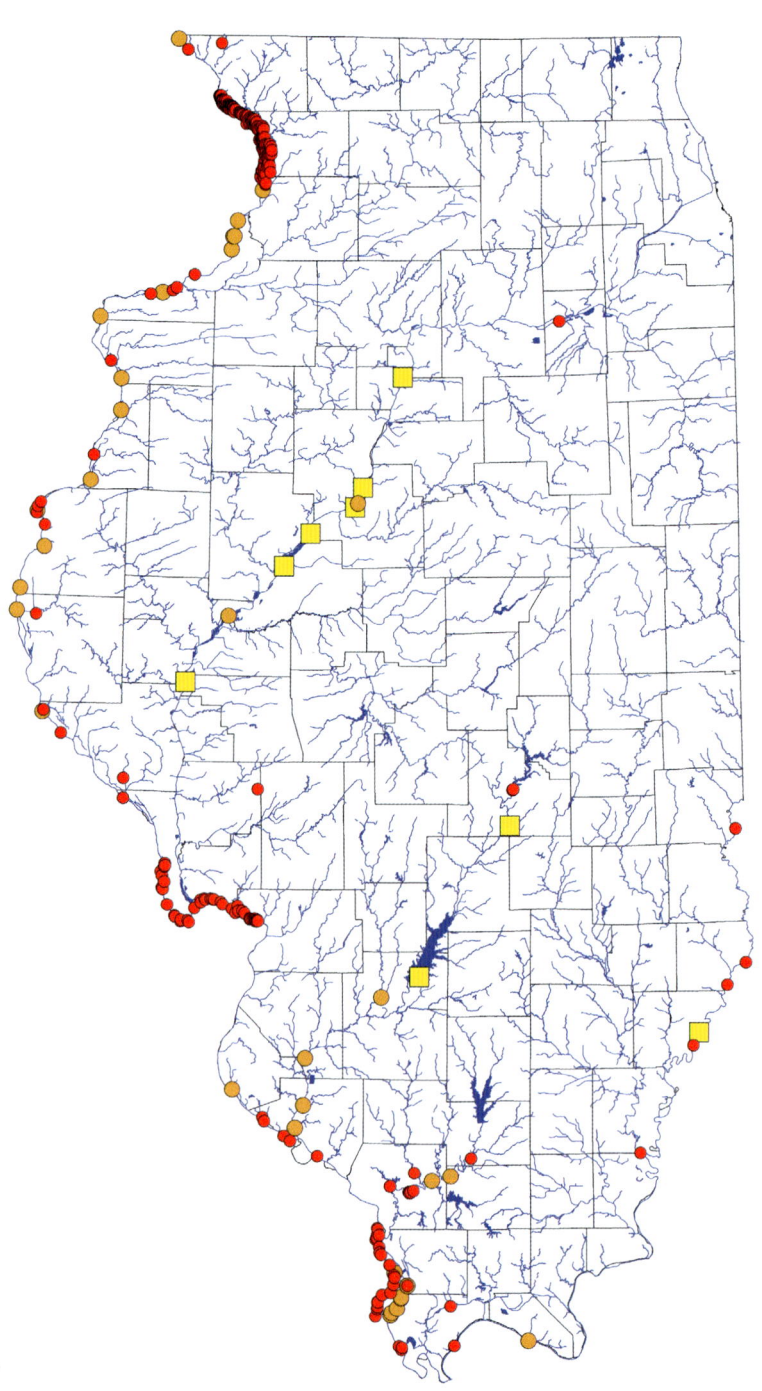

STARGAZING DARTER *Percina uranidea* (Jordan & Gilbert 1887)

IDENTIFICATION The Stargazing Darter has a dull red-brown back and upper side with 4 dark brown saddles extending down the side of the body to the lateral line. There are 9–12 dark brown blotches along the midside, and the lower side of the body is pale yellow. A small, black spot usually is present at the front of the first dorsal fin, and there is a large, black teardrop. The body is slender, and the snout is moderately blunt. The premaxillary frenum is absent or narrow. Large males have a long anal fin that extends to the caudal fin. The lateral line is complete with 46–58 pored scales, there are 2 anal spines, and the head canals are uninterrupted. Scutes are present on the breast and on the male in a row along the midline of the belly. To 3 in. (7.8 cm).

SIMILAR SPECIES The River Darter, *P. shumardi*, is dusky olive-green on the back, lacks bold dark saddles, and has a large, black spot at the rear of the first dorsal fin and 8–15 black bars along the side of the body. The Gilt Darter, *P. evides*, has large, white or yellow spots on the caudal-fin base and a wide premaxillary frenum.

HABITAT The Stargazing Darter lives in fast gravel runs of clear medium to large rivers.

DISTRIBUTION IN ILLINOIS The Stargazing Darter has not been collected in the Wabash River since the 1880s, when it was collected at Vincennes and New Harmony, Indiana. It no doubt also occurred on the Illinois side of the river at that time, as noted by Smith (1979), although he showed only the collection at New Harmony on his distribution map. The population of the Stargazing Darter in the Wabash River was far north of other populations of the species, which are in the St. Francis, White, and Ouachita River basins of Missouri, Arkansas, and northern Louisiana.

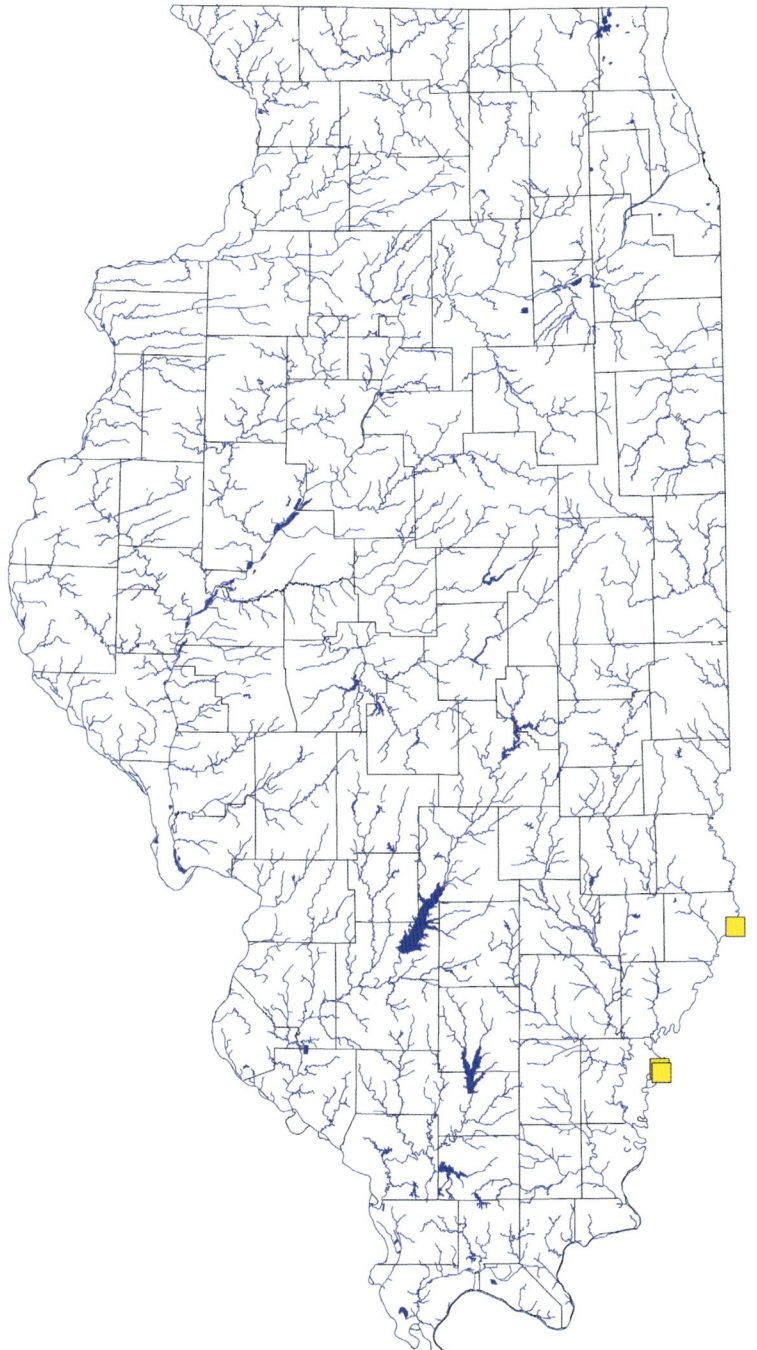

DARTERS AND PERCHES (Percidae)

373

SAUGER *Sander canadense* (Griffith & Smith 1834)

IDENTIFICATION The Sauger is a large fish with an elongate body that is deepest under the first dorsal fin and tapers to a slender caudal peduncle. The snout is long and pointed, and the large mouth extends beneath the middle of the eye and contains large, canine teeth. The caudal fin is forked. The yellow-brown back has dark green to black mottling and 3–4 dark brown saddles that extend onto the brassy yellow-blue side of the body as broad bars. Wavy dark brown bands are present on the clear to yellow fins. The 2 dorsal fins are well separated, and the first dorsal fin has black half-moon–shaped spots. The lateral line is complete with 79–100 lateral scales, usually 17–19 dorsal rays, 11–13 anal rays, and 5–8 pyloric caeca. To 30 in. (76 cm).

SIMILAR SPECIES The Walleye, *S. vitreum*, has a white tip on the lower lobe of the caudal fin and a large, black spot—but no black half-moons—on the first dorsal fin. The Yellow Perch, *Perca flavescens*, is deeper bodied, lacks large, canine teeth, and has dark triangular bars on the side of the body, 52–61 lateral scales, 12–14 dorsal rays, and 6–8 anal rays.

HABITAT The Sauger lives in sandy and gravelly runs and pools and backwaters of small to large rivers. It is most frequently encountered in large rivers and occasionally is found in lakes.

DISTRIBUTION IN ILLINOIS The Sauger is state-wide in distribution as it was in the earlier eras and appears to be at least as widespread as it was during the Smith era. The species has been stocked in Illinois as a sport fish for decades.

REMARKS Eyes of the Sauger and Walleye look silvery and opaque because of the tapetum lucidum, a layer of light-gathering tissue that enables these fishes to be active in low light.

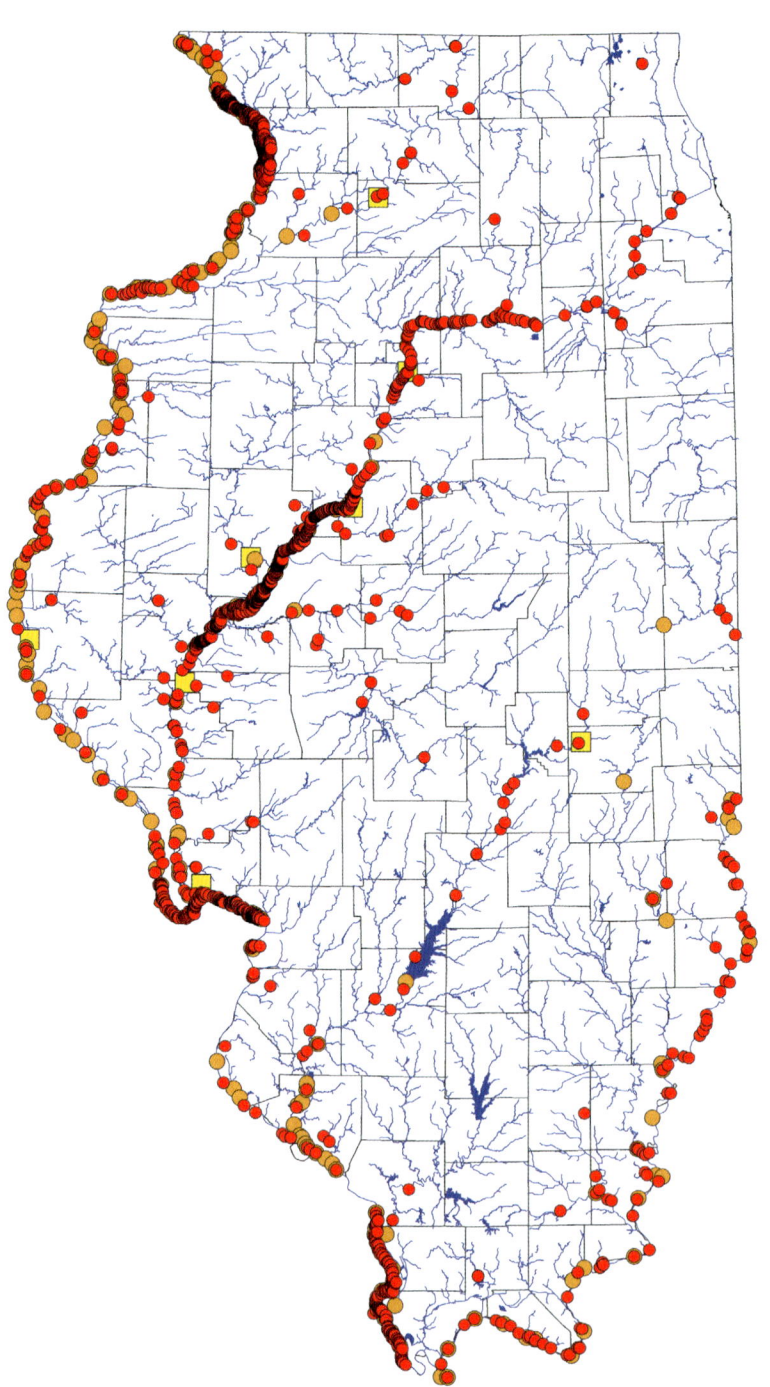

WALLEYE *Sander vitreum* (Mitchill 1818)

IDENTIFICATION The Walleye is a large fish with an elongate body that is deepest under the first dorsal fin and tapers to a slender caudal peduncle. The snout is long and pointed, and the large mouth extends beneath the middle of the eye and contains large, canine teeth. The caudal fin is forked. It has dark green mottling on the yellow-brown back and 5–12 brown saddles that extend onto the side of the brassy yellow-blue body as short bars; the bars are less obvious in large individuals. Wavy dark brown bands are present on the yellow second dorsal and caudal fins, and there is a white edge on the anal and pelvic fins and a white tip on the lower lobe of the caudal fin. The 2 dorsal fins are well separated, and a large, black spot is present posteriorly on the first dorsal fin. The lateral line is complete with 77–108 lateral scales, usually 19–22 dorsal rays, 11–14 anal rays, and 3 pyloric caeca. To 36 in. (91 cm).

SIMILAR SPECIES The Sauger, *S. canadense*, lacks the white tip on the lower lobe of the caudal fin and the large, black spot at the rear of the first dorsal fin and has black half-moon–shaped spots on the first dorsal fin. The Yellow Perch, *Perca flavescens*, is deeper bodied, lacks large, canine teeth, and has dark triangular bars on the side of the body, 52–61 lateral scales, 12–14 dorsal rays, 6–8 anal rays.

HABITAT The Walleye inhabits lakes and pools, backwaters, and runs of medium to large rivers. It often is found near brush.

DISTRIBUTION IN ILLINOIS The Walleye is state-wide in distribution and appears to be at least as frequently encountered as it was during the Smith era. Its preference for cooler water is evident in the greater abundance of records in the northern part of the state. The species has been stocked in Illinois as a sport fish for decades.

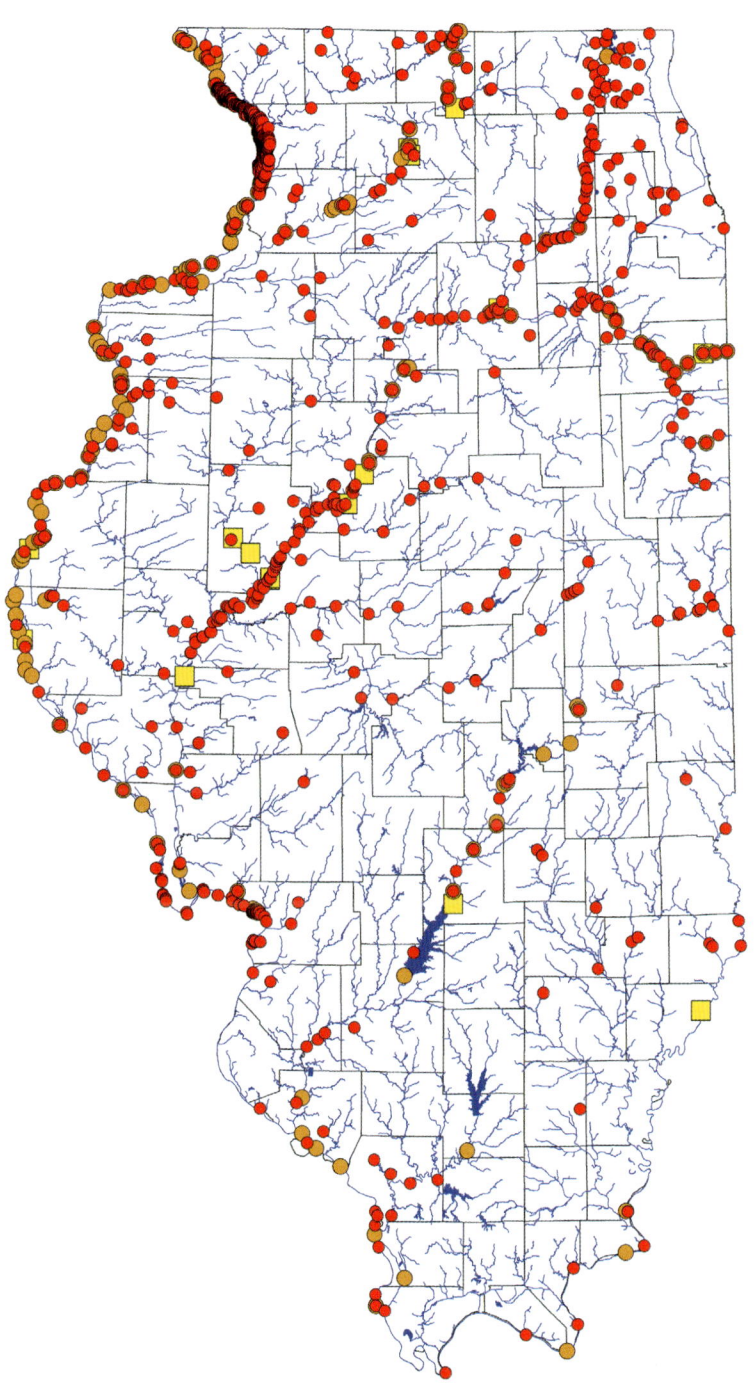

STICKLEBACKS

Family Gasterosteidae Bonaparte 1831

Sticklebacks are small fishes with 3–16 isolated dorsal spines ("stickles" on the back) followed by a dorsal fin with 14–16 rays. They have a deep body followed by an extremely narrow caudal peduncle. They lack scales but often have large, bony plates on the side of the body. The pelvic fin is thoracic with 1 spine and 1–2 rays. Sticklebacks inhabit both marine and fresh waters of North America and Eurasia. Through complex courting maneuvers of the male, the female is enticed to enter the nest (a tube constructed of plant material and held together with a sticky kidney secretion from the male) and deposit eggs. After she leaves the nest, the male enters and fertilizes the eggs. The male guards the eggs and young. Only 9 species are recognized in this family, but numerous taxonomic problems remain. Three of the 9 species are found in Illinois.

KEY TO THE STICKLEBACKS (Gasterosteidae)

1a Isolated dorsal spines 7–11, some inclined to the right and some to the left; caudal peduncle wider than deep, with a lateral keel; caudal fin slightly emarginated . . . **Page 380**

Ninespine Stickleback (*Pungitius pungitius*)

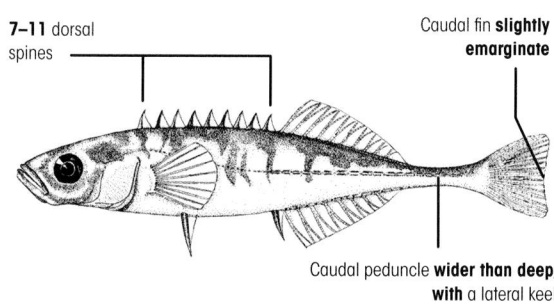

7–11 dorsal spines

Caudal fin **slightly emarginate**

Caudal peduncle **wider than deep, with** a lateral keel

1b Isolated dorsal spines 3–6, in a vertical line; caudal peduncle deeper than wide, lateral keel present or absent; caudal fin rounded to truncate **2**

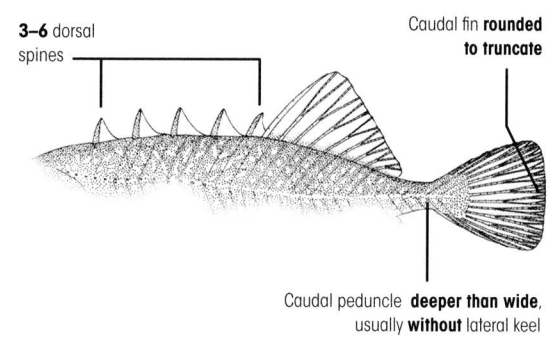

3–6 dorsal spines

Caudal fin **rounded to truncate**

Caudal peduncle **deeper than wide**, usually **without** lateral keel

2a Isolated dorsal spines 4–6; spines shorter than eye diameter; there are usually small, bony plates on side of body . . . .**Page 378**

Brook Stickleback (*Culaea inconstans*)

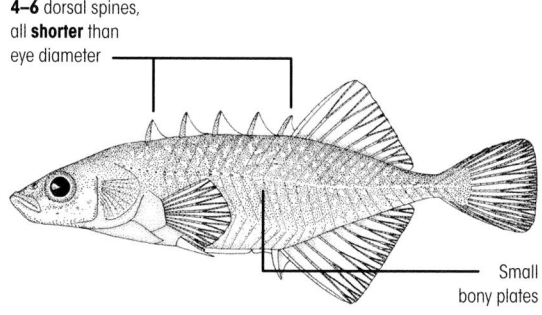

4–6 dorsal spines, all **shorter** than eye diameter

Small bony plates

2b Isolated dorsal spines 3; longest spine longer than eye diameter; conspicuous bony plates on side of body . . . .**Page 379**

Threespine Stickleback (*Gasterosteus aculeatus*)

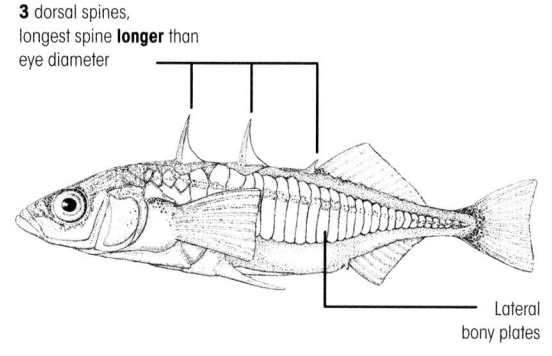

3 dorsal spines, longest spine **longer** than eye diameter

Lateral bony plates

BROOK STICKLEBACK *Culaea inconstans* (Kirtland 1840)

IDENTIFICATION The Brook Stickleback has a large head and a strongly compressed body that is deep anteriorly but tapers to a slender caudal peduncle. There are 4–6 short, isolated dorsal spines followed by a second dorsal fin with 9–11 rays and opposite an anal fin with 9–11 rays. The pelvic fin is extremely small with 1 spine and 1 ray. The head and body are olive-green with dark green to black mottling. The back and upper side are covered with pale green flecks, and the lower side and belly are light green to silver-white. During the breeding season the male is dark green to almost black and sometimes has red on the pelvic fins. There are usually small, bony plates on the lateral-line pores but no large, bony plates on the side of the body and no keel on the caudal peduncle. The branchiostegal membranes are joined to one another but are not attached to the isthmus. To 3½ in. (8.7 cm).

SIMILAR SPECIES The Ninespine Stickleback, *Pungitius pungitius*, has 7–12, usually 9 dorsal spines angled alternately to the left and right and a keel on the caudal peduncle. The Threespine Stickleback, *Gasterosteus aculeatus*, has 3 dorsal spines, a keel on the caudal peduncle, and branchiostegal membranes that are broadly attached to the isthmus.

HABITAT The Brook Stickleback lives in lakes, ponds, and quiet to slightly flowing pools of headwaters, creeks, and small rivers. It usually is found among vegetation over sand or mud.

DISTRIBUTION IN ILLINOIS The Brook Stickleback is restricted to the northern one-fourth of the state where it appears to be as widespread as it was in previous eras.

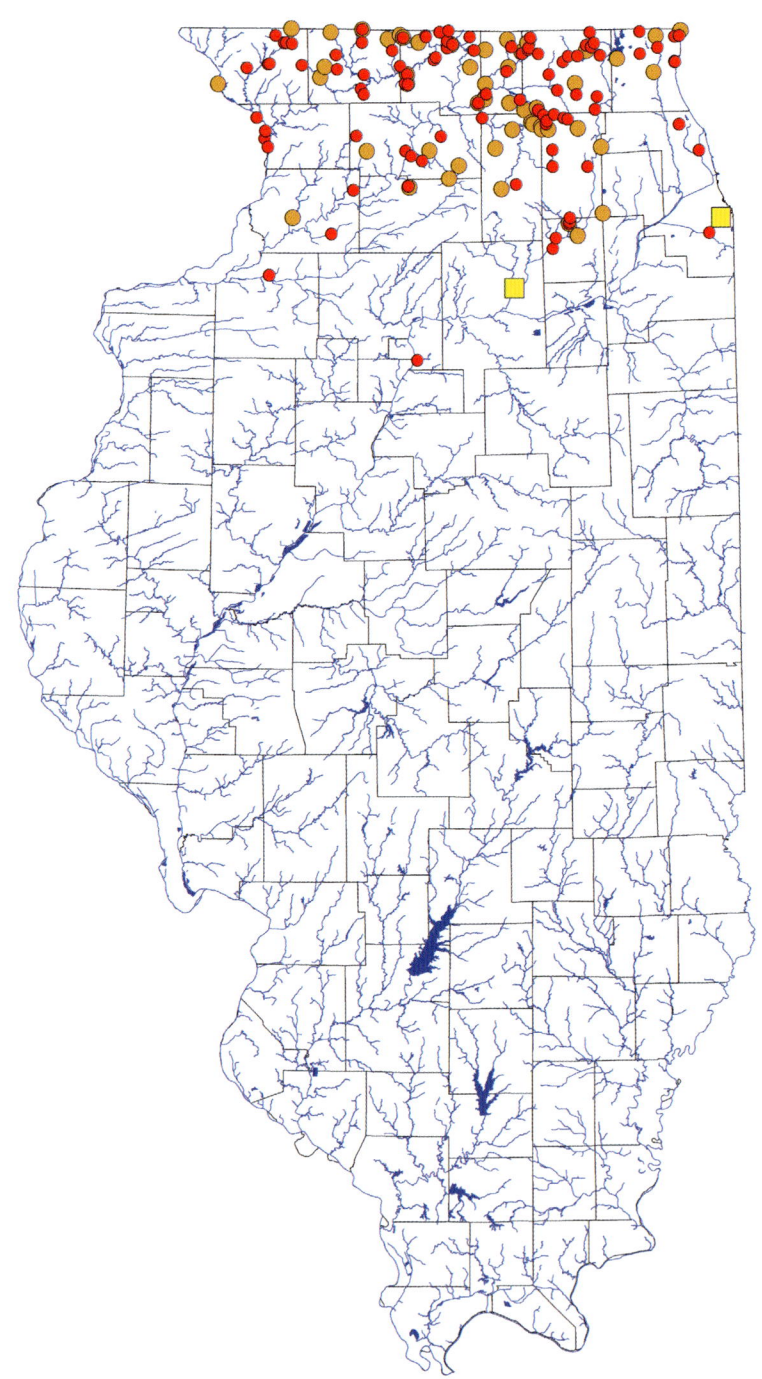

THREESPINE STICKLEBACK *Gasterosteus aculeatus* Linnaeus 1758

IDENTIFICATION The Threespine Stickleback has a large head and a strongly compressed body that is deep anteriorly but tapers to an extremely slender caudal peduncle. There are usually 3 dorsal spines, the last of which is much shorter than the other 2. The pelvic fin is extremely small with 1 long spine and 1 ray. The head and body are green to brown with a silver sheen and dark green mottling on the back and upper side of the body. The lower side and belly are silver-white. During the breeding season, the male has a blue back and upper side of the body and a bright blue eye and has red on the fins, lower side, and belly. Large females have a pink throat and belly. There is a bony keel along the side of the caudal peduncle and often bony plates on the side of the body. The branchiostegal membranes are broadly attached to the isthmus. To 4 in. (10 cm).

SIMILAR SPECIES The Ninespine Stickleback, *Pungitius pungitius*, has 7–12, usually 9, dorsal spines angled alternately to the left and right and branchiostegal membranes that are joined to one another but are not attached to the isthmus. The Brook Stickleback, *Culaea inconstans*, has 4–6 short dorsal spines, no keel on the caudal peduncle, and branchiostegal membranes that are joined to one another but not attached to the isthmus.

HABITAT The Threespine Stickleback lives among vegetated areas, usually over mud or sand in large lakes and channels of rivers.

DISTRIBUTION IN ILLINOIS The Threespine Stickleback first appeared in the Illinois portion of Lake Michigan in 1988. It has become more widespread in the contemporary era and has entered the upper Illinois River (although the species has not been recorded in Illinois for more than 20 years).

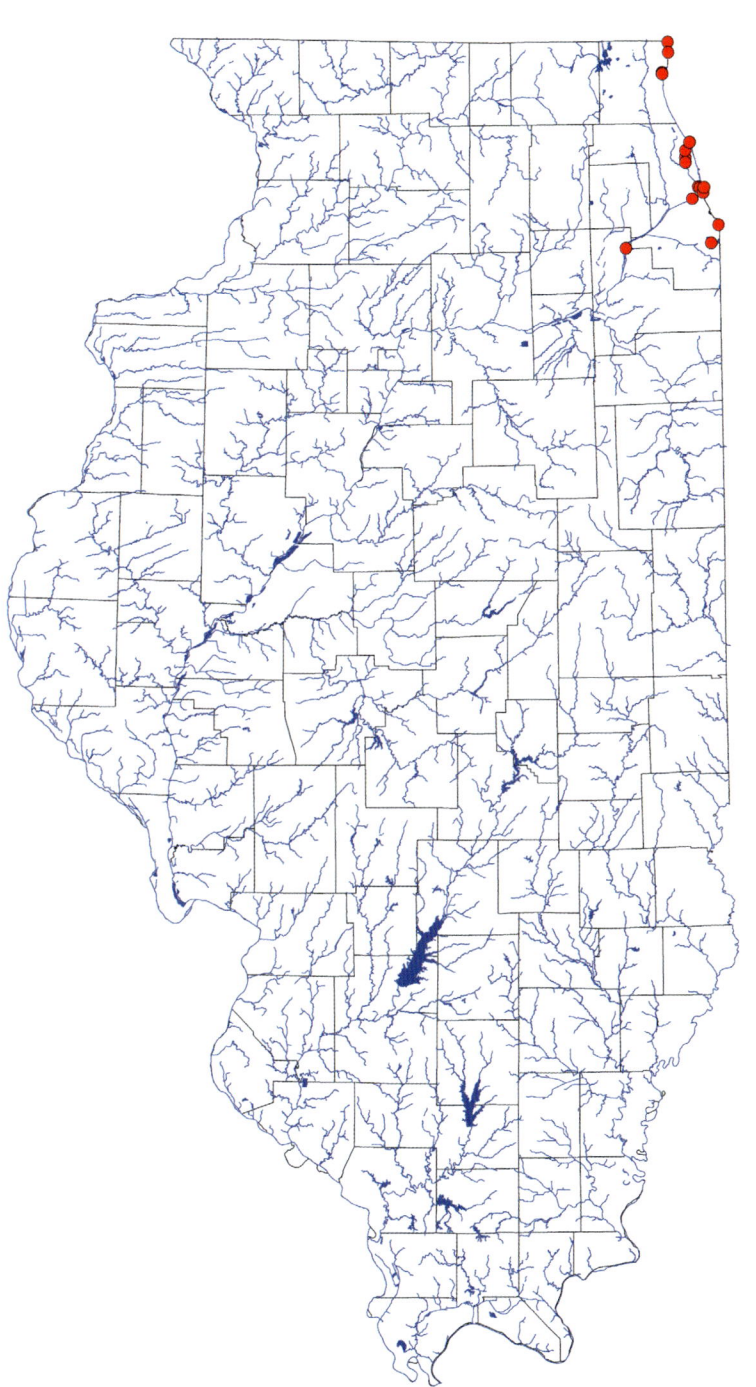

STICKLEBACKS (Gasterosteidae)

NINESPINE STICKLEBACK *Pungitius pungitius* (Linnaeus 1758)

IDENTIFICATION The Ninespine Stickleback has a strongly compressed, slender body that is deepest anteriorly but tapers to an extremely slender caudal peduncle. There are 7–12, usually 9, short dorsal spines that are angled alternately to the left and right. The pelvic fin is extremely small with 1 spine and 1 ray. The head and body are olive-green with dark gray mottling on the back and upper side. The lower side and belly are silver-white. During the breeding season large males have a black belly, white pelvic fins, and often red on the head. There are 0–8 small plates on the lateral-line pores on the front half of the body and a keel on the thin caudal peduncle. Branchiostegal membranes are joined to one another but are not attached to the isthmus. To 3½ in. (9 cm).

SIMILAR SPECIES The Brook Stickleback, *Culaea inconstans*, has 4–6 short dorsal spines and no keel on the caudal peduncle. The Threespine Stickleback, *Gasterosteus aculeatus*, has 3 dorsal spines and branchiostegal membranes that are broadly attached to the isthmus.

HABITAT The Ninespine Stickleback usually is found among vegetation along the shores of lakes, ponds, and pools of slow-moving streams. Less often it is found over sand in open water areas of lakes and ponds.

DISTRIBUTION IN ILLINOIS The Ninespine Stickleback occurs along the shores and harbors of Lake Michigan. Although Smith (1979) recorded 1 specimen collected in 1976 from the Illinois River, he surmised that the record was based on an individual that entered the river from Lake Michigan during a flood event. No contemporary-era records exist for the Illinois River, but the species appears to be as widespread in Lake Michigan as it was in the mid-20th century.

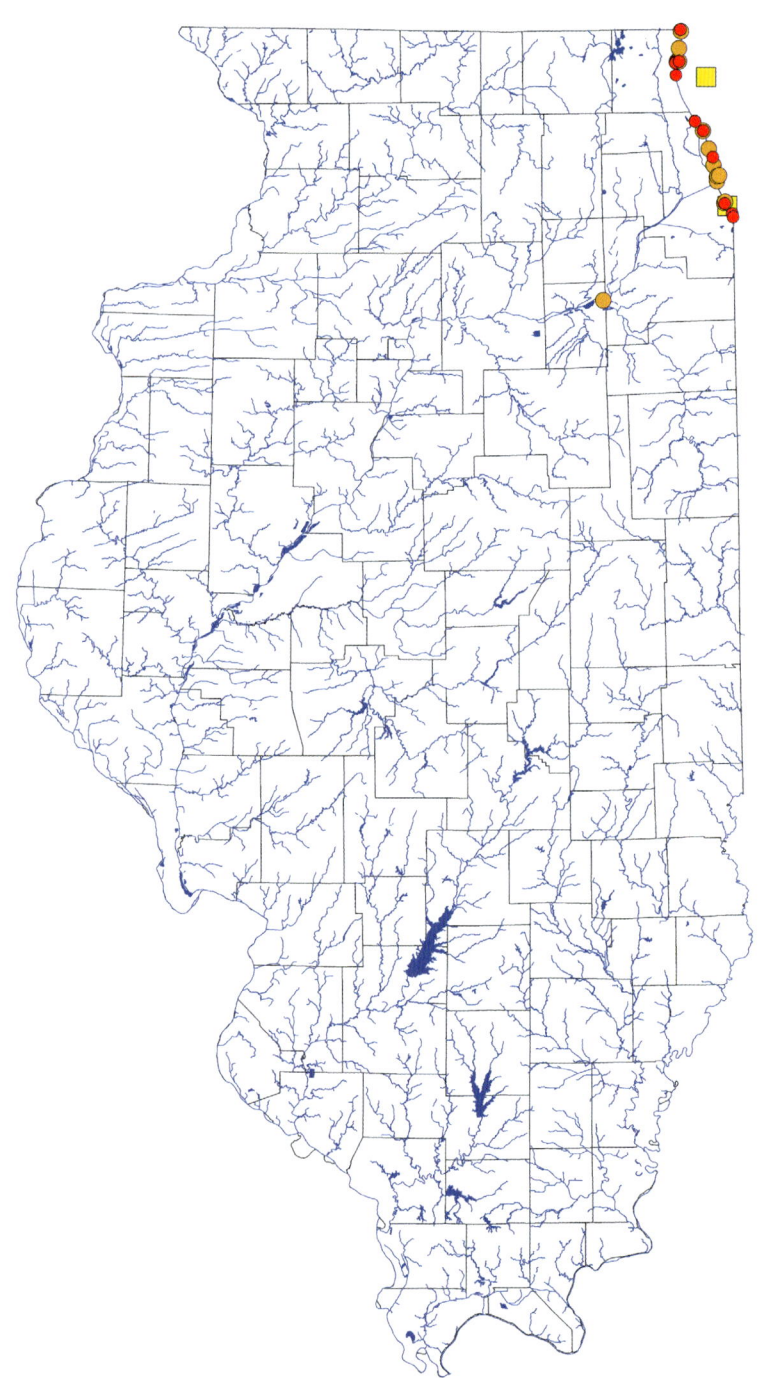

SCULPINS

Family Cottidae Bonaparte 1831

Sculpins have a large head and large mouth, a body that is wide at the front and tapers to a slender caudal peduncle, no scales on the body, large pectoral fins, 1–4 preopercular spines, thoracic pelvic fins, a long anal fin that lacks spines, and 2 dorsal fins, the first with spines and the second longer than the first and with flexible branched rays. Sculpins have a suborbital stay—a bony connection across the cheek uniting bones under the eye with the preopercle—the anterior part of the gill cover. Many species have small spines or prickles on the head and body. Most of the more than 280 species of sculpins are marine, but several inhabit fresh waters of North America and northern Eurasia. Five species occur in Illinois.

Sculpins are often difficult to identify. Most are mottled, and color pattern is of less use than in other groups of fishes. Development of prickles and preopercular spines, often useful in identifying species, can vary with the habitat. Individuals in lakes and other quiet waters have large prickles and spines, but those in small, cold streams tend to have poorly developed prickles and spines.

1a Gill covers not attached to isthmus; dorsal fins widely separated, space between fins much greater than eye length . . . .**Page 385**

1b Gill covers broadly attached to isthmus; dorsal fins touching or distance between fins much less than eye length **2**

Deepwater Sculpin
(*Myoxocephalus thompsonii*)

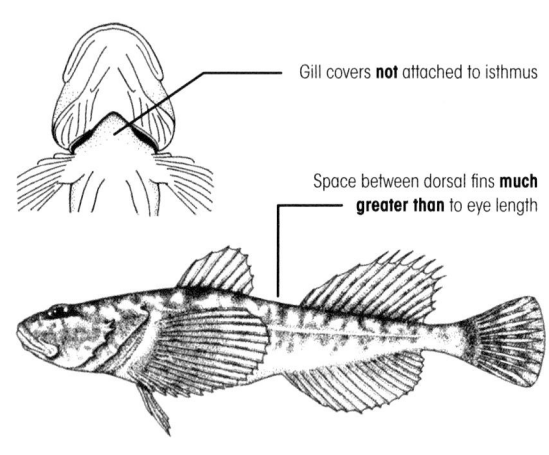

Gill covers **not** attached to isthmus

Space between dorsal fins **much greater than** to eye length

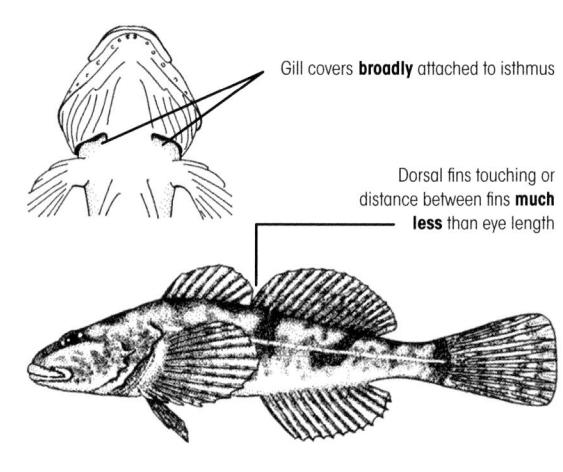

Gill covers **broadly** attached to isthmus

Dorsal fins touching or distance between fins **much less** than eye length

2a Lateral line complete to base of caudal fin **3**

2b Lateral line incomplete, usually ending below base of soft dorsal fin **4**

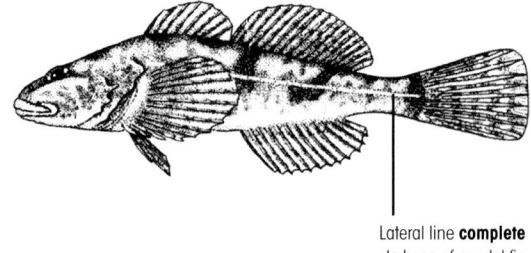

Lateral line **complete** to base of caudal fin

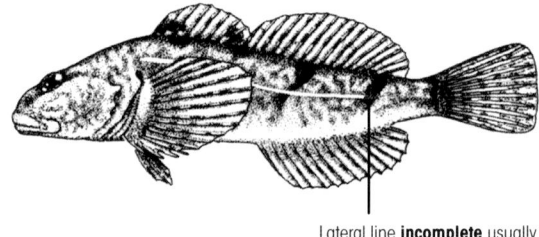

Lateral line **incomplete** usually ending below base of soft dorsal fin

3a Head wide and flat; upper head and body covered with prickles; upper spine on preopercle bone curved inwardly, its length about 2/3 length of eye; 4 dark saddles on back not extending down side of body as sharply defined bars; pores in lateral line usually 33–36; 1 pore at midline of chin . . . . **Page 384**

Spoonhead Sculpin (*Cottus ricei*)

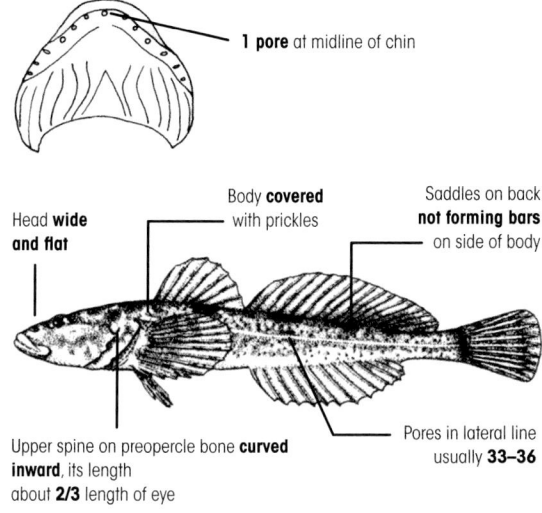

1 **pore** at midline of chin

Body **covered** with prickles

Saddles on back **not forming bars** on side of body

Head **wide and flat**

Upper spine on preopercle bone **curved inward**, its length about **2/3** length of eye

Pores in lateral line usually **33–36**

3b Head rounded and not obviously flattened; head and body without prickles (if present only behind pectoral fin); upper spine on preopercle bone straight or only slightly curved inward, its length about 1/2 length of eye; 4–5 dark saddles on back, last 3 extending down side of body as sharply defined bars; pores in lateral line usually 29–34; 2 pores at midline of chin . . . . **Page 387**

Banded Sculpin (*Uranidea carolinae*)

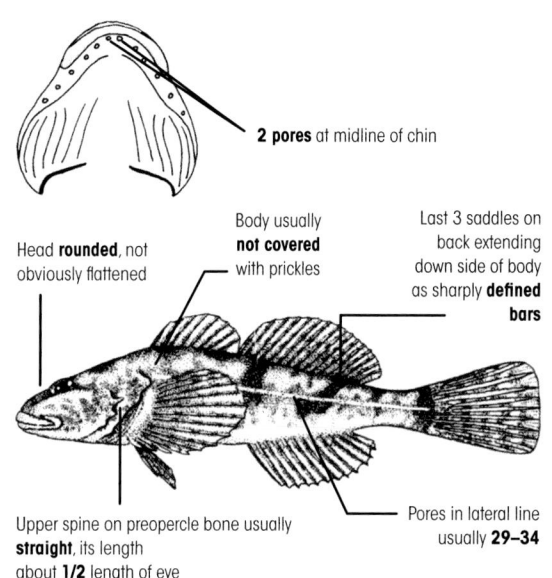

2 **pores** at midline of chin

Head **rounded**, not obviously flattened

Body usually **not covered** with prickles

Last 3 saddles on back extending down side of body as sharply **defined bars**

Upper spine on preopercle bone usually **straight**, its length about **1/2** length of eye

Pores in lateral line usually **29–34**

4a Pelvic fin rays 4, last a little shorter than third; length of caudal peduncle (A) less than distance from back edge of eye to edge of gill cover (B); last 2 rays of dorsal and anal fins arising from same base; palatine teeth usually present. . . . . **Page 386**

Mottled Sculpin (*Uranidea bairdii*)

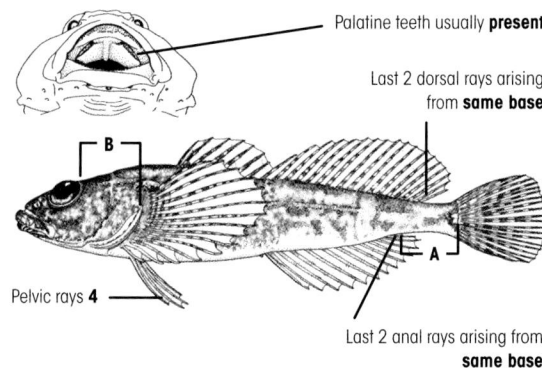

Palatine teeth usually **present**

Last 2 dorsal rays arising from **same base**

Pelvic rays **4**

Last 2 anal rays arising from **same base**

4b Pelvic fin rays usually 3, if 4, last much shorter than third; length of caudal peduncle (A) greater than distance from back edge of eye to edge of gill cover (B); last 2 rays of dorsal and anal fins arising from separate bases; palatine teeth usually absent . . . . **Page 388**

Slimy Sculpin (*Uranidea cognata*)

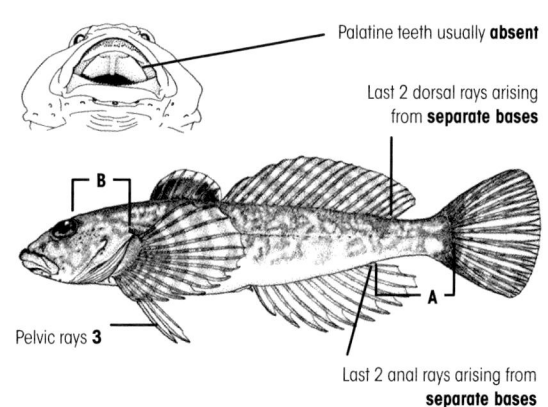

Palatine teeth usually **absent**

Last 2 dorsal rays arising from **separate bases**

Pelvic rays **3**

Last 2 anal rays arising from **separate bases**

SCULPINS (Cottidae)

SPOONHEAD SCULPIN *Cottus ricei* (Nelson 1876)

IDENTIFICATION The Spoonhead Sculpin has a large head and mouth, large pectoral fins, thoracic pelvic fins, a long anal fin that lacks spines, 2 dorsal fins—the first with spines and the second longer than the first and with flexible branched rays—and no scales on the body. The head is wide and flat, and the relatively wide body tapers to an extremely slender caudal peduncle. The back and upper side are light brown with dark brown specks and mottling. Up to 4 dark brown saddles may be present on the back. There are no large, black spots on the dorsal fins. The 2 dorsal fins are separate at the base. The body is covered with fleshy prickles. The lateral line is complete with 33–36 pores. There are 3 preopercular spines; the upper 1 is long and curved inward. There are 4 pelvic rays, 1 pore at the tip of the chin, and no palatine teeth. To 5 in. (13 cm).

SIMILAR SPECIES Species of *Uranidea* in Illinois lack fleshy prickles on the body, have 2 pores at the tip of the chin, a narrower and deeper head, and a deeper caudal peduncle. The Mottled Sculpin, *U. bairdii*, and Slimy Sculpin, *U. cognata*, also have an incomplete lateral line. The Deepwater Sculpin, *Myoxocephalus thompsonii*, has a large gap between the 2 dorsal fins.

HABITAT The Spoonhead Sculpin lives in rocky areas of swift creeks and rivers and along rocky shores of lakes. It also is found in lakes to depths of 450 ft. (137 m).

DISTRIBUTION IN ILLINOIS The Spoonhead Sculpin was described from Lake Michigan near Evanston, Illinois, in 1876 and was collected a few times along the shoreline of Lake Michigan in the earlier eras. However, it has not been collected in Lake Michigan or elsewhere in Illinois since 1939.

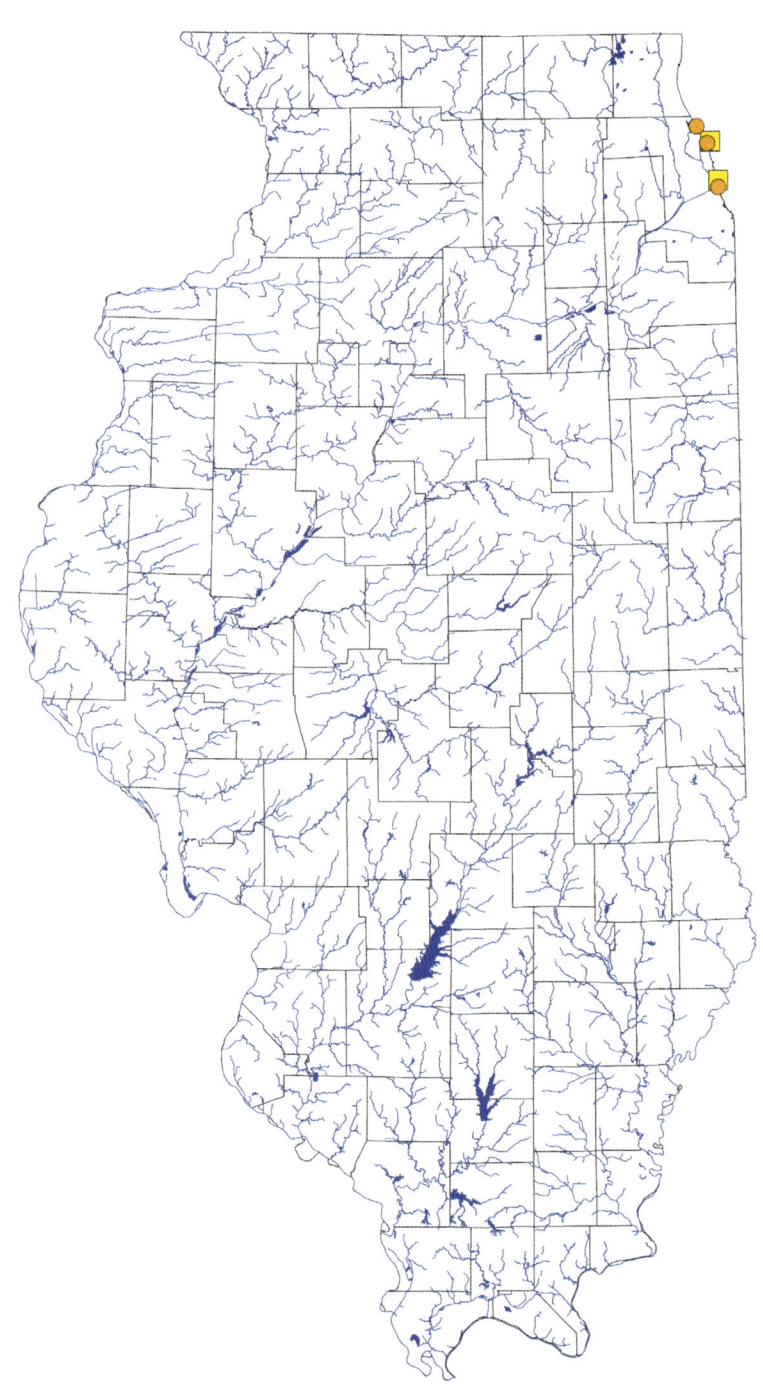

DEEPWATER SCULPIN *Myoxocephalus thompsonii* (Girard 1851)

IDENTIFICATION The Deepwater Sculpin has large pectoral fins, thoracic pelvic fins, a long anal fin that lacks spines, and 2 dorsal fins—the first with spines and the second longer than the first and with flexible branched rays. There is a large gap between the 2 dorsal fins. The head is extremely wide and flat, and the large mouth extends to beneath the eye. The body is wide anteriorly and tapers to an extremely slender caudal peduncle. The second dorsal and pectoral fins are large, especially on large males. There are no scales on the body and bony plates along the complete lateral line. The back and upper side are gray to light brown with dark brown to green mottling and 4–7 dark brown saddles. The underside is white, and the fins are mostly clear with dark brown bands. There are 4 preopercular spines; the upper 2 spines are large and directed upward, often appearing as 1 divided spine; the lower 2 spines are small and directed downward. There are 3 (rarely, 4) pelvic rays. To 9 in. (23 cm).

SIMILAR SPECIES Other sculpins (*Cottus* and *Uranidea*) in Illinois lack the large gap between the dorsal fins and bony plates along the lateral line.

HABITAT The Deepwater Sculpin lives on rocky substrates in deep and cold lakes. It is found to depths of 1,200 ft. (370 m).

DISTRIBUTION IN ILLINOIS The Deepwater Sculpin is restricted in Illinois to Lake Michigan. Although never frequently collected, it probably remains as widespread in the contemporary era as in the Forbes and Richardson and Smith eras.

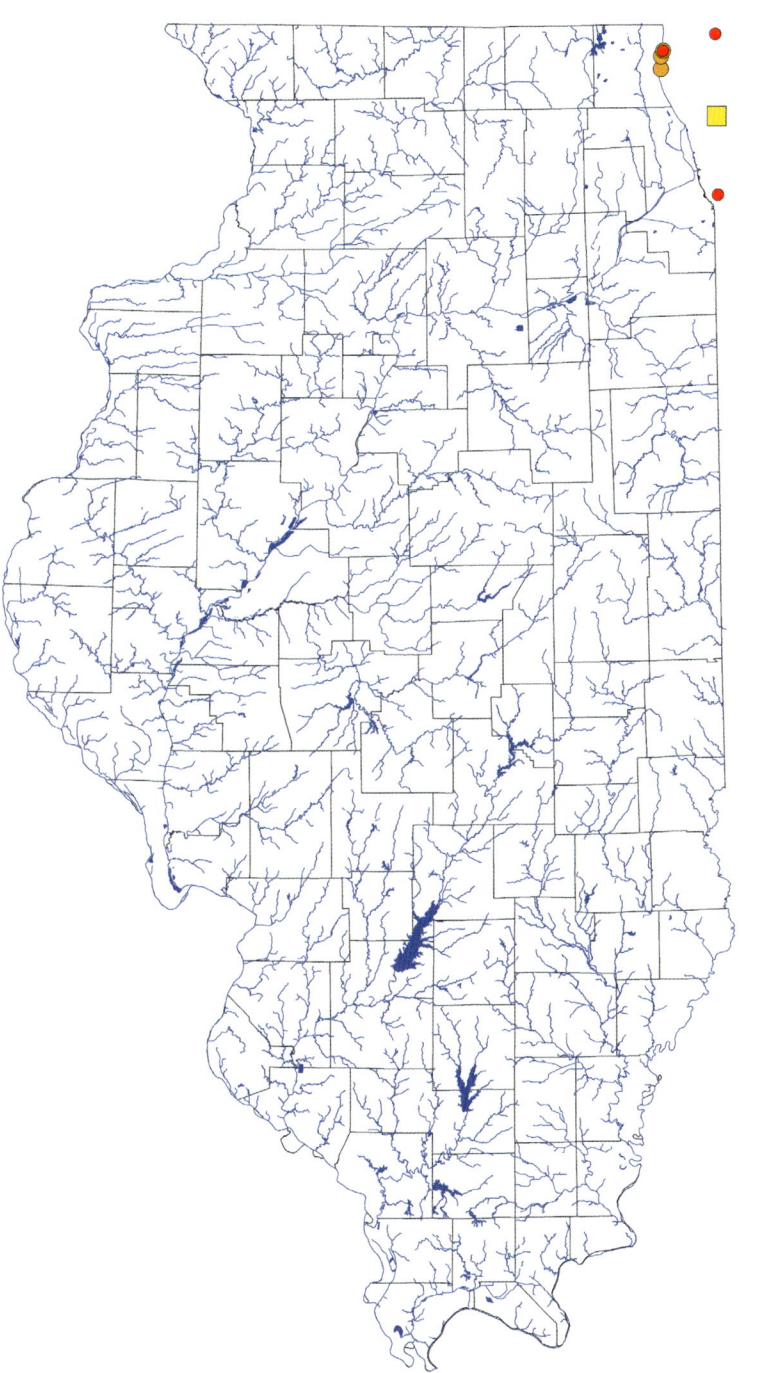

MOTTLED SCULPIN *Uranidea bairdii* (Girard 1850)

IDENTIFICATION The Mottled Sculpin has a large head and mouth, large pectoral fins, thoracic pelvic fins, a long anal fin that lacks spines, 2 dorsal fins—the first with spines and the second longer than the first and with flexible branched rays—and no scales on the body. The body is wide at front and tapers to a slender caudal peduncle. The back is brown with dark brown to black mottling and 2–4 dark brown-black saddles that extend onto the side of the body. There are a large, black spot anteriorly and another large, black spot posteriorly on the first dorsal fin. Large males have a black band and an orange edge on the first dorsal fin. The dorsal fins are joined at the base, and the lateral line is incomplete with 14–27 (usually 21–23) pores. There are 3 preopercular spines, 4 pelvic rays, and usually 2 pores at the tip of the chin. Palatine teeth usually are present. To 6 in. (15 cm).

SIMILAR SPECIES The Banded Sculpin, *U. carolinae*, has no large, black spots on the first dorsal fin, more-distinct bars with bold black edges on the side of the body, a complete lateral line, and dorsal fins that are separate at the base. Large males lack an orange edge on the first dorsal fin. The Slimy Sculpin, *U. cognata*, has 3 pelvic rays, a more slender body, lighter saddles and bars, dorsal fins separate at the base, and no palatine teeth. The Spoonhead Sculpin, *Cottus ricei*, has the body covered with fleshy prickles, a complete lateral line, 1 pore at tip of chin, and no palatine teeth. The Deepwater Sculpin, *Myoxocephalus thompsonii*, has a large gap between the 2 dorsal fins.

HABITAT The Mottled Sculpin lives in rocky riffles of headwaters, creeks, and small rivers and on rocky shores of lakes. It is frequently encountered in spring effluents.

DISTRIBUTION IN ILLINOIS The Mottled Sculpin is restricted to northern Illinois where it occurs in Lake Michigan, the upper Illinois River and Rock River basins, and the Vermilion River (Wabash) basin in Vermilion County. Neither Forbes and Richardson (1909) nor Smith (1979) recorded the species from the Vermilion River (Wabash) basin, but it was known in Smith's time to occur in the Vermilion River (Wabash) basin in adjacent Indiana. The species appears to be as widespread as it was in the Smith era.

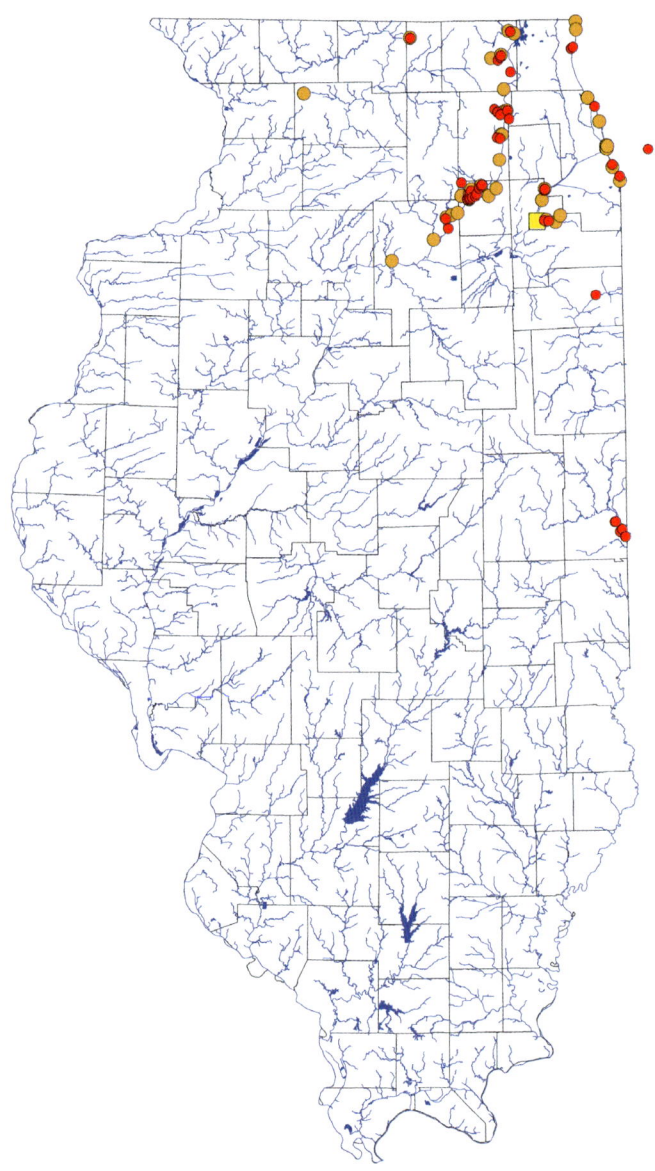

BANDED SCULPIN *Uranidea carolinae* (Gill 1861)

IDENTIFICATION The Banded Sculpin has a large head and mouth, large pectoral fins, thoracic pelvic fins, a long anal fin that lacks spines, and 2 dorsal fins—the first with spines and the second longer than the first and with flexible branched rays—and no scales on the body. The body is wide at front and tapers to a slender caudal peduncle. The back is red-brown or light brown with 4–5 dark brown saddles. The 3 posterior saddles have well-defined black edges and continue onto the upper side. There are no large, black spots on the dorsal fins and no prickles on the body. The dorsal fins are separate at the base. The lateral line is usually complete with 29–34 pores. There are 3 preopercular spines, 4 pelvic rays, and 2 pores at the tip of the chin. Palatine teeth are present. To 7¼ in. (18 cm).

SIMILAR SPECIES The Mottled Sculpin, *U. bairdii* has large, black spots on the first dorsal fin, less-distinct bars without bold black edges on the side of the body, an incomplete lateral line, dorsal fins that are joined at the base, and an orange edge on the first dorsal fin of large males. The Slimy Sculpin, *U. cognata*, has large, black spots on the first dorsal fin, 3 pelvic rays, a more slender body, incomplete lateral line, lighter saddles and bars, and no palatine teeth. The Spoonhead Sculpin, *Cottus ricei*, has the body covered with fleshy prickles, 1 pore at tip of chin, and no palatine teeth. The Deepwater Sculpin, *Myoxocephalus thompsonii*, has a large gap between the 2 dorsal fins.

HABITAT The Banded Sculpin lives in rocky riffles of headwaters, creeks, and small rivers. It is most common in rocky spring effluents.

DISTRIBUTION IN ILLINOIS The Banded Sculpin appears to be as widespread as it was in past eras. Smith described the distribution in Illinois as including extreme southern Illinois and streams flowing from the bluffs in the lower Illinois and Mississippi basins.

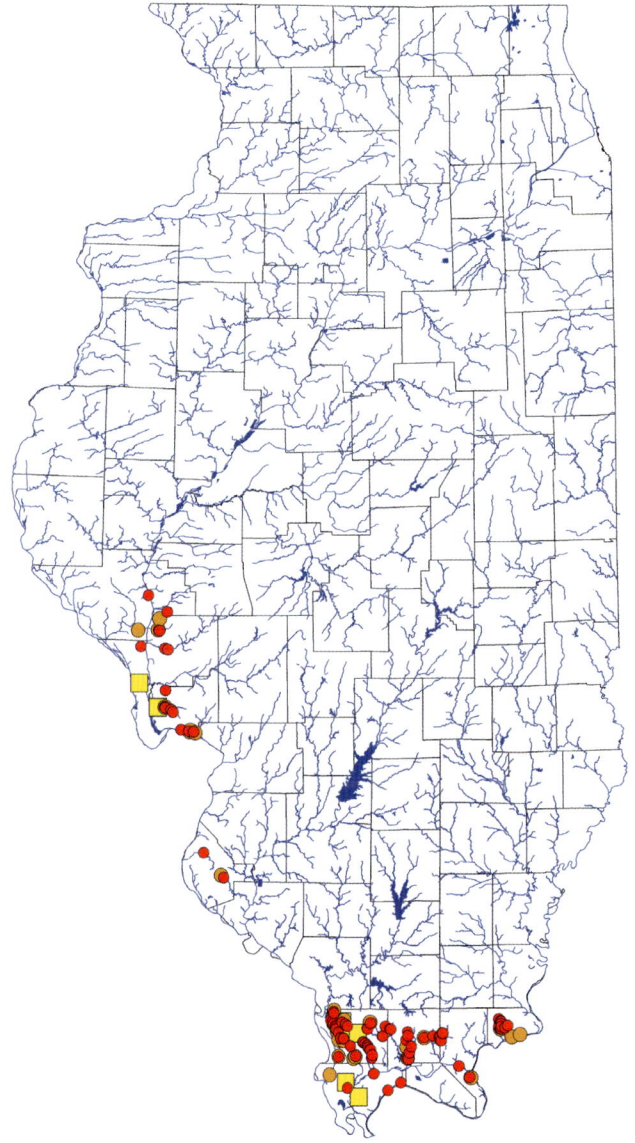

SLIMY SCULPIN *Uranidea cognata* (Richardson 1836)

IDENTIFICATION The Slimy Sculpin has a large head and mouth, large pectoral fins, thoracic pelvic fins, a long anal fin that lacks spines, and 2 dorsal fins—the first with spines and the second longer than the first and with flexible branched rays—and no scales on the body. The body is wide at front and tapers to a slender caudal peduncle. The back and side of the body are dark green-brown or gray and have dark gray mottling. There are 2 dark gray saddles under the second dorsal fin and a large, black spot anteriorly, and another large, black spot posteriorly, on the first dorsal fin. Large males are dark gray and have an orange edge on the first dorsal fin. The head is covered with fleshy prickles. The dorsal fins are separate at the base, and the lateral line usually is incomplete with 12–26 pores. There are 2–3 preopercular spines with the upper 1 being larger, usually 3 pelvic rays (if a fourth is present, it is greatly reduced), 2 pores at the tip of the chin, and no palatine teeth. To 4½ in. (12 cm).

SIMILAR SPECIES The Mottled Sculpin, *U. bairdii*, has 4 pelvic rays, a deeper body, darker saddles and bars, and dorsal fins joined at the base; palatine teeth usually are present. The Banded Sculpin, *U. carolinae*, lacks large, black spots on the first dorsal fin, has darker saddles and bars, palatine teeth, 4 pelvic rays, a deeper body, and a complete lateral line. The Spoonhead Sculpin, *Cottus ricei*, has the body covered with fleshy prickles, a complete lateral line, and 1 pore at the tip of the chin. The Deepwater Sculpin, *Myoxocephalus thompsonii*, has a large gap between the 2 dorsal fins.

HABITAT The Slimy Sculpin lives in rocky areas of lakes and rocky riffles of cold streams.

DISTRIBUTION IN ILLINOIS Forbes and Richardson (1909) and Smith (1979) reported the Slimy Sculpin from Lake Michigan, which is on the southern edge of the species's range. Contemporary-era records indicate that it remains in the Illinois portion of the lake.

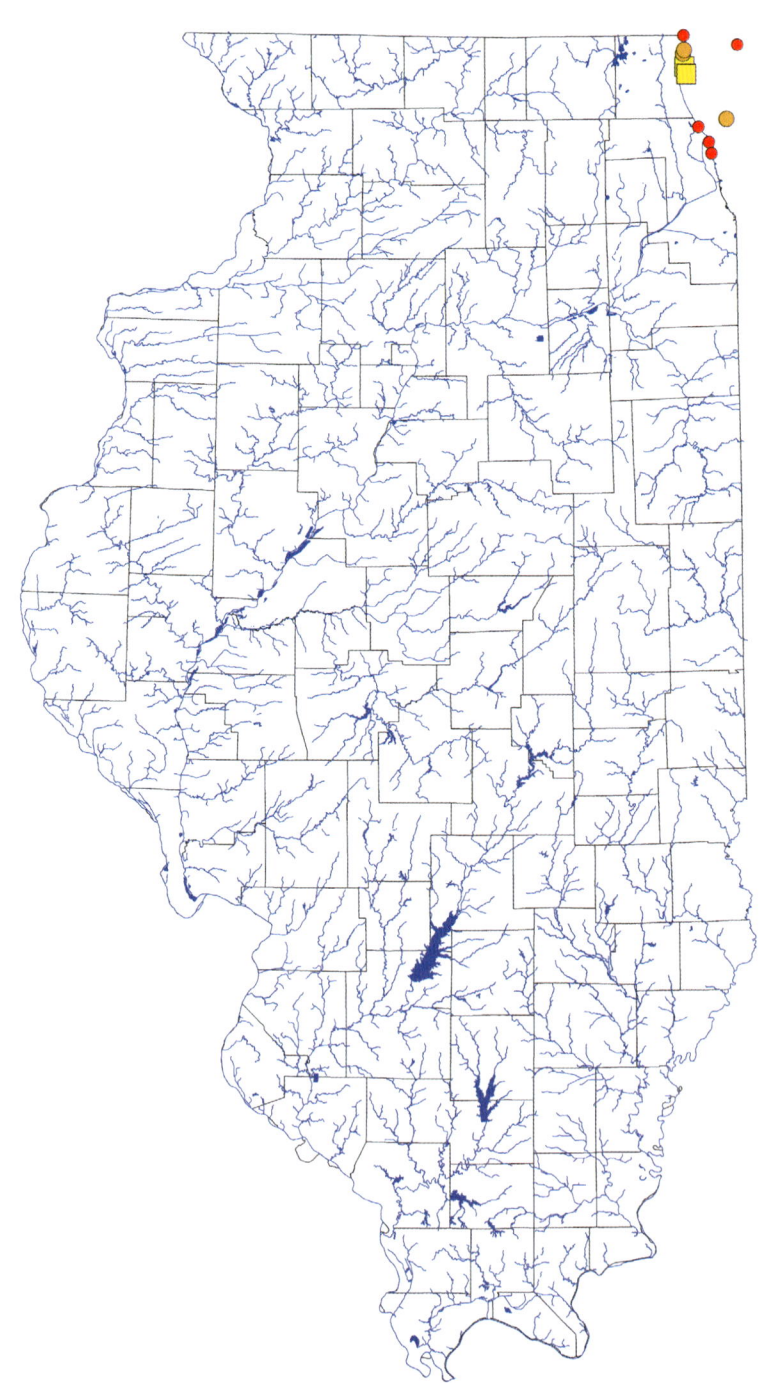

PHOTO CREDITS AND LOCATIONS

Photos are in order of their appearance in the family and species accounts.

Ohio Lamprey, *Ichthyomyzon bdellium*, from Green River, Green Co., KY. Photo by M. Thomas.

Chestnut Lamprey, *Ichthyomyzon castaneus*, from Ohio River, unknown co., KY. Photo by M. Thomas.

Northern Brook Lamprey, *Ichthyomyzon fossor*, from Middle Fork Red River, Powell Co., KY. Photo by M. Thomas.

Silver Lamprey, *Ichthyomyzon unicuspis*, from Missouri River, unknown co., MO. Photo by L. Merry.

Least Brook Lamprey, *Lampetra aepyptera*, from Elisha Branch, McCreary Co., KY. Photo by M. Thomas.

American Brook Lamprey, *Lethenteron appendix*, from White Water River, Cape Girardeau Co., MO. Photo by L. Merry.

Sea Lamprey, *Petromyzon marinus*, adult from Trail Creek, LaPorte Co., IN. Photo by B. Fisher; juvenile from Atlantic Ocean drainage, unknown co., DE. Photo by Z. Randall.

Lake Sturgeon, *Acipenser fulvescens*, from Mississippi River, Lewis Co., MO. Photo by L. Merry.

Pallid Sturgeon, *Scaphirhynchus albus*, captively propagated specimen. Photo by L. Merry.

Shovelnose Sturgeon, *Scaphirhynchus platorynchus*, from Mississippi River, Mississippi Co., AR. Photo by U. Thomas.

Paddlefish, *Polyodon spathula*, from Mississippi River, Mississippi Co., AR. Photo by U. Thomas.

Alligator Gar, *Atractosteus spatula*, juvenile from John Allen Fish Hatchery, MS. Photo by M. Thomas.

Alligator Gar, *Atractosteus spatula*, adult from John Allen Fish Hatchery, MS. Photo by M. Thomas.

Spotted Gar, *Lepisosteus oculatus*, from Illinois River, Grundy Co., IL. Photo by U. Thomas.

Longnose Gar, *Lepisosteus osseus*, from Mississippi River, Mississippi Co., AR. Photo by U. Thomas.

Shortnose Gar, *Lepisosteus platostomus*, from Illinois River, LaSalle Co., IL. Photo by U. Thomas.

Bowfin, *Amia calva*, juvenile from Ditch 5 spillway, Stoddard Co., MO. Photo by L. Merry.

Bowfin, *Amia calva*, adult from St. James Ditch, Mississippi Co., MO. Photo by L. Merry.

American Eel, *Anguilla rostrata*, from Caddo River, Clark Co., AR. Photo by U. Thomas.

Goldeye, *Hiodon alosoides*, from Mississippi River, Mississippi Co., AR. Photo by U. Thomas.

Mooneye, *Hiodon tergisus*, from Vermilion River, LaSalle Co., IL. Photo by U. Thomas.

Alabama Shad, *Alosa alabamae*, from Sante Fe River, Columbia Co., FL. Photo by Z. Randall.

Skipjack Herring, *Alosa chrysochloris*, from Illinois River, LaSalle Co., IL. Photo by U. Thomas.

Alewife, *Alosa pseudoharengus*, from Buck Creek, Pulaski Co., KY. Photo by M. Thomas.

Gizzard Shad, *Dorosoma cepedianum*, from Buck Creek, Pulaski Co., KY. Photo by M. Thomas.

Threadfin Shad, *Dorosoma petenense*, from Shawnee Creek, Ballard Co., KY. Photo by M. Thomas.

River Carpsucker, *Carpiodes carpio*, from Crooked Creek, Marion Co., AR. Photo by U. Thomas.

Quillback, *Carpiodes cyprinus*, from Big South Fork River, McCreary Co., KY. Photo by M. Thomas.

Highfin Carpsucker, *Carpiodes velifer*, from Ouachita River, Ouachita Co., AR. Photo by U. Thomas.

Longnose Sucker, *Catostomus catostomus*, from Flaugherty Creek, Somerset Co., PA. Photo by B. Zimmerman.

White Sucker, *Catostomus commersonii*, from Big Willis Creek, Clinton Co., KY. Photo by M. Thomas.

Blue Sucker, *Cycleptus elongatus*, from Missouri River, Boone Co., MO. Photo by L. Merry.

Western Creek Chubsucker, *Erimyzon claviformis*, from Humphrey Creek, Ballard Co., KY. Photo by M. Thomas.

Lake Chubsucker, *Erimyzon sucetta*, from Little Beaver Creek, Kankakee Co., IL. Photo by U. Thomas.

Northern Hog Sucker, *Hypentelium nigricans*, from Aux Sable Creek, Grundy Co., IL. Photo by U. Thomas.

Smallmouth Buffalo, *Ictiobus bubalus*, from Deer Creek, Livingston Co., KY. Photo by M. Thomas.

Bigmouth Buffalo, *Ictiobus cyprinellus*, from Clarks River, McCracken Co., KY. Photo by M. Thomas.

Black Buffalo, *Ictiobus niger*, from Clarks River, McCracken Co., KY. Photo by M. Thomas.

Spotted Sucker, *Minytrema melanops*, from Little Joe Creek, Hickman Co., KY. Photo by M. Thomas.

Silver Redhorse, *Moxostoma anisurum*, from Gasconade River, Maries Co., MO. Photo by L. Merry.

River Redhorse, *Moxostoma carinatum*, from Buck Creek, Pulaski Co., KY. Photo by M. Thomas.

Black Redhorse, *Moxostoma duquesnei*, from Gasconade River, Maries Co., MO. Photo by L. Merry.

Golden Redhorse, *Moxostoma erythrurum*, from Buck Creek, Pulaski Co., KY. Photo by M. Thomas.

Shorthead Redhorse, *Moxostoma macrolepidotum*, from Mississippi River, Ralls Co., MO. Photo by L. Merry.

Greater Redhorse, *Moxostoma valenciennesi*, from Aux Sable Creek, Grundy Co., IL. Photo by U. Thomas.

Goldfish, *Carassius auratus*, from Thorn Creek, Cook Co., IL. Photo by U. Thomas.

Common Carp, *Cyprinus carpio*, from Obion Creek, Hickman Co., KY. Photo by M. Thomas.

Grass Carp, *Ctenopharyngodon idella*, from private fish farm, AR. Photo by U. Thomas.

Silver Carp, *Hypophthalmichthys molitrix*, from Mississippi River, Cape Girardeau Co., MO. Photo by L. Merry.

Bighead Carp, *Hypophthalmichthys nobilis*, from Sangamon River, Macon Co., IL. Photo by L. Merry.

Black Carp, *Mylopharyngodon piceus*, from private fish farm, AR. Photo by U. Thomas.

Central Stoneroller, *Campostoma anomalum*, breeding male from Forked Creek, Will Co., IL. Photo by U. Thomas.

Central Stoneroller, *Campostoma anomalum*, female from Leaf River, Ogle Co., IL. Photo by U. Thomas.

Largescale Stoneroller, *Campostoma oligolepis*, breeding male from Mississippi River, Jo Daviess Co., IL. Photo by U. Thomas.

Largescale Stoneroller, *Campostoma oligolepis*, female from Rock Creek, McCreary Co., KY. Photo by M. Thomas.

Southern Redbelly Dace, *Chrosomus erythrogaster*, from Tower Creek, Kankakee Co., IL. Photo by U. Thomas.

Redside Dace, *Clinostomus elongatus*, female from Licking River drainage, unknown co., KY. Photo by M. Thomas.

Redside Dace, *Clinostomus elongatus*, breeding male from Licking River drainage, unknown co., KY. Photo by M. Thomas.

Lake Chub, *Couesius plumbeus*, from Horserace Brook, Piscataquis Co., ME. Photo by D. Neely.

Red Shiner, *Cyprinella lutrensis*, breeding male from Mayfield Creek, Graves Co., KY. Photo by M. Thomas.

Red Shiner, *Cyprinella lutrensis*, nonbreeding individual from Brush Creek, Hickman Co., KY. Photo by M. Thomas.

Spotfin Shiner, *Cyprinella spiloptera*, breeding male from Aux Sable Creek, Grundy Co., IL. Photo by U. Thomas.

Spotfin Shiner, *Cyprinella spiloptera*, nonbreeding individual from Sharps Creek, Simpson Co., KY. Photo by M. Thomas.

Blacktail Shiner, *Cyprinella venusta*, from Obion Creek, Hickman Co., KY. Photo by M. Thomas.

Steelcolor Shiner, *Cyprinella whipplei*, from South Fork Licking River, Pendleton Co., KY. Photo by M. Thomas.

Silverjaw Minnow, *Ericymba buccata*, from Bull Creek, Kankakee Co., IL. Photo by U. Thomas.

Gravel Chub, *Erimystax x-punctatus*, from Dry Fork Creek, Phelps Co., MO. Photo by L. Merry.

Western Silvery Minnow, *Hybognathus argyritis*, from Little White River, Mellette Co., SD. Photo by L. Merry.

Brassy Minnow, *Hybognathus hankinsoni*, from Sand Draw Creek, Brown Co., NE. Photo by L. Merry.

Cypress Minnow, *Hybognathus hayi*, from Locust Bayou, Calhoun Co., AR. Photo by U. Thomas.

Mississippi Silvery Minnow, *Hybognathus nuchalis*, from Caney Branch, Crittenden Co., KY. Photo by M. Thomas.

Plains Minnow, *Hybognathus placitus*, from North Fork Red River, Jackson Co., OK. Photo by U. Thomas.

Bigeye Chub, *Hybopsis amblops*, from Vermilion River, Vermilion Co., IL. Photo by U. Thomas.

Pallid Shiner, *Hybopsis amnis*, from Kankakee River, Will Co., IL. Photo by U. Thomas.

Striped Shiner, *Luxilus chrysocephalus*, breeding male from Bullskin Creek, Clay Co., KY. Photo by M. Thomas.

Striped Shiner, *Luxilus chrysocephalus*, nonbreeding individual from Little Beaver Creek, Kankakee Co., IL. Photo by U. Thomas.

Common Shiner, *Luxilus cornutus*, from Apple River, Jo Daviess Co., IL. Photo by U. Thomas.

Bleeding Shiner, *Luxilus zonatus*, breeding male from Jacks Fork River, Shannon Co., MO. Photo by U. Thomas.

Scarlet Shiner, *Lythrurus fasciolaris*, female from Buck Creek, Simpson Co., KY. Photo by M. Thomas.

Scarlet Shiner, *Lythrurus fasciolaris*, breeding male from Middle Fork Drakes Creek, Allen Co., KY. Photo by M. Thomas.

Ribbon Shiner, *Lythrurus fumeus*, from Beaverdam Slough, Marshall Co., KY. Photo by M. Thomas.

Redfin Shiner, *Lythrurus umbratilis*, female from Forked Deer Creek, Will Co., IL. Photo by U. Thomas.

Redfin Shiner, *Lythrurus umbratilis*, breeding male from Forked Deer Creek, Will Co., IL. Photo by U. Thomas.

Sturgeon Chub, *Macrhybopsis gelida*, from Mississippi River, Alexander Co., IL. Photo by L. Merry.

Shoal Chub, *Macrhybopsis hyostoma*, from Green River, Warren Co., KY. Photo by M. Thomas.

Sicklefin Chub, *Macrhybopsis meeki*, from Missouri River, Boone Co., MO. Photo by L. Merry.

Silver Chub, *Macrhybopsis storeriana*, from Vermilion River, LaSalle Co., IL. Photo by U. Thomas.

Hornyhead Chub, *Nocomis biguttatus*, from Vermilion River, Vermilion Co., IL. Photo by U. Thomas.

River Chub, *Nocomis micropogon*, breeding male from Little Sturgeon Creek, Owsley Co., KY. Photo by M. Thomas.

River Chub, *Nocomis micropogon*, female from Cumberland River, Whitley Co., KY. Photo by M. Thomas.

Golder Shiner, *Notemigonus crysoleucas*, from tributary of Little Beaver Creek, Kankakee Co., IL. Photo by U. Thomas.

Pugnose Shiner, *Notropis anogenus*, from St. Francis River, Wayne Co., MO. Photo by L. Merry.

Emerald Shiner, *Notropis atherinoides*, from Illinois River, Grundy Co., IL. Photo by U. Thomas.

River Shiner, *Notropis blennius*, from Mississippi River, Marion Co., MO. Photo by L. Merry.

Bigeye Shiner, *Notropis boops*, from Eassis Creek, Randolph Co., AR. Photo by U. Thomas.

Ghost Shiner, *Notropis buchanani*, from Marais des Cygnes River, Bates Co., MO. Photo by L. Merry.

Ironcolor Shiner, *Notropis chalybaeus*, from Spring Creek, Kankakee Co., IL. Photo by U. Thomas.

Bigmouth Shiner, *Notropis dorsalis*, from Thorn Creek, Cook Co., IL. Photo by U. Thomas.

Blackchin Shiner, *Notropis heterodon*, from Lake Wawasee, Kosciusko Co., IN. Photo by U. Thomas.

Blacknose Shiner, *Notropis heterolepis*, from Lake Wawasee, Kosciusko Co., IN. Photo by U. Thomas.

Spottail Shiner, *Notropis hudsonius*, from Lake Superior, Ashland Co., WI. Photo by L. Merry.

Taillight Shiner, *Notropis maculatus*, breeding male from Goose Pond, Ballard Co., KY. Photo by M. Thomas

Taillight Shiner, *Notropis maculatus*, female from Goose Pond, Ballard Co., KY. Photo by M. Thomas.

Ozark Minnow, *Notropis nubilus*, from Apple River, Jo Daviess Co., IL. Photo by U. Thomas.

Carmine Shiner, *Notropis percobromus*, from Apple Creek, Cape Girardeau Co., MO. Photo by L. Merry.

Silver Shiner, *Notropis photogenis*, from Middle Fork Drakes Creek, Allen Co., KY. Photo by M. Thomas.

Rosyface Shiner, *Notropis rubellus*, from Vermilion River, Vermilion Co., IL. Photo by L. Merry.

Silverband Shiner, *Notropis shumardi*, from Mississippi River, Scott Co., MO. Photo by L. Merry.

Sand Shiner, *Notropis stramineus*, from Thorn Creek, Cook Co., IL. Photo by U. Thomas.

Weed Shiner, *Notropis texanus*, from Little Beaver Creek, Iroquois Co., IL. Photo by U. Thomas.

Mimic Shiner, *Notropis volucellus*, from Black River, Butler Co., MO. Photo by L. Merry.

Channel Shiner, *Notropis wickliffi*, from Mississippi River, Cape Girardeau Co., MO. Photo by L. Merry.

Pugnose Minnow, *Opsopoeodus emiliae*, from Terre Noir Creek, Clark Co., AR. Photo by U. Thomas.

Suckermouth Minnow, *Phenacobius mirabilis*, from Mayfield Creek, Graves Co., KY. Photo by M. Thomas.

Bluntnose Minnow, *Pimephales notatus*, breeding male from North Fork Vermilion River, Livingston Co., IL. Photo by U. Thomas.

Bluntnose Minnow, *Pimephales notatus*, nonbreeding individual from Harmons Lick, Lincoln Co., KY. Photo by M. Thomas.

Fathead Minnow, *Pimephales promelas*, from Thorn Creek, Cook Co., IL. Photo by U. Thomas.

Bullhead Minnow, *Pimephales vigilax*, from Kankakee River, Will Co., IL. Photo by U. Thomas.

Flathead Chub, *Platygobio gracilis*, from Niobrara River, Cherry Co., NE. Photo by L. Merry.

Bluehead Shiner, *Pteronotropis hubbsi*, from Locust Bayou, Calhoun Co., AR. Photo by U. Thomas.

Longnose Dace, *Rhinichthys cataractae*, from Carroll Creek, Jo Daviess Co., IL. Photo by U. Thomas.

Western Blacknose Dace, *Rhinichthys obtusus*, from Stonecoal Branch, Rowan Co., KY. Photo by M. Thomas.

Creek Chub, *Semotilus atromaculatus*, from Bay Creek, Pope Co., IL. Photo by M. Thomas.

Oriental Weatherfish, *Misgurnus anguillicaudatus*, from tributary of Des Plaines River, Cook Co., IL. Photo by U. Thomas.

White Catfish, *Ameiurus catus*, from Guist Creek Lake, Shelby Co., KY. Photo by M. Thomas.

Black Bullhead, *Ameiurus melas*, from Deer Creek, Will Co., IL. Photo by U. Thomas.

Yellow Bullhead, *Ameiurus natalis*, from Pitman Creek, Pulaski Co., KY. Photo by M. Thomas.

Brown Bullhead, *Ameiurus nebulosus*, from Hazel Creek, Ballard Co., KY. Photo by M. Thomas.

Blue Catfish, *Ictalurus furcatus*, from Mississippi River, Mississippi Co., AR. Photo by U. Thomas.

Channel Catfish, *Ictalurus punctatus*, juvenile from Embarras River, Cumberland Co., IL. Photo by U. Thomas.

Channel Catfish, *Ictalurus punctatus*, adult from Aux Sable Creek, Grundy Co., IL. Photo by U. Thomas.

Mountain Madtom, *Noturus eleutherus*, from Tippecanoe River, Tippecanoe Co., IN. Photo by U. Thomas.

Slender Madtom, *Noturus exilis*, from Horse Creek, Kankakee Co., IL. Photo by U. Thomas.

Stonecat, *Noturus flavus*, from Horse Creek, Kankakee Co., IL. Photo by U. Thomas.

Tadpole Madtom, *Noturus gyrinus*, from Little Beaver Creek, Kankakee Co., IL. Photo by U. Thomas.

Brindled Madtom, *Noturus miurus*, from Hackett Creek, Douglas Co., IL. Photo by U. Thomas.

Freckled Madtom, *Noturus nocturnus*, from Embarras River, Cumberland Co., IL. Photo by U. Thomas.

Northern Madtom, *Noturus stigmosus*, from Ohio River, McCracken Co., KY. Photo by M. Thomas.

Flathead Catfish, *Pylodictis olivaris*, from Mississippi River, Mississippi Co., AR. Photo by U. Thomas.

Grass Pickerel, *Esox americanus*, from Spring Creek, Kankakee Co., IL. Photo by U. Thomas.

Northern Pike, *Esox lucius*, from Kankakee River, Kankakee Co., IL. Photo by U. Thomas.

Muskellunge, *Esox masquinongy*, from Mackinaw River, McLean Co., IL. Photo by L. Merry.

Central Mudminnow, *Umbra limi*, from Spring Creek, Kankakee Co., IL. Photo by U. Thomas.

Cisco, *Coregonus artedi*, from St. Mary's River, Chippewa Co., MI. Photo by T. Harmon.

Lake Whitefish, *Coregonus clupeaformis*, from Lake Superior, Ashland Co., WI. Photo by L. Merry.

Bloater, *Coregonus hoyi*, from Lake Superior, Ontonagon Co., MI. Photo by J. Lyons.

Blackfin Cisco, *Coregonus nigripinnis*, from Lake Michigan, Door Co., WI. Photo by J. Lyons.

Coho Salmon, *Oncorhynchus kisutch*, female from Root River, Racine Co., WI. Photo by N. Tessler.

Coho Salmon, *Oncorhynchus kisutch*, male from Sheboygan River, Sheboygan Co., WI. Photo by N. Tessler.

Rainbow Trout, *Oncorhynchus mykiss*, from Big Double Creek, Clay Co., KY. Photo by M. Thomas.

Chinook Salmon, *Oncorhynchus tshawytscha*, from French River, unknown co., MN. Photo by K. Schmidt.

Round Whitefish, *Prosopium cylindraceum*, from Shingle Point, Yukon Territory, Canada. Photo by C. Gallagher.

Brown Trout, *Salmo trutta*, from Shepherd of the Hills Fish Hatchery, Taney Co., MO. Photo by L. Merry.

Brook Trout, *Salvelinus fontinalis*, from Spring Brook, Geauga Co., OH. Photo by B. Zimmerman.

Lake Trout, *Salvelinus namaycush*, from Raystown Lake, Huntingdon Co., PA. Photo by R. Criswell.

Rainbow Smelt, *Osmerus mordax*, from Lake Superior, Ashland Co., WI. Photo by L. Merry.

Trout-perch, *Percopsis omiscomaycus*, from Locust Creek, Sullivan Co., MO. Photo by L. Merry.

Pirate Perch, *Aphredoderus sayanus*, from Hayden Ditch, Kankakee Co., IL. Photo by U. Thomas.

Shawnee Hills Cavefish, *Forbesichthys papilliferus*, from unnamed spring-head, Scott Co., MO. Photo by L. Merry.

Burbot, *Lota lota*, from Snake River, Kanabec Co., MN. Photo by L. Merry.

Round Goby, *Neogobius melanostomus*, from Thorn Creek, Cook Co., IL. Photo by U. Thomas.

Brook Silverside, *Labidesthes sicculus*, from Kankakee River, Will Co., IL. Photo by U. Thomas.

Inland Silverside, *Menidia beryllina*, from Hamby Pond, Fulton Co., KY. Photo by M. Thomas.

Northern Studfish, *Fundulus catenatus*, from Little South Fork of Cumberland River, Wayne Co., KY. Photo by M. Thomas.

Banded Killifish, *Fundulus diaphanus*, from Lake George, Lake Co., IN. Photo by U. Thomas.

Starhead Topminnow, *Fundulus dispar*, female from Spring Creek, Kankakee Co., IL. Photo by U. Thomas.

Starhead Topminnow, *Fundulus dispar*, male from Hamby Pond, Fulton Co., KY. Photo by M. Thomas.

Blackstripe Topminnow, *Fundulus notatus*, from Horse Creek, Kankakee Co., IL. Photo by U. Thomas.

Blackspotted Topminnow, *Fundulus olivaceus*, male from Brush Creek, Hickman Co., KY. Photo by M. Thomas.

Blackspotted Topminnow, *Fundulus olivaceus*, female from Brush Creek, Hickman Co., KY. Photo by M. Thomas.

Western Mosquitofish, *Gambusia affinis*, male from Brush Creek, Adams Co., OH. Photo by B. Zimmerman.

Western Mosquitofish, *Gambusia affinis*, female from Brush Creek, Hickman Co., KY. Photo by M. Thomas.

Striped Mullet, *Mugil cephalus*, left from Indian River, Brevard Co., FL; right from Indian River, Volusia Co., FL. Photos by Z. Randall.

Rock Bass, *Ambloplites rupestris*, from Stinking Creek, Knox Co., KY. Photo by M. Thomas.

Flier, *Centrarchus macropterus*, from Saline River, Howard Co., AR. Photo by U. Thomas.

Green Sunfish, *Lepomis cyanellus*, from Hayden Ditch, Kankakee Co., IL. Photo by U. Thomas.

Pumpkinseed, *Lepomis gibbosus*, from backwater of Kankakee River, Kankakee Co., IL. Photo by U. Thomas.

Warmouth, *Lepomis gulosus*, from Bayou Creek, McCracken Co., KY. Photo by M. Thomas.

Orangespotted Sunfish, *Lepomis humilis*, from Sangamon River, Macon Co., IL. Photo by L. Merry.

Bluegill, *Lepomis macrochirus*, from Salt River, Spencer Co., KY. Photo by M. Thomas.

Longear Sunfish, *Lepomis megalotis*, from Little Beaver Creek, Kankakee Co., IL. Photo by U. Thomas.

Redear Sunfish, *Lepomis microlophus*, from Little River, McCurtain Co., OK. Photo by U. Thomas.

Northern Sunfish, *Lepomis peltastes*, from tributary of Little Beaver Creek, Kankakee Co., IL. Photo by U. Thomas.

Spotted Sunfish, *Lepomis punctatus*, from an oxbow lake of Saline River, Drew Co., AR. Photo by U. Thomas.

Bantam Sunfish, *Lepomis symmetricus*, from Bayou Dorcheat, Lafayette Co., AR. Photo by U. Thomas.

Smallmouth Bass, *Micropterus dolomieu*, from Buck Creek, Pulaski Co., KY. Photo by. M. Thomas.

Spotted Bass, *Micropterus punctulatus*, from Brushy Creek, Pulaski Co., KY. Photo by M. Thomas.

Largemouth Bass, *Micropterus salmoides*, from White River, Independence Co., AR. Photo by U. Thomas.

White Crappie, *Pomoxis annularis*, from Guist Creek Lake, Shelby Co., KY. Photo by M. Thomas.

Black Crappie, *Pomoxis nigromaculatus*, from Kankakee River, Will Co., IL. Photo by U. Thomas.

Banded Pygmy Sunfish, *Elassoma zonatum*, from Terrapin Creek, Graves Co., KY. Photo by M. Thomas.

White Perch, *Morone americana*, from Branched Oak Lake, Lancaster Co., NE. Photo by L. Merry.

White Bass, *Morone chrysops*, from Illinois River, LaSalle Co., IL. Photo by U. Thomas.

Yellow Bass, *Morone mississippiensis*, from Guist Creek Lake, Shelby Co., KY. Photo by M. Thomas.

Striped Bass, *Morone saxatilis*, juvenile from Dennis Wildlife Center Fish Hatchery, Bonneau, SC. Photos by Z. Randall.

Striped Bass, *Morone saxatilis*, adult from Welaka National Fish Hatchery, Welaka, FL. Photo by Z. Randall.

Freshwater Drum, *Aplodinotus grunniens*, from Mississippi River, Mississippi Co., AR. Photo by U. Thomas.

Western Sand Darter, *Ammocrypta clara*, from Black River, Butler Co., MO. Photo by L. Merry.

Eastern Sand Darter, *Ammocrypta pellucida*, from Vermilion River, Vermilion Co., IL. Photo by U. Thomas.

Crystal Darter, *Crystallaria asprella*, from Black River, Butler Co., MO. Photo by L. Merry.

Mud Darter, *Etheostoma asprigene*, female from Quiver Creek, Mason Co., IL. Photo by U. Thomas.

Mud Darter, *Etheostoma asprigene*, breeding male from Quiver Creek, Mason Co., IL. Photo by U. Thomas.

Greenside Darter, *Etheostoma blennioides*, female from Knights Creek, Vermilion Co., IL. Photo by U. Thomas.

Greenside Darter, *Etheostoma blennioides*, breeding male from Knights Creek, Vermilion Co., IL. Photo by U. Thomas.

Rainbow Darter, *Etheostoma caeruleum*, female from Buck Creek, Lincoln Co., KY. Photo by M. Thomas.

Rainbow Darter, *Etheostoma caeruleum*, breeding male from Tower Creek, Kankakee Co., IL. Photo by U. Thomas.

Bluebreast Darter, *Etheostoma camurum*, female from Buck Creek, Pulaski Co., KY. Photo by M. Thomas.

Bluebreast Darter, *Etheostoma camurum*, breeding male from Kinniconick Creek, Lewis Co., KY. Photo by M. Thomas.

Bluntnose Darter, *Etheostoma chlorosoma*, from Johns Creek, Marshall Co., KY. Photo by M. Thomas.

Fringed Darter, *Etheostoma crossopterum*, female from Mill Creek, Union Co., IL. Photo by U. Thomas.

Fringed Darter, *Etheostoma crossopterum*, breeding male from Mill Creek, Union Co., IL. Photo by U. Thomas.

Iowa Darter, *Etheostoma exile*, female from Lake Wawasee, Kosciusko Co., IN. Photo by U. Thomas.

Iowa Darter, *Etheostoma exile*, breeding male from Sanctuary Pond, Lake Co., IL. Photo by L. Merry.

Fantail Darter, *Etheostoma flabellare*, female from Tower Creek, Kankakee Co., IL. Photo by U. Thomas.

Fantail Darter, *Etheostoma flabellare*, breeding male from Tower Creek, Kankakee Co., IL. Photo by U. Thomas.

Slough Darter, *Etheostoma gracile*, from Humphrey Creek, Ballard Co., KY. Photo by M. Thomas.

Harlequin Darter, *Etheostoma histrio*, from Little River, McCurtain Co., OK. Photo by U. Thomas.

Stripetail Darter, *Etheostoma kennicotti*, from Big Creek, Hardin Co., IL. Photo by U. Thomas.

Least Darter, *Etheostoma microperca*, from Tower Creek, Kankakee Co., IL. Photo by U. Thomas.

Johnny Darter, *Etheostoma nigrum*, from Quiver Creek, Mason Co., IL. Photo by U. Thomas.

Cypress Darter, *Etheostoma proeliare*, female from Goose Creek, Calloway Co., KY. Photo by M. Thomas.

Cypress Darter, *Etheostoma proeliare*, male from Beechy Creek, Calloway Co., KY. Photo by M. Thomas.

Orangethroat Darter, *Etheostoma spectabile*, breeding male from Stevens Creek, Mason Co., IL. (left) Photo by L. Merry.

Orangethroat Darter, *Etheostoma spectabile*, breeding male from Glenburn Creek, Vermilion Co., IL. (center) Photo by L. Merry.

Orangethroat Darter, *Etheostoma spectabile*, female from Forked Deer Creek, Will Co., IL. (right) Photo by U. Thomas.

Spottail Darter, *Etheostoma squamiceps*, female from Big Creek, Hardin Co., IL. Photo by U. Thomas.

Spottail Darter, *Etheostoma squamiceps*, male from Big Creek, Hardin Co., IL. Photo by U. Thomas.

Banded Darter, *Etheostoma zonale*, breeding male from Quiver Creek, Mason Co., IL. Photo by U. Thomas.

Yellow Perch, *Perca flavescens*, from Wolf Lake, Cook Co., IL. Photo by U. Thomas.

Logperch, *Percina caprodes*, from Cumberland River, Whitley Co., KY. Photo by M. Thomas.

Channel Darter, *Percina copelandi*, from Licking River, Bath Co., KY. Photo by M. Thomas.

Gilt Darter, *Percina evides*, from White River, Independence Co., AR. Photo by U. Thomas.

Blackside Darter, *Percina maculata*, from Forked Creek, Will Co., IL. Photo U. Thomas.

Slenderhead Darter, *Percina phoxocephala*, from Kankakee River, Kankakee Co., IL. Photo by U. Thomas.

Dusky Darter, *Percina sciera*, from Tippecanoe River, Pulaski Co., IN. Photo by U. Thomas.

River Darter, *Percina shumardi*, from Green River, Warren Co., KY. Photo by M. Thomas.

Stargazing Darter, *Percina uranidea*, from Ouachita River, Shelby Co., AR. Photo by U. Thomas.

Sauger, *Sander canadense*, from Kaskaskia River, Shelby Co., IL. Photo by L. Merry.

Walleye, *Sander vitreum*, from Kaskaskia River, Shelby Co., IL. Photo by L. Merry.

Brook Stickleback, *Culaea inconstans*, from Niobrara River, Cherry Co., NE. Photo by L. Merry.

Threespine Stickleback, *Gasterosteus aculeatus*, from Pacific Ocean, San Mateo Co., CA. Photo by D. Neely.

Ninespine Stickleback, *Pungitius pungitius*, from tributary of Pollack Brook, Lincoln Co., ME. Photo by D. Neely.

Spoonhead Sculpin, *Cottus ricei*, from Lake Superior, unknown co., WI. Photo by K. Schmidt.

Deepwater Sculpin, *Myoxocephalus thompsonii*, from Lake Saganaga, Cook Co., MN. Photo by K. Schmidt.

Mottled Sculpin, *Uranidea bairdii*, from Tippecanoe River, Fulton Co., IN. Photo by U. Thomas.

Banded Sculpin, *Uranidea carolinae*, from Pleasant Run, Logan Co., KY. Photo by M. Thomas.

Slimy Sculpin, *Uranidea cognata*, from Indian Stream, Coos Co., NH. Photo by D. Neely.

GLOSSARY

Some terms below have other or more encompassing meanings but are defined as applied to fishes in this book. Adapted from *A Field Guide to Freshwater Fishes of North America North of Mexico* (Page & Burr 2011) with some modifications from *Fishes in the Freshwaters of Florida* (Robins et al. 2018).

Adipose eyelid. Translucent tissue that partially covers eyeball in some fishes.

Adipose fin. Small, fleshy fin, usually without spines or rays, on back between dorsal fin and caudal fin (figure 18).

Adnate. Joined by having grown together.

Allopatric. Occurring in different geographic areas.

Ammocoete. Blind larva of lamprey.

Anadromous. Moving from ocean into fresh water to spawn (e.g., sturgeons).

Anal fin. Median fin located on undersurface, usually just behind anus (figure 18).

Axillary process. Fleshy flap (actually modified scale), usually narrow and projecting to rear, just above pectoral or pelvic fins.

Backwater. Quiet pool on side of stream channel. See *Pool.*

Band. Bar or stripe on fin. See *Bar, Stripe.*

Bar. (1) Vertical band of color. See *Band, Stripe.* (2) Ridge of sand or gravel in stream or along shore, formed by water currents.

Barbel. Fleshy sensory projections on head (sometimes whisker-like, usually near mouth; for example, chin, maxillary, and nasal barbels) (figure 18).

Basin. The land area draining to a body of water.

Basioccipital process. Extension of bone at rear edge of skull.

Belly. Abdomen; lower surface of a fish between pelvic fins and anus (figure 18).

Benthic. Living on or near the bottom.

Bifid. Divided into two parts.

Blotch. Irregularly shaped patch of color. See *Speck, Spot.*

Body depth. Straight-line measure from back to belly at the maximum distance (in most fishes, usually from dorsal fin origin to belly).

Branchiostegal membrane. Membrane connecting branchiostegal rays.

Branchiostegal ray. Ray-like bony support of branchiostegal membranes (figure 19).

Breast. Chest; lower surface of body between gill openings and pelvic fins (figure 18).

Bridle. Color mark across snout that suggests a bridle.

Canal. An artificial channel of water made for boat travel or for movement of water from one area to another.

Canine. Pointed, conical tooth, usually larger than surrounding teeth.

Cartilage. Material that forms skeleton of young fishes and that persists in adults of some species (notably, sharks and rays) but is largely converted to bone in most fishes.

Caudal fin. Median fin at posterior end of body; "tail" (figure 18).

Caudal peduncle. Part of body between base of last dorsal and anal fin rays and caudal fin base (figure 18).

Caudal spot. Spot at caudal fin base (origin).

Channel. Main course of a stream.

Cheek. Area below the eye (figure 19).

Coastal plain. Physiographic province extending inland along Atlantic (Atlantic Coastal Plain) and Gulf of Mexico coasts; extends north to southern Illinois on Former Mississippi Embayment.

Compressed. Flattened from side to side.

Concave. Bowed or curved inward.

Creek. A body of running water generally between 3 ft. (1 m) and 15 ft. (5 m) wide. See *Headwater, River, Stream.*

Ctenoid. Type of scale with toothed rear edge, making scale rough to touch.

Cusp. Principal projecting point of tooth; pointed anatomical feature.

Cycloid. Type of scale with smooth rear edge, making scale smooth to touch.

Decurved. Curved downward.

Depressed. Flattened from top to bottom.

Disc. See *Oral disc.*

Distal. Farthest from the point of attachment.

Dorsal. Upper part of fish; above.

Dorsal fin. Median rayed fin on back, often notched or subdivided; sometimes fully divided into 2 fins (figure 18).

Ear flap. Fleshy or bony extension on rear edge of opercle, as on sunfishes.

Electrofishing. Use of electrical current to temporarily stun fish for capture. Power is supplied by battery or generator, and current flows between submerged anodes and cathodes.

Embedded. Covered by skin (usually refers to scales).

Emarginate. Having a distal margin notched, indented, or slightly forked.

Endemic. Restricted to a particular basin, lake, state, and the like.

Falcate. Deeply indented or sickle-shaped (e.g., edge of fin).

Fall line. Boundary between Coastal plain and upland physiographic provinces, typically marked on rivers by waterfalls or large rapids.

Fin base. Part of fin that is attached to the body.

Floodplain. Land between top of stream bank and valley wall and floods during periods of high discharge.

Forked. In reference to caudal fin, used when rear edge is distinctly indented.

Frenum. Fleshy bridge or connection, as between snout and upper lip.

Fusiform. Cylindrical and tapering at both ends (usually refers to body shape).

Ganoid. Type of scale covered with hard enamel (e.g., on gars).

Gas bladder. Sac located between spinal column and gut cavity; also called air bladder or swim bladder.

Genus (plural: genera). Taxonomic category including 1 species or a closely related (i.e., all species sharing a common ancestor) group of species.

Gill. Breathing organ in fishes, including highly vascularized filaments used to extract oxygen from water (figure 19).

Gill arch. Bony or cartilaginous support to which gill filaments and gill rakers are attached. Area above maximum bend in arch is upper gill arch, below bend is lower gill arch (figure 19).

Gill cover. Bony flap covering outside of gill chamber (figure 19).

Gill filament. Feathery projection on rear of gill arch and for exchange of respiratory gases and excretion of certain physiological wastes.

Gill opening. Opening at rear of head, from gill chamber to outside (most fishes have 1 on each side).

Gill raker. Toothlike projection on front of gill arch; often used to trap food items between gill arches (figure 19).

Gonopodium. Front rays of anal fin of male livebearers modified to serve as an intromittent organ.

Gular plate. Bony plate on throat.

Habitat. Place where a fish lives; usually defined in terms of substrate, current, and stream size.

Halo. Circle of color around spot of another color.

Head length. Distance from tip of snout, lip, or chin—whichever is farthest forward—to rear edge of gill cover (figure 19).

Headwater. A body of running water generally less than 3 ft. (1 m) wide. See *Creek, River, Stream.*

Hermaphrodite. Having both male and female gonads in one individual.

Herringbone lines. Pattern of parallel, slanting lines (e.g., as caused by ribs of a herring or by blood vessels on upper side of a minnow).

Heterocercal. Type of caudal fin in which vertebral column extends into upper lobe, which is usually longer (e.g., sturgeons).

Holarctic. Northern parts of the Old and the New Worlds; the Nearctic and Palearctic Regions.

Humeral spot. Large spot above and usually slightly posterior to upper edge of pectoral fin base.

Hypural plate. Expanded terminal vertebral process that forms a fan-like plate onto which the caudal fin rays are attached.

Ichthyologist. Person who studies fishes.

Infraorbital. Below eye (e.g., infraorbital canal or pores).

Intergrade. Individual with characteristics intermediate between those of 2 subspecies.

Interorbital. Between eyes (orbits).

Invertebrate. Animal lacking backbone (e.g., insect, crayfish, worm).

Isthmus. Triangular, front-most part of underside of body; largely separated from head, in most bony fishes, by gill openings.

Juvenile. Young; usually small version of adult.

Keel. Shelflike fleshy or bony ridge.

Larva (plural: larvae). Newborn; developmental stage of a fish before it becomes a juvenile.

Lateral. On the side.

Lateral line. Rearward extension of sensory canal system (lateralis system) on head; usually single and located on midside of body; may be branched and variously placed, complete, incomplete, or interrupted; contains sense organs that detect pressure change (figure 19).

Lateral scale count. Number of scales along lateral line if present and complete or along midside if lateral line absent or incomplete.

Lateral scale. Row of scales along midside (usually along lateral line) from rear edge of gill cover to base of caudal fin (figure 19). Often called lateral line scales. See *Lateral scale count.*

Leptocephalus. Transparent, ribbon-like larva of eels, ladyfish, and tarpons.

Levee. A ridge or wall, usually parallel to a stream, designed to prevent lateral movement of water from a stream to its floodplain.

Medial. In middle plane or axis of body.

Median fin. Unpaired fin located on median plane of body; dorsal, caudal, and anal fins.

Melanophore. Cell containing melanin, a dark brown or black pigment. When contracted, these cells appear as pepper-like dots; when expanded, large areas of fish may become dark.

Midwater. In or near middle water stratum, as opposed to at surface or on bottom. See *Benthic.*

Myomere. Body segment. In lampreys, trunk myomeres extend from first segment after last gill pore to and including segment before anus.

Nape. Part of back immediately behind head; in spiny-rayed fishes, part between head and point where first (spiny) dorsal fin begins (figure 18).

Nocturnal. Active at night.

Nostril. Nasal opening (most fishes have 2 on each side).

Ocellus (plural: ocelli). Eyelike spot; usually dark, bordered by ring of light pigment.

Opercle. Uppermost and largest of bones forming gill cover (figure 19).

Oral disc. Fleshy circular structure surrounding mouth of lamprey.

Orbital. Related to the eye.

Origin. Point where fin begins, that is, point at which most anterior ray is inserted.

Oxbow. Lake formed in abandoned channel of stream meander after stream has cut through land at a narrow point in its meander.

Papilla (plural: papillae). Small nipple-like protuberance of connective tissue.

Papillose. With papillae.

Paired fins. Collectively, pectoral and pelvic fins.

Palatine. One of pair of bones on roof of mouth, one on each side, between jaw and midline; often used to describe teeth on this bone.

Palatine teeth. Teeth on palatine bones.

Parr. A juvenile Salmonid between the fry and smolt stages.

Pectoral fin. One of pair of fins (1 fin on each side) attached to shoulder girdle and just behind head (figure 18).

Peduncle. See *Caudal peduncle.*

Pelagic. Living in open water away from bottom.

Pelvic fin. One of pair of fins on lower part of body (figure 18); position ranges from on belly just in front of anal fin to under pectoral fin to isthmus.

Peritoneum. Lining of abdominal cavity; may be pigmented and visible externally.

Pharyngeal. Of or near the throat.

Plankton. Small plants (phytoplankton) and animals (zooplankton); mostly free-floating.

Plica (plural: plicae). Small fold of skin.

Plicate. With folds of skin.

Pool. Quiet, often relatively deep, segment of stream. See *Backwater, Riffle, Run*.

Pore. Tiny opening in skin; usually involved with sensory perception.

Pored scale. Scale with a pore (e.g., a lateral line scale).

Predorsal scales. Row of scales along middle of back between head and dorsal fin.

Premaxilla. One of a pair of bones at front of upper jaw (figure 19).

Premaxillary frenum. Bridge of flesh connecting upper lip and snout.

Preopercle. Bone at rear of cheek and in front of gill cover, often separated from gill cover by groove (figure 19).

Preopercular spine. Any spine along rear or lower edges of preopercle.

Preoperculomandibular canal pores. Pores of the lateral line that extend along the preopercle and mandible.

Preorbital. In front of eye (e.g., preorbital bar).

Protrusible. In reference to mouth with upper lip not attached to snout; may be extended far forward to catch prey.

Punctate lateral line. Pattern of rows of black spots above and below the lateral line.

Pyloric caecum (plural: caeca). Fingerlike tubes at junction of stomach and intestine.

Ray. Flexible, segmented fin ray; often branched (figure 18); also refers to bony element that supports and spreads branchiostegal membranes. See *Branchiostegal ray*.

Reticulate. Color markings in chainlike pattern or network.

Riffle. Fast-flowing, shallow segment of stream where surface of water is broken over rocks or debris. See *Pool, Run*.

River. A large body of running water; >15 ft. (5 m) wide. See *Creek, Headwater, Stream*.

Rudiment. Small, incompletely developed eye, fin ray, or gill raker (figure 19).

Run. Smooth-flowing transitional segment of stream between riffle and pool habitats, with moderate current and depth. See *Pool, Riffle*.

Saddle. Color mark, more or less rectangular, on back.

Scute. Enlarged scale, often with one or more bony projections.

Serrae. Sawtooth-like notches along an edge.

Serrate. With serrae.

Snout. Portion of head in front of eyes and above mouth (figure 19).

Snout length. Straight-line measure from tip of snout to anterior margin of eye (figure 19).

Soft ray. See *Ray*.

Speck. Very small blotch. See *Blotch, Spot*.

Spine. (1) Bony projection, usually on head (e.g., opercular spine). (2) Hard, unsegmented, and unbranched ray in fin—sometimes called a spinous ray (figure 18).

Spiracle. Opening (behind eye) of separate duct or canal leading to branchial chamber in sharks, rays, and certain ancient bony fishes (not the gill opening).

Spiral valve. Fold of tissue spiraling through intestine.

Spot. Circular color mark. See *Blotch, Speck*.

Standard length. Straight-line distance from tip of snout, lip, or chin—whichever is farthest forward—to rear end of vertebral column (that is, end of hypural plate—locate by lifting caudal fin and noting crease at caudal fin base). Used as standard measure of length of fish (figure 19).

Stream. A body of running water. See *Creek, Headwater, River*.

Stripe. Horizontal band of color. See *Band, Bar*.

Submandibular. Along the lower margin of the mandible.

Submarginal. In reference to fins, area along but not including edge of fin.

Suborbital. Below eye.

Subspecies. Geographically diagnosable population of a species. Subspecies name consists of 3 parts: genus, species, subspecies (e.g., *Lepomis punctatus miniatus*).

Subterminal. Position of mouth that describes when mouth opens below foremost point of head. See *Terminal, Upturned*.

Symphyseal frenum. Fleshy bridge or connection at junction of 2 bones (e.g., halves of upper jaw).

Teardrop. Vertical color mark under eye. Also called suborbital bar.

Terminal. Position of mouth that describes when mouth opens at foremost point of head, upper and lower jaws equal in length. See *Subterminal, Upturned*.

Territorial. Defending a particular area.

Thoracic. On breast (e.g., in reference to pelvic fin).

Tile drainage. Network of subsurface, porous, cylindrical tiles that removes excess groundwater and transports it to a body of surface water.

Total length. Straight-line distance from tip of snout, lip, or chin—whichever is farthest forward—to farthest rearward projection of tail (figure 19).

Truncated. Straight edged, squared off, as opposed to pointed or forked (e.g., edge of fin).

Tubercle. Small, white, hard (keratinized) protuberances on skin; usually seasonal in occurrence and only on breeding males.

Upturned. Position of mouth that describes when mouth opens above foremost point of head. See *Subterminal, Terminal*.

Urogenital opening. Opening through which bodily waste or eggs and sperm are expelled and into which intromittent organ of male is inserted in fishes with internal fertilization.

Vent. The opening of the anus on the ventral surface of a fish.

Ventral. Lower part of fish; below.

Vermiculation. Color pattern of wavy (wormlike) lines.

Villiform. Consisting of numerous, short, thin, hairlike structures grouped closely together.

Vomer. A median bone in front of roof of mouth; often used to describe teeth on this bone.

LITERATURE CITED

Bailey, R. M., W. C. Latta, and G. R. Smith. 2004. An atlas of Michigan fishes with keys and illustrations for their identification. Miscellaneous Publications of Museum of Zoology, University of Michigan 192.

Baldwin, N. A., R. W. Saalfeld, M. R. Dochoda, H. J. Buettner, and R. L. Eshenroder. 2006. Commercial fish production in the Great Lakes 1867–2006. Available: http://www.glfc.org/databases/commercial/commerc.php.

Barnhardt, M. L. 2010. Soils. p. 373–384. In: Geology of Illinois. Kolata, D. R., and C. K. Nimz (eds.). Illinois State Geological Survey, Champaign, Illinois.

Bart, H. L., Jr., M. D. Clements, R. E. Blanton, K. R. Piller, and D. L. Hurley. 2010. Discordant molecular and morphological evolution in buffalo-fishes (Actinopterygii: Catostomidae). Molecular Phylogenetics and Evolution 56:808–820.

Bayley, P. B., and D. C. Dowling. 1990. Gear efficiency calibrations for stream and river sampling. Illinois Natural History Survey Aquatic Ecology Technical Report 90/8.

Becker, G. C. 1983. Fishes of Wisconsin. University of Wisconsin Press, Madison, Wisconsin.

Bennett, G. W. 1958. Aquatic biology. p. 163–178. In: A century of biological research. Mills, H. B., G. W. Bennett, G. C. Decker, T. G. Scott, H. H. Ross, J. S. Ayars, C. J. Cedric, R. R. Warrick, and B. B. East (eds.). Illinois Natural History Survey Bulletin 27:85–234.

Berg, R. C. 2010. Protecting groundwater resources from contamination. p. 351–360. In: Geology of Illinois. Kolata, D. R., and C. K. Nimz (eds.). Illinois State Geological Survey, Champaign, Illinois..

Blum, M. J., D. A. Neely, P. M. Harris, and R. L. Mayden. 2008. Molecular systematics of the cyprinid genus Campostoma (Actinopterygii: Cypriniformes): disassociation between morphological and mitochondrial differentiation. Copeia 2008:360–369.

Bogue, M. B. 2000. Fishing the Great Lakes: an environmental history 1783–1933. The University of Wisconsin Press, Madison, Wisconsin.

Boltz, J. M., and J. R. Stauffer Jr. 1993. Systematics of Aphredoderus sayanus (Teleostei: Aphredoderidae). Copeia 1993:81–98.

Burr, B. M. 1976. Distribution and taxonomic status of the Stoneroller, Campostoma anomalum, in Illinois. Chicago Academy of Sciences, Natural History Miscellanea 194:1–8.

Burr, B. M. 1991. The fishes of Illinois: an overview of a dynamic fauna. p. 417–427. In: Our living heritage: the biological resources of Illinois. Page, L. M., and M. R. Jeffords (eds.). Illinois Natural History Survey Bulletin 34:357–477.

Burr, B. M. 1997. Stephen A. Forbes (1844–1930) and Philip W. Smith (1921–1986): ichthyological collection builders of the Illinois Natural History Survey and authors of The Fishes of Illinois. p. 517–523. In: Pietsch, T. W., and W. D. Anderson (eds.). Collection building in ichthy-

ology and herpetology, American Society of Ichthyologists and Herpetologists Special Publication 3.

Burr, B. M., K. M. Cook, D. J. Eisenhour, K. R. Piller, W. J. Poly, R. W. Sauer, C. A. Taylor, E. R. Atwood, and G. L. Seegert. 1996. Selected Illinois fishes in jeopardy: new records and status evaluations. Transactions of the Illinois State Academy of Science 89:169–186.

Burr, B. M., and R. L. Mayden. 1980. Dispersal of rainbow smelt, Osmerus mordax, in the upper Mississippi River (Pisces: Osmeridae). American Midland Naturalist 104:198–201.

Burr, B. M., and L. M. Page. 2009. Illinois fish communities: more than a century of change. p. 147–162. In: Canaries in the catbird seat: the past, present, and future of biological resources in a changing environment. Taylor, C. A., J. B. Taft, and C. E. Warwick (eds.). Illinois Natural History Survey Special Publication 30.

Burr, B. M., and M. L. Warren Jr. 1986. A distributional atlas of Kentucky fishes. Kentucky Nature Preserves Commission, Frankfort, Kentucky.

Burr, B. M., M. L. Warren Jr., and K. S. Cummings. 1988. New distributional records of Illinois fishes with additions to the known fauna. Transactions of the Illinois State Academy of Science 81:163–170.

Carlander, H. B. 1954. A history of fish and fishing in the upper Mississippi River. Sponsored by the Upper Mississippi River Conservation Committee.

Chick, J. H., D. K. Gibson-Reinemer, L Soeken-Gittinger, and A. F. Casper. 2019. Invasive Silver Carp is empirically linked to declines of native sport fish in the upper Mississippi River system. Biological Invasions. Available: https://doi.org/10.1007/s10530-019-02124-4.

Coker, R. E. 1930. Studies of common fishes of the Mississippi River at Keokuk. U.S. Department of Commerce, Bureau of Fisheries Document 1072:141–225.

Courtenay, W. R., Jr., D. A. Hensley, J. N. Taylor, and J. A. McCann. 1984. Distribution of exotic fishes in the continental United States. p. 41–77. In: Distribution, biology, and management of exotic fishes. Courtenay, W. R., Jr., and J. R. Stauffer Jr. (eds.). Johns Hopkins University Press, Baltimore, Maryland.

Croker, R. A. 2001. Stephen Forbes and the rise of American ecology. Smithsonian Institution Press, Washington, D.C.

Daxobeck, C., D. K. Barnard, and D. J. Randall. 1981. Functional morphology of the gills of the bowfin, Amia calva L., with special reference to their significance during air exposure. Respiration Physiology 43:349–364.

Duvernell, D. D., J. F. Schaefer, D. C. Hancks, J. A. Fonoti, and A. M. Ravanelli. 2007. Hybridization and reproductive isolation among syntopic populations of the topminnows Fundulus notatus and F. olivaceus. Journal of Evolutionary Biology 20:152–164.

Eisenhour, D. J. 1997. Distribution and systematics of Notropis wickliffi

(Cypriniformes: Cyprinidae) in Illinois. Transactions of the Illinois State Academy of Science 90:65–78.

Environmental Systems Research Institute (ESRI). 2017. ArcGIS release 10.6. Redlands, California.

Eshenroder, R. L., P. Vecsei, O. T. Gorman, D. L. Yule, T. C. Pratt, N. E. Mandrak, D. B. Bunnell, and A. M. Muir. 2016. Ciscoes (*Coregonus*, subgenus *Leucichthys*) of the Laurentian Great Lakes and Lake Nipigon. Great Lakes Fishery Commission, Miscellaneous Publication 2016-01.

Etnier, D. A., and W. C. Starnes. 1993. The fishes of Tennessee. University of Tennessee Press, Knoxville, Tennessee.

Fehrenbacher, J. B., J. D. Alexander, I. J. Jansen, R. G. Darmody, R. A. Pope, M. A. Flock, E. E. Voss, J. W. Scott, W. F. Andrews, and L. J. Bushue. 1984. Soils of Illinois. Soil Conservation Service Bulletin 778. University of Illinois at Urbana-Champaign, Illinois, College of Agriculture, Agricultural Experiment Station, and U.S. Department of Agriculture.

Field Museum of Natural History. 2019. Illinois fossil fishes. Retrieved September 17, 2019.

Forbes, S. A. 1882. The blind cavefishes and their allies. American Naturalist 16:1–5.

Forbes, S. A. 1884. A catalog of the native fishes of Illinois. Report of the Illinois State Fish Commissioner for 1884:60–89.

Forbes, S. A. 1885. Description of new Illinois fishes. Illinois State Laboratory of Natural History Bulletin 2:135–139.

Forbes, S. A. 1887. The lake as a microcosm. Bulletin of the Scientific Association (Peoria, Illinois), 1887:77–87. Reprinted in the Illinois Natural History Survey Bulletin 15:537–550, 1925.

Forbes, S. A. 1907. On the local distribution of certain Illinois fishes: an essay in statistical ecology. Illinois State Laboratory of Natural History 7:273–303.

Forbes, S. A., and R. E. Richardson. 1905. On a new shovelnose sturgeon from the Mississippi River. Illinois State Laboratory of Natural History Bulletin 7:37–44.

Forbes, S. A., and R. E. Richardson. 1909. The fishes of Illinois. Illinois Natural History Survey, Urbana, Illinois.

Forbes, S. A., and R. E. Richardson. 1920. The fishes of Illinois, 2nd ed. Illinois Natural History Survey, Urbana, Illinois.

Fricke, R., W. N. Eschmeyer, and R. van der Laan (eds.). 2020. Eschmeyer's catalog of fishes: genera, species, references. Available: http://research archive.calacademy.org/research/ichthyology/catalog/fishcatmain.asp.

Gilbert, C. R. 1961. Hybridization versus intergradation: an inquiry into the relationship of two cyprinid fishes. Copeia 1961:181–192.

Gilbert, C. R. 1964. The American cyprinid fishes of the subgenus *Luxilus* (genus *Notropis*). Bulletin of the Florida State Museum 8:95–194.

Grande, L., and W. E. Bemis. 1999. Historical biogeography and historical paleoecology of Amiidae and other halecomorph fishes. p. 413–424. *In*: Mesozoic fishes 2: systematics and fossil record. Arratia, G., and H. P. Schultze (eds.). Verlag, Dr. Friedrich Pfeil, München, Germany.

Hansel, A. K., and W. H. McKay. 2010. Quaternary Period. p. 216–247. *In*: Geology of Illinois. Kolata, D. R., and C. K. Nimz (eds.). Illinois State Geological Survey, Champaign, Illinois.

Havera, S. P. 1989. Ecological changes in the Illinois River Valley. p. 5. *In*: Proceedings of the Illinois Nature Preserves Commission 25th Anniversary Symposium on Preserving the Aquatic Biodiversity of Illinois: Inventory, Research, Regulation, and Protection. Philippi, M. A., and B. D. Anderson (eds.).

Hiland, T. M., and W. J. Poly. 2000. The Bleeding Shiner, *Luxilus zonatus*, in Illinois (Cyprinidae). Transactions of the Illinois State Academy of Science 93:145–146.

Hubbs, C. L., and K. F. Lagler. 2004. Fishes of the Great Lakes Region. Revised ed. by G. R. Smith. University of Michigan, Ann Arbor, Michigan.

Illinois Endangered Species Protection Board. 2020. Checklist of endangered and threatened animals and plants of Illinois. Illinois Endangered Species Protection Board, Springfield, Illinois. Available: Illinois Endangered and Threatened Species.pdf.

Illinois State Climatologist Office. 2020. Climate change in Illinois—temperature and precipitation. Available: https://www.isws.illinois.edu/statecli/climate-change/iltrends.htm.

Illinois State Fish Commission. 1884. Report of the Illinois State Fish Commission to the Governor of Illinois. H. W. Rokker, State Printer and Binder, Springfield, Illinois.

Illinois State Water Survey. 2011. Illinois climate normals. Data provided by the Illinois State Climatologist's Office, a part of the Illinois State Water Survey (ISWS), Champaign and Peoria, Illinois. Available: https://stateclimatologist.web.illinois.edu.

Iverson, L. R. 1991. Forest resources of Illinois: what do we have and what are they doing for us? p. 361–374. *In*: Our living heritage: the biological resources of Illinois. Page, L. M., and M. R. Jeffords (eds.). Illinois Natural History Survey Bulletin 34:357–477.

Janssen, J., and D. J. Jude. 2001. Recruitment failure of Mottled Sculpin *Cottus bairdi* in Calumet Harbor, southern Lake Michigan, induced by the newly introduced Round Goby *Neogobius melanostomus*. Journal of Great Lakes Research 27:319–328.

Jordan, D. S. 1878. A catalogue of the fishes of Illinois. Illinois Laboratory of Natural History Bulletin 1:37–70.

Jordan, D. S., and C. H. Gilbert. 1883. Synopsis of the fishes of North America. Bulletin of the U.S. National Museum 16.

Kelly, L. S. 1997. Patterns of faunal exploitation at Cahokia. p. 68–123. *In*: Cahokia: domination and ideology in the Mississippian world. Pauketat, T. R., and T. E. Emerson (eds.). University of Nebraska Press, Lincoln, Nebraska.

Kelly, L. S., and P. G. Cross. 1984. Zooarchaeology. p. 215–232. *In*: American bottom archaeology. Bareis, C. J., and J. W. Porter (eds.). University of Illinois Press, Urbana, Illinois.

Kennicott, R. 1855. Catalogue of animals observed in Cook County, Illinois. Illinois State Agricultural Society Transactions 1:577–595.

Kinziger, A. P., R. M. Wood, and D. A. Neely. 2005. Molecular systematics of the genus *Cottus* (Scorpaeniformes: Cottidae). Copeia 2005:303–311.

Kolata, D. R. 2010. Overview of Illinois geology. p. 59–73. *In*: Geology of Illinois. Kolata, D. R., and C. K. Nimz (eds.). Illinois State Geological Survey, Champaign, Illinois.

Kolata, D. R., and C. K. Nimz (eds.). 2010. Geology of Illinois. Illinois State Geological Survey, Champaign, Illinois.

Large, T. [1903]. A list of the native fishes of Illinois, with keys. Appendix to Report of the State Board of Fish Commissioners for September 30, 1900, to October 1, 1902.

Larimore, R. W., and P. W. Smith. 1963. The fishes of Champaign County, Illinois, as affected by 60 years of stream changes. Illinois Natural History Survey Bulletin 28:299–382.

Larson, D. R., and B. L. Herzog. 2010. Aquifers. p. 325–336. *In*: Geology of Illinois. Kolata, D. R., and C. K. Nimz (eds.). Illinois State Geological Survey, Champaign, Illinois.

Lewis, W. M. 1955. The fish population of the main stream of the Big Muddy River. Illinois Academy of Science Transactions 47:20–24.

Lyons, J., P. A. Cochran, and D. Fago. 2000. Wisconsin fishes 2000: status and distribution. University of Wisconsin Sea Grant Publication WISCU-B-00-001.

Mather, F. 1889. Brown Trout in America. Bulletin of the U.S. Fish Commission 7 (for 1887):21–22.

Mayden, R. L., R. B. Cross, and O. T. Gorman. 1987. Distributional history of the rainbow smelt, *Osmerus mordax*, (Salmoniformes: Osmeridae), in the Mississippi River Basin. Copeia 1987:1051–1055.

McClelland, M. A., G. G. Sass, T. R. Cook, K. S. Irons, N. N. Michaels, T. M. O'Hara, and C. S. Smith. 2012. The long-term Illinois River fish population monitoring program. Fisheries 37:340–350.

McCorvie, M. R., and C. L. Lant. 1993. Drainage district formation and the loss of midwestern wetlands, 1850–1930. Agricultural History 67:13–39.

McElrath, D. L., T. E. Emerson, A. C. Fortier, and J. L. Phillips. 1984. Late Archaic Period. p. 34–58. *In*: American bottom archaeology. Bareis, C. J., and J. W. Porter (eds.). University of Illinois Press, Urbana, Illinois.

Meek, S. E., and S. F. Hildebrand. 1910. A synoptic list of the fishes known to occur within fifty miles of Chicago. Field Museum of Natural History 142. Zoological Series 7:223–338.

Michaels, N. N., S. M. Tyszko, M. A. McClelland, and G. G. Sass. 2011. The long-term Illinois, Mississippi, Ohio, and Wabash Rivers Fish Population Monitoring Program. Illinois Natural History Survey Technical Report 2011(23).

Miner, J. J., and M. V. Miller. 2010. Wetlands geology. p. 361–369. *In*: Kolata, D. R., and C. K. Nimz (eds.). Geology of Illinois. Illinois State Geological Survey, Champaign, Illinois.

Morris, M. A., and L. M. Page. 1981. Variation in western logperches (Pisces: Percidae), with description of a new subspecies from the Ozarks. Copeia 1981:95–108.

Nelson, E. W. 1876. A partial catalogue of the fishes of Illinois. Illinois Museum of Natural History Bulletin 1(1):33–52; also Illinois State Laboratory of Natural History Bulletin 1:33–52.

Niemiller, M. L., J. R. McCandless, R. G. Reynolds, J. Caddle, T. J. Near, C. R. Tillquist, W. D. Pearson, and B. M. Fitzpatrick. 2013. Effects of climatic and geological processes during the Pleistocene on the evolutionary history of the northern cavefish, *Amblyopsis spelaea* (Teleostei: Amblyopsidae). Evolution 67:1011–1025.

O'Donnell, J. D. 1935. Annotated list of the fishes of Illinois. Illinois Natural History Survey Bulletin 20:473–500.

Page, L. M. 1985. The crayfishes and shrimps (Decapoda) of Illinois. Illinois Natural History Survey Bulletin 3:335–448.

Page, L. M. 1991. Streams of Illinois. p. 439–446. *In*: Our living heritage: the biological resources of Illinois. Page, L. M., and M. R. Jeffords (eds.). Illinois Natural History Survey Bulletin 34:357–477.

Page, L. M., and B. M. Burr. 2011. A field guide to freshwater fishes of North America north of Mexico. 2nd ed. The Peterson Field Guide Series, Houghton-Mifflin-Harcourt, Boston, Massachusetts.

Page, L. M., and M. R. Jeffords (eds.). 1991. Our living heritage: the biological resources of Illinois. Illinois Natural History Survey Bulletin 34:357–477.

Page, L. M., and R. L. Smith. 1970. Recent range adjustments and hybridization of *Notropis lutrensis* and *Notropis spilopterus* in Illinois. Transactions of the Illinois State Academy of Science 63:264–272.

Parmalee, P. W., A. A. Paloumpis, and N. Wilson. 1972. Animals utilized by woodland peoples occupying the Apple Creek Site, Illinois. Illinois Valley Archaeological Program, Research Papers 5.

Pegg, M. A., and M. A. McClelland. 2004. Spatial and temporal patterns in fish communities along the Illinois River. Ecology of Freshwater Fish 13:125–135.

Pflieger, W. L. 1997. The fishes of Missouri. Revised ed. Missouri Department of Conservation, Jefferson City, Missouri.

Pierce, R. L., B. M. Kahler, and G. J. Soulliere. 2014. State x BCR assessment: Illinois 22-Eastern Tallgrass Prairie. Upper Mississippi River/Great Lakes Region Joint Venture, U.S. Fish and Wildlife Service, Bloomington, Minnesota. Available: www.UpperMissGreatLakesJV.org.

Piller, K. R., H. L. Bart Jr., and D. L. Hurley. 2008. Phylogeography of the greenside darter complex, *Etheostoma blennioides* (Teleostomi: Percidae): a wide-ranging polytypic taxon. Molecular Phylogenetics and Evolution 46:974–985.

Prince, H. C. 1997. Wetlands of the American Midwest: a historical geography of changing attitudes. University of Chicago Press, Chicago, Illinois.

Ratcliff, E. N., E. J. Gittinger, T. M. O'Hara, and B. S. Ickes. 2014. Long Term Resource Monitoring Program procedures: Fish Monitoring, 2nd ed. U.S. Army Corps of Engineer Upper Mississippi River Restoration–Environmental Management Program. June 2014. Program Report LTRMP 2014-P001.

Retzer, M. E. 2005. Changes in the diversity of native fishes in seven basins in Illinois, USA. American Midland Naturalist 153:121–134.

Retzer, M. E., and C. R. Kowalik. 2002. Recent changes in the distribution of the River Redhorse (*Moxostoma carinatum*) and Greater Redhorse (*Moxostoma valenciennesi*) (Cypriniformes: Catostomidae) in Illinois and comments on their natural history. Transactions of the Illinois State Academy of Science 95:327–333.

Rice, D., and B. Zimmerman. 2019. A naturalist's guide to the fishes of Ohio. Special publication. Ohio Biological Survey, Columbus, Ohio.

Rios, N. E., and H. L. Bart. 2010. GEOLocate. Version 3.22. Tulane University Museum of Natural History, Belle Chasse, Louisiana.

Robins, R. H., L. M. Page, J. D. Williams, Z. S. Randall, and G. E. Sheehy. 2018. Fishes in the fresh waters of Florida. University of Florida Press, Gainesville, Florida.

Sabaj, M. H., K. S. Cummings, and L. M. Page. 1997. Annotated catalog of type specimens in the Illinois Natural History Survey fish collection. Illinois Natural History Survey Bulletin 35:253–300.

Savitz, J., C. Aiello, and L. G. Bardygula. 1989. The first record of the White Perch (*Morone americana*) in Illinois waters of Lake Michigan. Transactions of the Illinois Academy of Science 82:57–58.

Scott, R. J., P. W. Willink, and B. J. Norton. 2018. Biogeography and distribution of the cryptic species Rosyface Shiner *Notropis rubellus* and Carmine Shiner *Notropis percobromus* in Illinois. Copeia 106:524–531.

Smith, P. W. 1961. The amphibians and reptiles of Illinois. Illinois Natural History Survey Bulletin 28:1–298.

Smith, P. W. 1965. A preliminary annotated list of the lampreys and fishes of Illinois. Illinois Natural History Survey Biological Notes 54:1–12.

Smith, P. W. 1971. Illinois streams: a classification based on their fishes and an analysis of factors responsible for disappearance of native species. Illinois Natural History Survey Biological Notes 76:1–14.

Smith, P. W. 1973. A key to the fishes of Illinois. Illinois Department of Conservation Division of Fisheries, Fishery Bulletin 6.

Smith, P. W. 1979. The fishes of Illinois. University of Illinois Press, Urbana, Illinois.

Smith, P. W. 1986. A naturalist in the environmental crisis. Carlton Press, New York, New York.

Sparks, R. E., and W. C. Starrett. 1975. An electrofishing survey of the Illinois River, 1959–1974. Illinois Natural History Survey Bulletin 31:317–380.

Styles, B. W. 1981. Faunal exploitation and resource selection: Early Late Woodland subsistence in the lower Illinois Valley. Northwestern University Archeological Program, Scientific Papers 3.

Tan, M., and J. W. Armbruster. 2018. Phylogenetic classification of extant genera of fishes of the order Cypriniformes (Teleostei: Ostariophysi). Zootaxa 4476:6–39.

Tanner, H. H. (ed.). 1987. Atlas of the Great Lakes Indian history. University of Oklahoma Press, Norman, Oklahoma.

Taylor, C. A., J. B. Taft, and C. E. Warwick (eds.). 2009. Canaries in the catbird seat: the past, present, and future of biological resources in a changing environment. Illinois Natural History Survey Special Publication 30.

Thompson, J. 2002. Wetlands drainage, river modification, and sectoral conflict in the Lower Illinois Valley, 1890–1930. Southern Illinois University Press, Carbondale, Illinois.

Tranah, G., D. E. Campton, and B. May. 2004. Genetic evidence for hybridization of Pallid and Shovelnose Sturgeon. Journal of Heredity 95:474–480.

Trautman, M. L. 1931. *Notropis volucellus wickliffi*, a new subspecies of cyprinid fish from the Ohio and upper Mississippi Rivers. Ohio Journal of Science 31:467–474.

United States Army Corps of Engineers. 2020a. National inventory of dams. Available: https://nid.sec.usace.army.mil/ords/f?p=105:1.

United States Army Corps of Engineers. 2020b. Operations Division: Locks and Dams Project Office, Louisville District. Available: https://web.archive.org/web/20080610061205/http://www.lrl.usace.army.mil/opl/.

United States Census Bureau. 1900 Census Statistics. Available: https://www.biggestuscities.com/il/1900.

United States Fish and Wildlife Service. 2021. Endangered Species. Available: https://www.fws.gov/endangered/.

Wahl, D., et al. 2009. 150 years of fishery science: changes, progress, and the role of the Illinois Natural History Survey. p. 177–198. *In*: Canaries in the catbird seat: the past, present, and future of biological resources in a changing environment. Taylor, C. A., J. B. Taft, and C. E. Warwick (eds.). Illinois Natural History Survey Special Publication 30.

Warren, M. L., Jr., and B. M. Burr. 1988. Reassessment of the Illinois ranges of the bigeye chub, *Hybopsis amblops*, and the pallid shiner, *Notropis amnis*. Ohio Journal of Science 88:181–183.

Webb, D. W. 2002. The winter stoneflies of Illinois (Insecta: Plecoptera): 100 years of change. Illinois Natural History Survey Bulletin 36:195–274.

White, J. 1978. Illinois natural areas inventory technical report. Vol. 1: Survey methods and results. Illinois Natural Areas Inventory, Urbana, Illinois.

Wieczorek, J., Q. Guo, and R. J. Hijmans. 2004. The point-radius method for georeferencing locality descriptions and calculating associated uncertainty. International Journal of Geographical Information Science 18:745–767.

Wiggers, R. 1997. Geology underfoot in Illinois. Mountain Press, Missoula, Montana.

Willink, P. W., J. S. Tiemann, J. L. Sherwood, E. R. Larson, A. Otten, and B. Zimmerman. 2019. The mystery of the Banded Killifish *Fundulus diaphanus* population explosion: where did they all come from? American Currents 44:3–6.

Willink, P. W., T. A. Widloe, V. J. Santucci Jr., D. Makauskas, J. S. Tiemann, S. D. Hertel, J. T. Lamer, and J. L. Sherwood. 2018. Rapid expansion of Banded Killifish *Fundulus diaphanus* across northern Illinois: dramatic recovery or invasive species? American Midland Naturalist 179:179–190.

Willman, H. B., and J. C. Frye. 1970. Pleistocene stratigraphy of Illinois. Illinois State Geological Survey Bulletin 94:1–204.

INDEX OF COMMON
AND SCIENTIFIC NAMES

Brian A. Metzke is State Aquatic Ecologist for the Illinois Department of Natural Resources. **Brooks M. Burr** is an emeritus professor of zoology at Southern Illinois University Carbondale and a coeditor of *Freshwater Fishes of North America*. **Leon C. Hinz Jr.** is the Wildlife Action Plan Coordinator for the Illinois Department of Natural Resources. **Lawrence M. Page** is Curator of Fishes at the Florida Museum of Natural History. He is coauthor of *Peterson Field Guide to Freshwater Fishes of North America North of Mexico*. **Christopher A. Taylor** is an aquatic biologist and Curator of Fishes and Crustaceans with the Illinois Natural History Survey at the Prairie Research Institute at the University of Illinois Urbana-Champaign.

The University of Illinois Press
is a founding member of the
Association of University Presses.

———————————————

Text designed by Dustin Hubbart
and Jim Proefrock
Composed in 9.5/13 ITC Avant Garde
Book Condensed by Jim Proefrock
at the University of Illinois Press
Manufactured by Versa Press, Inc.

University of Illinois Press
1325 South Oak Street
Champaign, IL 61820-6903
www.press.uillinois.edu